Calibration/Validation of Visible Infrared Imaging Radiometers and Applications

Special Issue Editor

Changyong Cao

Image credit of Bin Zhang and Yan Bai

Special Issue Editor
Changyong Cao
NOAA/NESDIS/STAR, College Park
USA

Editorial Office
MDPI AG
St. Alban-Anlage 66
Basel, Switzerland

This edition is a reprint of the Special Issue published online in the open access journal, *Remote Sensing* (ISSN 2072-4292) from 2015–2016, available at:

http://www.mdpi.com/journal/remotesensing/special_issues/VIIRS

For citation purposes, cite each article independently as indicated on the article page online and as indicated below:

Author 1; Author 2; Author 3 etc. Article title. *Journalname*. **Year**. Article number/page range.

ISBN 978-3-03842-318-8 (Pbk)
ISBN 978-3-03842-319-5 (PDF)

Table of Contents

Chapter 1: Overview of Calibration/Validation

Chapter 2: Instrument Onboard Calibration and Prelaunch Characterization

Chapter 3: Sensor Data Records Intercomparison and Monitoring

Chapter 4: Environmental Data Records Product Calibration/Validation

About the Guest Editor

Changyong Cao, Ph.D., is a research physical scientist specializing in the calibration of radiometers onboard NOAA's (National Oceanic and Atmospheric Administration) operational environmental satellites. In addition to the operational pre- and post-launch calibration support, he is responsible for developing and refining the methodology for inter-satellite calibration using the Simultaneous Nadir Overpass (SNO) method, which has been used for the long-term on-orbit instrument performance monitoring of all radiometers on NOAA's polar orbiting satellites, and is being used by scientists for quantifying and correcting inter-satellite calibration biases in developing long-term time series for climate change detection studies. This methodology has been adopted by the World Meteorological Organization (WMO) as one of the cornerstones for the Global Space-based Inter-Calibration System (GSICS). Changyong's primary research interests are optical sensor calibration/validation, SI (International System of Units) traceability from pre-launch to post-launch, and calibration reanalysis for long-term time series and climate change detection. He was the instrument scientist for the legacy instrument High Resolution Infrared Radiation Sounder (HIRS) which has provided four decades of observations of the Earth and atmosphere. He is currently the JPSS (Joint Polar Satellite System) VIIRS sensor science team lead, and co-chair for the GOES-R (Geostationary Operational Environmental Satellite—R Series) calibration working group. He was former chair of the CEOS/WGCV (Committee on Earth Observation Satellites/Working Group on Calibration/Validation)—the international committee for space agencies. Before joining NOAA in 1999, Changyong was a senior scientist in remote sensing with a major aerospace company at NASA Stennis Space Center, where he supported a number of NASA projects, from hyperspectral spaceborne/airborne instrument preflight calibration, inflight radiometric and spectral calibration, validation and verification, to advanced remote sensing applications. He was an assistant professor and laboratory manager at Southern Illinois University in the early 1990s. Changyong received his Ph.D. degree in geography specializing in remote sensing and geographic information systems from Louisiana State University, and Bachelor of Science degree in geography from Peking (Beijing) University. He is the recipient of two gold and one silver medals honored by the U.S. Department of Commerce for his scientific and professional achievements, and has over one hundred peer reviewed publications.

Preface to "Calibration/Validation of Visible Infrared Imaging Radiometers and Applications"

The success of the Suomi National Polar-orbiting Partnership (NPP) brings us into a new era of global daily Earth observations, ranging from the faintest light of human settlements and air glows, to the dramatic events of hurricanes and forest fires, as well as the subtle changes in the planet Earth which we call home. At the heart of all satellite applications, calibration/validation of the instrument measurements and derived products is essential. Satellite product calibration and validation have become increasingly more important and challenging, in order to meet the stringent requirements for accurate and quantitative data for numerical weather prediction, climate change detection, and environmental intelligence. Validation is required not only for the satellite measurements, but also for all geophysical retrievals, including aerosols, cloud properties, radiation budget, sea surface temperature, ocean color, active fire, albedo, snow and ice, vegetation, as well as nightlights from human settlements. Active validation research includes, without being limited to, comparisons with similar products from other satellites, with in situ, aircraft measurements, or observations from other platforms. Validation results not only help users and decision-makers but also serve as feedback to calibration, which in turn improves the operational products.

This Special Issue of *Remote Sensing* explores recent results in the calibration and validation of the Suomi National Polar-orbiting Partnership satellite (Suomi NPP)/Joint Polar Satellite System (JPSS) radiometers. Studies involving the Suomi NPP/JPSS instruments in general, and VIIRS in particular, are included in this volume which consists of 30 papers covering a wide range of topics involving calibration and validation. This Special Issue consists of four Chapters: 1. Overview of calibration/validation; 2. Instrument onboard and prelaunch calibration; 3. Sensor data records, intercomparison and monitoring; 4. Environmental data records, product calibration and validation.

Finally, I would like to take this opportunity to thank all authors, co-authors, editors, reviewers, other contributors and supporters for the hard work and dedication that made this Special Issue possible.

Changyong Cao

Guest Editor

Chapter 1: Overview of Calibration/Validation

Article

Assessment of S-NPP VIIRS On-Orbit Radiometric Calibration and Performance

Xiaoxiong Xiong [1], James Butler [1], Kwofu Chiang [2], Boryana Efremova [2,†], Jon Fulbright [2,‡], Ning Lei [2], Jeff McIntire [2,*], Hassan Oudrari [2], Zhipeng Wang [2] and Aisheng Wu [2]

1 Sciences and Exploration Directorate, NASA/GSFC, Greenbelt, MD 20771, USA; xiaoxiong.xiong-1@nasa.gov (X.X.); james.j.butler@nasa.gov (J.B.)
2 Science Systems and Applications, Inc., Lanham, MD 20706, USA; kwofu.chiang@ssaihq.com (K.C.); boryana.efremova@noaa.gov (B.E.); jon.p.fulbright@nasa.gov (J.F.); ning.lei@ssaihq.com (N.L.); hassan.oudrari-1@nasa.gov (H.O.); zhipeng.wang@ssaihq.com (Z.W.); Aisheng.Wu@ssaihq.com (A.W.)
* Correspondence: jeffrey.mcintire@ssaihq.com; Tel.: +1-301-867-2073; Fax: +1-301-867-2151
† Current affiliation: Earth Resources Technology, Inc., Silver Spring, MD 20707, USA.
‡ Current affiliation: Columbus Technologies and Services, Inc., Greenbelt, MD 20770, USA.

Academic Editors: Xiaofeng Li and Prasad S. Thenkabail
Received: 25 November 2015; Accepted: 16 January 2016; Published: 23 January 2016

Abstract: The VIIRS instrument on board the S-NPP spacecraft has successfully operated for more than four years since its launch in October 2011. Many VIIRS environmental data records (EDR) have been continuously generated from its sensor data records (SDR) with improved quality, enabling a wide range of applications in support of users in both the operational and research communities. This paper provides a brief review of sensor on-orbit calibration methodologies for both the reflective solar bands (RSB) and the thermal emissive bands (TEB) and an overall assessment of their on-orbit radiometric performance using measurements from instrument on-board calibrators (OBC), as well as regularly scheduled lunar observations. It describes and illustrates changes made and to be made for calibration and data quality improvements. Throughout the mission, all of the OBC have continued to operate and function normally, allowing critical calibration parameters used in the data production systems to be derived and updated. The temperatures of the on-board blackbody (BB) and the cold focal plane assemblies are controlled with excellent stability. Despite large optical throughput degradation discovered shortly after launch in several near- and short-wave infrared spectral bands and strong wavelength-dependent solar diffuser degradation, the VIIRS overall performance has continued to meet its design requirements. Also discussed in this paper are challenging issues identified and efforts to be made to further enhance the sensor calibration and characterization, thereby maintaining or improving data quality.

Keywords: S-NPP; VIIRS; on-orbit; radiometric; performance; calibration

1. Introduction

The first Visible Infrared Imaging Radiometer Suite (VIIRS) sensor on board the Suomi National Polar Orbiting Partnership (S-NPP) satellite has been successfully operated for four years since its launch in October 2011. Designed with a strong MODIS heritage, VIIRS has 22 spectral bands spanning visible and infrared wavelengths from 0.4 μm–12.5 μm. These bands are designed to support the generation of a number of environmental data records (EDR) that benefit users in the land, ocean and atmospheric science disciplines [1–6]. The VIIRS instrument is a cross-track scanning (whiskbroom) radiometer. It uses a rotating telescope assembly (RTA) to collect data continuously from the Earth view (EV) and the calibration views every 1.78 s. In combination with the RTA, a half-angle mirror (HAM) rotates at half the rate of the RTA to direct light into stationary optics and onto different focal plane assemblies (FPAs). The S-NPP satellite is operated in a near Sun-synchronous polar orbit with a nominal altitude of 828 km and at an inclination angle of approximately 98 degrees relative to the Equator (the equatorial crossing time is 1:30 PM) [7]. With an EV scan angle range of about ±56 degrees, the VIIRS sensor is capable of making continuous global observations twice daily. The VIIRS spectral bands and detectors are located on three FPAs, the visible and near-infrared (VIS/NIR), the short- and mid-wave infrared (S/MWIR) and the long-wave infrared (LWIR). The S/MWIR and LWIR FPAs are temperature controlled at 80 K.

The fourteen reflective solar bands (RSB) are calibrated by observing solar radiance reflected off a solar diffuser (SD) and by observing a dark reference through a space view (SV) port. A solar diffuser stability monitor (SDSM) is used to track the SD on-orbit degradation. The RSB consist of three imaging bands (I1–I3) and eleven moderate resolution bands (M1–M11). Of these, six bands (M1–M5 and M7) are optimized using dual gain electronics, such that the high gain stage is used over low radiance scenes (e.g., oceans), while low gain is used over mid/high radiance scenes (e.g., land and clouds). VIIRS also has a reflective solar, panchromatic day-night band (DNB) on a separate FPA, used not only for imagery, but also for science studies of nighttime scenes with high radiometric quality. The seven thermal emissive bands (TEB) are calibrated using an on-board blackbody (BB) and dark offset signals from the SV. The TEB consist of two imaging bands (I4–I5) and five moderate resolution bands (M12–M16). M13 is the only dual gain band in the TEB, designed for measurements of high scene temperatures needed for fire products. The imaging and moderate resolution bands have nominal nadir spatial resolutions of 375 and 750 m, respectively, and the ground swath is approximately 3040 km in the cross-track direction. Some of the key characteristics of VIIRS spectral bands are shown in Table 1, including their wavelength ranges, focal plane location, typical and maximum scene spectral radiances or temperatures and specified signal-to-noise ratios (SNR) or noise-equivalent temperature differences (NEdT) at their corresponding typical radiances or temperatures. In this table, radiance and SNR are used for the RSB, while temperature and NEdT are used for the TEB.

Listed in Table 2 are some of the key events for S-NPP VIIRS on-orbit operation and calibration. Prior to opening the nadir aperture door, a series of sensor and on-board calibrator (OBC) functional tests were conducted. The "first light" images on 21 November 2011 were produced by the spectral bands in the visible (VIS) and near-infrared (NIR) spectral regions. Observations by S/MWIR and LWIR spectral bands were not scientifically useful before the cryo-cooler door was opened on 18 January 2012. It took about two days before the S/MWIR and LWIR FPAs reached their operational temperatures. Key calibration events performed during the sensor's initial intensive calibration and validation phase included routine SDSM and BB operations, as well as special calibration maneuvers. Only the first event of each routine calibration activity is listed in Table 2.

The VIIRS sensor data records (SDR) generated from its EV observations include calibrated and geolocated radiance, as well as the reflectance and brightness temperature for the RSB and TEB, respectively [8]. Since launch, processing of the SDR products has been under continuous enhancement either from the identification of and correction for mistakes in the operational processing algorithm or due to better understanding of the sensor operations and generation of improved and consistent calibration look-up tables (LUTs) [9–11]. Currently, S-NPP VIIRS is normally operated with all product and intermediate product files being generated routinely and the LUTs updated on a regular basis in the operational processing system, leading to very stable and high quality instrument performance.

The list of activities performed to generate VIIRS SDR includes: new RSB calibration coefficients (in LUT form) developed every week to generate the radiance and reflectance in the SDR products [12]; the DNB detector offsets and gain ratios generated on a monthly basis [13]; and the LUTs needed for the DNB stray light correction are updated every month in the operational processing [13]. Other activities are also routinely performed to monitor the instrument calibration and data quality, which include monthly lunar views to track the quality of the SD-based calibration for the RSB [14] and vicarious calibrations to track the quality of the SDR data using stable and well-characterized Earth view targets [15].

Table 1. VIIRS spectral band design specifications including typical scene readiances or temperatures (Ltyp or Ttyp) and maximum scene radiances or temperatures (Lmax or Tmax). VG denotes variable gain (low gain, middle gain and high gain). Units are $Wm^{-2} \cdot sr^{-1} \cdot \mu m^{-1}$ for the RSB and K for the TEB; DNB radiance units are $Wm^{-2} \cdot sr^{-1}$. SNR are listed at Ltyp, and NEdT are listed at Ttyp. M and I indicate moderate and imaging resolution.

	Band	Spectral Range (μm)	Band Gain	Ltyp or Ttyp	Lmax or Tmax	SNR or NEdT
				VIS/NIR		
Reflective Bands	DNB	0.500–0.900	VG	0.00003	200	6
	M1	0.402–0.422	High	44.9	135	352
			Low	155	615	316
	M2	0.436–0.454	High	40	127	380
			Low	146	687	409
	M3	0.478–0.498	High	32	107	416
			Low	123	702	414
	M4	0.545–0.565	High	21	78	362
			Low	90	667	315
	I1	0.600–0.680	Single	22	718	119
	M5	0.662–682	High	10	59	242
			Low	68	651	360
	M6	0.739–0.754	Single	9.6	41	199
	I2	0.846–0.885	Single	25	349	150
	M7	0.846–0.885	High	6.4	29	215
			Low	33.4	349	340
				S/WMIR		
	M8	1.230–1.250	Single	5.4	165	74
	M9	1.371–1.386	Single	6	77.1	83
	I3	1.580–1.640	Single	7.3	72.5	6
	M10	1.580–1.640	Single	7.3	71.2	342
	M11	2.225–2.275	Single	0.12	31.8	10
Emissive Bands	I4	3.550–3.930	Single	270	353	2.5
	M12	3.660–3.840	Single	270	353	0.396
	M13	3.973–4.128	High	300	343	0.107
			Low	380	634	0.423
				LWIR		
	M14	8.400–8.700	Single	270	336	0.091
	M15	10.263–11.263	Single	300	343	0.07
	I5	10.500–12.400	Single	210	340	1.5
	M16	11.538–12.488	Single	300	340	0.072

Table 2. Key events for S-NPP VIIRS on-orbit operation and calibration. SDSM, solar diffuser stability monitor; RTA, rotating telescope assembly; HAM, half-angle mirror; BB, blackbody.

Date	Event Description
28/10/2011	Suomi-NPP launch
08/11/2011	VIIRS turned on
08/11/2011	First SDSM operation (initially every orbit)
18/11/2011	First RTA/HAM sync loss reported
21/11/2011	Nadir door open
25/11/2011	First VIIRS recommended operating procedure for DNB calibration
25/11/2011	First VIIRS safe mode due to 1394 data bus anomaly that caused single board computer lock-up
04/01/2012	First planned lunar calibration (with roll maneuver)
18/01/2012	Cryo-cooler door open
19/01/2012	SDSM calibration frequency changed to once per day
06/02/2012	First BB warm-up and cool-down
15/02/2012	Yaw maneuver (fourteen orbits)
20/02/2012	Pitch maneuver
24/03/2012	Spacecraft anomaly: Sun point mode
16/05/2014	SDSM calibration frequency changed to three times a week

The SD on-orbit degradation continues to exhibit consistent wavelength dependence, as reported previously [9,16], with more degradation towards the shorter wavelengths. The largest SD degradation is at 0.41 μm and is currently about 31%. The large sensor responsivity degradation discovered shortly after S-NPP's launch in some of the NIR and SWIR spectral bands is approaching its limit as predicted by the sensor degradation model [17,18]. The TEB performance in terms of detector response and noise characteristics remains extremely stable, as reported in previous studies [19]. The largest change in TEB spectral band radiometric response has been less than 1.3% (I5) since launch. The telemetry trending of the VIIRS instrument has also exhibited the high stability of various instrument temperatures, showing well-controlled cold FPAs and BB, with temperature variations being less than 6 mK and 25 mK, respectively.

This paper provides the status of VIIRS instrument operations and calibration activities that are crucial to the SDR and EDR data quality. To some extent, this paper is an update to our previous study of VIIRS initial on-orbit calibration and performance [9]. In addition, it describes key changes that have been made in support of the SDR data processing system, including the offline processing and generation of calibration LUTs, to either enhance data quality or to mitigate issues affecting the sensor performance. Section 2 of this paper will provide a brief overview of VIIRS on-orbit calibration methodologies and activities, including the lunar calibration scheduling and implementation strategies. Section 3 will present the on-orbit calibration performance results based on OBC and telemetry data, including on-orbit changes in spectral band radiometric responses and sensor characterization, as well as calibration improvements. A list of lessons learned and future work to mitigate concerns identified in the operational processing will be discussed in Section 4, followed by a conclusion and summary in Section 5.

2. On-Orbit Calibration Methodologies and Activities

The VIIRS solar calibration system designed for the RSB consists of an on-board SD, a permanent solar attenuation screen (SAS) and an on-board SDSM. The SDSM is a ratioing device used to track on-orbit changes in the SD bidirectional reflectance distribution function (BRDF) via alternate measurements of direct sunlight through a fixed attenuation screen and the sunlight reflected off the SD panel. The DNB low gain stage is also calibrated by the SD/SDSM system. In addition to solar calibration, regularly-scheduled lunar observations made through the instrument SV port are used to support RSB on-orbit calibration. For the TEB, an on-board V-grooved BB panel is used as the calibration target. Illustrated in Figure 1 are the VIIRS instrument scan cavity and the OBC, including its extended SV port for lunar acquisitions and measurements of instrument background and offset reference.

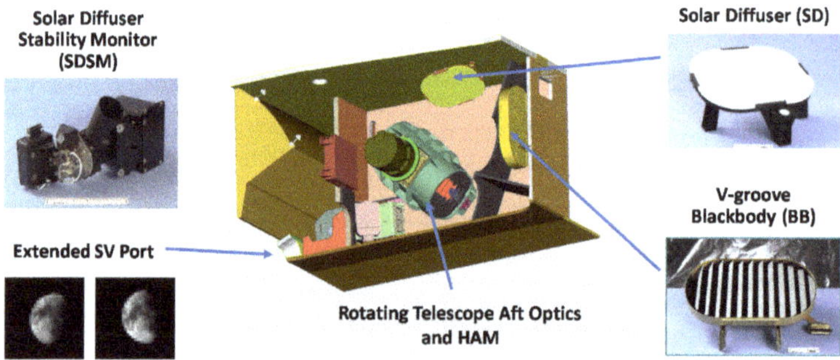

Figure 1. VIIRS sensor showing the positions of the SD, SDSM, BB and space view (SV).

Both VIIRS RSB and TEB apply a quadratic polynomial algorithm to retrieve their EV scene spectral radiance using their background subtracted detector response (dn_{EV}),

$$L_{EV} = F \times (c_0 + c_1 \times dn_{EV} + c_2 \times dn_{EV}^2)/RVS_{EV} \quad (1)$$

where c_0, c_1 and c_2 are the instrument temperature-dependent calibration coefficients derived from pre-launch characterization, RVS_{EV} is the detector's response *versus* scan angle at the EV HAM angle of incidence (AOI), also derived from pre-launch measurements, and F is a calibration scaling factor derived from on-orbit measurements of the SD or BB, known as the F-factor [8].

Specifically, the F-factor is determined by comparing the known calibration source spectral radiance (L_{CS}) with that retrieved by the sensor (L_{RET}) using the pre-launch calibration coefficients,

$$F = L_{CS}/L_{RET} \tag{2}$$

where:

$$L_{RET} = (c_0 + c_1 \times dn_{CS} + c_2 \times dn_{CS}^2)/RVS_{CS} \tag{3}$$

Similarly to Equation (1), the dn_{CS} and RVS_{CS} in Equation (3) are the detector response to the known calibration source and the RVS at the calibration source view HAM AOI, respectively.

2.1. Solar Calibration for the RSB

For the RSB, the on-board SD provides a known calibration source when it is fully illuminated by the Sun. In this case, the L_{RET} is computed using Equation (3) with the subscript CS replaced by SD. The calibration source spectral radiance can be computed from the solar spectral irradiance at the spacecraft, $E_{SUN}(\lambda)$, and the SD BRDF(λ) using the following expression,

$$L_{CS} = \tau_{SAS} \times \cos(\theta_{SD}) \times \int (RSR(\lambda) \times E_{SUN}(\lambda) \times BRDF(\lambda) \times d\lambda) / \int (RSR(\lambda) \times d\lambda) \tag{4}$$

where τ_{SAS} is the SAS transmission function, θ_{SD} is the SD solar zenith angle and λ is the wavelength. Both τ_{SAS} and SD BRDF are functions of the solar illumination angle. The E_{SUN} and SD BRDF in Equation (4) are weighted by the detector's relative spectral response, RSR(λ). The RSB F-factor for each calibration event is the average of multiple scan-by-scan computations when the SD is fully illuminated by the Sun and is band, detector and HAM side dependent [9]. The official RSR and E_{SUN} functions used for VIIRS SDR calibration can be found on the NOAA website [20]. The SD BRDF is based on the pre-launch $BRDF_0$ with corrections applied to account for its on-orbit degradation.

Algorithm details for using the SDSM to track the on-orbit SD degradation can be found in a number of references [16,21,22]. For VIIRS, an H-factor is used to represent the SD BRDF degradation. It is determined from the time series of ratios of the SDSM's SD view response (dc_{SD}) to its Sun view response (dc_{SUN}),

$$H \propto (dc_{SD}/(\tau_{SAS} \times \cos(\theta_{SD})))/(dc_{SUN}/\tau_{SUN}) \tag{5}$$

where τ_{SUN} is the transmission function for the SDSM Sun view screen. τ_{SUN} is both detector dependent and a function of solar illumination angle. The SDSM has eight detectors (D1–D8) with their center wavelengths located at 0.41, 0.44, 0.49, 0.56, 0.67, 0.75, 0.86 and 0.93 μm, respectively. The H-factor is computed separately using measurements made by each SDSM detector. For wavelengths greater than 0.93 μm, it is assumed that the SD degradation is negligible; this assumption will be revisited in Section 4.

2.2. Lunar Calibration for the RSB

The Moon is an extremely stable radiometric calibration target in the reflective solar spectral region [23]. Like MODIS, VIIRS lunar observations have been regularly scheduled and implemented in support of its RSB on-orbit calibration. Similar to the solar calibration, a lunar calibration F-factor (F_{MOON}) for the RSB is derived using the following expression,

$$F_{MOON} = I_{MODEL}/I_{RET} \tag{6}$$

where I_{MODEL} and I_{RET} are the model-predicted lunar irradiance, integrated over the entire lunar disk, and sensor-retrieved lunar irradiance, respectively. Currently, the USGS Robotic Lunar Observatory (ROLO) lunar model is used as the VIIRS lunar calibration reference. The ROLO model provides the predicted lunar irradiance values (I_{MODEL}) that depend on lunar viewing parameters, such as the Sun-Earth and sensor-Moon distances, the lunar phase angle and lunar libration. The lunar irradiance retrieved by the sensor (I_{RET}) is computed by integrating the radiances from individual detectors over the lunar disk using their pre-launch calibration coefficients. Details of VIIRS lunar calibration methodologies are found in a number of references [14,24–27].

It should be noted that F_{MOON} is determined for each spectral band and detector at the HAM AOI of 60.2 degrees, which is nearly identical to the SD HAM AOI. As a result, temporal changes in F_{MOON} reflect on-orbit changes in spectral band or detector response (or gain) and can be independently used to validate, and correct if necessary, the temporal changes in the SD F-factor. F_{MOON} is currently calculated using the "center-scans" approach by integrating the radiance of all detectors for each scan with complete lunar images captured by the FPA. In addition, detector-dependent F_{MOON} factors can be derived using the "all-scans" approach by integrating the radiance of all scans with lunar images for each detector [27].

2.3. DNB Calibration

The VIIRS DNB is designed to make global observations during both day and nighttime with a large dynamic range implemented through three different gain stages: low gain stage (LGS), mid-gain stage (MGS) and high gain stage (HGS). Each gain stage is a separate CCD (HGS has two redundant CCD arrays, HGA and HGB). It has 32 aggregation modes, implemented to achieve a nearly constant footprint across the entire scan. The DNB calibration is performed separately for each gain stage and aggregation mode. The DNB on-orbit calibration requires three pieces of information: the dark offset, the LGS linear gain and the gain ratios (MGS/LGS and HGS/MGS) [13].

The DNB dark offsets were determined early in the mission using data collected during the spacecraft pitch maneuver. Their on-orbit changes are tracked using on-board BB data collected under the darkest conditions (nighttime over the Pacific Ocean during new moon) as a dark reference. The DNB LGS calibration coefficients are determined from on-orbit SD observations when the SD is fully illuminated by sunlight, with the exception that the DNB radiance retrieved and used in the calibration is the spectral band integrated radiance (units: $Wm^{-2} \cdot sr^{-1}$). The LGS calibration is transferred to the MGS and HGS via gain ratios determined from SD view data collected before and after the SD is fully illuminated.

2.4. BB Calibration for the TEB

The VIIRS TEB on-orbit calibration is performed with reference to the on-board BB. The calibration source spectral radiance (L_{CS}) is modeled as the radiance difference between the BB and SV paths or:

$$L_{CS} = L_{BB} + (1 - RVS_{SV}/RVS_{BB}) \times ((1 - \rho_{RTA}) \times L_{RTA} - L_{HAM})/\rho_{RTA} \quad (7)$$

where L_{BB}, L_{RTA} and L_{HAM} are the radiances of the BB, RTA and HAM averaged over the spectral response of each band, RVS_{SV} and RVS_{BB} are the RVS at the SV and BB HAM AOI, respectively, and ρ_{RTA} is the reflectivity of the RTA mirrors. The TEB calibration source radiance is the sum of emitted radiance and reflected radiance (emission from thermal sources around the BB, reflected off the BB into the optical path). The emitted BB contribution is the dominant term, as the BB has an emissivity above 99.6%. The other terms in Equation (7) account for the RTA and HAM emission that do not cancel in the path difference (due to differences in the RVS at the two view angles). Similar contributions from RTA and HAM emission must also be included in Equation (1) for the TEB EV radiance retrieval [9,19].

Again, Equation (3) is used to determine the retrieved spectral radiance (L_{RET}) for the BB view by replacing the subscript CS with BB. The TEB F-factors are then estimated by substituting Equations (3) and (7) into Equation (2). This is performed every other scan for all TEB, because of the two HAM sides, except for M13 high gain, where the F-factors are calculated every fourth scan (M13 low gain F-factors are set to one, as the BB radiance is too low for accurate trending).

2.5. On-Orbit Calibration Activities

The VIIRS instrument does not have an aperture door for the SD port, except for the permanently fixed SAS placed in front of the SD panel. As a result, on-orbit SD calibration is performed for each orbit. The SDSM is used to track the SD on-orbit degradation and is operated via uploaded commands. As expected, the SDSM was operated more frequently early in the mission, initially every orbit and then daily (see Table 2). Starting from 16 May 2014, the SDSM has been operated only three times per week.

Table 3 is a summary of lunar observations successfully scheduled and implemented for S-NPP VIIRS RSB calibration. It contains spacecraft roll angles and lunar phase angles. If the predicted roll angle is within ±1 degree, no spacecraft roll maneuver is needed. In order to capture lunar images for all spectral bands at the same time, a sector rotation is implemented, such that the EV data sector is centered on the SV view angle. Most of the lunar phase angles are centered at −51 degrees, where the negative phase angle indicates a waxing Moon. Only a few lunar observations performed at the beginning of the mission have their phase angles centered at −55 degrees. The −55 degrees lunar phase angle was selected to match the Aqua MODIS lunar observations. The change from −55 degrees to −51 degrees was made in response to additional operational constraints on the S-NPP spacecraft and to optimize lunar observation opportunities. The lunar observations, as well as SD/SDSM measurements are primarily used for RSB calibration.

The BB is nominally controlled at about 292.7 K, and its warm-up and cool-down (WUCD) operation has been performed quarterly since launch (starting from February 2012) for a total of 15 times thus far. In order to minimize thermal effects during BB WUCD that have shown a small, but undesirable impact on data quality, the frequency of WUCD is set to be reduced to semi-annually starting in the future. During the WUCD, the BB temperature cycles through a series of discrete temperatures and varies between an instrument ambient temperature of approximately 267 K and 315 K. The WUCD operation is performed regularly to characterize and track changes in TEB detector offset and nonlinear response. It also allows TEB detector noise characterization to be examined over a range of source temperatures.

Table 3. S-NPP VIIRS lunar calibrations implemented (launch to September 2015). Angle units are degrees. No satellite roll maneuver was performed if the predicted roll angle is within ±1 degree (denoted with *).

Date	H:M:S	Roll Angle	Phase Angle	Date	H:M:S	Roll Angle	Phase Angle
04/01/2012	08:48:53	−9.490	−55.41	11/01/2014	09:59:45	−6.727	−51.30
03/02/2012	04:21:32	−5.445	−56.19	10/02/2014	05:34:12	−3.714	−51.03
03/02/2012	06:03:34	−5.279	−55.38	12/03/2014	01:11:43	−3.944	−51.05
02/04/2012	23:05:11	−3.989	−51.24	10/04/2014	20:53:17	−4.977	−50.60
02/05/2012	10:20:06	−3.228	−50.92	10/05/2014	13:13:00	−4.177	−50.91
31/05/2012	14:47:14	−0.081 *	−52.97	09/06/2014	03:48:42	+0.301 *	−51.05
25/10/2012	06:58:15	−4.048	−51.02	04/10/2014	17:29:10	+0.696 *	−50.81
23/11/2012	21:18:20	−9.429	−50.74	03/11/2014	01:07:35	−0.609	−50.53
23/12/2012	15:00:50	−7.767	−50.90	02/12/2014	08:41:44	−10.841	−50.83
22/01/2013	12:13:35	−3.383	−50.81	31/12/2014	19:38:07	−8.981	−50.73
21/02/2013	09:31:25	−1.712	−50.71	30/01/2015	08:22:14	−5.674	−51.16
23/03/2013	03:29:00	−3.320	−51.15	01/03/2015	00:34:22	−4.048	−50.91
21/04/2013	19:47:54	−3.882	−50.82	30/03/2015	16:49:09	−5.236	−51.29
21/05/2013	08:43:15	−0.809 *	−50.67	29/04/2015	12:29:27	−4.701	−50.43
14/10/2013	21:39:19	−1.305	−50.95	29/05/2015	04:47:10	−2.304	−51.07
13/11/2013	06:57:41	−7.981	−50.66	27/06/2015	14:17:10	0.314 *	−54.43
12/12/2013	19:35:46	−9.438	−50.39				

On-orbit changes in VIIRS DNB offsets are determined approximately monthly via a VIIRS recommended operating procedure. This procedure is implemented at new moon over dark oceans.

An additional VIIRS recommended operating procedure is performed monthly near the terminator to determine the gain ratios needed to transfer the LGS gain to the MGS and HGS.

3. On-Orbit Performance

3.1. RSB Performance

The H-factor is calculated using Equation (5). In this equation, τSUN was determined from both the yaw maneuver data and a small portion of regular on-orbit data [28]. The τSAS*BRDF0 was derived from the yaw maneuver data. Corrected solar vectors were used in the screen transmittance determinations and in the calculation of the H-factor. It should be noted that due to mishandling of the solar vector in different coordinate systems, an error in the solar vector occurred in the operation processing prior to 22 November 2014. The SDSM system on VIIRS has tracked the SD on-orbit degradation well. Estimated H-factor errors on a per satellite orbit basis range from 0.0002 (D6)–0.0007 (D1) (after removing a bias in the calculated τSAS*BRDF0, due to the angular dependence of the H-factor at the time of the yaw maneuvers, and using both the yaw maneuver and regular on-orbit data to calculate τSAS*BRDF0 [29]). Figure 2 displays the per-orbit H-factors measured by the eight SDSM detectors over the lifetime of the mission normalized to their respective extrapolated values at satellite launch. The H-factors are determined from data within a solar vertical angle range of ± 1.0 degree in SDSM screen coordinates [9]. The bluest detector (D1) indicates that the SD has degraded by close to 31% since launch, whereas the degradation is only slightly larger than 1% at 926 nm (D8). There are a few undulations with magnitudes around 0.001 in the H-factor curves that are most likely due to unresolved features in the calculated transmission functions of the attenuation screens. The H-factors in the figure were calculated without considering SDSM detector RSR out-of-band (OOB) effects. The RSR OOB is spectral filter orientation dependent, and the orientation is unknown for the S-NPP VIIRS SDSM. The impact on the H-factor due to the RSR OOB response is negligible for SDSM D5–D8 [30]. Radiometric calibration through lunar observation has shown that the dependence of the H-factor on view angle off the SD and the OOB response effect on the H-factor almost cancel (about a 1% residual effect remains at the central wavelength of M1; see Figure 3).

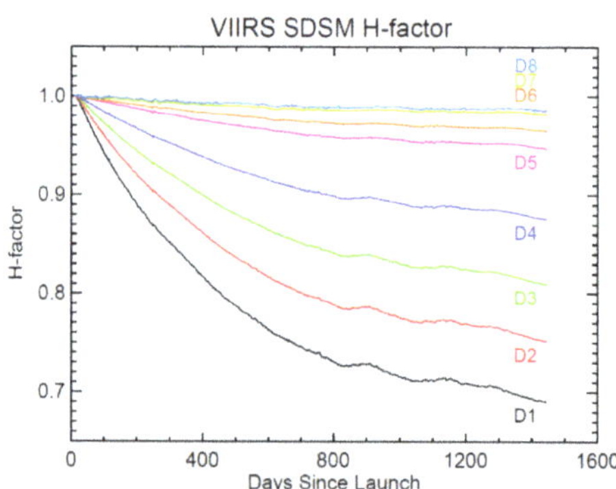

Figure 2. H-factors for SDSM D1–D8 determined by Equation (5). The H-factors are normalized to the satellite launch date.

The normalized RSB 1/F-factors (or gains) derived from the SD calibration are shown by the solid lines in Figure 3a–d. The normalization is made to the first available date for the VIS/NIR bands: 8 November 2011. For the SWIR bands (M8–M11 and I3), the normalization is made to the modeled values on 8 November 2011. The 1/F-factors shown in Figure 3 are daily averaged values over all detectors and HAM sides for each spectral band. No additional fitting or smoothing is made to the 1/F-factors. A noticeable feature is the sudden decrease in the 1/F-factors for NIR and SWIR bands.

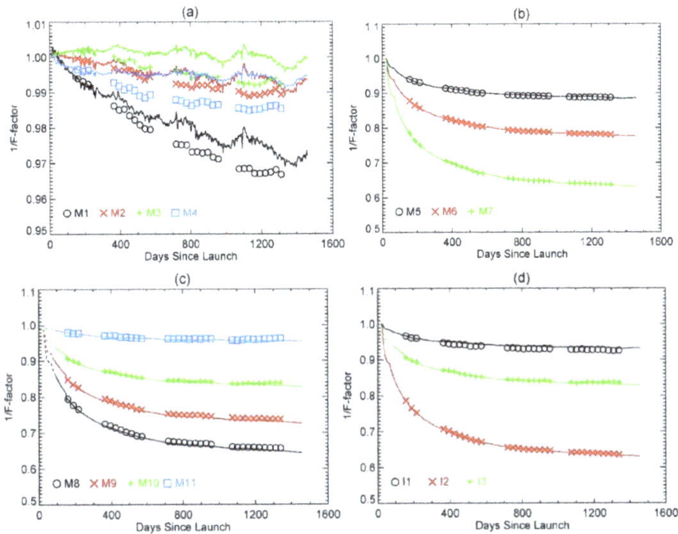

Figure 3. 1/F-factors (gains) trends for the RSB: (**a**) for bands M1–M4; (**b**) for bands M5–M7; (**c**) for bands M8–M11; and (**d**) for bands I1–I3. The solid lines represent the SD observations, and the symbols denote the lunar observations. The lunar 1/F-factors are scaled so that their values on 2 April 2012 (157 days after launch) exactly match their corresponding SD 1/F-factors.

For the bands I2 and M7, the 1/F-factors have decreased about 37% since launch. The root cause of this sudden decrease is the contamination of the RTA mirrors by tungsten oxides that become light absorbent when exposed to UV light [17,18,31,32]. The optical throughput decrease due to this mirror contamination has slowed significantly since launch. Despite the RTA throughput degradation, the projected signal-to-noise ratio for each RSB will continue to meet the design requirements at the end of seven years of operation. About 140 days after launch, the SWIR band 1/F-factors dipped out of trend. Right before the dip, the spacecraft was briefly in Sun point mode (see Table 2), resulting in the spacecraft control computer reset likely due to over-heating. The root cause for the dip remains to be understood.

The RSB detector gains (1/F-factors) derived from VIIRS lunar calibration for the RSB are also plotted in Figure 3 as symbols. For comparison purposes, the lunar calibration results are normalized to their respective SD calibration gains on 2 April 2012; after this date, all subsequent scheduled VIIRS lunar observations have been conducted with a consistent system setup, including a sector rotation and fixed high gain. There are seasonal oscillations in both the SD and lunar trending, although they are not directly related. For the lunar data, the oscillation is mainly caused by systematic effects within the ROLO model that are related to the uncompensated libration effect. Empirical fitting has been applied to the lunar F-factor to reduce the oscillation [25].

Overall, the normalized SD and lunar gains match very well in terms of their long-term trends, in line with expectations, since they both reflect the detector gain change for nearly the same optical path and AOI. However, there are noticeable band-dependent temporal drifts between the two trends of up to 0.8%. For bands M1–M4, the gain changes derived from lunar calibration are larger than from SD calibration. For bands M8 and M9, the gain changes derived from lunar calibration are smaller than from the SD. The Moon is viewed by the instrument optics directly. Therefore, it is likely that the lunar trending better reflects the gain change than the SD calibration. The drift between the two trends reflects systematic uncertainty in the SD calibration, especially in the SD degradation monitored by the SDSM.

The on-orbit RSB SNR is estimated from partial views of the SD during a solar calibration event. The SNR at the spectral radiance level L (calculated from Equation (3)) is determined by dividing the average SD response by its standard deviation. A fit of the calculated SNR to the estimated radiance,

$$\mathrm{SNR} = \mathrm{L}/\sqrt{(\mathrm{k}_0 + \mathrm{k}_1 \times \mathrm{L})} \tag{8}$$

determines the coefficients k_0 and k_1. From Equation (8), the SNR at the specified typical radiance is calculated. The SNRs at the typical radiance divided by their respective required SNRs are shown in Table 4. The optical throughput degradation described earlier has caused the SNR of affected bands to decrease by approximately the square root of the radiance; otherwise, the RSB SNR trends have been generally stable.

Table 4. VIIRS noise characterization results expressed using ratios between measured SNR (or NEdT) and specified SNR (or NEdT) for RSB (or TEB). The requirements are satisfied if the ratio is greater than one for the RSB and less than one for the TEB.

Band	Band Gain	Pre-Launch	06/02/12	10/09/12	18/03/13	16/09/13	15/03/14	18/09/14	20/03/15	19/09/15
M1	High	1.75	1.69	1.66	1.72	1.67	1.67	1.69	1.66	1.66
	Low	3.46	3.37	3.32	3.54	3.37	3.49	3.31	3.33	3.33
M2	High	1.64	1.54	1.53	1.56	1.54	1.53	1.56	1.53	1.56
	Low	2.73	2.58	2.53	2.63	2.74	2.34	2.62	2.54	2.50
M3	High	1.66	1.53	1.54	1.54	1.57	1.52	1.54	1.54	1.57
	Low	2.68	2.49	2.45	2.54	2.55	2.36	2.42	2.44	2.48
M4	High	1.61	1.50	1.49	1.50	1.53	1.47	1.52	1.49	1.53
	Low	3.06	2.72	2.78	2.77	2.98	2.70	2.77	2.76	2.90
I1	Single	2.02	1.79	1.76	1.76	1.75	1.74	1.73	1.73	1.73
M5	High	1.51	1.37	1.35	1.30	1.33	1.30	1.29	1.26	1.29
	Low	2.30	1.74	1.69	1.65	1.68	1.62	1.62	1.66	1.61
M6	Single	2.09	1.82	1.74	1.69	1.68	1.69	1.65	1.68	1.65
I2	Single	2.03	1.72	1.55	1.50	1.47	1.45	1.43	1.43	1.42
M7	High	2.42	2.08	1.87	1.80	1.77	1.75	1.73	1.72	1.72
	Low	2.49	1.83	1.61	1.53	1.50	1.50	1.45	1.48	1.47
M8	Single	3.69	3.10	2.73	2.60	2.52	2.49	2.48	2.45	2.41
M9	Single	3.05	2.79	2.54	2.47	2.43	2.39	2.39	2.38	2.33
I3	Single	28.67	25.48	24.12	23.58	23.53	23.20	23.27	23.18	22.95
M10	Single	2.09	1.74	1.68	1.60	1.63	1.62	1.60	1.60	1.59
M11	Single	2.50	2.21	2.16	2.14	2.13	2.15	2.15	2.14	2.10
I4	Single	0.16	0.16	0.16	0.16	0.16	0.16	0.16	0.16	0.16
M12	Single	0.33	0.32	0.31	0.30	0.29	0.29	0.30	0.30	0.30
M13	High	0.37	0.39	0.39	0.37	0.37	0.37	0.37	0.37	0.37
	Low	-	-	-	-	-	-	-	-	-
M14	Single	0.66	0.58	0.62	0.60	0.60	0.60	0.60	0.60	0.60
M15	Single	0.43	0.41	0.41	0.39	0.39	0.39	0.39	0.39	0.39
I5	Single	0.27	0.26	0.26	0.26	0.26	0.26	0.26	0.26	0.26
M16	Single	0.53	0.39	0.39	0.39	0.39	0.39	0.39	0.40	0.40

3.2. *TEB Performance*

Overall, the VIIRS TEB calibration has been very stable since launch. VIIRS telemetry, TEB F-factor, NEdT and WUCD results have all been monitored over the mission life, as will be discussed below. Note that all TEB trends begin when the cold FPAs reached their operational temperatures (20 January 2012).

Key telemetry points that are used in the thermal model and for monitoring the health of the instrument (such as the cold FPA temperatures) were trended both over the short term and over the entire mission. Most telemetry parameters exhibit an orbital cycle in their temperature variations. In particular, the BB temperature varies with the orbital cycle with an amplitude of about 25 mK (highest during daytime and lowest at night); Figure 4a plots the average BB temperature over approximately two orbits. This behavior has been reported earlier in the mission [19,33] and has continued to the present. The orbital cycle of the six individual thermistors shows that Thermistors 3 and 6 (the thermistors farthest from the EV port) are driving the average BB temperature variation; the likely cause of these variations is heating of one side of the BB due to Earth illumination. Note that the BB is still operating within its design requirements in terms of both stability and uniformity. Figure 4b shows the variation in the S/MWIR FPA temperature; the variation within an orbit is under 6 mK.

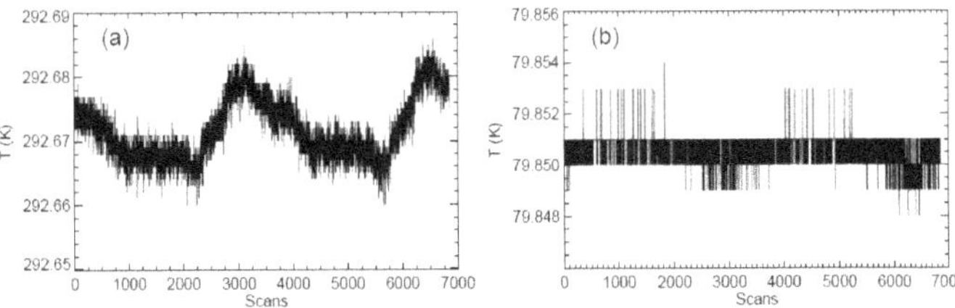

Figure 4. VIIRS BB (**a**) and S/MWIR focal plane assembly (FPA) (**b**) temperature trending over Orbits 20,326–20,327 on 30 September 2015.

The long-term trending of the principle telemetry is shown in Figure 5a–d. In Figure 5a, the average BB temperature is shown; the BB temperature has been stable to within ~25 mK for the entire mission, excluding WUCD events and instrument anomalies. The small discontinuities of 15 mK observed were due to the implementation of two slightly different BB settings (one uploaded after a WUCD event and the other after an instrument anomaly) [19]; the operational processing has been updated to avoid this feature. After this update, the BB has been stable to within ~10 mK. The other temperatures used in the thermal model were trended in Figure 5b. All temperatures show some slight yearly variation of less than 2 K; the local maximums correspond to the Earth at perihelion. The electronics and instrument temperatures shown in Figure 5c exhibit a slight increase of about 1 K over a four-year period. Figure 5d graphs the cold FPA temperatures. The LWIR FPA temperature has been very steady at 79.94 K since it reached its operational temperature; the S/MWIR FPA temperature (which is tied to the LWIR FPA temperature) has drifted upward over three years from ~79.83 K–~79.85 K.

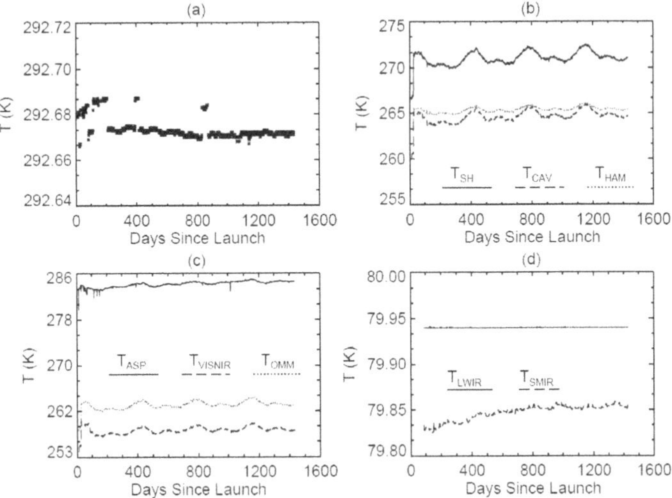

Figure 5. Daily averaged VIIRS sensor telemetry trended over the entire mission. (**a**) The trending of the BB temperature; (**b**) graph of the temperature trend for the inputs to the thermal model: HAM, scan cavity (CAV) and BB shield temperatures; (**c**) plot of the temperature trends for electronics (ASP), instrument (OMM) and VIS/NIR FPA temperature; and (**d**) the S/MWIR and LWIR FPA temperature trends.

The TEB 1/F-factors were trended over on-orbit operations using two granules collected at the solar observation of every orbit. The band average results (HAM side A) are plotted in Figure 6. The MWIR bands all showed a slight decrease in the beginning of the mission, but subsequently leveled off (a recent discontinuity in band I4 was observed to cause an additional slight decrease). Band I5 has shown a steady decrease of its 1/F-factor over time and has now changed by ~1.3% since the spacecraft anomaly (see Table 2). The other LWIR bands have shown little long-term trend, but they do show a slight annual cycle (including band I5); the days when Earth is at perihelion are visible as local minimums. These annual cycles in the 1/F-factor are related to the annual cycles in the temperatures used in the sensor thermal calibration model; however, since these temperatures are highly correlated, it is difficult to attribute the cycle to any of them. There were some discontinuities in the early mission due to instrument and spacecraft anomalies (see Table 2), which resulted in the cold FPA temperature spiking. Short-term trending was also performed, and small orbital variations were observed on the order of ~0.1% or less using a granule average; these variations are partially the result of the orbital variation in the BB temperature [19,33].

The trending for the NEdT at the BB operational temperature of ~292 K has been very stable over the entire mission; the I bands have shown NEdT between 0.14 K and 0.18 K, while the M bands have shown NEdT between 0.03 K and 0.06 K. Detector 2 in band I5, detector 16 in band M12 and detector 9 in band M16A are higher than other detectors in their respective bands and have been consistently over the entire mission.

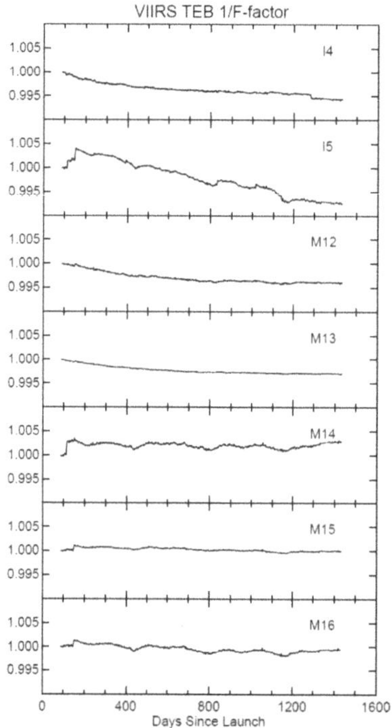

Figure 6. VIIRS TEB band average 1/F-factor trending over the entire mission, normalized to 20 January 2012.

The offset and nonlinear coefficients derived during each WUCD have generally been stable over the fifteen WUCDs so far; moreover, they are consistent with the LUTs values (derived during prelaunch testing) currently used by the SDR algorithm. There is some difference between the WU and CD results, where the WU generally provides larger offsets and nonlinear terms. The WUCD is also used to check the NEdT at the specified typical temperature (see Table 4) and to assess the response nonlinearity; both have been very stable since the mission began and are well within the sensor design specifications.

3.3. DNB Performance

The DNB LGS gains derived from SD observations have been trended over the mission. The daily average of the LGS gain for detector 8, aggregation mode 5 has steadily increased by approximately 11% since launch, as shown in Figure 7a. All detectors show a similar trend for all aggregation modes. The MGS/LGS and HGA/MGS gain ratios, derived via the method described in Section 2.3, are shown in Figure 7b,c, respectively, for the same detector and aggregation mode. The MGS/LGS ratio trend is stable with small seasonal variations throughout the mission, indicating that the MGS gain has increased by about the same amount as the LGS gain. Overall, the HGA/MGS ratio has more fluctuations in addition to seasonal variation; over the entire mission, this ratio decreases about 2% on average. The behavior of the gain ratios for other detectors and aggregation modes is similar.

The on-orbit DNB detector SNRs are computed by using BB and SV data at 32 different aggregation modes during the nighttime orbits. An example of the SNR trending is graphed in Figure 7d. The results from daily trending show that the SNR performance of all of the aggregation modes is better than the design specification (six for scan angles less than 53 degrees and five for scan angles greater than 53 degrees).

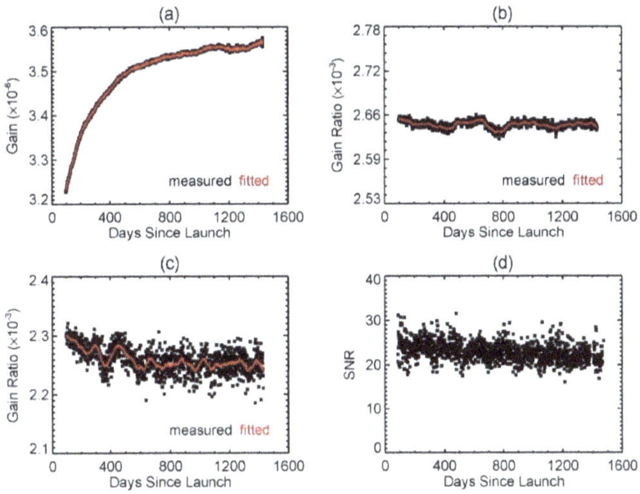

Figure 7. (**a**) DNB low gain stage (LGS) gain; (**b**) mid-gain stage (MGS)/LGS gain ratio; (**c**) high gain stage A (HGA)/MGS gain ratio and (**d**) HGA SNR trended for detector 8, aggregation mode 5, HAM side A over the entire mission. Fits of the LGS gains and gain ratios are shown by the red curves.

3.4. Changes and Improvements (Made Since Launch for Online and Offline Processing)

An improvement on the calculated SDSM screen transmission function has been made. Both the yaw maneuver and a small portion (~3 months) of regular on-orbit data were used to determine the screen transmission function at a finer scale in both the vertical and horizontal solar angles. Figure 8 shows the SDSM screen transmission functions derived from the prelaunch data (Figure 8a), the yaw maneuver data (Figure 8b) [34] and the combination of yaw maneuver and regular on-orbit data (Figure 8c). The effects of the transmission functions on the calculated H-factors are shown in Figure 9. The yearly undulations in the figure are from the solar vector issue mentioned in Section 3.1.

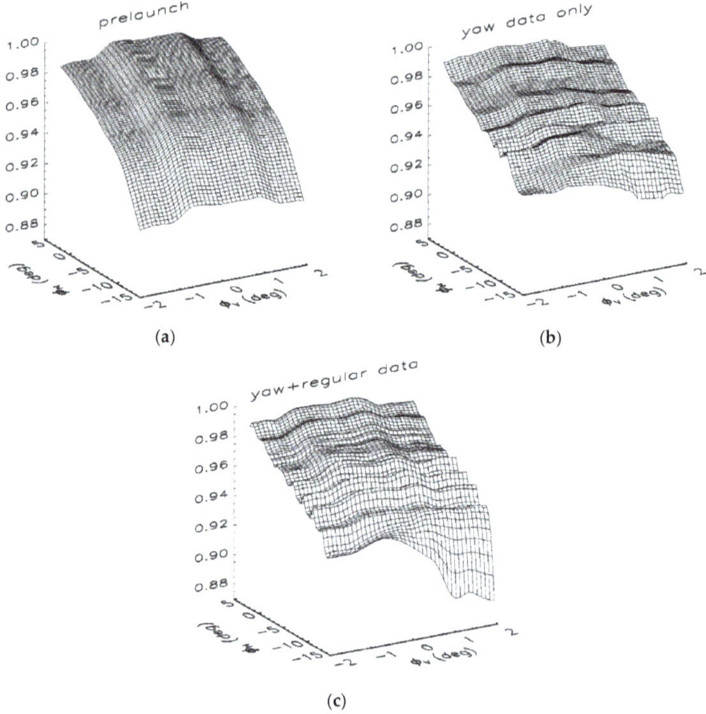

Figure 8. Plots of the relative transmittance of the SDSM screen for D8 using prelaunch data (**a**); yaw maneuver data (**b**) and yaw maneuver and a small portion of regular on-orbit data (**c**).

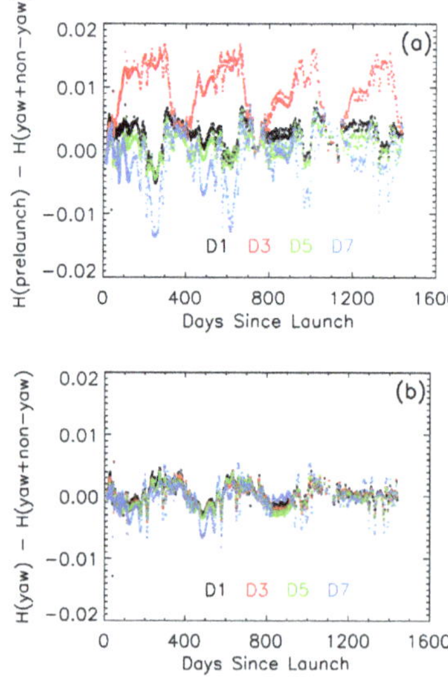

Figure 9. The difference in the H-factors using the SDSM screen transmission functions for the odd-numbered detectors derived from prelaunch data (**a**) and from the yaw maneuver data only (**b**) relative to the H-factor using the SDSM screen transmission functions derived from both yaw maneuver and regular on-orbit data.

In early 2014, with the assistance of the Aerospace Corporation, the NASA S-NPP VIIRS geo-calibration group found that there was an error in the solar vectors due to a mishandling of the coordinate systems. The uncorrected solar vectors resulted in as much as a 0.2 degree error in the solar angle and, thus, an H-factor error as large as 0.005 [35]. Since then, corrected solar vectors have been produced and used for the screen transmission function determinations and the computations of the F- and H-factors.

Since launch, the RTA has experienced optical throughput degradation. The degradation is wavelength dependent and therefore results in modulation of the RSB detector RSRs. The modeled RTA degradation was used to compute the modulated RSRs that are used in the calculation of the F-factors (see Equation (4)). Figure 10a,b shows the prelaunch RSR and the modulated RSR at 3.7 years after launch for band M1 and the DNB, respectively.

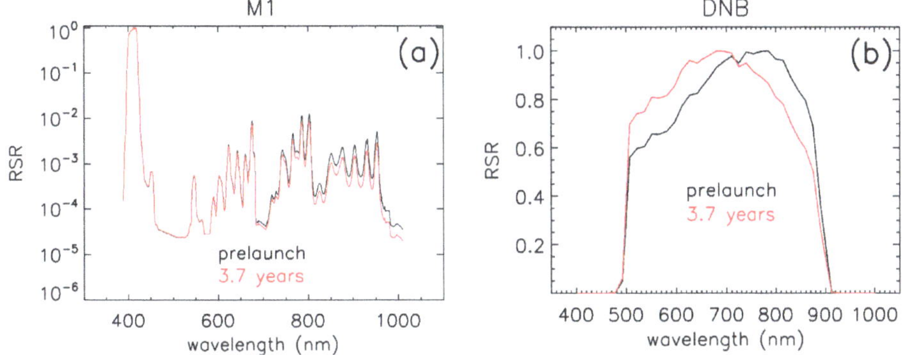

Figure 10. Comparison of the prelaunch RSR (black) and the modulated RSR at 3.7 years after launch (red) for band M1 (**a**) and the DNB (**b**).

Stray light contamination of DNB has been observed in nighttime images near the transitions to and from daytime [13]. Figure 11 displays two DNB nighttime image pairs, which are three years apart from 2012 and 2015, over the eastern U.S. Each pair shows the images before (Figure 11a,c) and after (Figure 11b,d) the stray light correction. The same algorithm used to compute the amount of stray light was used to derive the monthly correction LUTs; this has demonstrated the effectiveness of the time-dependent correction applied to the SDR product. In Figure 11d, there is still a small amount of under-corrected stray light in the twilight region, which is near the minimum of the detectable light by the DNB. This is the area for future improvement.

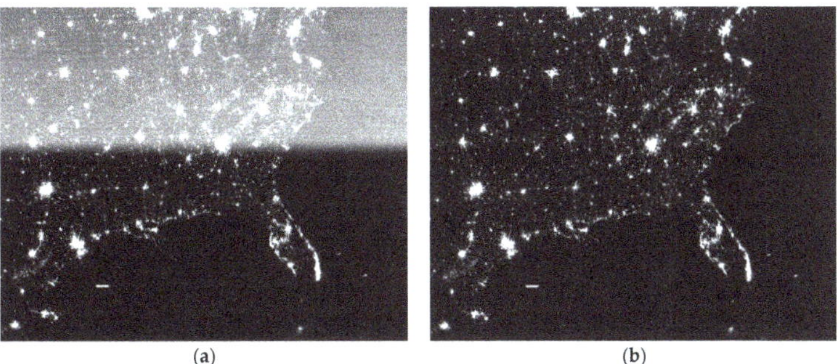

(**a**) (**b**)

Figure 11. *Cont.*

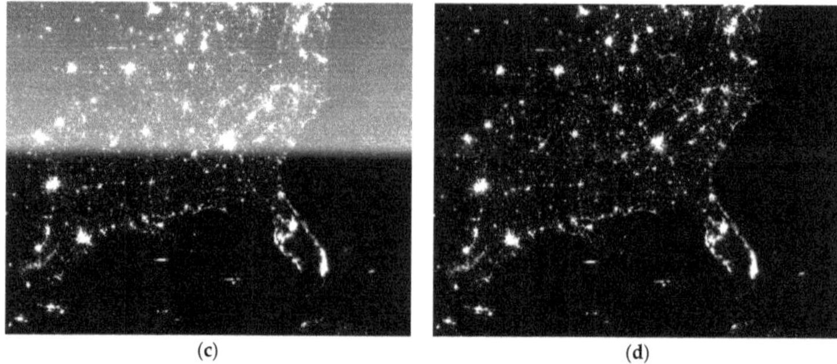

Figure 11. DNB nighttime SDR images over the eastern U.S. on (**a**) 19 June 2012 before the stray light correction; (**b**) 19 June 2012 after the correction; (**c**) 12 June 2015 before the correction and (**d**) 12 June 2015 after the correction.

The first on-orbit emissive band calibration BB WUCD was performed on 6 February 2012, which lasted for more than four days, as shown in the BB temperature profile in Figure 12a. During the WU period, two of the BB temperature set points, 297.5 K and 312.5 K, were maintained for about 16 hours each, much longer than necessary for the TEB calibration. After the first WUCD was completed, the BB WU cycle procedure was reviewed by the SDR team, which recommended that the duration of those two temperature plateaus be shortened. The second WUCD took three days to complete, as shown in Figure 12b. To track the on-orbit change in the TEB detector offset and non-linear coefficients, the calibration analysis is performed separately on the WU and CD parts of the WUCD. After further review of the procedure, an updated WU cycle was introduced for the third WUCD, as shown in Figure 12c, which further reduced the duration by another day. This new procedure only takes two days to complete the WUCD by removing a redundant part of the CD process while keeping all of the temperature set points. This new procedure has been adopted and has been used for the WUCD since September 2012.

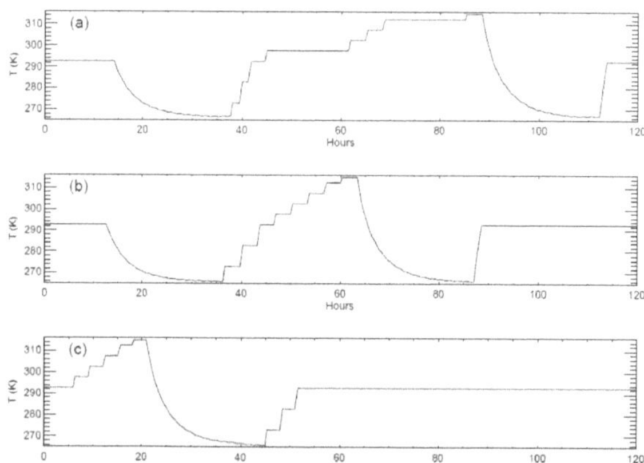

Figure 12. Blackbody temperature profiles during warm-up and cool-down (WUCD) calibration activities: (**a**) 6 February 2012 (first WUCD); (**b**) 22 May 2012 (second WUCD); and (**c**) 10 September 2012 (third WUCD).

In addition to OBC and lunar observations, a number of pseudo-invariant calibration sites and a series of simultaneous nadir overpasses have been used as part of the effort to assess the VIIRS RSB on-orbit performance. Extensive calibration and validation efforts using either similar or different approaches have also been made by other calibration groups and the user community. Results from using the Libya-4 desert and deep convective clouds show that the RSB temporal stability is within ±0.6% per year based on the desert and deep convective cloud pixels collected during the first two years of the mission [36,37]. Results from inter-comparison with Aqua MODIS based on simultaneous nadir overpasses and two invariant targets (the Libya-4 desert and Dome C) indicate that the VIIRS VIS/NIR spectral bands' calibration has been stable to within 1% over the first three years of the mission [38–40]. These studies show that after correcting for the spectral response differences, the absolute calibration differences between VIIRS and MODIS are within 2%, with the exception of the 0.865-μm band (M7), where they differ by more than 3%.

The VIIRS TEB performance has also been under evaluation using on-orbit comparisons between VIIRS and the Cross-track Infrared Sounder (CrIS) instrument on S-NPP, as well as between VIIRS and the Infrared Atmospheric Sounding Interferometer (IASI) on MetOp-A. The VIIRS-IASI and VIIRS-CrIS findings closely agree for the spectrally-matched TEB for warm scenes, but small offsets exist for cold scenes [41]. The calibration stability of VIIRS TEB was evaluated by reference to stable ground-based temperature measurements at the Dome C site [42]. No significant evidence of any temporal drift was observed due to the fact that there is a large variability in the temperature difference trends, and it is expected that at least 10 years of cumulative datasets are required in order to detect any calibration trends. The calibration differences between VIIRS and Aqua MODIS spectrally-matched TEB were also assessed by comparing the brightness temperatures retrieved from their simultaneous nadir overpasses [43]. The S-NPP CrIS data taken simultaneously with the VIIRS data are used to derive a correction for the slightly different spectral coverage of VIIRS and MODIS TEB. After the spectral correction, the agreement is well within 0.10 K over a scene temperature range of 220 K–290 K.

4. Lessons Learned and Future Improvements

For the RSB on-orbit calibration, the calculated F-factor should use the H-factor derived from the RTA SD view, or H(RTA). Determining H(RTA) from the H-factor derived from the SDSM SD view, or H(SDSM), by comparing the F-factors from the SD and lunar observations shows that H(RTA) differs from H(SDSM) at the band M1 central wavelength by about 5% at present, with H(SDSM) computed with an SDSM detector RSR averaged over the four possible filter orientations [44]. A mathematical formula was developed to relate H(RTA) to H(SDSM). Due to the angular dependence of the H-factor [45] and to avoid biasing the calculated H- and F-factors, the H- and F-factors should be calculated with a fixed angle between the incident solar radiative energy and the SD surface. The bias in the computed τ_{SAS}*$BRDF_0$ for both the RTA and SDSM views due to the angular dependence of the H-factor should be removed [45]. Additionally, to improve the precision of the computed τ_{SAS}*$BRDF_0$ for both the RTA and SDSM views, both yaw maneuver and regular on-orbit data should be used. The SD degradation at wavelengths above 1.0 μm is very small, so in this work, the H-factor was set to 1.0 for the SWIR bands [46]. However, recent analysis indicates that at present, the H-factor is about 0.996 at 1.26 μm (equivalent to band M8) [22].

The uncertainty in the lunar calibration is likely introduced by the lunar irradiance model itself. Therefore, further improvement of lunar calibration results depends greatly on the improvement of the model. The lunar gain coefficients can also be calculated for each detector and HAM side, allowing an independent validation of the detector difference and HAM side difference of the gain coefficients characterized by the SD calibration. In addition to the scheduled lunar calibrations discussed above, VIIRS can view the Moon through its SV port occasionally at various lunar phase angles without a roll maneuver or sector rotation. The lunar images captured during these unscheduled observations can be used to track the detector gain change and help to investigate how the uncertainty of the lunar model changes with lunar phase angle and libration angle.

For TEB on-orbit calibration, the sea surface temperature EDR shows a bias during the WUCD cycles. When the BB temperature is away from its nominal temperature, the sea surface temperature results show an upward trend (particularly when the BB temperature is lower than its nominal temperature). One approach to resolve this is to use a parametric model to determine if any biases exist using prelaunch data; prelaunch data had shown some BB temperature-dependent differences between the temperature retrievals and the known external blackbody temperature. In addition, overlapping orbits have shown a slight mismatch in temperature retrievals at the beginning and end of the scan. One possible cause is RVS error; for the LWIR bands, the RVS changes by between 3% and 10% from the beginning to end of the scan. Reanalysis of prelaunch data, as well as pitch maneuver data may shed some light on this phenomenon.

For DNB on-orbit calibration, the current SDR stray light correction LUT consists of a large table to cover offsets in both solar zenith angle and scan angle. The correction algorithm is being revised in order to reduce the size of the stray light correction LUT. Another possible future improvement is prediction of the stray light correction. The current method for the SDR forward process is to reuse the correction LUT from the same month of the previous year. The Earth-Sun-spacecraft relative geometry roughly repeats at an annual cycle and so does the stray light pattern. However, there are other factors that impact the amount of stray light entering the VIIRS instrument that differ from year to year. Future work to improve the algorithm and the accuracy of the stray light prediction are being investigated.

5. Conclusions

This paper has presented the status of S-NPP VIIRS on-orbit radiometric performance. All of the OBC have been consistently providing high quality calibration data needed to maintain SDR calibration for both RSB and TEB. Improved understanding and processing of the SD and SDSM data have led to better characterization of the screen transmissions and their uncertainties and, hence, better calibration of the RSB SDR products. Analyses of the SD/SDSM derived F-factors have shown that the wavelength-dependent optics degradation due to RTA mirror contamination has been gradually slowing down, which is consistent with the modeling results. As expected, the SD degradation is larger at shorter wavelengths, and the rate of degradation has gradually decreased. The TEB calibration performance continues to be extremely stable over the mission, and a minor feature observed in the sea surface temperature trending (occurring only during BB WUCD) is being addressed by the SDR team. A series of planned on-orbit operations have been successfully executed to monitor and enhance the sensor performance and characterization, including the monthly spacecraft roll maneuvers to view the Moon for verification of the SD/SDSM-based calibration for the RSB, as well as the WUCD events to characterize and calibrate the TEB. Results from ground-based vicarious calibration and validation, as well as inter-comparison with other sensors have shown good performance for the VIIRS RSB and TEB in general. Currently, there are further enhancements that the SDR team is working on that will either reinforce the quality of the current calibration or correct some of the features observed in the calibration trending. These calibration enhancements are expected to be part of the future reprocessing effort and should lead to significantly better SDR quality needed by the science and climate research communities.

Acknowledgments: The authors of this paper would like to thank the VIIRS team members from NOAA, Aerospace Corporation, NASA Science Team, University of Wisconsin and Raytheon for their valuable contributions to VIIRS testing and performance verifications. We also want to thank the MODIS Characterization and Support Team (MCST), as well as previous VIIRS Characterization Support Team (VCST) members, who provided valuable support to the VIIRS program and to the pre-launch calibration and characterization effort.

Author Contributions: Xiaxiong Xiong and Jeff McIntire compiled and edited the manuscript; Hassan Oudrari, Ning Lei, Zhipeng Wang, Kwofu Chiang, and Jeff McIntire each wrote sections of the manuscript and produced the figures; Ning Lei, Zhipeng Wang, Kwofu Chiang, Boryana Efremova, Jon Fulbright, and Jeff McIntire conducted the analysis that the results presented in the manuscript were based on; and Xiaoxiong Xiong, James Butler, and Hassan Oudrari provided oversight, coordination, and technical assistance in the performance of the analysis underlying this work.

Conflicts of Interest: The authors declare no conflict of interest.

References

1. Schueler, C.F.; Clement, J.E.; Ardanuy, P.E.; Welsch, C.; de Luccia, F.; Swenson, H. NPOESS VIIRS sensor design overview. *Proc. SPIE* **2002**, *4483*. [CrossRef]
2. Lee, T.F.; Miller, S.D.; Schueler, C.; Miller, S. NASA MODIS previews NPOESS VIIRS capabilities. *Weather Forecast.* **2006**, *21*, 649–655. [CrossRef]
3. Ardanuy, P.E.; Schueler, C.F.; Miller, S.W.; Kealy, P.M.; Cota, S.A.; Haas, M.; Welsch, C. NPOESS VIIRS design process. *Proc. SPIE* **2002**, *4483*. [CrossRef]
4. Jackson, J.M.; Liu, H.; Laszlo, I.; Kondragunta, S.; Remer, L.A.; Huang, J.; Huang, H.-C. Suomi-NPP VIIRS aerosol algorithms and data products. *J. Geophys. Res. Atmos.* **2013**, *118*, 12–673. [CrossRef]
5. Justice, C.O.; Román, M.O.; Csiszar, I.; Vermote, E.F.; Wolfe, R.E.; Hook, S.J.; Friedl, M.; Wang, Z.; Schaaf, C.B.; Miura, T.; *et al.* Land and cryosphere products from Suomi NPP VIIRS: Overview and status. *J. Geophys. Res. Atmos.* **2013**, *118*, 9753–9765. [CrossRef] [PubMed]
6. Wang, M.; Liu, X.; Tan, L.; Jiang, L.; Son, S.; Shi, W.; Rausch, K.; Voss, K. Impacts of VIIRS SDR performance on ocean color products. *J. Geophys. Res. Atmos.* **2013**, *118*, 10–347. [CrossRef]
7. Cao, C.; Xiong, X.; Wolfe, R.; De Luccia, F.; Liu, Q.; Blonski, S.; Lin, G.; Nishihama, M.; Pogorzala, D.; Oudrari, H.; *et al. Visible Infrared Imaging Radiometer Suite (VIIRS) Sensor Data Record (SDR) User's Guide;* NOAA Technical Report NESDIS: College Park, MD, USA, 2013.
8. *Joint Polar Satellite System (JPSS) VIIRS Radiometric Calibration Algorithm Theoretical Basis Document (ATBD);* NASA Goddard Space Flight Center: Greenbelt, MD, USA, 2013.
9. Xiong, X.; Butler, J.; Chiang, K.; Efremova, B.; Fulbright, J.; Lei, N.; McIntire, J.; Oudrari, H.; Sun, J.; Wang, Z.; *et al.* VIIRS on-orbit calibration methodology and performance. *J. Geophys. Res. Atmos.* **2014**, *119*, 5065–5078. [CrossRef]
10. Eplee, R.E.; Turpie, K.R.; Meister, G.; Patt, F.S.; Franz, B.A. Updates to the on-orbit calibration of SNPP VIIRS for ocean color applications. *Proc. SPIE* **2015**, *9607*. [CrossRef]
11. Wolfe, R.E.; Lin, G.; Nishihama, M.; Tewari, K.P.; Tilton, J.C.; Isaacman, A.R. Suomi NPP prelaunch and on-orbit geometric calibration and characterization. *J. Geophys. Res. Atmos.* **2013**, *118*, 11–508. [CrossRef]
12. Moy, G.; Rausch, K.; Haas, E.; Wilkinson, T.; Cardema, J.; de Luccia, F. Mission history of reflective solar band calibration performance of VIIRS. *Proc. SPIE* **2015**, *9607*. [CrossRef]
13. Lee, S.; Chiang, K.; Xiong, X.; Sun, C.; Anderson, S. The S-NPP VIIRS day-night band on-orbit calibration/characterization and current state of SDR products. *Remote Sens.* **2014**, *6*, 12427–12446. [CrossRef]
14. Xiong, X.; Sun, J.; Fulbright, J.; Wang, Z. Lunar Calibration and Performance for S-NPP VIIRS Reflective Solar Bands. *IEEE Tran. Geosci. Remote Sens.* **2015**. [CrossRef]
15. Uprety, S.; Cao, C.; Xiong, X.; Blonski, S.; Wu, A.; Shao, X. Radiometric intercomparison between Suomi-NPP VIIRS and Aqua MODIS reflective solar bands using simultaneous nadir overpass in the low latitudes. *J. Atmos. Ocean. Technol.* **2013**, *30*, 2720–2736. [CrossRef]
16. Xiong, X.; Fulbright, J.; Angal, A.; Wang, Z.; Geng, X.; Butler, J. Assessment of MODIS and VIIRS solar diffuser on-orbit degradation. *Proc. SPIE* **2015**, *9607*. [CrossRef]
17. De Luccia, F.; Moyer, D.; Johnson, E.; Rausch, K.; Lei, N.; Chiang, K.; Xiong, X.; Fulbright, J.; Haas, E.; Iona, G. Discovery and characterization of on-orbit degradation of the visible infrared imaging radiometer suite (VIIRS) rotating telescope assembly (RTA). *Proc. SPIE* **2012**, *8510*, 85101A.
18. Lei, N.; Xiong, X.; Guenther, B. Modeling the detector radiometric gains of the Suomi NPP VIIRS reflective solar bands. *IEEE Trans. Geosci. Remote Sens.* **2015**, *53*, 1565–1573. [CrossRef]
19. Efremova, B.; McIntire, J.; Moyer, D.; Wu, A.; Xiong, X. S-NPP VIIRS thermal emissive bands on-orbit calibration and performance. *J. Geophys. Res.-Atmos.* **2014**, *119*. [CrossRef]
20. Standardized Calibration Parameters. Available online: https://cs.star.nesdis.noaa.gov/NCC/Standardized CalibrationParameters (accessed on 21 January 2016).
21. Fulbright, J.; Lei, N.; McIntire, J.; Efremova, B.; Chen, X.; Xiong, X. Improving the characterization and performance of the Suomi-NPP VIIRS solar diffuser stability monitor. *Proc. SPIE* **2013**, *8866*, 88661J.
22. Lei, N.; Xiong, X. Determination of the SNPP VIIRS solar diffuser BRDF degradation factor over wavelengths longer than 1 m. *Proc. SPIE* **2015**, *9607*, 96071W.
23. Kieffer, H.H.; Stone, T.C. The spectral irradiance of the Moon. *Astronom. J.* **2005**, *129*, 2887–2889. [CrossRef]
24. Fulbright, J.; Wang, Z.; Xiong, X. Suomi-NPP VIIRS lunar radiometric calibration observations. *Proc. SPIE* **2014**, *9218*. [CrossRef]

25. Eplee, R.E.; Turpie, K.R.; Meister, G.; Patt, F.S.; Franz, B.A.; Bailey, S.W. On-orbit calibration of the Suomi national polar-orbiting partnership visible infrared imaging radiometer suite for ocean color applications. *Appl. Opt.* **2015**, *54*, 1984–2006. [CrossRef] [PubMed]
26. Sun, J.; Xiong, X.; Barnes, W.L.; Guenther, B. MODIS reflective solar bands on-orbit lunar calibration. *IEEE Trans. Geosci. Remote Sens.* **2007**, *45*, 2383–2393. [CrossRef]
27. Wang, Z.; Fulbright, J.; Xiong, X. Update on the performance of Suomi-NPP VIIRS lunar calibration. *Proc. SPIE* **2015**, *9607*. [CrossRef]
28. Lei, N.; Chen, X.; Xiong, X. Determination of the SNPP VIIRS SDSM screen relative transmittance from both yaw maneuver and regular on-orbit data. *IEEE Trans. Geosci. Remote Sens.* **2015**. [CrossRef]
29. Lei, N.; Xiong, X. Estimation of the accuracy of the SNPP VIIRS SD BRDF degradation factor determined by the solar diffuser stability monitor. *Proc. SPIE* **2015**, *9607*, 96071V.
30. Lei, N.; Wang, Z.; Fulbright, J.; Xiong, X. Effect of the SDSM detector relative spectral response in determining the degradation coefficient of the SNPP VIIRS solar diffuser reflectance. *Proc. SPIE* **2013**, *8866*. [CrossRef]
31. Iona, G.; Butler, J.; Guenther, B.; Graziani, L.; Johnson, E.; Kennedy, B.; Kent, C.; Lambeck, R.; Waluschka, E.; Xiong, X. VIIRS on-orbit optical anomaly: investigation, analysis, root cause determination and lessons learned. *Proc. SPIE* **2012**, *8510*, 85101C.
32. Barrie, J.D.; Fuqua, P.D.; Meshishnek, M.J.; Ciofalo, M.R.; Chu, C.T.; Chaney, J.A.; Moision, R.M.; Graziani, L. Root cause determination of on-orbit degradation of the VIIRS rotating telescope assembly. *Proc. SPIE* **2012**, *8510*, 8510IB.
33. Moyer, D.; McIntire, J.; de Luccia, F.; Efremova, B.; Chiang, K.; Xiong, X. VIIRS thermal emissive bands calibration algorithm and on-orbit performance. *Proc. SPIE* **2012**, *8510*, 85101D.
34. McIntire, J.; Moyer, D.; Efremova, B.; Oudrari, H.; Xiong, X. On-orbit Characterization of S-NPP VIIRS Transmission Functions. *IEEE Trans. Geosci. Remote Sens.* **2015**, *53*, 2354–2363. [CrossRef]
35. Fulbright, J.; Anderson, S.; Lei, N.; Efremova, B.; Wang, Z.; McIntire, J.; Chiang, K.; Xiong, X. The solar vector error within the SNPP Common GEO code, the correction, and the effects on the VIIRS SDR RSB calibration. *Proc. SPIE* **2014**, *9264*, 92641T.
36. Bhatt, R.; Doelling, D.R.; Wu, A.; Xiong, X.; Scarino, B.R.; Haney, C.O.; Gopalan, A. Initial stability assessment of S-NPP VIIRS reflective solar band calibration using invariant desert and deep convective cloud targets. *Remote Sens.* **2014**, *6*, 2809–2826. [CrossRef]
37. Wang, W.; Cao, C. Assessing the VIIRS RSB calibration stability using deep convective clouds. *Proc. SPIE* **2014**, *9264*. [CrossRef]
38. Uprety, S.; Cao, C. Suomi NPP VIIRS reflective solar band on-orbit radiometric stability and accuracy assessment using desert and Antarctica Dome C sites. *Remote Sens. Environ.* **2015**, *166*, 106–115. [CrossRef]
39. Wu, A.; Xiong, X.; Cao, C.; Sun, C. Monitoring NPP VIIRS on-orbit radiometric performance from TOA reflectance time series. *Proc. SPIE* **2013**, *8866*, 88660Q.
40. Wu, A.; Xiong, X.; Cao, C.; Chiang, K. Assessment of SNPP VIIRS VIS/NIR radiometric calibration stability using Aqua MODIS and invariant surface targets. *IEEE Trans. Geosci. Remote Sens.* **2015**. [CrossRef]
41. Moeller, C.; Tobin, D.; Quinn, G. S-NPP VIIRS thermal band spectral radiance performance through 18 months of operation on-orbit. *Proc. SPIE* **2013**, *8866*, 88661N.
42. Madhavan, S.; Wu, A.; Brinkmann, J.; Wenny, B.; Xiong, X. Evaluation of VIIRS and MODIS thermal emissive band calibration consistency using Dome C. *Proc. SPIE* **2015**, *9639*. [CrossRef]
43. Efremova, B.; Wu, A.; Xiong, X. Relative spectral response corrected calibration inter-comparison of S-NPP VIIRS and Aqua MODIS thermal emissive bands. *Proc. SPIE* **2014**, *9218*, 92180G.
44. Lei, N.; Xiong, X. Impact of the angular dependence of the SNPP VIIRS solar diffuser BRDF degradation factor on the radiometric calibration of the reflective solar bands. *Proc. SPIE* **2015**, *9607*, 96071Y.
45. Lei, N.; Xiong, X. Examination of the angular dependence of the SNPP VIIRS Solar diffuser bidirectional reflectance distribution function degradation factor. *Proc. SPIE* **2014**, *9218*, 92181N.
46. Fulbright, J.; Lei, N.; Efremova, B.; Xiong, X. Suomi-NPP VIIRS Solar diffuser stability monitor performance. *IEEE Trans. Geosci. Remote Sens.* **2015**. [CrossRef]

Review

VIIRS Reflective Solar Bands Calibration Progress and Its Impact on Ocean Color Products

Junqiang Sun [1,2,*] and Menghua Wang [1]

1 NOAA National Environmental Satellite, Data, and Information Service, Center for Satellite Applications and Research, E/RA3, 5830 University Research Ct., College Park, MD 20740, USA; menghua.wang@noaa.gov

2 Global Science and Technology, 7855 Walker Drive, Suite 200, Greenbelt, MD 20770, USA

* Correspondence: junqiang.sun@noaa.gov; Tel.: +1-301-863-3338; Fax: +1-301-863-3301

Academic Editors: Changyong Cao, Xiaofeng Li and Prasad S. Thenkabail

Received: 28 November 2015; Accepted: 23 February 2016; Published: 27 February 2016

Abstract: The radiometric calibration for the reflective solar bands (RSB) of the Visible Infrared Imaging Radiometer Suite (VIIRS) on board the Suomi National Polar-orbiting Partnership (SNPP) platform has reached a mature stage after four years since its launch. The characterization of the vignetting effect of the attenuation screens, the bidirectional reflectance factor of the solar diffuser, the degradation performance of the solar diffuser, and the calibration coefficient of the RSB have all been made robust. Additional investigations into the time-dependent out-of-band relative spectral response and the solar diffuser degradation non-uniformity effect have led to newer insights. In particular, it has been demonstrated that the solar diffuser (SD) degradation non-uniformity effect induces long-term bias in the SD-calibration result. A mitigation approach, the so-called Hybrid Method, incorporating lunar-based calibration results, successfully restores the calibration to achieve ~0.2% level accuracy. The successfully calibrated RSB data record significantly impacts the ocean color products, whose stringent requirements are especially sensitive to calibration accuracy, and helps the ocean color products to reach maturity.

Keywords: VIIRS; reflective solar bands; radiance; solar diffuser; moon; calibration; ocean color remote sensing

---

1. Introduction

The Visible Infrared Imaging Radiometer Suite (VIIRS) is one of five instruments housed by the Suomi National Polar-orbiting Partnership (SNPP) satellite launched on 28 October 2011 [1,2]. It is a scanning radiometer, which collects visible and infrared imagery and radiometric measurements of the land, atmosphere, cryosphere, and oceans. With an ascending sun-synchronous orbit passing the equator around 1:30 p.m. local time, VIIRS covers almost the entire Earth surface every day. VIIRS data are used to measure cloud and aerosol properties, ocean color, sea and land surface temperature, ice motion and temperature, fires, and Earth's albedo [1]. They are also used to improve our understanding of global climate change.

VIIRS has 22 spectral bands covering a spectral range from 0.410–12.013 μm, which include 14 reflective solar bands (RSB), 7 thermal emissive bands (TEB), and a panchromatic day/night band (DNB). The VIIRS RSB are calibrated on-orbit with an on-board solar diffuser (SD) [3–5], whose performance and degradation is itself tracked by a solar diffuser stability monitor (SDSM) [6–8]. The VIIRS RSB on-orbit changes are also monitored with scheduled monthly lunar observations through the instrument's space view (SV) port [9–12], which is also used to provide the instrument's dark scene response. Figure 1 is a schematic diagram for VIIRS and its on-board calibrators [13]. VIIRS views the

SV, earth view (EV), blackbody (BB), and SD, respectively, via a rotating telescope assembly (RTA) and half-angle mirror (HAM). Figure 2 shows the scan angles as well as the Angle of Incidence (AOI) for each view [13,14].

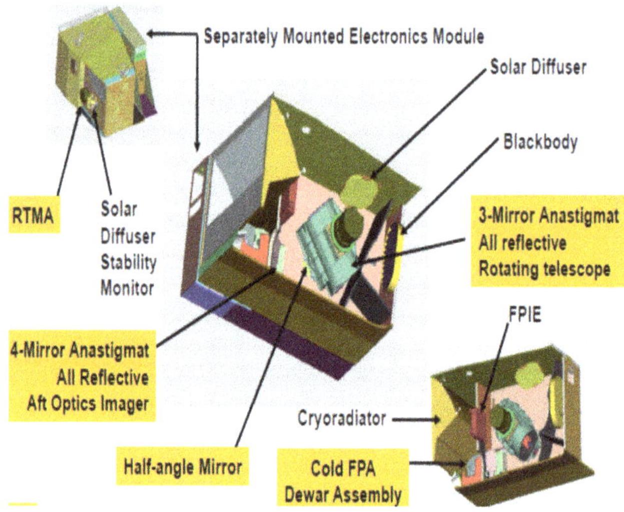

Figure 1. VIIRS instrument and its on-board calibrators.

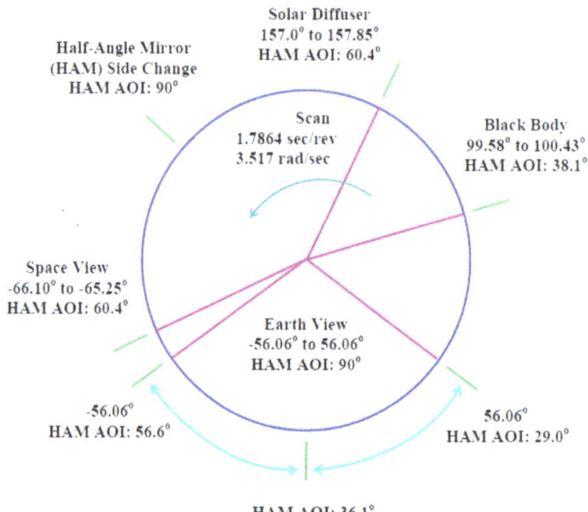

Figure 2. Scan angles of VIIRS view sectors and their corresponding AOIs on the HAM.

The accuracy of the RSB calibration using the SD strongly depends on that of the bidirectional reflectance factor (BRF) of the SD and that of the vignetting function (VF) of the screen placed in the front of the SD port, through which the SD is illuminated by the Sun [15,16]. The BRF changes on-orbit with time due to the degradation of the SD caused by solar exposure. Any inaccuracy in the BRF or the VF brings out an annual oscillation as well as a long-term drift in the derived calibration coefficients, also called F-factors in the VIIRS community [3–5]. We have carefully derived the BRF and the VF [16], improved the accuracy of the SDSM calibration [8], and refined other related algorithms [5]. With these improvements, both short-term and long-term stability and high accuracy of the F-factors are obtained [5]. However, it is found that the SD degrades non-uniformly with respect to both incident and outgoing directions, thus invalidates the key assumption in the SD/SDSM calibration methodology that the SD degradation in the outgoing direction towards the SDSM can be used interchangeably with the result for the outgoing direction towards the RTA. The actual discrepancy results in a long-term bias in the calibration coefficients derived from the SD calibration, especially for short wavelength bands [5,8].

The lunar calibration does not have the long-term bias issue since the reflectance of the lunar surface is very stable in the VISible (VIS) and Near-InfraRed (NIR) spectral range [17]. However, the accuracy of the calibration strongly depends on that of view geometric effect correction [9–12,18]. Any error in the correction induces a seasonal oscillation in the calibration coefficients derived from the lunar calibration [10–12]. In addition, lunar calibration can only be implemented about nine months each year [9]. With careful correction of the geometric effect, stable and clean RSB calibration coefficients can be derived from the lunar calibration [12,16]. In general, the derived lunar F-factors are consistent with those derived from the SD/SDSM observations [10–12,16]. However, there are observable differences between the lunar and the SD F-factors [10–12,16]. For VIIRS, the AOI of the SD exactly coincides with that of the SV, as illustrated in Figure 2. Thus, the SD/SDSM calibration and the lunar calibration should in principle provide the same on-orbit changes over time. Considering the potential bias of the SD F-factors due to the temporal non-uniformity of the SD degradation, which is a primary reason for the differences of the two sets of F-factors, and the stability of the lunar surface's reflectance, the lunar calibration should provide a more reliable long-term calibration baseline [16]. Nevertheless, the monthly lunar observations are infrequent and unavailable for several months of the year due to a spacecraft roll angle safety constraint preventing a lunar view through the space view port [9–11]. A hybrid approach has been developed to appropriately combine SD-based and lunar-based calibration coefficients to generate a set of *hybrid calibration coefficients*, leading to overall stable short- and long-term calibrated VIIRS RSB Sensor Data Records (SDR) [14,19]. This is especially important for VIIRS ocean color Environmental Data Records (EDR).

The ocean color EDR products [20–24] are highly sensitive to the accuracy of the RSB calibration and can also be used to check the RSB calibration accuracy itself [25]. Significant long-term drifts and unexpected features have indeed been discovered in the VIIRS normalized water-leaving radiances from the short wavelength bands and the VIIRS chlorophyll-a calculated using the NOAA Interface Data Processing Segment (IDPS) SDR [25,26], which are the current official SDR products produced using the SD F-factors derived by the VIIRS SDR team [4]. The application of the reprocessed SDR with our best and latest improved SD F-factors does significantly improve the quality of the ocean color EDR products due to removal of the seasonal oscillations and other errors in the SD calibration coefficients [26], but still fails to address the long-term drifts present in the products. The hybrid F-factors finally remove the long-term drifts observed in the ocean color EDR and meet the ocean color EDR's stringent requirement for the RSB calibration [14]. Preliminary results of the performance of the ocean color EDR with the hybrid F-factors-implemented SDR have been reported in our previous works [14,26].

In this paper, we review the algorithms of the VIIRS SDR from Raw Data Records (RDR), the RSB on-orbit calibration algorithms and performance, the ocean color EDR improvements, and discuss the challenging issues related to the VIIRS RSB calibration. In Section 2, the SDR algorithms are reviewed and related issues are discussed. In Section 3, the algorithms for SD calibration, lunar calibration, and hybrid calibration coefficients are reviewed, calibration results are displayed, and challenge issues are discussed. In Section 4, ocean color EDR processed using the SDR generated with IDPS and hybrid F-factors, respectively, are demonstrated and compared. They are also compared with the *in situ* measurements. It is shown that the ocean color EDR processed using the SDR generated with the hybrid F-factors match very well. This work completes the standard core of the methodology and analyses, and enables VIIRS to further many science products to reach higher data quality, especially for ocean color products. Section 5 summaries and concludes the work.

2. VIIRS Sensor Data Records

The 14 VIIRS RSB cover a spectral range from 410–2250 nm. Table 1 lists the center wavelengths of the RSB and their specification. The background response for a detector of an RSB is provided by the detector reading of the SV, as mentioned previously. In the VIIRS RSB radiance retrieval methodology, a quadratic approximation is applied to establish the relationship between the radiance of the incident sunlight at the center wavelength of a RSB detector and the background-subtracted digital number *dn* of the detector [13]. This is different from the methodology for the Moderate-resolution Imaging Spectroradiometer (MODIS), where a simple linear approximation is used to establish the relationship [27,28].

Table 1. VIIRS RSB and SDSM Specification.

SDSM		VIIRS RSB			
Detector	CW (nm)	Band	CW (nm)	BW(nm)	Gain
D1	412	M1	410	20	DG
D2	450	M2	443	18	DG
D3	488	M3	486	20	DG
D4	555	M4	551	20	DG
NA	NA	I1	640	80	SG
D5	672	M5	671	20	DG
D6	746	M6	745	15	SG
D7	865	M7	862	39	DG
D7	865	I2	862	39	SG
D8	935	NA	NA	NA	NA
NA	NA	M8	1238	20	SG
NA	NA	M9	1378	15	SG
NA	NA	M10	1610	60	SG
NA	NA	I3	1610	60	SG
NA	NA	M11	2250	50	SG

CW: Center Wavelength; BW: Bandwidth; DG: Dual Gain; SG: Single Gain.

The top-of-atmosphere (TOA) radiance, called SDR in the VIIRS community, is the baseline for all science products. A remote sensor can only provide its responses to a target at selected wavelengths. To convert the instrument responses to the corresponding TOA radiance, a relationship between them has to be established first. For SNPP VIIRS RSB, a quadratic form is applied to relate the at-aperture radiance and instrument background-subtracted response, as mentioned above [13,14], that is:

$$L_B(P,S,D,t) = \frac{F(B,D,M,G,t)\sum_{j=0}^{2} c_j\,(B,D,M,G)\,dn_B^j(P,S,D)}{RVS\,(\vartheta,B)} \tag{1}$$

where B, D, P, S, M, and t, denote band, detector, pixel, scan, and HAM side, and time, respectively. $L_B(P,S,D,t)$ is the radiance at wavelength of band B, $RVS(\vartheta,B)$ is the response-versus-scan angle (RVS) at the angle of incidence (AOI), ϑ, of the HAM. $c_0(B,D,M,G)$, $c_1(B,D,M,G)$, and $c_2(B,D,M,G)$ are the temperature-effect-corrected prelaunch measured offset, linear, and nonlinear coefficients of the quadratic form [29]. $dn_B(P,S,D)$ is the background-subtracted instrument response, and $F(B,D,M,G,t)$ is the scale factor, called F-factor, as mentioned previously, for the coefficients of the quadratic form at time t relative to those of the quadratic form of the prelaunch [13]. In Equation (1), it is assumed that the three coefficients of the quadratic form change at the same rate for all time. This assumption enables a robust ratio approach to be developed for the SDR reprocessing in cases when only the F-factors are updated [30]. From Equation (1), it is clearly seen that the offset should be zero since for the dark scene on both sides it should be zero. The prelaunch-measured calibration coefficients were originally fitted with prelaunch measurements without forcing the offset to be zero, but they were refitted later with the offset set to zero after the instrument had been on orbit for more than a year. Even though the non-zero nonlinear term is considered in VIIRS RSB SDR algorithm, the impact of the non-linear effect is small. In fact, it is smaller than 1% for all RSB as long as the radiance does not exceed the specified maximum radiance, L_{max}.

F(B,D,M,G,t) is the time-dependent F-factor and needs to be updated constantly. The RVS was measured prelaunch but may change on-orbit due to the degradation of the scan mirror as demonstrated by both Aqua and Terra MODIS scan mirrors [31]. Since the HAM is inside the instrument, it degrades much more slowly than the scan mirrors of the two MODIS instruments, where the scan mirror directly faces the Earth's surface. The slower degradation of the VIIRS HAM can be seen from the much slower degradation of the VIIRS short wavelength VIS bands [5]. Nevertheless, it degrades with time and its degradation is wavelength-dependent. Its degradation should also be AOI-dependent, as demonstrated also by the time-dependent RVS of MODIS [31]. The degradation of the HAM at the AOI of the SV is tracked by the F-factors and it is one of the contributions to the variations of the F-factors with time [5]. However, the HAM degradation attributing to AOI difference, or time-dependent RVS, is not considered in this paper and it is beyond the need of this work due to its small effect.

The current official forward SDR products are produced by the NOAA IDPS using the F-factors derived from the SD/SDSM calibration [4]. The VIIRS Ocean Color Team at the NOAA Center for Satellite Applications and Research (STAR) has reprocessed the SDR with the hybrid F-factors using the Application Development Library (ADL) for the early mission. For the later mission, the ratio approach [30] was used, since new prelaunch calibration coefficients, $c_0(B,D,M,G)$, $c_1(B,D,M,G)$, and $c_2(B,D,M,G)$ have already been applied in IDPS SDR. The reprocess of SDR using the ADL is a time-consuming process. The ratio approach dramatically improves the efficiency by two orders of magnitude [30]. The VIIRS Ocean Color Team has started, since September 2014, to produce near real-time SDR on a daily basis using our hybrid F-factors for scientific quality of the ocean color EDR products.

3. RSB Calibration Methodology and Performance

In this section, we review the SD and lunar calibration algorithms and the RSB on-orbit performance. To distinguish the F-factor derived from the SD/SDSM calibration and the F-factors obtained from the lunar calibration, they will be called as SD F-factors and lunar F-factors in this and later sections, respectively. We will discuss the advantages and disadvantages of the approaches. We will also review the hybrid approach methodology and show the hybrid calibration coefficients.

3.1. SD and SDSM Calibration

The SNPP VIIRS SD is a flat panel which is made of Spectralon® and is installed inside the VIIRS instrument. Figure 3 is a schematic diagram for the VIIRS SD/SDSM calibration. The BRF of the SD was measured prelaunch and validated on-orbit using the measurements with yaw maneuvers [9,15,16,32]. The on-orbit degradation of the SD BRF is tracked by an SDSM [6–8]. To prevent saturation of the RSB, a screen is placed in the front of the SD port to reduce the intensity of the sunlight. The transmittance of the screen, described by a vignetting function (VF), was also characterized prelaunch and validated on-orbit using the yaw measurements [15,16,33]. For on-orbit calibration, the SD is fully illuminated only in a short window of time in each orbit when the satellite crosses the terminator from the nightside to the dayside of the Earth and only the measurements in this short full-illumination interval can be used to derive the RSB calibration coefficients, or F-factors [3–5].

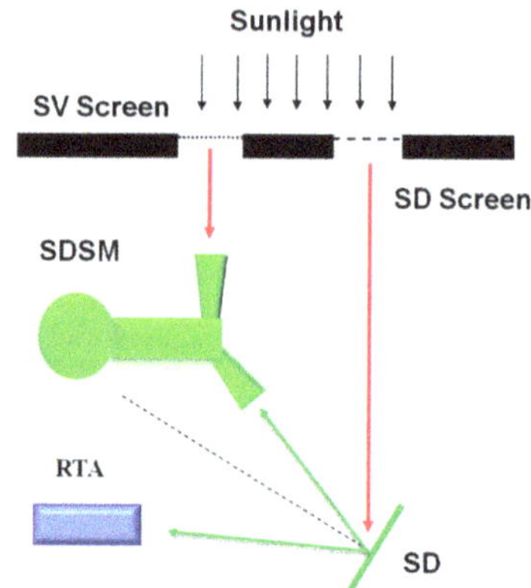

Figure 3. Schematic diagram for SD/SDSM calibration.

The RSB calibration coefficients, or the F-factors, using the SD (solar diffuser) observation can be calculated by [5]:

$$F(B,D,M,G,t) = \left\langle \frac{RVS\,(\vartheta_{SD},B)\int RSR_B(\lambda,t)\cdot L_{SD}(\lambda)d\lambda}{\left[\sum_{j=0}^{2} c_j\,(B,D,M,G)\,dn_B^j(P,S,D)\right]\int RSR_B(\lambda,t)d\lambda} \right\rangle_{P,S} \quad (2)$$

where θ_{SD} is the AOI of the SD, RSR_B is relative spectral response of band B, and $< \ldots >_{P,S}$ indicates the average over the selected scans, which will be discussed in more detail later, and pixels in each scan. In Equation (2), $L_{SD}(\lambda)$ is the radiance of the sunlight reflected by the SD and can be expressed as:

$$L_{SD}(\lambda) = I_{Sun}(\lambda)\cdot BRF_{RTA}(\lambda)\cdot \tau_{SDS}\cdot \cos(\theta_{SD})\cdot h(\lambda)/d_{VS}^2 \quad (3)$$

where I_{Sun} is solar irradiance, BRF_{RTA} is the BRF for the outgoing direction of the RTA through which the RSB views the EV, SV, SD, and other on-board calibrators. τ_{SDS} is the VF of the SD screen, $h(\lambda)$ is the on-orbit SD degradation, called H-factors [6–8], tracked by the SDSM and is normalized to 28 October 2011, and d_{VS} is the VIIRS-Sun distance.

Both the BRF and the VF vary with the incident direction of the sunlight, which can be described by two independent solar angles. The solar declination and azimuth angles in the instrument coordinate system are usually selected to describe the incident direction. As mentioned previously, the BRF and VF were measured prelaunch [32,33] and their incident direction dependence was validated on-orbit using the on-orbit measurements from the planned yaw maneuvers [9,15,16]. Errors in the obtained BRF or VF induce seasonal oscillations in the derived F-factors, as seen in early literatures [3,4]. We have carefully re-derived BRF and VF from the yaw measurements and demonstrated that the re-derived BRF and VF significantly reduce the seasonal oscillations in the derived F-factors [5,16]. The declination angle changes 360° every orbit and the SD is fully illuminated in a small range of the declination, which corresponds to a short time window just before the instrument passes the south pole from nightside to dayside of the Earth, as mentioned previously [5]. Since the full-illumination interval changes with the seasons, a smaller window, called "sweet spot", is selected to guarantee that the SD is fully illuminated in the selected window in all seasons, and that only the data in the window are used to derive the F-factors. An improperly selected "sweet spot" may induce seasonal oscillation as well. We have carefully selected the "sweet spot" to be in the declination angle range from 13°–17° for VIIRS SD calibration [5,16]. Our work uses the much-improved BRF and VF as well as the carefully selected "sweet spot".

Figure 4 shows the actual solar azimuth angle for the full illumination time periods in the last four years since the instrument launch. The vertical line around day 110 corresponds to 15 and 16 February 2012, when the yaw maneuvers were implemented. The solar azimuth angle shows a time-varying annual pattern, as expected. It is also noticeable that the solar azimuth angle has shown a long-term increase, which implies that the SNPP satellite orbit is drifting. From the plot, it is seen that the solar azimuth angle is still within the range covered by the yaw measurement, but it can be a concern that the solar azimuth angle may drift beyond the selected range, after which the BRF and the VF may have larger errors in the range where they are not as well characterized.

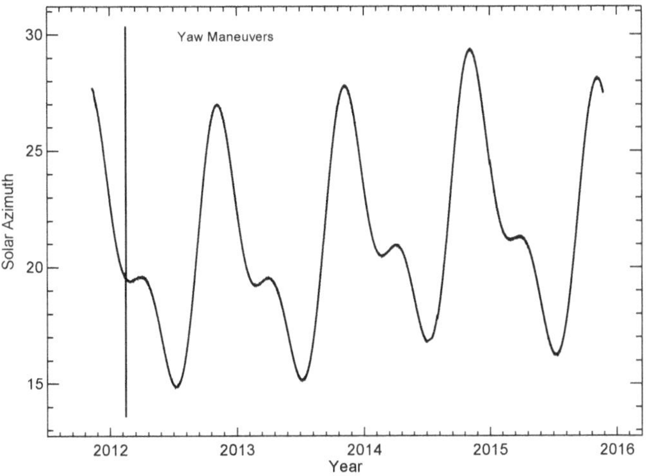

Figure 4. SNPP VIIRS solar-azimuth angle. The vertical line indicates the yaw maneuvers.

The VIIRS SD on-orbit degradation, $h(\lambda)$, is tracked by the on-board SDSM calibration, as mentioned previously [6–8]. VIIRS SDSM calibration was performed for every orbit in the first few months on orbit, and then the operational rate was reduced to once *per* day, and further reduced to once every two days after 16 May 2014. So far, more than 1730 SDSM measurements have been performed for VIIRS. The SDSM has eight detectors with different center wavelengths tracking the SD degradation at the eight wavelengths [6–8]. Figure 5 shows the SD degradation derived from the SDSM measurements in symbols. They change smoothly over time except for those measured by the SDSM detectors D7 and D8, showing more noise [8]. Solid lines in Figure 5 are either the exponential functions of time fitted to the measured data (detectors D7 and D8), or the simply linear connections of the measured data (all other detectors). Both the symbols and solid lines are normalized at the time of launch, 28 October 2011. The solid lines for the SD degradation at eight different wavelengths and their linear interpolations for other wavelengths within the spectral coverage are used in Equation (2) to derive the F-factors [5,8]. As expected, the SD degrades faster at shorter wavelengths. In the past four years since VIIRS first light, the SD has degraded about 31.8%, 25.4%, 19.7%, 11.8%, 5.2%, 3.4%, 1.9%, and 1.3% at wavelengths of 412, 450, 488, 555, 672, 746, 865, and 935 nm, respectively. The SD degradation for the shortwave infrared (SWIR) bands, or wavelengths longer than 935 nm, is beyond the spectral coverage of the SDSM, as seen in Table 1, and typically is assumed to be null for wavelengths beyond this range. The SD may have non-negligible SD degradation at the wavelength of the shortest SWIR band, M8 (1238 nm), and the degradation is estimated to be around 0.5% [8,34,35].

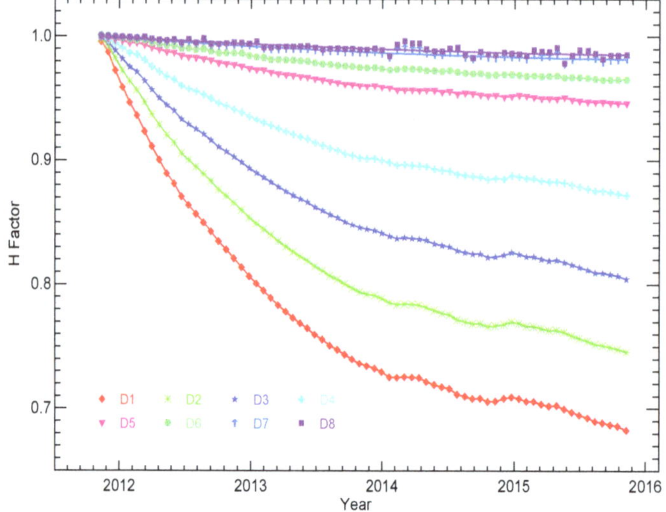

Figure 5. SD degradation derived from the SDSM measurements.

The SD degraded smoothly before February 2014, as expected, but started to behave abnormally since then, especially at short wavelengths. The SD reflectance started to increase in February 2014 for about the next 75 days before reverting back. Furthermore, the subsequent trend did not return to the previously declining pattern expected [8]. The reflectance also started to increase again at about the same time in following years and resulted in an extra seasonal repeating pattern on top of the speculated SD degradations. This unexpected behavior is considered to be a real SD reflectance change according to its impact on the calculated RSB F-factors using the SD calibration [5]. The real physical or chemical phenomena for this behavior are not known yet. Attention needs to be paid to this abnormal behavior and further investigation of the cause of the behavior can be valuable.

As seen in Figure 3, the view directions of the SDSM and the RTA with respect to the SD surface are very different. The SD degradation, $h(\lambda)$, derived from the SDSM calibration describes the SD degradation for the outgoing direction towards the SDSM view. We have demonstrated that the SD may degrade non-uniformly with respect to incident and outgoing directions [5,8]. This means that the SD degradation observed by the SDSM may differ from the actual degradation of the SD for the outgoing direction towards the RTA view. This issue has been discussed in our previous work and will be discussed further in later sections [5,8,14].

The relative spectral response (RSR) may change on-orbit, especially when the out-of-band (OOB) RSR contribution is large. The RSR on-orbit changes can be induced by the degradation of the instruments' optical system, the degradation of the detectors and other possible reasons. If the degradation of optical system is the main cause of the on-orbit changes of the SNPP VIIRS RSB, they can be derived from the time-dependent F-factors [14,36]. An iterative approach can be used to calculate both the F-factors and the RSR with the prelaunch-measured RSR. It has been shown that the rate of RSR change is comparatively larger in the early mission but then slows afterward until becoming almost negligible after about August of 2013 [14]. It has also been shown that the on-orbit RSR changes impact most of the RSB at the 0.1% level or less, except for band M1, with 0.14% at its maximum value, and the effect mainly occurs early in the mission [14]. Generally speaking, the impacts of the on-orbit RSR changes on RSB F-factors derived from the SD calibration are not significant. Nevertheless, the F-factors shown in this paper are all calculated with the time-dependent RSR from our iterative calculation.

The SD is fully illuminated whenever the instrument passes the southern pole from the nightside of the Earth to dayside, as mentioned previously, generating about 14 SD calibration events daily, from which 14 sets of RSB calibration coefficients can be derived. The F-factors are calculated for each RSB, detector, gain status, and HAM side using Equation (2). Figures 6 and 7 show the F-factors for bands M1 and M4, respectively, with high gain and HAM side 1, where D1, D2, . . . , and, D16 denote detector 1, 2, . . . , and 16, respectively, of each band. For both bands, F-factors are clearly detector- and time-dependent. In the last four years, they have changed about 3% for band M1 and less than ~0.5% for band M4. Compared to those for the corresponding MODIS bands (with similar center wavelengths), their changes are much smaller. The F-factors displayed in Figures 6 and 7 for both bands M1 and M4 are stable and the level of noises or fluctuations in them are within ~0.2%. Figure 8 shows the detector-averaged F-factors for band M1 for four combinations of the gain status and HAM side. The F-factors increase smoothly with noises being within ~0.2% for all four combinations. The differences among the four combinations have maintained at about the same level in the last four years since the VIIRS launch.

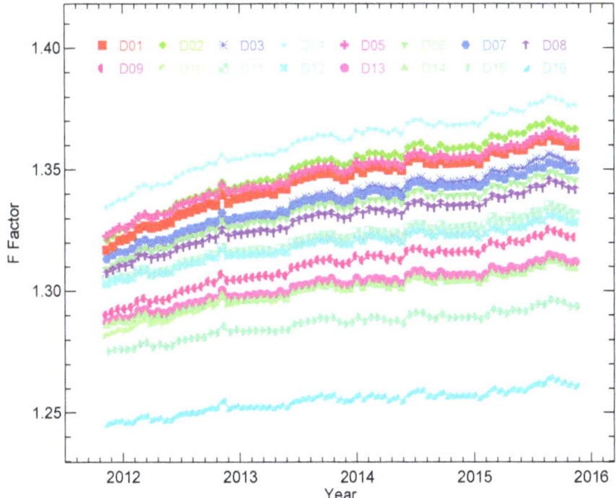

Figure 6. VIIRS band M1 HAM 1 HG SD F-factors.

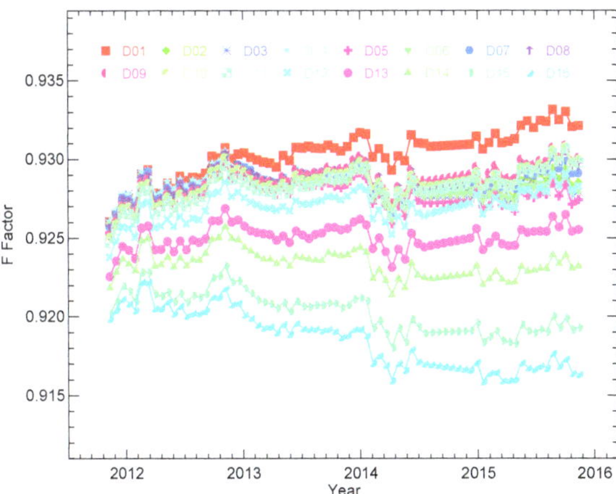

Figure 7. VIIRS band M4 HAM 1 HG SD F-factors.

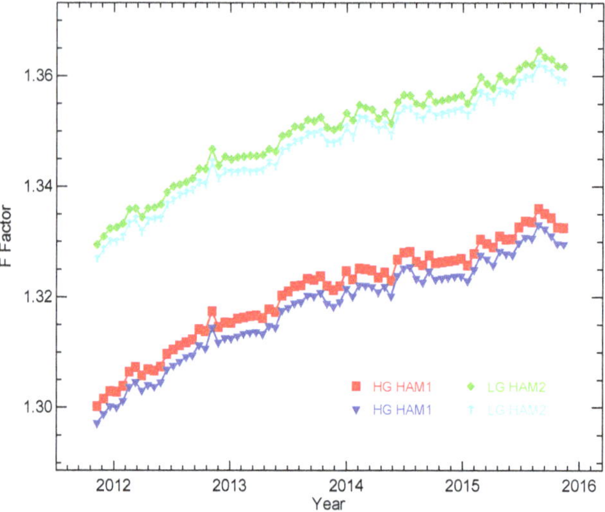

Figure 8. VIIRS band M1 detector-averaged SD F-factors.

The detector-averaged F-factors for all RSB with high-gain status and HAM 1 are displayed in Figure 9. The F-factors for non-SWIR bands are normalized at the first measurement. The SWIR bands were normalized on 20 January 2012 when the temperature of the S/MWIR FPA started to be controlled. The SWIR bands were not stable and usable before 20 January 2012. The F-factors shown in all figures are the calculated values from each individually selected SD measurement without any average over different events. From Figure 9, it is seen that the F-factors of bands I2 and M7 have the largest increase, of about 58.5%. M8 has the second largest F-Factor increase of about 35.0%. The F-factor of band M3 has the least increase and is actually a modest decrease of about 0.3% in the last four years. The F-factor of band M1, which has the shortest wavelength among all RSB, has only increased ~2.6% in the last four years since launch. Since VIIRS on-orbit, the F-factors of bands M2, M4, I1, M5, I3, M6, M9, M10, and M11 have increased about 0.5%, 0.4%, 7.0%, 13.0%, 13.5%, 28.5%, 24.0%, 13.5%, and 3.6%, respectively. The gain of a band is inversely proportional to the F-factor of the band. Thus, the increase of a band's F-factor means the degradation of the band and a larger increase of the F-factor indicates a larger degradation of the band's gain. Generally, the NIR and SWIR bands have degraded much faster than VIS bands. The degradation of the NIR and SWIR bands is mainly due to the degradation of the RTA, which has the largest degradation, around a wavelength of 1000 nm [5,37–39]. Among bands M1-M3, which have the shortest wavelengths, the band with the shorter wavelength has the larger degradation. The degradations of these bands are mainly due to the degradation of the HAM, which degrades faster at a shorter wavelength [5].

The HAM side differences provide important information about the degradation of the HAM. To examine the HAM side difference, we need the actual calibration coefficients, which are calculated by the multiplication of F-factors and the prelaunch calibration coefficients. The HAM side ratios of the actual calibration coefficients are calculated and shown in Figure 10 for all RSBs. They are very stable and close to 1, within ±0.8%, for all RSBs, which is much smaller than those observed in MODIS instruments [27,28,31]. It is also seen from Figure 10 that the HAM side differences are wavelength-dependent and have had no observable changes in the last four years. The largest HAM side differences occur in bands I1 and M4. This is consistent with the polarization sensitivity on HAM side differences observed in prelaunch measurements only in bands I1 and M4 [40].

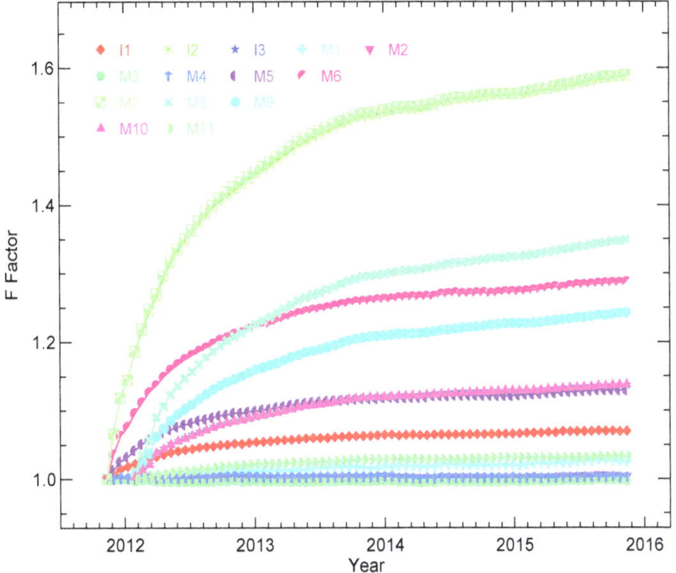

Figure 9. VIIRS RSB HAM 1 HG detector-averaged SD F-factors.

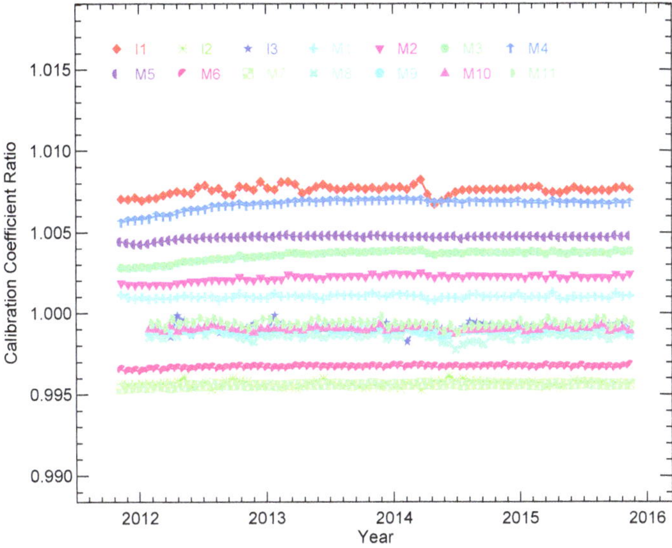

Figure 10. VIIRS RSB calibration coefficients mirror side ratios.

3.2. Lunar Calibration

The Moon is known to have a very stable reflectance in the VIS and NIR spectral regions [18] and has been widely used to track the RSB on-orbit gain changes [41–43]. Since the lunar surface is not smooth, the lunar irradiance instead of the lunar radiance is used for the RSB calibration. The lunar irradiance strongly depends on the viewing geometry, especially the lunar phase angle. To reduce the calibration uncertainty, the lunar phase angles of all lunar calibration events for a remote sensor are typically kept in a small selected range [9]. To keep the phase angle in the selected small range, an instrument roll maneuver, which increases the lunar observation opportunity, is usually required. For SNPP VIIRS, the phase angle range was selected as (−56°, −55°) early in the mission and then changed to (−51.5°, −50.5°) to minimize the roll angles needed for the lunar observations [10,11]. For the safety of the instrument, the roll angles for the roll maneuvers are required to be in the range of (−14°, 0°), where the angle is defined in the instrument coordinate system [10,11]. SNPP VIIRS always views a waxing Moon with its designed orbit and the confined roll angle range. The phase angle for a SNPP VIIRS lunar observation is negative because the instrument always views a waxing Moon and the phase angle is defined as negative for the Moon [9]. Due to the confinement of the roll angle, the Moon can be observed only in about nine months every year. SNPP VIIRS is scheduled to view the Moon once each month in the nine months. VIIRS views the Moon from the SV port. It was found early in the mission that the scene co-registration was not applied in the SV sector, thus the SV sector cannot provide a full lunar image for all bands [10]. The solution is to apply sector rotation during a lunar observation to store lunar observation data in the EV sector, and this has been done since April 2012.

The lunar irradiance can be calculated for each detector of a RSB, considering the oversampling effect and then the F-factor can be derived from the lunar observation for the detector [42]. However, the F-factors derived from a lunar observation for individual detectors are noisier than those derived from the SD calibration due to the roughness of the lunar surface. For lunar calibration, the detector-averaged F-factors are more reliable and adequate for the need of this analysis. During a lunar observation, the Moon can be seen in many scans and in most of the scans the instrument sees a partial Moon. The RSB F-factors can be derived, using all the scans with correction of the oversampling effect, or using a few scans around the center of the lunar observation in which the instrument sees a full Moon [42]. In this analysis, the latter approach is used. With the approximation that the difference between the detector on-orbit changes is negligible, the detector-averaged lunar F-factor can be expressed as [10]:

$$F(B, D, M, G, t) = \frac{g(B) N_M}{\sum\limits_{D,P,S} L_{pl}(B, D, P, S)\delta(M, M_S)} \tag{4}$$

where $g(B)$ corrects the relative geometric effects for band B, N_M is the number of scans, with HAM side M. In the summation, $L_{pl}(B,D,P,S)$ is the lunar irradiance calculated using the prelaunch calibration coefficients, and M_S is the mirror side for the scan S. $g(B)$ depends on the lunar view geometry, which is described by Sun-Earth distance, sensor-Moon distance, lunar phase angle, and lunar librations. The impacts of the first two factors can be described analytically and that of the last term is relatively small. The impact of the lunar phase angle is much larger than that of the lunar librations and thus lunar observations are scheduled for phase angles to be as identical as possible [9–11]. For SNPP VIIRS, the lunar phase angle, defined as negative for a waxing moon, is confined in (−56°, −55°) in first few months on-orbit and then changed to (−51°, −50°) due to instrument operational safety considerations. The confinement may occasionally be relaxed due to various other considerations. The viewing geometry can never be identical and the geometric effect has to be corrected to get accurate lunar F-factors. The Robotic Lunar Observatory (ROLO) model [18,41] can provide the predicted lunar irradiance, which can be used as $g(B)$ to account for the geometric effect in Equation (4). The absolute uncertainty of the ROLO model can be as large as ~5% for short-wavelength bands and the uncertainty is even larger for the NIR bands. Thus, the lunar calibration cannot provide accurate absolute F-factors for the RSB. It can only be used to track the RSB on-orbit change. However, absolute lunar F-factors can be obtained by normalizing the lunar F-factors to the SD F-factors at the time of the first lunar calibration. The ROLO model uncertainty may vary with the viewing geometry. The relative uncertainty of the irradiance predicted by the ROLO model over the entire view geometry is about 1% [41]. Since the lunar phase angle is confined to a selected and restricted region, as mentioned above [9–11], the relative uncertainty of the lunar irradiance predicted by the ROLO model for VIIRS-scheduled lunar observations should be much smaller than ~1%. Thus, the lunar calibration can provide relative F-factors or the on-orbit gain change within ~1% accuracy. The relative uncertainty of the ROLO model prediction induces a seasonal oscillation, which is indeed smaller than 1%, as expected [11], in the derived F-factors. We have introduced a correction based on the viewing geometry to reduce the oscillation pattern and further reduce the uncertainty of geometric effect correction.

The F-factors, derived for SNPP VIIRS VIS and NIR bands from the scheduled lunar observations, are shown in Figure 11 as symbols. Similar to SD F-factors, the lunar F-factors are strongly wavelength-dependent and increase with time [5]. As expected, the largest increase occurs in the F-factors for bands I2 and M7, and changes in those for short wavelength bands are small. Among the three shortest wavelength bands M1-M3, band M1 has had the largest increase in lunar F-factors in the last four years. Compared to those reported in literature [10,11], the lunar F-factors in Figure 11 are much smoother and less noisy. The SD F-factors are also drawn in Figure 11 for VIS and NIR bands with solid lines for comparison. The lunar F-factors are normalized to the SD F-factors corresponding to the April 2012 lunar observation. It can be seen that the F-factors from the two calibrations are in general agreement but with observable differences. The differences between the two sets of F-factors also increase with time. For a more clear demonstration of the differences, Figure 12 shows only the F-factors for bands M1-M4. Among the four bands, the largest difference occurs in the band M4, which is about 1.3%, and band M2 has the second largest difference, which is about 1.0%. The differences for bands M1 and M2 are about 0.6% and 0.7%, respectively. The exact root causes of the differences between the SD/SDSM calibration and the lunar calibration are difficult to identify. However, in our previous works using both SDSM and RSB measurements from SD observations, it has been demonstrated that the SD degrades non-uniformly with respect to incident direction, especially for short wavelength bands [5,8]. This indicates that the SD degradation derived from the SDSM calibration may not be exactly the same as the SD degradation for the RTA view direction, resulting in errors featured as a long-term drift in the SD F-factors. The lunar calibration based on irradiance, however, faces no degradation issue and, in principle, should provide more reliable and accurate long-term RSB gain on-orbit changes. Thus, the non-uniformity of the SD degradation should be one of the main reasons for the discrepancy of the two sets of F-factors and should be the primary one for short wavelength bands.

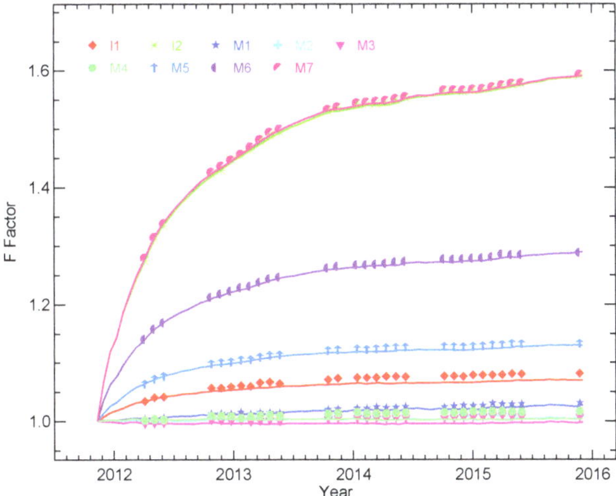

Figure 11. SNPP VIIRS RSB SD F-factors (lines) and lunar F-factors (symbols).

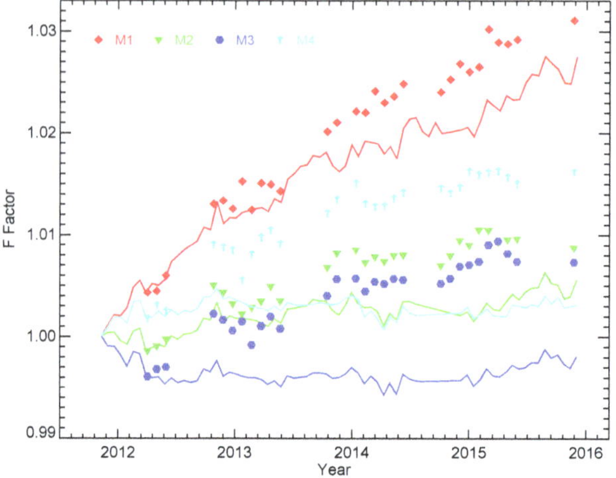

Figure 12. SNPP VIIRS RSB SD F-factors (lines) and lunar F-factors (symbols).

3.3. Hybrid Approach

SD/SDSM or lunar calibration alone has its advantages and disadvantages. Lunar calibration provides more accurate and reliable long-term VIIRS RSB on-orbit gain changes but is less frequent, only about nine times each year, and may have greater measurement uncertainty due to difficulty in accurately correcting the viewing geometry effect on the lunar irradiance. On the other hand, the SD/SDSM calibration can provide VIIRS RSB on-orbit change for each orbit and is smooth and stable in a short time frame, although the derived F-factors demonstrate long-term bias due to the degradation non-uniformity effect [5,8]. A hybrid approach has been proposed and applied to combine the two sets of F-factors by using the lunar F-factors as the long-term baseline and the SD F-factors for short-term gain variation [14].

In the hybrid approach, the ratios of the lunar factors and the SD F-factors for an RSB are calculated first, and then the ratios are fitted to a quadratic form of time. Figure 13 shows the band-averaged ratios of the two sets of F-factors and fitted functions for VIS and NIR bands, where symbols are the measured ratios and solid lines are fitted functions. It is clearly shown that the ratios are band-dependent and increase with time. The non-negligible differences between the SD and lunar F-factors for short wavelength bands are as expected according to the non-uniformity degradation of the SD. It is also noticeable that there are non-negligible differences between the two sets of F-factors for other bands with longer wavelengths. This indicates additional and unknown mechanisms contributing to the differences of the F-factors besides the non-uniformity of the SD degradation. It is also worth noting that the ratios can be approximately classified into two groups, with bands M3, M4, and I1 as one group with larger ratios, and the remaining bands as another. The fitted smooth functions are normalized at the time of the first lunar observation, April 2012, used in this analysis.

The hybrid F-factors are calculated by multiplying the SD F-factors and the quadratic form. The new set of F-factors averts the errors of the SD F-factors caused by the SD degradation non-uniformity effect and other unknown reasons, but keeps the frequency and smoothness of the SD F-factors. Figure 14 displays the band-averaged hybrid F-factors and the SD F-factors. As expected, the hybrid F-factors are larger than the SD F-factors and the differences between the two sets of F-factors increase with time. It is also seen that the differences between the two sets of F-factors for the band M4 are larger than those for the other three bands. This is consistent with the ratios of the lunar and SD F-factors displayed in Figure 14. The hybrid approach is not applied to the time period before April 2012 due to the inaccuracy of the lunar F-factors induced by the partial observations of the lunar surface [10,11].

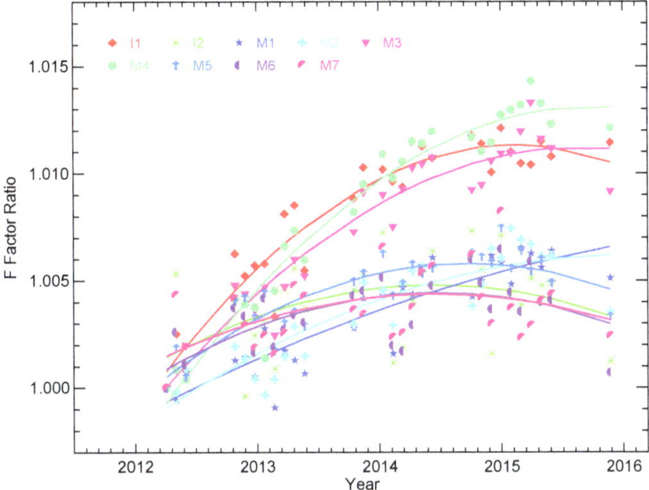

Figure 13. Ratios of lunar F-factors over SD F-factors.

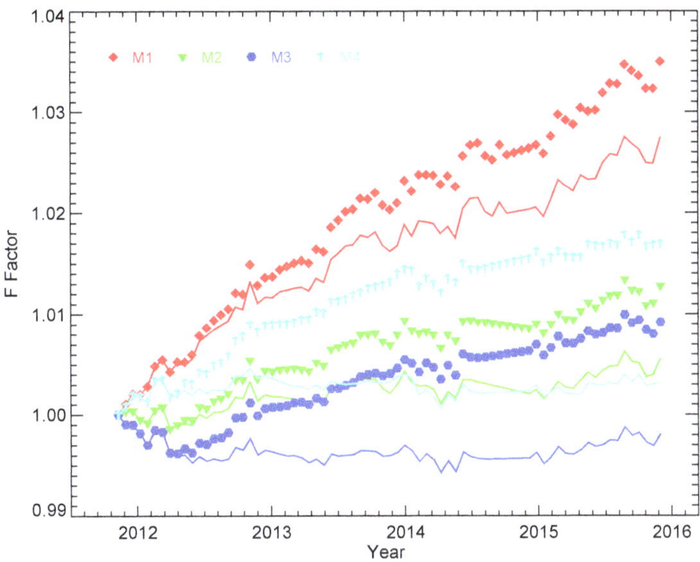

Figure 14. Band-averaged hybrid F-factors (symbols) and band-averaged SD F-factors (lines).

4. Ocean Color EDR Improvements

The main ocean color EDR products are the normalized water-leaving radiance spectra, $nL_w(\lambda)$, which is the radiance that would exit the ocean in the absence of atmosphere and with the Sun at the zenith [20]. It is wavelength-dependent and can be computed for the wavelength of each of the instrument's ocean bands designed for ocean color applications. From the radiances at the wavelengths of the ocean bands, ocean optical, biological, and biogeochemical properties can be derived. Since the atmosphere and ocean surface radiances contribute about more than 90% in the TOA radiance in the visible wavelength range, water-leaving radiance only contributes about <10% to the TOA [44]. Since the atmosphere and ocean surface contributions to the TOA radiances (*i.e.*, atmospheric correction) are simulated using a theoretical model, all instrument calibration errors are passed on to satellite-derived water-leaving radiance spectra [20,44]. Thus the percentage of the uncertainty in the SDR radiance (or calibration uncertainty) will be amplified approximately by an order of magnitude in the normalized water-leaving radiance (or ocean color EDR products) [20,21,44]. Therefore, the radiometric accuracies of the RSBs are critical to the quality of ocean color products [25]. All other ocean color biological and biogeochemical EDR products are derived from the normalized water-leaving radiance. Chlorophyll-*a* concentration, *Chl-a*, is an estimate of the phytoplankton biomass and can be derived from the $nL_w(\lambda)$ at the wavelengths of 443 (or 486) and 551 nm. For VIIRS, they correspond to the center wavelengths of bands M2 (or M3) and M4, respectively.

The NOAA Ocean Color Team has developed a global near real-time VIIRS ocean color data processing system, which automatically downloads global VIIRS RDR (Level-0), SDR (Level-1B), and ancillary data in near real-time, and then processes them into ocean color EDR (Level-2) data [19] using the NOAA Multi-Sensor Level-1 to Level-2 (NOAA-MSL12) software package [44–46]. They also routinely generate the global Level-3 binned products (daily, 8-day, monthly, and climatology) for evaluation purposes [25,26]. Using the system, the NOAA Ocean Color Team has been routinely processing and evaluating VIIRS ocean color products from the start of the VIIRS mission using the IDPS SDR products. They have also reprocessed ocean color EDR using the SDR processed with the improved SD F-factors and the hybrid F-factors described in our previous work and also in this analysis. As mentioned previously, they will also routinely process the SDR using the hybrid F-factors and then produce the ocean color science quality EDR with high data quality and accuracy using the improved SDR for forward daily processing, utilizing the previously-mentioned ratio-approach method, which reduces the computational effort by at least two magnitudes [30]. The method can be directly applied in ocean color EDR processing and then the computational effort is further reduced and disk storage space for improved SDR is not needed. In fact, the NOAA Ocean Color Team uses the latter approach for VIIRS ocean color EDR reprocessing and forward daily processing with scientific quality using the hybrid F-factors.

The preliminary evaluation of the reprocessed ocean color EDR using the SDR generated with the hybrid F-factors has been reported in our previous papers [14,26]. The detailed evaluation of the improvements of new hybrid LUTs on ocean color products, as well as improvements of the ocean color algorithms, will be discussed elsewhere. Here we briefly show the improvements of the further updated hybrid F-factors on the VIIRS ocean color products. Figures 15 and 16 show the time series of VIIRS-derived $nL_w(\lambda)$ at wavelengths of 443 nm (M2) and 551 nm (M4), respectively, over the Hawaii region (oligotrophic waters). $nL_w(\lambda)$ spectra derived with the IDPS SDR processed with standard operational F-factors are represented by solid diamonds. The $nL_w(\lambda)$ derived using the SDR reprocessed with our new hybrid F-factors at the two wavelengths are shown by solid squares. The two figures show that $nL_w(\lambda)$ data derived with the IDPS SDR have a large anomaly before 20 April 2012 and a long-term drift for both bands. The newly derived $nL_w(\lambda)$ spectra with hybrid F-factors are much improved and the long-term drifts are significantly reduced. This is also true for 412 nm (M1) and 488 nm (M3). The trend from the Marine Optical Buoy (MOBY) data, the direct *in situ* measurements of $nL_w(\lambda)$ using a system of buoys [47,48], are also shown in Figures 15 and 16 (in solid triangles). The comparison with the MOBY result further confirms that the SDR using hybrid F-factors significantly elevates the accuracy of the ocean color products. Figure 17 shows VIIRS *Chl-a* derived from the newly reprocessed SDR with the hybrid F-factors and IDPS SDR in the same region. *Chl-a* data based on IDPS SDR (solid diamonds) show a clear long-term drift of about 15%, while the new *Chl-a* results show a clearly reduced long-term trend (solid squares). It is worth mentioning that there is a large spike in the *Chl-a* shown in Figure 17 at the end of 2014. The spike is mainly due to the big dip of the $nL_w(\lambda)$ at wavelengths of 443 nm (M2) shown in Figure 15 at the same time. The dip was observed by both SNPP VIIRS B2 and the MOBY and therefore it is a real feature of the $nL_w(\lambda)$. This indicates that the aforementioned spike in Figure 17 is the actual performance of the *Chl-a*.

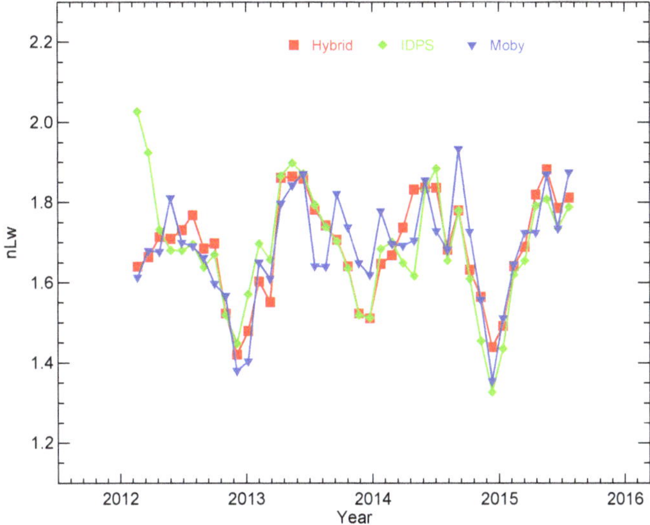

Figure 15. Ocean $nL_w(\lambda)$ trending for the band M2 along with MOBY *in situ* data.

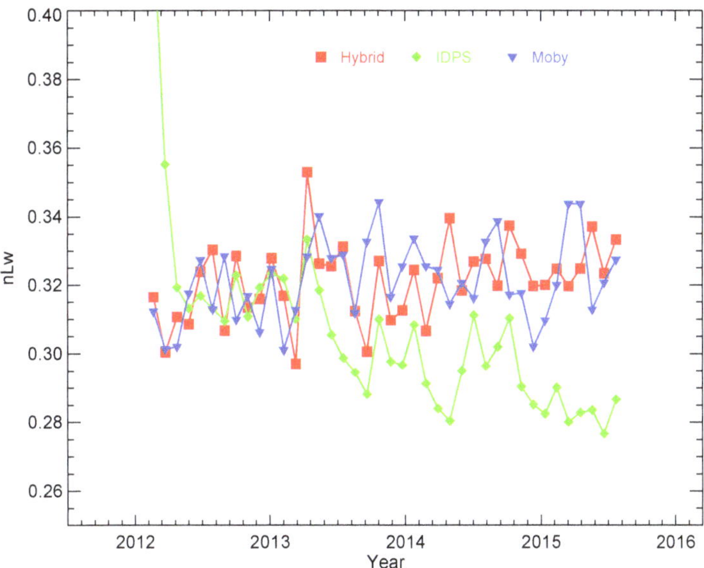

Figure 16. Ocean $nL_w(\lambda)$ trending for the band M4 along with MOBY *in situ* data.

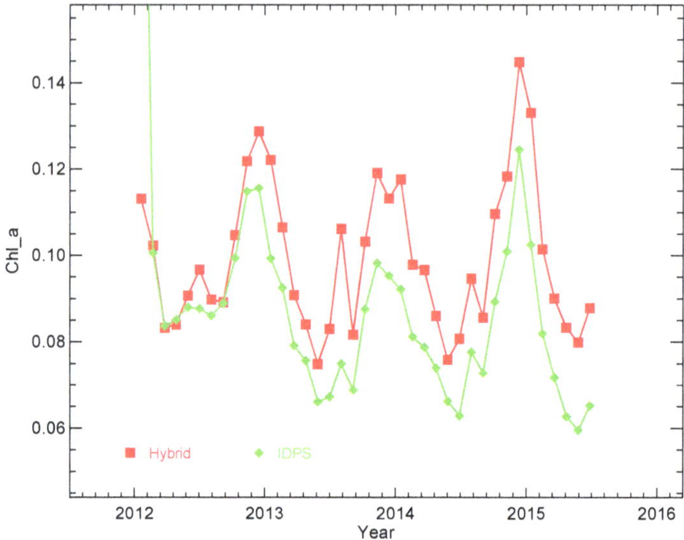

Figure 17. Ocean *Chl-a* trending.

5. Summary

We have made a significant progress in the calibration of the SNPP VIIRS RSB. All key components in the calibration pipeline have been carefully examined and robustly characterized. This includes the VFs, the BRF, the H-factors, and the F-factors. The newly built hybrid F-factors that incorporate lunar-based calibration results, motivated by the failure of the standard methodology sourced to the effect of the SD degradation non-uniformity, is now the *de facto* standard. The results are clean, robust, without previously observed artifacts such as seasonal variations, and achieve an accuracy level estimated at ~0.2% and better. The ocean color products, which are sensitively dependent on RSB calibration, demonstrate significant improvement. These investigative efforts have shown that, with proper treatment of the physical or optical effects, such as the SD degradation non-uniformity, in the calibration methodology, the SNPP VIIRS is a splendid instrument functioning as well as can be expected in its first four years on orbit.

Acknowledgments: The work was supported by the Joint Polar Satellite System (JPSS) funding. We thank the MOBY team for providing the *in situ* data. The views, opinions, and findings contained in this paper are those of the authors and should not be construed as an official NOAA or U.S. Government position, policy, or decision.

Author Contributions: Junqiang Sun is the chief investigator of this topic and is fully responsible for the formulation, technical advancement, and all materials presented in this work. Menghua Wang lead the NOAA VIIRS ocean color effort, provided technical direction and guidance for VIIRS ocean color products, and supported this work.

Conflicts of Interest: The authors declare no conflict of interest.

References

1. Cao, C.; Deluccia, F.; Xiong, X.; Wolfe, R.; Weng, F. Early on-orbit performance of the Visible Infrared Imaging Radiometer Suite (VIIRS) onboard the Suomi National Polar-orbiting Partnership (S-NPP) satellite. *IEEE Trans. Geosci. Remote Sens.* **2014**, *52*, 1142–1156. [CrossRef]
2. Xiong, X.; Butler, J.; Chiang, K.; Efremova, B.; Fulbright, J.; Lei, N.; McIntire, J.; Oudrari, H.; Sun, J.; Wang, Z.; *et al.* VIIRS on-orbit calibration methodology and performance. *J. Geophys. Res. Atmos.* **2014**, *119*, 5065–5078. [CrossRef]

3. Lei, N.; Wang, Z.; Fulbright, J.; Lee, S.; McIntire, J.; Chiang, K.; Xiong, X. Initial on-orbit radiometric calibration of the Suomi NPP VIIRS reflective solar bands. *Proc. SPIE* **2012**, *8510*. [CrossRef]
4. Cardema, J.C.; Rausch, K.; Lei, N.; Moyer, D.I.; DeLuccia, F. Operational calibration of VIIRS reflective solar band sensor data records. *Proc. SPIE* **2012**, *8510*, 851019.
5. Sun, J.; Wang, M. On-orbit calibration of the Visible Infrared Imaging Radiometer Suite reflective solar bands and its challenges using a solar diffuser. *Appl. Opt.* **2015**, *54*, 7210–7223. [CrossRef] [PubMed]
6. Hass, E.; Moyer, D.; DeLuccia, F.; Rausch, K.; Fulbright, J. VIIRS solar diffuser bidirectional reflectance distribution function (BRDF) degradation factor operational trending and update. *Proc. SPIE* **2012**, *8510*, 851016.
7. Fulbright, J.; Lei, N.K.; Chiang, K.; Xiong, X. Characterization and performance of the Suomi-NPP VIIRS solar diffuser stability monitor. *Proc. SPIE* **2012**, *8510*, 851015.
8. Sun, J.; Wang, M. Visible infrared image radiometer suite solar diffuser calibration and its challenges using solar diffuser stability monitor. *Appl. Opt.* **2014**, *36*, 8571–8584. [CrossRef] [PubMed]
9. Sun, J.; Xiong, X. Solar and lunar observation planning for Earth-observing sensor. *Proc. SPIE* **2011**, *8176*, 817610.
10. Sun, J.; Xiong, X.; Butler, J. NPP VIIRS on-orbit calibration and characterization using the Moon. *Proc. SPIE* **2012**, *8510*, 85101I.
11. Xiong, X.; Sun, J.; Fulbright, J.; Z. Wang, Z.; Butler, J. Lunar Calibration and Performance for S-NPP VIIRS Reflective Solar Bands. *IEEE Trans. Geosci. Remote Sens.* **2016**. [CrossRef]
12. Eplee, R.E., Jr.; Turpie, K.R.; Meister, G.; Patt, F.S.; Franz, B.A.; Bailey, S.W. On-orbit calibration of the Suomi National Polar-Orbiting Partnership Visible Infrared Imaging Radiometer Suite for ocean color applications. *Appl. Opt.* **2015**, *54*, 1984–2006. [CrossRef] [PubMed]
13. Baker, N. *Joint Polar Satellite System (JPSS) VIIRS Radiometric Calibration Algorithm Theoretical Basis Document (ATBD)*; Goddard Space Flight Center: Greenbelt, MA, USA, 2013.
14. Sun, J.; Wang, M. Radiometric Calibration of the Visible Infrared Imaging Radiometer Suite Reflective Solar Bands with Robust Characterizations and Hybrid Calibration Coefficients. *Appl. Opt.* **2015**, *54*, 9331–9342. [CrossRef] [PubMed]
15. McIntire, J.; Moyer, D.; Efremova, B.; Oudrari, H.; Xiong, X. On-Orbit Characterization of S-NPP VIIRS Transmission Functions. *IEEE Trans. Geosci. Remote Sens.* **2015**, *53*, 2354–2365. [CrossRef]
16. Sun, J.; Wang, M. On-orbit characterization of the VIIRS solar diffuser and solar diffuser screen. *Appl. Opt.* **2015**, *54*, 236–252. [CrossRef] [PubMed]
17. Kieffer, H.H. Photometric Stability of the Lunar Surface. *Icarus* **1997**, *130*, 323–327. [CrossRef]
18. Kieffer, H.H.; Stone, T.C. The Spectral Irradiance of the Moon. *Astronom. J.* **2005**, *129*, 2887–2901. [CrossRef]
19. Sun, J.; Wang, M. VIIRS Reflective Solar Bands On-Orbit Calibration and Performance: A Three-Year Update. *Proc. SPIE* **2014**, *9264*, 92640L.
20. Gordon, H.R.; Wang, M. Retrieval of water-leaving radiance and aerosol optical thickness over the oceans with SeaWiFS: A Preliminary Algorithm. *Appl. Opt.* **1994**, *33*, 443–452. [CrossRef] [PubMed]
21. Wang, M. Remote sensing of the ocean contributions from ultraviolet to near-infrared using the shortwave infrared bands: Simulations. *Appl. Opt.* **2007**, *46*, 1535–1547. [CrossRef] [PubMed]
22. Wang, M.; Shi, W. The NIR-SWIR combined atmospheric correction approach for MODIS ocean color data processing. *Opt. Express* **2007**, *15*, 15722–15733. [CrossRef] [PubMed]
23. Wang, M.; Tang, J.; Shi, W. MODIS-derived ocean color products along the China east coastal region. *Geophys. Res. Lett.* **2007**, *34*, L06611. [CrossRef]
24. Wang, M.; Shi, W.; Tang, J. Water property monitoring and assessment for China's inland Lake Taihu from MODIS-Aqua measurements. *Remote Sens. Environ.* **2011**, *115*, 841–845. [CrossRef]
25. Wang, M.; Liu, X.; Tan, L.; Jiang, L.; Son, S.; Shi, W.; Rausch, K.; Voss, K. Impact of VIIRS SDR performance on ocean color products. *J. Geophys. Res. Atmos.* **2013**, *118*, 10347–10360. [CrossRef]
26. Wang, M.; Liu, X.; Jiang, L.; Son, S.; Sun, J.; Shi, W.; Tan, L.; Naik, P.; Mikelsons, K.; Wang, X.; Lance, V. Evaluation of VIIRS ocean color products. *Proc. SPIE* **2014**, *9261*, 92610E.
27. Xiong, X.; Sun, J.; Barnes, W.; Salomonson, V.; Esposito, J.; Erives, H.; Guenther, B. Multiyear on-orbit calibration and performance of Terra MODIS reflective solar bands. *IEEE Trans. Geosci. Remote Sens.* **2007**, *45*, 879–889. [CrossRef]

28. Xiong, X.; Sun, J.; Xie, X.; Barnes, W.L.; Salomonson, V.V. On-orbit calibration and performance of Aqua MODIS reflective solar bands. *IEEE Trans. Geosci. Remote Sens.* **2010**, *48*, 535–545. [CrossRef]
29. Oudrari, H.; McIntire, J.; Xiong, X.; Butler, J.; Lee, S.; Lei, N.; Schwarting, T.; Sun, J. Prelaunch radiometric characterization and calibration of THE S-NPP VIIRS sensor. *IEEE Trans. Geosci. Remote Sens.* **2015**, *53*, 2195–2210. [CrossRef]
30. Sun, J.; Wang, M.; Tan, L.; Jiang, L. An Efficient Approach for VIIRS RDR to SDR Data Processing. *IEEE. Trans. Geosci. Remote Sens. Lett.* **2014**, *11*, 2037–2041.
31. Sun, J.; Xiong, X.; Angal, A.; Chen, H.; Wu, A.; Geng, X. Time dependent response *versus* scan angle for MODIS reflective solar bands. *IEEE Trans. Geosci. Remote Sens.* **2014**, *52*, 3159–3174. [CrossRef]
32. Lessel, K.; McClain, S. Low uncertainty measurements of bidirectional reflectance factor on the NPOESS/VIIRS solar diffuser. *Proc. SPIE* **2007**, *6677*, 66771O.
33. JPSS. *Joint Polar Satellite System (JPSS) VIIRS Reflective Solar Bands—Performance Verification Report (PVR)*; NASA Goddard Space Flight Center: Greenbelt, MD, USA, 2011.
34. Lei, N.; Xiong, X. Determination of the SNPP VIIRS solar diffuser BRDF degradation factor over wavelengths longer than 1 μm. *Proc. SPIE* **2015**, *9607*, 96071W.
35. Xiong, X.; Angal, A.; Fulbright, J.; Lei, N.; Mu, Q.; Wang, Z.; Wu, A. Calibration improvements for MODIS and VIIRS SWIR spectral bands. *Proc. SPIE* **2015**, *9607*, 96071Z.
36. Lei, N.; Guenther, B.; Wang, Z.; Xiong, X. Modeling SNPP VIIRS reflective solar bands optical throughput degradation and its impacts on the relative spectral response. *Proc. SPIE* **2013**, *8866*, 88661H.
37. De Luccia, F.; Moyer, D.; Johnson, E.; Rausch, K.; Lei, N.; Chiang, K.; Xiong, X.; Fulbright, J.; Haas, E.; Iona, G. Discovery and characterization of on-orbit degradation of the Visible Infrared Imaging Radiometer Suite (VIIRS) Rotating Telescope Assembly (RTA). *Proc. SPIE* **2012**, *8510*, 85101A.
38. Barrie, J.D.; Fugua, P.D.; Meshishnek, M.J.; Ciofalo, M.R.; Chu, C.T.; Chaney, J.A.; Moision, R.M.; Graziani, L. Root cause determination of on-orbit degradation of the VIIRS rotating telescope assembly. *Proc. SPIE* **2012**, *8510*, 8510B.
39. Iona, G.; Butler, J.; Guenther, B.; Graziani, L.; Johnson, E.; Kenedy, B.; Kent, C.; Lambeck, R.; Waluschka, E.; Xiong, X. VIIRS on-orbit optical anomaly—Investigation, analysis, root cause determination and lessons learned. *Proc. SPIE* **2014**, *8510*, 85101C.
40. Sun, J. Sigma Space Corporation: Lanham, MD, USA, Unpublished; 2010.
41. Stone, T.C.; Kieffer, H.H. Assessment of Uncertainty in ROLO Lunar Irradiance for On-orbit Calibration. *Proc. SPIE* **2004**, *5542*, 300–310.
42. Sun, J.; Xiong, X.; Barnes, W.L.; Guenther, B. MODIS Reflective Solar Bands On-Orbit Lunar Calibration. *IEEE Trans. Geosci. Remote Sens.* **2007**, *43*, 2383–2393. [CrossRef]
43. Barnes, R.A.; Eplee, R.E., Jr.; Patt, F.S.; McClain, C.R. Changes in the radiometric stability of SeaWiFS determined from lunar and solar measurements. *Appl. Opt.* **1999**, *38*, 4649–4664. [CrossRef] [PubMed]
44. Wang, M. A sensitivity study of SeaWiFS atmospheric correction algorithm: Effects of Spectral Band Variations. *Remote Sens. Environ.* **1999**, *67*, 348–359. [CrossRef]
45. Wang, M.; Franz, B.A. Comparing the ocean color measurements between MOS and SeaWiFS: A Vicarious Intercalibration Approach for MOS. IEEE Trans. *Geosci. Remote Sens.* **2000**, *38*, 184–197. [CrossRef]
46. Wang, M.; Isaacman, A.; Franz, A.B.; McClain, C.R. Ocean color optical property data derived from the Japanese Ocean Color and Temperature Scanner and the French Polarization and Directionality of the Earth's Reflectances: A Comparison Study. *Appl. Opt.* **2002**, *41*, 974–990. [CrossRef] [PubMed]
47. Clark, D.; Gordon, H.; Voss, K.; Broenkow, W.; Trees, C. Validation of atmospheric correction over the oceans. *J. Geophys. Res.* **1997**, *102*, 17209–17217. [CrossRef]
48. Clark, D.; Yarbrough, M.; Feinholz, M.; Flora, S.; Broenkow, W.; Kim, Y.S.; Johnson, B.C.; Brown, S.W.; Yuen, M.; Mueller, J.W. MOBY, A Radiometric Buoy for Performance Monitoring and Vicarious Calibration of Satellite Ocean Color Sensors: Measurement and Data Analysis Protocols. In *Ocean. Optics Protocols for Satellite Ocean Color Sensor Validation*; Mueller, J., Fargion, G., McClain, C., Eds.; National Aeronautics and Space Administration Goddard Space Flight Center: Greenbelt, MD, USA, 2003; pp. 3–34.

Review

Comparison of the Calibration Algorithms and SI Traceability of MODIS, VIIRS, GOES, and GOES-R ABI Sensors

Raju Datla [1,*], Xi Shao [1,2], Changyong Cao [3] and Xiangqian Wu [3]

1 NOAA Affiliate, ERT, Inc., Laurel, MD 20707, USA; xi.shao@noaa.gov
2 Department of Astronomy, University of Maryland, College Park, MD 20742, USA
3 NOAA/NESDIS/STAR, College Park, MD 20737, USA; changyong.cao@noaa.gov (C.C.); Xiangqian.wu@noaa.gov (X.W.)
* Correspondence: raju.datla@noaa.gov; Tel.: +1-240-305-3484

Academic Editors: Dongdong Wang, Richard Müller and Prasad S. Thenkabail
Received: 20 November 2015; Accepted: 27 January 2016; Published: 6 February 2016

Abstract: The radiometric calibration equations for the thermal emissive bands (TEB) and the reflective solar bands (RSB) measurements of the earth scenes by the polar satellite sensors, (Terra and Aqua) MODIS and Suomi NPP (VIIRS), and geostationary sensors, GOES Imager and the GOES-R Advanced Baseline Imager (ABI) are analyzed towards calibration algorithm harmonization on the basis of SI traceability which is one of the goals of the NOAA National Calibration Center (NCC). One of the overarching goals of NCC is to provide knowledge base on the NOAA operational satellite sensors and recommend best practices for achieving SI traceability for the radiance measurements on-orbit. As such, the calibration methodologies of these satellite optical sensors are reviewed in light of the recommended practice for radiometric calibration at the National Institute of Standards and Technology (NIST). The equivalence of some of the spectral bands in these sensors for their end products is presented. The operational and calibration features of the sensors for on-orbit observation of radiance are also compared in tabular form. This review is also to serve as a quick cross reference to researchers and analysts on how the observed signals from these sensors in space are converted to radiances.

Keywords: remote sensing; calibration algorithm; calibration equations; SI traceability; LEO and GEO optical sensors

---

1. Introduction

The current activity at National Oceanic and Atmospheric Administration (NOAA) National Calibration Center (NCC) for calibration algorithm harmonization and establishing SI traceability for satellite optical sensor measurements can be traced back to late 1980s. The American Institute of Aeronautics and Astronautics (AIAA) initiated a special task group to specifically address the concern of the radiometric measurement community for the widely diverse approaches being taken in definition, performance and evaluation of sensor systems in space. As a result of their effort, the NIST Handbook on "Recommended Practice; Symbols, Terms, Units and Uncertainty analysis for Radiometric Sensor Calibration" was published [1]. The recommended practice for enabling harmonization and establishing SI traceability gathered more impetus for meeting the need for high accuracy observations across the globe for monitoring climate change. A series of workshops sponsored by NOAA and supported by National Aeronautics and Space Administration (NASA) and National Institute of Standards and Technology (NIST) over the last decade culminated in the establishment of NOAA NCC in 2011 as a virtual center to provide a knowledge base for calibration algorithm harmonization following best practices for achieving SI traceability of operational sensors [2–7].

The center's knowledge base helps the Calibration Working Groups (CWG) and teams on each operational sensor at NOAA/NESDIS/STAR to work with NASA and the instrument vendors towards SI traceability from pre-launch testing to on-orbit operations.

Based on the International Vocabulary of Metrology (VIM) SI traceability can be defined as the result of a measurement that can be related to a reference standard through a documented unbroken chain of calibrations, each contributing to the measurement uncertainty [8]. The emphasis is to create a common basis so that the measurement result can be related to other measurements through their common SI reference and the uncertainty budget accounts all the known components of uncertainly, Type A and Type B, following the Guide to the expression of uncertainty in measurement (GUM) [9]. The best practice guideline for radiometric measurements from space is to compare results obtained through different measurement approaches such as simultaneous observations from different sensors in space through the Global Space-based Inter-calibration System (GSICS) [10] and to use standards such as the moon for RSB radiometric validation. The calibration algorithm and its implementation on orbit to deduce radiances should be flexible enough to study independently the anomalies observed compared to expectations. In all this the calibration equation plays an important role to take into account all possible effects contributing to measured radiance. Both Pre- and Post-launch calibration activity is critical for continuous maintenance of SI traceability of the sensor through its life time [11].

In this article the basic equations that describe radiance measurements by a multiband sensor are first introduced drawing from the recommended practice for radiometric sensor calibration [1]. These equations are recast to the commonly used forms in the Calibration Algorithm Theoretical Basis Documents (C-ATBD) for measuring radiance at the aperture stop of each sensor considered in this article. The optical sensors considered in this article are filter radiometers and the channels defined by the filters are divided as reflective solar bands (RSB) for wavelengths range between 400 nm and 2500 nm and as thermal emissive bands (TEB) for wavelengths in the range 2.5 μm and 100 μm. However the longest wavelength TEB among the sensors in this review is in MODIS at 14.4 μm.

Measurement Equation and Calibration Equations

The general form of the measurement equation illustrates that the response in digital counts of the detector in the sensor focal plane array for flux at the entrance aperture of a band pass filter radiometer is obtained by integration over the appropriate variables:

$$\mathrm{DN} = Gain \iiint \int L(\lambda)\ R_{\mathrm{I}}(\lambda) \cos\theta \, \mathrm{d}A_{\mathrm{s}} \, \mathrm{d}\omega_{\mathrm{s}} \, \mathrm{d}\lambda \, \mathrm{d}t \tag{1}$$

where DN is the digital output by a detector in the instrument assuming linear response for simplicity, *Gain* is gain of the instrument detector plus digitization electronics, $L(\lambda)$ is the source spectral radiance, $R_{\mathrm{I}}(\lambda)$ is the sensor absolute (bandpass) spectral responsivity, θ is the angle the sensor subtends at the source, A_{s} is the area of the source, ω_{s} is the solid angle subtended by the sensor entrance aperture at the source, $\mathrm{d}\lambda$ is the wavelength interval and $\mathrm{d}t$ is the time interval [1]. Instrument response non-linearity and background are not shown in Equation (1). As a first step, the spectral and spatial domains can be considered independent and also the radiance can be considered spatially uniform and stable in time so that the variables can be separated in Equation (1), and Δt is the integration time assuming stable response.

$$\mathrm{DN} = Gain\,\Delta t \int L(\lambda)\, R_{\mathrm{I}}(\lambda)\, \mathrm{d}\lambda \int \cos\theta\ \mathrm{d}\omega_{\mathrm{s}} \int \mathrm{d}A_{\mathrm{s}} \tag{2}$$

The spatial integral:

$$\int \cos\theta\ \mathrm{d}\omega_{\mathrm{s}} \int \mathrm{d}A_{\mathrm{s}} = A_{\mathrm{s}}\,\Omega_{\mathrm{s}} = A_{\mathrm{c}}\,\Omega_{\mathrm{c}} \tag{3}$$

is the source throughput and by the reciprocity theorem equal to the sensor throughput $A_{\mathrm{c}}\,\Omega_{\mathrm{c}}$ where A_{c} is the sensor aperture area Ω_{c} is the projected solid angle subtended by the source at the sensor aperture. It is the sensor field-of-view for a uniform extended area source. The sensor absolute responsivity $R_I(\lambda)$ is determined by the quantum efficiency (QE (λ)) of the detector for the incident photon energy and the transmittance (ρ) of the sensor optics. The optical transmittance ρ of the instrument optics (refractory and reflective) can be represented by:

$$\rho(\lambda) = \prod_{1}^{N_{opt}} \rho_i(\lambda) \tag{4}$$

where ρ_i is the transmittance of optical elements and N_{opt} is the number of optical elements. Therefore:

$$R_I(\lambda) = QE(\lambda)\frac{\lambda}{hc}\rho \tag{5}$$

In general the relative spectral responsivity (RSR) of the sensor is defined by using the pre-launch laboratory data of the sensor for each spectral band.

$$\text{RSR}(\lambda) = \frac{\text{QE}(\lambda)\lambda\ \rho(\lambda)}{\max\{\text{QE}(\lambda)\lambda\ \rho(\lambda)\}} \tag{6}$$

Therefore Equation (2) becomes the following measurement equation

$$\text{DN} = \frac{Gain\,\Delta t\, A_c\, \Omega_c \max\{\text{QE}(\lambda)\lambda\ \rho(\lambda)\}}{hc}\int L(\lambda)\,\text{RSR}(\lambda)\,\text{d}\lambda \tag{7}$$

By rewriting Equation (7) as follows, the band averaged radiance, $\overline{L(\lambda)}$ can be obtained.

$$\overline{L(\lambda)} = \frac{\int L(\lambda)\,\text{RSR}(\lambda)\,\text{d}\lambda}{\int \text{RSR}(\lambda)\,\text{d}\lambda} = m\,\text{DN}\ and\ m = \frac{hc}{Gain\ \Delta t\ A_c\,\Omega_c \max\{\text{QE}(\lambda)\,\lambda\,\rho\,(\lambda)\}\int \text{RSR}(\lambda)\,\text{d}\lambda} \tag{8}$$

Equation (8) is called the Calibration Equation where the quantity m is the calibration coefficient. To simplify Equation (8), we can assume that the variables in the integral do not have strong wavelength dependence. This is valid for an instrument with narrow bandwidth channels. Equation (2) can be rewritten using Equations (3)–(5) as below [12].

$$\text{DN} = Gain\ A_c\,\Omega_c\,L(\lambda)\,\Delta\lambda\,\text{QE}\frac{\lambda}{hc}\,\rho\,\Delta t \tag{9}$$

Again instrument response non-linearity and background are not shown in Equations (7) and (9). These quantities are determined in pre-launch instrument characterization tests and are incorporated in instrument radiometric models and in the production of measured radiances.

Equation (9) can be re-written as the calibration equation equivalent to Equation (8) for obtaining radiance from the detector response in digital counts

$$L(\lambda) = m\ \text{DN} \tag{10}$$

where

$$m = \frac{hc}{Gain\ A_c\,\Omega_c\,\Delta\lambda\,\text{QE}\,\lambda\ \rho\,\Delta t} \tag{11}$$

The calibration coefficient m (in Equation (8) or (11)) is determined pre-launch by viewing uniform sources of known radiance, such as blackbodies for TEB, and well characterized integrating sphere sources for RSB. It is also determined from characterization of individual components such as mirror reflectance, polarization responsivity, spectral radiance responsivity based on the sensor specifications and operational requirements. The individual component level measurements are also used as input to sensor radiometric mathematical models to predict sensor performance and the calculation of band by band measurement uncertainty. Both sensor performance predictions and the system level measurements of m help to establish its pre-launch SI traceable calibration uncertainty. On-orbit the calibration coefficient m is monitored by viewing on-board sources of known radiance. For TEB the on-board blackbody maintained at constant known temperatures serves as a SI traceable standard. However, the background from the emission of the scanning optics in front of its aperture adds to the signal and is to be accounted and subtracted. In general, to accomplish this as a second point of calibration, the sensor on orbit is provided to have a space view of zero radiance.

For RSB, the radiance in Equation (1) can be written for observing the earth as solar reflected radiance,

$$\mathrm{DN} = Gain \iiiint E_{sun}(\lambda)\, \mathrm{BRDF}\,(\phi_h, \phi_v)\, R_{\mathrm{I}}(\lambda) \cos\theta \, \mathrm{d}A_{\mathrm{s}}\, \mathrm{d}\omega_{\mathrm{s}}\, \mathrm{d}\lambda\, \mathrm{d}t \tag{12}$$

where E_{sun} is the solar spectral irradiance, BRDF is the bi-directional reflectance distribution function and φ_h, φ_v are the horizontal and the vertical incidence angles of solar illumination on the reflectance standard or the earth scene. The instrument calibration for RSB is performed on orbit using a reflectance standard such as Spectralon™ to observe reflected solar radiation in comparison to the reflected light from the earth scene. Again, space view helps to discriminate any background contribution.

In Sections 2 and 3 below the radiometric calibration methods of the sensors, the Moderate Resolution Imaging Spectroradiometer (MODIS) and the Visible Infrared Imaging Radiometer Suite (VIIRS) in polar orbit, *i.e.*, low earth orbit (LEO) and the Geostationary Operational Environmental (GOES) Imager and the GOES-R Advanced Baseline Imager (GOES-R ABI) in geostationary orbit (GEO) are analyzed. However, the symbols and terminology used in their Algorithm Theoretical Basis Documents (ATBD) are followed without changing for this article as there will be less confusion to refer back to the literature on these sensors. The calibration equations for TEB and RSB are separately discussed for these sensors in each section due to the difference in calibration instrumentation and methodology for these two spectral regions. Both equations and the SI traceability are analyzed for each sensor. In the discussion Section 4 the wavelength bands and the calibration features of MODIS and VIIRS, and GOES Imager and GOES-R ABI are compared. The conclusion is given in Section 5. The authors note given as Appendix shows Table A1 comparing the few terms and symbols used differently across the various sensor algorithms.

2. Methods of Radiometric Calibration—Polar Satellite Sensors

2.1. MODIS

2.1.1. MODIS TEB Radiometric Calibration

The MODIS Level 1B ATBD describes the sensor instrumentation and develops the calibration equation [13]. The MODIS scan cavity and on-board calibrators and the schematic of the optical system are shown in Figures 1 and 2 of Reference [13]. The double sided Primary mirror (scan mirror) scans the earth scene and the two calibrators, the on-board blackbody (BB) and the space view for zero radiance. It is set to scan the on board calibrators at different angles of incidence compared to the earth observations. As the mirror rotates the scan angles for earth observations on each side is accompanied by the calibrator observations on the same side. These observations alternate between sides. The scan angles are given in Figure 4a of Reference [13].

The TEB radiometric calibration for MODIS is performed by using the digital counts from the two point observations of the BB and Space.

In BB Look, the radiance observed by the sensor through the optics in the beam path is given by the following radiometric equation and it holds for each mirror side, spectral band and detector (Equation (3.1) in [13]),

$$L_{BB\_Path} = RVS_{BB}\ \varepsilon_{BB}\ L_{BB} + (1 - RVS_{BB})\ L_{SM} + RVS_{BB}\ (1 - \varepsilon_{BB})\,\varepsilon_{CAV}\ L_{CAV} + L_{BKG} \tag{13}$$

where L_{BB_Path} is the radiance of all elements in the beam path when the sensor views the blackbody (BB), RVS_{BB} is the normalized system response *versus* scan angle when the sensor views the BB. It is essentially the reflectance of the scan mirror for the appropriate angle. The important pre-launch characterization is the *RVS* measurement for each band and the mirror side. It is normalized for each viewing angle to its value at the BB viewing angle of $-152.5°$.

ε_{BB} is the emissivity of the BB (Pre-launch characterization).
L_{BB} is the radiance from the BB at its set temperature. It is the averaged over the *RSR*.
L_{SM} is the radiance of the scan mirror at its temperature.
ε_{CAV} is the emissivity of the cavity formed by the housing of all the optics (Pre-launch Characterization).
L_{CAV} is the radiance of the scan housing that acts as a cavity emitting radiation based on its temperature.
RVS_{BB} is the normalized system response *versus* scan angle when the sensor views the BB.
L_{BKG} is the radiance from the common path background.

Equation (13) essentially shows the total radiance observed by the sensor while viewing the BB as the sum of radiance from BB emission, the radiance due to the scan mirror emission, the radiance of the optics housing cavity reflected by the blackbody and the scan mirror, and any other unaccounted back ground.

In the Space Look, the sensor views deep space through the Space View (SV) port at the scan angle of $-99°$ on each mirror side and the sensor response establishes the "zero radiance" for each band. The radiance at Space view (*SV*), L_{SV_Path} is given by

$$L_{SV\_Path} = (1 - RVS_{SV}) \times L_{SM} + L_{BKG} \quad (14)$$

where RVS_{SV} is the normalized sensor response *versus* scan angle when the sensor views deep-space. Space itself is considered providing zero radiance.

The difference in counts between the BB view and the space view, $\mathrm{DN}_{BB} - \mathrm{DN}_{SV}$ is measured as dn_{BB}. The calibration equation with instrument response in counts written as a quadratic function of dn_{BB} is as follows,

$$RVS_{BB}\,\varepsilon_{BB}\,L_{BB} + (RVS_{SV} - RVS_{BB})\,L_{SM} + RVS_{BB}\,(1-\varepsilon_{BB})\,\varepsilon_{CAV}\,L_{CAV} = a_0 + b_1 dn_{BB} + a_2 dn_{BB}^2 \quad (15)$$

The calibration coefficients a_0, b_1 and a_2 are the quadratic polynomial coefficients. The linear coefficient b_1 is determined scan by scan using on–orbit data of the sensor response to the BB in its view. In fact Equation (15) allows determining the dominant Calibration Coefficient, b_1 whereas the offset, a_0 and the non-linearity term, a_2 available from pre-launch calibration are updated by the fitting process to reduce the size of the residuals.

In the Earth Scene Look, the sensor views the earth scene over the scan angles $\pm 55°$ and corresponding Earth radiance, L_{EV} is determined by the following calibration equation similar to Equation (15) above accounting for the background for the earth view. The term $RVS_{BB}\ (1-\varepsilon_{BB})\,\varepsilon_{CAV}\,L_{CAV}$ in Equation (15) drops out as there is no reflected IR radiance in the earth scene. Therefore the calibration equation for the earth scene radiance L_{EV} is:

$$RVS_{EV} \cdot L_{EV} + (RVS_{SV} - RVS_{EV}) \cdot L_{SM} = a_0 + b_1 \cdot dn_{EV} + a_2 \cdot dn_{EV}^2 \quad (16)$$

where dn_{EV} is the difference in counts between the earth view and the space view, $\mathrm{DN}_{EV} - \mathrm{DN}_{SV}$. The earth scene radiance L_{EV} is obtained from Equation (16) as all the other quantities are known from Equation (15) evaluated in the same scan.

2.1.2. MODIS RSB Radiometric Calibration

The calibration equation for the bands in the reflective solar wavelength region (RSB) is described in Section 4 of Reference [13]. For RSB, Equation (12) describes the measured digital counts in terms of the reflected radiance due to the solar irradiance of the earth scene. The On-board calibrator for RSB is the Solar Diffuser (SD) panel made from space-grade Spectralon™. Also MODIS is equipped with on-board Solar Diffuser Stability Monitor (SDSM) to monitor and correct for SD degradation. Extensive pre-launch calibration and characterization was performed for RSB on the Response *vs.* Scan Angle (RVS) of the rotating half angle mirror, the Focal Plane Array Detectors, the SD and the SDSM. The MODIS Radiometric Calibration Equation for the RSB Earth view radiance is given below based on earth view and space view digital counts.

The earth view reflectance factor is related to the digital counts by:

$$\rho_{EV} \cos\left(\theta_{EV}\right) = m_1 dn_{EV}^* d_{ES}^2 \tag{17}$$

where ρ_{EV} earth view scene reflectance, θ_{EV} the the solar Zenith angle of the earth view pixel, m_1 the calibration coefficient is determined from the on orbit measurements of the solar diffuser (SD) and SD Stability Monitor (SDSM) and updated regularly, d_{ES} the earth-sun distance at the time of the earth view scene observation and dn_{EV}^* the background subtracted, earth view angle difference adjusted, and instrumental temperature effect corrected digital signal. It is evaluated by:

$$dn_{EV}^* = \left(\mathrm{DN}_{EV} - \langle \mathrm{DN}_{SV} \rangle\right)\left(1 + k_{inst}\Delta T\right) / RVS_{EV} \tag{18}$$

where DN_{EV} and $\langle \mathrm{DN}_{SV} \rangle$ are earth view and space view raw digital counts, respectively, k_{inst} represents the relative dependence of the digital count on the instrument temperature, ΔT is the difference of the instrument temperature from its reference value, and RVS is the response *versus* scan angle normalized at the angle of incidence of the SD. For each band and detector, k_{inst} was calculated from pre-launch measurements using the response of detector at different instrument temperatures. RVS was measured pre-launch and updated from on-orbit characterization and monitoring.

The linear calibration coefficient m_1 is determined from the measurements of the SD and SDSM through

$$m_1 = \frac{\rho_{SD} \cos\left(\theta_{SD}\right)}{dn_{SD}^* d_{ES\_SD}^2} \Gamma_{SDS} \Delta_{SD} \tag{19}$$

where ρ_{SD} is the SD reflectance measured pre launch and dn_{SD}^* is the corrected digital signal given by Equation (18) when measuring SD. The d_{ES_SD} is the Earth- sun distance in AU at the time of SD measurement. The Δ_{SD} is the SD degradation factor determined from the SDSM. For high gain bands, a solar diffuser screen (SDS) is closed to attenuate direct Sun light and Γ_{SDS} is the vignetting (transmission) function. It is unity when the SDS is open and not attenuating the sunlight.

The calibration equation for Earth view radiance can be written using Equation (16) for the reflectance factor.

$$L_{EV} = E_{Sun} \rho_{EV} \cos\left(\theta_{EV}\right) / \pi d_{ES}^2 = m_1 dn_{EV}^* \frac{E_{Sun}}{\pi} \tag{20}$$

where E_{Sun} is the solar irradiance normalized with π at $d_{ES} = 1$ AU and dn_{EV}^* and m_1 are obtained from Equations (18) and (19) respectively.

2.1.3. Analysis of the MODIS Calibration Algorithm and SI Traceability

The development of Equation (16) for TEB and Equation (20) for RSB for MODIS reflects the best practice as the radiometric calibration methodology accounts for various possible contributions to the radiance and is amenable to be modified to implement corrections or improvements based on observations on-orbit.

For example, the case of MODIS sensor in Terra satellite is discussed below [14]. Due to various limitations during pre-launch the scan angle dependence of the scan mirror contribution at large angles for the 2nd term in Equation (16) could not be characterized well and the task was left for post launch. A satellite maneuver was carried out to orient the earth view port to view cold space during the eclipse part of an orbit. Such a satellite maneuver is called the deep space maneuver (DSM). The measured radiances from the earth view and the normal space view in such a configuration are both due to the mirror radiance except its dependence on the mirror angle as shown below.

$$(RVS_{SV} - RVS_{EV})\ L_{SM} = a_0 + b_1 dn_{EV} + a_2 dn_{EV}^2 \tag{21}$$

The RVS_{EV} for the entire Earth view was determined using Equation (21), by knowing dn_{EV} from DSM. The scan mirror emitted radiance, L_{SM} is calculated from its on orbit measured temperature. The offset term a_0 and the nonlinear term a_2 are provided in the Look up Tables (LUT) from pre-launch calibration or updated from blackbody warm-up and cool-down cycles (WUCD) performed quarterly. The WUCD cycle allows the measurement of TEB detectors responses over a range of radiances corresponding to the blackbody temperature variation from 270 K to 315 K. The linear coefficient b_1 of each detector is determined every scan knowing the blackbody temperature and other variables in the Equation (15). This new *RVS* improved the Terra imagery and reduced the calibration errors [14]. Based on the lesson learned from Terra MODIS Aqua MODIS pre-launch *RVS* characterization was much more comprehensive and the performance of the sensor was much more stable with fewer corrections to be made [15]. The degradation of the Solar Diffuser tracking using Solar Diffuser stability Monitor (SDSM) and lunar observations is another example of on-orbit flexibility of the calibration algorithm to track and maintain calibration of RSB for both Terra and Aqua sensors. The spacecraft roll maneuver was performed 9 to 10 times per year for lunar observations through space view port and special SD/SDSM operations during spacecraft yaw maneuvers allowed to derive the SD screen vignetting function [16]. Similarly using on-orbit data sets and the calibration equations, the MODIS team was able to track potential changes to be made to the calibration algorithm to monitor and maintain the on-orbit calibration of TEB and RSB. The uncertainty evaluation of TEB and RSB radiances was done based on the analysis of respective calibration equations [17,18]. It stands as an exemplary effort for the uncertainty analysis for follow up sensors. The analysis followed the important feature of GUM identifying component uncertainties and combining them by square root of sum of squares to arrive at total uncertainty. However, the GUM terminology and procedure of identifying the component uncertainties with uncertainty budget distinguishing as Type A or Type B was not followed. The best practice guideline of using GUM will allow evaluation of the relative contributions from individual variables in the calibration equation. It enables optimization of the calibration equation in a transparent way by removing negligible effects from consideration in meeting the requirements.

2.2. VIIRS

The VIIRS optics has the Rotating Telescope Assembly (RTA), the Half Angle Mirror (HAM) and all of the optics past HAM called *aft*. Reflected and emitted radiation from the earth enters the sensor through the RTA and is reflected from HAM into the *aft* optics sub system. The HAM is a two sided mirror and derives its name as half angle mirror because it rotates at half the angular speed of RTA as both sides of the mirror become active one after another in its full revolution to reflect the radiation from the RTA via a fold mirror into the aft optics. The VIIRS ATBD Section 2.2.2 describes the Opto-Mechanical Module in full detail [19]. The on-board calibrators are the black body (OBCBB) for TEB and the Solar Diffuser (SD) for RSB with a Solar Diffuser Stability Monitor (SDSM) to monitor SD degradation just as in MODIS. The radiometric equations are developed in Section 2.3 in Reference [19].

The transmittance through the RTA follows Equation (4) and an assumption is made that the scan angle dependence and the wavelength dependence of the transmittance (ρ) of the optics are separable as the spectral wavelength is narrow for each band. The scan angle dependence of the response is due to the rotation of HAM presenting different angles during its scan and is denoted as

Response Versus Scan (RVS (θ,B)). It is measured in pre-launch testing and is an important parameter in the development of calibration equation. The term in the brackets in the equation below is the product of the reflectance of HAM wavelength dependence part and the angular dependence part of HAM reflectance.

$$\rho_{sys}(\lambda) = \prod_{1}^{N_{op}} \rho_j(\lambda) = \rho_{rta}(\lambda)\ \rho_{aft}(\lambda)\left[\rho_{ham}(\lambda)\ \mathrm{RVS}(\theta, B)\right] \tag{22}$$

The angle independent transmittance of the system is combined as $\rho_{fix}(\lambda)$ given by

$$\rho_{fix}(\lambda) = \rho_{rta}(\lambda)\ \rho_{ham}(\lambda)\ \rho_{aft}(\lambda) \tag{23}$$

where $\rho_{rta}(\lambda)$, $\rho_{ham}(\lambda)$, $\rho_{aft}(\lambda)$ denote the transmittance through rotating telescope assembly, the HAM and the aft assembly respectively.

The VIIRS (RTA) has three views, the space, the blackbody and the earth. They are abbreviated as sv, obc and ev.

The measurement equation is given by Equation (8) in Section 2.3 of Reference [18],

$$N_e = \frac{\Omega_{stop}\cdot \Delta t \cdot A}{hc} \int \mathrm{QE}(\lambda)\cdot\lambda\cdot\left[L_{ap}(\lambda,\theta)\cdot \rho_{fix}(\lambda)\cdot \mathrm{RVS}(\theta,B) + \frac{E_{bkg}(\lambda,\theta)}{\Omega_{stop}}\right] d\lambda \tag{24}$$

where N_e is the number of photo electrons per detection, Ω_{stop} is the solid angle of the aperture stop as seen from the field stop and A is the area of the field stop, $L_{ap}(\lambda,\theta)$ is the spectral radiance at the aperture at angle θ, and $E_{bkg}(\lambda,\theta)$ is the spectral irradiance at the field stop due to the self-emissive background detected at the scan angle θ.

Equation (24) is in integral form to account for wavelength dependence within the band. Also, in Reference [4] the relative spectral response, RSR is defined as in Equation (6) in Section 1.

$$\mathrm{RSR}(\lambda) = \frac{QE(\lambda)\cdot\lambda\cdot\rho_{fix}(\lambda)}{max\left(QE(\lambda)\cdot\lambda\cdot\rho_{fix}(\lambda)\right)} \tag{25}$$

The following notation was used in Reference [18] for any quantity to be band averaged. For example, the quantity $F(\lambda)$ is band averaged as follows

$$\overline{F(\lambda)} = \frac{\int \mathrm{RSR}(\lambda)\cdot F(\lambda)\, d\lambda}{\int \mathrm{RSR}(\lambda)\, d\lambda} \tag{26}$$

Using Equations (25) and (26), the quantities in the bracket of Equation (24) are band averaged and the resulting quantity is called the band-averaged detectable radiance, $\overline{L_{det}}(\theta, B)$.

$$\overline{L_{det}}(\theta,B) = \mathrm{RVS}(\theta,B)\cdot\overline{L_{ap}}(\lambda,\theta) + \overline{L_{\mathrm{det\_}bkg}}(\theta,B) \tag{27}$$

Equation (27) is transformed into the calibration equation as shown below

$$\overline{L_{det}}(\theta,B) = G\ N_e \tag{28}$$

where the quantity G relates N_e to the radiance at the detector.

Equation (27) relates the radiance at the detector as due to radiance entering the aperture modified by the response *vs.* scan angle of the HAM and the background radiance.

The space view (*sv*) provides zero radiance entering the aperture. It is a calibration of zero external radiance input for the sensor output, *i.e.*, $L_{ap}(\lambda,\theta_{SV}) = 0$ and in the notation B representing each band can be replaced with corresponding λ for that band; *i.e.*, $\overline{L_{ap}}(\theta,B) = \overline{L_{ap}(\lambda,\theta)}$.

By defining

$$\overline{\Delta L_{det}}(\theta, B) = \overline{L_{det}}(\theta, B) - \overline{L_{\text{det\_}bkg}}(\theta_{SV}, B) \tag{29}$$

$$\overline{\Delta L_{det\_bkg}}(\theta, B) = \overline{L_{\text{det\_}bkg}}(\theta, B) - \overline{L_{\text{det\_}bkg}}(\theta_{SV}, B) \tag{30}$$

Substituting from Equations (27) and (30), the net radiance detected, in Equation (29) is given by

$$\overline{\Delta L_{det}}(\theta, B) = \text{RVS}(\theta, B) \cdot \overline{L_{ap}}(\lambda, \theta) + \overline{\Delta L_{\text{det\_}bkg}}(\theta, B) \tag{31}$$

Basically, we have the radiance causing the detector to respond $\overline{\Delta L_{det}}(\theta, B)$ given by Equation (31) that excludes the space view background. The non-linearity effects and the temperature effects of the detectors in the FPA and the electronics are parameterized by coefficients C_i and expresses $\overline{\Delta L_{det}}(\theta, B)$ as $\sum_{i=0}^{2} C_i dn^i$ where dn is the detector output signal in digital counts after subtracting the space view counts accounting for the space view background. The detector output is expressed as a second order polynomial and coefficients C_i which are theoretically analyzed as combination of individual components a_i for the detector and b_i for the electronics and shown in Tables 11–13 in Section 2.3.1 in Reference [19] for various possible scenarios of VIIRS performance. The C_i determined pre-launch are changed to $C_i{}'$ post launch and are being tracked and calibrated. A scale factor F is introduced to account for the change quantitatively and to be dynamically calibrated on orbit. It is assumed that all three coefficients change by the same factor F and $c_i' = F \,.\, c_i$.

Therefore,

$$\overline{\Delta L_{det}}\ (\theta, B) = \overline{\Delta L_{det}}\ (dn) = \sum_{i=0}^{2} C'_i dn^i = F \ldots\ldots \sum_{i=0}^{2} C_i dn^i \tag{32}$$

where $= DN - \overline{DN_{sv}}$, is the background subtracted counts.

2.2.1. VIIRS TEB Radiometric Calibration

The RTA views the emissive sources in all three of its views, the space, the earth and the on-board blackbody and always having the background contribution from the optics and any reflected radiation. In Reference [19] all background sources within the solid angle of the aperture stop are analyzed and it is assumed that all components of the RTA are at same temperature T_{rta} and the temperature of the HAM mirror T_{ham} will be different.

The band averaged residual background for any view subtracting the space view is given in Reference [19] Equation (44).

$$\overline{\Delta L_{\text{det\_}bkg}}(\theta, B) = (RVS(\theta, B) - RVS(\theta_{SV}, B)) \cdot \left(\frac{\{(1 - \overline{\rho_{rta}(\lambda)}) \cdot \overline{L(T_{rta}, \lambda)} - \overline{L(T_{ham}, \lambda)}\}}{\overline{\rho_{rta}}(\lambda)} \right) \tag{33}$$

So Equation (31) transforms to the following equation which is essentially Equation (45) in Reference [18].

$$\begin{gathered} \text{RVS}(\theta, B)\, \overline{L_{ap}}(\theta, B) = \overline{\Delta L_{det}}(\theta, B) - \overline{\Delta L_{\text{det\_}bkg}}(\theta, B) = \overline{\Delta L_{det}}(dn) - \overline{\Delta L_{\text{det\_}bkg}}(\theta, B) = \\ \sum_{j=0}^{2} C\prime_j dn^j - \overline{\Delta L_{\text{det\_}bkg}}(\theta, B) \end{gathered} \tag{34}$$

The OBCBB look is used for calibration as the RTA views it in each scan. The contribution to the radiance from other possible sources that add to the OBC radiance is analyzed. Other sources are the reflections from the blackbody shield, the cavity and the telescope. The band averaged radiance at the aperture stop for OBC look is given by

$$\overline{L_{ap}}\ (\theta_{obc}\ ,\ B) = \overline{\left[\epsilon_{obc}\ (\lambda) \cdot L\ (T_{obc}\ ,\ \lambda) + \ L_{obc\_rfl}\ (T_{sh}\ ,\ T_{cav}\ , T_{tele}\ ,\lambda)\right]} \quad (35)$$

So the measurement equation while viewing the OBC is given as essentially Equation (111) in Reference [18] and can be written with the Cal factor F in Equation (32) as below.

$$\begin{aligned} F \cdot \sum_{j=0}^{2} C_j dn^j &= \text{RVS}\ (\theta_{obc}, B) \cdot \overline{\left[\epsilon_{obc}\ (\lambda) \cdot L\ (T_{obc}\ ,\ \lambda) + \ L_{obc_{rfl}}\ (T_{sh}\ ,\ T_{cav}\ , T_{tele}\ ,\lambda)\right]} + \\ &(\text{RVS}\ (\theta_{obc}, B) - \ \text{RVS}\ (\theta_{SV}, B)) \cdot \left(\frac{\left\{ \left(1 - \overline{\rho_{rta}(\lambda)}\right) \cdot \overline{L\,(T_{rta},\ \lambda)} - \overline{L\,(T_{ham},\ \lambda)} \right\}}{\overline{\rho_{rta}}\ (\lambda)} \right) \end{aligned} \quad (36)$$

F can be determined as part of On Board calibration. The RVS is arbitrarily normalized to one scan angle (Space view angle) RVS $(\theta_{SV}, B) = 1$. The calibration factor F is determined from Equation (36).

$$F = \frac{\text{RVS}\ (\theta_{obc}, B) \cdot \left[\overline{\epsilon_{obc}\ (\lambda)} \cdot \overline{L\ (T_{obc}\ (t),\ \lambda)} + \overline{L_{obc_{rfl}}\ (T_{sh}\ (t),\ T_{cav}\ (t), T_{tele}\ (t),\lambda)}\right]}{\sum_{j=0}^{2} C_j dn_{obc}{}^j} + \frac{(\text{RVS}\ (\theta_{obc}, B) - \ 1) \cdot \left(\frac{\left\{ \left(1 - \overline{\rho_{rta}(\lambda)}\right) \cdot \overline{L\,(T_{rta}\ (t),\ \lambda)} - \overline{L\,(T_{ham}\ (t),\ \lambda)} \right\}}{\overline{\rho_{rta}}\ (\lambda)} \right)}{\sum_{j=0}^{2} C_j dn_{obc}{}^j} \quad (37)$$

The band averaged OBC blackbody reflected radiance is:

$$\overline{L_{obc\_rfl}\ (T_{sh}\ ,\ T_{cav}\ , T_{tele}\ ,\lambda)} = \left(1 - \ \overline{\varepsilon_{obc}\ (\lambda)}\right) \cdot \ \left(F_{sh} \cdot\ \overline{L\ (T_{sh},\lambda)} + \ F_{cav} \cdot\ \overline{L\ (T_{cav},\lambda)} + \ F_{tele} \cdot\ \overline{L\ (T_{tele},\lambda\ )}\right) \quad (38)$$

where the factors F_{sh}, F_{cav} , F_{tele} represent the fraction of the reflectance off the OBC blackbody originating from the blackbody shield, cavity and telescope. Assuming the emissivity of these three sources to be 1, it follows $F_{sh} + F_{cav} + F_{tele} = 1$. The OBC blackbody reflected radiance is routinely updated knowing the temperatures of all relevant components in Equation (38).

The scene radiance $\overline{L_{ap}}\ (\theta, B)$ is obtained for Earth view angle θ using the Equations (34)–(37) and is given by the following calibration equation

$$\overline{L_{ap}}\ (\theta_{ev}, B) = \frac{F\ \sum_{j=0}^{2} C_i\ (T_{det}, T_{ele})\ dn_{ev}{}^j + (1 - \ \text{RVS}\ (\theta_{ev}, B))\ \left(\frac{\left\{ \left(1 - \overline{\rho_{rta}(\lambda)}\right)\ \overline{L\,(T_{rta},\ \lambda)} - \overline{L\,(T_{ham},\ \lambda)} \right\}}{\overline{\rho_{rta}}\ (\lambda)} \right)}{\text{RVS}\ (\theta_{ev},\ B)} \quad (39)$$

The above equation essentially gives the radiance (Earth view) as the sum of Calibrated FPA signal converted to radiance accounting for HAM scan angle dependence and the residual background from the RTA and the HAM.

In summary, for evaluating Equation (39), the pre-launch data provide the HAM scan angle dependence of the RVS (response vs scan angle) as LUTs. The background is evaluated from the emitted radiance determined at the temperature of the optics dynamically measured on-orbit. The $\overline{\rho_{rta}}\ (\lambda)$ is determined pre-launch and stored in the LUTs. The gain Coefficient F is determined from the blackbody view and Equation (37).

2.2.2. VIIRS Radiometric RSB Calibration

The calibration equation shown below for the RSB is developed from the general calibration algorithm developed earlier *i.e.*, Equations (31) and (32).

$$\overline{L_{ap}}\ (\theta, B) = \overline{L_{ap}\ (\theta, \lambda)} = \frac{\overline{\Delta L_{det}}\ (\theta, B)}{RVS\ (\theta, B)} = \frac{F\ .\ \sum_{i=0}^{2} C_i\ .\ dn^i}{RVS\ \ (\theta, B)} \quad (40)$$

The calibration source for RSB is the solar diffuser (SD) for which the reflectance factor, ρ_{sd}, is determined as discussed for MODIS sensor, by the pre-launch measurements of the SD Bi-directional

Reflectance Distribution Function (BRDF) and corrected to account for the SD degradation on-orbit based on the trending of the solar diffuser stability monitor (SDSM) output. There is also a SD screen (SDS) at the entrance to the instrument aperture to attenuate the solar irradiance when pointing to the SD. The radiance at the entrance aperture viewing the SD can be written as:

$$L_{ap}\ (\theta_{sd},\lambda) = \tau_{sds}\ (\phi_h,\phi_v,\ \lambda,d).E_{sun}(\ \lambda,d_{se}).\cos{(\theta_{inc})}\ .\mathrm{BRDF}\ (\phi_h,\phi_v,\ \lambda) \tag{41}$$

where ϕ_v and ϕ_h are the vertical and horizontal incidence angles of solar illumination upon the SD, θ_{inc} is the incidence angle onto the SD relative to normal, d_{se} is the distance from the sun to the earth, $\tau_{sds}\ (\phi_h, \phi_v,\ \lambda,\ d)$ is the transmittance of the SDS, d is detector index and $E_{sun}(\lambda,d_{se})$ is the irradiance from the sun upon a surface with its normal pointing toward the sun. Integrating Equation (41) over the spectral band "B" and substituting it into Equation (40), we obtain the measurement equation.

$$F = \frac{RVS\ (\theta_{sd},B)\ .\cos{(\theta_{inc})}\ .\left[\overline{\tau_{sds}\ (\phi_h,\phi_v,\ \lambda,\ d)\ .E_{sun}\ (\lambda,d_{se})\ \mathrm{BRDF}\ (\phi_h,\phi_v,\ \lambda)}\right]}{\sum_{i=0}^{2} c_i\ .\ dn_{sd}^{i}} \tag{42}$$

where the average denotes the averaging over the spectral band "B". It is further assumed that τ and BRDF are invariant with wavelength within the narrow band and are taken out of the integral. All the variables on the right of Equation (42) are based on preflight measurements and on angles that can be determined from the geometry. All values on the right are known. The three c coefficients are determined pre-launch and Equation (40) allows the scale factor F to be determined for RSB from the solar diffuser measurements. The determination of F for RSB resembles the case of IR bands as discussed in relation to Equation (32).

After determining the F factor from the solar diffuser measurements, the calibrated earth view at-aperture radiance for RSB is calculated using the calibration equation *i.e.*, Equation (40),

$$\overline{L}_{ap}\ (\theta_{ev},\ B) = \frac{F\ .\sum_{i=0}^{2} c_i\ .dn_{ev}^{i}}{\mathrm{RVS}\ (\ \theta_{ev},\ B)} \tag{43}$$

where $\overline{L}_{ap}\ (\theta_{ev},\ B)$, the band-averaged spectral radiance at the aperture for earth view scan angle θ_{ev}, $\overline{L}_{ap}\ (\theta_{ev},\ B)$, is the response *versus* scan function at earth view scan angle θ_{ev} for band B and $dn_{ev} = DN_{ev} - DN_{sv}$, the difference between total digital output for earth view angle θ_{ev} and digital counts for space view.

The spectral earth-view reflectance for VIIRS RSB can be written as discussed for MODIS sensor,

$$\rho_{ev}\ (\theta_{ev},\lambda) = \frac{\pi.\ L_{ap}\ (\theta_{ev},\lambda)}{\cos{(\theta_{sun\_earth})}\ .\ E_{sun}\ (\lambda,\ d_{se})} \tag{44}$$

Applying band-averaging for Equation (44) over spectral band "B" and using Equation (40) to substitute for $\overline{L}_{ap}\ (\theta_{ev},\ B)$, we obtain the band- averaged earth-view reflectance as

$$\overline{\rho}_{ev}\ (\theta_{ev},\ B) = \frac{\pi\ .\ F\ .\sum_{i=0}^{2} c_i\ .dn_{ev}^{i}}{\mathrm{RVS}\ (\ \theta_{ev},\ B)\ .\ \cos{(\theta_{sun\_earth})}\ .\ \overline{E_{sun}\ (\lambda,\ d_{se})}} \tag{45}$$

2.2.3. Analysis of the VIIRS Calibration Algorithm and SI Traceability

The VIIRS sensor was launched into orbit in October 2011 and much experience on its On-orbit performance is reported in literature. The calibration equations, Equation (39) for TEB and Equation (45) for RSB have been developed from first principles of Radiometry as shown in Section 1 of this paper. The calibration algorithm differs from the simplicity of MODIS due to new Cal factor "F" introduced as a scaling factor in the calibration equations as discussed in Sections 2.2.1 and 2.2.2 for monitoring on-orbit performance compared to the results of extensive pre-launch characterization and calibration of the sensor. The pre-launch test data was comprehensive over the full range of instrument operating conditions guided by stringent uncertainty requirements that provided LUTs for on-orbit data analysis. The F factor is evaluated at every scan, at every detector of every band, for every HAM mirror side which is a good practice to identify anomalies. However there is an intrinsic difficulty in determining the F factor for TEB due to the RVS variation with angle of incidence as can be seen in the Equations (37) and (39). The space view does not cancel the background because of the RVS difference between space view and the blackbody view. As such one needs accurate RVS data to evaluate F using the blackbody radiance as input in Equation (37). The other way out is to have RVS for blackbody view same as for the space view by design and provide an accurate measurement of this RVS in LUT. This will eliminate background contribution and F can be determined from Equation (37) without getting coupled to RVS and Equation (38) can be used for analyzing earth view radiances knowing the other parameters. However, at the time of this writing the long wave TEB radiances for Sea Surface Temperature (SST) Environmental data Record (EDR) for the blackbody Warm Up and Cool Down (WUCD) time periods showed anomalous values. David Moyer *et al.*, are analyzing the WUCD data to resolve the issue [20].

The radiometric performance and stability of VIIRS sensor TEB during normal operations is considered excellent compared to expectations [21]. The large degradation of the NIR and SWIR of the RSB bands soon after launch was addressed and resolved as caused by the tungsten contamination in the RTA mirror coatings. This degradation is currently reported to have considerably leveled off [22]. As part of good practice, spacecraft maneuvers are being performed to verify and update key parameters. At the beginning of the mission yaw maneuver was performed to validate and update the transmission of the SD and SDSM screen [23]. Roll maneuvers are performed on nearly monthly-basis for lunar observations for independent validation of RSB calibration using the SD and SDSM. Pitch maneuver was performed and the data validated the relative TEB RVS values [24]. However, a pitch maneuver coupled to WUCD may provide independent data to address the SST EDR anomaly. The pitch maneuver will point the earth view to another space view and thus provide another zero radiance reference. The path difference background signal can be analyzed independently as a function of the temperature of the HAM mirror and the angle dependent RVS. This independent experiment could validate the parameters evaluated by Reference [19] for SST anomaly resolution. If there are limitations of time for the complete WUCD cycle, even a part of the cycle during the useful time of the pitch maneuver may help to get TEB data for the anomaly resolution.

The SI traceability is addressed in the ATBD comprehensively as part of stringent uncertainty requirements [19]. The Reference [19] and published literature followed GUM to a large extent to determine uncertainty by using the calibration equations [23,25]. However, the usage of terminology of GUM such as "Standard uncertainty" for individual components and "Combined Standard uncertainty" for the total are yet to be introduced in to the common practice for VIIRS calibration.

3. Methods of Radiometric Calibration—Geostationary Satellite Sensors

3.1. GOES Imager

The current GOES imagers trace back to GOES-8 launched in 1994 as the start of 2nd generation GOES satellites (also called as the 3rd generation by Tim Schmit *et al.*, [26]. Currently GOES-13 (N) is covering the eastern part of the United States and GOES-15 (P) is covering the West. They belong to extended 2nd generation N-O-P satellite series (Tim Schmit *et al.*, called it as the 4th generation that is equipped with improved image navigation and registration performance [26]. All satellites in the series carried a five channel (one visible band (RSB) and four infrared bands (TEB)) imaging radiometer designed to sense the earth both in emitted and reflected energy and a warm blackbody for the calibration of the TEB. The space view served to provide zero radiance reference for calibration. There was no on-board visible reference standard to calibrate the visible band on-orbit and the pre-launch calibration provided the calibration coefficients. The radiometer had a two axis gimbaled scan mirror, a Cassegrain telescope to focus the energy into after telescope (aft) optics that separates the visible and four infrared bands through beam splitters and filters to produce the image by the detector array at each band [27]. The gimbaled scan mirror produces a bidirectional scan (East/West and North/South) of the earth scene and provides a space look at an extreme East West coordinate and a blackbody look when it is rotated in the N-S direction through an angle of approximately 180 degrees.

3.1.1. GOES TEB Radiometric Calibration

Weinreb *et al.* described in detail the operational calibration algorithm based on GOES-8 and GOES-9 on orbit observations [28]. This algorithm is being followed to date on all sensors in the series. The calibration equation for the TEB was given originally at launch by the vendor as

$$R = q\,X^2 + m\,X + b \tag{46}$$

where R is the radiance from the scene, X is the signal output for each band, q, m, and b are the coefficients [28]. The radiance R for a band and detector is the average of spectral radiance over the spectral response function as shown in Section 1 for the calculation of band average. The value of the nonlinear coefficient q was determined from the prelaunch measurements to account for instrument and detector operating temperature dependence. The slope m and the intercept b are given by the following equations:

$$m = \frac{\left[R_{\rm bb} - q\left(X_{\rm bb}^2 - X_{\rm sp}^2\right)\right]}{\left(X_{bb} - X_{sp}\right)} \tag{47}$$

$$b = -m\,X_{sp} - q\,X_{\rm sp}^2 \tag{48}$$

where the subscripts bb and sp refer to data taken from blackbody and space views respectively.

However, the above equations were found to be insufficient to describe the observed data as there was significant dependence of the radiance R on the emissivity of the scan mirror as a function of its east-west scan angle. Weinreb *et al.* modified Equation (46) as follows:

$$[1 - \epsilon(\theta)]\,R + \epsilon(\theta)\,R_M = q\,X^2 + m\,X + b \tag{49}$$

where $\epsilon(\theta)$ is the emissivity of the scan mirror as a function of its angle θ. The parameter R_M is the radiance of the scan mirror that can be computed from its monitored temperature using Planck's law and the spectral response function of the band. Weinreb *et al.* developed an algorithm to derive $\epsilon(\theta)$ using the laboratory data on all bands for the emissivity of witness samples of the scan mirror at 45°, $\epsilon(45)$ that corresponds to the viewing angle at the center of the east-west scan [28]. It is implemented through special data collection of the instrument's output during east-west scans of space across the entire east-west field of regard above and below the earth and is described in detail in Reference [28]. Knowing $\epsilon(\theta)$, Equation (49) can be solved for m and b using the two point calibration of blackbody-look and space-look instrument output, X_{bb} and X_{sp} respectively as given below.

$$m = \frac{\left[[1 - \epsilon(45)]\,R_{\rm bb} + [\epsilon(45) - \epsilon(\rm sp)]\,R_{M,bb} - q\left(X_{\rm bb}^2 - X_{\rm sp}^2\right)\right]}{\left(X_{bb} - X_{sp}\right)} \tag{50}$$

$$b = -m\,X_{sp} - q\,X_{\rm sp}^2 + \epsilon(sp)\,R_{M,sp} \tag{51}$$

where $R_{M,bb}$ and $R_{M,sp}$ are the radiances of the scan mirror computed from its temperature at the time of blackbody-look and space-look respectively. So from the observed detector output X for the earth-view, the radiance R is computed for each band from Equation (49) using Equations (50) and (51) and the detector output X_{sp} collected for the space-look and interpolated as needed. The expression for R is given below.

$$R = \frac{\left[q\left[X^2 - X_{sp}^2\right] + m\left[X - X_{sp}\right] - \left[\epsilon\left(\theta\right) - \epsilon\left(sp\right)\right] R_{M,sp}\right]}{\left[1 - \epsilon\left(\theta\right)\right]} \tag{52}$$

3.1.2. GOES Visible Band (0.55–0.75 μm) Radiometric Calibration

Weinreb *et al.* showed the calibration equation for the visible band radiance expressed in two ways [28].

$$R = m\,X + b \tag{53}$$

$$R = m\,\left(X - X_{sp}\right) \tag{54}$$

where m and b are the calibration coefficients determined prelaunch, X and X_{sp} are the observed signals at the earth view and space view respectively and the radiance R is the band averaged radiance computed with the spectral response function. Their recommendation is to use Equation (54) as the b from prelaunch calibration in Equation (53) may not be valid on orbit because of drifts and noise spikes. However for GOES-8 and GOES-9 imagers, Equation (53) is used with $b = m\,X_0$ where X_0 is chosen to be 29.

The reflectance factor of the earth scene ρ is obtained from the radiance R as:

$$\rho = \frac{\pi\,R}{E_{sun}} \tag{55}$$

where E_{sun} is the band averaged solar spectral irradiance computed with the spectral response function of the visible band. The pre-launch calibration coefficients determined as above have been used post-launch until 2005. Since then those post launch coefficients are being updated monthly based on the collocated Terra MODIS cloud observations using GSICS inter comparison methodology [29].

3.1.3. Analysis of the GOES Calibration Algorithm and SI Traceability

The GOES Imager was in orbit prior to the LEO sensors discussed earlier and it provided a learning experience for the establishment of SI traceable observations from satellite sensors that followed as its main anomalies listed below for the TEB required critical examination of instrument design and assumptions of pre-launch characterization.

As discussed in Section 3.1.1 the calibration algorithm had to be modified on-orbit to account for the scan angle dependence of the reflectivity of the scan-mirror that was not observed until the launch of GOES-8.

In addition to the normal, expected diurnal variation of the calibration slope, m in Equation (50) (responsivity), caused primarily by the diurnal changes in background flux reaching the detectors, there was another effect occurring around midnight called the "midnight calibration anomaly". It was attributed to the extra heating of the instrument surfaces around midnight when the sun directly shines into the scan-mirror cavity in the front of the instrument. The emitted radiation from the rapidly heated surfaces gets reflected by the blackbody (emissivity < 1) and reaches the detector when the blackbody look is performed for calibration. This extra flux adding to the blackbody flux erroneously reduces the calculated slope during the heating phase, which leads to erroneously low measurements of the scene temperature. This effect is not noticed when the blackbody and the instrument cavity are close to the same temperature. In addition to this effect it was noticed that the space view gets contaminated with scattered solar flux during eclipse seasons resulting in a larger subtraction and increased slope. These effects were well studied and a correction algorithm was implemented [30]. The GSICS methodology based comparisons to well studied SI-traceable radiometers in LEO orbits, Aqua MODIS for visible band and Infrared Sounding Radiometer (IASI) on Metop-A for TEB were made to reduce the uncertainties in the corrections [31].

There were erroneous SRF induced scene dependent biases noticed in brightness temperature (T_b) in GOES series and the GSICS methodology based comparisons with Atmospheric Infrared Sounder (AIRS) and IASI allowed to correct the Spectral Response Functions of GOES TEB especially for Band 6 (13.3 μm) [32,33].

Various vicarious calibration methods were suggested and used based on Empirical Distribution Function (EDF) of earth view data, on measurements from stable earth target such as desert, on star observations, and on MODIS data [34]. As noted earlier Collocated Terra MODIS cloud observations are being used for operationally updating the calibration coefficients [29]. However, it is difficult to establish SI traceable uncertainty because of lack of quantitative measurements of environmental parameters and lack of sufficient validation of model assumptions to arrive at Top of the Atmosphere (TOA) reflected radiances. Based on the experience gained on the GOES Imager and other innovations in imaging technology, the ABI sensor is built as the next generation GEO sensor and will be discussed in the next section.

3.2. GOES-R Advanced Baseline Imager (ABI)

GOES-R forms the beginning of the 3rd generation series (also called as the 5th generation by Tim Schmit *et al.*, [26]) of geostationary satellites with the ABI instrument replacing the GOES Imager for weather, oceanographic, climate, and environmental applications. ABI is an advanced imager with sixteen spectral bands spanning visible to long wave infrared (0.45 μm–13.6 μm), three times more spectral information, four times the spatial resolution, and more than five times faster temporal coverage than the current GOES imager [35]. It has a unique 2-mirror scanner design with independent East/West and North/South scan mirrors allowing the system to be repositioned quickly while using less power. This design allows flexible, custom scanning configurable on orbit [36,37].

There are on-board SI traceable standards for radiometric calibration consisting of a three bounce cavity design blackbody with emissivity guaranteed to be better than 0.995 called the Internal Calibration Target (ICT) for TEB calibration and a Spectralon™ diffuser called the Solar Calibration Target (SCT) for the calibration of RSB. The ICT is observed at least every 15 min depending on the scan mode selected. The space look is collected every 30 s. The solar calibration can be scheduled as needed and ABI interrupts operational image collection and performs the solar calibration using SCT by opening the solar calibration cover (SCC) for reflecting sunlight by the diffuser in to the sensor through its solar viewing port. The SCC is closed the rest of the time to protect it from unnecessary exposure and degradation. The ABI can collect lunar images when the moon appears in its field of view for RSB calibration stability validation through intercomparison with the SD calibration

The GOES-R is yet to be launched and much of the ABI instrument specific measurements and calibration details are yet to be released to the public. However, the hardware configuration is similar to the Advanced Himawari Imager (AHI) in the geostationary meteorological satellite of the Japan Meteorological Agency (JMA), Himawari-8, that was launched in October 2014 and entered into operation in July 2015. The information for the review below is drawn from the JMA presentation [38] and presentations of ABI calibration scientists at technical meetings [39–41].

The general formulation of the calibration algorithm is same as of the GOES Imager except it is now adapted to the ABI optical setup. The radiance observed by GOES-R ABI viewing an object is determined from the measured counts (C) of the object view (*ov*) and the space view (*sv*) as follows

$$\Delta L = L^{ov} - L^{sv} = m\,\Delta C_{ov} + q\Delta C_{ov}^2 \tag{56}$$

where,

$$\Delta C_{ov} = C_{ov} - C_{sv} \tag{57}$$

The quantity m is determined pre-launch and up dated on orbit. The quantity q is derived from pre-launch measurements to account for any detector non linearity. Again the radiance L is used in

Equation (56) as a short hand and it represents the band averaged spectral radiance with the spectral response function of each band.

3.2.1. GOES-R ABI TEB Radiometric Calibration

As noted earlier the ABI instrument has two mirrors, North/South and the East/West. The North/South mirror projects the earth, the ICT and the space views to the East/West mirror which raster scans and reflects it to the telescope aperture. The angular dependence of the mirror reflectance and the emissivity of the mirrors are measured pre-launch and provided as LUTs for on-orbit radiometric data analysis to produce radiances. The mirror emission and reflection contribute to the total radiance observed. So these contributions are accounted and for convenience the following notation is used to represent their contributions as they are in the optical path to the telescope. The effective radiance L_x^{eff} at the telescope aperture from an object x is the product of its emissivity and its radiance with unity emissivity, and the reflectance of the North/South mirror, ρ_N^x and the East/West mirror, ρ_E^x at the angles they are set to view the object x.

$$L_x^{eff} = \varepsilon_x L^x \,.\, \rho_N^x \rho_E^x \tag{58}$$

Similarly the effective radiance $L_{N@x}^{eff}$ of the North/South mirror while viewing the object x is represented as the product of its emitted radiance at its temperature T_N and the reflectance of the East/West mirror.

$$L_{N@x}^{eff} = \varepsilon_N^x L^{T_N,\,x} \rho_E^x \tag{59}$$

The effective radiance $L_{E@x}^{eff}$ of the East/West mirror while viewing the object x is simply its emitted radiance at its temperature because it is the final optical element in the optical path to the telescope.

$$L_{E@x}^{eff} = \varepsilon_E^x L^{T_E,\,x} \tag{60}$$

Using the above notation the radiance at the Telescope aperture, L^x by accounting for the radiance contributions of various optical elements in the beam path for the space view (*sv*), ICT view (*ICTv*) and the Scene on the earth view (*ev*) can be written as follows.

For the space view:

$$L^{sv} = L_{N@sv}^{eff} + L_{E@sv}^{eff} \tag{61}$$

In Equation (59) the terms in the right hand side are the radiance contributions from the North/South and East/West mirrors given by Equations (57) and (58) respectively and the radiance contribution of space is dropped off as the radiance of space is assumed to be zero for all practical purposes.

Similarly, we can write using Equation (56), the radiance for the ICT view,

$$L^{ICTv} = L_{N@ICT}^{eff} + L_{E@ICT}^{eff} + L_{ICT}^{eff} \tag{62}$$

and the radiance for the scene on the earth view,

$$L^{ev} = L_{N@ev}^{eff} + L_{E@ev}^{eff} + L_{ev}^{eff} \tag{63}$$

Using Equations (59) and (60) and the calibration equation; Equation (54) the linear calibration coefficient m can be determined from

$$m = \frac{L_{ICT}^{eff} - q\,\Delta C_{ICT}^2 + \left(L_{E@ICT}^{eff} + L_{N@ICT}^{eff}\right) - \left(L_{E@sv}^{eff} + L_{N@sv}^{eff}\right)}{\Delta C_{ICT}} \tag{64}$$

Therefore, m is updated on orbit at least viewing ICT every 15 min depending on the scan mode as is being done in AHI every 10 min [38]. The rest of the quantities are measured from the ICT view and the space view or deduced from measurements using the LUTs.

Using m obtained from Equation (64), the scene radiance on the earth view can be determined from the calibration Equation (56) and using Equations (63) and (58), as follows:

$$\langle L_{ev}\rangle = \frac{m\ \Delta C_{ev} + q\Delta C_{ev}^2 - (L_{N@ev}^{eff} - L_{N@sv}^{eff}) - \left(L_{E@ev}^{eff} - L_{E@sv}^{eff}\right)}{\rho_N^{ev}\rho_E^{ev}} \quad (65)$$

where $\langle L_{ev}\rangle$ is band averaged radiance of the earth scene, *i.e.*, mW/(m^2-sr-cm^{-1}) for each band that can be converted to corresponding brightness temperature using Planck's law (emissivity = 1). It should be noted that in Equation (64) any reflected radiation from the blackbody due to its finite emissivity is neglected as is the case for GOES sensor where as it is accounted in the calibration equations of MODIS and VIIRS sensors.

3.2.2. GOES-R ABI RSB Radiometric Calibration

The radiance observed for the RSB bands follows from Equation (65) except that the mirror contributions for radiance in this spectral region are negligible and the offset due to the radiance from mirrors is set to zero. However, the space view counts are still subtracted in determining ΔC_{ev} to eliminate any other background effect and biases for the RSB. Therefore, for RSB,

$$\langle L_{ev}\rangle = \frac{m\ \Delta C_{ev} + q\Delta C_{ev}^2}{\rho_N^{ev}\rho_E^{ev}} \quad (66)$$

where $\langle L_{ev}\rangle$ is the band-averaged spectral radiance for the scene in the earth view, *i.e.*, W/(m^2-sr-µm) for the RSB, m and q are the linear and quadratic coefficients, respectively. Again the quadratic coefficient q is determined pre-launch per band and per detector element as noted in Section 3.2. The linear coefficient m for RSB is determined on-orbit viewing the Solar Calibration Target (SCT), which is a solar diffuser. Essentially Equation (64) holds for determining m except the mirror radiance contributions drop off as noted earlier for RSB and

$$m = \frac{f_{int,ch}\ L_{SCT}^{eff} - q\ \Delta C_{SCT}^2}{\Delta C_{SCT}} \quad (67)$$

where $f_{int,ch}$ is the integration factor for each band for viewing *SCT* determined pre-launch. The integration factor accounts for the change in integration time to measure a higher signal in viewing the SCT. L_{SCT}^{eff} is the effective band average calibrated radiance defined for viewing *SCT* as

$$L_{SCT}^{eff} = \langle L_{SCT}\rangle\ \rho_N^{sct}\rho_E^{sct} \quad (68)$$

Here, ρ_E^{SCT} and ρ_N^{SCT} are the reflectance of the East/West and North/South scan mirrors respectively, when viewing *SCT*. In Equation (66), $\langle L_{SCT}\rangle$ is the SCT band-averaged on-orbit calibrator radiance and is computed from the equation

$$\langle L_{SCT}\rangle = K_{\beta_{eff}}^{detector\ row\#} cos\left(\theta_{sun}\right)\left[\pi\langle L_{100\%\alpha}\rangle\left(\frac{R_{sun}}{r_{sun}}\right)^2\right] \quad (69)$$

In Equation (69), R_{sun} is the average radial distance from Earth to the Sun (*i.e.*, 1 AU) and r_{sun} is the actual distance between the Earth and the Sun at the time of the calibration. $\pi\langle L_{100\%\alpha}\rangle$ is the solar irradiance at 1 AU over a Lambertian surface with 100% albedo. The factor

$cos\left(\theta_{sun}\right)\left[\pi\langle L_{100\%\alpha}\rangle\left(\frac{R_{sun}}{r_{sun}}\right)^2\right]$ is solar irradiance normal to the solar diffuser surface when the solar incident angle is θ_{sun} and earth-sun distance is r_{sun}. The factor $K_{\beta_{eff}}^{detector\ row\#}$ is related to the effective BRDF of the *SCT*. It is an instrument property parameter that is determined pre-launch.

3.2.3. Analysis of the GOES-R ABI Calibration Algorithm and SI Traceability

The ABI with its advanced imaging capabilities and on-board SI traceable standards built to meet stringent uncertainty requirements opens a new era of SI traceable radiometric observations from GEO. The calibration algorithm is transparent in describing all the terms considered for converting the scene signal into radiance. The reflection of background from the blackbody in the calibration is not considered in the calibration algorithm. One of the reasons to neglect this effect is that the emissivity of the blackbody is high. Also the instrument is designed to maintain the temperature of the blackbody at a constant temperature and the background is well controlled by thermal stabilization of the environment. However, including the term would have been better just in case to account for anomalies that may happen due to uncontrolled background as happened with GOES Imager.

There was extensive pre-launch calibration using transfer standards from NIST that adds solid support for pre-launch SI traceability. The vendor measured spectral response functions (SRF) were validated through additional band pass measurements at NIST on witness samples for the sensor geometry of incidence angles and operating temperature. The sensor was tested in a cryo-chamber at LN_2 background for validating the TEB radiance calibration algorithm with a blackbody called the External Calibration Target (ECT) and was found to meet the specifications [42]. The ECT was calibrated with the NIST Thermal IR transfer standard radiometer (TXR) for the radiance observed in its field of view and predicted from the ECT thermal model [43]. GOES-R ABI has large number of detectors (hundreds) per band and the SRF differences between detectors in each band are relatively small and met the sensor specifications. In order to maintain computational efficiency to meet the needs of operational processing in a timely manner, it is decided to use a band averaged single SRF for each band. However, the individual detector level SRFs are available for GOES-R project to reprocess on-orbit data for research and analysis. Studies are done on the impacts for on-orbit analyses of using simulated average of the SRFs of all detectors in each band *vs.* individual measured detector SRFs for TEB and the radiometric biases in the RSB in a simulated comparison between GOES-R ABI and VIIRS expected radiances from Sonoran Desert and White Sands National Monument desert sites. The knowledge base built in these studies helps assess uncertainties in the forthcoming GOES-R ABI on-orbit data analysis [40,41]. The size of the solar diffuser in GOES-R ABI chosen by design limitations only fills partially the full aperture of the telescope and as such the uncertainties estimated in prelaunch testing need to be validated post launch on orbit. Also, there is no solar diffuser stability monitor for RSB. So there is extensive post launch testing planned to monitor and validate the individual detector spectral responsivity uniformity band by band using the North/South mirror scanning a vicarious calibration scene of uniform radiance while the East/West scan mirror points to each band systematically. The GOES-R ABI CWG at NOAA, NASA program office and the Vendor have been working together in implementing such best practice guidelines for SI traceability [39]. Also the GOES-R ABI CWG and NIST researchers worked together and created methodology to translate the sensor radiometric requirements written in error analysis framework to the modern approach based on the GUM that allows implementation of the guideline for SI traceability [44]. It helps to adopt the best practice in analyzing and reporting ABI radiance results and comparisons to other sensors on-orbit in the GSICS frame-work.

4. Discussion—Comparison of the Spectral Bands and Their Calibration Features of the Four Sensors, MODIS, VIIRS, GOES and GOES-R ABI

4.1. Comparison of Spectral Bands

The spectral data on the bands of VIIRS and MODIS that are equivalent for their primary purpose are shown in Tables 1 and 2. Their horizontal spatial resolution (HSR) specification is also given in the tables. There are 22 bands in VIIRS. The calibration uncertainty specification in spectral reflectance for a scene at typical radiance of VIIRS RSB both M and I bands is less than 2%. The specifications on the absolute uncertainty of emissive M bands varied based on the scene temperature and typically for 270 K it is 0.7% for M12 and M13, 0.6% for M14 and 0.4% for M15 and M16.

The VIIRS imaging bands are compared with equivalent MODIS bands in Table 2 in the same way as Table 1. The absolute uncertainty on the VIIRS imaging TEB is 5% for I4 and 2.5% for I5. The polychromatic Day-Night Band (DNB) in VIIRS is a special addition and MODIS did not carry its equivalent. The DNB uncertainty requirement is based on its gain state in the range of 5% to 100%. More details on these specifications of all VIIRS bands can be found in Reference [19]. These specifications are currently being validated and reported [21,22].

There are 36 bands in MODIS in total 15 bands go beyond the VIIRS equivalent 21 bands. The full details on all the MODIS bands can be seen in Reference [13]. The uncertainty requirement for MODIS RSB bands was 2% in reflectance and 5% in radiance. The MODIS TEB uncertainty specification was 1% in radiance except 0.5% for bands 31 and 32 for SST measurements, 0.75% for Band 20 and 10% for band 21 for fire detection [45]. These specifications are being continuously validated operationally and reported [17,18].

Table 1. Visible Infrared Imaging Radiometer Suite (VIIRS) Spectral Bands (M) and Moderate Resolution Imaging Spectroradiometer (MODIS) Equivalent Bands (M for Moderate spatial resolution; HSR for Horizontal Spatial Resolution).

VIIRS Band	Central Wavelength (μm)	Wavelength Range (μm)	Nadir HSR (m)	Primary Use	MODIS Equal Band(s)	Central Wavelength (μm)	Wavelength Range (μm)	Nadir HSR (m)
M1	0.412	0.402–0.422	750	Ocean color Aerosals	8	0.4113	0.405–0.420	1000
M2	0.445	0.436–0.454	750	Ocean color Aerosals	9	0.442	0.438–0.448	1000
M3	0.488	0.478–0.488	750	Ocean color Aerosals	3 10	0.4656 0.4869	0.459–0.479 0.483–0.493	500 1000
M4	0.555	0.545–0.565	750	Ocean color Aerosals	4 12	0.5536 0.5468	0.545–0.565 0.546–0.556	500 1000
M5	0.672	0.662–0.682	750	Ocean color Aerosals	13 14	0.6655 0.6768	0.662–0.672 0.673–0.683	1000 1000
M6	0.746	0.739–0.754	750	Atmospheric Correction	15	0.7464	0.743–0.753	1000
M7	0.865	0.846–0.885	750	Ocean color Aerosals	16	0.8662	0.862–0.877	1000
M8	1.240	1.23–1.25	750	Cloud Particle size	5	1.2416	1.23–1.25	500
M9	1.378	1.371–1.386	750	Cirrus/Cloud Cover	26	1.38	1.36–1.39	1000
M10	1.61	1.58–1.64	750	Snow Fraction	6	1.629	1.628–1.652	500
M11	2.25	2.23–2.28	750	Clouds	7	2.114	2.105–2.155	500
M12	3.7	3.61–3.79	750	Sea Surface Temperature (SST)	20	3.79	3.66–3.84	1000
M13	4.05	3.97–4.13	750	SST/Fires	21 22 23	3.96 3.96 4.06	3.929–3.989 3.929–3.989 4.02–4.08	1000 1000 1000
M14	8.55	8.4–8.7	750	Cloud Top Properties	29	8.52	8.4–8.7	1000
M15	10.763	10.26–11.26	750	SST	31	11.02	10.78–11.28	1000
M16	12.013	11.54–12.49	750	SST	32	12.03	11.77–12.27	1000

Table 2. VIIRS Spectral Bands (I) and MODIS Equivalent Bands (I for Imaging).

VIIRS Band	Central Wavelength (μm)	Wavelength Range (μm)	Nadir HSR (m)	Primary Use	MODIS Equal Band (s)	Central Wavelength (μm)	Wavelength Range (μm)	Nadir HSR (m)
DNB	0.7	0.5–0.9	750 (across full scan)	Imagery				
I1	0.64	0.6–0.68	375	Imagery	1	0.6455	0.62–0.67	250
I2	0.865	0.85–0.88	375	NDVI	2	0.8565	0.841–0.876	250
I3	1.61	1.58–1.64	375	Binary Snow Map	6	1.6291	1.628–1.652	500
I4	3.74	3.55–3.93	375	Imagery of Clouds	20	3.79	3.66–3.84	1000
I5	11.45	10.5–12.4	375	Imagery of Clouds	31 32	11.02 12.03	10.78–11.28 11.77–12.27	1000 1000

The GOES-R ABI imager has 16 bands and their wavelength range, central wavelength, Instantaneous Geometric Field of View (IGFOV), primary use, comparable bands in MODIS, VIIRS and GOES are shown in Table 3 [35]. As noted earlier the GOES-R series carry only the imager ABI and do not have a sounder. However some of the bands of the GOES sounder are carried into the ABI as shown Table 3.

Table 3. Advanced Baseline Imager (ABI) Spectoral band data and equivalent bands in heritage instruments.

ABI Band	Wavelength Range (μm)	Central Wavelength (μm)	IGFOV (km)	Primary Use	Heritage Instrument (s)
1	0.45–0.49	0.47	1	Daytime aerosol land, coastal water mapping	MODIS B 3 and B 10; VIIRS M3
2	0.59–0.69	0.64	0.5	Daytime Clouds, fog, Insolation, winds	Current GOES Imager; Sounder
3	0.846–0.885	0.865	1	Vegetation/ burn scar and aerosol over water, winds	VIIRS I2; MODIS B 2
4	1.371–1.386	1.378	2	Daytime Cirrus Cloud	VIIRS M9; MODIS B 26
5	1.58–1.64	1.61	1	Daytime Cloud-top Phase and particle size, snow	VIIRS I3 and M10; MODIS B 6
6	2.225–2.275	2.25	2	Daytime land/cloud properties, particle size, vegetation, snow	VIIRS M11; MODIS B 7
7	3.80–4.00	3.9	2	Surface and Cloud, fog at night, fire and winds	Current GOES Imager
8	5.77–6.6	6.19	2	High-level atmospheric water vapor, winds, rainfall	Current GOES Imager
9	6.75–7.15	6.95	2	Mid-level atmospheric water vapor, winds and rainfall	Current GOES Sounder
10	7.24–7.44	7.34	2	Lower-level water vapor, winds, rainfall	Current GOES Sounder
11	8.3–8.7	8.5	2	Total water for stability, cloud phase, dust, SO_2	VIIRS M14; MODIS B 29
12	9.42–9.8	9.61	2	Total Ozone, Turbulence and winds	Current GOES Sounder
13	10.1–10.6	10.35	2	Surface and Cloud	VIIRS M15; MODIS B 31
14	10.8–11.6	11.2	2	Imagery, SST, Clouds and rainfall	Current GOES Sounder
15	11.8–12.8	12.3	2	Total water, ash, and SST	Current GOES Sounder
16	13.0–13.6	13.3	2	Air temperature, cloud heights and amounts; CO_2	Current GOES Imager; Current GOES Sounder

4.2. Comparison of Sensor Features for Calibration and On-Orbit Observations

In Table 4 below the scanning optics hardware of the four sensors is compared. The VIIRS design is an advancement based upon the experience from heritage sensors. The VIIRS has RTA that is chosen because of the need to have better scattered light control than MODIS because of VIIRS higher altitude orbit to cover the earth in a day. Also the HAM rotates at half the angle in VIIRS and reduces the scan angle effects on the sensor response compared to the paddle mirror in MODIS as discussed in Reference [19].

Table 4. Comparison of the radiometer optics.

MODIS	VIIRS	GOES Imager	GOES-R ABI
Passive Cross-track imaging radiometer	Passive Cross-track imaging radiometer	Passive staring imaging radiometer	Passive staring imaging radiometer
Two sided beryllium paddle wheel scan mirror continually rotates at 20.3 rpm (1.478 s for each mirror side).	Scanning telescope: RTA (Rotating telescope assembly) projecting on to a half angle mirror (HAM). The RTA continuously rotates at constant speed taking 1.7864 s per revolution and is synchronized with the HAM once per scan.	Scan mirror that alternately sweeps east to west and west to east perpendicular to a north south path to direct the beam to a Casegrain telescope. Full disk scan interval 30 min	N/S scanning mirror reflection scanned by E/W mirror to the fore-optics (Off-Axis Four Mirror Assembly (FMA) Telescope. Independent N/S and E/W scanners. E/W raster scan rate 1.4 degree/Section Full disk scan interval 5 min.

The GOES-R ABI has 2-mirror scanner design that is unique with its capability to scan many ways that is not possible in the GOES heritage sensor architecture. The major difference is GOES-R ABI uses a raster scan that allows collection of all data in one scan direction. The scan from one point to another point is called a swath and the image collection is swath based. The time interval across swath boundary is constant. The swaths can be chosen to cover the full disk or a commanded area to be observed (CONUS) or meso scale scene or other choices.

Tables 5 and 6 show the comparison of various other features of the sensor calibration algorithms of TEB and RSB respectively.

Table 5. Comparison of various features of Emissive band calibration of the four sensors.

Feature	MODIS	VIIRS	GOES Imager	GOES-R ABI
Instrumentation temperature dependency	Yes	Yes	No (Calibration algorithm). Yes (for corrections)	No
Scan mirror Radiance considered	Yes	Yes	Yes	Yes (Both East/West and North/South mirrors)
Scan mirror housing cavity	Yes	Yes	No	No
Telescope Radiance considered	No Need	Yes	No Need	No Need
Calibration function	Quadratic with offset	Quadratic with offset	Quadratic with offset (before space view subtraction)	Quadratic (Offset cancelled by space view)
On-orbit Calibration Coefficients updated	Linear coefficient (Offset and non-linear terms from LUTs updated from on orbit BB warm up and cool down cycles)	Scale Factor F to update calibration coefficients provided in the LUTs (Offset and non-linear terms from LUTs updated from on orbit BB warm up and cool down cycles)	Calibration coefficients determined on orbit	Linear Coefficient is updated and LUT provides quadratic coefficient and no offset

Table 5. *Cont.*

Feature	MODIS	VIIRS	GOES Imager	GOES-R ABI
Spectral Response Function (SRF)	Pre-launch	Pre-launch	Prelaunch (Corrections using GSICS LEO comparison [31,32]	Pre-launch
Band averaged spectral radiance	LUT generated using SRF	LUT generated using SRF	Approximate formulas generated using Planck and SRF.	LUT generated using SRF
Response *versus* mirror scan angle	Pre-launch Post-launch verification–Update once done for TERRA Deep Space maneuver)	Pre-launch and on-orbit verification (Pitch Maneuver)	Determined using on orbit space look data at various angles and laboratory witness sample data for 45° scan angle.	Pre-launch (potential on-orbit verification)
Calibration Interval	Scan-by scan	Scan-by-scan for single Gain bands. Dual gain M13 band Low gain LUT on-orbit update.	Blackbody 30 min; Space view 2.2 s, or 36.6 s	5–15 min

Table 6. Comparison of various features of RSB calibration of the four sensors.

Feature	MODIS	VIIRS	GOES Imager	GOES-R ABI
Instrumentation temperature dependency	Yes	Yes	No	No
Detector temperature dependency	Yes (detector temperatures are correlated with instrument temperatures)	Yes	No	No
Calibration function	Linear without offset	Quadratic with offset	Linear with offset	Quadratic (Offset cancelled by space view)
On-orbit Calibration Coefficient	Linear coefficient based on calibration by Solar Diffuser	Scale Factor F using Solar Diffuser calibration to update Calibration coefficients provided in the LUTs	Calibration coefficients are being updated monthly by vicarious calibration using collocated Terra MODIS cloud observations since 2005.	Linear Coefficient using solar Diffuser Calibration and also using LUT
Spectroradiometric Calibration assembly Sector (SRCA)	Yes	No	No	No
Spectral Response Function (SRF)	Pre-launch and on orbit verification using SRCA	Pre-launch	Prelaunch	Pre-launch
Response *versus* mirror scan angle	Post Launch on–orbit using moon or earth view	Pre-launch (Pitch maneuver—on-orbit verification)	No	Pre-launch (potential on-orbit verification)
Lunar Calibration	Yes	Yes	Yes	Yes
Solar Diffuser Stability Monitor (SDSM)	Yes	Yes	Not Applicable as No Solar Diffuser	No
Solar Diffuser (SD) Calibration Interval	Weekly to tri-weekly	Daily	No SD	Weekly to monthly
SD–Vignette Function (VF) of Partial Aperture (PA)	Averaged VF Pre-launch, and on-orbit verification	Band-dependent VF Pre-launch (Updated on orbit)	No SD	Partial Aperture Pre-launch
Solar Irradiance data	3 data sets; (0.4 to 0.8 µm [46]; 0.8 to 1.1 µm [47]; Above 1.1 µm [48].	MODTRAN 4.3 [49]	Bishop and Rossow [50]	4 data sets; 0.4 to 1.2 µm [47]), 033 to 1.25 µm [50]; 1.2 to 100 µm [51]; 0.2 to 10.1 µm, [52].

5. Conclusions

The review shows that the calibration algorithm harmonization facilitated by NOAA NCC, accounting for all the contributions to the radiometric signal based on SI traceable measurements from pre-launch to post launch is being implemented at NOAA/NESDIS/STAR following the best practice guidelines going from MODIS to VIIRS and GOES to GOES-R ABI. The new sensors VIIRS and GOES-R ABI have SI traceable standards for on-orbit radiometric calibration having the methodology evolved from lessons learned from the legacy sensors MODIS and GOES. However, it is an open question to see how the assumption made in neglecting the background reflection from the blackbody in GOES-R ABI holds on orbit.

The critical component that leads to uncertainty in on-orbit measurements is the scanning mirror temperature and its reflectance. In Terra MODIS the RVS (scan angle dependence of the scanning mirror reflectance) had to be measured on-orbit with a deep space maneuver of the spacecraft as discussed in Section 2.1.3. The follow up Aqua MODIS pre-launch characterization based on Terra experience was more thorough and the RVS problem was thus mitigated. Current S-NPP VIIRS SST data during WUCD of the on-board blackbody is being shown as anomalous compared to GSICS based comparisons to other sensors. The WUCD of the blackbody is carried out quarterly and the discrepancy noticed is consistent in the time series. Currently this issue is under investigation by the VIIRS SDR team. Lessons learned in resolving this issue could help the upcoming VIIRS operational sensor on Joint Polar Satellite System (JPSS) satellite missions JPSS -1 and JPSS-2 to eliminate the problem.

The GOES-R ABI optical setup has dual scanning mirrors with advanced scanning capabilities and their RVS was measured pre-launch to meet stringent requirements to provide reliable data as LUTs and should provide accurate data base to mitigate anomalies. The legacy experience of GOES sensor in this regard as discussed in Section 3.1.3 gave a solid base in the design of GOES-R ABI and its calibration algorithm. The blackbody temperature is controlled to one operating temperature and the scanning environment is effectively shielded from spurious reflections and heating effects, thus reducing the possibility of diurnal variation and midnight anomaly (Section 3.1.3) experienced in GOES. The limited data available from Himawari-8 AHI the equivalent of GOES-R ABI largely support the expected on-orbit performance of GOES-R ABI. However, including the possibility of background reflection from the blackbody in the calibration algorithm would have been better to start with for possible use if needed.

The best practice guidelines give equal importance to both pre-launch and post launch testing and monitoring of SI traceability of measurements. Both VIIRS and GOES-R ABI are following the best practice guidelines. For RSB, the VIIRS calibration algorithm is similar to MODIS and the VIIRS SD degradation issue was closely monitored using the MODIS data for inter comparison under GSICS methodology. There is no SDSM in GOES-R ABI and the calibration algorithm uses lunar observations for SD stability monitoring besides taking the precaution that the SD is exposed to Sun only during RSB calibration. The RSB observations of Himawari-8 AHI show thus far that the sensor is largely following expectations; although anomalies due to some scattered light issues are being discussed in private meetings.

Acknowledgments: We thank Boryana Effermova, Fangfang Yu, Mike Weinreb, Aaron Pearlman and Frank Padula at NOAA reading the manuscript and giving us valuable comments. The views, opinions, and findings contained in this paper are those of the authors and should not be construed as official positions, policy, or decisions of the NOAA or the U. S. Government. Commercial companies identified in this paper are only to foster understanding. Such identification does not imply recommendation or endorsement by the NOAA, nor does it imply that they are the best available for the purpose.

Author Contributions: All authors contributed equally to this work.

Conflicts of Interest: The authors declare no conflict of interest.

Appendix

The radiometric symbols, terminology and methodology in the calibration algorithms of sensors reviewed in this article mostly follow self explanatory common usage. However, Table A1 below shows a few specific exceptions.

Table A1. Few differences in symbols, terminology and methodology across sensors.

Terminology	Symbols			
	MODIS	VIIRS	GOES/Imager	GOES-R ABI
Radiance	*L*	*L*	*R*	*L*
Detector output (counts) used in the Calibration equation	The difference in counts between sensor view of an object and space view.	The difference in counts between sensor view of an object and space view.	Detector response in Counts.	The difference in counts between sensor view of an object and space view.
(*SV* = *Space View*)	$dn = \mathrm{DN} - \mathrm{DN}_{SV}$	$dn = DN - \overline{DN_{sv}}$	X	$\Delta C = C - C_{sv}$.

References

1. Wyatt, C.L.; Privalsky, V.; Datla, R. Symbols, terms, units and uncertainty analysis for radiometric sensor calibration, NIST handbook 152. In *Recommended Practice*; National Technical Information Service, U.S. Department of Commerce: Springfield, VA, USA, 1998; pp. 1–91.
2. Ohring, G.; Wielicki, B.; Spencer, R.; Emery, W.J.; Datla, R. *Satellite Instrument Calibration for Measuring Global Climate Change, NISTIR 7047*; National Institute of Standards and Technology: Gaithersburg, MD, USA, 2004; pp. 1–101.
3. Ohring, G.; Tansock, J.; Emery, W.; Butler, J.; Flynn, L.; Weng, F.; Germain, K.S.; Wielicki, B.; Cao, C.; Goldberg, M.; *et al.* Achieving satellite instrument calibration for global climate change. *EOS Trans. Am. Geophys. Union* **2007**, *88*, 136–136. [CrossRef]
4. Datla, R.U.; Rice, J.P.; Lykke, K.R.; Johnson, B.C.; Butler, J.J.; Xiong, X. Best practice guidelines for pre-launch characterization and calibration of instruments for passive optical remote sensing. *J. Res. Natl. Inst. Stand. Technol.* **2011**, *116*, 621–646. [CrossRef]
5. Cooksey, C.; Datla, R. Workshop on bridging satellite climate data gaps. *J. Res. Natl. Stand. Technol.* **2011**, *116*, 505–516. [CrossRef]
6. Datla, R.; Weinreb, M.; Rice, J.; Johnson, B.C.; Shirley, E.; Cao, C. Optical passive sensor calibration for satellite remote sensing and the legacy of NOAA and NIST cooperation. *J. Res. Natl. Inst. Stand. Technol.* **2014**, *119*, 235–255. [CrossRef] [PubMed]
7. NOAA National Calibration Center (NCC). Available online: http://ncc.nesdis.noaa.gov/ (accessed on 27 January 2016).
8. Borzyminski, J.; Buzoianu, M.M.; Bievre, P.D.; Imai, H.; Karshenboim, S.; Kool, W.; Krystek, M.; Mari, L.; Muller, M.M.; Narduzzi, C.; *et al.* Joint Committee for Guides in Metrology (JCGM). In *International Vocabulary of Metrology—Basic and General Concepts and Associated Terms (VIM Third Edition), JCGM 200:201*; BIPM: Paris, France, 2012.
9. Bich, W.; Cox, M.; Ehrlich, C.D.; Elster, C.; Estler, W.T.; Fischer, N.; Hibbert, D.B.; Imai, H.; Mussio, L.; Nielsen, L.; Pendrill, L.R.; *et al.* Joint Committee for Guides in Metrology (JCGM). In *Evaluation of Measurement Data—Guide to the Expression of Uncertainty in Measurement (GUM) JCGM 100:2008*; BIPM, IEC, IFCC, ISO, IUPAC, IUPAP, OIML: Paris, France, 1995.
10. Goldberg, M.; Ohring, G.; Butler, J.; Cao, C.; Datla, R.; Doelling, D.; Gärtner, V.; Hewison, T.; Iacovazzi, B.; Kim, D.; *et al.* The global space-based inter-calibration system. *Bull. Amer. Meteor. Soc.* **2011**, *92*, 467–475. [CrossRef]
11. Tansock, J.; Bancroft, D.; Butler, J.; Cao, C.; Datla, R.; Hanse, S.; Helder, D.; Kacker, R.; Latvakoski, H.; Mlynczak, M.; *et al.* *Guidelines for Radiometric Calibration of Electro-Optical Instruments for Remote Sensing, NISTHB 157*; National Institute of Standards and Technology: Gaithersburg, MD, USA, 2015; pp. 1–131.
12. Butler, J.; Johnson, B.C.; Barnes, R.A. The calibration and characterization of earth remote sensing and environmental monitoring instruments. In *Optical Radiometry*; Parr, A., Datla, R.U., Gardner, J.L., Eds.; Elsevier Academic Press: San Diego, CA, USA, 2005; Volume 41, pp. 453–534.
13. Xiong, J.; Toller, G.; Chiang, V.; Sun, J.; Esposito, J.; Barnes, W. *MODIS Level 1-B Algorithm Theoretical Basis Document*; Version 4; National Aeronautic and Space Administration (NASA), Goddard Space Flight Center (GSFC): Greenbelt, MD, USA, 2013; pp. 1–40.

14. Xiong, X.; Salomonson, V.; Chiang, K.; Wu, A.; Guenther, B.; Barnes, W. On-orbit characterization of RVS for MODIS thermal emissive bands. *Proc. SPIE* **2004**, *5652*, 210–218.
15. Xiong, X.; Wenny, B.N.; Barnes, W.L. Overview of NASA earth observing systems Terra and Aqua moderate resolution imaging spectroradiometer instrument calibration algorithms and on-orbit performance. *J. Appl. Remote Sens.* **2009**, *3*, 1–25.
16. Xiong, X.; Barnes, W.; Chiang, K.; Erives, H.; Che, N.; Sun, J. Status of aqua MODIS on-orbit calibration and characterization. *Proc. SPIE* **2004**, *5570*, 317–327.
17. Chiang, K.; Xiong, X.; Wu, A.; Barnes, W. MODIS Thermal emissive bands calibration uncertainty analysis. *Proc. SPIE* **2004**, *5542*, 437–447.
18. Esposito, J.; Xiong, X.; Wu, A.; Sun, J.; Barnes, W. MODIS reflective solar bands uncertainty analysis. *Proc. SPIE* **2004**, *5542*, 448–458.
19. Baker, N. *Joint Polar Satellite System (JPSS) VIIRS Radiometric Calibration Algorithm Theoretical Base Document (ATBD)*; National Aeronautic and Space Administration (NASA), Goddard Space Flight Center (GSFC): Greenbelt, MD, USA, 2014; pp. 1–195.
20. Moyer, D.; Luccia, F.; Moy, G.; Haas, E.; Wallisch, C. *Personal Communication*; The Aerospace Corporation: Los Angeles, CA, USA, 2015.
21. Efremova, B.; McIntire, J.; Moyer, D.; Wu, A.; Xiong, X. S-NPP VIIRS thermal emissive bands on-orbit calibration and performance. *J. Geophys. Res. Atmos.* **2014**. [CrossRef]
22. Xiong, X.; Butler, J.; Chiang, K.; Efremova, B.; Fulbright, J.; Ler, N. VIIRS on-orbt calibration methodology and performance. *J. Geophys. Res. Atmos.* **2013**. [CrossRef]
23. McIntire, J.; Moyer, D.; Efremova, B.; Oudari, H.; Xiong, X. On-orbt characterization of S-NPP VIIRS transmission functions. *IEEE Trans. Geosci. Remote Sens.* **2015**, *53*, 2354–2365. [CrossRef]
24. Wu, A.; Xiong, X.; Chiang, K.; Sun, C. Assessment of the NPP VIIRS RVS for the thermal emissive bands using the first pitch maneuver observations. *Proc. SPIE* **2012**, *8510*, 85101Q.
25. Johnson, E.; Galang, K.; Ranshaw, C.; Robinson, B. NPP visible/infrared imager radiometer suite (VIIRS) radiance uncertainty, emissive bands–tested performance. *Proc. SPIE* **2010**, *7808*. [CrossRef]
26. Fact Checking GOES Current and Future. Available online: http://www.ssec.wisc.edu/media/images/january2013/cimss_slide_shows/menzel_schmit.pdf (accessed on 27 January 2016).
27. Imager–GOES Project Science. Available online: https://www.yumpu.com/en/document/view/7736366/goes-n-databook-goes-project-science-nasa/7 (accessed on 27 January 2016).
28. Weinreb, M.P.; Jamieson, M.; Fulton, N.; Chen, Y.; Johnson, J.X.; Smith, C.; Bremer, J.; Baucom, J. Operational calibration of geostationary operational environmental satellite-8 and -9 imagers and sounders. *App. Opt.* **1997**, *36*, 6895–6904. [CrossRef]
29. Wu, X.; Sun, F. Post-launch calibration of GOES imager visible channel using MODIS. *Proc. SPIE* **2005**, *5882*. [CrossRef]
30. Johnson, R.X.; Weinreb, M. GOES-8 Imager midnight effects and slope correction. *Proc. SPIE* **1996**, *2812*, 596–607.
31. Yu, F.; Wu, X.; Rama Varma Raja, M.K.; Li, Y.; Wang, L.; Goldberg, M. Diurnal and scan angle variations in the calibration of GOES imager infrared channels. *IEEE Trans. Geosci. Remote Sens.* **2013**, *51*, 2354–2365. [CrossRef]
32. Yu, F.; Wu, X. Correction for GOES Imager spectral response function using GSICS. Part II: Applications. *IEEE Trans. Geosci. Remote Sens.* **2013**, *51*, 1200–1214. [CrossRef]
33. Wu, X.; Yu, F. Correction for GOES imager spectral response function using GSICS. Part I: Theory. *IEEE Trans. Geosci. Remote Sens.* **2013**, *51*, 1215–1223. [CrossRef]
34. Wu, X.; Weinreb, M.; Chang, I.-L.; Crosby, D.; Dean, C.; Sun, F.; Han, D. Calibration of GOES Imager visible channels. *IEEE Int. Geosci. Remote Sens. Symp.* **2005**, *5*, 3432–3435.
35. Schmit, T.J.; Gunshor, M.M.; Menzel, W.P.; Gurka, J.J.; Li, J.; Bachmeier, A.S. Introducing the next generation advanced baseline imager on GOES-R. *BAMS* **2005**, *86*, 1079–1096. [CrossRef]
36. Exelis–Brochure. Available online: http://www.exelisinc.com/solutions/ABI/Documents/ABI_Brochure.pdf (accessed on 27 January 2016).
37. ABI Delivers Significantly Increased Capabilities over Current Imagers. Available online: http://www.goes-r.gov/ downloads/GOES_Users_ConferenceIV/Complete%20Posters/GUC4_poster_Griffith.pdf (accessed on 27 January 2016).

38. Okuyama, A.; Andou, A.; Date, K.; Hoasaka, K.; Mori, N.; Murata, H.; Tabata, T.; Takahashi, M.; Yoshino, R.; Bessho, K. Preliminary validation of Himawari-8/AHI navigation and calibration. *Proc. SPIE* **2015**, *9607*, 96072E.
39. Slack, K. ABI Calibration. Available online: http://www.google.com/url?sa=t&rct=j&q=&esrc=s&source=web&cd=1&ved=0ahUKEwjrx5H5yvLJAhXBJx4KHausCUwQFggcMAA&url=http%3A%2F%2Fwww.goes-r.gov%2Fdownloads%2FGOES-R_Series_Program%2F2014%2F11-Slack-pres.pptx&usg= AFQjCNE9QxMvtTaNsJLMM9IgfzX5cwcF1A&bvm=bv.110151844,d.eWE (accessed on 27 January 2016).
40. Pearlman, A.; Pogorzala, D.; Cao, C. Goes-R advanced baseline imager: Spectral response functions aqnd radiometric biases with the NPP visible infrared imaging radiometer suite evaluated for desert calibration sites. *App. Opt.* **2013**, *52*, 7660–7668. [CrossRef] [PubMed]
41. Pearlman, A.; Padula, F.; Cao, C.; Wu, X. The GOES-R advanced baseline imager: Detector spectral response effects on thermal emissive band calibration. *Proc. SPIE* **2015**, *9639*, 963917.
42. Pearlman, A.; Datla, R.; Cao, C.; Wu, X. Multichannel IR sensor calibration validation using Planck's law for next generation environmental geostationary systems. In Proceedings of the Calcon Technical Meeting: Meeting on Characterization and Radiometric Calibration for Remote Sensing, Logan, UT, USA, 26 August 2015.
43. Datla, R.U.; Rice, J.P. *NIST TXR Calibration of the GOES-R External Calibration Target (ECT) for the GOES-R Advanced Baseline Imager (ABI) Program, NISTIR 7797*; National Institute of Standards and Technology: Gaithersburg, MD, USA, 2011; pp. 1–32.
44. Pearlman, A.; Datla, R.; Kacker, R.; Cao, C. Translating radiometric requirements for satellite sensors to match international standards. *J. Res. Natl. Inst. Stand. Technol.* **2014**, *119*, 272–276. [CrossRef] [PubMed]
45. Xiong, X. MODIS On-Orbit Calibration and Lessons Learned. In Proceedings of the Calcon Technical Conference: Pre-Conference Tutorial, Logan, UT, USA, 27 August 2012.
46. Thuillier, G.; Herse, M.; Simon, P.C.; Labs, D.; Mandel, H.; Gillotay, D.; Foujols, T. The visible solar spectral irradiance from 350 to 850 nm as measured by the SOLSPEC spectrometer during the ATLAS-1 mission. *Sol. Phys.* **1998**, *177*, 41–61. [CrossRef]
47. Neckel, H.; Labs, D. The solar radiation between 3300 and 12500 A. *Solar Phys.* **1984**, *90*, 205–258. [CrossRef]
48. Smith, E.V.P.; Gottlieb, D.M. Solar flux and its variations. *Space Sci. Rev.* **1974**, *16*, 771–802. [CrossRef]
49. Kurucz, R.L. The Solar irradiance by computation. In Proceedings of the 17th Annual Conference on Atmospheric Transmission Models, PL-TR-95-2060, Hanscom AFB, MA, USA, 8–9 June 1994; pp. 333–334.
50. Bishop, J.K.B.; Rossow, W.B. Spatial and temporal variability of global surface solar irradiance. *J. Geophys. Res.* **1991**, *96*, 16839–16858. [CrossRef]
51. Thekaekara, M.P. Extraterrestial solar spectrum. *App. Opt.* **1974**, *13*, 518–522. [CrossRef] [PubMed]
52. Wehrli, C. *Extraterrestial Solar Spectrum, Publication No. 615*; Physikalisch-Metereologisches Observatorium (PMO) + World Radiation Center (WRC): Davos, Switzerland, 1985.

Article

An Overview of the Joint Polar Satellite System (JPSS) Science Data Product Calibration and Validation

Lihang Zhou [1,*], Murty Divakarla [2,†] and Xingpin Liu [2,†]

1 NOAA/NESDIS Center for Satellite Applications and Research (STAR), 5830 University Research Court, MD 20740, USA

2 IM Systems Group, Inc., 3206 Tower Oaks, Blvd., Suite 300, Rockville, MD 20852, USA; Murty.Divakarla@noaa.gov (M.D.); Xingpin.Liu@noaa.gov (X.L.)

* Correspondence: Lihang.Zhou@noaa.gov; Tel.: +1-301-638-3595; Fax: +1-301-683-3612

† These authors contributed equally to this work.

Academic Editors: Changyong Cao, Alfredo R. Huete and Prasad S. Thenkabail

Received: 22 December 2015; Accepted: 25 January 2016; Published: 8 February 2016

Abstract: The Joint Polar Satellite System (JPSS) will launch its first JPSS-1 satellite in early 2017. The JPSS-1 and follow-on satellites will carry aboard an array of instruments including the Visible Infrared Imaging Radiometer Suite (VIIRS), the Cross-track Infrared Sounder (CrIS), the Advanced Technology Microwave Sounder (ATMS), and the Ozone Mapping and Profiler Suite (OMPS). These instruments are similar to the instruments currently operating on the Suomi National Polar-orbiting Partnership (S-NPP) satellite. In preparation for the JPSS-1 launch, the JPSS program at the Center for Satellite Applications and Research (JSTAR) Calibration/Validation (Cal/Val) teams, have laid out the Cal/Val plans to oversee JPSS-1 science products' algorithm development efforts, verification and characterization of these algorithms during the pre-launch period, calibration and validation of the products during post-launch, and long-term science maintenance (LTSM). In addition, the team has developed the necessary schedules, deliverables and infrastructure for routing JPSS-1 science product algorithms for operational implementation. This paper presents an overview of these efforts. In addition, this paper will provide insight into the processes of both adapting S-NPP science products for JPSS-1 and performing upgrades for enterprise solutions, and will discuss Cal/Val processes and quality assurance procedures.

Keywords: JPSS; S-NPP; calibration; validation; AIT; AMP; NJO; IDPS; NDE; JSTAR; L1RDS

---

1. Introduction

The Joint Polar Satellite System (JPSS) is the National Oceanic and Atmospheric Administration's (NOAA) operational program that provides continuity of global environmental data from multiple polar-orbiting satellites for operational remote sensing of weather, climate and other environmental applications. The Suomi National Polar-orbiting Partnership (S-NPP) satellite launched in October 2011 was the first satellite designed to bridge into the future JPSS constellation. The S-NPP satellite carried aboard the following five instruments: Visible Infrared Imaging Radiometer Suite (VIIRS), Cross-track Infrared Sounder (CrIS), Advanced Technology Microwave Sounder (ATMS), Ozone Mapping and Profiler Suite (OMPS) and Clouds and the Earth's Radiant Energy System (CERES). Details of these instruments, channel characteristics, calibration, and the science data products derivable from these instruments are discussed in a compendium of scientific papers published as a special issue in the *Journal of Geophysical Research* [1]. The JPSS-1 (planned for launch in early 2017) and follow-on satellites will carry aboard similar instruments that are currently operating on-board the S-NPP with additional improvements [2].

The NOAA Center for Satellite Applications and Research (STAR) leads the efforts to develop, test, validate and refine the science algorithms to process S-NPP and JPSS instruments data from VIIRS,

CrIS, ATMS, and OMPS into user required data products such as Temperature Data Records (TDRs), Sensor Data Records (SDRs), and Environmental Data Records (EDRs) (here after referred to as xDRs collectively). Raw Data Records (RDRs) refer to the raw data generated by sensors on the satellites. Science algorithms that perform calibration and geo-location are applied on the RDRs to produce SDRs and TDRs. EDRs are geophysical parameters that are derived by applying retrieval algorithms on the SDRs. EDR products provide global measurements of such quantities as sea surface temperature, ocean color, ozone and trace gases, aerosols, clouds, temperature and moisture profiles, wind speeds, land surface properties, and snow and ice cover, *etc.* The data products produced from the S-NPP/JPSS satellites provide continuity of critical observations for accurate weather forecasting, reliable severe storm outlooks, and ocean ecosystem dynamics [3]. Examples of ocean applications include detecting and predicting harmful algal blooms [4], tracking runoff plumes and sediment [5], monitoring water quality [6], and hazards detection (e.g., oil spill detection, [7]).

JPSS provides data products to the primary NOAA User Community, which includes all centers of NOAA and worldwide weather agencies. The National Weather Service (NWS) and the European Center for Medium Range Weather Forecasting (ECMWF) assimilate JPSS products into their Numerical Weather Prediction (NWP) models. JPSS also partners with the users from National Aeronautics and Space Administration (NASA), Department of Commerce (DOC), Department of Agriculture, and Environmental Protection Agency (EPA) to utilize JPSS data products to support their research and operations. Direct readout users receive live S-NPP/JPSS data using direct downlink capabilities and the Community Satellite Processing Package (CSPP) allows creation of many SDR and imagery products in realtime [8]. S-NPP/JPSS data products are also accessible to public users worldwide through the NOAA Comprehensive Large Array-data Stewardship System (CLASS, [9]).

2. Science Data Products and Key Performance Parameters (KPPs)

Figure 1 provides a list of xDR products that are operationally produced from the current S-NPP suite of instruments. The S-NPP SDRs and associated EDR products (Figure 1) generated from the VIIRS, CrIS, ATMS, and OMPS instruments, and the Cal/Val efforts validating the products, are published in many journal articles [1,10–14]. The VIIRS SDRs are used to produce more than 20 EDR products (e.g., imagery, aerosol optical thickness (AOT), sea surface temperature, and many others as shown in Figure 1) and provide critical data for environmental assessments, forecasts and warnings [10]. The CrIS/ATMS SDR products are assimilated into NWP models worldwide and have been shown to provide considerable impact in reducing medium range forecast errors [15]. A variety of CrIS/ATMS EDR sounding products derived from the NOAA Unique CrIS/ATMS Processing System (NUCAPS) are currently ingested into the Advanced Weather Interactive Processing System (AWIPS-2, [16]) for their utility by many Weather Forecasting Offices (WFOs) nationwide for analyzing atmospheric instabilities, potential outbreaks of severe weather, and now-casting applications [17]. The OMPS Nadir Mapper and Nadir Profiler measurements are used to create global ozone maps and UV index forecasts [18].

The JPSS-1 and subsequent series of satellites will continue to produce these base-line products, along with upgrades planned as part of science improvements, and new and additional products derivable as a direct result of those improvements. Based on the utility of these science data products by customers and the user community, these products are prioritized as Key Performance Parameters (KPPs), and priority 2, 3, and 4 products. The requirements and priorities of the JPSS products are defined in the Level 1 Requirement Document (L1RD, [19]) and L1RD Supplement (L1RDS, [20]). The designations of the xDR priorities also dictate the priorities of the Cal/Val efforts towards evaluation of threshold or objective attributes. The S-NPP xDR product algorithms are in operations through either the Interface Data Processing Segment (IDPS) implemented by Raytheon or through S-NPP Data Exploration (NDE) at Environmental Satellite Processing Center (ESPC). Operationalization of the JPSS-1 xDR products (either in IDPS or through NDE/ESPC) will follow the S-NPP operational protocols and time-to-time directives from the NOAA JPSS Program office (NJO) on the implementation strategies for the S-NPP and JPSS data products. Most of the S-NPP products are operationally available to all of the user agencies worldwide through the NOAA CLASS. In preparation for the JPSS-1 launch, the JPSS STAR (JSTAR) Calibration/Validation (Cal/Val) teams have laid out the Cal/Val plans to oversee JPSS-1 science products algorithm development efforts, verification and characterization of these algorithms. The team also has developed schedules for deliverables and infrastructure for routing JPSS-1 xDR algorithm(s), improvements and updates in compliance with the Algorithm Change Process (ACP, [21]).

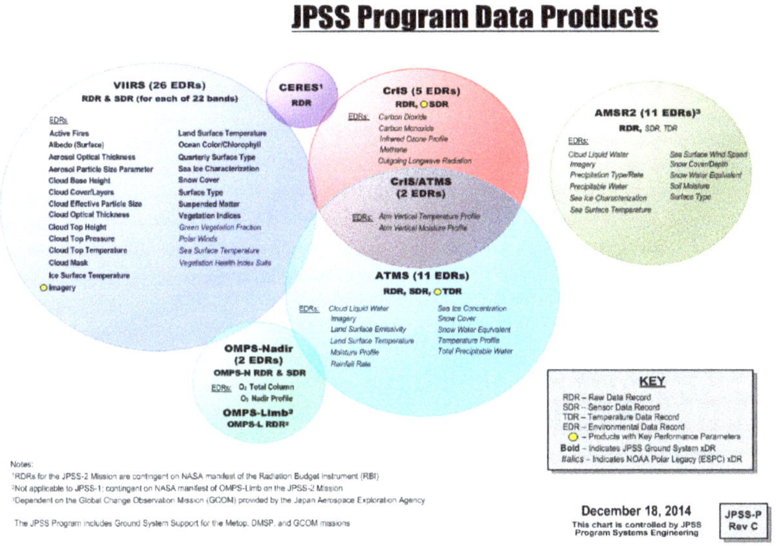

Figure 1. Sensor and Environmental Data (xDR) products from the S-NPP/JPSS-1 instruments suite.

3. JSTAR Cal/Val Teams

Two primary goals of the JSTAR Cal/Val Program are: (1) providing robust, affordable, and flexible state-of-the-art scientific solutions to meet JPSS requirements and (2) assuring operational viability of the data products meeting JPSS mission objectives. To achieve these goals, the JSTAR Cal/Val Program formed the Cal/Val teams consisting of highly distinguished scientists and engineers from an array of government agencies (NOAA, NASA; Department of Defense, DOD), NOAA's Cooperative Institutes (University partners), and industry partners. The Cal/Val team members have first-hand knowledge of algorithms developed for a variety of satellite systems (NOAA Polar orbiting Environmental Satellites, POES; Geostationary Operational Environmental Satellites, GOES; Defense Meteorological Satellite Program, DMSP; Earth Observing Systems, EOS; Meteorological Operational satellites (MetOp) program; and GOES-R) to define robust, affordable, and flexible state-of-art scientific solutions meeting JPSS requirements.

The JSTAR Cal/Val teams consist of four SDR teams to generate high quality ATMS, CrIS, VIIRS and OMPS SDRs. In addition, twelve EDR teams cover all of the JPSS EDR products related to the Atmosphere, Land and Oceans. All of these SDR/EDR teams coordinate with the JSTAR program management, other government agencies and industry partners to define various Cal/Val activities and execute them, thereby meeting schedules and deliverables. The SDR/EDR teams lay out different tasks to be accomplished during the pre-launch characterization, post-launch validation, and subsequent reactive maintenance and sustainment activities to realize and maintain high quality SDR and EDR products. Details of various Cal/Val tasks for each xDR product, as well as the roles and responsibilities of team members, are discussed in the respective Cal/Val plan documents [22]. The Cal/Val teams envision improvements to the existing products, investigate feasibility of new products development, and test algorithm updates accordingly in their offline systems. The teams coordinate with the STAR Algorithm Integration Team (AIT) in delivering Algorithm Change Packages (ACPs) for IDPS [21] and NDE [23] operations. The JSTAR management, in association with the Algorithm Management Program (AMP) team members, and the JPSS Ground Segment engineers, ensures that the Cal/Val plans, processing and analysis approach are robust for realization and in maintaining data throughput and integrity.

4. Cal/Val Processes

4.1. Pre-Launch to Post-Launch and Cal/Val Maturity

Declaring SDR/EDR product maturity is the result of a specific review of artifacts that document that the products meet a series of criteria defined for each maturity stage. The Cal/Val plan documents describe in detail the methodologies on how xDR products specified in the JPSS Program L1RD and L1RDS will be evaluated and validated for each of the maturity status. The planned Cal/Val activities during the (a) Pre-Launch; (b) Early Orbit Check-out (EOC); (c) Intensive Cal/Val (ICV), and (d) Long-term Monitoring (LTM) phases are also included in the Cal/Val plans. This is schematically depicted in Figure 2. During pre-launch, the emphasis is on characterizing the product performance utilizing the data sets available for the specific instrument (e.g., Thermal Vacuum Chamber (TVAC) test data sets; proxy or synthetic data sets generated from previous missions) and leveraging on the expertise gained from the previous missions. The Cal/Val teams also develop tools during the pre-launch to exercise post-launch activities. During the post-launch EOC phase, the emphasis is on sensor characterization and calibration, and quick-look analysis of the products derived from real satellite observations. This is followed by post-launch ICV with emphasis on characterizing the product performance (Accuracy, Precision and Uncertainty, APU) for different scenarios defined in the L1RDS using a variety of correlative data sets.

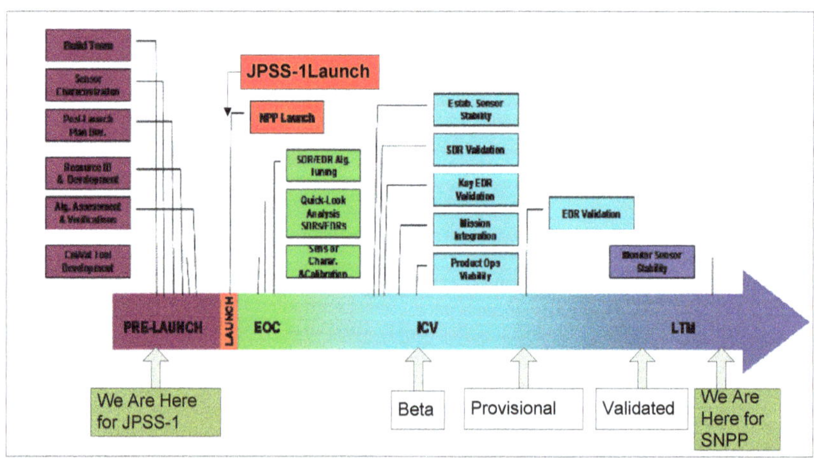

Figure 2. Components of Cal/Val Process from Pre-Launch to Post-Launch.

The Cal/Val teams identify any algorithm issues and make updates necessary to meet the required criteria as defined in product maturity definitions as shown in Table 1. The critical path during the post-launch Cal/Val phase is to ensure that the operational xDRs meet the beta and provisional maturity requirements and are ready for validated maturity, and to transition the products towards LTM. The JPSS Cal/Val teams have followed this process for S-NPP and currently most of the S-NPP products have reached validated maturity and are rolling towards the LTM phase. For JPSS-1, all of the Cal/Val teams have described in detail their Cal/Val plans and schedules starting from pre-launch to post-launch in their respective Cal/Val plan documents [22].

Table 1. Algorithm Maturity Definitions for S-NPP/JPSS-1 SDR/EDR Products.

- **Beta**
  - Product is minimally validated, and may still contain significant identified and unidentified errors.
  - Information/data from validation efforts can be used to make initial qualitative or very limited quantitative assessments regarding product fitness-for-purpose.
  - Documentation of product performance and identified product performance anomalies, including recommended remediation strategies, exists.
- **Provisional**
  - Product performance has been demonstrated through analysis of a large, but still limited (*i.e.*, not necessarily globally or seasonally representative) number of independent measurements obtained from selected locations, time periods, or field campaign efforts.
  - Product analyses are sufficient for qualitative, and limited quantitative, determination of product fitness-for-purpose.
  - Documentation of product performance, testing involving product fixes, identified product performance anomalies, including recommended remediation strategies, exists.
  - Product is recommended for potential operational use (user decision) and in scientific publications after consulting product status documents.
- **Validated**
  - Product performance has been demonstrated over a large and wide range of representative conditions (*i.e.*, global, seasonal).
  - Comprehensive documentation of product performance exists that includes all known product anomalies and their recommended remediation strategies for a full range of retrieval conditions and severity level.
  - Product analyses are sufficient for full qualitative and quantitative determination of product fitness-for-purpose.
  - Product is ready for operational use based on documented validation findings and user feedback.
  - Product validation, quality assurance, and algorithm stewardship continue through the lifetime of the instrument.

4.2. Cal/Val Review Process

The JSTAR team facilitates the Cal/Val maturity review process for all of the xDR products. Each product's Cal/Val maturity status is determined by evaluating the algorithm performance with the truth data sets and comparing the performance with the requirements. The review panel is comprised of the JPSS Program and Project Scientists, NOAA customers (National Weather Service, NWS; National Ocean Service, NOS; National Marine Fisheries Services, NMFS; Office of Oceanic and Atmospheric Research, OAR) and external users, Low-earth Orbiting Working Group (LORWG) advisors, product development managers (NDE and IDPS), and the JSTAR program manager. Following the Cal/Val maturity science reviews presented by the Cal/Val teams, the review board assesses the product performance and makes an overall assessment of each algorithm and approves the maturity status according to the criteria for Beta, Provisional, and Validated Maturity (Table 1). Based on the assessment, the review panel may also recommend further actions to the respective Cal/Val teams to achieve the desired product maturity status.

4.3. Quality Assurance

To ensure efficient research-to-operations transitions for science product algorithm updates, the JSTAR Program has implemented Quality Assurance (QA) procedures. The QA procedures include coordination activities between the algorithm teams on Cal/Val activities, supporting the algorithm change process (ACP), identifying potential shortfalls (risks) and mitigation, Configuration Management (CM) for documentation, schedules and milestones, and Earned Value Management (EVM). The J-STAR QA also facilitates technical reviews and coordination among STAR Cal/Val teams and the AIT to bring consistency to algorithm development and the delivery processes for operations. As part of the quality assurance and users engagement, JSTAR also developed a website [24] that provides detailed information on the JPSS instruments, science data products and documents, product maturity status, long-term validation and science monitoring of SDR/EDR products.

5. S-NPP Science Data Products Cal/Val Maturity Status

The Algorithm Theoretical Basis Documents (ATBDs) provide detailed descriptions of the science algorithms, and the Cal/Val documents associated with each product provide details of the Cal/Val efforts implemented during various phases of the S-NPP Cal/Val process. All of these documents are accessible from the JSTAR document website [22]. Most of the S-NPP data products have been validated with truth data sets and the product performances have been verified with the L1RDS requirements for progression from Beta, Provisional, and Validated Maturity stages at the times specified in Table 2. These SDR/EDR products and the product utility "readme" files are available through the NOAA CLASS and the JSTAR website [25] to global users for end-user applications. The review panel assessment reports on the Cal/Val maturity status presentations on each product are available on the STAR website [24] to assess the utility of the data product for various applications. The JPSS STAR Annual Science Team Meetings [26,27] have numerous presentations on the utility of the data products for many end-user applications. As of this publication, all of the SDRs and most of the EDRs had been declared to have reached the validated maturity. CrIS and ATMS radiances have been operationally assimilated in the National Weather Prediction (NWP) Centers and provided continuity of essential atmospheric sounding information for weather forecasting. Figure 3 demonstrated that CrIS and ATMS are among the top two contributors for reducing forecast errors.

Table 2. S-NPP SDR/EDR products, Cal/Val Maturity Status.

Sensor	Algorithm	Beta	Provisional	Validated
ATMS	ATMS SDR	January-2012	October-2012	December-2013
CrIS	CrIS SDR	April-2012	October-2012	December-2013
VIIRS	VIIRS SDR	April-2012	October-2012	December-2013
OMPS	OMPS SDR: NTC & NP	February-2012	October-2012	August-2015
VIIRS	Imagery (Not Near-Constant Contrast)	May-2012	January-2013	January-2014
VIIRS	NCC Imagery	October-2012	Aug-2013	January-2014
VIIRS	Cloud Mask	June-2012	January-2013	January-2014
VIIRS	Cloud Property Algorithms	June-2013	January-2014	September-2014 *
VIIRS	Aerosol Optical Thickness and Particle Size	September-2012	April-2013	August-2014
VIIRS	Aerosols-Suspended Matter	June-2013	**	**
VIIRS	Ice Surface Temperature	May-2013	August-2013	January-2014
VIIRS	Sea Ice Concentration and Ice Thickness	May-2013	November-2013	**
VIIRS	Binary Snow Cover	May-2013	November-2013	January-2014
VIIRS	Fraction Snow Cover	May-2013	November-2013	**
VIIRS	Active Fires	October-2012	August-2013	September-2014
VIIRS	Land Surface Temperature	December-2012	April-2013	December-2014
VIIRS	Land Surface Albedo	January-2013	April-2014	December-2014
VIIRS	Surface Type	February-2013	January-2014	December-2014
VIIRS	Land Surface Reflectance	February-2013	August-2013	September-2014
VIIRS	Vegetation Index	February-2013	August-2013	September-2014
VIIRS	Ocean Color	January-2013	January-2014	March-2015
VIIRS	Sea Surface Temperature	February-2013	January-2014	September-2014
CrIS	Soundings	August-2012	January-2013	September-2014
OMPS	Total Column Ozone EDR	July-2012	January-2013	August-2015
OMPS	Nadir Profiler Ozone EDR	August-2012	January-2013	August-2015

* Within the Cloud Properties Algorithms, all products except Cloud Cover Layer and Nighttime Cloud Optical Properties have been validated. ** Enterprise Algorithms are being implemented.

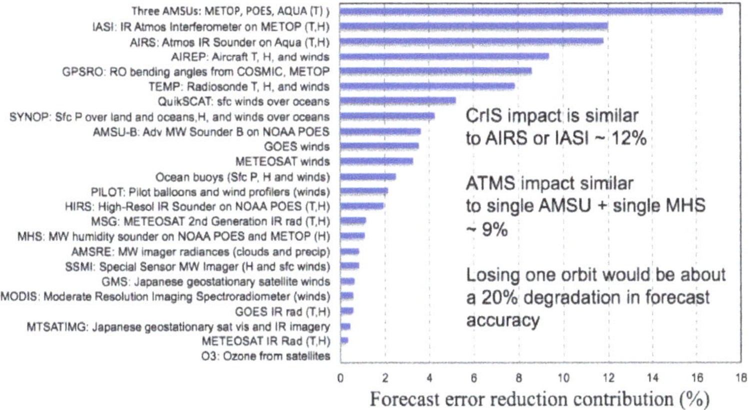

Figure 3. Satellite observations contribution to reducing forecast errors (plot courtesy of Carla Cardinali and Sean Healy, ECMWF).

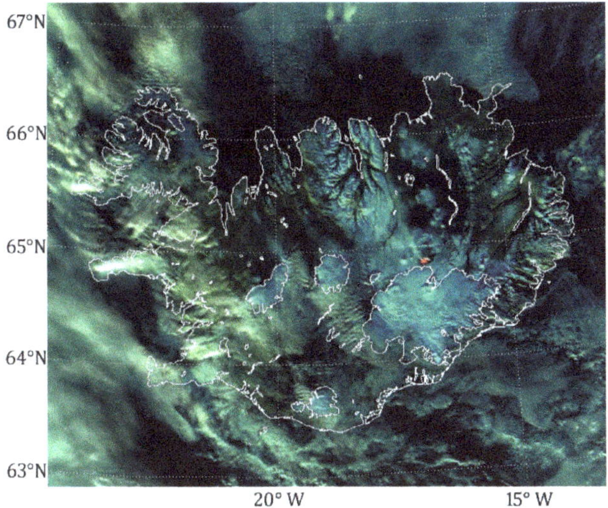

Figure 4. The multi-spectral capabilities of VIIRS are useful for investigating the volcano: VIIRS RGB composite of channels M4, M7, and M11 (red is 2.15 um, green is 0.86 um, and the blue is 0.55 um reflectance values). This combination makes vegetation appear green, ice appears dark cyan, clouds appear a light cyan color and the hot spot from the volcano appears red. The image is showing the eruption of Bárðarbunga in Iceland, as it appeared at 13:42 UTC 17 November 2014 (Image courtesy of C. Seaman, Cooperative Institute for Research in the Atmosphere, CIRA, Colorado State University).

Another example of EDR impacts on weather forecasting is the VIIRS Imagery EDR, one of the Key Performance Parameter (KPP) products [14], which reached the validated maturity in January 2014. Both Non-Near Constant Contrast (NCC) and NCC Imagery attained this level simultaneously. Non-NCC imagery has been well validated and ahead of NCC Imagery for months/years. NCC Imagery recently made significant advances in full coverage and stray light suppression (an SDR issue). The VIIRS imagery products have been widely used and proved to be critical for weather forecasts and environmental monitoring applications. Figure 4 is a VIIRS RGB composite imagery of channels M4, M7, and M11, showing the eruption of Bárðarbunga in Iceland, as it appeared at 13:42 UTC 17 November 2014. The VIIRS Imagery EDR User's guide is now available on the JSTAR website [22]. The Users' Guide is a required part of the Validation Maturity process and it has been released for many S-NPP SDR and EDR products that reached the validated maturity.

5.1. Product Performance Monitoring

Product performance monitoring is an integral element of Cal/Val and essential for anomaly detection and mitigation, and science maintenance of the long-term stability of the data products. The JSTAR teams have developed the Intensive Cal/Val System (ICVS, [28]) to monitor in near real-time the S-NPP spacecraft and onboard instruments' health status, performance, and SDR product data quality. The ICVS aids the SDR/EDR Cal/Val teams with corrective actions and in making occasional algorithm upgrades for reactive maintenance and long-term sustainment. The ICVS has demonstrated its tremendous value in monitoring satellite instrument health for JPSS as well as other national/international operational weather satellites. It has become a powerful tool that is being widely used by operational and research users for timely monitoring of the sensors' performance. It also ensures that qualities of the satellite observations are intercomparable and are tied to international standards for weather, climate, ocean and other environmental applications. Leveraging on the experience of ICVS, an EDR product long-term science monitoring system (LTSM, [29]) is also in the developmental phase to integrate science product monitoring and routine validations and to aid the operational users and product scientists in monitoring the EDR product quality in near real time, both routinely as well as on a long-term basis. Figure 5 shows an example of the CrIS radiance standard deviation per scan as monitored by the ICVS system. Figure 6a,b shows the S-NPP VIIRS and Aqua Moderate-Resolution Imaging Spectroradiometer (MODIS) global Land Surface Temperature (LST) maps as depicted by the EDR product monitoring system.

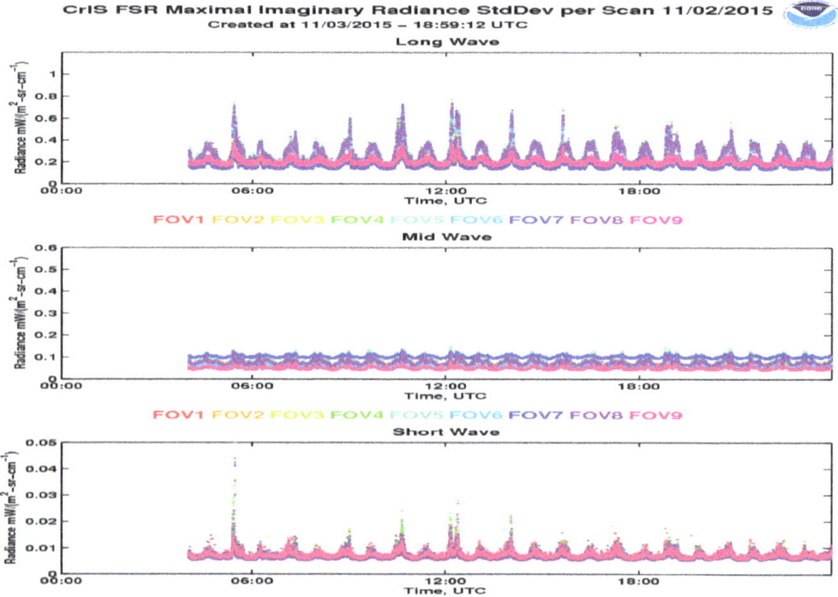

Figure 5. CrIS radiance standard deviation per scan as monitored by the ICVS system.

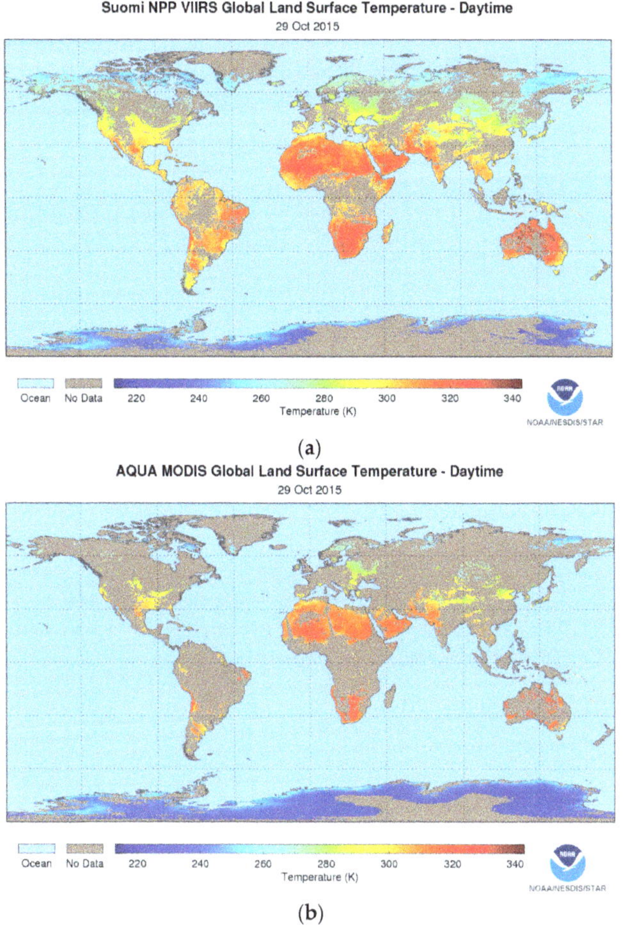

Figure 6. (**a**) S-NPP VIIRS and (**b**) Aqua MODIS global Land Surface Temperature (LST) maps as depicted by the EDR product monitoring system. The VIIRS instrument's wider swath and the cloud flag criteria used by the LST EDR product results in much better spatial coverage.

5.2. Enterprise Algorithm Developments and Other Future Improvements

Based on feedback from science panels and end-users acquired during JSTAR Annual Science Meetings, the NJO has directed STAR to work on NOAA enterprise systems for JPSS Priority 3 and 4 products. Enterprise solutions use the same scientific methodology and software base to create the same classification of product from differing input data (e.g., satellite, *in situ*, ancillary) [30]. This architecture provides a wide variety of benefits alleviating the challenges of ever increasing satellite and ancillary data volumes, cost effectiveness, and user demands for enhanced product quality and consistency. Enterprise solutions provide optimal and reliable data products across multiple satellite platforms through consistent algorithms, blended products, and longevity in conjunction with legacy and next generation systems (POES to JPSS, GOES to GOES-R, and NASA EOS to the Decadal Survey Mission). Enterprise solutions offer cost effectiveness by maximizing code reuse, better maintenance and quality control, facilitated integration of observations from multiple data streams, coordinated programs for inter-comparisons with other instruments, and risk identification and management.

JSTAR teams are currently working towards developing and implementing the Enterprise algorithms for both S-NPP and JPSS-1. The Cal/Val process described earlier in this article will be applied to the S-NPP/JPSS enterprise algorithms.

To further enhance the utilization of S-NPP/JPSS data products and to bridge the gaps between data products and users' applications, JSTAR teams are also working on setting up the environments to process/reprocess SDR/EDR products with the most advanced algorithms in a demonstration testbed mode. The testbed approach would provide a mechanism to evaluate alternative algorithms requested by the users and speed up the transition from research to operation.

6. Plans and Preparations for JPSS-1

The JPSS-1 and S-NPP xDR algorithms may be similar for those instruments that have remained essentially the same between the two platforms (e.g., ATMS, VIIRS), while the algorithms could be different for instruments with updated characteristics (e.g., OMPS and CrIS). In addition to pre-launch configuration files and Processing Coefficient Table (PCT) updates, the JPSS-1 SDR algorithms also have to accommodate upstream modifications and mitigations due to instrument waivers. Furthermore, based on the experience gained through the S-NPP science product development and end-user utility/feedback, the Cal/Val teams have identified many new products and associated algorithm refinements and improvements. The JPSS-1 science data product Cal/Val thus starts with the pre-launch utilization of a variety of data sets, including: (1) test data sets provided by various instrument vendors; (2) proxy data sets generated by the SDR and EDR Cal/Val teams for functional testing; and (3) data sets specifically required to evaluate whether the performance of SDR and EDR algorithms (Cal/Val) will meet the requirements slated by the L1RDS. These algorithms and the upgrades tested off-line have to be optimized for Research to Operations (R2O). The process thus calls for a coordinated activity among JPSS teams starting with the Flight team, the STAR xDR Cal/Val teams, the STAR AIT, and the JPSS Ground Segment team. All of these teams have defined their roles and responsibilities, and the Cal/Val teams have updated the ATBDs and laid-out JPSS-1 Cal/Val process plans spanning from pre-launch characterization to post-launch validations (documents accessible through STAR JPSS website, [22]. Based on specific algorithm readiness levels, the JSTAR teams established a schedule of anticipated dates for the algorithms to achieve Beta, Provisional and Validated statuses. Figure 7 shows the projected Cal/Val timeline for the JPSS-1 xDR products. Figure 8a–c shows a comparison of the projected JPSS-1 timelines to that of S-NPP in achieving Beta, Provisional and Validated maturity status. These projections clearly indicate that the S-NPP Cal/Val experience aids in expediting the JPSS-1 Cal/Val maturity process.

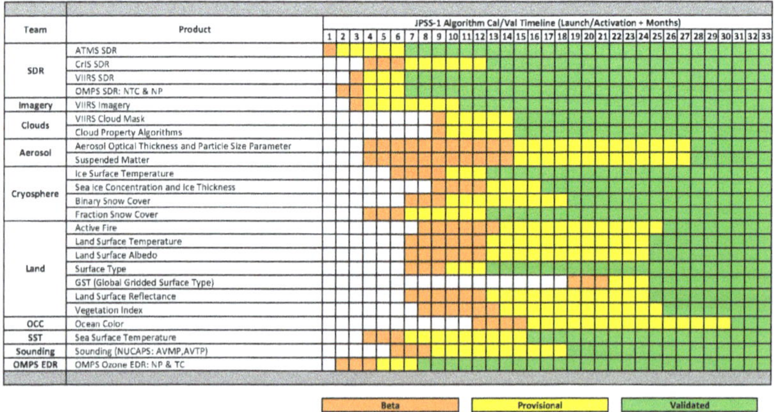

Figure 7. JPSS-1 Algorithm Cal/Val Timelines.

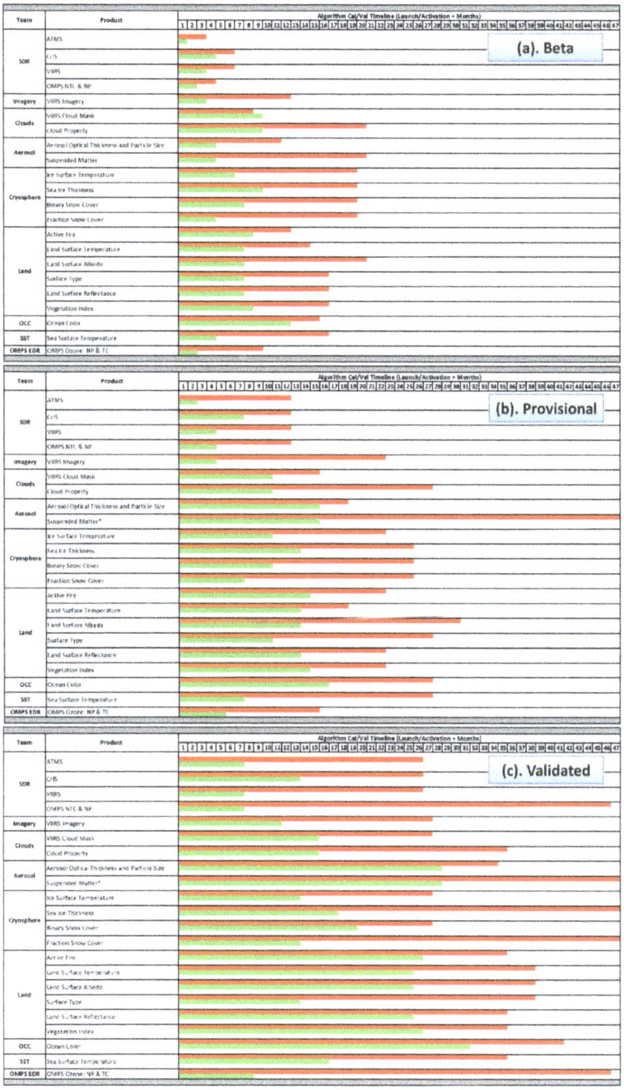

Figure 8. (**a–c**) Projected JPSS-1 Timelines (green bars) and comparison with the S-NPP timelines (red bars) in achieving (**a**) beta (top); (**b**) provisional (middle); and (**c**) validated (bottom) maturity status. The Cal/Val activities for JPSS-1 are expected to be much more accelerated than those for S-NPP, and JPSS-1 data products will be provided to decision makers/users with a much improved latency.

7. Summary and Conclusions

Most of the operational S-NPP data products have reached the Validated maturity level. The S-NPP data products are easily accessible via NOAA operations, direct readout, and NOAA CLASS. The scientific maturity of these products is well documented and the Cal/Val artifacts are available on the JSTAR website. S-NPP Product evaluation and updates are continuing and are currently in long-term monitoring and reactive maintenance phase. Replacement and upgrade of current S-NPP algorithms with NOAA enterprise algorithms are ongoing.

With improved knowledge of the pre-launch characterization of the J1 instruments and by leveraging the S-NPP Cal/Val experience, the JSTAR teams work with JPSS Algorithm Management Project (AMP) and JPSS Program Science, and are ready to accelerate the Cal/Val activities for JPSS-1 sensor and data products. Detailed Cal/Val activities and schedules are described in the JPSS-1 Cal/Val plans, which have been delivered to the JPSS program and are also available to the public on the JSTAR webpage. The Cal/Val activities for JPSS-1 are expected to be much more accelerated than those for S-NPP, and JPSS-1 data products will be provided to decision makers/users with a much-improved latency.

Acknowledgments: The work of JSTAR program is funded by NOAA JPSS Office (NJO). This paper describes the collective work of government industry and academic teams over the course of many years. The authors would like to acknowledge the hard work and dedication of all contributing individual companies and organizations. The manuscript contents are solely the opinions of the authors and do not constitute a statement of policy, decision, or position on behalf of NOAA or the U.S. government.

Author Contributions: Lihang Zhou is the Program Manager for the JSTAR, deputy for JPSS AMP, and is instrumental in the planning, execution, and oversight of the S-NPP/JPSS-1 Cal/Val program. Murty Divakarla coordinates and facilitates JSTAR Cal/Val processes, aids the algorithm change process, and acts as a liaison to the Algorithm Management Program and operational transition. Xingpin Liu is the quality assurance manager for the JSTAR and oversees schedules, milestones, and website maintenance.

Conflicts of Interest: The authors declare no conflict of interest.

References

1. JGR special issue of Suomi NPP Calibration and Validation Scientific Results. Available online: http://onlinelibrary.wiley.com/journal/10.1002/(ISSN)2169-8996/specialsection/SNPPCVSR1 (accessed on 19 December 2015).
2. Joint Polar Satellite System: A Collaborative Mission between NOAA and NASA. Available online: http://www.jpss.noaa.gov (accessed on 19 December 2015).
3. Krasnopolsky, V.; Nadiga, S.; Mehra, A.; Bayler, E.; Behringer, D. Neural networks technique for filling gaps in satellite measurements: Application to ocean color observations. *Comput. Intell. Neurosci.* **2016**, *9*. [CrossRef] [PubMed]
4. Qi, L.; Hu, C.; Cannizzaro, J.; Corcoran, A.A.; English, D.; Le, C. VIIRS Observations of a Karenia brevis Bloom in the Northeastern Gulf of Mexico in the Absence of a Fluorescence Band. *IEEE Geosci. Remote Sens. Lett.* **2015**, *12*, 2213–2217. [CrossRef]
5. Vandermeulen, R.A.; Arnone, R.; Ladner, S.; Martinolich, P. Enhanced satellite remote sensing of coastal waters using spatially improved bio-optical products from SNPP–VIIRS. *Remote Sens. Environ.* **2015**, *165*, 53–63. [CrossRef]
6. Son, S.; Wang, M. Diffuse attenuation coefficient of the photosynthetically available radiation Kd (PAR) for global open ocean and coastal waters. *Remote Sens. Environ.* **2015**, *159*, 250–258. [CrossRef]
7. Hu, C.; Chen, S.; Wang, M.; Murch, B.; Taylor, J. Detecting surface oil slicks using VIIRS nighttime imagery under moon glint: A case study in the Gulf of Mexico. *Remote Sens. Lett.* **2015**, *6*, 295–301. [CrossRef]
8. Gumley, L.; Huang, A.; Strabala, K.; Mindock, S.; Garcia, R.; Martin, G.; Cureton, G.; Weisz, E.; Smitha, N.; Bearson, N.; *et al.* Community Satellite Processing Package (CSPP) Polar-Orbiting Satellite Software and Products. Available online: http://www.ssec.wisc.edu/meetings/cspp/2015/Agenda%20PDF/Wednesday/Gumley_CSPP_Darmstadt.pdf (accessed on 19 December 2015).
9. NOAA Comprehensive Large-Array Data Stewardship System (CLASS), NOAA. Available online: http://www.nsof.class.noaa.gov (accessed on 19 December 2015).
10. Cao, C.; Xiong, J.; Blonski, S.; Liu, Q.; Uprety, S.; Shao, X.; Bai, Y.; Weng, F. Suomi NPP VIIRS sensor data record verification, validation, and long-term performance monitoring. *J. Geophys. Res. Atmos.* **2013**, *118*, 11664–11678. [CrossRef]
11. Han, Y.; Revercomb, H.; Cromp, M.; Gu, D.; Johnson, D.; Mooney, D.; Scott, D.; Strow, L.; Bingham, G.; Borg, L.; *et al.* Suomi NPP CrIS measurements, sensor data record algorithm, calibration and validation activities, and record data quality. *J. Geophys. Res. Atmos.* **2013**, *118*, 12734–12748. [CrossRef]

12. Weng, F.; Zou, X.; Sun, N.; Yang, H.; Tian, M.; Blackwell, W.J.; Wang, X.; Lin, L.; Anderson, K. Calibration of Suomi national polar-orbiting partnership advanced technology microwave sounder. *J. Geophys. Res. Atmos.* **2013**, *118*, 11187–11200. [CrossRef]
13. Pan, C.; Kowalewski, M.; Buss, R.; Flynn, L.; Wu, X.; Caponi, M.; Weng, F. Performance and calibration of the nadir Suomi-NPP ozone mapping profiler suite from early-orbit images. *IEEE J. Sel. Top. Appl. Earth Obs. Remote Sens.* **2013**, *6*, 1539–1551. [CrossRef]
14. Hillger, D.; Kopp, T.; Lee, T.; Lindsey, D.; Seaman, C.; Miller, S.; Solbrig, J.; Kidder, S.; Bachmeier, S.; Jasmin, T.; *et al.* First-light imagery from Suomi NPP VIIRS. *Bull. Am. Meteorol. Soc.* **2013**, *94*, 1019–1029. [CrossRef]
15. Goldberg, M.D.; Kilcoyne, H.; Cikanek, H.; Mehta, A. Joint Polar Satellite System: The United States next generation civilian polar-orbiting environmental satellite system. *J. Geophys. Res. Atmos.* **2013**, *118*, 13463–13475. [CrossRef]
16. Advanced Weather Interactive Processing System, AWIPS II Forecasting Software. Available online: http://www.unidata.ucar.edu/software/awips2 (accessed on 19 December 2015).
17. Line, W.; Calhoun, K. *GOES-R and JPSS Proving Ground Demonstration at the 2015 Spring Experiment—Experimental Warning Program (EWP) and Experimental Forecast Program (EFP)*; NOAA Hazardous Weather Testbed (HWT): Norman, OK, USA, 2015.
18. Long, C.; Wild, J.; Zhou, S.; Yang, S.; Flynn, L.; Beach, E. Application of OMPS Ozone Products. NOAA/NESDIS/STAR JPSS annual meeting. 2014. Available online: http://www.star.nesdis.noaa.gov/star/documents/meetings/2014JPSSAnnual/dayThree/07_Session5e_Long_Application%20of%20OMPS%20Ozone%20Products.pdf (accessed on 19 December 2015).
19. Joint Polar Satellite System (JPSS): Level 1 Requirements Document. Version 1.8. 2014. Available online: http://www.jpss.noaa.gov/pdf/L1RD_JPSS_REQ_1001_final_v1.8-1.pdf (accessed on 19 December 2015).
20. Joint Polar Satellite System (JPSS): Program Level 1 Requirements Supplement. Version 2.10. June 2014. Available online: http://www.jpss.noaa.gov/pdf/L1RDS_JPSS_REQ_1002_NJO_v2.10_100914_final-1.pdf (accessed on 19 December 2015).
21. Dafoe, L. Joint Polar Satellite System (JPSS) Algorithm Change Management Plan. Available online: https://jpssmis.gsfc.nasa.gov/documentation/doc_view_dsp.cfm?DTTM=20150721101511&Request Timeout=5000&ri=1678&pv=0 (accessed on 19 December 2015).
22. STAR S-NPP/JPSS Science Documents. Available online: http://www.star.nesdis.noaa.gov/jpss/Docs.php (accessed on 19 December 2015).
23. Roy, P.; Wolf, W.; Schott, T.; Guch, I. Improving Implementation Efficiency with Process Reviews of NESDIS Satellite Product Development Projects from Research to Operations. In Proceedings of the American Meteorological Society (AMS) Annual Meeting, Phoenix, AZ, USA, 4–8 January 2015.
24. STAR Joint Polar Satellite System Website. Available online: http://www.star.nesdis.noaa.gov/jpss/index.php (accessed on 19 December 2015).
25. STAR JPSS Algorithm Maturity Matrix. Available online: http://www.star.nesdis.noaa.gov/jpss/AlgorithmMaturity.php (accessed on 19 December 2015).
26. STAR JPSS Annual Science Team Meeting. 2014. Available online: http://www.star.nesdis.noaa.gov/star/meeting_2014JPSSAnnual_agenda.php (accessed on 19 December 2015).
27. STAR JPSS Annual Science Team Meeting. 2015. Available online: http://www.star.nesdis.noaa.gov/star/meeting_2015JPSSAnnual_agenda.php (accessed on 19 December 2015).
28. STAR ICVS Integrated Calibration/Validation System Long Term Monitoring. Available online: http://www.star.nesdis.noaa.gov/icvs/index.php (accessed on 19 December 2015).
29. Suomi-NPP/JPSS EDR Product Quality and Performance Monitoring. Available online: http://www.star.nesdis.noaa.gov/jpss/EDRs/index.php (accessed on 19 December 2015).
30. Wolf, W.; Li, A.; Wang, N.; Sampson, S.; Roy, P. NOAA/NESDIS ground enterprise architecture system (GEARS) algorithm prototyping whitepaper. *Proc. SPIE* **2014**. [CrossRef]

Article

User Validation of VIIRS Satellite Imagery

Don Hillger [1,*], Tom Kopp [2], Curtis Seaman [3], Steven Miller [3], Dan Lindsey [1], Eric Stevens [4], Jeremy Solbrig [3], William Straka III [5], Melissa Kreller [6], Arunas Kuciauskas [7] and Amanda Terborg [8]

1. NOAA/NESDIS Center for Satellite Applications and Research (StAR), Fort Collins, CO 80523, USA; dan.lindsey@noaa.gov
2. The Aerospace Corporation, El Segundo, CA 90245, USA; Thomas.J.Kopp@aero.org
3. CIRA, Colorado State University, Fort Collins, CO 80523, USA; Curtis.Seaman@colostate.edu (C.S.); Steven.Miller@colostate.edu (S.M.); Jeremy.Solbrig@colostate.edu (J.S.)
4. Geographic Information Network of Alaska (GINA), Fairbanks, AK 99775, USA; eric@gina.alaska.edu
5. CIMSS, University of Wisconsin, Madison, WI 53706, USA; wstraka@ssec.wisc.edu
6. NWS, Fairbanks, AK 99775, USA; melissa.kreller@noaa.gov
7. NRL, Marine Meteorology Division, Monterey, CA 93943, USA; Arunas.Kuciauskas@nrlmry.navy.mil
8. Aviation Weather Center, NWS, Kansas, MO 64153, USA; amanda.terborg@noaa.gov

* Correspondence: don.hillger@noaa.gov; Tel.: +1-970-491-8498; Fax: +1-970-491-8241

Academic Editors: Changyong Cao, Xiaofeng Li and Prasad S. Thenkabail
Received: 30 October 2015; Accepted: 21 December 2015; Published: 24 December 2015

Abstract: Visible/Infrared Imaging Radiometer Suite (VIIRS) Imagery from the Suomi National Polar-orbiting Partnership (S-NPP) satellite is the finest spatial resolution (375 m) multi-spectral imagery of any operational meteorological satellite to date. The Imagery environmental data record (EDR) has been designated as a Key Performance Parameter (KPP) for VIIRS, meaning that its performance is vital to the success of a series of Joint Polar Satellite System (JPSS) satellites that will carry this instrument. Because VIIRS covers the high-latitude and Polar Regions especially well via overlapping swaths from adjacent orbits, the Alaska theatre in particular benefits from VIIRS more than lower-latitude regions. While there are no requirements that specifically address the quality of the EDR Imagery aside from the VIIRS SDR performance requirements, the value of VIIRS Imagery to operational users is an important consideration in the Cal/Val process. As such, engaging a wide diversity of users constitutes a vital part of the Imagery validation strategy. The best possible image quality is of utmost importance. This paper summarizes the Imagery Cal/Val Team's quality assessment in this context. Since users are a vital component to the validation of VIIRS Imagery, specific examples of VIIRS imagery applied to operational needs are presented as an integral part of the post-checkout Imagery validation.

Keywords: VIIRS; DNB; NCC; imagery; validation; Alaska; KPP

1. Introduction

A major component in the overall strategy for the Imagery calibration and validation (Cal/Val) effort for the Visible Infrared Imaging Radiometer Suite (VIIRS) is to ensure that the Imagery is of suitable quality for effective operational use. Imagery of sufficient quality is often determined by the ability of human users to easily locate and discriminate atmospheric and ground features of interest. Such features include clouds and their type, especially convection and low clouds/fog, sea/lake ice edges, snow cover, volcanic eruptions, tropical cyclone structure, and dust storms [1–3]. In many of these cases, multi-spectral algorithms presented as false-color imagery are required to best identify land and atmospheric features over a given location. Pursuant to the Cal/Val strategy, VIIRS Imagery are analyzed to determine if it can be used to interpret the features noted above, considering both single and multi-spectral applications as appropriate.

The primary measurement obtained by VIIRS is digital counts (also known as digital numbers, or DN), designed to respond proportionally to the photons received by the detectors in the 22 bands that comprise the instrument. These DN values are converted into calibrated radiance, reflectance, and/or brightness temperatures based on the methodology of Cao *et al.* [4]. These calibrated data are distributed to the scientific community as Sensor Data Records (SDRs). A subset of the 22 bands is remapped to a satellite-relative Mercator projection and is distributed to the user community as Environmental Data Records (EDRs). Imagery produced from these data (both SDR and EDR) have been examined by the VIIRS EDR Imagery Team and collaboratively with the general user community to gauge performance and identify artifacts that characterize the overall quality of the imagery.

During the post-launch checkout, both single and multi-spectral images are analyzed to look for artifacts such as striping, banding, noise, geolocation errors, and collocation differences between bands that may plague multi-spectral imagery. In many cases, the root cause for such issues lies with the sensor and, as such, repairing or mitigating these artifacts are primarily the responsibility of the VIIRS SDR Cal/Val Team. The benefits of these ad-hoc software corrections are then inherited by the imagery EDRs produced from the SDRs. Much of this activity occurs before Imagery is declared to be an officially "validated" EDR by the Imagery Team. After this commissioning, VIIRS Imagery is made available to a wide audience of users who continue the Imagery validation process via assessments of operational utility, some examples of which are detailed herein.

Although S-NPP VIIRS visible and infrared capabilities are found to be clearly superior to heritage operational instrumentation (the Advanced Very High Resolution Radiometer, AVHRR), user validation is also tied to the unique features of VIIRS. These include the Day/Night Band (DNB) [5] that is not available from any current or near-future geostationary platform. The DNB, and the Near Constant Contrast (NCC) [6], a more user-friendly product derived from DNB, have found widespread use across NOAA, National Weather Service (NWS) and the U.S. Navy. NCC is capable of providing visual images at night, even under no moon conditions. While the quality varies with the amount of moonlight, day-night images have proven useful to numerous users [5,7,8].

2. VIIRS EDR Imagery

The Suomi National Polar-orbiting Partnership (S-NPP) satellite, which includes VIIRS as its principal imaging radiometer, was launched into a sun-synchronous, 1330 local time ascending node polar orbit on 28 October 2011. The VIIRS Imagery EDR is in fact comprised of three different sets of Imagery products. These are five Imaging-resolution or I-band products, six Moderate-resolution or M-band products, and Near Constant Contrast (NCC) Imagery derived from the DNB sensor. Each of these is remapped to a Ground Track Mercator (GTM) projection, the I-band with 400 m resolution and the others at 800 m resolution [9]. Details on the GTM mapping may be found in the VIIRS Imagery Products Algorithm Theoretical Basis Document (ATBD) [10] and are not discussed here.

The VIIRS Imagery EDR benefits from the many programs which preceded it, and many of the bands chosen to be created as Imagery products are based on heritage from such sensors as the Moderate Resolution Imaging Spectroradiometer (MODIS), AVHRR, and the Operational Line Scanner (OLS). Details on the VIIRS bands transformed into Imagery products by the JPSS ground system are provided in Tables 1 and 2. The Imagery EDR products in Table 1 include both the VIIRS bands explicitly spelled out in the Level 1 Requirements Documents (L1RD) [11] as well as VIIRS I3 band (1.61 μm, excellent in identifying snow/ice locations and ice-topped clouds) and the DNB, both of which became new Key Performance Parameter (KPP) bands in mid-2015. The products in Table 2 are the remaining EDR Imagery products created at present but not classified as KPPs at this time.

Table 1. Required Imagery Environmental Data Records (EDRs).

Imagery EDR Product	VIIRS Band	Wavelength (μm)	SDR Spatial Resolution Nadir/Edge-of-Scan (km)
Daytime Visible	I1	0.60–0.68	0.4/0.8
Short Wave IR (SWIR)	I3	1.58–1.64	0.4/0.8
Mid-Wave IR (MWIR)	I4	3.55–3.93	0.4/0.8
Long-Wave IR (LWIR)	I5	10.5–12.4	0.4/0.8
LWIR	M14	8.4–8.7	0.8/1.6
LWIR	M15	10.263–11.263	0.8/1.6
LWIR	M16	11.538–12.488	0.8/1.6
NCC	DNB	0.5–0.9	0.8/1.6

Table 2. Other Imagery EDRs.

Imagery EDR Product	VIIRS Band	Wavelength (μm)	Spatial Resolution Nadir/Edge-of-Scan (km)
Near Infrared (NIR)	I2	0.846–0.885	0.4/0.8
Visual	M1	0.402–0.422	0.8/1.6
Visual	M4	0.545–0.565	0.8/1.6
SWIR	M9	1.371–1.386	0.8/1.6

The required Imagery products shown in Table 1 primarily reflect user needs in the Alaskan weather theatre. Since no geostationary satellite properly covers the entire Alaskan region, user dependency on polar-orbiting products is much greater than any other US forecasting region. Of the explicit EDRs, bands I1 (0.64 μm), I4 (3.74 μm), and I5 (11.45 μm) comprise the basic building blocks for standard Imagery applications in the visible, MWIR, and LWIR portions of the radiative spectrum. The other three Moderate-resolution band images are LWIR bands that assist in locating clouds and in determining their composition (water or ice).

Use of the remaining bands in Table 2 varies according to the specific atmospheric and surface features of interest at a particular location. The I2 (0.86 μm) band complements the I1 (0.64 μm) band in the visible spectrum. Bands M1 (0.41 μm) and M4 (0.55 μm) assist in the creation of "true-color" and "natural-color" Imagery, while band M9 (1.378 μm) is superior to any other band at identifying thin cirrus during the day. These Imagery products are of practical use to various subsets of the user community.

The final EDR, the NCC as derived from the DNB, has found widespread use across NOAA, the NWS, and the U.S. Navy. NCC is capable of providing visible-wavelength images at night, even under no moon conditions [6]. While the quality varies with the amount of moonlight, NCC has proven useful at night in locating clouds, ice edge, snow cover, tropical cyclone centers (eyes), fires and gas flares, lightning, dust storms, and volcanic eruptions [5,12,13].

While Tables 1 and 2 encompass all of the EDR Imagery products created by the Ground System, they include only 6 of the 16 available M bands, implying that some of the VIIRS SDRs are not currently made available as Imagery EDRs. Some users work with the VIIRS SDRs from the 10 non-EDR rendered bands for their particular imagery needs. The Imagery Cal/Val Team is aware of these needs, and works with users to ensure that all spectral bands derived from VIIRS are of operational quality. DNB radiances, in particular, are used heavily in quantitative applications at the National Environmental Information Center (NEIC), Boulder CO [14] and at the National/Navy Ice Center (NIC).

3. VIIRS Imagery as a Key Performance Parameter (KPP)

Officially, only a subset of the global VIIRS Imagery products is designated as KPP. The specific verbiage of the VIIRS KPP reads as follows (updated to include the two new KPP bands): "VIIRS Imagery EDR at 0.64 μm (I1), 1.61 μm (I3), 3.74 μm (I4), 11.45 μm (I5), 8.55 μm (M14), 10.763 μm

(M15), 12.03 μm (M16), and Near Constant Contrast EDR for latitudes greater than 60°N in the Alaskan region."

The Cal/Val efforts place emphasis on high latitudes, as articulated by this verbiage. The overarching position taken by the Imagery Cal/Val Team is that the imagery requirements must be met first and foremost by the bands in Table 1 in the Alaskan region. Furthermore, the application of VIIRS Imagery must meet the user's expectations. This assessment relies heavily on user engagement and identification of representative use-cases.

Because Imagery is a KPP for VIIRS, it also is required to meet "Minimum Mission Success" in the Post Launch Test (PLT) time frame. How this will be accomplished for JPSS-1 is now spelled out in the newly-revised JPSS Cal/Val Plan [15] for the VIIRS Imagery Product. For JPSS-1 the PLT time frame ends at launch +85 days (L + 85). The objective of the Imagery Cal/Val Team is to show Imagery indeed meets the KPP criteria at L + 85, with the caveat understood that the time frame and season may limit the completeness of the Alaskan data set used for this assessment. For example, if the L + 85 period for JPSS-1 occurs primarily during the Northern Hemisphere winter, little visual data (Imagery) would be available to analyze north of 60° latitude. In contrast, if the period occurs primarily during the Northern Hemisphere summer season, when areas north of 60° latitude are bathed in sunlight, there will be few opportunities to assess the nocturnal component of the DNB's NCC product. Furthermore, certain key atmospheric events that drove the KPP for Alaska, such as volcanic ash, may not occur in the PLT time frame over Alaska. In these cases, the Imagery Cal/Val Team will use appropriate Imagery from other locations to show VIIRS Imagery products is sufficient or better than heritage imagery. Such use of alternative non-high-latitude locations complements the assessment over Alaska, and in most cases is adequate to show by proxy that VIIRS Imagery will achieve Minimum Mission Success.

The requirements for Imagery, as stated in the L1RD-Supplement, are simply spatial resolution requirements, as opposed to requirements that quantitatively address the quality of the EDR Imagery radiances and reflectances. Because of the lack of such specifications, Imagery validation rides heavily upon user feedback, such that the Cal/Val Team's work is complemented by users who are well versed in the use of Imagery applications.

4. Imagery Validation

During the early orbit instrument checkout phase, Imagery products were created as soon as the SDRs became available. In these early stages, VIIRS data are retrieved and evaluated from several sources: the Government Resource for Algorithm Verification Independent Testing and Evaluation (GRAVITE), the Comprehensive Large Array-data Stewardship System (CLASS), the Product and Evaluation and Test Element (PEATE), or from direct-broadcast line-of-sight reception sites. Post-checkout operational users have access to VIIRS data and derived products via NOAA's NPP Data Exploitation (NDE), as well as the NWS Advanced Weather Interactive Processing System (AWIPS) satellite data distribution system.

4.1. Validation Tools

There are many tools available designed to exploit the VIIRS Imagery. These tools, employed by the Cal/Val Team, are the Man-computer Interactive Data Access System (McIDAS-V in particular) [16], TeraScan processing software from SeaSpace Corporation [17], and various data processing and display tools using the Interactive Data Language (IDL) [18]. These tools are made available to various users during the Cal/Val process for Imagery. The primary operational user, the NWS, has its own display tools that are integrated into AWIPS [19]. Direct broadcast reception sites may display Imagery using the Community Satellite Processing Package (CSPP) [20] or the International Polar Orbiting Processing Package (IPOPP) [21]. Feedback from users of these systems is considered part of the extended validation process.

4.2. Intensive Calibration/Validation Phase

Among the basic qualities of Imagery are its spectral, spatial, temporal, and radiometric resolution. Spectral resolution is determined by the bands and bandwidths of VIIRS, as well as optical effects of the reflecting mirrors that direct light into the focal plane array. The spatial and radiometric resolutions of VIIRS are determined at the SDR level. The SDR and EDR radiances are unchanged, except for the special processing that goes into the DNB/NCC pair [6]. Table 3 compares the VIIRS SDRs and Imagery EDRs.

Table 3. Similarities and Differences between Visible/Infrared Imaging Radiometer Suite (VIIRS) Sensor Data Records (SDRs) *vs.* Imagery EDRs.

Characteristic	SDR	EDR
Solar reflective (visible) bands	Radiances and reflectances	Radiances and reflectances (same as SDR)
Infrared (thermal) bands	Radiances and brightness temperatures	Radiances and brightness temperatures (same as SDR)
Geo-spatial mapping	Satellite projection (with bowtie deletions and overlapping pixels)	Ground Track Mercator (GTM) projection (rectangular grid, no pixel deletions or pixel overlap)
Day/Night Band (DNB) imagery	DNB radiances (may vary by up to 7 orders of magnitude, depending on lunar and/or solar illumination)	NCC pseudo-albedos [6] (may vary by up to 3 orders of magnitude, to display features under conditions ranging from no moon to full solar illumination, as well as artificial lights)

The primary objective of the intensive Cal/Val phase is establishing in official sequence beta, provisional, and validated status for the Imagery products. The Imagery Cal/Val Team works with the VIIRS SDR Cal/Val Team to verify those requirements for Imagery that are tied to the SDR quality (e.g., radiance calibration accuracy, detector noise and striping, geolocation accuracy, *etc.*). The spatial resolution requirements are tied to the GTM projection, and are straightforward to verify. However, Imagery only passes through these validation stages as it is determined to be of quality for operational users.

The second component of Cal/Val user-oriented validation is operational user analysis of the Imagery. Band combinations targeting the characterization of clouds and cloud types, clouds phase, snow/ice on the surface, and dust storms are emphasized. Given the KPP emphasis on the Alaska region, this user base has a direct say in determining when validation has been reached. This determination is made via evaluation of VIIRS Imagery applications in the day-to-day operational environment.

The evaluation of NCC Imagery is a special case, due to its unique nature and its growing use for multiple purposes. DNB/NCC is used for such features ranging from ice edge at night to tropical cyclone center fixing. Some of the capabilities demonstrated at lower latitudes have analogous applications at the high latitudes. For example, the ability to peer through thin cirrus and reveal low-level circulation in tropical cyclones via moonlight also holds utility for peering through frontal cirrus and detecting the distribution of clouds or sea ice below (e.g., [5]). The Imagery Cal/Val Team helps users evaluate DNB/NCC for specialized applications such as gas flares, fishing boats, and auroras, while the operational applications of this nighttime imagery are the ultimate validation of its usefulness.

Mitigation of Non-Linearity for DNB on JPSS-1

The only significant difference with the input SDRs between S-NPP and JPSS-1 is with the DNB. It was discovered during routine laboratory testing that the DNB for JPSS-1 contains an anomalous non-linear response at high scan angles. The anomalies necessitated the design of DNB SDR post-processing software to mitigate the associated imagery artifacts. The method chosen impacts the spacing between DNB pixels towards the edge of the scan, as well as the location of nadir within the DNB SDR itself. The mitigation procedure is asymmetric, such that the DNB will actually extend in one direction (referred to as an "extended scene"). Furthermore, instead of preserving constant pixel size, the resolution of the DNB on JPSS-1 will degrade to approximately 1.2 km at the edge of the scan, and the spacing between pixels will not be as constant across the scan line as it is with S-NPP. The degradation in spatial resolution with the JPSS-1 DNB does not occur until a 49° viewing angle is reached.

Figure 1 shows the impact of these mitigation steps. These figures were produced using S-NPP DNB SDRs and executing the mitigation intended for JPSS-1. Figure 1A shows how the DNB SDR would look for JPSS-1. The blank space (blue) on the right side is the extended scene component. Nadir actually lies on the left side of the SDR image, as noted by the dashed line. In Figure 1B, which is the resulting NCC image, the NCC process shifts nadir (dashed line) to the center of the image, and truncates the extended scene portion of the DNB. This is intended, so NCC imagery from JPSS-1 will be similar to that from S-NPP, as well as match the GTM mapping for the VIIRS M-bands.

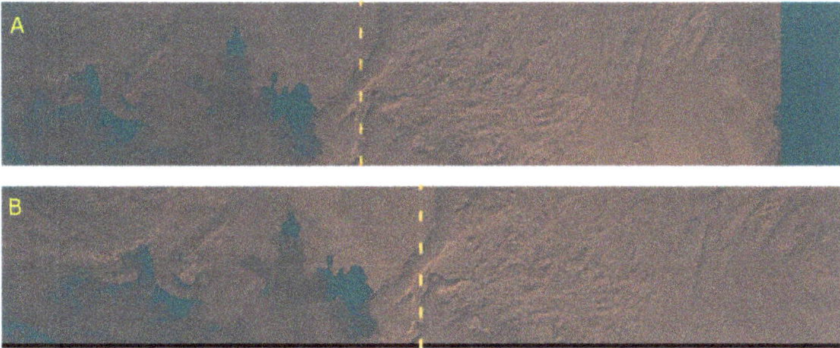

Figure 1. (**A**) DNB from Suomi National Polar-orbiting Partnership (S-NPP) used to display how DNB will look from JPSS-1, with the blue area on the right filled with extended scene imagery (currently missing in this simulation); (**B**) The DNB remapped into the GTM mapping used for Near Constant Contrast (NCC), showing that the NCC shifts the DNB imagery to the right, placing nadir at the center and ignoring the extended scene data on the right. In each image, the dashed line shows the approximate location of nadir.

4.3. Long-Term Monitoring Phase

Long-term monitoring of VIIRS Imagery is the responsibility of the VIIRS SDR and Imagery EDR Teams collectively. The task extends to the Imagery user base as well, who at any time may help the Teams to identify an imagery anomaly that may be either transient or recurring. Hence, long-term monitoring focuses on the ongoing value of the Imagery product quality as it applies to users.

With those Cal/Val basics as background, the rest of this article presents key examples of the uses of VIIRS Imagery that show its quality as being highly useful (even exceeding expectations) for many analysis and forecasting applications. In the cases presented, the Imagery or image products proved beneficial to the users.

5. Applications of VIIRS Imagery in Meteorological Operations

5.1. Alaska Examples

As noted in previous sections, Alaska users of VIIRS are specifically spelled out as primary users of VIIRS imagery since they are located on the northern edge of most geostationary satellite views and polar-orbiting data is best utilized in Polar Regions where there is a high frequency of satellite overpasses. The KPP explicitly spells out the Alaskan region as the most critical area for Imagery coverage and quality.

5.1.1. Visible and Longwave Imagery

VIIRS Imagery bands are displayable on NWS Alaska Region AWIPS as single-band products. Figure 2A shows an AWIPS screen capture of I1 (0.64 μm) visible imagery from 29 July 2015 at 0043 UTC (4:43 pm Alaska Daylight Time), 28 July 2015. Figure 2B shows the I5 (11.45 μm) longwave IR imagery. Each of these single-band products has its strengths and weaknesses. For example, consider the deck of low stratus over the ocean near the Bering Strait. This stratus is obvious in the VIIRS I1 visible imagery. But clouds like these typically develop in the lower troposphere and have temperatures similar to nearby sea surface temperatures, with the result that the distinction between the stratus deck and clear skies over the ocean (in the yellow circle) cannot be made using the VIIRS I5 (11.45 μm) longwave IR imagery alone. The I5 imagery helps forecasters identify colder convective clouds more easily than can be done in the I1 visible imagery. The colder clouds inland on the right side of these figures (noted by yellow arrows) represent typical summertime afternoon convection over the rough terrain of the Nulato Hills. These convective clouds are easy to identify in the I5 imagery because of their colder temperatures, but these clouds might not be immediately identified as convective in the I1 visible imagery alone, since they have the same white color as the low stratus over the nearby marine areas. Figure 2C is the result of combining the different advantages of the VIIRS I1 visible and I5 longwave IR imagery into a single product, thereby allowing the forecaster to gain the meteorological insights contained in two different bands by looking at just one product. In this case, the two Imagery bands are simply overlaid and blended together. This approach is simplistic, but effective. A more sophisticated and much more helpful approach is not to overlay different products but to combine two or more single bands into a multi-spectral product as described in the following section.

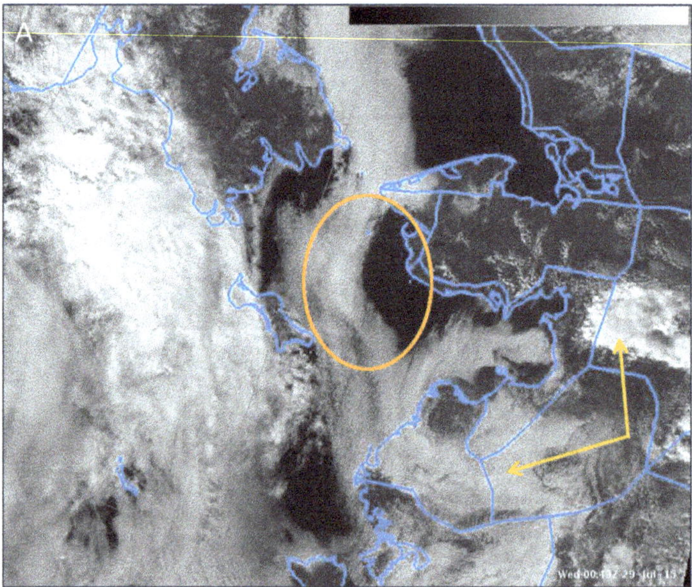

Figure 2. *Cont.*

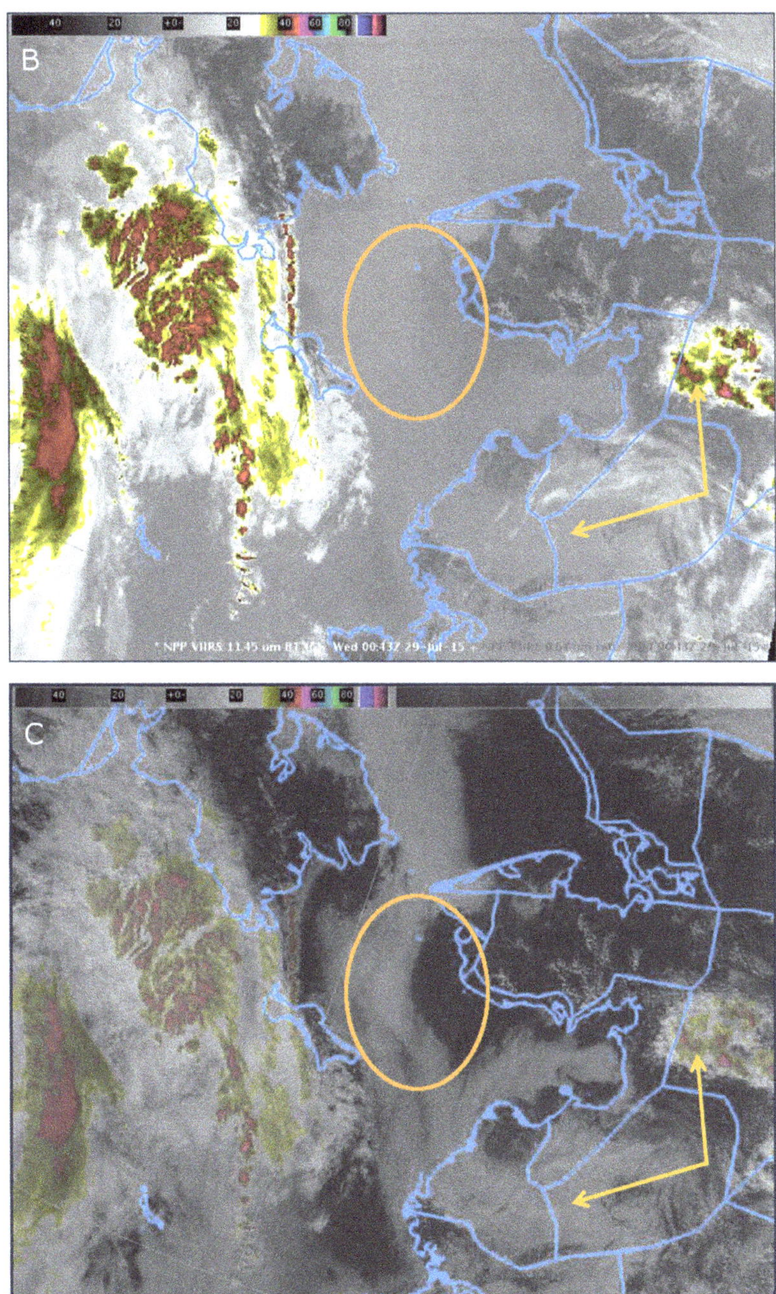

Figure 2. Annotated Advanced Weather Interactive Processing System (AWIPS) screen capture on 29 July 2015 at 0043 UTC centered over the Bering Strait of (**A**) VIIRS I1 (0.64 μm) band visible imagery; (**B**) VIIRS I5 (11.45 μm) band longwave infrared imagery; and (**C**) the two bands overlaid in AWIPS into a VIIRS Imagery product.

5.1.2. Fog/Low Cloud

Beginning with the GOES-8 imager, when the 3.9 μm band first became available, the 11 μm–3.9 μm brightness temperature difference (BTD) has traditionally been used for fog/low cloud detection [22]. At night, the 3.9 μm and I5 (11.45 μm) bands detect differences in radiometric temperature (due to spectral emissivity differences between the two bands) rather than thermodynamic temperature. These emissivity differences are related to the size of the particles, meaning that small droplets (such as fog) can be distinguished from larger droplets as well as cloud free surfaces. While these emissivity effects are present during the day, they are overwhelmed by the solar signal in the 3.9 μm band. Thus, this product is more useful during the night time. Fog is of critical importance in the Alaska region due to the large amount of private and commercial aviation and maritime traffic. During the northern hemisphere winter months, much of the Alaska region is in "night" or near-terminator illumination conditions, allowing for the application of the 11 μm–3.9 μm BTD for fog detection. Owing to the availability of timely direct broadcast S-NPP data, provided by the Geographic Information Network of Alaska (GINA, located at the University of Alaska, Fairbanks), there have been several instances where the NWS Weather Forecasting Office (WFO) in Fairbanks, Alaska, has been able to gain information on fog/stratus along the North Slope of Alaska in near real-time.

Figure 3 shows a color-enhanced 11 μm–3.9 μm BTD for the northern Alaska forecast area including the North Slope. The orange colors indicate areas of low cloud/fog, the light grays indicate higher clouds, and the black areas indicate thin cirrus. To accompany this figure, portions of the NWS Northern Alaska Forecast Discussion (AFD) for 11 March 2013 issued at 1258 pm (which follows) mention the use of VIIRS imagery (with yellow highlighting emphasizing the S-NPP VIIRS Imagery fog product). The AFD points out that the fog was evident in the VIIRS Imagery, surface observations, and MODIS products; but the higher spatial resolution of the VIIRS Imagery captured the fog in more detail than MODIS, and certainly more detail than surface observation; as well as VIIRS providing additional imagery at times other than MODIS overpasses.

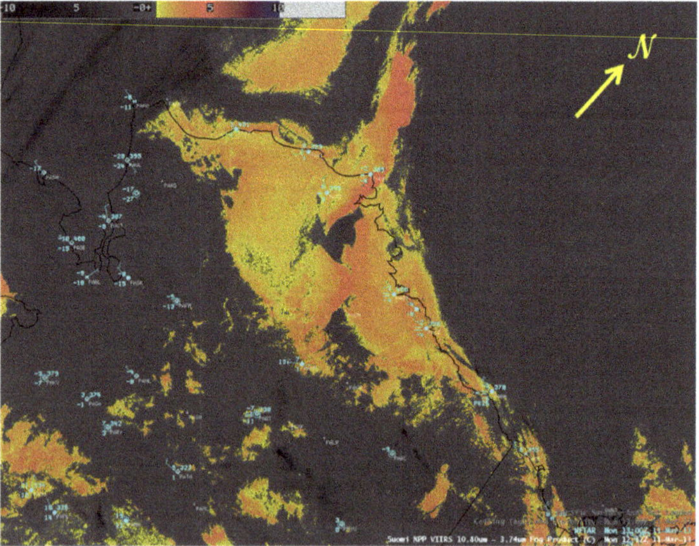

Figure 3. Color-enhanced 11 μm–3.9 μm BTD for 11 March 2013 at 1252 UTC, along with METAR observations at 1300 UTC. The oranges indicate areas of low cloud/fog, where the light grays indicate higher clouds and the black indicates thin cirrus.

"The Suomi NPP VIIRS satellite fog product was indicating a decent layer of stratus along the North Slope. Observations across the area generally indicated 1 to 2 miles (1.5 to 3 km) in visibility with flurries and fog. The IFR conditions align very well with the higher probabilities of MODIS IFR product. There are some very isolated pockets of higher probabilities of the MODIS LIFR conditions. These conditions should remain through Tuesday evening or Wednesday morning as the surface high pressure remains within the area. By Wednesday morning the surface pressure gradient begins to tighten providing an increase in winds and perhaps a break in some of the fog."

Fog and stratus have large impacts year round along the North Slope, particularly to the aviation community. The fog/stratus example in Figure 4 highlights the critical utilization the VIIRS imagery of the I3 (1.61 μm) band at the NWS Fairbanks WFO in the preparation of the Terminal Aerodrome Forecast (TAFs) issuance. The timely high-resolution imagery at 0004 UTC on 26 April 2015 allowed forecasters to identify and forecast the mesoscale circulation feature moving to the northwest of the Nuiqsut AK and showed improving visibility conditions for air flights. Meanwhile it was apparent that the visibility at the Prudhoe Bay airport was still ¼ mile (400 m) in freezing fog and would remain IFR conditions as well as along the Arctic Coast.

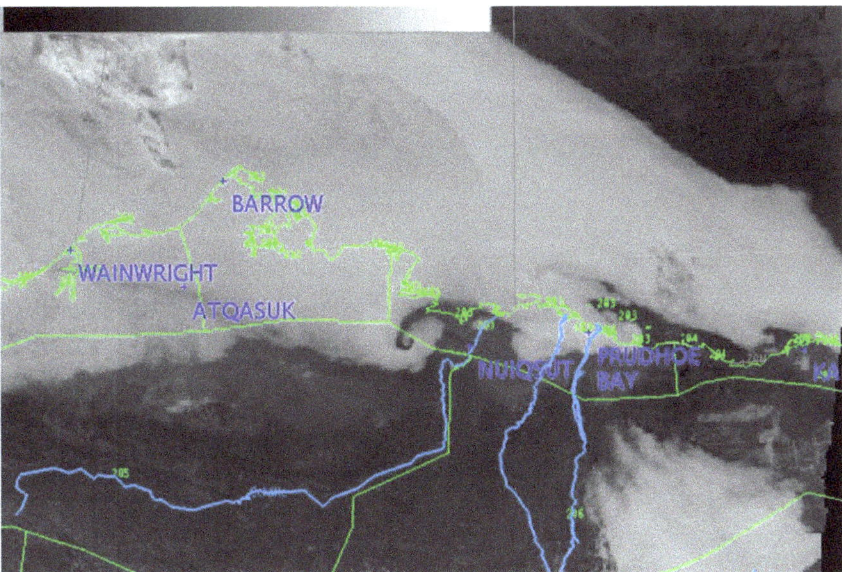

Figure 4. The Suomi National Polar-orbiting Partnership (S-NPP) VIIRS I3 (1.61 μm) band along the North Slope on 26 April 2015 at 0004 UTC. The white colors indicate areas of low clouds and fog, where the darker gray indicates the clear conditions. The light blue lines are the rivers with the green lings indicating the boundaries for the zone forecast areas.

5.1.3. Multi-Spectral Imagery

The large number of VIIRS I-bands and M-bands (5 and 16 bands, respectively) offers a wide variety of possible combinations into multi-spectral, or Red, Green, Blue (RGB) composite, image products. Figure 5A shows a true-color RGB from 9 July 2015 at 2302 UTC (3:02 pm Alaska Daylight Time), that combines the M5 (0.67 μm), M4 (0.55 μm), and M3 (0.49 μm) bands into a true-color image that represents what the human eye would see if we could ride along on the S-NPP satellite. A true-color image has great advantages for meteorological surveillance, in part because it is comparatively intuitive to interpret. Every multi-spectral Imagery product has its strengths and weaknesses, and while a strength of the true-color RGB is its ease of interpretation, a particular weakness (from the Alaskan perspective) is its inability to offer a distinction between clouds and sea ice over the Arctic Ocean because both clouds and ice appear white in true-color imagery. Figure 5B is a natural-color RGB composite image that assigns the VIIRS I3 (1.61 μm) band to the red component, the I2 (0.86 μm) band to the green component, and the I1 (0.64 μm) band to the blue component. This product takes advantage of the fact that incoming sunlight is absorbed or reflected to various degrees by different surfaces at these three wavelengths. Most helpfully, the resulting multi-spectral product depicts sea ice and ice clouds as cyan and liquid-based clouds as pink, making the separation of low clouds from ice comparatively straightforward. (Sea ice and ice clouds can normally be easily discriminated by texture and context.) The number of multi-spectral products available to NWS meteorologists in Alaska on their AWIPS workstations has increased substantially over recent years, with RGB composite images now being built routinely from VIIRS and MODIS data received locally in Alaska via direct broadcast antennas.

Figure 5. (**A**) VIIRS True-color Red, Green, Blue (RGB) combining the I1 (0.64 μm) band with the M3 (0.49 μm) and M4 (0.55 μm) bands; and (**B**) VIIRS natural-color RGB composite image combining the VIIRS I3 (1.61 μm), I2 (0.86 μm), and I1 bands, both from 9 July 2015 at 2302 UTC.

5.1.4. The Day/Night Band at Night

The VIIRS DNB has proven particularly useful to NWS forecasters during the extended periods of darkness during the Alaskan winter, because the DNB offers forecasters the ability to analyze meteorological and terrain features in the visible portion of the spectrum when ambient light levels are too low for conventional visible satellite imagery to be helpful. Figure 6 is an AWIPS screen capture of DNB imagery from 24 January 2013 at 1354 UTC (4:54 am Alaska Standard Time), a period of total darkness over all of Alaska. The scaling of DNB imagery needs to be appropriate to the lunar illumination at the time of the image, as the DNB radiances can vary by 7 orders of magnitude, from full daylight to nearly total darkness under new-moon conditions. Terrain and cloud features are very evident, as are the lights from the cities of Anchorage and Fairbanks. The raw S-NPP data used to produce this image were received via a direct broadcast antenna at GINA on the campus of the University of Alaska Fairbanks, then processed at GINA with CSPP software, and finally delivered to the NWS in Alaska for display in AWIPS via the Local Data Manager (LDM).

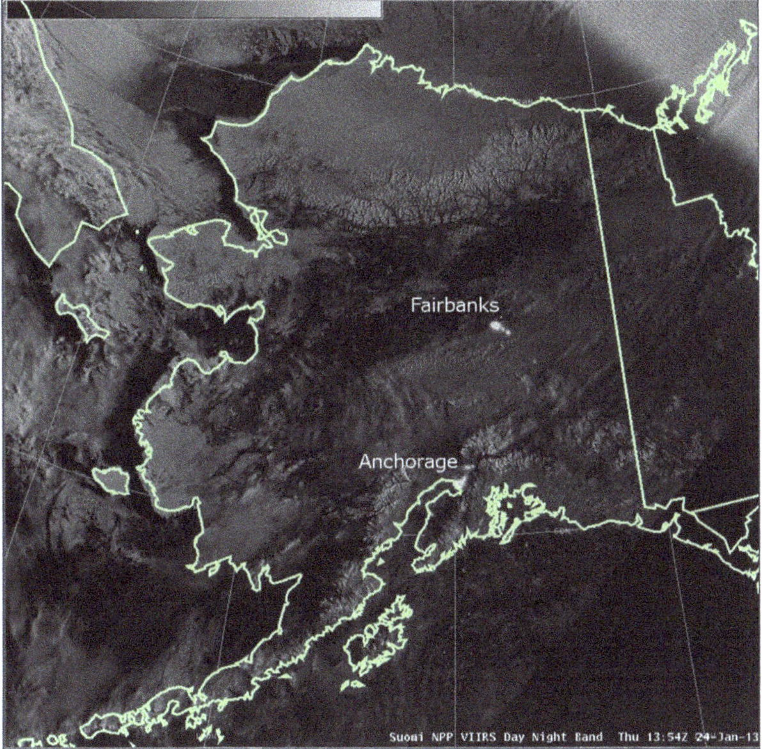

Figure 6. Example of VIIRS DNB imagery for 24 January 2013 as displayed on a National Weather Service (NWS) AWIPS workstation. An adjustable gray-scale appropriate to the lunar (or solar) illumination at the time is used to enhance the DNB radiances to best reveal cloud and surface features.

5.1.5. Volcanic Ash

Volcanic ash is also a common hazard in Alaska, not only from current eruptions, but even for ash from long-dormant eruptions as in the following example.

The 6–8 June 1912 eruption of Novarupta, one of the largest volcanic eruptions in recorded history, expelled ash with an estimated column height exceeding 23 km and deposited a layer of ash exceeding 100 m thick in the nearby Valley of Ten Thousand Smokes on the Alaska Peninsula [23]. During periods of high winds, this volcanic ash may become re-suspended in the atmosphere, posing a hazard to aviation [24] and human health [25]. VIIRS imagery was able to capture one such incident that occurred on 30–31 October 2012. According to local media reports [26], ash was lofted up to 1.2 km in altitude and resulted in the diversion and cancellation of flights in the vicinity of Kodiak Island during the high wind event. The NWS Alaska Aviation Weather Unit (AAWU) issued a Significant Meteorological event warning (SIGMET) for the re-suspended volcanic ash. Portions of the initial SIGMET [27] issued at 2216 UTC on 30 October 2012 is included below.

"Volcano: Novarupta 1102-18

Eruption details: NO Eruption. Resuspended ASH.

RMK: Resuspended ash due to high winds in area. Not from eruption."

Figure 7 shows the VIIRS DNB image (collected at night) along with the true-color RGB (Red/Green/Blue) composite image (collected during the following afternoon) from 30 October 2012. The plume of volcanic ash is visible in both the DNB and true-color images. In each image, the location of Novarupta is indicated by the yellow arrows. The ash plume extends from Novarupta to the southeast, across the Shelikof Strait and Kodiak Island. While the initial SIGMET was based on analysis of MODIS imagery, this case highlights the importance of satellite imagery for volcanic ash detection and demonstrates the utility of VIIRS imagery for this purpose.

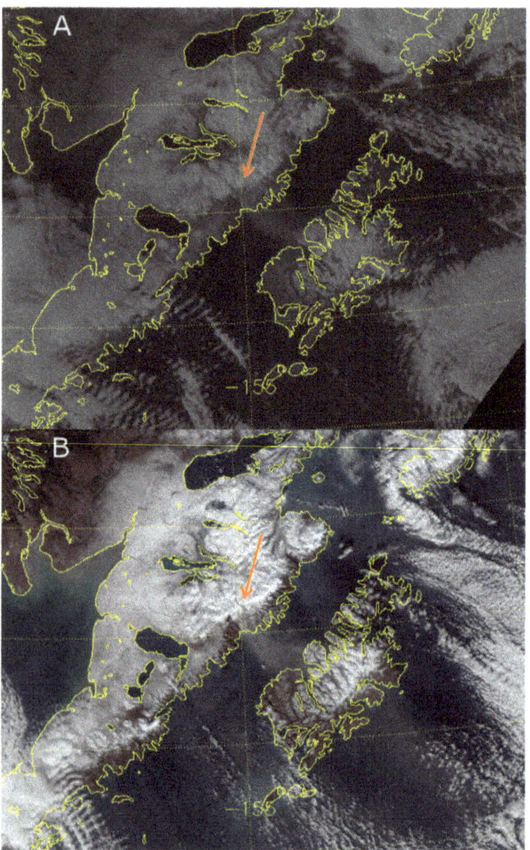

Figure 7. (**A**) VIIRS DNB image showing re-suspended ash from the 1912 Novarupta eruption (1411 UTC 30 October 2012) (**B**) VIIRS true-color image of the ash plume (2223 UTC 30 October 2012). In each image, the location of Novarupta is indicated by a red arrow. The ash plume extends from the volcano to the southeast across the Shelikof Strait and Kodiak Island.

5.1.6. Fire Weather

The VIIRS imagery with the I4 (3.74 μm) and I1 (0.64 μm) band is extremely beneficial to NWS forecasters for identifying hotspots and areas of smoke. The I4 brightness temperatures allow for the identification of the fire "hotspots", with the I1 visible band overlaid identifying the smoke plumes. The combination of these two products as in Figure 8 is heavily utilized during the Alaska fire weather season from late June through August. The 2015 fire season was extreme, and a record number of acres were burned. The area burned, 5.148 million acres (20.8 km^2), puts the season in second place (out of 66 years) behind the extreme 2004 season. During the active fire season three Incident meteorologists were dispatched to the wildfire complexes in the Fairbanks warning area [28] and one IMET was dispatched to the office to help with the number of wildfire spot forecasts (909 Fire Weather Spot forecasts were issued for the 2015 Fire weather season).

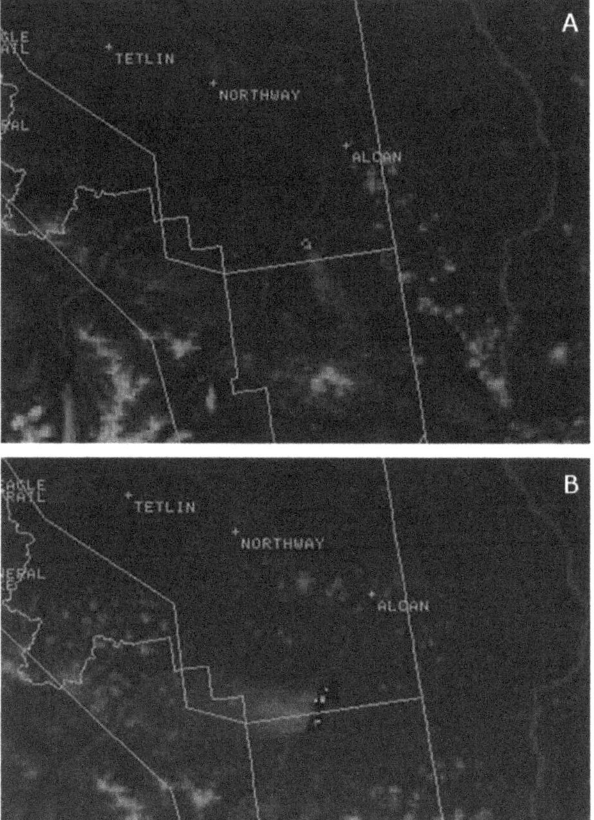

Figure 8. (**A**) The VIIRS I4 (3.74 μm) band brightness temperature and I1 (0.64 μm) band visible satellite imagery overlaid at 2211 UTC on 17 June 2013 in which you can see the smoke plume moving towards the southeast and a fire "hotspot" from the Chisana River, Eagle Creek, and Bruin Creek wildfires; (**B**) Image from 18 June at 2013 UTC after the winds shifted from the northwest to the southeast and moved the smoke plume.

With the active fire weather period, it was also important not only to identify these hot spots but also to inform the public on smoke hazards. The NWS Fairbanks WFO typically informs the public via social media, as in Figure 9 for 6 July 2015 regarding the Aggie Creek wildfire located northwest of Fairbanks. The reports show (top) the pyrocumulus images from the Aggie Creek fire, and (bottom) a side-by-side comparison of the VIIRS false-color imagery at 2312 UTC with the radar reflectivity image from the Pedro Dome (PAPD) radar on 6 July 2015. In this case the VIIRS Imagery supports the radar, webcam, and other fire-related information, adding to the validity of the information presented to the public, which benefits from analyses such as this one compiled from multiple sources by NWS forecasters.

(a)

(b)

Figure 9. Two social media reports from 6 July 2015 about the Aggie Creek fire located northwest of Fairbanks: (**a**) as seen from the NWS Fairbanks office on the University of Fairbanks campus; (**b**) the VIIRS False-Color satellite imagery at 2312 UTC and the radar reflectivity image from the Pedro Dome radar are shown side by side.

5.1.7. Sea Ice

The determination of the sea ice state is important in Alaska in all seasons. VIIRS Imagery, when cloud-free, can provide a good view of sea ice, which is valuable input for a sea ice analysis. These ice analyses are particularly valuable when lives and properly are at stake, such as in the following example when VIIRS imagery allowed the NWS to help the US Coast Guard (USCG) rescue a mariner in need.

On 10 July 2014 the USCG was called upon to rescue a solo mariner in a small boat attempting to sail the Northwest Passage. The mariner had become stuck in the pack ice north of Barrow, Alaska, and the USCG coordinated with the NWS Ice Program based in Anchorage, Alaska. The NWS used multi-spectral VIIRS imagery, to analyze the sea ice in the area of concern and provide the USGC guidance regarding the best path to approach the mariner. The annotated VIIRS-based sea ice analysis for this case is shown in Figure 10. As a result, the Coast Guard successfully rescued the stranded sailor [29].

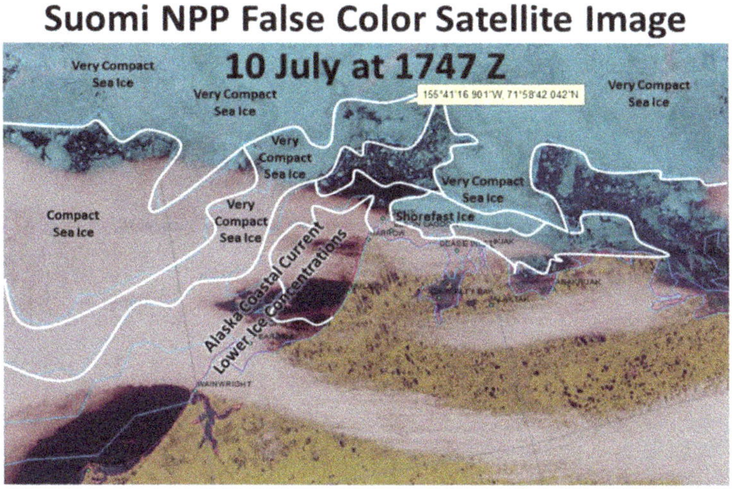

Figure 10. Annotated VIIRS RGB composite image from the NWS Ice Program in Anchorage AK, including the position of the boat stranded in the ice. (Image courtesy of Mary-Beth Schreck.)

5.2. *Examples Outside of Alaska*

Because S-NPP is a polar-orbiting satellite with worldwide coverage every 12 h, there are many opportunities for the use of VIIRS Imagery around the world, not just in the polar regions that specifically define VIIRS as a KPP.

5.2.1. Tropical Storm Centering at Night

It has become common practice at the Joint Typhoon Warning Center to consult an RGB composite image product that combines the VIIRS DNB with the I5 (11.45 μm) band during analysis of tropical cyclones. This imagery, made available by Naval Research Laboratory in Monterey CA (NRL-MRY) via the Automated Tropical Cyclone Forecasting System (ATCF) and the NRL Tropical Cyclone Webpage [30], has proven useful for distinguishing between high and low clouds in subjective analysis. Additionally, this band combination can allow analysts to discern situations where low cloud can be seen through optically-thin high cloud. RGB composite images have aided analysts in improving their estimates of storm center on multiple occasions.

To create the RGB composite image, the DNB lunar reflectance (following the model of Miller and Turner [31], or via NCC pseudo-albedo following Liang *et al.* [6]) is assigned to the red and green components, and inverted I5 band brightness temperatures are assigned to the blue component. This results in a false-color image product where: open ocean and optically-thin low cloud appear black; optically-thick low cloud appears yellow; optically-thin high cloud appears blue; optically-thick high cloud appears white; and optically-thin high cloud overlaying optically-thick low cloud also appears white, but can often be discerned from optically-thick high cloud by context.

In cases of extremely sheared storms, the RGB composite image can enable an analyst to observe the exposed low-altitude circulation center; however, even in less sheared cases, the imagery can provide useful information about the low-level circulation. An example of this is shown in Figure 11 where the left-hand panel shows I5 band brightness temperatures and the right-hand panel shows an RGB composed of the DNB and I5 band brightness temperatures. Both images depict Typhoon Linfa on 3 June 2015 at 1752 UTC. The I5 band image provides information about the storm's convective tops, but provides little information about the low-altitude circulations. Based solely on this I5 band image, analysts at the Joint Typhoon Warning Center (JTWC) would likely have determined that the storm center was at the center of the convective region (marked with a §).

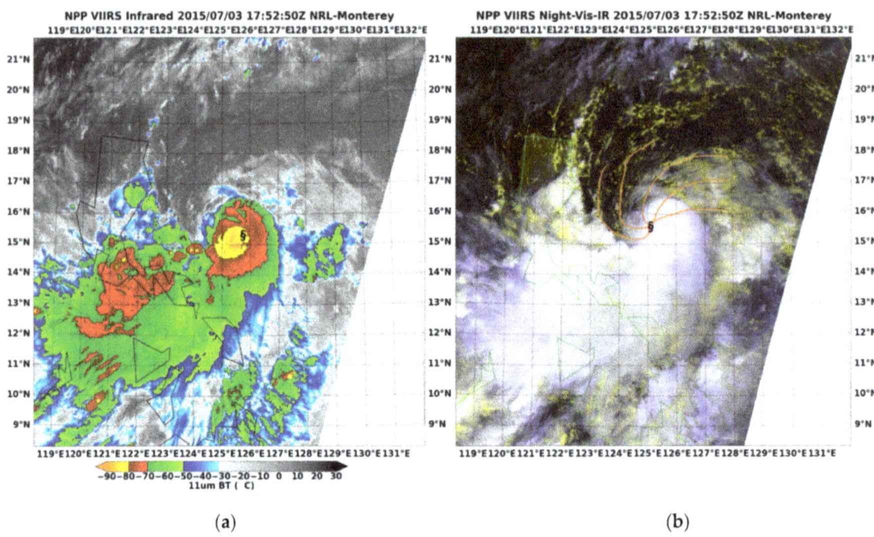

Figure 11. VIIRS (**a**) I5 (11.45 μm) band brightness temperature and (**b**) RGB composite imagery of Typhoon Linfa from 3 July 2015 at 1752 UTC. The RGB composite is composed of the DNB in the red and green bands and inverted I5 brightness temperatures in the blue. The RGB composite was used by Joint Typhoon Warning Center (JTWC) analysts to correctly determine the center of the typhoon by tracing the low-altitude cloud lines (orange) to the center of circulation.

The RGB composite image product provides a significant amount of additional information. The imagery allows analysts to readily discriminate low altitude patterns from high altitude patterns. JTWC stated that "without the VIIRS DNB image, the TC position would have been derived from IR only, which would have placed the center further southeast about 20 miles (30 km) with the assistance of the VIIRS DNB image, the forecaster correctly placed the best track position, which in turn improved model initialization and subsequent forecast accuracy."

In another example, the VIIRS DNB proved useful for identifying low cloud features in the presence of higher overriding cirrus clouds. The event occurred east of Hawaii on 29 July 2013. Tropical Storm Flossie was east of the Big Island of Hawaii, moving generally to the west-northwest. The Central Pacific Hurricane Center (CPHC) faced the challenge of issuing guidance on possible landfall over the nighttime hours when infrared data gave misleading information on the circulation center. After appealing to DNB imagery, the CPHC issued the following statement (with yellow highlighting noting the use of VIIRS nighttime visible imagery): "The center of Flossie was hidden by high clouds most of the night before VIIRS nighttime visual satellite imagery revealed an exposed low level circulation center farther north than expected. We re-bested the 0600 UTC position based on the visible data" [32].

Figure 12A shows the VIIRS DNB image at 1103 UTC that was being referred to, and Figure 12B shows the corresponding I5 (11.45 μm) infrared image. High cirrus clouds with brightness temperatures around −30 °C can be seen in the infrared image to the northwest of the deepest convection, but in the DNB some low clouds can be seen underneath. CPHC inferred the center of circulation based on the spiral structure of these low level clouds, and their 1200 UTC analysis of the center location is denoted in the figure with a maroon dot. First-light daytime visible imagery from GOES-W confirmed the DNB moonlight-based guidance. The infrared imagery alone would not have been useful in locating the center.

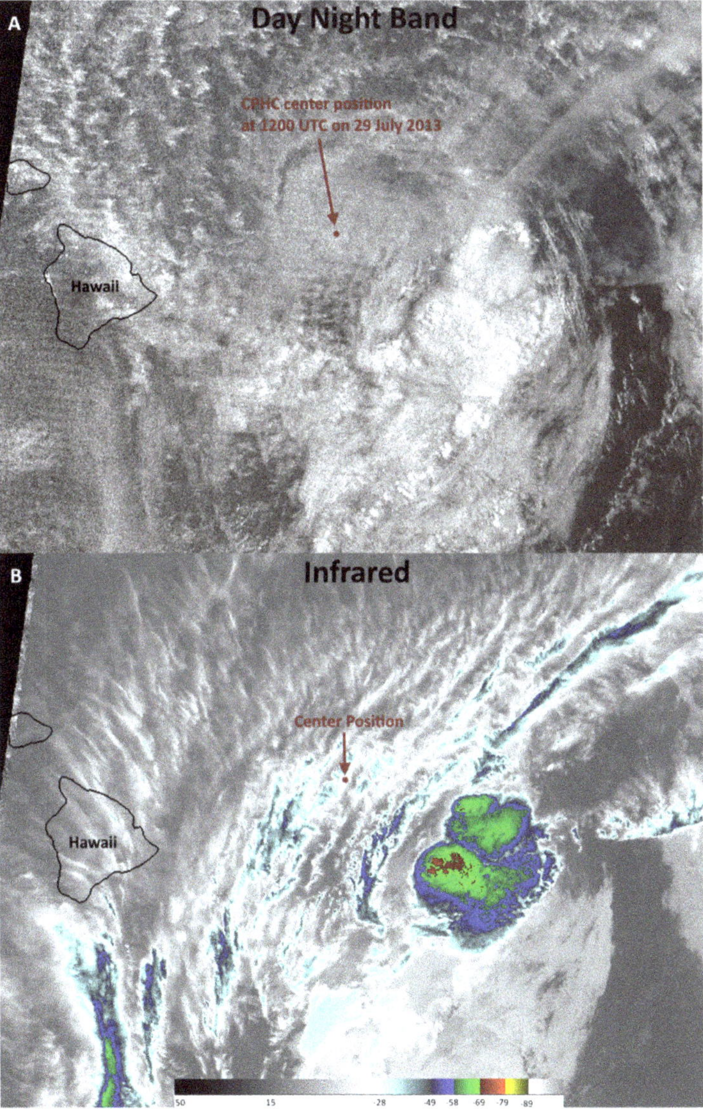

Figure 12. VIIRS (**A**) DNB and (**B**) I5 (11.45 μm) image showing Tropical Storm Flossie east of Hawaii on 29 July 2013 at 1103 UTC. The analyzed position by the Central Pacific Hurricane Center of the center of the storm at 1200 UTC is denoted by a maroon dot in both images. The units of brightness temperature in (b) are degrees C. Striping in the DNB becomes more apparent as the signal level decreases under lowlight illumination.

5.2.2. Puerto Rico Dust from Saharan Air Layer

During the spring through autumn months, the greater Caribbean region is susceptible to major dust outbreaks due to passages of dust-laden Saharan Air Layer (SAL) that can significantly degrade air quality and suppress local convection. SAL occurrence increases the potential for massive wildfire outbreaks throughout Puerto Rico and the surrounding West Indies. Additionally, the public in the greater Caribbean region suffers some of the world's highest rates of asthma, which has been postulated by health officials to be attributable to long-term exposure to high dust concentrations and the bio-chemical content of this dust [33–35]. The NWS in San Juan, Puerto Rico (NWS-PR) is responsible for issuing hazardous warning alerts to other agencies and the public. The resources available to the NWS-PR are limited with respect to making accurate and effective assessment and predictions of SAL events.

S-NPP VIIRS products are actively being used by the NWS-PR in monitoring the SAL via the NexSat website [36] from NRL-MRY. Through a wealth of remote sensing datasets and image products, in-situ observations, and an operational global dust model, NRL-MRY provides NWS-PR with enhanced capabilities to detect, assess, and predict SAL events as they propagate across the forecast area of responsibility (AOR). Included among the suite of products provided to NWS-PR through NRL-MRY's NexSat website are products derived from the S-NPP VIIRS sensor. These products include visible, infrared, true-color, dust, Aerosol Optical Depth (AOD), and DNB imagery.

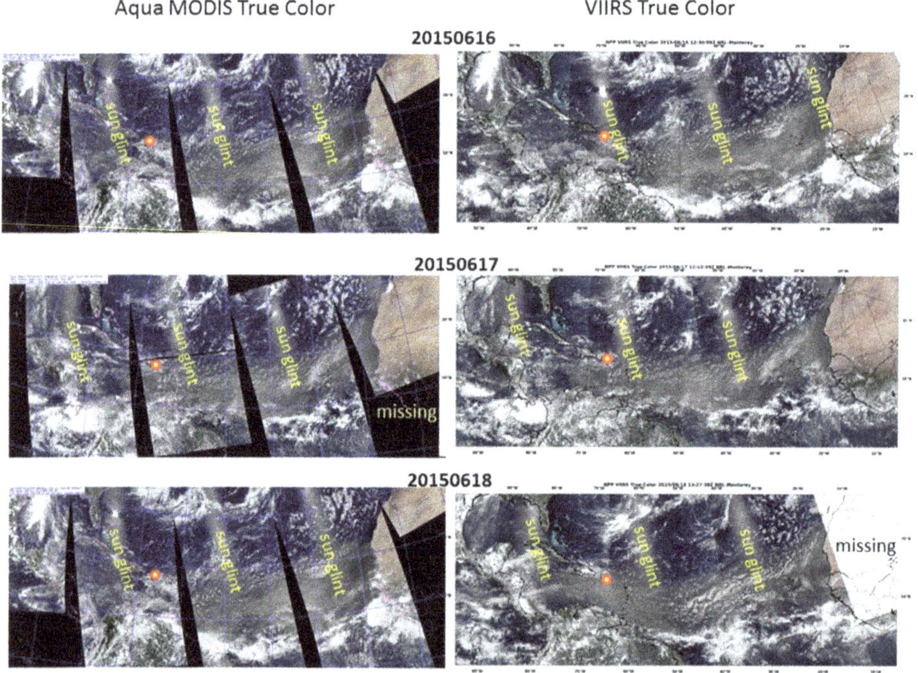

Figure 13. Comparison of Aqua MODIS (left panels) *vs.* S-NPP VIIRS (right panels) true-color image products while monitoring a SAL event across the north tropical Atlantic basin during 16–18 June 2015. The increased swath within VIIRS reduces the "guesswork" in tracking dust. Sun glint regions indicate enhanced levels of reflection from the ocean surface.

During the May–July 2015 timeframe, an almost continuous stream of strong African dust events propagated across the north tropical Atlantic basin, eventually impacting the greater Caribbean region.

These events produced high counts of particulate matter ($PM_{2.5}$ and PM_{10}), high AOD values, and a rash of wildfires due to dry subsiding air, particularly within the forested regions of Puerto Rico. Figure 13 compares true-color imagery from Aqua MODIS and S-NPP VIIRS at the height of a SAL event that occurred during the period of 16–18 June 2015. The wider swath width and increased resolution at scan edge allows VIIRS (right-hand column) to provide a more continuous and more easily-analyzed product than MODIS (left-hand column), reducing the "guesswork" in tracking dust across the Atlantic basin. Figure 14 presents VIIRS-derived blue-light dust products [37] (left-hand column) during the approach of the same SAL-borne dust plume towards Barbados (marked with a red/yellow dot) for the period of 16–18 June. In this imagery, dust appears in shades of pink, while clouds are shown in cyan.

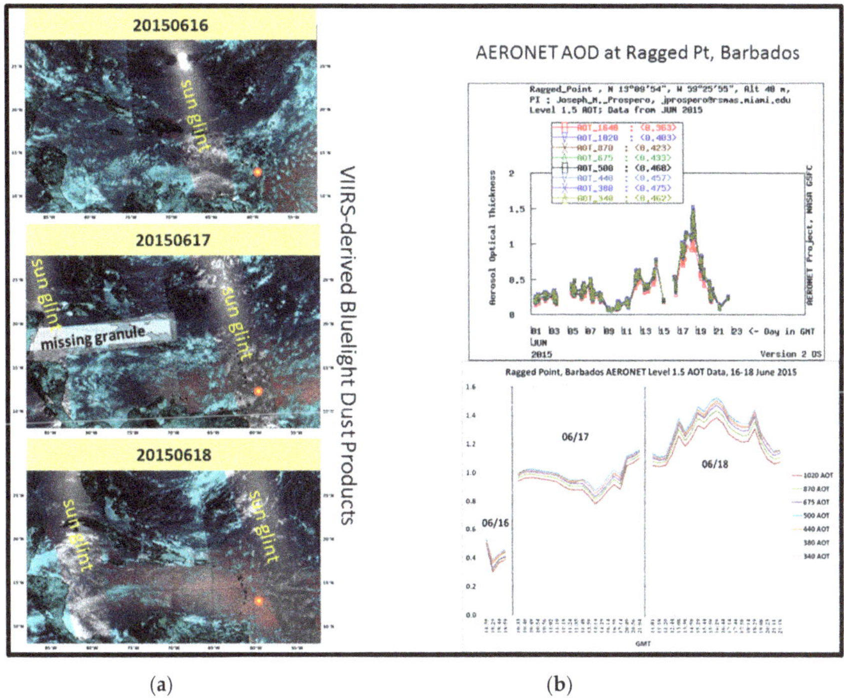

(a) (b)

Figure 14. (**a**) A series of VIIRS-derived bluelight dust products during 16–18 June 2015 showing the approach and eventual impact of the SAL plume covering the greater Caribbean region. Dust appears in shades of pink, whereas clouds are in shades of cyan. The island of Barbados is annotated in red/yellow circles; (**b**) AERONET AOD measurements taken at Ragged Point, Barbados. The top-right panel indicates the monthly profile, with a spike during the SAL event (17–19 June). The bottom-right panel highlights the SAL event peaking during 14–17 UTC on 18 June.

Surface-based measurements of AOD indicate that the VIIRS imagery correctly identifies dust for subjective analysis. The plot in the top right of Figure 14 shows AOD measurements from the Aerosol Robotic Network (AERONET) site at Ragged Point, Barbados over the period from 1–23 June 2015. AOD values spike between the 17th and 19th, indicating the impact of the SAL-borne dust on the region. The figure in the lower right shows the same AOD data, but focusing on the period of 16–18 June, showing that the event peaks between 14 and 17 UTC on 18 June. These observations of elevated AOD values coincide with the approach of the dust plume as observed by VIIRS. In this case, VIIRS was able to provide forecasters with advanced warning of the SAL event and allowed forecasters to disseminate that information to the public.

5.2.3. Blowing Dust in the Great Plains

Another use of VIIRS multi-spectral Imagery is exemplified by the 29 April 2014 dust case over the Texas Panhandle. That day ended a three-day-long dust event associated with an intense upper-level system that lumbered across the western and central U.S. The placement and movement of the system kept a very strong jet over western Colorado and the Texas Panhandle, resulting in widespread severe winds gusting from 55 mph (90 km/h) to even greater than 65 mph (105 km/h) at times. Sustained drought over the region provided an ample supply of dust, which was immediately lofted by strong surface winds. Smaller regional airports in the Texas Panhandle reported visibilities at or near zero for the duration of this event. These conditions occurred on three consecutive afternoons from the 27 to the 29 April, and necessitated the issuance of multiple blowing dust SIGMET advisories.

On the afternoon of the 29 April, Aviation Weather Center (AWC) forecasters examined the VIIRS and MODIS dust enhancements as they issued several SIGMETs related to this event. Figure 15a is a VIIRS true-color image from 2000 UTC. There were no obvious signs of dust in the scene. However, the corresponding dust enhancement (Figure 15b), based on the multi-spectral algorithm [37], highlights in pink a large swath of blowing dust crossing into the Texas Panhandle from southwestern Colorado.

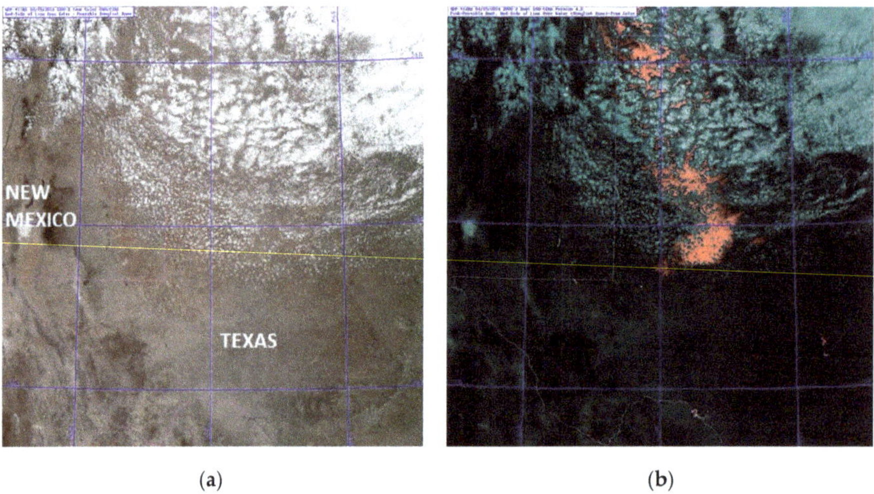

(a) (b)

Figure 15. (**a**) VIIRS true-color imagery for the Texas Panhandle at 2000 UTC on 29 April 2014 with no obvious signs of dust in the scene; (**b**) VIIRS dust-enhancement imagery at the same time, highlighting the blowing dust in pink below the overlying clouds in cyan.

A blowing dust SIGMET was issued at 1700 UTC, valid until 2100 UTC. Because of the limited temporal resolution of polar imagery, this SIGMET was issued without the use of the dust enhancement. However, an extension of this SIGMET was issued just after 2100 UTC after forecasters examined the VIIRS dust enhancement imagery, and it was extended further south and east into central Texas. In this case the dust enhancement proved a very useful situational awareness tool for amending the SIGMET, which will only improve when geostationary assets begin carrying the bands required for implementing this algorithm.

6. Discussion

Although the initial post-launch Cal/Val checkout of VIIRS Imagery is shared between the EDR Imagery Team and the VIIRS SDR Team, many Imagery applications are validated during the intensive Cal/Val period ending with an 'official' validation status, signifying that Imagery is ready for use by a wide range of users worldwide. Each step involves users, but not to the extent that users become involved in the post-checkout timeframe. The post-launch checkout is followed by the long-term monitoring phase that continues for the life of VIIRS on each satellite. It is during this stage that users become heavily involved in VIIRS validation. Users are sought who benefit from the advantages that VIIRS has over heritage satellite instrumentation, particularly the DNB/NCC imagery not available from geostationary orbit.

The validation examples presented, most with direct and specific feedback from operational users, are the crux of the post-launch validation of VIIRS Imagery. Emphasis has been on the Alaska operations, since the Alaska region is specifically mentioned in the KPP statement for VIIRS. Alaska examples include identification of fog, convection, and the ability to distinguish between sea ice and clouds over the Arctic Ocean and Bering Sea. However, not all VIIRS validation necessarily takes place in the Polar Regions. Examples of tropical cyclone re-centering and Saharan dust in the Caribbean and dust on the Great Plains were presented as applications of VIIRS DNB imagery that has potential for worldwide application.

7. Conclusions

These cases present an overall picture of wide use and multiple-user benefits gained from VIIRS Imagery, both as individual bands and as band combinations and RGB composite image products suitable to specific atmospheric and land phenomena. The fact that numerous users are pleased with VIIRS is the "user" validation that goes beyond the "official" validation accomplished by the Imagery Team alone.

Future plans for VIIRS Imagery validation include continuing/routine checkout as operational ground systems (hardware and software) change, to make sure Imagery remains at consistent high quality. Then, the validation process begins over with follow-on satellites, JPSS-1 to be launched in 2016 or 2017, and similarly for JPSS-2 and beyond.

Acknowledgments: The authors would like to thank other on the VIIRS Imagery and Visualization Team [38] who are not listed as co-authors for their contributions: Stan Kidder, Debra Molenar, Steve Finley, Renate Brummer, Chris Elvidge, Kim Richardson, and Bill Thomas. The authors would also like to thank Capt. Brian Decicco and TSgt. Ricky Frye of the Joint Typhoon Warning Center for their contributions. Funding for this work was provided by the JPSS Program Office, NOAA/NESDIS/StAR, and the Naval Research Laboratory (Grant #N00173-14-G902), the Oceanographer of the Navy through office at the PEO C4I & Space/PMW-120 under program element PE-0603207N. The views, opinions, and findings contained in this article are those of the authors and should not be construed as an official National Oceanic and Atmospheric Administration (NOAA) or U.S. Government position, policy, or decision.

Author Contributions: Don Hillger and Tom Kopp, as VIIRS Imagery Team co-leads, wrote most of the text. The rest of the authors contributed the Imagery examples and the text to accompany those examples.

Conflicts of Interest: The authors declare no conflict of interest.

References

1. Hillger, D.; Kopp, T.; Lee, T.; Lindsey, D.; Seaman, C.; Miller, S.; Solbrig, J.; Kidder, S.; Bachmeier, S.; Jasmin, T.; *et al.* First-light imagery from Suomi NPP VIIRS. *Bull. Am. Meteor. Soc.* **2013**. [CrossRef]
2. Hillger, D.; Seaman, C.; Liang, C.; Miller, S.D.; Lindsey, D.; Kopp, T. Suomi NPP VIIRS imagery evaluation. *J. Geophys. Res.* **2014**, *119*, 6440–6455. [CrossRef]
3. Kuciauskas, A.; Solbrig, J.; Lee, T.; Hawkins, J.; Miller, S.; Surratt, M.; Richardson, K.; Bankert, R.; Kent, J. Next-generation satellite meteorology technology unveiled. *Bull. Amer. Meteor. Soc.* **2013**, *94*, 1824–1825. [CrossRef]

4. Cao, C.; Shao, X.; Xiong, X.; Blonski, S.; Liu, Q.; Uprety, S.; Shao, X.; Bai, Y.; Weng, F. Suomi NPP VIIRS sensor data record verification, validation, and long-term performance monitoring. *J. Geophys. Res. Atmos.* **2013**, *118*, 11664–11678. [CrossRef]
5. Miller, S.; DStraka, W.; Mills, S.P.; Elvidge, C.D.; Lee, T.F.; Solbrig, J.; Walther, A.; Heidinger, A.K.; Weiss, S.C. Illuminating the capabilities of the Suomi NPP VIIRS Day/Night Band. *Remote Sens.* **2013**, *5*, 6717–6766. [CrossRef]
6. Liang, C.K.; Hauss, B.I.; Mills, S.; Miller, S.D. Improved VIIRS Day/Night band imagery with near constant contrast. *IEEE TGRS* **2014**, *52*, 6964–6971. [CrossRef]
7. Miller, S.D.; Mills, S.P.; Elvidge, C.D.; Lindsey, D.T.; Lee, T.F.; Hawkins, J.D. Suomi satellite brings to light a unique frontier of environmental imaging capabilities. *Proc. Nat. Acad. Sci. USA* **2012**, *109*, 15706–15711. [CrossRef] [PubMed]
8. Solbrig, J.E.; Lee, T.E.; Miller, S.D. Advances in remote sensing: Imaging the Earth by moonlight. *Eos* **2013**, *94*, 349–350. [CrossRef]
9. Seaman, C.; Hillger, D.; Kopp, T.; Williams, R.; Miller, S.; Lindsey, D. *Visible Infrared Imaging Radiometer Suite (VIIRS) Imagery Environmental Data Record (EDR) User's Guide*; NOAA Technical Report NESDIS: Washington, WA, USA, 2015.
10. Algorithm Theoretical Basis Documents (ATBD). Available online: http://www.star.nesdis.noaa.gov/jpss/Docs.php (accessed on 21 December 2015).
11. JPSS Level 1 Requirements Document Final. Available online: http://www.jpss.noaa.gov/pdf/L1RD_JPSS_REQ_1001_final_v1.8–1.pdf (accessed on 21 December 2015).
12. Miller, S. A satellite sensor that can see in the dark is revealing new information for meteorologists, firefighters, search teams and researchers worldwide. *Sci. Am.* **2015**, *312*, 78–81. [CrossRef] [PubMed]
13. Straka, W.J.; Seaman, C.J.; Baugh, K.; Cole, K.; Stevens, E.; Miller, S.D. Utilization of the Suomi National Polar-Orbiting Partnership (NPP) Visible Infrared Imaging Radiometer Suite (VIIRS) Day/Night band for arctic ship tracking and fisheries management. *Remote Sens.* **2015**, *7*, 971–989. [CrossRef]
14. Elvidge, C.D.; Zhizhin, M.; Hsu, F.-C.; Baugh, K.E. VIIRS nightfire: Satellite pyrometry at night. *Remote Sens.* **2013**, *5*, 4423–4449. [CrossRef]
15. JPSS Imagery CVP. *JPSS Cal/Val Plan for Imagery Product*; JPSS Program Office: Washington, WA, USA, 2015.
16. McIDAS-V. Available online: http://www.ssec.wisc.edu/mcidas/software/v/ (accessed on 21 December 2015).
17. SeaSpace Corporation. Available online: http://www.seaspace.com/software.php (accessed on 21 December 2015).
18. Interactive Data Language. Available online: http://exelisvis.com/ProductsServices/IDL.aspx (accessed on 21 December 2015).
19. Advanced Weather Interactive Processing System. Available online: http://www.nws.noaa.gov/ops2/ops24/awips.htm (accessed on 21 December 2015).
20. Community Satellite Processing Package. Available online: http://cimss.ssec.wisc.edu/cspp/ (accessed on 21 December 2015).
21. International Polar Orbiting Processing Package. Available online: https://directreadout.sci.gsfc.nasa.gov/?id=dspContent&cid=68 (accessed on 21 December 2015).
22. Ellrod, G.P. Advances in the detection and analysis of fog at night using GOES multispectral infrared imagery. *Wea. Forecast.* **1995**, *10*, 606–619. [CrossRef]
23. Fierstein, J.; Hildreth, W. The plinian eruptions of 1912 at Novarupta, Katmai National Park, Alaska. *Bull. Volcanol.* **1992**, *54*, 646–684. [CrossRef]
24. Casadevall, T. The 1989–1990 eruption of redoubt volcano, Alaska—Impacts on aircraft operations. *J. Volcanol. Geotherm. Res.* **1994**, *62*, 301–316. [CrossRef]
25. Horwell, C.J.; Baxter, P.J. The respiratory health hazards of volcanic ash: A review for volcanic risk mitigation. *Bull. Volcanol.* **2006**, *69*, 1–24. [CrossRef]
26. Ash from century-old Novarupta Volcanic Eruption Sweeps over Kodiak Island. Available online: http://www.adn.com/article/ash-century-old-novarupta-volcanic-eruption-sweeps-over-kodiak-island (accessed on 21 December 2015).
27. SIGMET. Available online: http://www.volcanodiscovery.com/archive/vaac/latest-reports-2012.html (accessed on 21 December 2015).

28. Alaska Wildfire Season Worst on Record So Far: NOAA Providing On-The-Scene Assistance. Available online: http://www.noaa.gov/features/03_protecting/070215-alaska-wildfire-season-worst-on-record-so-far.html (accessed on 21 December 2015).
29. USCG Rescues Man Trapped in Arctic Ice. Available online: http://navaltoday.com/2014/07/14/uscg-rescues-man-trapped-in-arctic-ice/ (accessed on 21 December 2015).
30. NRL Tropical Cyclone Webpage. Available online: http://www.nrlmry.navy.mil/TC.html (accessed on 21 December 2015).
31. Miller, S.D.; Turner, R.E. A dynamic lunar spectral irradiance dataset for NPOESS/VIIRS Day/Night band nighttime environmental applications. *IEEE Trans. Geosci. Remote Sens.* **2009**, *47*, 2316–2329. [CrossRef]
32. CPHC Statement. Available online: http://www.prh.noaa.gov/cphc/tcpages/archive/2013/TCDCP1.EP062013.019.1307291511 (accessed on 21 December 2015).
33. Akinbami, O.J.; Moorman, J.E.; Liu, X. *Asthma Prevalence, Health Care Use, and Mortality: United States, 2005–2009*; US Department of Health and Human Services, Centers for Disease Control and Prevention, National Center for Health Statistics: Centers for Disease Control and Premention: Washington, WA, USA, 2011.
34. Anstey, M.H. Climate change and health—What's the problem? *Global Health* **2013**, *9*. [CrossRef] [PubMed]
35. NexSat. Available online: http://www.nrlmry.navy.mil/NEXSAT.html (accessed on 21 December 2015).
36. Cadelis, G.; Molinie, J. Short-term effects of the particulate pollutants contained in Saharan Dust on the visits of children to the emergency department due to asthmatic conditions in Guadeloupe (French Archipelago of the Caribbean). *PloS ONE* **2014**, *6*. [CrossRef] [PubMed]
37. Miller, S.D. A consolidated technique for enhancing desert dust storms with MODIS. *Geophys. Res. Lett.* **2003**, *30*, 2071–2074. [CrossRef]
38. VIIRS Imagery and Visualization Team. Available online: http://rammb.cira.colostate.edu/projects/npp/ (accessed on 21 December 2015).

Chapter 2:
Instrument Onboard Calibration and Prelaunch Characterization

Article

Spectral Dependent Degradation of the Solar Diffuser on Suomi-NPP VIIRS Due to Surface Roughness-Induced Rayleigh Scattering

Xi Shao [1,*], Changyong Cao [2] and Tung-Chang Liu [3]

1 Department of Astronomy, University of Maryland, College Park, MD 20742, USA
2 NOAA (National Oceanic and Atmospheric Administration)/NESDIS (National Environmental Satellite, Data, and Information Service)/STAR (Center for Satellite Applications and Research), NCWCP, E/RA2, 5830 University Research Ct., College Park, MD 20740, USA; changyong.cao@noaa.gov
3 Department of Physics, University of Maryland, College Park, MD 20742, USA; tcliu@umd.edu
* Correspondence: xshao@umd.edu; Tel.: +1-301-405-7936; Fax: +1-301-405-2929

Academic Editors: Richard Müller and Prasad S. Thenkabail
Received: 20 November 2015; Accepted: 11 March 2016; Published: 17 March 2016

Abstract: The Visible Infrared Imaging Radiometer Suite (VIIRS) onboard Suomi National Polar Orbiting Partnership (SNPP) uses a solar diffuser (SD) as its radiometric calibrator for the reflective solar band calibration. The SD is made of Spectralon™ (one type of fluoropolymer) and was chosen because of its controlled reflectance in the Visible/Near-Infrared/Shortwave-Infrared region and its near-Lambertian reflectance property. On-orbit changes in VIIRS SD reflectance as monitored by the Solar Diffuser Stability Monitor showed faster degradation of SD reflectance for 0.4 to 0.6 μm channels than the longer wavelength channels. Analysis of VIIRS SD reflectance data show that the spectral dependent degradation of SD reflectance in short wavelength can be explained with a SD Surface Roughness (length scale << wavelength) based Rayleigh Scattering (SRRS) model due to exposure to solar UV radiation and energetic particles. The characteristic length parameter of the SD surface roughness is derived from the long term reflectance data of the VIIRS SD and it changes at approximately the tens of nanometers level over the operational period of VIIRS. This estimated roughness length scale is consistent with the experimental result from radiation exposure of a fluoropolymer sample and validates the applicability of the Rayleigh scattering-based model. The model is also applicable to explaining the spectral dependent degradation of the SDs on other satellites. This novel approach allows us to better understand the physical processes of the SD degradation, and is complementary to previous mathematics based models.

Keywords: solar diffuser; spectral dependent degradation; surface roughness; VIIRS; Rayleigh scattering

---

1. Introduction

The Suomi-NPP satellite was successfully launched on 28 October 2011. VIIRS (Visible Infrared Imager Radiometer Suite) is one of five instruments onboard the Suomi-NPP (SNPP) satellite and acquired its first measurements in November 2011 [1–3]. The VIIRS is a scanning radiometer and has 22 spectral bands covering the spectrum between 0.412 μm and 11.5 μm, including 14 reflective solar bands (RSB), seven thermal emissive bands (TEB), and one day–night band (DNB). It primarily focuses on clouds, Earth surface variables, surface temperature and imagery, and provides moderate-resolution, radiometrically accurate images of the globe once per day for the RSBs and twice daily for the TEBs and DNB. It has wide-swath (3000 km) with spatial resolutions of 375 and 750 m at nadir for the imaging bands (I-bands) and moderate resolution bands (M-bands), respectively.

For the VIIRS RSBs, the radiometric calibration uncertainty in spectral reflectance for a scene at typical radiance is expected to be less than 2%. Calibration methods such as using the onboard solar diffuser (SD), inter-comparisons with instruments on other satellites, vicarious calibration at desert and ocean sites and lunar calibration [2–12] have been routinely performed to trend and validate the on-orbit radiometric performance of VIIRS RSBs. As a result of continuous calibration efforts, both radiometric and signal-to-noise ratio performances of the VIIRS RSBs continue to meet its requirements.

In operations, the VIIRS uses onboard SD and space view data to perform radiometric calibration of RSBs for Sensor Data Record (SDR) generation. The space view data are used to determine the background offset and the reflected solar light data from the SD is used to determine the gain for RSBs. The solar diffuser is made of Spectralon™ material and degrades in reflectance (especially at the blue end of the spectrum) due to exposure to space radiation such as solar UV light and energetic particles in space [13–20]. To mitigate this effect, VIIRS uses a Solar Diffuser Stability Monitor (SDSM) in the 0.4–0.94 μm wavelength to monitor changes of the SD reflectance over time and provides a correction factor (H factor) to the calibration coefficients of VIIRS RSBs [9–12]. The H factor monitored by the VIIRS SDSM revealed that reflectance of the 0.4 to 0.6 μm channels of VIIRS SD degraded faster than the SD reflectance of longer wavelength RSB channels. Over ~2.5 years, the degradation of SD reflectance of VIIRS M1 (412 nm) channel reached ~29%. Similar degradation of the SD reflectance, but with a much slower rate, also occurred to Moderate Resolution Imaging Spectroradiometer (MODIS) on Aqua and Terra. For comparison, the degradation of the VIIRS SD reflectance over 1.5 years is comparable to 11 years' degradation of SD on Aqua MODIS [21,22]. It is assumed that the difference is mainly due to the limited exposure of the Aqua MODIS SD to sunlight with the use of a shutter door when not calibrating. However, other possibilities such as accelerated SD reflectance degradation due to contamination with Pennzane are also being investigated [20].

Radiometric calibration of RSB using onboard SD is quite common for modern-day satellite instruments such as MODIS on AQUA and TERRA [21,22], Thermal And Near-infrared Sensor for carbon Observation-Fourier Transform Spectrometer (TANSO-FTS) onboard GoSat [23], Operational Land Imager on LandSat-8 [24] and VIIRS on SNPP [2,3]. Instruments from other upcoming missions such as the Advanced Baseline Imager (ABI) on GOES-R, TANSO-FTS on GoSat2 and VIIRS on JPSS series will also use onboard SD for radiometric calibration. The degradation of SD reflectance in space occurs inevitably when it is used for RSB calibration and exposed to space radiation. Efforts in instrument design have been devoted to track and/or reduce the SD degradation. For instruments such as MODIS and VIIRS, SDSM has been used in tracking the SD degradation and for TANSO on GoSat, double-sided SD with different solar exposure frequency has been used.

There have also been continuous efforts in laboratory experiments [13–20] to characterize the reflectance degradation of SD material by exposing samples to radiations such as UV and energetic particles. While these experiments connect the spectral reflectance degradation of the SD material with radiation exposure, there has been an insufficient understanding of the relationship between the observed spectral dependent degradation and the changes in material surface properties previously.

In this paper, we introduce a SD Surface Roughness based Rayleigh Scattering (SRRS) physical model to explain the spectral dependent degradation of SD material under space radiation. Characteristic parameters of surface roughness can be derived from the long term spectral dependent degradation of VIIRS SD reflectance and used to monitor the surface roughness change over its operation period. In Section 2, we introduce the onboard calibration of VIIRS RSBs using SD and SDSM. Section 3 presents the model of spectral dependent degradation due to surface roughness and specifies the modeling parameters for characterizing surface roughness. In Section 4, the model is applied to spectral reflectance data of VIIRS SD, and an explanation of the spectral dependent degradation of VIIRS SD in terms of surface roughness change is also given. In Appendix, general effects of space radiation such as UV light and energetic particles on the SD material are presented. Experimental evidence of spectral dependent degradation of SD material and surface roughness changes are reviewed.

2. Onboard Calibration of VIIRS RSBs with SD and SDSM

2.1. VIIRS SD for Onboard Calibration of RSBs

VIIRS is a conventional differencing radiometer that uses a space view to determine zero radiance and observations of reflected solar light from SD (Figure 1) to determine gain for RSBs. When the Suomi-NPP satellite moves from the night side toward day side of the Earth near the South Pole, the Sun illuminates SD panel and sun-view port of the SDSM through attenuation screens. The attenuation screen consists of many small holes packed closely together and has about 11% transmission. It is mounted in the SD view port to limit the solar radiance reflected off the SD and to make it comparable to typical Earth scenes. Since the geometric locations of light source, *i.e.*, azimuth and elevation of the Sun, keep changing for each scan, both the transmission screen and the SD need to be characterized as a function of solar incidence angles. The bidirectional reflectance distribution function (BRDF) measurements were performed during the pre-launch testing and updated with the on-orbit yaw maneuver measurements.

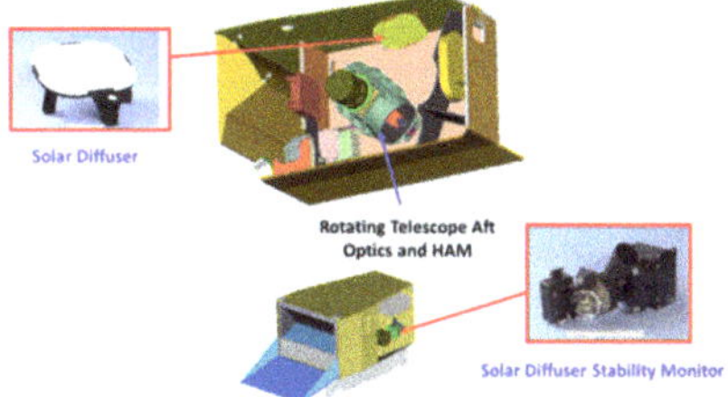

Figure 1. Solar diffuser (SD) and solar diffuser stability monitor (SDSM) used for onboard calibration for VIIRS RSBs (from JPSS VIIRS SDR ATBD).

Figure 2a shows the signals of the reflected solar light from the SD detected by the detectors of the I1 channel (0.64 μm wavelength) of VIIRS together with the spacecraft solar zenith angle variation over multiple-orbits around the Earth. The I1 view of the SD shows spikes of solar illumination when the satellite passes from the dark side to the sunlit side of the Earth in the high latitudes of the southern hemisphere. These spikes indicate the time interval when the SD is illuminated by the sunlight passing through the SD screen. Figure 2b shows the zoom-in of the I1 view of SD during one orbit. The main spike in the I1 view of SD lasts ~10 min. There are remnant radiation side lobes with amplitudes of about 1/10 of the main spike appearing in the I1 view of SD as the satellite transits to the dayside. The side lobes in the I1 view of SD is likely due to the reflected sun light from Earth which enters through the VIIRS nadir door, contaminating the SD and observed by I1. Due to the exposure of the SD to sunlight over time, it degrades and this degradation is tracked using the SDSM to calculate the H-factors.

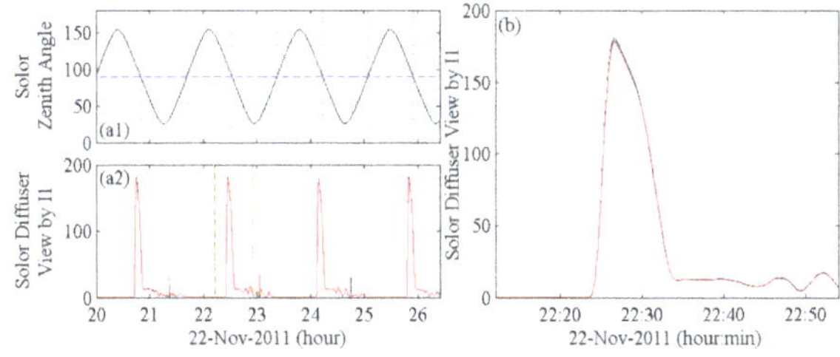

Figure 2. (**a1**) Spacecraft solar zenith angle and (**a2**) solar light illumination of VIIRS SD as seen by I1 channel detectors (overlaid together) over multiple orbits; and (**b**) zoom-in view of I1 view of SD.

2.2. Degradation of VIIRS SD as Monitored by the SDSM

The VIIRS SDSM monitors the SD reflectance degradation using a three-position fold mirror to perform a three-scan cycle, which sequentially views a dark scene, screened sunlight, and illumination from the SD. The direct sunlight to the SDSM is attenuated through a pinhole-filter to keep the radiance within the dynamic range of the SDSM's detector/amplifier combination. The SDSM is basically a ratioing radiometer and monitors the SD degradation rate by the ratio between signals from screened sunlight and reflected light from the SD. Proper correction of the attenuation filter for the Sun view as function of incident angles is applied at the time of observation. In operations, during each SDSM calibration event, the SDSM acquires data using eight individual detectors over an approximately one minute period, which is ~33 VIIRS scans. These measurements are accumulated, trended and projected as spectral H factors in order to provide an update on SD reflectance change for the corresponding VIIRS RSBs. The VIIRS H factors were updated daily before 14 May 2014 and later the operation frequency was switched to three times a week. Table 1 lists the center wavelengths of VIIRS RSB that use the H factors as monitored by the SDSM to trend SD degradation.

Table 1. Wavelength of VIIRS RSBs used in monitoring SD degradation.

VIIRS Band	M1	M2	M3	M4	M5	M6	M7
Center λ (μm)	0.412	0.445	0.488	0.555	0.672	0.746	0.865

Figure 3a shows the SD degradation trend, *i.e.*, evolution of H factor for M1–M7 band of VIIRS. The SD reflectance degradation is faster at shorter visible wavelengths. Over ~2.5 years, the degradation of SD reflectance of VIIRS M1 (412 nm) channel reached ~29%, while the degradation for M7 (865 nm) channel is 1.3%. Figure 3b also shows that the degradation of SD reflectance is spectrally dependent, *i.e.*, maintaining a smooth spectral shape during the degradation, which will be further investigated in this paper.

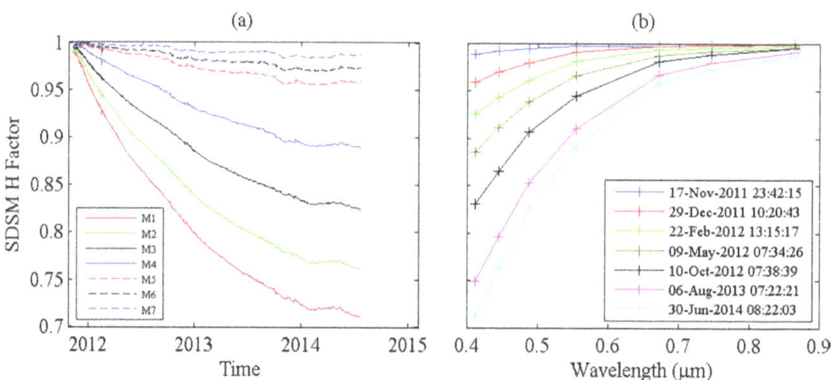

Figure 3. (**a**) VIIRS SD degradation over time as revealed by H factor for M1–M7 band; and (**b**) spectral dependence of VIIRS SD reflectance change over time.

The larger SD degradation at shorter wavelengths is very similar to what has been observed on Terra and Aqua MODIS, which were launched in December 1999 and May 2002, respectively. Different from VIIRS, the MODIS design includes a SD door [21,22]. The SD door is commanded to an open position only during each scheduled SD/SDSM calibration event in order to prevent unnecessary solar exposure on to the SD panel and the SD door is closed after the calibration event. For MODIS on Terra, a SD door operation anomaly occurred on 6 May 2003 during a scheduled SD/SDSM calibration event.

The SD door on MODIS Terra was permanently set at the "open" position since 2 July 2003 and the SD was under solar exposure every orbit afterwards. For Aqua MODIS, there has been no schedule change, and the SD/SDSM system was operated according to the scheduled frequencies. VIIRS does not have a SD door and the SD calibration is obtained every-orbit.

After more than 12 years of on-orbit operation, reflectance degradation of Aqua MODIS SD varies from 0.6% at 0.94 μm to 19.0% at 0.41 μm [22]. Due to more frequent solar exposure, the Terra MODIS SD has experienced a much larger degradation, varying from 2.3% at 0.94 μm to 48.0% at 0.41 μm. However, the VIIRS SD degradation is faster than MODIS on Aqua. Degradation of VIIRS SD over 2.5 years has exceeded 11 years' degradation of Aqua MODIS SD. The difference in the magnitude of spectral dependent degradation among VIIRS and MODIS on Aqua and Terra is mainly due to the difference in the solar exposure time of SD.

3. Model on Spectrally-Dependent Scattering over Rough Surfaces

In this section, we provide a physics-based SRRS theoretical model to interpret the spectral dependent degradation of solar diffuser as shown in Figure 3, and in particular, the wavelength dependence that is shown in Figure 3b. The portion of light being scattered from the SD is dominantly determined by two factors: the wavelength of the light and the surface structure of the SD. The SD is made of Spectralon^{TM} (Fluorocarbon-based polymer) and designed to have the property of Lambertian reflectance (see Appendix A.1.). The spectral properties of SD reflectance inevitably degraded due to UV exposure and particle radiations in space [13–20], which are summarized in Appendixes A.2. and A.3. Laboratory experiment also showed that there are surface roughness change on the order of tens of nanometers for Fluoropolymer samples after combined UV and particle exposures [25] (see Appendix A.4.). While these experiments provide direct evidence of increased polymer surface roughness after UV and particle irradiations, there were no measurements or models on the spectral reflectance changes of the SD samples to connect surface roughness change to spectral reflectance change. In the following, we present a physics-based SRRS model to link the spectral dependent degradation of SD material, e.g., Spectralon panel on VIIRS, with the surface roughness change due to space radiation exposure.

Due to dielectric property changes of the local irregularities formed within the surface, the light irradiated on a rough surface can be scattered. Spectral scattering due to surface roughness has been extensively studied in theory and by experiments [26–33]. Scattering of light depends on the wavelength of the light being scattered and the scale length of surface roughness. In this section, we discuss the model of how surface roughness of length scale << wavelength changes the spectral reflectance of a SD material. Before being illuminated by space radiation, the diffuse reflectance $R_0(\lambda)$ of porous SD material for the wavelength range between 0.4 μm and 1.0 μm is almost flat and can be as high as 99% in the case of Spectralon. Under illumination of UV radiation and/or particle radiation, fine scale surface roughness increases as observed in the experiment by [25]. The length scale of these surface roughness is of tens nanometer and is much smaller than wavelength of interest. Therefore, the light-scattering from the surface roughness is of Rayleigh-type. Rayleigh scattering from a sphere of diameter d and refractive index n from a beam of light of wavelength $\lambda(d << \lambda)$ and intensity I_0 is given by

$$I = I_0 \frac{1+\cos^2\theta}{2R^2} \left(\frac{2\pi}{\lambda}\right)^4 \left(\frac{n^2-1}{n^2+2}\right)^2 \left(\frac{d}{2}\right)^6 \quad (1)$$

where R is the distance to the particle, θ is the scattering angle, and I is the intensity of scattered light. Two important aspects of Rayleigh scattering are worth noting.

(a) For $d << \lambda$, the angular pattern of scattering is symmetric between the forward and backward direction.

(b) The dependence of scattering on wavelength follows λ^{-4}, so light of shorter wavelength suffers more scattering.

Therefore, the Rayleigh-type scattering from the fine scale surface roughness on a SD material causes more scattering at shorter wavelengths than at longer wavelength and the symmetric scattering pattern can cause short wavelength light to be scattered back to the material, trapped or transmitted through. This is the origin of the degradation of reflectance in the short wavelength range for SD material after exposure to space radiation. It should be noted that although previous studies have used $H = 1 - \alpha/\lambda^{\eta}$ equations to mathematically model the SD degradation [34], our study is the first to establish the physical model based on SD Surface Roughness induced Rayleigh scattering on the SD surface and related degradation. To quantify the effect of Rayleigh-type scattering on the spectral dependent degradation of SD material due to surface roughness change, we first introduce parameters to characterize the surface roughness in Section 3.1, and the model for surface roughness-induced degradation is presented in Section 3.2.

3.1. Statistical Parameters Characterizing Surface Roughness

To introduce parameters characterizing surface roughness, we set $h = h(x, y)$ as the height of the surface at the coordinate point (x, y) and conveniently choose the coordinate system such that it has a zero mean. Two relevant statistical surface characteristics are the surface height distribution $SHD(h)$ and the surface auto-covariance function $ACF(x, y)$, defined, respectively, as

$$SHD(H)\mathrm{d}H = \mathrm{Area}(\{(x, y)|H < h(x, y) < H + \mathrm{d}H\}) \tag{2}$$

and

$$ACF(x, y) =< h(x_0, y_0)h(x_0 + x, y_0 + y) > \tag{3}$$

where $<\bullet>$ is the ensemble average over all regions of coordinates (x_0, y_0). An illustration of the surface roughness characteristic parameters is shown in Figure 4, where the root-mean-square (RMS) surface roughness σ_s is the standard deviation of $SHD(h)$, and the auto-correlation length l is the half-width of the $ACF(x, y)$ at $1/e$ peak value.

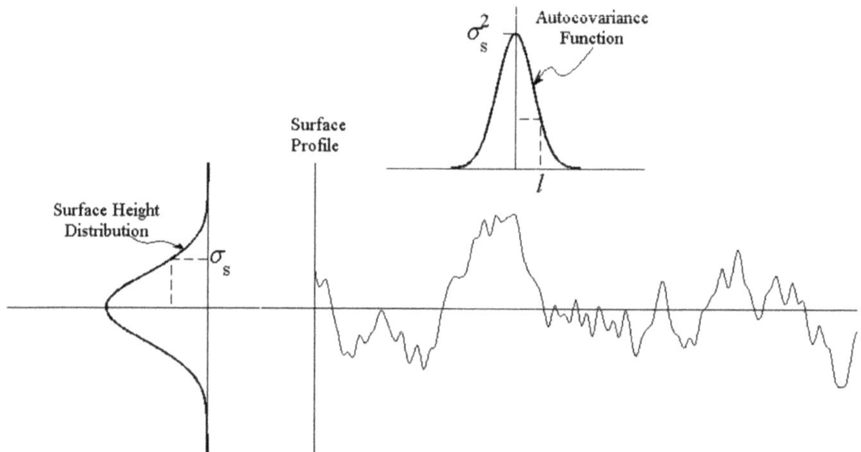

Figure 4. Illustration of statistical parameters used in characterizing surface roughness.

For many cases of interest, the surface heights are normally distributed, *i.e.*, $SHD(h)$ is Gaussian. However, in most instances, $ACF(x, y)$ is material and process dependent. To simplify the problem, both of them are assumed Gaussian in this paper.

3.2. SRRS Model of Reflectance Change Due to Surface Roughness

The local fine surface roughness on a SD material causes the scattering of light, and accordingly the reflectance to be modified as $R_m(\lambda) = R_0(\lambda)[1 - S_T(\lambda)]$ [30], where the reduction factor $S_T(\lambda)$ represents the fraction of scattered light that is trapped in the material or transmitted through. Here we assume that the original SD is Lambertian and the diffusely scattered light is uniformly distributed in all directions. This allows us to focus on studying the physical origin of the spectral dependent degradation of SD reflectance.

The total integrated scattering from surface roughness, *i.e.*, $TIS(\lambda)$, has been studied in [26,27,29,32]. For surface roughness with length scale $l >> \lambda$, in [26], it is derived that the fraction of diffusively scattered light could be expressed as

$$TIS(\lambda) = 1 - \exp\left[-\left(\frac{4\pi\cos\theta_i\sigma_s}{\lambda}\right)^2\right] \tag{4}$$

where θ_i is the angle of incidence. In [29], it is further suggested that scattering should be significantly dependent on the roughness correlation length l if it is much smaller than the wavelength ($l << \lambda$) and quantitatively derived

$$TIS(\lambda) = \frac{64}{3}\frac{\pi^4\sigma_s^2 l^2}{\lambda^4} \tag{5}$$

for normal incident light. The relationship $TIS(\lambda)\propto\lambda^{-4}$ for $l << \lambda$ is due to Rayleigh scattering, which is dominantly elastic scattering. For oblique incident light with an angle of incidence θ_i, the scattering can be expressed as [31,33]

$$TIS(\lambda) = \frac{64}{3}\frac{\pi^4\sigma_s^2 l^2\cos^2\theta_i}{\lambda^4} \tag{6}$$

In this case, the modification of *TIS* due to oblique light incidence is applied to the root-mean-square surface roughness σ_s as a factor of $\sigma_s\cos\theta_i$ [31]. In the case of scattering from surface roughness on SD material, Equation (6) is applicable and the fraction of scattered light that is trapped in the material or transmitted through can be estimated as

$$S_T(\lambda) = \alpha \times TIS(\lambda) = \alpha\frac{64}{3}\frac{\pi^4\sigma_s^2 l^2\cos^2\theta_i}{\lambda^4} \tag{7}$$

where α is fraction of light that is not reflected. The resulting modified reflectance due to light scattering from surface roughness can be expressed as

$$R_m(\lambda) = R_0(\lambda)[1 - S_T(\lambda)] = R_0(\lambda)\left(1 - \alpha\frac{64}{3}\frac{\pi^4\sigma_s^2 l^2\cos^2\theta_i}{\lambda^4}\right) \tag{8}$$

4. Analysis of Spectral Dependent Degradation of VIIRS SD with SRRS Model

We can apply the model of surface roughness-induced spectral dependent degradation presented in Section 4 to estimate the length scale of surface roughness grown on SD using the spectral reflectance data of VIIRS SD and check the validity of the model. In the case of SD material, initial reflectance $R_0(\lambda)$ before illumination by space radiation can be approximated as being independent of wavelength over the wavelength of interest, *i.e.*, $R_0(\lambda) \approx 1$, and the dependence of $R_m(\lambda)$ on wavelength can be attributed to the factor $[1 - (64/3)\alpha\pi^4\sigma_s^2 l^2\cos^2\theta_i\lambda^{-4}]$ in Equation (8). Therefore, by fitting the spectral reflectance $R_{SD}(\lambda)$ of VIIRS SD, *i.e.*, *H* factor, using Equation (8), we can estimate the SD surface roughness length parameter $\sigma_s l$.

Figure 5 shows the fitting of the VIIRS SD spectral reflectance data on 6 August 2013 with the SRRS model (Equation (8)). In the fitting, we use $\alpha = 0.5$ to account for that the scattering is symmetric between forward and backward direction in the regime of Rayleigh scattering and focus on the dependence on wavelength. The solar light incident angle to SD is approximately taken as

$\theta_i = 52.4$ degree. For comparison, the model based on Equation (4) [26,27] for surface roughness length $l >> \lambda$ is also plotted. It can be seen that the Rayleigh scattering-based reflectance correction model (Equation (8)) matches the VIIRS SD data very well while the model based on [26] does not match the spectral trend in the SD reflectance data. On the other hand, surface roughness length parameter $\sqrt{\sigma_s l}$ can be estimated using Equation (8) which gives $\sqrt{\sigma_s l} = 66.5$nm for this case. Both facts confirm that the SD surface roughness length is much less than the wavelength and Rayleigh scattering is the dominant mechanism responsible for the observed SD spectral dependent degradation. In [34], it is first empirically determined that the degradation factor is proportional to $1/\lambda^{4.07}$, which is consistent with our SRRS theoretical model.

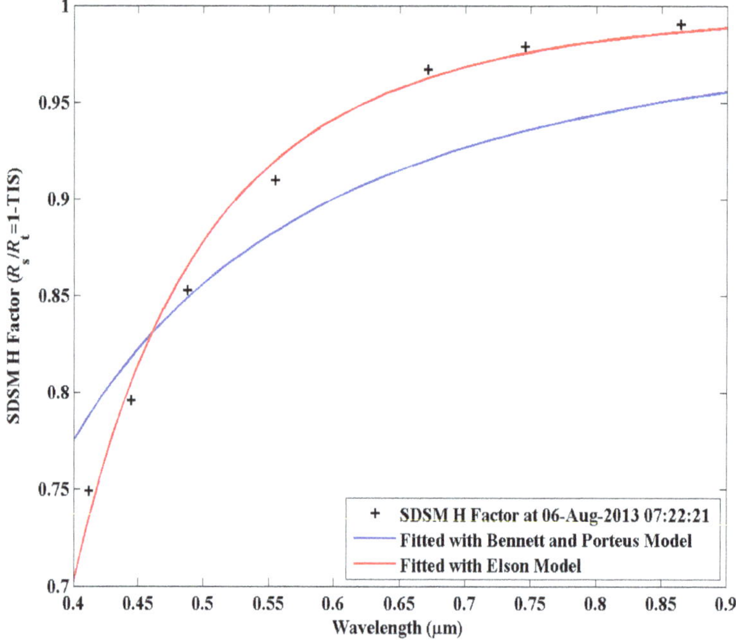

Figure 5. Applying Bennett and Porteus (Equation (4)) and Surface Roughness based Rayleigh Scattering (SRRS) model to fit for spectral dependent degradation of VIIRS SD.

Figure 6 shows a fitting of the SD spectral reflectance with the SRRS reflectance correction model (Equation (8)) for several cases during the 2.5 years operation of VIIRS. In general, the matching between the modeled and observed spectral reflectance of the SD for these cases is quite good as the SD operation time accumulates and the degradation becomes more severe.

Figure 7 shows the evolution of SD surface roughness length parameter $\sigma_s l$ derived from fitting of the ~2.5 year time series of VIIRS SD spectral reflectance data using Equation (8). The SD surface roughness appears to be growing with a slowing down growth rate as the radiation exposure of SD persists during each orbit. In year 2012, the $\sigma_s l$ increases fastest at an average rate of 2147 nm^2/year. In 2013, $\sigma_s l$ increases at 736 nm^2/year and in 2014, $\sigma_s l$ appears flat for early half year. Over 2.5 years, $\sqrt{\sigma_s l}$ changed ~2.3 times from 30.7 nm to 69.1 nm. Given the average exposure time of the SD of ~5 min with ~14 orbits per day, the corresponding VIIRS SD exposure time of solar light for such a roughness increase is ~1000 h. Since there is a SD screen, which attenuates the solar light on VIIRS, an estimation of exact UV and VUV exposure time requires knowledge of the transmittance in UV wavelength range. We also note that the radiation exposure experiment for Teflon fluorinated ethylene propylene (FEP, similar Fluorocarbon-based polymer as Spectralon™) samples by [25] showed that surface roughness length scale of Teflon FEP changed from 8 nm to 35 nm after combined VUV and oxygen ion exposure. Our estimation of the surface roughness length parameter for VIIRS SD is consistent with the experimental results, *i.e.*, both are in tens nanometer scale and are much less than the wavelength. This further validates the application of SRRS model to explain the spectral dependent degradation of SD reflectance.

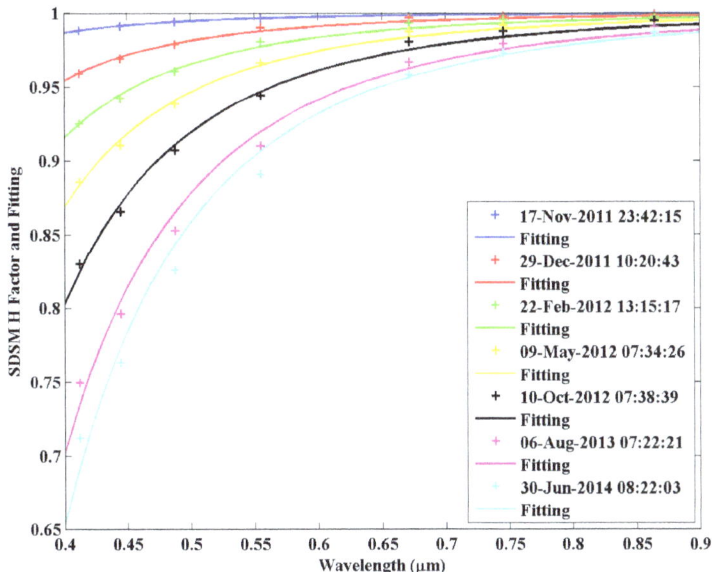

Figure 6. Fitting of VIIRS SD spectral reflectance with the SRRS reflectance correction model (Equation (8)) at several instants during 2.5 years operation of VIIRS.

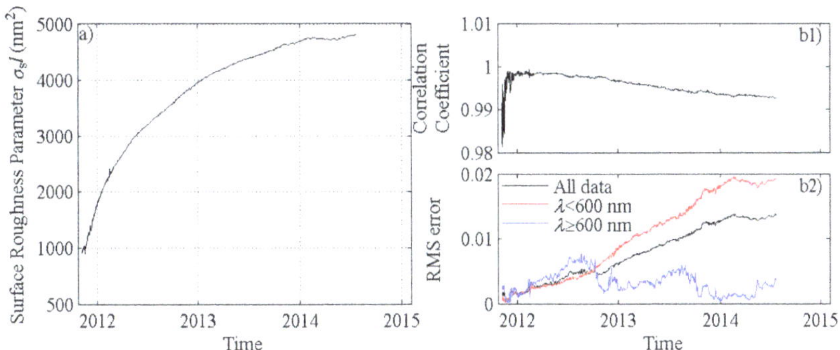

Figure 7. (**a**) Trending of VIIRS SD surface roughness characteristic parameter $\sigma_s l$; and (**b**) evolution of correlation coefficient (**b1**) and Root-Mean-Square (RMS) error (**b2**) of the spectral fitting using Equation (8). The RMS errors are calculated for short wavelengths (<600 nm), long wavelengths (>600 nm) and whole spectral range combined, respectively.

Figure 7 also shows the evolution of the correlation coefficient and fitting residual error between the SRRS reflectance model and observed SD spectral reflectance data to quantify the matching between the model and data. In general, the model matches the observed SD data well with cross correlation coefficient >0.99 except for the period before 2012 which maintains correlation coefficient >0.98. The RMS error between model and observation is <0.014 (black line in Figure 7b2) for the whole spectral range combined and there is a trend of increase in the RMS error as the SD operation time accumulates. To further analyze the wavelength dependence of the RMS error, we calculated the RMS errors with wavelengths above and below 600 nm, respectively. The RMS error for long wavelengths (black line in Figure 7b2) grows faster to reach a maximum of 0.0077 in August 2012 and then decreases to stay within a region with small value (<0.005). On the other hand, the error with short wavelength (red line in Figure 7b2) grows continuously and gradually approaches a value of about 0.020.

The above analysis of the VIIRS SD reflectance data shows that the modeled and observed reflectance matches well and the estimated surface roughness length parameter is much less than wavelength and is consistent with the length scale obtained from lab measurements [25]. Therefore, the spectral dependent degradation of SD reflectance in short wavelength can be explained with the SRRS model ($\propto\lambda^{-4}$). The growth of SD surface roughness is due to the exposure of fluoropolymer-based SD material to space radiations such as solar UV light and energetic particles.

5. Discussion

Recently, there have been several studies [34–37] to improve the modeling accuracy of the VIIRS SD reflectance degradation and the radiometric calibration accuracy of VIIRS RSB sensor. For example, Lei and Xiong presented a method in [34] to determine the SNPP VIIRS SD BRDF degradation factor over wavelengths longer than 1 μm. They used a phenomenological power law model with fitting function $1 - H = \alpha/\lambda^{\eta}$ to simulate the spectral dependent degradation of SD monitored by SDSM. This formulation is similar to ours presented in this study. However, our model fixed the exponent as 4 since the dominant origin of SD spectral degradation is due to surface roughness-induced Rayleigh scattering in our SRRS model. With two fitting parameters, it is determined in [34] that the best fitting exponent $\eta = 4.07$ for SDSM data over four wavelengths: 675, 746, 874 and 929 nm. This fitting exponent is consistent with our SRRS ($\propto 1/\lambda^4$)-model.

In [34], it is also shown that with two fitting parameters, the average (over the entire mission) of the absolute values of the differences between modeled and measured degradation factors can be as low as 0.00037 over the four wavelengths from 675 nm to 929 nm. Our model shows deviation of SDSM data from the model for short wavelength (Figure 6) when only one fitting parameter, *i.e.*, $\sigma_s l$, is used to fit for all of the M bands. The deviation can be due to the impacts of SDSM detector relative spectral responses [35]. The SDSM detector response degradation is wavelength-dependent with larger changes at longer (NIR) wavelengths. This type of degradation is most likely due to displacement damage to the SDSM detector system by the high-energy protons in the space operation environment. Since the SDSM is essentially a ratioing radiometer, its degradation has minimal impact on the determination of the SD reflectance degradation [35]. While using high order fitting parameters improves the modeling accuracy, our work focused on revealing the dominant physical origin of the SD BRDF degradation, which suggests that the degradation process is dominated by surface roughness-induced Rayleigh scattering.

In [34], the power-law model is extended to the wavelength region of the SWIR bands from 1.238 to 2.257 μm (bands M8 to M11 of VIIRS) to determine the corresponding degradation factors. These bands are not monitored by SDSM and it is predicted in [34] that on day 1300 the degradation can vary from ~0.04% to ~0.4% for wavelengths from 1.238 to 2.257 microns. The physics-based model presented in this paper can be used to explain the phenomenological power law model presented in [34]. The SRRS model provides physical foundation for extrapolating SD reflectance degradation to longer wavelength with the power law model such as in [34].

In [36,37], Lei and Xiong analyzed the impact of the angular dependence of the VIIRS SD BRDF degradation factor on the radiometric calibration of the RSBs. It was found that the H factors of the SD are angle-dependent. They examined the H factors with different solar vertical angle ϕ_V. It was shown that during orbit 5447 and 10,838, for short wavelengths ($\lambda < 500$ nm), the H factor could increase by about 2% when ϕ_V increases from 25° to 42°. It was also found that the factor of SD transmittance multiplied by the BRDF depends on the solar horizontal angle ϕ_H by a relative difference of no more than 0.14%, which has a negligible impact on the SD BRDF degradation. Since the SDSM and the VIIRS telescope SD views have different angles, angular dependence of BRDF impacts the determination of the SD screen transmittance viewed by both the SDSM detectors and the VIIRS telescope. Therefore, taking the dependence of SD BRDF degradation on solar vertical angles ϕ_V into account can improve the accuracy in operational radiometric calibration of VIIRS RSBs.

Since the focus of this paper is to explore the physical mechanism, and explain the spectral dependent degradation of the SD in terms of the SRRS model and estimate the surface roughness length scale, we only took into account of the nominal solar light incident angle on the SD and did not consider the angular dependence of the SD degradation. For the operational calibration of VIIRS RSB, a more comprehensive model that incorporates angular dependence of SD BRDF degradation with more fitting parameters should be used, such as presented in [36,37].

6. Summary

A physics-based surface roughness Rayleigh scattering (SRRS) model (Equation (8)) is presented in this paper to explain the spectral reflectance degradation of VIIRS SD in terms of Rayleigh scattering from surface roughness caused by space radiation. Space radiations such as UV light and energetic particles can modify the SD material through breaking C-C and C-F bonds, scissioning or cross linking, ionizing the polymer, which causes the growth of surface roughness. It is found that the model matches well with the spectrally structured degradation of VIIRS SD reflectance data collected over 2.5 years, which indicates that the SD reflectance decrease in shorter wavelength is due to enhanced scattering ($\propto\lambda^{-4}$) from surface roughness. The surface roughness length parameter is trended using SD reflectance data over 2.5 years. The derived SD surface roughness length parameter is of tens of nanometer scale (much less than wavelength) and consistent with the length scale found from previous radiation exposure experiments of fluoropolymer samples [25]. This further verifies that the SRRS model is self-consistent and could explain the spectral-dependent degradation of SD material due to surface roughness change under space radiation.

SD plays an important role in onboard radiometric calibration of RSB sensor for existing missions such as MODIS, GoSat and SNPP VIIRS as well as for the future mission such as GOES-R ABI, JPSS VIIRS and GoSat2. Understanding and characterizing the SD reflectance degradation can help improve the instrument design and calibration performance and is of particular importance for missions such as ABI onboard GOES-R which uses a SD for calibration, but has no SDSM for tracking its degradation.

The model presented in this paper provides insights in understanding the physics in the spectral dependent degradation of SD after radiation exposure in space and can be used for other missions. Future efforts will be focused on cross-comparing the evolution of surface roughness length parameter using SD spectral reflectance data from different missions and investigating its dependence on the accumulated solar light exposure time. The model presented in this paper also calls for lab experiments to investigate surface roughness change of sample SD material after radiation illumination and explore the connection between surface property change and spectral dependent degradation.

Acknowledgments: We thank Slawomir Blonski for providing Suomi-NPP VIIRS H-factor data, and Jason Choi, Bin Zhang, Bengt Eliasson and Gennady Milikh for helpful discussion. The views, opinions, and findings contained in this paper are those of the authors and should not be construed as official positions, policy, or decisions of the NOAA or the U. S. Government. Commercial companies identified in this paper are only to foster understanding. Such identification does not imply recommendation or endorsement by the NOAA, nor does it imply that they are the best available for the purpose.

Author Contributions: All authors contributed equally to this work.

Conflicts of Interest: The authors declare no conflict of interest.

Appendix: Radiation-Induced Spectral Dependent Degradation of SD Material

A.1. SD Material

The SD on VIIRS is made of Spectralon™ material, which is produced by Labsphere. It is composed of pure polytetrafluoroethylene (PTFE) polymer resin that is compressed into a hard porous white material in a proprietary procedure. The fluoropolymer is a fluorocarbon-based polymer with strong C-F bonds. The diffuse reflectance of Spectralon is generally >99% for 400 nm to 1500 nm wavelength range and >95% for a wider wavelength range from 250 nm to 2.5 μm. The porous network of thermoplastic on the Spectralon surface produces multiple reflections in the first few tenths of a millimeter depth. This makes the surface and immediate subsurface structure of Spectralon exhibit highly Lambertian reflection behavior. The Spectralon is thermally stable up to >350° C. It is chemically inert, environmentally stable and its reflectance has been characterized with NIST traceable calibration. Spectralon's optical properties make it ideal as a reference surface in remote sensing and spectroscopy. Space-grade Spectralon combines high-reflectance with an extremely Lambertian reflectance profile. Therefore, it has been used as a SD material for terrestrial remote sensing applications on MODIS, SNPP-VIIRS, GoSat, LansSat-8, and will be used on upcoming J1-VIIRS and GOES-R ABI.

A.2. Space Environment Effects on SD Material

Spectralon, as a high density fluorocarbon-based polymer, is subject to space environmental effects such as UV radiation, high-energy protons, electrons, and energetic ions [25]. The total energy of solar UV radiation (100–400 nm) carried by solar light is ~8% of the solar constant (1366 W/m^2). The photon energy at the near UV (200–400 nm) range is >3 eV and >5 eV for Vacuum UV (VUV, wavelength <200 nm) range. The intensity of VUV part of solar radiation fluctuates in large amplitude when the solar activity varies from minimum to maximum during a solar cycle. In space, the UV radiation illuminates the VIIRS SD directly after passing through the pinhole SD screen and is energetic enough to break polymer bonds such as C-C, C-F and functional groups or excite bonds with no direct atomic displacements. The modification of long chain polymers such as SD material by UV radiation occurs through scissioning, cross linking without mass loss, and creation of volatile polymer fragments with mass loss. These modifications can result in SD material surface roughness changes.

The radiation belt particles trapped by the Earth's magnetic field experience combined motion of gyration, bounce and drifting around the Earth, forming stable doughnut-shaped zones of highly energetic charged particles around Earth. The belts contain energetic electrons (up to several MeV) that form the outer belt and a combination of protons (up to several 100 of MeV) and electrons that form the inner belt. Most of the particles that form the belts come from solar wind and proton energies exceeding 50 MeV in the inner belts are the results of the beta decay of neutrons created by cosmic ray collisions with nuclei of the upper atmosphere. Polar-orbiting satellites such as SNPP transit through both the electron and proton radiation belt along each orbit around the earth. In particular, when the satellite goes through the South Atlantic anomaly (SAA) region, the energetic proton and electron fluxes increase. These radiation belt particles can lead to gradual degradation of SD performance through ionizing radiation, phonon excitations and atomic displacement on the polymer.

Atomic oxygen and energetic ions such as O^+ of plasma thermal energy (from space or from outgas of instrument material induced by high vacuum in space) can impact polymers through contamination or collisionally-induced scission of fluorocarbon-based polymers, and result in surface erosion, changes in chemical composition, and formation of particulate and molecular contamination of the material surfaces.

In short, these above-mentioned space environment effects will modify the surface properties of a polymer-based SD panel. The consequent degradation of reflectance is investigated in this paper.

A.3. Radiation Exposure Experiments on SD Material

Extensive experiments of particle and UV irradiation on SD material were performed [13–20] to study changes in directional-hemispheric spectral reflectance, the bidirectional spectral reflectance factor and the polarization properties of diffuser material after the radiation exposure. In early experiment of [13], optical-grade Spectralon material underwent proton exposure and UV-irradiation. A 10-cm-diameter proton beam generated by proton-accelerator facilities at Jet Propulsion Laboratory was used to irradiate the Spectralon sample with 10^{10} particle/cm^2 at energy levels of 1 keV, 1 MeV and 10 MeV, respectively. This fluence level is representative of accumulated proton radiation exposure during a five-year Earth orbiting satellite mission. Under proton radiation, the sample reflectance had little degradation at a wavelength of 0.4 μm wavelength and exhibited ~15% decrease in its diffuse reflectance at 0.2 μm wavelength. The Spectralon sample was next irradiated with UV light with an intensity of 1.5 equivalent suns generated from a 5 kilowatt short arc Xenon lamp for 333 h, yielding 500 equivalent hours exposure. The sample diffuse reflectance decreased from 1.0 μm wavelength with a steady decrease to 0.2 μm. The maximum reflectance decrease was 65% for Spectralon. Overall, UV light exposure degrades the Spectralon reflectance more than the proton exposure for wavelength greater than 0.4 μm. Other experiments [14–19] also show similar magnitude of reflectance degradation after exposure to UV radiation. Recent experiment in [20] studied changes in the reflectance properties of pressed and sintered PTFE diffusers induced by exposure to VUV irradiation before and after controlled contamination with Pennzane. Their results show that the 8° directional hemispherical reflectance of vacuum-baked sample degraded about 70% after 40 h of VUV exposure. The reflectance of Pannzance-contaminated sample degrades for an additional 6%–8% after the same VUV exposure time.

Although these previous UV exposure experiments consistently show structured reflectance degradation with faster degradation at shorter wavelengths, there has not been any further analysis to relate the spectral dependent degradation of SD material with the changes in sample surface properties, especially surface roughness, as investigated in this paper.

A.4. Experimental Evidence of Surface Roughness Change after VUV Radiation on Fluoropolymer Material

Separate from the Spectralon studies, in [25], experiments were performed to study the surface property changes of fluoropolymers after exposure of VUV radiation and/or energetic oxygen ion. Fluoropolymer samples such as Teflon fluorinated ethylene propylene (FEP), Tefzel, Tedlar and Polyethylene were exposed to low fluences of VUV radiation generated by a 30-W deuterium lamp with a MgF_2 window and/or 50 eV oxygen ions generated by a Kaufman source. The total VUV exposure time is 20 equivalent Sun hours (ESH) and the total exposure to oxygen ion is 5×10^{17} O^+/cm^2. While Tefzel, Tedlar and Polyethylene are all H-containing polymers and have C-H bond, Teflon FEP consists purely of C-C and C-F bond and is copolymer of hexafluoropropylene and tetrafluoroethylene. Teflon FEP's composition is very similar to the PTFE-based Spectralon. The main difference is in the processing techniques and Teflon FEP is softer than PTFE, more easily formable, and melts at 260 °C. Therefore, surface property changes of Teflon FEP after radiation exposure can provide implications on surface changes of Spectralon material under similar exposures.

Surface roughness of Teflon FEP was measured using Atomic Force Microscopy in [25]. The results showed that both VUV irradiation and oxygen ions caused significant mass loss for Teflon FEP. Surface roughness length scale of Teflon FEP changed from 8 nm to 14 nm after 20 ESH VUV exposure and to 35 nm after combined VUV and oxygen ion exposure. This indicates that the irradiation of Teflon FEP causes scissoring (breaking of C-C and C-F bonds), cross linking and erosion, and it results in surface roughening.

References

1. Miller, S.D.; Mills, S.P.; Elvidge, C.D.; Lindsey, D.T.; Lee, T.F.; Hawkins, J.D. Suomi satellite brings to light a unique frontier of nighttime environmental sensing capabilities. *Proc. Natl. Acad. Sci. USA* **2012**, *109*, 15706–15711. [CrossRef] [PubMed]
2. Cao, C.; Xiong, J.; Blonski, S.; Liu, Q.; Uprety, S.; Shao, X.; Bai, Y.; Weng, F. Suomi NPP VIIRS sensor data record verification, validation, and long-term performance monitoring. *J. Geophys. Res. Atmos.* **2013**, *118*, 11664–11678. [CrossRef]
3. Cao, C.; de Luccia, F.J.; Xiong, X.; Wolfe, R.; Weng, F. Early on-orbit performance of the visible infrared imaging radiometer suite onboard the Suomi National Polar-Orbiting Partnership (S-NPP) Satellite. *IEEE Trans. Geosci. Remote Sens.* **2014**, *52*, 1142–1156. [CrossRef]
4. Xiong, X.; Butler, J.; Chiang, K.; Efremova, B.; Fulbright, J.; Lei, N.; McIntire, J.; Oudrari, H.; Sun, J.; Wang, Z.; *et al.* VIIRS on-orbit calibration methodology and performance. *J. Geophys. Res. Atmos.* **2014**, *119*, 2013JD020423. [CrossRef]
5. Rausch, K.; Houchin, S.; Cardema, J.; Moy, G.; Haas, E.; de Luccia, F.J. Automated calibration of the Suomi National Polar-Orbiting Partnership (S-NPP) Visible Infrared Imaging Radiometer Suite (VIIRS) reflective solar bands. *J. Geophys. Res. Atmos.* **2013**, *118*. [CrossRef]
6. Uprety, S.; Cao, C.; Xiong, X.; Blonski, S.; Wu, A.; Shao, X. Radiometric intercomparison between Suomi-NPP VIIRS and Aqua MODIS Reflective solar bands using simultaneous nadir overpass in the low latitudes. *J. Atmos. Ocean. Technol.* **2013**, *30*, 2720–2736. [CrossRef]
7. Uprety, S.; Cao, C. Suomi NPP VIIRS reflective solar band on-orbit radiometric stability and accuracy assessment using desert and Antarctica Dome C sites. *Remote Sens. Environ.* **2015**, *166*, 106–115. [CrossRef]
8. Shao, X.; Choi, T.; Cao, C.; Blonski, S.; Wang, W.; Ban, Y. Trending of Suomi-NPP VIIRS radiometric performance with lunar band ratio. *Proc. SPIE* **2014**, *9264*. [CrossRef]
9. Lei, N.; Wang, Z.; Fulbright, J.; Lee, S.; McIntire, J.; Chiang, K.; Xiong, X. Initial on-orbit radiometric calibration of the Suomi NPP VIIRS reflective solar bands. *Proc. SPIE* **2012**, *8510*, 851018.
10. Fulbright, J.P.; Lei, N.; Chiang, K.; Xiong, X. Characterization and performance of the Suomi-NPP/VIIRS solar diffuser stability monitor. *Proc. SPIE* **2012**, *8510*. [CrossRef]
11. Fulbright, J.P.; Lei, N.; McIntire, J.; Efremova, B.; Chen, X.; Xiong, X. Improving the characterization and performance of the Suomi-NPP VIIRS solar diffuser stability monitor. *Proc. SPIE* **2013**, *8866*. [CrossRef]
12. Sun, J.; Wang, M. Visible infrared imaging radiometer suite solar diffuser calibration and its challenges using a solar diffuser stability monitor. *Appl. Opt.* **2014**, *53*, 8571–8584. [CrossRef] [PubMed]
13. Guzman, C.T.; Palmer, J.M.; Slater, P.N.; Bruegge, C.J.; Miller, E.A. Requirements of a solar diffuser and measurements of some candidate materials. *Proc. SPIE* **1991**, *1493*, 120–131.
14. Bruegge, C.J.; Stiegman, A.E.; Coulter, D.R.; Hale, R.R.; Diner, D.J.; Springsteen, A.W. Reflectance stability analysis of Spectralon diffuse calibration panels. *Proc. SPIE* **1991**, *1493*, 132–142.
15. Stiegman, A.E.; Bruegge, C.J.; Springsteen, A.W. Ultraviolet stability and contamination analysis of Spectralon diffuse reflectance material. *Opt. Eng.* **1993**, *32*, 799–804. [CrossRef]
16. Bruegge, C.J.; Stiegman, A.E.; Rainen, R.A.; Springsteen, A.W. Use of Spectralon as a diffuse reflectance standard for in-flight calibration of earth-orbiting sensors. *Opt. Eng.* **1993**, *32*, 805–814. [CrossRef]
17. Petroy, S.B.; Leland, J.E.; Chommeloux, B.; Bruegge, C.J.; Gourmelon, G. Phase 1: analysis of Spectralon material for use in on-board calibration systems for the medium resolution imaging spectrometer (MERIS). *Proc. SPIE* **1994**, *2210*, 616–624.
18. Leland, J.E.; Arecchi, A.V. Phase 2 analysis of Spectralon material for use in on-board calibration systems for the medium-resolution imaging spectrometer (MERIS). *Proc. SPIE* **1995**, *2475*, 384–392.
19. Georgiev, G.T.; Butler, J.J. Long-term calibration monitoring of Spectralon diffusers BRDF in the air-ultraviolet. *Appl. Opt.* **2007**, *46*, 7892–7899. [CrossRef] [PubMed]
20. Georgiev, G.T.; Butler, J.J.; Thome, K.J.; Ramos-Izquierdo, L.A.; Ding, L.; Graziani, L.J.; Meadows, G.A. Initial studies of the directional reflectance changes in pressed and sintered PTFE diffusers following exposure to contamination and ionizing radiation. *Metrologia* **2014**, *51*. [CrossRef]
21. Xiong, X.; Sun, J.; Xie, X.; Barnes, W.L.; Salomonson, V.V. On-orbit calibration and performance of aqua MODIS reflective solar bands. *IEEE Trans. Geosci. Remote Sens.* **2010**, *48*, 535–546. [CrossRef]

22. Xiong, X.; Angal, A.; Sun, J.; Choi, T.; Johnson, E. On-orbit performance of MODIS solar diffuser stability monitor. *J. Appl. Remote Sens.* **2014**, *8*. [CrossRef]
23. Markham, B.; Barsi, J.; Kvaran, G.; Ong, L.; Kaita, E.; Biggar, S.; Czapla-Myers, J.; Mishra, N.; Helder, D. Landsat-8 operational land imager radiometric calibration and stability. *Remote Sens.* **2014**, *6*, 12275–12308. [CrossRef]
24. Yoshida, Y.; Kikuchi, N.; Yokota, T. On-orbit radiometric calibration of SWIR bands of TANSO-FTS onboard GOSAT. *Atmos. Meas. Tech.* **2012**, *5*, 2515–2523. [CrossRef]
25. Grossman, E.; Gouzman, I. Space environment effects on polymers in low earth orbit. *Nucl. Instrum. Methods Phys. Res. Sect. B Beam Interact. Mater. Atoms* **2003**, *208*, 48–57. [CrossRef]
26. Bennett, H.E.; Porteus, J.O. Relation between surface roughness and specular reflectance at normal incidence. *J. Opt. Soc. Am.* **1961**, *51*. [CrossRef]
27. Bennett, H.E. Scattering characteristics of optical materials. *Opt. Eng.* **1978**, *17*. [CrossRef]
28. Sparks, M. Explanation of λ^{-2} optical scattering and λ^{-2} Strehl on-axis irradiance reduction. *J. Opt. Soc. Am.* **1983**, *73*. [CrossRef]
29. Elson, J.M. Theory of light scattering from a rough surface with an inhomogeneous dielectric permittivity. *Phys. Rev. B* **1984**, *30*, 5460–5480. [CrossRef]
30. Schröder, S.; Duparré, A.; Coriand, L.; Tünnermann, A.; Penalver, D.H.; Harvey, J.E. Modeling of light scattering in different regimes of surface roughness. *Opt. Express* **2011**, *19*, 9820–9835. [CrossRef] [PubMed]
31. Harvey, J.E.; Schröder, S.; Choi, N.; Duparré, A. Total integrated scatter from surfaces with arbitrary roughness, correlation widths, and incident angles. *Opt. Eng.* **2012**, *51*. [CrossRef]
32. Wolff, L.B. Diffuse-reflectance model for smooth dielectric surfaces. *J. Opt. Soc. Am. A* **1994**, *11*, 2956–2968. [CrossRef]
33. Elson, J.M.; Rahn, J.P.; Bennett, J.M. Relationship of the total integrated scattering from multilayer-coated optics to angle of incidence, polarization, correlation length, and roughness cross-correlation properties. *Appl. Opt.* **1983**, *22*, 3207–3219. [CrossRef] [PubMed]
34. Lei, N.; Xiong, X. Determination of the SNPP VIIRS solar diffuser BRDF degradation factor over wavelengths longer than 1 μm. *Proc. SPIE* **2015**, *9607*. [CrossRef]
35. Xiong, X.; Fulbright, J.; Angal, A.; Wang, Z.; Geng, X.; Butler, J. Assessment of MODIS and VIIRS solar diffuser on-orbit degradation. *Proc. SPIE* **2015**, *9607*. [CrossRef]
36. Lei, N.; Xiong, X. Estimation of the accuracy of the SNPP VIIRS SD BRDF degradation factor determined by the solar diffuser stability monitor. *Proc. SPIE* **2015**, *9607*. [CrossRef]
37. Lei, N.; Xiong, X. Impact of the angular dependence of the SNPP VIIRS solar diffuser BRDF degradation factor on the radiometric calibration of the reflective solar bands. *Proc. SPIE* **2015**, *9607*. [CrossRef]

Article

Suomi NPP VIIRS Reflective Solar Bands Operational Calibration Reprocessing

Slawomir Blonski [1,*] and Changyong Cao [2]

1 Earth System Science Interdisciplinary Center, University of Maryland, 5825 University Research Court, Suite 4001, College Park, MD 20740, USA

2 NOAA (National Oceanic and Atmospheric Administration)/NESDIS (National Environmental Satellite, Data, and Information Service)/STAR (Center for Satellite Applications and Research), NCWCP, E/RA2, 5830 University Research Ct., Suite 2838, College Park, MD 20740, USA; Changyong.Cao@noaa.gov

* Correspondence: Slawomir.Blonski@noaa.gov; Tel.: +1-301-683-3603; Fax: +1-301-683-3616

Academic Editors: Richard Müller and Prasad S. Thenkabail

Received: 20 October 2015; Accepted: 25 November 2015; Published: 2 December 2015

Abstract: Radiometric calibration coefficients for the VIIRS (Visible Infrared Imaging Radiometer Suite) reflective solar bands have been reprocessed from the beginning of the Suomi NPP (National Polar-orbiting Partnership) mission until present. An automated calibration procedure, implemented in the NOAA (National Oceanic and Atmospheric Administration) JPSS (Joint Polar Satellite System) operational data production system, was applied to reprocess onboard solar calibration data and solar diffuser degradation measurements. The latest processing parameters from the operational system were used to include corrected solar vectors, optimized directional dependence of attenuation screens transmittance and solar diffuser reflectance, updated prelaunch calibration coefficients without an offset term, and optimized Robust Holt-Winters filter parameters. The parameters were consistently used to generate a complete set of the radiometric calibration coefficients for the entire duration of the Suomi NPP mission. The reprocessing has demonstrated that the automated calibration procedure can be successfully applied to all solar measurements acquired from the beginning of the mission until the full deployment of the automated procedure in the operational processing system. The reprocessed calibration coefficients can be further used to reprocess VIIRS SDR (Sensor Data Record) and other data products. The reprocessing has also demonstrated how the automated calibration procedure can be used during activation of the VIIRS instruments on the future JPSS satellites.

Keywords: Suomi NPP; VIIRS; calibration; reprocessing

---

1. Introduction

VIIRS (Visible Infrared Imaging Radiometer Suite) instrument has been operating onboard the Suomi NPP (National Polar-orbiting Partnership) satellite since November 2011 [1]. Suomi NPP VIIRS is the first instrument in the series that is being deployed on the JPSS (Joint Polar Satellite System) spacecraft by NOAA (National Oceanic and Atmospheric Administration) and NASA (National Aeronautics and Space Administration). Although VIIRS is an integrated system with common telescope and electronics, it can also be seen as a suite of three scanning radiometers with distinct performance characteristics and calibration methods:

- TEB (Thermal Emissive Band) measurements (in seven channels) cover the spectral range from 3.5 to 12.5 μm and are calibrated using onboard blackbody and space-view data collected during every scan.
- RSB (Reflective Solar Band) measurements consist of 14 channels covering the range from 400 to 2280 nm: RSB calibration uses onboard solar diffuser data collected once per orbit and space-view data from every scan.

- DNB (Day/Night Band) covers the broad spectral range of 500–900 nm with high dynamic range measurements that are only partially calibrated by the solar diffuser data: complete DNB calibration requires special Earth observations conducted once per month (an alternative method that uses only the onboard calibrator data has been proposed [2], but its implementation in operational software is not fully tested yet).

Only RSB radiometric calibration is a subject of this paper. While DNB is a reflective band as well, its calibration methodology differs so much that it is beyond scope of the present work.

Since it was found early in the Suomi NPP mission that sensitivity of the VIIRS instrument in several spectral bands decreases with time much faster than expected [3], accuracy of the radiometric calibration has been maintained by updating processing coefficients with the weekly frequency (approximately every 100 orbits). The radiometric calibration procedure together with processing parameters has been improved several times throughout the lifetime of the mission. Initially, only values of the calibration coefficients were updated every week. Later, changes of the coefficients during each week were predicted as well. To further reduce calibration uncertainty between the updates, an automated procedure has been implemented that calculates the RSB radiometric calibration coefficients after every orbit. Rausch *et al.* [4] developed the automated calibration computer code, called RSBAutoCal, and delivered it to the JPSS program for implementation in the operational processing software. The RSBAutoCal code has enabled us to conduct long-term testing and reprocessing of the RSB calibration. Partial, preliminary results from the reprocessing have been presented previously [5], but this is the first description and discussion of the complete results that include all reprocessed spectral bands and cover the entire duration of the Suomi NPP mission.

This paper describes a long-term testing of the automated calibration procedure based on reprocessing VIIRS solar calibration data starting from the beginning through near four years of the Suomi NPP mission. The presented work had two goals:

1. Validation: To demonstrate that the automated calibration procedure can be successfully applied to all solar diffuser measurements acquired from the beginning of the mission until the full operational deployment of the automated procedure.
2. Reprocessing: To generate a complete set of the calibration coefficients while consistently using the latest processing parameters that are optimized based on instrument knowledge gained during the entire mission so far.

The reprocessed radiometric calibration coefficients derived by the automated procedure are compared with those previously applied in the SDR (Sensor Data Record) production at the NOAA operational IDPS (Interface Data Processing Segment) system.

IDPS generates VIIRS data products in near real time with latency currently limited to no more than few hours. The processing coefficients in IDPS are not based on interpolation between past calibration measurements, but rather an extrapolation of the calibration coefficients into the future is always required. Reprocessing that uses this approach may be less accurate than those based solely on the interpolation [6,7], but it provides the SDR products that are most consistent with the latest operational output.

2. Reprocessing Methodology

2.1. Calibration Equation

Radiometric calibration is applied for each VIIRS RSB Earth observation sample using the following quadratic equation [8], which is simplified here by omitting dependence on spectral band, detector, gain state, HAM (half-angle mirror) side, scan angle, and instrument temperature, but otherwise defines the main terms in the calculations:

$$L = \langle F \rangle \frac{c_0 + c_1 \cdot dn + c_2 \cdot dn^2}{RVS} \quad (1)$$

Radiance L is derived from the Earth measurements after a dark signal (Space view) subtraction (dn is the difference between the Earth and Space views digital counts). The c_i coefficients have been obtained from prelaunch tests and are functions of instrument temperature [9]. RVS (response *versus* scan-angle) corrects for dependence on the Earth viewing angle along the scan. RVS has been derived from the prelaunch test data acquired by the sensor manufacturer [10]. The F factors provide scaling between the prelaunch testing and on orbit performance. They are calculated from the onboard calibrator data as a ratio of radiance predicted from the solar spectrum to on-orbit measurements of radiance reflected from the solar diffuser:

$$F = \frac{RVS_{SD} \cdot \cos AOI}{c_0 + c_1 \cdot dn_{SD} + c_2 \cdot dn_{SD}^2} \frac{\int \tau_{SDS}(\lambda) \cdot BRDF_{SD}(\lambda) \cdot \frac{\Phi_{sun}(\lambda)}{4\pi d^2} \cdot RSR(\lambda) d\lambda}{\int RSR(\lambda) d\lambda} \quad (2)$$

A Space view signal is also subtracted from the solar diffuser measurements (dn_{SD}). Angle of incidence (AOI) of the solar illumination on the diffuser is calculated for each scan from spacecraft's coordinates and attitude. Transmittance of the solar diffuser attenuation screen (τ_{SDS}) is tabulated for different directions of the sunlight, as is the bidirectional reflectance of the diffuser ($BRDF_{SD}$). Solar spectrum (Φ_{sun}) is integrated over the relative spectral response (RSR) of each band and is corrected for the changes in the Sun-Earth distance (d). Although screen transmittance and diffuser reflectance do not change much within the spectral response range of each band, they are still included in the integration.

Since VIIRS solar calibration is based on measurements of light reflected from the onboard solar diffuser, knowledge of the diffuser reflectance is fundamental for calibration accuracy. Changes in the solar diffuser's reflectance with time are measured by a separate, onboard instrument: the SDSM (Solar Diffuser Stability Monitor). SDSM measurements are used to derive for each band an H factor: a ratio of the current $BRDF_{SD}$ to its value at the beginning of the mission [8,11]. The H factors are used to calculate the $BRDF_{SD}$ values used in Equation (2). In this way, the H factors are directly used to scale the calibration coefficients for the visible and near-infrared bands (M1–M7 and I1–I2). Degradation of the solar diffuser reflectance is considered negligible for the short-wave infrared bands (I3 and M8–M11) [12].

While the solar calibration is conducted once per orbit, stability of the diffuser is monitored less frequently. Only during the initial on-orbit checkout were the SDSM measurements conducted on every orbit. Later, SDSM was operated once per day. The frequency has been reduced to three times per week since May 2014. Linear extrapolation is used to calculate the H factor values for each orbit from the less frequent SDSM measurements.

2.2. Data Processing

VIIRS SDR products are generated by the IDPS system that implements the above calibration equation. The IDPS processing codes are also included in the ADL (Algorithm Development Library) software that can be executed outside of the IDPS system to support additional development of the algorithms (https://jpss.ssec.wisc.edu/). To conduct the long-term testing of the automated calibration procedure, the VIIRS RSB radiometric calibration coefficients were calculated using the ADL software that includes version Mx8.10 of the IDPS code. Figure 1 shows a flowchart of the conducted ADL processing. The reprocessing was started with the VIIRS RDRs (Raw Data Records) to mitigate earlier processing errors such as an incorrect solar vector rotation or omission of some SDSM data. Only granules that contain the solar calibration data were processed for each orbit. The datasets were obtained from the NOAA CLASS (Comprehensive Large Array-data Stewardship System) archive (http://www.class.noaa.gov/) or from the JPSS GRAVITE (Government Resources for Algorithm Verification, Independent Testing and Evaluation) system (https://gravite.jpss.noaa.gov/). When multiple RDR sets existed in the archives for the same data granule, only the most complete one or the most recent one (when equal in size) was selected for processing. OBC IP (Onboard Calibrator Intermediate Product) files, which have been generated by IDPS since 21 November 2014 (orbit 15,893), did not require reprocessing and were directly obtained from the above data archives.

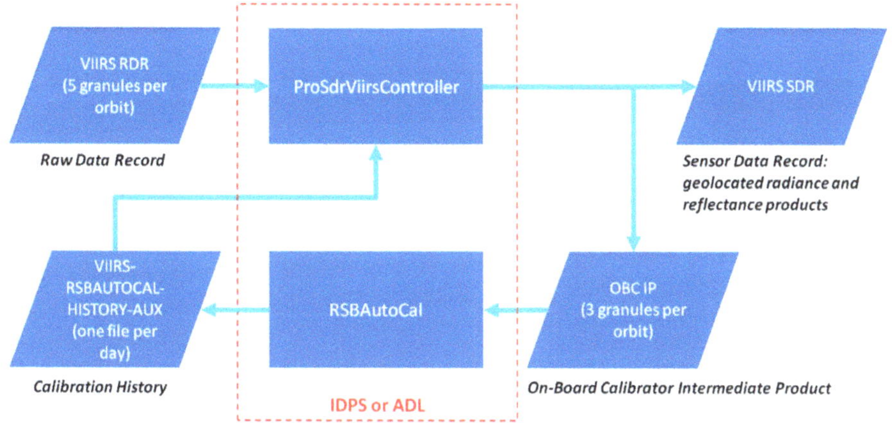

Figure 1. Flowchart of the VIIRS RSB (Reflective Solar Band) solar calibration reprocessing.

In IDPS, the program that generates the SDR and OBC IP files (ProSdrViirsController) is executed in parallel with the automated calibration program (RSBAutoCal). After a new OBC IP file is created by ProSdrViirsController, RSBAutoCal is invoked to process that file, together with two OBC IP files from the preceding and subsequent granules. When a new solar calibration event is detected in the OBC IP data, RSBAutoCal generates an updated VIIRS-RSBAUTOCAL-HISTORY-AUX file. Before producing SDR files in the automated calibration mode, ProSdrViirsController selects the VIIRS-RSBAUTOCAL-HISTORY-AUX file that contains the F factors for the orbit being processed. In our calibration reprocessing with ADL, these two processes were separated. All OBC IP files with solar calibration data were produced first using ProSdrViirsController, and then they were processed using RSBAutoCal to generate the VIIRS-RSBAUTOCAL-HISTORY-AUX files. To apply the reprocessed calibration coefficients in SDR production using ADL, the VIIRS-RSBAUTOCAL-HISTORY-AUX files should be provided, together with the RDR files required for a given SDR granule, as input to ProSdrViirsController executed in the automated calibration mode.

The calculations were based on the latest processing parameters from the operational system that incorporate the following major changes:

- Corrected solar vectors: After a processing code update, orientation of the Sun-satellite vectors has changed by as much as 0.2° [13].
- Improved solar attenuation screens transmittance and solar diffuser bidirectional reflectance tables: angular dependence was optimized based on reanalysis of the on-orbit yaw maneuver and the routine onboard calibrator measurements acquired during the first three years of the Suomi NPP mission [14].
- Updated prelaunch calibration coefficients: To improve consistency between bands with similar spectral response, offset terms were set to zero while only the linear and quadratic terms remained [9].
- Optimized RHW (Robust Holt-Winters) filter parameters: Smoothing of the calibration coefficients time series was improved by damping oscillations while maintaining sensitivity to trend changes [11].

The same set of the processing parameters was consistently used in the entire reprocessing of the VIIRS RSB calibration coefficients. Table 1 list versions of the LUT (Lookup Table) files applied during the calibration reprocessing. Effectivity start dates of these LUT files have been often extended to the beginning of the Suomi NPP mission. Only the spectral response LUT was changed during the reprocessed time period (as indicated in Table 1 by the effectivity dates), similarly to the changes that occurred in the operational SDR production. The calibration history file, VIIRS-RSBAUTOCAL-HISTORY-AUX, which was used to initialize the calculations, did set the H and F factor values to one and their trends to zero at time of the Suomi NPP launch.

Table 1. Lookup tables applied in the VIIRS RSB calibration reprocessing.

LUT Name	Version Identifier	Effectivity Start Date
VIIRS-RSBAUTOCAL-BRDF-SCREEN-TRANSMISSION-PRODUCT-RTA-VIEW-LUT	CCR-15-2496-JPSS-DPA-003	26 October 2011
VIIRS-RSBAUTOCAL-BRDF-SCREEN-TRANSMISSION-PRODUCT-SDSM-VIEW-LUT	CCR-15-2248-JPSS-DPA-002	26 October 2011
VIIRS-RSBAUTOCAL-H-AUTOMATE-LUT	CCR-15-2353-JPSS-DPA-009	26 October 2011
VIIRS-RSBAUTOCAL-H-LUT	CCR-15-2639-JPSS-DPA-003	26 October 2011
VIIRS-RSBAUTOCAL-ROT-MATRIX-LUT	CCR-11-226-JPSS-DPA-001	26 October 2011
VIIRS-RSBAUTOCAL-RSB-F-AUTOMATE-LUT	CCR-15-2249-JPSS-DPA-007	26 October 2011
VIIRS-RSBAUTOCAL-RVF-LUT	CCR-13-876-JPSS-DPA-001	26 October 2011
VIIRS-RSBAUTOCAL-SDSM-SOLAR-SCREEN-TRANS-LUT	CCR-15-2248-JPSS-DPA-002	26 October 2011
VIIRS-RSBAUTOCAL-SDSM-TIME-LUT	NPP-1	26 October 2011
VIIRS-SDR-DELTA-C-LUT	CCR-15-2253-temp-corrected-JPSS-DPA-004	26 October 2011
VIIRS-SDR-GEO-MOD-PARAM-LUT	CCR-13-1171-JPSS-DPA-009	26 October 2011
VIIRS-SDR-QA-LUT	NPP-1	26 October 2011
VIIRS-SDR-RADIOMETRIC-PARAM-V3-LUT	CCR-13-1220-JPSS-DPA-005	26 October 2011
VIIRS-SDR-RELATIVE-SPECTRAL-RESPONSE-LUT	CCR-13-876-JPSS-DPA-001	26 October 2011
VIIRS-SDR-RELATIVE-SPECTRAL-RESPONSE-LUT	CCR-14-1965-JPSS-DPA-002	5 April 2013
VIIRS-SDR-SOLAR-IRAD-LUT	NPP-1	26 October 2011
VIIRS-SDR-TELE-COEFFS-LUT	CCR-12-346-JPSS-DPA-002	26 October 2011

3. Results and Discussion

3.1. Solar Diffuser Degradation

Two graphs included in Figure 2 show time series of the H factors calculated during the current reprocessing with and without applying the RHW filter. While only the filtered (smoothed and extrapolated) H values are used in the reprocessed F factor calculations, the unfiltered values provide additional insights into uncertainties of the SDSM measurements.

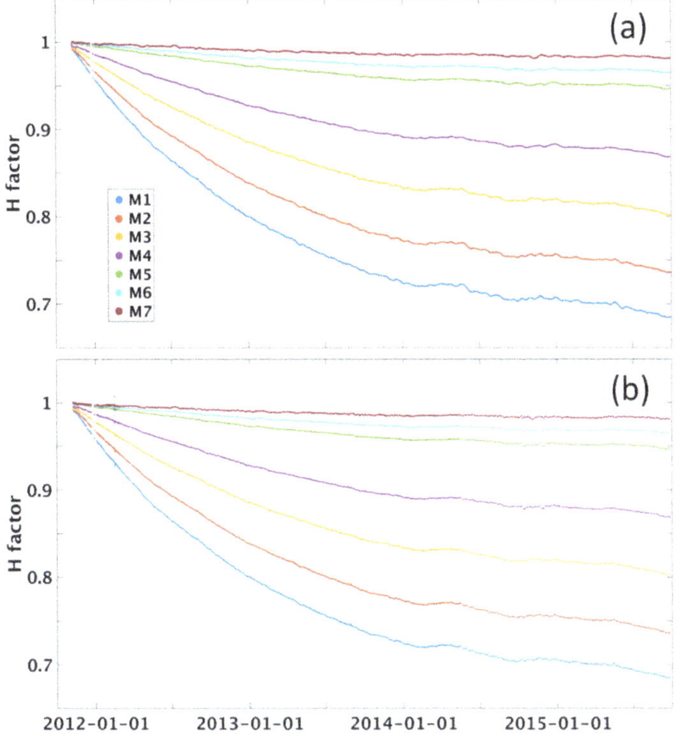

Figure 2. Degradation of the Suomi NPP VIIRS solar diffuser reflectance (the H factor) for bands M1–M7 as measured by the SDSM and analyzed by the automated calibration procedure with (**a**) and without (**b**) the time series filtering.

The H factor time series show that the measured solar diffuser reflectance appears to decrease exponentially from the beginning of the mission until February 2014. Since then, the pace of the diffuser degradation changes from time to time. Although the irregular trend changes in Figure 2 seem more pronounced for the shortest wavelength bands M1–M3, they are in fact similar for all reflective bands when compared with the degradation range for each band. The cause of these unexpected trend changes is still unknown, and it is not clear if these are real changes in the diffuser reflectance or artifacts in the SDSM data, either from the measurements themselves or from their analysis. Nevertheless, the H factor trend changes emphasize the need for using an automated calibration procedure that promptly responds to the observed variability.

Application of the RHW filter removes some artifacts from the SDSM data such as the incompletely corrected measurements acquired during the yaw maneuver in February 2012. Effectiveness of the filter occasionally appears less certain (Figure 3). RHW filter parameters have been optimized in a tradeoff between smoothness of the time series and sensitivity to temporal changes, but the choice of the RHW smoothing parameters is only briefly discussed in the literature [15]. A recent analysis has shown that the smoothing parameters for the measured values should be at least several times larger than the smoothing parameters for the measurement's trends [16]. Such RHW parameters have been used to filter both the *H* factors and the *F* factors in this study and in the operational processing of the VIIRS data [11].

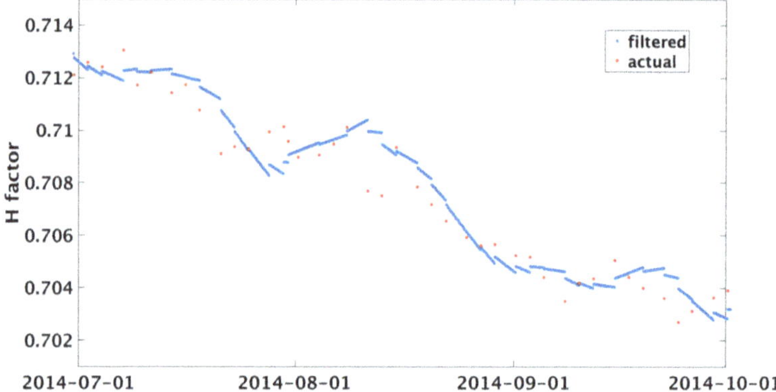

Figure 3. Comparison of the band M1 *H* factor values derived from the actual SDSM measurements without filtering and the values smoothed and interpolated with the RHW filter.

3.2. Radiometric Gain Changes

Figure 4 shows a comparison of the reprocessed *F* factors with those previously used in the operational SDR production for one of the spectral bands that is not significantly affected by the VIIRS telescope throughput degradation: band M3 [17]. The *F* factors calculated by the automated procedure not only extend to the beginning of the Earth observations from VIIRS, but also appear more stable throughout the duration of the mission. Variability visible in the time series of the reprocessed *F* factors is on the order of only 1% and reflects uncertainty of the calibration. The differences between the reprocessed coefficients and the original operational ones do not generally exceed $\pm$ 2%, in agreement with the radiometric calibration uncertainty requirements. During the first year of the mission, the changes in the operational *F* factors shown on the graph originate from improvements in characterization of the VIIRS radiometric response, with the most significant differences before May 2012. Since February 2014, the operational F factor variability for band M3 is mostly affected by the observed *H* factor trend changes. The automated calibration procedure appears to respond better to these changes. Small seasonal variability that can be seen in the reprocessed *F* factors coincides with changes in direction of solar illumination that make SDSM measurements more difficult during the months of October through December each year. Future improvements of the VIIRS SDR processing parameters are expected to further reduce this variability.

Figure 5 shows the *F* factor comparison for band M7 that is strongly affected by the rotating telescope mirror degradation [3]. Even with the *F* factor changes dominated by the throughput decrease, one can notice that the reprocessed calibration coefficients are free of the spurious variability that affected the earlier operational *F* factors (changes in December 2012 are only due to an investigation of the degradation anomaly). Several phases of the operational calibration improvements can be detected in the differences between the operational and reprocessed *F* factors. The largest differences existed until October 2012 when angular dependence of the attenuation screen transmittance and the solar diffuser reflectance were determined more accurately from the Suomi NPP orbital yaw maneuver [18]. The next change occurred in April 2014 when the offset terms in the calibration equation (the c_0 coefficients) were set to zero and the prelaunch c_1 and c_2 coefficients were recalculated from the prelaunch test data [9]. The differences diminished in 2015 when calculations of the operational *F* factors became similar to the automated procedure with the RHW filter [11]. This illustrates the tradeoff between the operational procedure and the automated one: by applying the automated calibration, at the cost of small variability observed in 2015, larger biases such as those from the first half of 2014 can be avoided.

Although some improvements of the calibration trending can be expected just from the consistent use of the same processing parameters for the whole mission period, Figures 4 and 5 show that after the reprocessing the solar diffuser degradation is better corrected for band M3, and the telescope degradation that is diminishing with time is better corrected after every orbit for band M7. Similar improvements in the reprocessed *F* factors can be seen for the other visible and near-infrared bands as well as for the short-wave infrared bands. The respective graphs are shown below and in the figures included in the Appendix.

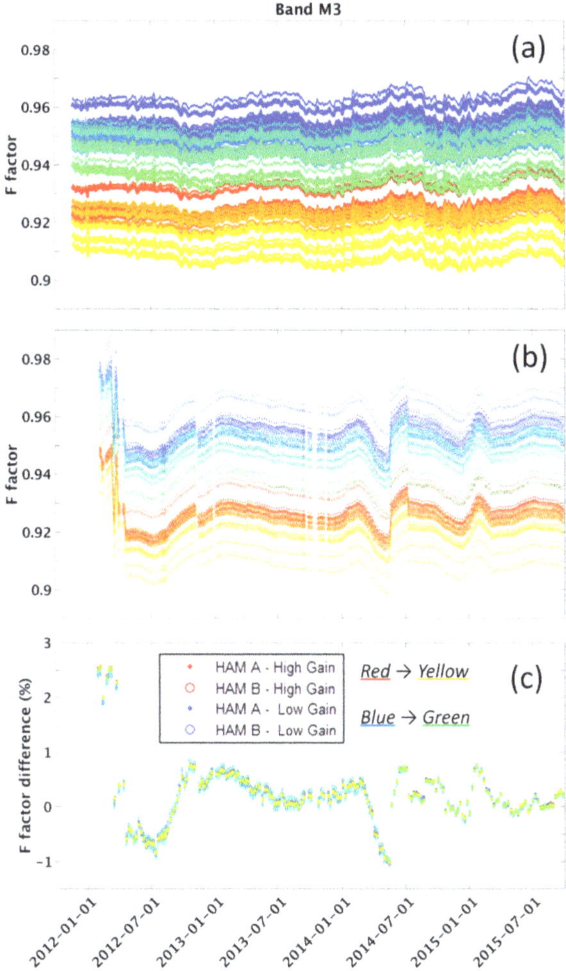

Figure 4. (**a**) VIIRS band M3 (488 nm, dual gain) radiometric calibration scaling coefficients (the *F* factors) derived by the automated calibration procedure; (**b**) the *F* factors used in the operational production of the VIIRS SDR; and (**c**) relative difference between the operational *F* factors and the reprocessed ones (used here as a baseline). The half-angle mirror sides and the gain states are indicated by different symbols (dots and circles), while the *F* factors are shown for each detector with a different color from the red-yellow (for high/single gain) and blue-green (for low gain) palettes.

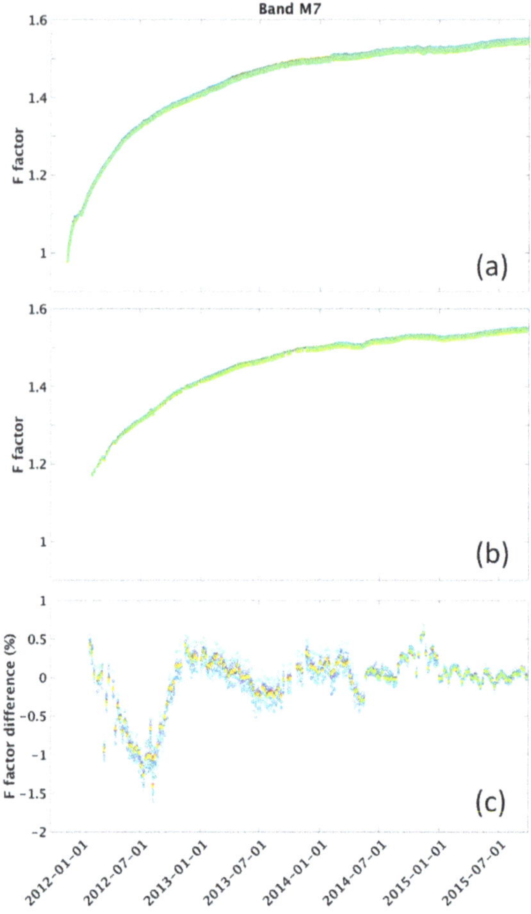

Figure 5. (**a**) VIIRS band M7 (865 nm, dual gain) radiometric calibration scaling coefficients (the *F* factors) derived by the automated calibration procedure; (**b**) the *F* factors used in the operational production of the VIIRS SDR; and (**c**) relative difference between the operational *F* factors and the reprocessed ones (used here as a baseline). Symbol legend is the same as in Figure 4.

Figure 6, for band M6, shows F factor values that are distinctly different for the two HAM sides. The difference compensates for the reciprocal difference that exists in the prelaunch *c* coefficients for the two HAM sides: the radiometric gains for band M6 are very similar for both HAM sides. This exemplifies that the *F* factors are a good estimate of the calibration coefficients (or the inverse of the *F* factors are a good estimate of the sensor radiometric gains) only when the *c* coefficients are virtually the same. If values of the *c* coefficients are different or have significantly changed, the product of the F factor and the c_1 coefficient is a better estimate, but it is still an approximation because of the non-linear form of the calibration equation. This is also illustrated in Figure 7 for band I3, which had the largest calibration change after the offset terms in the calibration equation were set to zero. The operational *F* factors have abruptly changed in April 2014 because the *c* coefficients were changed. The reprocessed *F* factors do not display such a change because the same set of the *c* coefficients was used throughout the entire reprocessing.

Since the F factors provide scaling between the effective radiometric gains determined from prelaunch testing and the gains measured on orbit during solar calibration events, values of the F factors can be either smaller (as for bands M2–M4) or larger than one (as for M1) depending on whether the prelaunch c coefficients overestimated or underestimated the actual radiometric response on orbit. The gains may have changed during launch or as a result of differences between the conditions during the thermal vacuum testing on the ground and in the real Space environment on orbit. The gain changes can also originate in instrument degradation occurring on orbit, such as in the case of bands M5–M11. The apparent radiometric response changes may also indicate uncertainties of the prelaunch testing created by such factors as the differences between irradiance spectra of the laboratory calibration lamps and the Sun.

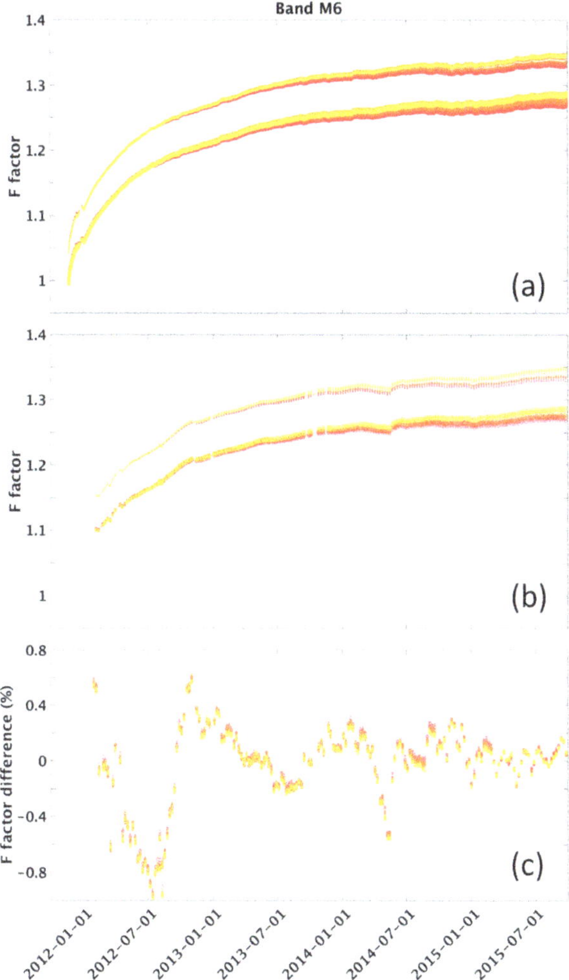

Figure 6. (**a**) VIIRS band M6 (746 nm, single gain) radiometric calibration scaling coefficients (the F factors) derived by the automated calibration procedure; (**b**) the F factors used in the operational production of the VIIRS SDR; and (**c**) relative difference between the operational F factors and the reprocessed ones (used here as a baseline). Symbol legend is the same as in Figure 4.

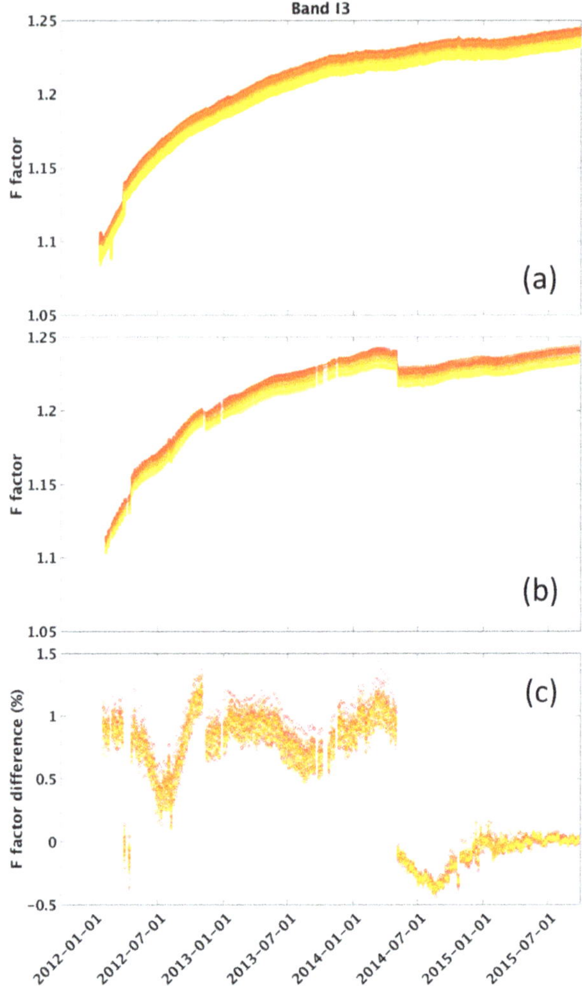

Figure 7. (**a**) VIIRS band I3 (1610 nm, single gain) radiometric calibration scaling coefficients (the *F* factors) derived by the automated calibration procedure; (**b**) the *F* factors used in the operational production of the VIIRS SDR; and (**c**) relative difference between the operational *F* factors and the reprocessed ones (used here as a baseline). Symbol legend is the same as in Figure 4.

The SWIR (short-wave infrared) bands (I3 and M8–M11) also show an *F* factor discontinuity in March 2012. This reflects an actual gain change that occurred after the Suomi NPP spacecraft was temporarily in safe mode during which the spacecraft was oriented toward the Sun. After this gain change, the reprocessed *F* factors are updated sooner by the automated calibration procedure than they were during the original operational processing. This creates the spikes in the differences between the reprocessed and operational *F* factors for the SWIR bands, and it also shows once more that the automated calibration procedure improves upon the operational processing.

A downward spike in the reprocessed SWIR-band *F* factors occurs in February 2012 during an orbital pitch maneuver of the Suomi NPP spacecraft. During the pitch maneuver, SWIR detectors temperature briefly increased above the nominal level, and the SWIR bands gains changed. Solar calibration data acquired during the pitch maneuver were excluded from the operational *F* factor calculations, because Earth observations during the maneuver were mostly invalid, but these solar data were processed by the automated procedure and are included on the presented graphs for completeness.

During the early on-orbit checkout of the Suomi NPP satellite, the VIIRS SWIR bands were activated later than the other reflective bands: the SWIR detectors were cooled down to an operational temperature after the RSB data acquisition has already started. The SWIR measurements acquired before the detectors reached the nominal temperature were dominated by random noise and therefore were processed by initially applying the automated calibration procedure without filtering. The RHW filter was applied only after the SWIR detector temperature has stabilized. Residual variability can still be seen at the beginning of the F factor time series for the SWIR bands, but magnitude of the transient changes is small and they diminish quickly. This approach demonstrates how the automated calibration procedure can be used during activation of the VIIRS instruments on the future JPSS satellites.

4. Conclusions

Reprocessing of the VIIRS RSB calibration coefficients using a new automated procedure has increased confidence in this approach by demonstrating that it can be applied to all solar calibration data previously acquired during the entire Suomi NPP mission. With optimized RHW filter parameters, the automated calculations of the *F* factors remained stable for the entire reprocessed time period. While the automated procedure improved upon calibration used in the operational processing of the VIIRS data, the differences between the reprocessed and operational *F* factors in general have not exceeded the VIIRS SDR radiometric calibration uncertainty and remain in the 1%–2% range, except for bands M1–M3 before May 2012. The improved calibration coefficients can be further used to reprocess VIIRS SDR and other data products. Since the SDR radiance and reflectance products are directly proportional to the *F* factors (Equation 1), the relative improvements between the original, operational products and the reprocessed ones will be the same as the differences between the *F* factors. Because the F factor differences appear cross-correlated for some band pairs, the differences for data products derived from SDR using reflective band ratios may be smaller. Such will be the case for NDVI (Normalized Difference Vegetation Index) EDR (Environmental Data Record) that is produced from bands I1 and I2 [19]. The improved calibration coefficients may especially benefit direct broadcast ground stations when reprocessing of previously acquired VIIRS datasets is needed. The reprocessed calibration history files are available though the VIIRS SDR home page at the NOAA National Calibration Center website (http://ncc.nesdis.noaa.gov/VIIRS/).

Acknowledgments: This study was partially funded by the JPSS program office and by the NOAA grant NA14NES4320003 for the Cooperative Institute for Climate and Satellites (CICS) at the University of Maryland/ESSIC. The manuscript contents are solely the opinions of the authors and do not constitute a statement of policy, decision, or position on behalf of NOAA or the U.S. government. The authors thank anonymous reviewers for constructive comments that greatly helped in improving the manuscript.

Author Contributions: Slawomir Blonski designed the study, developed methodology and software tools, performed the analysis and wrote the manuscript. Changyong Cao contributed to the design of the study and provided technical oversight of the project.

Conflicts of Interest: The authors declare no conflict of interest.

Appendix

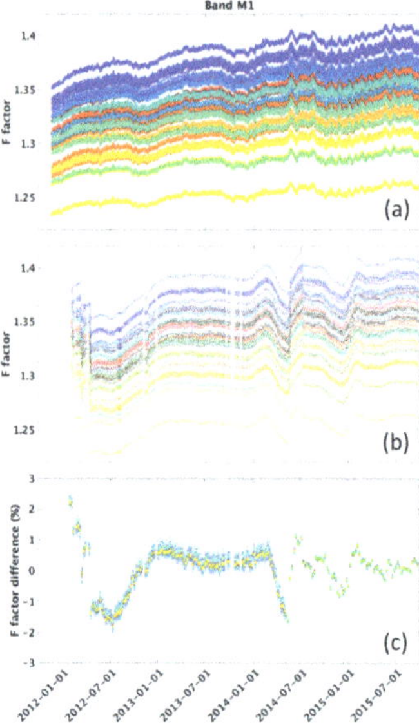

Figure A1. (**a**) VIIRS band M1 (412 nm, dual gain) radiometric calibration scaling coefficients (the *F* factors) derived by the automated calibration procedure; (**b**) the *F* factors used in the operational production of the VIIRS SDR; and (**c**) relative difference between the operational *F* factors and the reprocessed ones (used here as a baseline). Symbol legend is the same as in Figure 4.

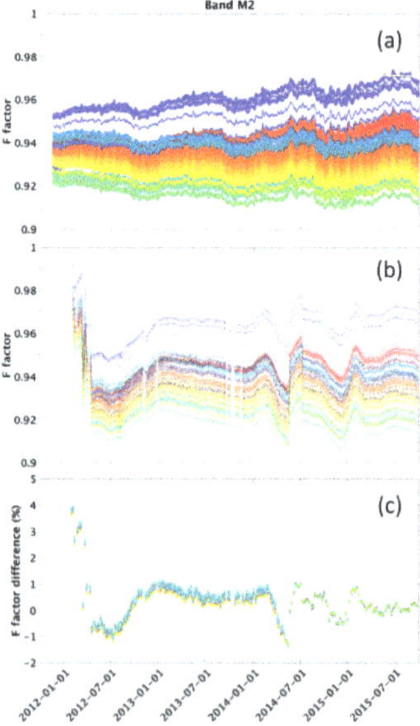

Figure A2. (**a**) VIIRS band M2 (445 nm, dual gain) radiometric calibration scaling coefficients (the *F* factors) derived by the automated calibration procedure; (**b**) the *F* factors used in the operational production of the VIIRS SDR; and (**c**) relative difference between the operational *F* factors and the reprocessed ones (used here as a baseline). Symbol legend is the same as in Figure 4.

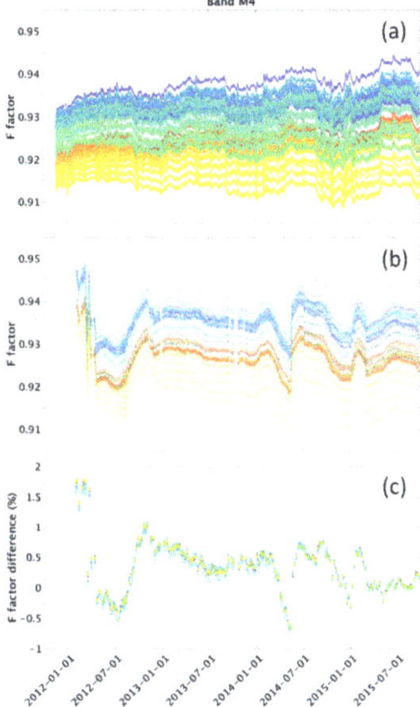

Figure A3. (**a**) VIIRS band M4 (555 nm, dual gain) radiometric calibration scaling coefficients (the *F* factors) derived by the automated calibration procedure; (**b**) the *F* factors used in the operational production of the VIIRS SDR; and (**c**) relative difference between the operational *F* factors and the reprocessed ones (used here as a baseline). Symbol legend is the same as in Figure 4.

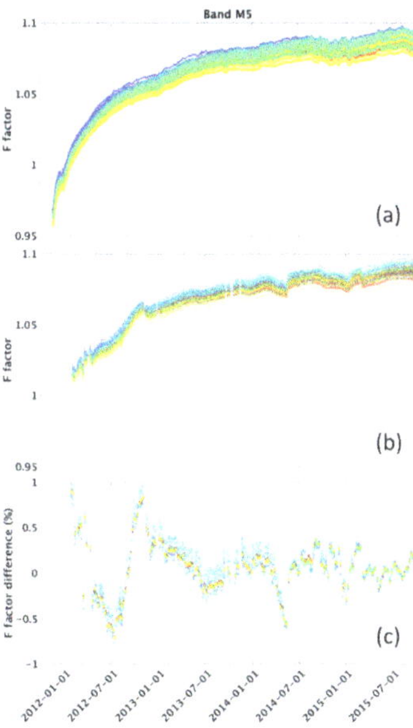

Figure A4. (**a**) VIIRS band M5 (672 nm, dual gain) radiometric calibration scaling coefficients (the *F* factors) derived by the automated calibration procedure; (**b**) the *F* factors used in the operational production of the VIIRS SDR; and (**c**) relative difference between the operational *F* factors and the reprocessed ones (used here as a baseline). Symbol legend is the same as in Figure 4.

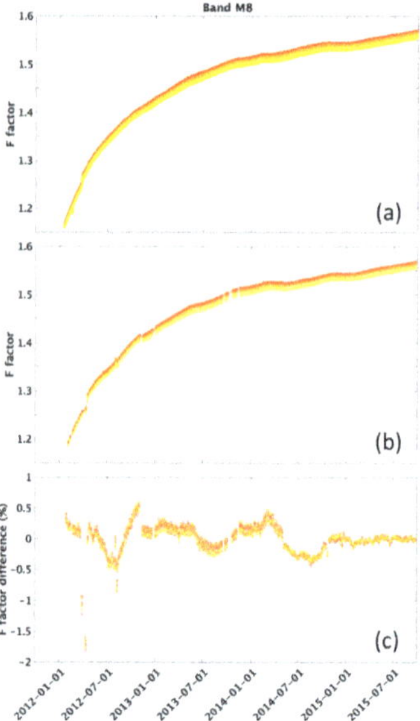

Figure A5. (**a**) VIIRS band M8 (1240 nm, single gain) radiometric calibration scaling coefficients (the *F* factors) derived by the automated calibration procedure; (**b**) the *F* factors used in the operational production of the VIIRS SDR; and (**c**) relative difference between the operational *F* factors and the reprocessed ones (used here as a baseline). Symbol legend is the same as in Figure 4.

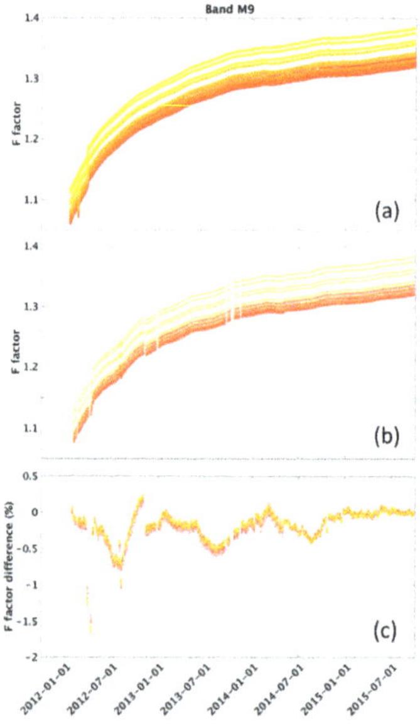

Figure A6. (**a**) VIIRS band M9 (1378 nm, single gain) radiometric calibration scaling coefficients (the *F* factors) derived by the automated calibration procedure; (**b**) the *F* factors used in the operational production of the VIIRS SDR; and (**c**) relative difference between the operational *F* factors and the reprocessed ones (used here as a baseline). Symbol legend is the same as in Figure 4.

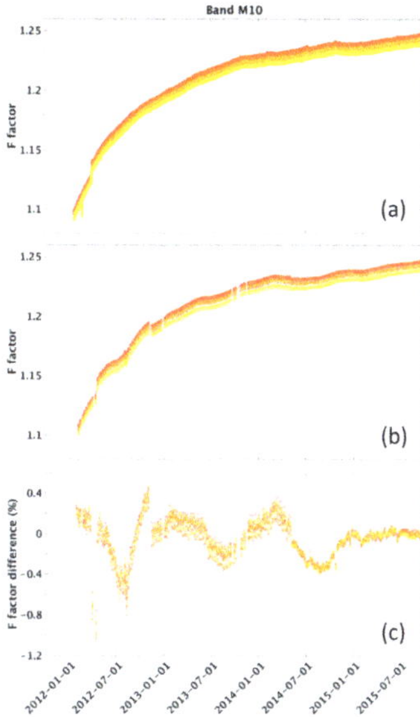

Figure A7. (**a**) VIIRS band M10 (1610 nm, single gain) radiometric calibration scaling coefficients (the *F* factors) derived by the automated calibration procedure; (**b**) the *F* factors used in the operational production of the VIIRS SDR; and (**c**) relative difference between the operational *F* factors and the reprocessed ones (used here as a baseline). Symbol legend is the same as in Figure 4.

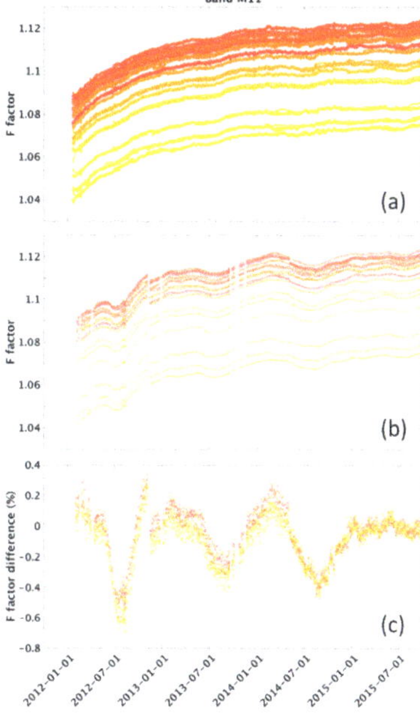

Figure A8. (**a**) VIIRS band M11 (2250 nm, single gain) radiometric calibration scaling coefficients (the *F* factors) derived by the automated calibration procedure; (**b**) the *F* factors used in the operational production of the VIIRS SDR; and (**c**) relative difference between the operational *F* factors and the reprocessed ones (used here as a baseline). Symbol legend is the same as in Figure 4.

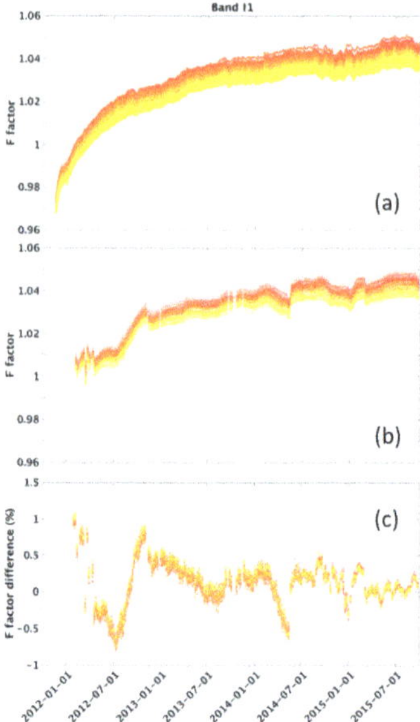

Figure A9. (**a**) VIIRS band I1 (640 nm, single gain) radiometric calibration scaling coefficients (the *F* factors) derived by the automated calibration procedure; (**b**) the *F* factors used in the operational production of the VIIRS SDR; and (**c**) relative difference between the operational *F* factors and the reprocessed ones (used here as a baseline). Symbol legend is the same as in Figure 4.

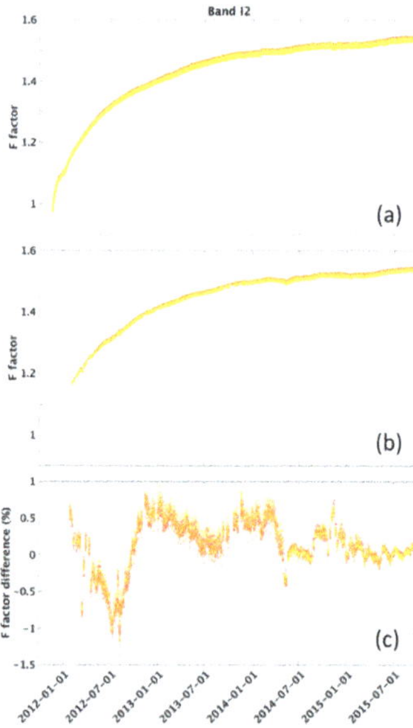

Figure A10. (**a**) VIIRS band I2 (865 nm, single gain) radiometric calibration scaling coefficients (the *F* factors) derived by the automated calibration procedure; (**b**) the *F* factors used in the operational production of the VIIRS SDR; and (**c**) relative difference between the operational *F* factors and the reprocessed ones (used here as a baseline). Symbol legend is the same as in Figure 4.

References

1. Cao, C.; De Luccia, F.J.; Xiong, X.; Wolfe, R.; Weng, F. Early on-orbit performance of the Visible Infrared Imaging Radiometer Suite onboard the Suomi National Polar-Orbiting Partnership (S-NPP) satellite. *IEEE Trans. Geosci. Remote Sens.* **2013**, *52*, 1142–1156. [CrossRef]
2. Lee, S.; McIntire, J.; Oudrari, H.; Schwarting, T.; Xiong, X. A new method for Suomi-NPP VIIRS day-night band on-orbit radiometric calibration. *IEEE Trans. Geosci. Remote Sens.* **2015**, *53*, 324–334.
3. Blonski, S.; Cao, C. Monitoring and predicting rate of VIIRS sensitivity degradation from telescope contamination by tungsten oxide. *Proc. SPIE* **2013**, *8739*. [CrossRef]
4. Rausch, K.; Houchin, S.; Cardema, J.; Moy, G.; Haas, E.; De Luccia, F.J. Automated calibration of the Suomi National Polar-Orbiting Partnership (S-NPP) Visible Infrared Imaging Radiometer Suite (VIIRS) reflective solar bands. *J. Geophys. Res. Atmos.* **2013**, *118*, 13434–13442. [CrossRef]
5. Blonski, S.; Cao, C. VIIRS reflective solar bands calibration reprocessing. In Proceedings of the 2015 IEEE International Geoscience and Remote Sensing Symposium (IGARSS), Milan, Italy, 26–31 July 2015; pp. 3906–3909.
6. Sun, J.; Wang, M. Visible Infrared Imaging Radiometer Suite solar diffuser calibration and its challenges using a solar diffuser stability monitor. *Appl. Opt.* **2014**, *53*, 8571–8584. [CrossRef] [PubMed]
7. Eplee, R.; Turpie, K.; Meister, G.; Patt, F.; Franz, B.; Bailey, S. On-orbit calibration of the Suomi National Polar-Orbiting Partnership Visible Infrared Imaging Radiometer Suite for ocean color applications. *Appl. Opt.* **2015**, *54*, 1984–2006. [CrossRef] [PubMed]
8. Lei, N.; Wang, Z.; Xiong, X. On-Orbit radiometric calibration of Suomi NPP VIIRS reflective solar bands through observations of a sunlit solar diffuser panel. *IEEE Trans. Geosci. Remote Sens.* **2015**, *53*, 5983–5990. [CrossRef]
9. Moyer, D.; Vandermierden, N.; Rausch, K.; De Luccia, F. VIIRS reflective solar bands on-orbit calibration coefficient performance using imagery and moderate band intercomparisons. *Proc. SPIE* **2014**, *9223*, 922305.
10. Wu, A.; McIntire, J.; Xiong, X.; De Luccia, F.J.; Oudrari, H.; Moyer, D.; Xiong, S.; Pan, C. Comparison of VIIRS pre-launch RVS performance using results from independent studies. *Proc. SPIE* **2011**, *8153*. [CrossRef]
11. Moy, G.; Rausch, K.; Haas, E.; Wilkinson, T.; Cardema, J.; De Luccia, F. Mission history of reflective solar band calibration performance of VIIRS. *Proc. SPIE* **2015**, *9607*. [CrossRef]
12. Xiong, X.; Angal, A.; Fulbright, J.; Lei, N.; Mu, Q.; Wang, Z.; Wu, A. Calibration improvements for MODIS and VIIRS SWIR spectral bands. *Proc. SPIE* **2015**, *9607*. [CrossRef]
13. Fulbright, J.; Anderson, S.; Lei, N.; Efremova, B.; Wang, Z.; McIntire, J.; Chiang, K.V.; Xiong, X. The solar vector error within the SNPP Common GEO code, the correction, and the effects on the VIIRS SDR RSB calibration. *Proc. SPIE* **2014**, *9264*. [CrossRef]
14. Haas, E.M.; De Luccia, F.J.; Moyer, D. Maintaining SNPP VIIRS reflective solar band sensor data record quality: On-orbit update of screen transmission and solar diffuser BRDF parameters. In Proceedings of the 11th Symposium on New Generation Operational Environmental Satellite Systems, Phoenix, AZ, USA, 4–8 January 2015; Available online: https://ams.confex.com/ams/95Annual/webprogram/Paper261791.html (accessed on 22 September 2015).
15. Gelper, S.; Fried, R.; Croux, C. Robust forecasting with exponential and Holt-Winters smoothing. *J. Forecast.* **2010**, *29*, 285–300. [CrossRef]
16. Anderson, S.; Chiang, K.V. VIIRS SDR RSBAutoCal Holt-Winters Filter Coefficients: NASA VCST Internal Memo 2014-16. > 2014. unpublished.
17. De Luccia, F.; Moyer, D.; Johnson, E.; Rausch, K.; Lei, N.; Chiang, K.; Xiong, X.; Fulbright, J.; Haas, E.; Iona, G. Discovery and characterization of on-orbit degradation of the Visible Infrared Imaging Radiometer Suite (VIIRS) Rotating Telescope Assembly (RTA). *Proc. SPIE* **2012**, *8510*. [CrossRef]
18. McIntire, J.; Moyer, D.; Efremova, B.; Oudrari, H.; Xiong, X. On-Orbit characterization of S-NPP VIIRS transmission functions. *IEEE Trans. Geosci. Remote Sens.* **2015**, *53*, 2354–2365. [CrossRef]

19. Song, C.; Chen, J.M.; Hwang, T.; Gonsamo, A.; Croft, H.; Zhang, Q.; Dannenberg, M.; Zhang, Y.; Hakkenberg, C.; Li, J. Ecological characterization of vegetation using multisensor remote sensing in the solar reflective spectrum. In *Land Resources Monitoring, Modeling, and Mapping with Remote Sensing*; Thenkabail, P.S., Ed.; CRC Press: Boca Raton, FL, USA, 2015; pp. 533–575.

Article

Soumi NPP VIIRS Day/Night Band Stray Light Characterization and Correction Using Calibration View Data

Shihyan Lee [1,2,*] and Changyong Cao [3]

[1] Science Applications International Corp., Goddard Space Flight Center Code 616, Greenbelt, MD 20771, USA
[2] ERT Corp., Laurel, MD 20740, USA
[3] NOAA/NESDIS/STAR, College Park, MD 20737, USA; changyong.cao@noaa.gov
* Correspondence: shihyanlee@yahoo.com; Tel.: +1-617-947-1605

Academic Editors: Richard Müller and Prasad S. Thenkabail
Received: 2 November 2015; Accepted: 29 January 2016; Published: 8 February 2016

Abstract: The Soumi NPP VIIRS Day/Night Band (DNB) nighttime imagery quality is affected by stray light contamination. In this study, we examined the relationship between the Earth scene stray light and the signals in VIIRS's calibrators to better understand stray light characteristics and to improve upon the current correction method. Our analyses showed the calibrator signal to be highly predictive of Earth scene stray light and can provide additional stray light characteristics that are difficult to obtain from Earth scene data alone. In the current stray light correction regions (mid-to-high latitude), the stray light onset angles can be tracked by calibration view data to reduce correction biases. In the southern hemisphere, it is possible to identify the angular extent of the additional stray light feature in the calibration view data and develop a revised correction method to remove the additional stray light occurring during the southern hemisphere springtime. Outside of current stray light correction region, the analysis of calibration view data indicated occasional stray light contamination at low latitude and possible background biases caused by Moon illumination. As stray light affects a significant portion of nighttime scenes, further refinement in characterization and correction is important to ensure VIIRS DNB imagery quality for Soumi NPP and future missions.

Keywords: remote sensing; nighttime lights; Day/Night Band; VIS/NIR; VIIRS; on-orbit calibration; stray light; moon

---

1. Introduction

One of the unique capabilities on Soumi National Polar-Orbiting Partnership (SNPP) is the Visible Infrared Imaging Radiometer Suite (VIIRS) Day/Night Band (DNB), a visible and near infrared panchromatic band (500–900 nm) that is capable of making observations during both day and night [1,2]. The VIIRS DNB is designed to improve upon the global nighttime lights observations initiated more than two decades ago by the Defense Meteorological Satellite Program's (DMSP) Operational Linescan System (OLS) [3–9]. The VIIRS DNB is sensitive enough to pinpoint the location of lights from bridges [10] or the faint air glow pattern during moonless nights [11]. Many other anthropogenic activities, e.g., city light power consumption and outages [12], seasonal light activities [13], and fishing and shipping tracks [14], have been observed and quantified by analysis of VIIRS DNB data. When there is sufficient moonlight, much of the Earth's surface features seen during the daytime, e.g., water, cloud, and land surfaces of different types, can also be observed by DNB at night. This improved capability reveals the Earth's nocturnal secrets in unprecedented detail for a new chapter in nighttime remote sensing [14].

The superior DNB nighttime data quality in SNPP is partly due to the fact that its on-orbit performance exceeds its original design specification [15]. On-orbit DNB calibration and characterization is a challenging undertaking because most useful observations have radiance values that are below the sensor's designed minimum observable radiance. Stray light contamination causes the most significant degradation in DNB nighttime image quality. The most persistent stray light contamination is in the mid-to-high latitude regions where the spacecraft is crossing the northern and southern day/night terminators [16]. The exact latitudes affected by stray light can be determined by the Earth-Sun-spacecraft geometry and orbital inclinations, which change over time. In these regions, the spacecraft is under direct solar illumination due to its elevated orbital track at ~830 km relative to the nighttime Earth surface. Due to the large difference between the Sun and nighttime Earth view radiances, the tiny fraction of Sunlight entering the optical system causes significant contamination of the images [17]. Without correction, large swaths of the nighttime scenes will have little use because the stray light dominates most nighttime imagery signals.

The current VIIRS operational calibration algorithm includes stray light correction to remove the mid-to-high latitude stray light contamination that occurs during the spacecraft's northern and southern day/night terminator crossings. The stray light is estimated based on Earth view data measured over dark surfaces during moonless nights, with an assumption that stray light and airglow are the only contributors to the observed signals [16,17]. Stray light intensity is estimated as a function of the satellite's zenith angle to account for the effects of Earth-Sun-spacecraft geometry. The estimated stray light is also detector- and scan angle-dependent, which indicates that there is a scattered light path that is sensitive to the positioning of the rotating telescope assembly (RTA) as well as to the minor differences in detector locations on the focal plane. Operationally, stray light is estimated once per month during the new moon to best approximate the stray light optical path. The monthly estimated stray light magnitudes are stored in a look-up table (LUT). The online calibration algorithm applies the correction derived from the LUT with the closest Earth-Sun-spacecraft geometries relative to the scene to minimize biases.

The accuracy of the stray light estimation depends on how well the dark surfaces are selected, and airglow approximated. Since a completely dark surface over the entire stray light-affected region is unobtainable, the signals over the dark surfaces at each detector and scan angle are estimated using many orbits of data, excluding data with potential light sources [16,17]. Since the dark surface selection process is imperfect, small uncertainties could occur due to residual data contamination. With the selected dark surfaces, the dark signals can be computed, taking into account the combined effect of stray light and airglow. To estimate the stray light, airglow is estimated from dark surfaces outside of the stray light affected region as an approximation. This airglow approximation is apt to be biased because airglow is not spatially and temporally uniform. Since airglow intensity is usually much smaller than the stray light, the error in stray light estimates is expected to be small. However, the uncertainty in airglow approximation indicates that current stray light-corrected images likely do not have accurate airglow patterns and magnitudes.

Additional uncertainty in stray light correction is due to the slight shift in the Earth-Sun-spacecraft geometry between the time stray light is estimated and the imaging time. The stray light correction biases due to the mismatch in Earth-Sun-spacecraft geometry are small in areas where stray light change, with respect to the satellite's zenith angle, is gradual. However, in regions with sharp stray light changes, *i.e.*, the penumbra and the solar diffuser onset angle, large biases could occur. The large biases result in a stray light-corrected image with an observable horizontal strip due to significant under/over correction.

In addition, the current operational method cannot resolve an additional stray light feature that occurs in the southern hemisphere during its springtime (October–December) [18]. The additional stray light could affect up to one-fourth of the total stray light region. The additional stray light usually exists near the onset of the twilight region, an area where stray light is most difficult to estimate [16,17]. The additional stray light cannot be predicted by current methods because it has features considerably different than the overall stray light pattern. Modification of the current methods is needed in order to remove the additional stray light features in the southern hemisphere.

Previous studies have suggested using calibration view data to improve stray light characterization and correction [16,17,19,20]. In this study, we performed detail analyses on DNB signal characteristics in the VIIRS on-board calibrators, and their corresponding Earth view stray light characteristics. We show the potential of using the DNB calibration view signals to improve stray light onset angle tracking and to refine the current stray light correction. In the southern hemisphere, the regions with additional stray light features can be identified in the calibrator signals, and the current correction algorithm can be modified to remove these features. Last, we will show the use of calibration view data to identify possible stray light contamination outside of the current mid-to-high latitude stray light regions, as well as other potential background contaminations.

2. DNB Calibration Framework

The SNPP VIIRS is a Sun-synchronized, polar-orbiting, scanning radiometer [21]. On orbit, VIIRS collects data in four separate view windows, Earth view (EV), blackbody (BB), solar diffuser (SV), and space view (SV), successively at each scan (Figure 1). The BB, SD, and SV are on-board calibrators (OBCs) designed to calibrate VIIRS's reflective solar bands (0.4–2.5 μm) and thermal emissive bands, and are also used as an alternative DNB calibration methodology [22–24]. The VIIRS BB has high emissivity and is temperature-controlled at ~292.5 K to calibrate the thermal emissive bands. The SD is a near Lambertian Spectralon® panel used to calibrate the reflective solar bands during orbital solar calibration events near the southern hemisphere day/night terminator crossings. The SV is located at a preselected scan angle near the start of EV scan angle, about three degrees off the edge of the Earth's limb, which is used to provide a deep space view for dark references.

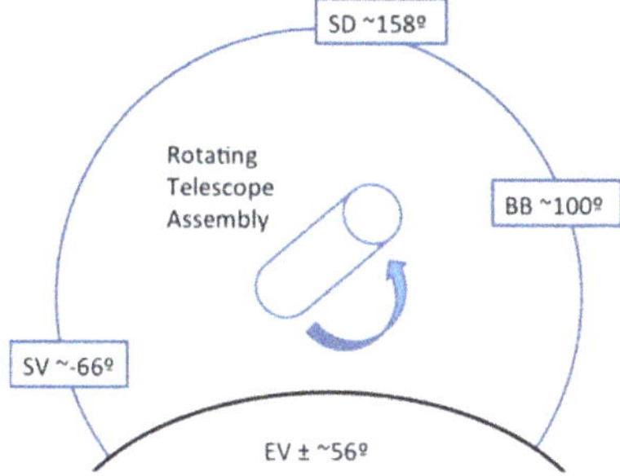

Figure 1. Schematic of VIIRS data collection windows and their approximated angles. The scan angle at the Earth's limb is ~62.5 degrees.

The DNB consists of four sectors of detector arrays: one is in the low-gain stage (LGS); one is in the intermediate-gain stage (MGS); and two redundant arrays, high-gain A (HGA) and high-gain B (HGB), are in the high-gain stage (HGS). In each scan, the DNB collects samples using the four detector arrays and aggregates them into 4064 EV samples and 16 calibration view samples. The output gain stage is selected at each EV sample to ensure its radiance level is within the dynamic range of that gain stage to reduce down-link bandwidth. The HGA and HGB are normally averaged on-board and down-linked as HGS. In calibration views, all four detector arrays are reported in each pixel.

Figure 2 shows typical DNB signal profiles in each of the calibration views for the entire orbit. In Figure 2, the DNB signals were plotted against solar declination angles to indicate the geometric relationship between satellite, Sun, and Earth. During daytime, the SD signals are several orders of magnitude higher than the SV and BB signals. The stronger SD signals are due to high reflectivity surfaces that are illuminated by the daytime Earthshine coming through VIIRS EV port. The BB's low reflectivity absorbs most of the incoming lights, and the recorded BB signals are likely dominated by scattered lights entering into the optical paths. During daytime, the SV signals are slightly higher than BB in general but share similar patterns, except when the spacecraft is near the day/night terminators.

Since the SV is about three degrees off the Earth's limb, the Earth scene stray light coming through the telescope could contribute additional signals added onto the stray light. The position of the telescope could also change the scattered light patterns; an effect that can be observed in EV, where the stray light signal varies by scan angles (see Section 3). Examining the calibration view signal characteristics could provide additional information on potential stray light paths since each calibrator has its own unique properties.

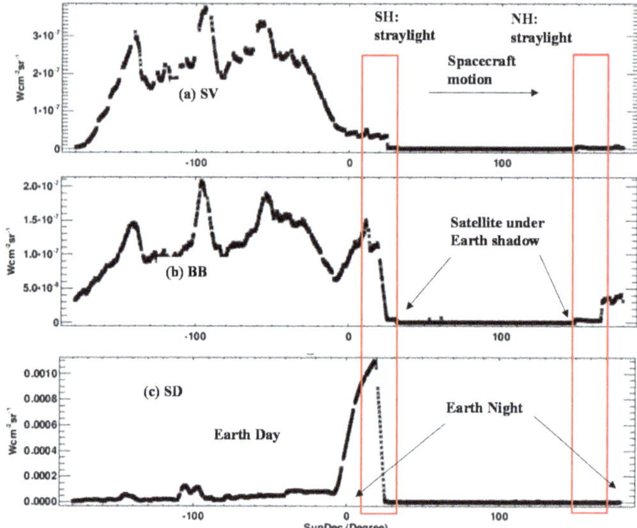

Figure 2. DNB calibration view signals for S-NPP orbit 15,758 on 12 November 2014. DNB detector 1 signals for SV, BB, and SD are shown in **(a–c)**, respectively. The y-axis are DNB signals in $W\cdot cm^{-2}\cdot sr^{-1}$. The x-axis is the VIIRS Sun declination angle, in degrees. NH: northern hemisphere, SH: southern hemisphere.

The DNB images have the most significant stray light problem (highlighted in Figure 2) near the day/night terminators. Although the stray light is stronger during the day than the day/night terminators, the effect is negligible because the stray light is several orders of magnitude smaller than the typical daytime radiance. In the stray light problem regions, the calibration view signals are elevated with patterns that are different between the northern and southern hemisphere. In the northern hemisphere, the bulk of the signals likely came through the EV port from the direct solar illumination. There is an additional scattered light through BB at the end of the DNB EV stray light problem region that caused sharp increase in BB signal. The timing of the sharp increase in BB signals corresponds to the onset of twilight on the Earth's surface. However, there is no corresponding increase in SV signals due to the increased Earth scene radiance in the twilight region. The lack of response in SV signals indicates that an RTA pointing angle might have provided a better shield from the scattered light paths from the Earth's twilight scene.

In the southern hemisphere, the onset of calibration view signal increase corresponds to the direct solar illumination entering the VIIRS's EV port. The sharp increase in the SD signal marked the beginning of solar calibration events when SD is being illuminated by the Sun. Both BB and SV showed increased signals, but in different patterns, due to additional lights entering from SD screen. The BB signals showed a rise and fall that largely corresponded to the SD signals, although several orders of magnitude smaller than SD. The SV signals showed a sharp increase at the onset of SD calibration events but remained more or less constant throughout the solar calibration event. The recorded calibration view signal levels during solar calibration could indicate the level of stray light contamination. Figure 3 shows that BB and SV have signals that are at least four orders of magnitude smaller than the SD signals during solar calibration. Unfortunately, we cannot directly measure how much of the SD signals are stray light, which will cause bias in calibration. The signal ratios of SV/SD and BB/SD imply the potential stray light contamination during solar calibration of less than 0.01%. Unless there is a large wavelength stray light dependency, the results indicate that the reflective solar bands calibration error due to stray light contamination should be negligible.

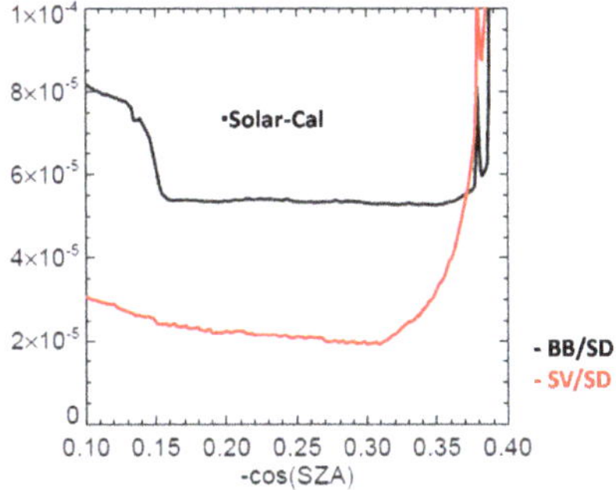

Figure 3. Mean ratios of BB/SD and SV/SD signals around solar calibration events on August 25, 2014. The dashed lines indicate the approximated solar calibration angle range. SZA = solar zenith angle.

3. Stray Light Characterization and Correction

The most prominent stray light contamination on DNB imagery occurs in the mid-to-high latitude of the northern and southern hemispheres when spacecraft is crossing the day/night terminators [16]. The stray light magnitude and its affected latitudes depend on the Earth-Sun-spacecraft geometric relationships, and the patterns usually reoccur at a yearly cycle when the similar Earth-Sun-spacecraft geometric relationships repeat. To remove the Earth scene stray light, the stray light is estimated based on the observed EV dark signals. Currently two different methods are implemented in the publicly-available DNB imagery products [16,17]. In this section, we examine the stray light characteristics based on the relationship of observed EV stray light and calibration view signals to explore ways to improve current stray light corrections. We will demonstrate that the information in calibration view data can be used to modify current stray light correction method to remove the additional stray light features that occur during southern hemisphere springtime.

3.1. Northern Hemisphere

Figure 4 shows the typical EV dark signals (d–f) in the northern hemisphere DNB EV stray light-affected regions and the corresponding calibration views signals (a–c). In the EV plots (d–f) the estimated stray light and mean signals after correction are also plotted in red and blue, respectively. Figure 4 shows that both EV and calibration view signals increased sharply in the penumbra region, when the spacecraft is transitioning from the dark to the bright part of the orbit. The recorded SV signals are similar to the EV dark signals both in patterns and magnitude before the Earth scene enters the twilight region. Since SNPP is on a descending orbit and has a declination angle of ~100 degrees, the Earth scene at end of scan (EOS) will observe the twilight region earlier than the beginning of the scan (BOS). The result is shown in Figure 4 as the sharp EV signal increase due to twilight happening at a lower solar zenith angle (SZA) at BOS (Figure 4d) than at EOS (Figure 4f). The BB signals also showed similar patterns to SV and EV from penumbra to SZA $\approx$ 108 degrees ($-\cos(\text{SZA}) \approx 0.32$), then the signals were slightly elevated before sharp increases at SZA $\approx$ 100 degrees ($-\cos(\text{SZA}) \approx 0.18$). The signal pause at SZA $\approx$ 108 degrees coincided with EOS twilight onset angle, indicating the increased Earth shine might have contributed slightly to BB signals. Similar to SV, BB, and EV, the

SD signals show the initial sharp increase in the penumbra region, then a gradual decrease before increasing again at around BOS twilight onset angles when the Earth view entered the twilight region. The SD signals are at least an order of magnitude stronger than SV, BB, or EV, indicating the bulk of the signal likely is the reflected stray light off the SD surface.

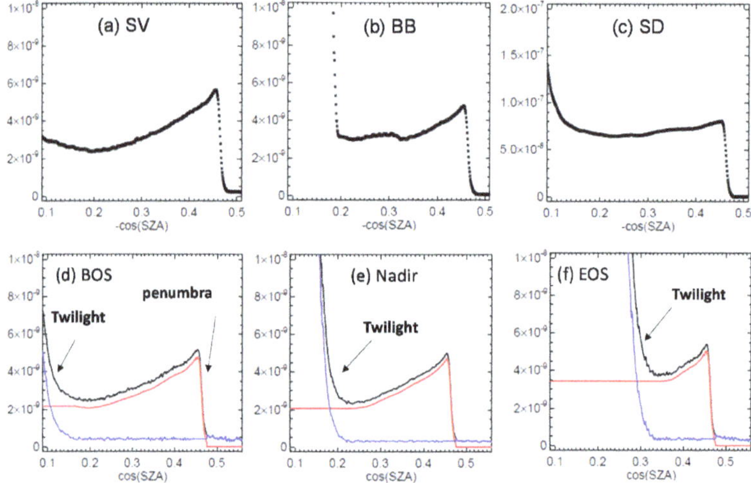

Figure 4. DNB detector 8 EV and calibration view dark signals near northern hemisphere stray light problem regions on October 5, 2013. (**a**–**c**) show calibration views. EV dark signals for beginning of scan (BOS), nadir, and end of scan (EOS) are shown in (**d**), (**e**), and (**f**), respectively. Black curves: dark signals; red curves: estimated stray light; blue curves: signals after stray light correction. SZA = solar zenith angle.

The current operational stray light correction methods approximate the stray lights as the EV dark signals minus the airglow (red curves in Figure 4d–f) approximated by the EV dark signals observed outside of the stray light affected regions [16,17]. In thre northern hemisphere, this method performed well in removing the bulk of the stray light, except for the penumbra regions. Since the sharp stray light increases with respect to SZA in the penumbra region, even a small mismatch in the penumbra onset angles between the imaging and LUT generation time could increase noticeable image artifacts after correction [17]. The effects can be demonstrated in Figure 5, where the same image is corrected using correction LUTs derived from three different months. Figure 5 shows the corrected image using the current month's LUT has the best quality (Figure 5a), and the worst image (Figure 5b) was corrected using the prior month's LUT. To show the penumbra angle mismatch was the cause, we estimated the median penumbra angles in SV signal profiles (Figure 6). Based on the estimated penumbra angles, the potential angle bias between the stray light correction LUTs and the scene in Figure 5 is about 0.07 degrees for (b) and 0.02 degrees for (c). The larger penumbra angle bias results in worse image quality after correction. Figure 6 shows that the estimated penumbra regions had an annual cycle with a magnitudes of ~0.2 degrees. The repeatable annual cycle indicates a prior year's correction LUT will usually have a better match in penumbra angle than the prior month's correction LUT. Without angular adjustment in correction LUT, a prior year's correction LUT should perform better in forward operational processing when the most current month's correction LUT is not available. The bias in penumbra angles could be further reduced if the yearly angular drifts are adjusted. For reprocessing when all historical correction LUTs are available, it is possible to temporally interpolate the penumbra onset angles to further reduce the correction residuals due to the minor penumbra angular drifts.

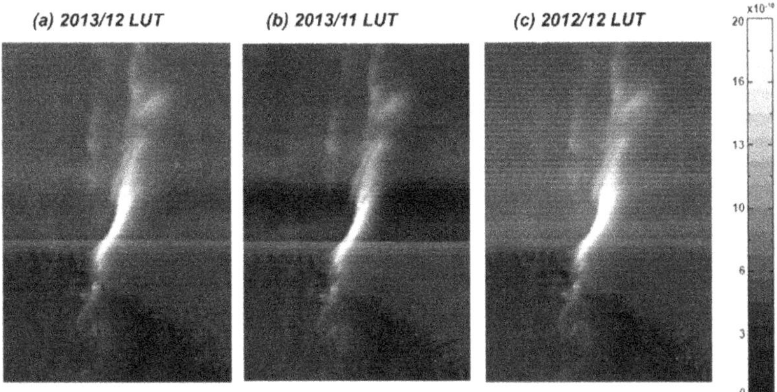

Figure 5. Stray light corrected images for December 3, 2013, 7:43 GMT, using LUTs derived from (**a**) 2013 December; (**b**) 2013 November; and (**c**) 2012 December data.

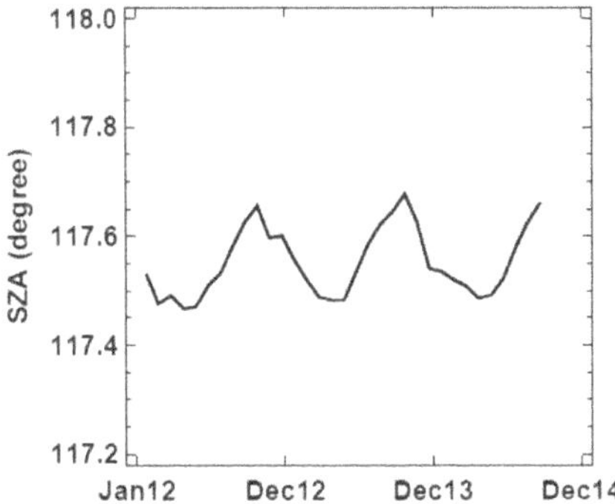

Figure 6. Median penumbra angle estimated using DNB SV data. The angle is estimated as the maximum slope in SV signal profile.

3.2. Southern Hemisphere

Figure 7 shows the EV and calibration view signal levels around the DNB EV stray light affected-regions in the southern hemisphere. In the penumbra region, EV and calibration view signals show an increase similar to what was observed in the northern hemisphere, but with a smaller magnitude. The BB and SV also showed similar patterns and magnitude as EV signals before the onset of solar calibration when additional solar illumination is coming through the SD screen. At the onset of solar calibration, both SV and EV signals showed a sudden jump that coincides with a signal jump in SD (Figure 7c). After the signal jump, the SV signal is about one order of magnitude larger than the EV before it transitions into the twilight region. At the beginning of the solar calibration event, the SD signal gradually increased to reflect the increase in solar illumination on SD. The BB signals show a similar pattern to SD with a magnitude of four to five orders smaller.

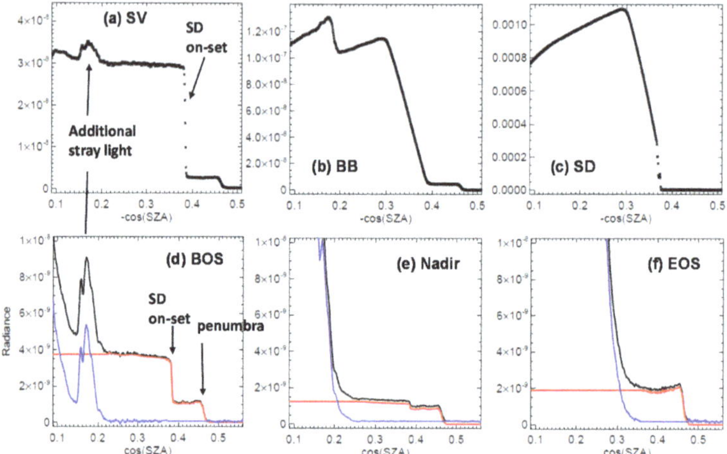

Figure 7. DNB detector 8 EV and calibration views dark signals near southern hemisphere stray light problem regions on October 5, 2013. (**a–c**) show calibration views. EV dark signals for beginning of scan (BOS), nadir, and end of scan (EOS) are shown in (**d–f**), respectively. Black curves are computed dark signals, red curves are estimated stray light, and blue curves are the signals after stray light correction. SZA = solar zenith angle.

Figure 7 data is taken when additional stray light features occurred in the southern hemisphere. Based on the current correction method, the large correction residual can be seen in Figure 8, which shows that the additional stray light obscures the underlying scene structure. The additional stray light occurred near the twilight region and can be traced back to the elevated SV, BB, and EV dark signals (Figure 7a,b,d). The mean corrected EV dark signal in Figure 7d (blue curve) shows the additional stray light remains within the corrected signals. For comparison, the SV, BB, and EV dark signals for September, 2013 (Figure 9) did not show the corresponding additional stray light features in SV. The current method relies on extrapolation to estimate the stray light in the twilight region due to the difficulty of separating stray light signals from Earth radiance [16,20]. The current method had been proven to work well roughly nine months of the year when there are no additional stray light features. The blue curve in Figure 9 shows no obvious artifact after correction. However, a different method would be required to remove this additional stray light feature.

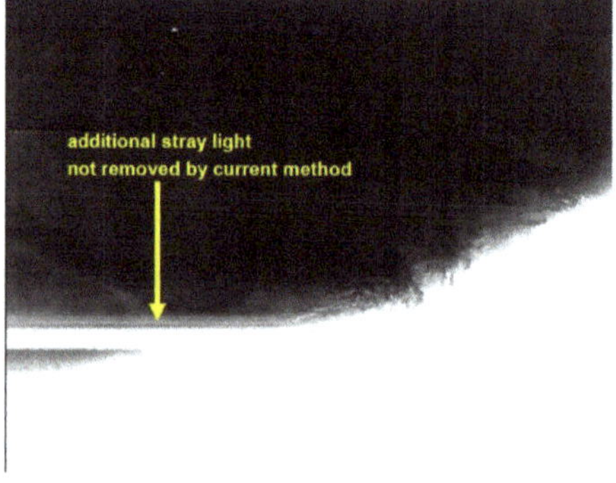

Figure 8. DNB image from October 5, 2013, 5:06 GMT, corrected by the current method. The residual artifact from additional stray light can be seen as a large swath across the image.

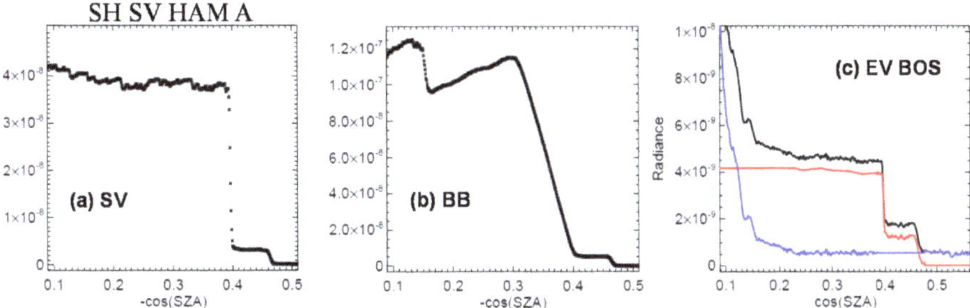

Figure 9. DNB detector of eight dark signals near the southern hemisphere stray light problem regions on September 5, 2013 for (**a**) SV; (**b**) BB; and (**c**) EV at BOS. Black curves are computed dark signals, red curves are estimated stray light, and blue curves are the signals after stray light correction. SZA = solar zenith angle.

To estimate the additional stray light contamination, the current correction method is modified in the twilight region. First, the angular extent of the additional stray light is estimated from SV signals (Figure 7). Then, the expected EV dark signals are estimated based on an exponential fit using data outside of the additional stray light region. The additional stray light is then approximated as the difference between the computed EV dark signals and the approximated EV dark signals. The stray light estimates outside of the new regions identified by SV data remain the same as the current method, in which extrapolated values were used. Based on this updated method, Figure 10 shows the computed EV dark signals, estimated EV stray light, and the expected EV dark signals after stray light correction. Compared with the current method (Figure 7), the updated method was able to estimate the bulk of the additional stray light even within the twilight region. Notice that the estimated additional stray light features were at the onset of twilight at BOS, completely within the twilight at nadir, and nonexistent at EOS, as the additional stay light becomes insignificant to EV signals.

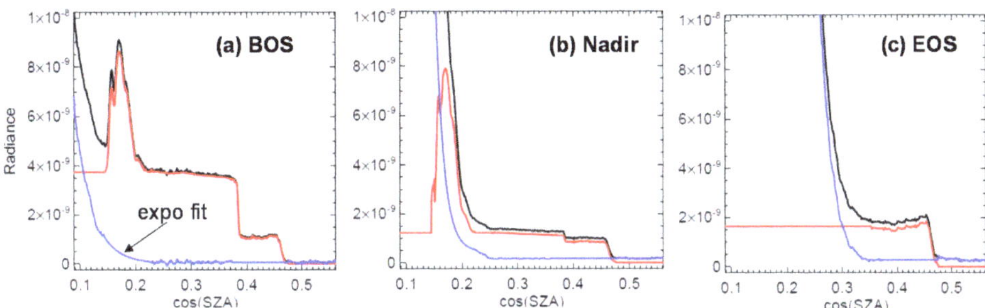

Figure 10. DNB EV dark signals near the southern hemisphere stray light problem regions on October 5, 2013. (**a–c**) show EV dark signals for beginning of scan (BOS), nadir, and end of scan (EOS), respectively. Black curves are computed dark signals, red curves are estimated stray light based on the updated method, and blue curves are the signals after stray light correction. SZA = solar zenith angle.

To test the updated method, we applied the updated stray light estimates to Figure 8 and generated the updated DNB stray light corrected image for comparison. The updated DNB stray light corrected image (Figure 11) shows that the bulk of the additional stray light is removed with some correction residuals. Compared with the current image (Figure 8), the updated image reveals additional features under the prior additional stray light contamination region with the well-defined twilight transition across the image. The additional stray light region slices through the twilight transition zones, which complicates the stray light estimation due to the overlapping of stray light and twilight signals. As shown in Figure 7, the additional stray light is clearly identified at BOS, but is partially embedded in the twilight at nadir. At EOS, the additional stray light becomes insignificant when compared with the Earth scene radiance. The result indicates that the updated correction method successfully entangled the stray light features from the twilight, as the corrected image (Figure 11) did not show strong scan angle-dependent residual biases.

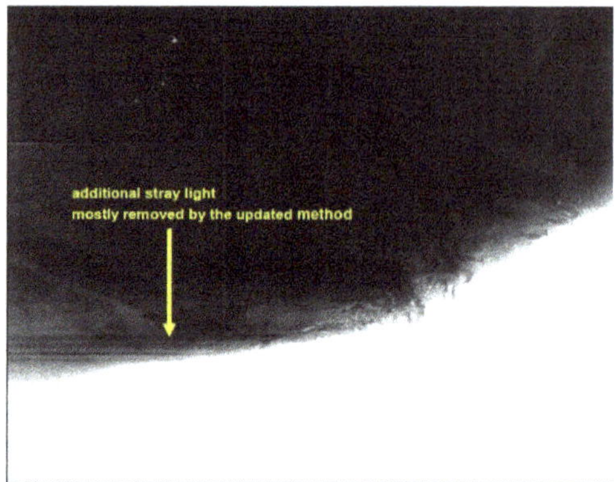

Figure 11. DNB image from October 5, 2013, 5:06 GMT, corrected by the updated method. The residual artifact from additional stray light is significantly reduced.

The southern hemisphere stray light correction experienced a similar penumbra angle mismatch issue discussed earlier in the northern hemisphere stray light correction. Furthermore, a much larger angle mismatch could come from the onset of the SD angle as its temporal drift is much larger. To track the temporal drift, we estimated the onset of the SD angle using the signal jumps in the SV data (see Figure 7a). The estimated SD onset angles (Figure 12) show an annual oscillation of about two degrees, an order of magnitude larger than the penumbra's angular cycle. The large angular drift indicates the importance in selecting stray light correction LUT to avoid a significant correction error due to bias in the SD onset angles. Figure 13 shows the image corrected using the prior month's LUT has significant overcorrection (black strip in Figure 13b) due to the bias in the SD onset angle. The image corrected using the prior year's LUT (Figure 13c) showed a more comparable result with correction performed with the current month's LUT (Figure 13a).

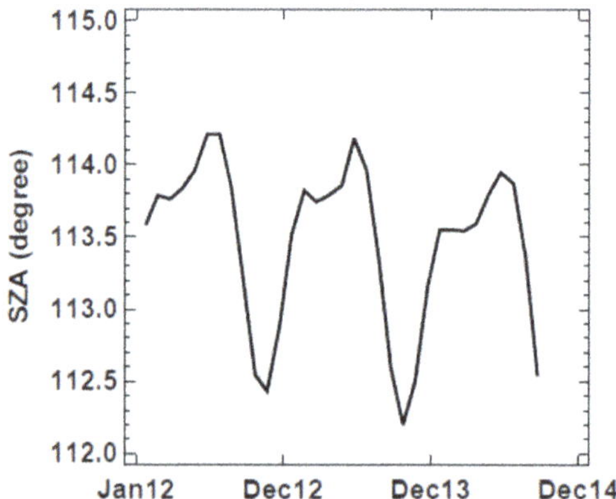

Figure 12. SD onset angle estimated using DNB SV data.

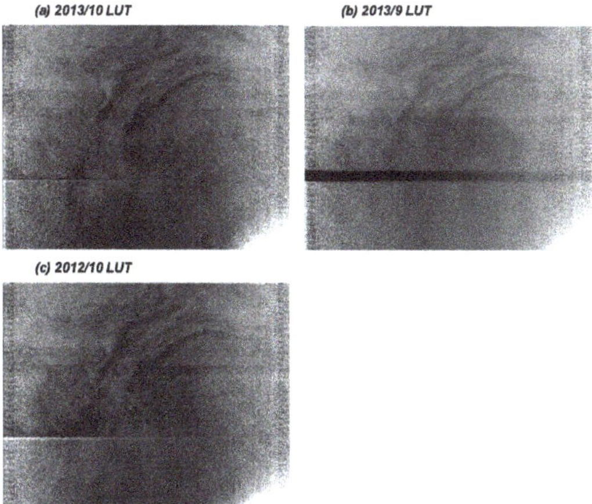

Figure 13. Stray light corrected images for October 5, 2013, 1:41 GMT, using LUTs derived from (**a**) 2013 October; (**b**) 2013 September; and (**c**) 2012 October data.

3.3. Low Latitude Stray Lights

Occasionally, the DNB images could also be contaminated by stray light in the lower latitude regions. Unlike the persistent and predictable mid-to-high latitude DNB stray light previously described, the low latitude stray light appears to occur without predictable periodicity. Figure 14 shows an image in Southeast Asia from December 28, 2014, and striping can be observed in the upper left part of the image. The striping occurred between 10 and 15 degrees north, which is far beyond the reach of the known mid-to-high latitude DNB stray light. Analysis of the calibration view signals showed an elevated SV signal that is several times higher during the time when the striping occurred in the Earth view image (Figure 15). The BB and SD signals are not shown here as no additional signals were observed. The correlation between SV signals and low latitude EV striping features were also found in many other instances (results not shown here). The result indicates a possible stray light source near SV as a cause of the striping feature. One probable source could be the Moon, which is at its first quarter phase and near the SV view angle when the elevated signals occurred. A source near SV could also explain the striping features being most prominent near the BOS, and no additional signals are observed in BB and SD data.

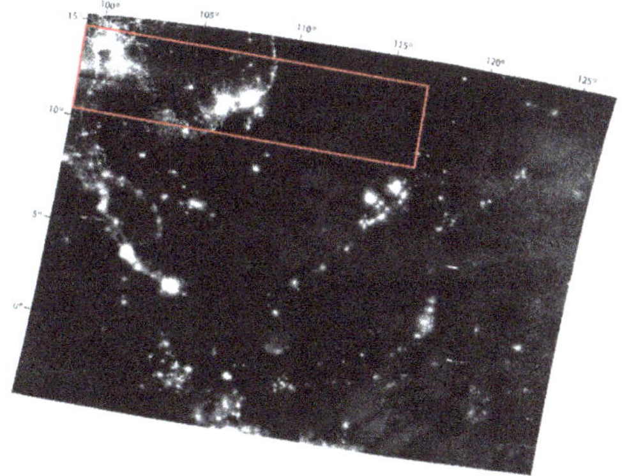

Figure 14. SNPP DNB image from December 28, 2014, 18:06 GMT. The red box indicates the approximated areas with contaminations.

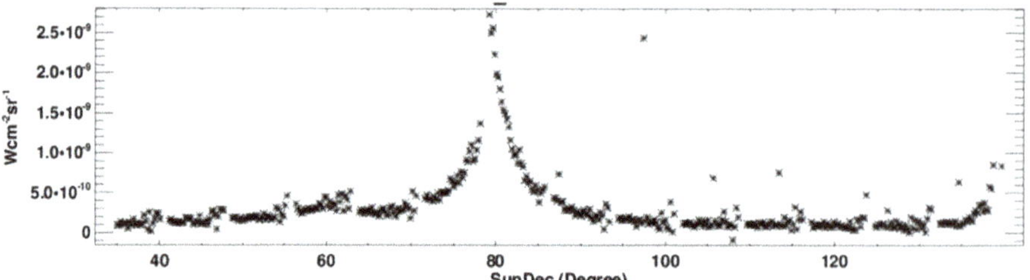

Figure 15. DNB HGA scan averaged SV signals plotted against spacecraft Sun declination angle (SunDec) for orbit 16,420 on December 28, 2014. The signal increase arounda Sun declination angle of 80 degrees corresponded to the timing when EV striping occurred at the beginning of the scan in Figure 14.

3.4. Background Stray Light from the Moon

The operational VIIRS DNB EV dark offset is determined once a month via a special operation [25]. The dark offset is determined using data collected over the dark ocean during each month's new moon. Studies have shown the dark ocean scenes are not sufficiently dark to set the DNB HGS offset for the nighttime due to airglow and the biases that could be up to 8×10^{-10} W$\cdot$ cm$^{-2}\cdot$ sr^{-1} [15,20]. One solution is to use an airglow-free dark reference derived from the spacecraft pitch maneuvers' deep space scene and track the dark offset drift from the BB signals [17]. However, the BB nighttime signal trending (Figure 16) showed periodic bumps that are found to be correlated with Moon illumination. To remove the periodic stray light from the Moon, the dark drifts are fitted using the BB data near new moon, and the fitted values are used to adjust dark offset derived from pitch maneuver data [17].

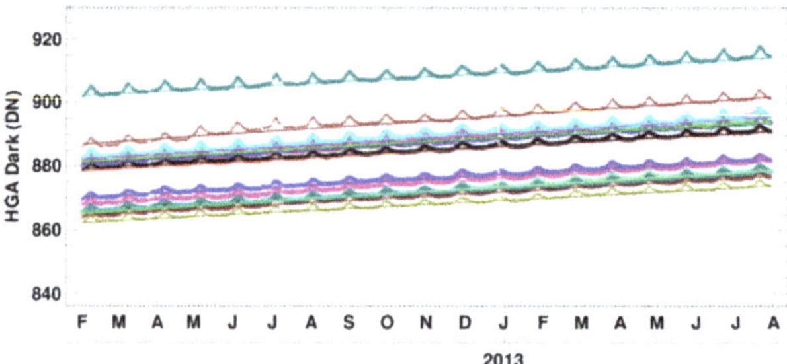

Figure 16. DNB BB daily mean dark signal DN for HGA, aggregation mode 1. The measured values for each detector are represented by symbols. The lines are fitted values using new moon data. The periodic increase in DN corresponded to the increase in lunar illumination. At full moon, the dark response is up to 5 dn (~ 8×10^{-11} W$\cdot$ cm$^{-2}\cdot$ sr^{-1}) higher than the new moon.

Although this method provided a dark offset that is free from Moon contamination, it is uncertain if, and how much of, the Moon stray light affects EV observation. Based on the stray light analysis and its corresponding EV and calibration views signal features presented earlier, the stray light signals in BB suggest similar effects in EV data. If the Moon stray light observed in BB is, indeed, elevating background signal levels for EV observations, then the EV dark offset should be adjusted based on Moon illumination. The EV and BB relationships presented earlier (Figures 4 and 7) indicated that the EV stray light is at a similar level of BB signal. In Figure 16, the Moon stray light observed in BB is less than 8×10^{-11} W$\cdot$ cm$^{-2}\cdot$ sr^{-1}, which is just below the usable DNB radiance of 10^{-10} W$\cdot$ cm$^{-2}\cdot$ sr^{-1} [15], and much smaller than the typical airglow. If the Moon stray light causes similar background signal increase in EV as in BB, the impact on DNB imagery quality will be minimal. The potential impact would only be in the airglow over dark surfaces, e.g., oceans, as most other surfaces will have radiance much higher than the Moon stray light.

4. Conclusions

In this paper we analyzed the Soumi NPP VIIRS DNB calibration view data to better understand DNB stray light characteristics. Our analyses show that although the calibration view data provided indirect measurements of Earth view stray light, the calibration view signal features contain useful information for improving current stray light correction and identifying additional stray light features.

In the current stray light correction, matching the onset angles in regions with sharp signal changes, *i.e.*, penumbra and solar diffuser calibration, are critical to avoid large errors. We found that the temporal changes in the penumbra and onset of solar diffuser calibration angles can be tracked by space view data. The angle differences between the creation of stray light correction LUT and the corrected scene can be adjusted to reduce image artifacts after correction. Both penumbra and solar diffuser calibration onset angles are affected by the Earth-Sun-spacecraft geometric relationship, as the long term trends show yearly cycles with a small long-term drift. The yearly repeated onset angles indicating a yearly recycled correction LUT should perform reasonably well with the current scene, thus improving stray light correction in forward operational correction when the current month's correction LUT is not available. However, some angle adjustment in the recycled LUT could be made to account for the long-term onset angular drifts, especially for the solar diffuser calibration onset region as its temporal change is much larger than the penumbras.

One limitation in the current stray light correction method is the inability to remove an additional stray light feature that occurs during the southern hemisphere's springtime. The additional stray light region overlaps with the Earth scene twilight signals but can be clearly identified in the calibration view data. Based on the calibration view data estimated angular extents of this additional feature, we modified how stray light is estimated when the additional stray light feature occurred. A comparison of DNB images generated by the current and updated methods showed the updated method removed most of the additional stray light and revealed portions of the Earth's surface previously obscured by stray light.

The analysis of calibration view signal characteristics also indicates possible stray light contamination outside of the currently-known stray light problem regions. The stray light-like features occasionally observed in low-latitude Earth scenes were linked to the elevated signals in the space view. The cause of this low-latitude stray light is unknown, although the Moon might be a possible source. Moon illumination was also found to cause a small increase in the calibration view's background signals which, potentially, could also affect the Earth view observations.

Finally, the VIIRS DNB stray light is a complex issue currently without clearly identified root causes. As the DNB stray light problem will continue for the follow-on VIIRS missions for the next couple of decades, the full potential of DNB will depend on how well the stray light can be characterized and corrected. In this study, we showed the potential of using calibration view data to improve Earth view stray light characterization and correction. We believe that further studies on operational methods to incorporate time-dependent onset angles in stray light correction, refinement of the additional stray light correction in the twilight regions, and quantifying other stray light sources are critical to continue improving DNB radiometric calibration accuracy in the future.

Acknowledgments: The authors would like to thank the anonymous reviewers for their constructive comments and suggestions, which helped improve our paper. Thanks to Samuel Anderson and Julia Gutin for their editorial assistance. The views, opinions, and findings contained in this paper are those of the authors and should not be construed as official positions, policy, or decisions of the NOAA or the U.S. Government.

Author Contributions: Shihyan Lee designed the study, developed the methodology, performed the analysis and wrote the manuscript. Changyong Cao provided the direction of the study, oversaw the progress and contributed to the writing and revising of the manuscript.

Conflicts of Interest: The authors declare no conflict of interest.

References

1. Mills, S. *VIIRS Radiometric Calibration Algorithm Theoretical Basis Document*; Doc. No.: D43777; Northrop Grumman Aerospace Systems: Redondo Beach, CA, USA, 2010.
2. Hillger, D.; Kopp, T.; Lee, T.; Lindsey, D.; Seaman, C.; Miller, S.; Solbrig, J.; Kidder, S.; Bachmeier, S.; Jasmin, T.; *et al.* First-light imagery from Suomi SNPP VIIRS. *Bull. Amer. Meteor. Soc.* **2013**, *94*, 1019–1029. [CrossRef]
3. Southwell, K. Remote sensing: Night lights. *Nature* **1997**, *390*. [CrossRef]
4. Imhoff, M.L.; Lawrence, W.T.; Elvidge, C.D.; Paul, T.; Levine, E.; Privalsky, M.V.; Brown, V. Using nighttime DMSP/OLS images of City lights to estimate the impact of urban land use on soil resources in the United States. *Remote Sens. Environ.* **1997**, *59*, 105–117. [CrossRef]
5. Imhoff, M.L.; Lawrence, W.T.; Stutzer, D.C.; Elvidge, C.D. A technique for using composite DMSP/OLS "City Lights" satellite data to map urban area. *Remote Sens. Environ.* **1997**, *61*, 361–370. [CrossRef]
6. Elvidge, C.; Imhoff, M.L.; Baugh, K.E.; Hobson, V.R.; Nelson, I.; Safran, J.; Dietz, J.B.; Tuttle, B.T. Nighttime lights of the world: 1994–1995. *ISPRS J. Photogramm. Remote Sens.* **2001**, *56*, 81–99. [CrossRef]
7. Lawrence, W.T.; Imhoff, M.L.; Kerle, N.; Stutzer, D. Quantifying urban land use and impact on soils in Egypt using diurnal satellite imagery of the earth surface. *Int. J. Remote Sens.* **2002**, *23*, 3921–3937. [CrossRef]
8. Elvidge, C.D.; Baugh, K.E.; Kihn, E.A.; Kroehl, H.W.; Davis, E.R. Relation between satellite observed visible-near infrared emissions, population, and energy consumption. *Int. J. Remote Sens.* **1997**, *18*, 1373–1379. [CrossRef]
9. Elvidge, C.D.; Baugh, K.E.; Kihn, E.A.; Kroehl, H.W.; Davis, E.R. Mapping of city lights using DMSP operational linescan system data. *Photogramm. Eng. Remote Sens.* **1997**, *63*, 727–734.
10. Cao, C.; Bai, Y. Quantitative analysis of VIIRS DNB nightlight point source for light power estimation and stability monitoring. *Remote Sens.* **2014**, *6*, 11915–11935. [CrossRef]
11. Miller, S.D.; Mills, S.P.; Elvidge, C.D.; Lindsey, D.T.; Lee, T.F.; Hawkins, J.D. Suomi satellite brings to light a unique frontier of nighttime environmental sensing capabilities. *Proc. Natl. Acad. Sci. USA* **2012**, *109*, 15707–15710. [CrossRef] [PubMed]
12. Cao, C.; Shao, X.; Uprety, S. Detecting light outages after severe storms using the S-SNPP/VIIRS day/night band radiances. *IEEE Trans. Geosci. Remote Sens.* **2013**, *10*, 1582–1586. [CrossRef]
13. Roman, M.; Stokes, E.C. Holidays in lights: Tracking cultural patterns in demand for energy services. *Earth's Future* **2015**. [CrossRef]
14. Millers, S. Satellite Sensor Reveals Earth's Nocturnal Secrets. Available online: http://www.scientificamerican.com/article/miller-satellite-sensor-reveals-earth-s-nocturnal-secrets/ (accessed on 1 February 2016).
15. Liao, L.B.; Weiss, S.; Mills, S.; Hauss, B. Suomi SNPP CIIRS day-night band on-orbit performance. *J. Geophys. Res. Atmos.* **2013**, *118*, 12707–12718. [CrossRef]
16. Mills, S.; Weiss, S.; Liang, K. VIIRS day/night band (DNB) stray light characterization and correction. *Proc. SPIE* **2013**. [CrossRef]
17. Lee, S.; Sun, C.; Chiang, K.F.; Xiong, X. An overview of NASA VIIRS Day-Night Band (DNB) on-orbit radiometric calibration. *Proc. SPIE* **2014**. [CrossRef]
18. Lian, L.; Weiss, S.; Liang, C. DNB Performance. 2013 December SNPP SDR Review. Available online: http://www.star.nesdis.noaa.gov/star/documents/meetings/SNPPSDR2013/dayTwo/Liao_DNBPerf.pdf (accessed on 1 February 2016).
19. Mills, S.; Miller, S.D. VIIRS Day-Night Band (DNB) calibration methods for improved uniformity. *Proc. SPIE* **2014**. [CrossRef]
20. Lee, S.; Chiang, K.F.; Xiong, X.; Sun, C.; Samuel, A. The S-SNPP VIIRS day-night band on-orbit calibration/characterization and current state of SDR products. *Remote Sens.* **2014**, *6*, 12427–12446. [CrossRef]
21. Lee, T.F.; Nelson, S.C.; Dills, P.; Riishojgaard, L.P.; Jones, A.; Li, L.; Miller, S.; Flynn, L.E.; Jedlovec, G.; McCarty, W.; *et al.* NPOESS: Next-generation operational global earth observations. *Bull. Amer. Meteor. Soc.* **2010**, *91*, 727–740. [CrossRef]
22. Lee, S.; McIntire, J.; Oudrari, H.; Schwarting, T.; Xiong, X. *A New Method for Suomi-NPP VIIRS Day Night Band (DNB) On-Orbit Radiometric Calibration*; Calcon: Logan, UT, USA, 2013.
23. Lee, S.; McIntire, J.; Oudrari, H.; Schwarting, T.; Xiong, X. A new method for Suomi-NPP VIIRS day-night band on-orbit radiometric calibration. *IEEE Trans. Geosci. Remote Sens.* **2015**, *53*, 324–334.

24. Rausch, K.; Houchin, S.; Cardema, J.; Moy, G.; Hass, E.; De Luccia, F.J. Automated calibration of the Suomi National Polar-Orbiting Partnership (S-SNPP) Visible Infrared Imaging Radiometer Suite (VIIRS) reflective solar bands. *J. Geophys. Res. Atmos.* **2013**, *118*, 13434–13442. [CrossRef]
25. Geis, J.; Florio, C.; Moyer, D.; Rausch, K.; De Luccia, F.J. VIIRS day-night band gain and offset determination and performance. *Proc. SPIE* **2012**, *8510*, 1–12.

Article

Assessing the Effects of Suomi NPP VIIRS M15/M16 Detector Radiometric Stability and Relative Spectral Response Variation on Striping

Zhuo Wang [1,*] **and Changyong Cao** [2]

1 CICS (Cooperative Institute for Climate and Satellites), University of Maryland, 5825 University Research Court, Suite 4001, Room 3045, M-Square, College Park, MD 20740, USA

2 NOAA (National Oceanic and Atmospheric Administration)/NESDIS (National Environmental Satellite, Data, and Information Service)/STAR (Center for Satellite Applications and Research), NCWCP, E/RA2, 5830 University Research Ct., College Park, MD 20740, USA; changyong.cao@noaa.gov

* Correspondence: Zhuo.Wang@noaa.gov; Tel.: +1-301-683-3553

Academic Editors: Jose Moreno, Richard Müller and Prasad S. Thenkabail
Received: 20 October 2015; Accepted: 2 February 2016; Published: 16 February 2015

Abstract: Modern satellite radiometers have many detectors with different relative spectral response (RSR). Effect of RSR differences on striping and the root cause of striping in sensor data record (SDR) radiance and brightness temperature products have not been well studied. A previous study used MODTRAN radiative transfer model (RTM) to analyze striping. In this study, we make efforts to find the possible root causes of striping. Line-by-Line RTM (LBLRTM) is used to evaluate the effect of RSR difference on striping and the atmospheric dependency for VIIRS bands M15 and M16. The results show that previous study using MODTRAN is repeatable: the striping is related to the difference between band-averaged and detector-level RSR, and the BT difference has some atmospheric dependency. We also analyzed VIIRS earth view (EV) data with several striping index methods. Since the EV data is complex, we further analyze the onboard calibration data. Analysis of Variance (ANOVA) test shows that the noise along track direction is the major reason for striping. We also found evidence of correlation between solar diffuser (SD) and blackbody (BB) for detector 1 in M15. Digital Count Restoration (DCR) and detector instability are possibly related to the striping in SD and EV data, but further analysis is needed. These findings can potentially lead to further SDR processing improvements.

Keywords: Suomi NPP; VIIRS; striping; LBLRTM; detector level Relative Spectral Response (RSR); striping index; atmospheric dependency; noise; stability

---

1. Introduction

Suomi National Polar Orbiting Partnership (S-NPP) spacecraft was successfully launched on 28 October 2011 with the Visible Infrared Imaging Radiometer Suite (VIIRS), which provides capacities for operational environmental remote sensing for weather, climate and other environmental applications. In contrast to conventional imaging radiometers [1], a single scan for VIIRS M bands (16 detectors aligned in the along-track direction) has a slower scan rate than that of the traditional single detector, therefore the spatial resolution is enhanced without losing the signal-to-noise (SNR). However, "bow tie" deletion and imaging striping occurs as a trade-off to this multi-detector arrangement that must be dealt with [1]. An anomalous striping pattern has been observed in SNPP VIIRS sea surface temperature products [2–4]. These striping are assumed to be caused by differences in the detector-level RSR. Currently SST EDR team developed an adaptive destriping algorithm to improve the operational SST imagery, but this destriping algorithm does not solve the problem at the root cause. The VIIRS SDR team performed studies to investigate possible root causes.

Padula and Cao [4] used MODTRAN to evaluate the brightness temperature difference caused by the difference between band averaged (which is used in the operations) and detector level SRF. However, some studies [5] indicate that the forward model MODTRAN does not reproduce spectral, angular, and water vapor dependencies with accuracies acceptable for SST analyses. In the current study, we analyze the striping using atmospheric radiative transfer models LBLRTM to support assessment of VIIRS TEB calibration variations due to atmospheric effects.

The SST EDR group reported that some striping patterns were observed in SST product, which is possibly caused by the difference between detector-level and band-averaged SRF for some thermal emissive bands (TEB). Since the radiance in the SDR operational process is retrieved using the band-averaged instead of the detector level SRF, the difference will affect the derived brightness temperature and therefore SST. The level 2 daytime VIIRS SST environmental data is retrieved from a Non-linear Split Window algorithm from VIIRS bands M15 and M16 [2,5]:

$$SST = a_0 + a_1 \cdot BT_{15} + a_2\,(BT_{15} - BT_{16})\;SST_{guess} + a_3\,(BT_{15} - BT_{16})\;(\sec\theta - 1) \tag{1}$$

where a_i (i = 0, 1, 2, 3) is the coefficient derived from regression, BT_{15} and BT_{16} are the observed brightness temperature in VIIRS bands M15 and M16. SST_{guess} is the simulated first guess SST, and θ is the sensor zenith angle. This algorithm will amplify the striping caused by $BT_{15} - BT_{16}$ due to the multiplication with the factor $a_2\ SST_{guess}$ greater than 1.0 over the tropical ocean, but not always the case for the ocean over the polar region, and propagates it to the SST products [2,5]. Coincidently, most SST striping are reported for the tropical regions. From the water vapor absorption spectrum [6], we can see that the water vapor absorption in band centered at 12 μm (VIIRS M16) is larger than that in band centered at 10.8 μm (VIIRS M15), so M16 is more sensitive to water vapor absorption, *i.e.*, the atmospheric impact on the BT variation is more obvious than that in M15. The brightness temperature in M16 is colder than that in M15 due to more water vapor absorption in M16.

In this study, Section 2 introduces the methodology. Section 3 describes the results from radiative transfer model. Section 4 analyzes satellite earth view observation and onboard calibration data. Conclusion and discussion are given in Section 5.

2. Methodology

The detector-level radiance simulated from a Radiative Transfer Model, measured Suomi NPP hyperspectral radiance data and onboard calibration data are analyzed to investigate the impacts of detector-level SRF difference on striping, possibility of atmospheric dependency, as well as detector stability.

2.1. Radiative Transfer Model

In a previous study [4], the detector-level radiance and brightness temperature were simulated using MODTRAN [7], which is a "narrow band model" atmospheric radiative transfer with a spectral resolution of 1 cm^{-1} for TEB bands. In comparison, the Line-By-Line Radiative Transfer Model (LBLRTM) is a more accurate and flexible radiative transfer model that can be used over the full spectral range from the microwave to the ultraviolet, providing the foundation for many radiative transfer applications [8,9]. LBLRTM calculations in the thermal infrared bands are recognized as a reference standard for intercomparisons of radiative transfer models [10]. LBLRTM has been widely used for a number of years as the foundation for retrieval algorithms. It is also used to train fast radiative transfer models used in Numerical Weather Prediction (NWP) assimilation systems and the Optimal Spectral Sampling model [11] implemented in the Joint Center for Satellite Data Assimilation (JCSDA) Community Radiative Transfer Model (CRTM).

LBLRTM can provide much finer spectral resolution such as at 0.01 cm^{-1}, which is more accurate for validation purposes. High spectral resolution is critical to identify and separate the causes of RSR effects from other possible causes. In this work, we use the LBLRTM v12.2, which was released in October 2012. It uses the AER line parameter database (hereafter AER3.2) on the basis of the HITRAN 2008 line parameter [12]. The Voigt line shape is used at all atmospheric levels with an algorithm according to a linear combination of approximating functions. Line coupling in LBLRTM is modeled using the first-order perturbation approach [13]. LBLRTM incorporates the continuum model MT_CKD [14], which includes self- and foreign-broadened water vapor continua as well as continua for CO_2, O_2, N_2, O_3, and extinction due to Rayleigh scattering.

To cover a range of environment conditions for evaluating TEB calibration and SST striping, we selected sites under various atmospheric conditions to perform simulations using LBLRTM, which can characterize atmospheric effects on SNPP VIIRS TEB calibration due to variability in atmospheric conditions. The results from LBLRTM will be compared with that from MODTRAN, which can serve as cross-check among models and help better determine the VIIRS detector level calibration biases due to effects of atmospheric radiative transfer.

2.2. Detector Level RSR

VIIRS is a whiskbroom scanning radiometer with a large scan angle which covers 112.56° at the nominal altitude of 829 km. It has many detectors with slightly different Relative Spectral Response (RSR). However, the impact of RSR difference among detectors on imagery artifacts, as well as geographical retrieval uncertainties has not been well studied until recently. Figure 1 shows detector-level and band averaged RSR function in M15. Band M16 includes two detector arrays A and B with time delay integration (VIIRS Radiometric Calibration ATBD, 2014). M16A and M16B figures are similar to that of M15, so they are not shown here. From the figure, we can see that RSR is slightly different among 16 detectors, which will affect the radiance and hence the brightness temperature. The impact of the detector level variation on the imagery artifacts will be analyzed in Sections 3 and 4.

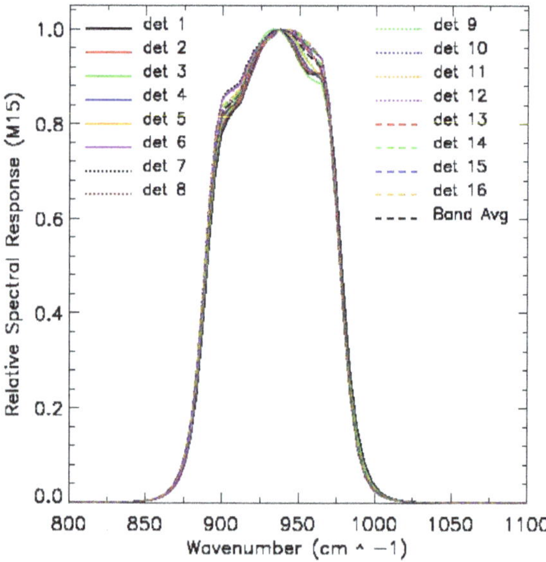

Figure 1. Detector-level and band-averaged relative spectral response (RSR) in M15. Note: M16A and M16B figures are similar, not shown here.

2.3. Data Processing

LBLRTM v12.2 is used in this study to simulate the top of the atmosphere (TOA) radiance in the spectral range (722, 2650.0) cm^{-1} with spectral resolution of 0.01 cm^{-1}. Six standard LBLRTM atmospheric profiles, including Tropical, Mid-Latitude Summer (MLS), Mid-Latitude Winter (MLW), Sub-Arctic Summer (SAS), Sub-Arctic Winter (SAW), and U.S. standard 1976 were used to ensure a broad range of simulated atmospheric conditions. The surface temperatures of 300 K, 290 K, 273 K, 280 K, 263 K, and 290 K were used for the above six atmospheres respectively.

The LBLRTM output TOA spectral radiance is then convolved with S-NPP VIIRS relative spectral response (RSR) to get the in-band averaged radiance mW/($m^2 \cdot sr \cdot cm^{-1}$):

$$L_{avg} = L(\nu_0, T) = \frac{\int_{\nu 1}^{\nu 2} L(\nu) \cdot RSR(\nu)\, d\nu}{\int_{\nu 1}^{\nu 2} RSR(\nu)\, d\nu} \tag{2}$$

where ν is the wavenumber, $L(\nu)$ is the at sensor radiance and $RSR(\nu)$ is the RSR at a given band. The computed in-band radiance (L_{avg}) can be converted to effective brightness temperature (T_{eff}) by using the radiance to brightness temperature Look Up Tables. The original VIIRS LUT converts radiance in wavelength units to temperature. However, the spectral radiance output from LBLRTM is in the unit of W/($m^2 \cdot sr \cdot cm^{-1}$), so we used the temperature from VIIRS LUT to compute the spectral radiance in wavenumber unit using the Planck function and then convolve with spectral RSR to get the in-band radiance in wavenumber unit in order to match the unit from the model output which is in per wavenumber. By doing so, the interpolation is used to get finer spectral radiance *versus* BT LUT.

The blackbody radiance is computed from the Planck function:

$$L(\nu, T) = \frac{c_1 \nu^3}{e^{c_2 \nu / T} - 1} \tag{3}$$

where $L(v,T)$ is the blackbody radiance (W/$m^2 \cdot sr \cdot cm^{-1}$), c_1 = 1.191042×10^{-8} (W/$m^2 \cdot sr \cdot cm^{-1}$), c_2 = 1.4387752 (K cm), v is the wavenumber (cm^{-1}) and T is the blackbody temperature (K).

In this study, we can see that the expected striping is caused by the difference in effective brightness temperature (ΔT_{eff}) between the detector-level and band averaged RSR:

$$\Delta T_{eff} = T_{eff\,(\mathrm{det}\,RSR)} - T_{eff\,(avg\,RSR)} \tag{4}$$

where $T_{eff(\mathrm{det}\,RSR)}$ and $T_{eff(avg\,RSR)}$ are the effective brightness temperature computed using detector-level and band averaged RSR, respectively.

The striping due to detector level RSR difference can be defined as the difference in ΔT_{eff} between two contiguous detectors i and i+1:

$$striping\ (i, i+1) = \Delta\ T_{eff}\ (i) - \Delta\ T_{eff}\ (i+1) \tag{5}$$

To evaluate the atmospheric dependencies, the simulated clear-sky radiance was processed using Equation (2) and detector level RSR to obtain the in-band radiance in M15 and M16. The radiance was then converted to effective brightness temperature for detector-level ($T_{eff(\mathrm{det}\,RSR)}$) and band averaged ($T_{eff(avg\,RSR)}$) by using Look Up Tables. The brightness temperature difference ((ΔT_{eff}) was then computed from Equation (4) and the results are analyzed in next section.

3. Model Results and Discussion

Based on the above equations, Figure 2 shows the effective temperature difference between detector level and band averaged RSR for six LBLRTM atmospheres in M15, M16, and M15 − M16. As shown on the top left panel of Figure 2, in M15, we found that there is a small but obvious atmospheric dependency. The odd/even detector pattern is observed, especially for detectors 1 to 8. The smallest BT difference is at detector 5. The magnitude of variation is 0.01 K for tropical atmosphere, and 0.025 K for subarctic atmosphere. In order to see how the out-of-band response affects the temperature difference, we extend the spectral range from (800, 1100) cm^{-1} to the entire range (800, 1333.33) cm^{-1} in M15 and from (769, 950) cm^{-1} to (769, 1250) cm^{-1} in M16. After including the out-of-band response, the pattern in the temperature difference does not change much. The results (not shown) are similar to those of Figure 2. It is noted that the results from LBLRTM is similar to those in a previous study with MODTRAN [4].

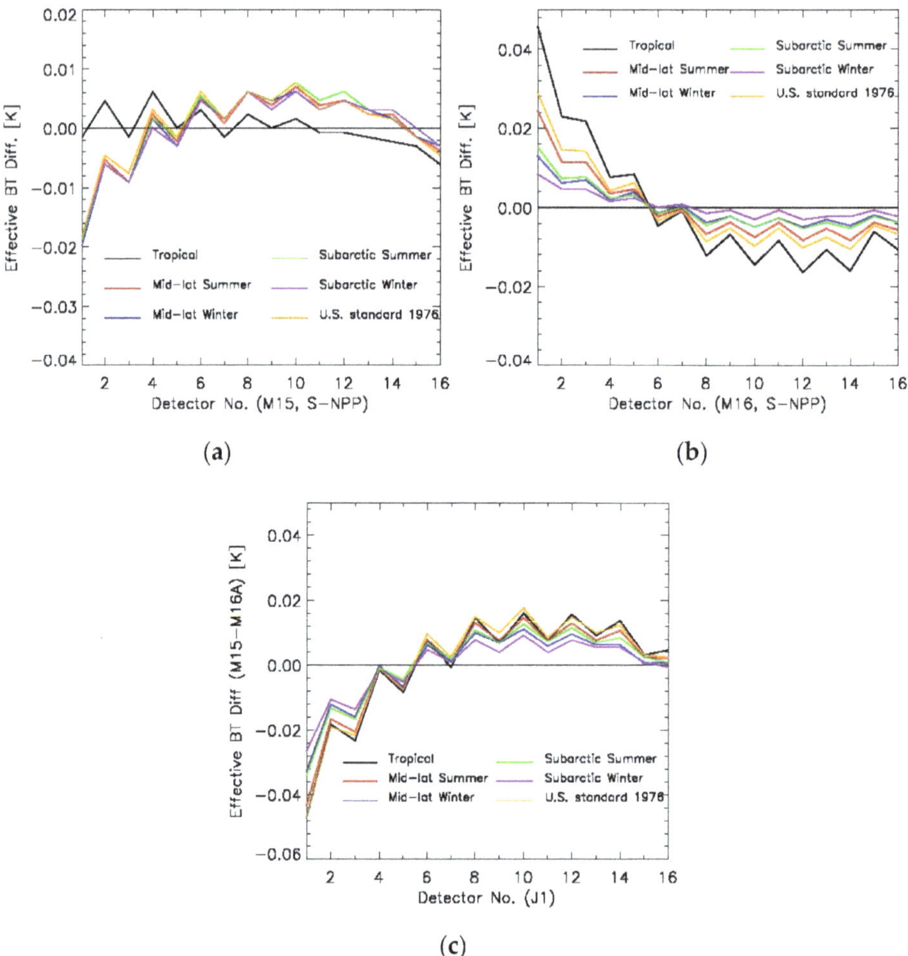

Figure 2. Effective temperature difference between detector level and band averaged RSR for six line-by-line radiative transfer model (LBLRTM) atmospheres in M15 (**a**); M16 (**b**, the average of M16A and M16B); and M15 − M16 (**c**).

M16 is computed from the average of M16A and M16B and is shown in top right panel of Figure 2. There is more obvious atmospheric impact on BT difference than in M15, and the tropical atmosphere pattern has the largest variation. The magnitude of variation is 0.063 K for tropical atmosphere, and 0.022 K for subarctic atmosphere. We also observed apparent odd/even detector pattern. Figure 2b shows that in M16 band, detectors 1 and 2 deviate most from the band average than other detectors. The BTs for detectors 1–5 are larger than the band averaged value, but BTs for detectors 6, and 8–16 are smaller than the band average value. In Figure 2c, the BT differences ($BT_{15} - BT_{16}$) from detectors 1–5 are smaller than the band average value, but they are larger than the band average for detectors 6–16. For detector 4 to 16, even if Sub-arctic Summer has higher temperature and more water vapor than Mid-latitude winter (see Table 1 [15]), but it doesn't have higher variation, so striping is not always atmospheric dependent. Therefore, water vapor and temperature may not be the only cause for striping, and other instrument factors may also be involved, which was further explored later in the paper. After extending to the entire spectral range (800, 1333.33) cm^{-1} in M15 and (769, 1250) cm^{-1} in M16 to include the out-of-band response, the pattern is similar. Since the term ($BT_{15} - BT_{16}$) is used in the VIIRS SST retrieval algorithm (Equation (1)), the difference between detector-level and band averaged BT in M15 − M16 should also be analyzed. The bottom panel of Figure 2 shows that the magnitude of variation in M15 − M16 is larger than that in single band, for example, they are 0.066 K and 0.063 K for tropical atmosphere in M15 − M16 and M16, respectively.

Table 1. The dependence of atmospheric water vapor in the MODerate resolution atmospheric TRANsmission (MODTRAN) standard cases [15].

Name	Columnar Liquid Water Equivalent (mm)
1976 Standard Atmosphere	17
Tropical	48
Mid-latitude summer (MLS)	35
Mid-latitude winter (MLW)	11
Sub-arctic summer (SAS)	25
Sub-arctic winter (SAW)	5

The ΔT in tropical atmosphere has larger magnitude (0.066 K) than subarctic (0.051 K). Compared to the band average, detectors 1–3 show larger atmospheric effect, with detector 1 showing the largest difference up to 0.046 K for tropical case, which is close to the 0.05 K in previous study [4] using MODTRAN model.

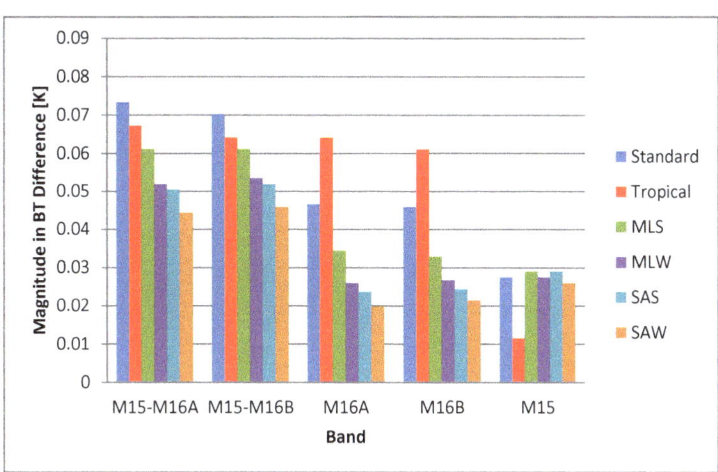

Figure 3. The magnitude of brightness temperature difference (from Equation (4)) between using detector level and band averaged RSR for six LBLRTM atmospheres.

Figure 3 provides a summary of the magnitude in brightness temperature difference (Equation (4)) between using detector level and band averaged RSR from LBLRTM output for six atmospheres. It is shown that the magnitude in M15 − M16 is larger than that in single band. Comparing with M16 and M15 − M16, M15 is much less affected by atmosphere. Except for M15 and standard atmosphere, the magnitude has obvious atmospheric dependency: tropical region has much larger magnitude than subarctic region. We also noticed that although with larger temperature and liquid water content than MLW, SAS has a little bit smaller magnitude than MLW in M15 − M16 and M16. Therefore, besides water vapor and temperature, other instrument effects may also play a role in striping.

In general, the result from radiative transfer model indicates that LBLRTM does show atmospheric dependency although the difference is very small and is hard to validate. The water vapor has impact on the striping pattern in the satellite images. For example, in M16, the BT differences between using the detector-level and band averaged RSR are the largest for tropical atmosphere, and is the smallest for subarctic winter. The warm and moist atmosphere shows larger atmospheric impact on BT difference than the cold and dry atmosphere. The atmospheric impact is more obvious in band difference (M15 − M16) than for a single band. For example, the magnitude of ΔT_{eff} for tropical atmosphere is 0.066 K, 0.063 K, 0.010 K in M15 − M16, M16, and M15, respectively. The magnitudes of ΔT_{eff} for tropical and subarctic atmosphere are 0.066 K and 0.051 K in M15 − M16, and are 0.063 K and 0.024 K in M16. Compared to the band average, detector 1−3 shows large atmospheric effect, with detector 1 showing the largest difference up to 0.046 K for tropical case, which is close to the 0.05 K in previous study by Padula and Cao [4] using MODTRAN model. The impact of water vapor on the BT difference variation can be observed. In addition, other instrument effects may exist. The real satellite observation data are further analyzed in the next section to see whether the water vapor is a dominant factor affecting the striping pattern in the following sections.

4. Striping in VIIRS Earth Observation Data

In order to investigate the relationship between water vapor and striping in VIIRS brightness temperature images, The VIIRS SDR brightness temperature data for bands M15 and M16 are analyzed in sample cases from 2012 to 2014. Table 2 shows analysis of six cases over the "uniform" clear sky ocean surface near tropical and polar region. A major step forward in this study is that we experimented quantifying striping using different methods with more data samples.

Table 2. Six cases used in this study.

Cases	Granule
Tropical case 1: Bay of Bengal	d20130619_t0746444
Tropical case 2: Bay of Bengal	d20140622_t0753301
Tropical case 3: Bay of Bengal	d20140703_t0748470
Polar case 1: Alaska	d20140520_t2158272
Polar case 2: Alaska	d20140603_t2237573
Polar case 3: Polar	d20150421_t1802552

4.1. Earth View Striping Quantification Using Cumulative Histogram Method

For each case, we selected a small uniform region with size of 60 pixels along scan direction, and 16 scan × 16 detector along track direction under clear sky condition according to VIIRS Cloud Mask Intermediate Product. Although the striping patterns can be observed from these images, it needs to be quantified with an index (named VSI). We used the cumulative histogram defined in previous studies [16,17] for striping quantification:

$$H_{i,\,HAM}(k) = \frac{1}{N_{i,\,HAM}} \sum_{l=0}^{k} \left(\sum l \in (l,\, i,\, HAM) \right) \quad (6)$$

Where the first sum is to count the number of pixels with the value l (for detector i and HAM side A or B), and the second sum is over the pixel value l. In Equation (5), l can be BT or BT difference

($BT_{15} - BT_{16}$), which depends on the striping index whether you want to quantify a single band or the band difference. The parameter $N_{i,\ HAM}$ in Equation (6) is the total number of the pixels in an image for the detector i and HAM side A or B. Here we didn't separate the HAM side because we focus on the detector level RSR difference instead of HAM side difference in this study. H_i refers to the percentage of the pixels with value less than k in an image. If there is striping pattern in an image, the histogram H_i diverges for different detectors. The divergence of the histogram can be represented as the horizontal distances among the different histograms:

$$g_{i,\ i''}(P) = k - k' \quad (7)$$

where P is the percentage of the pixels with the value less than the value in X-axis. X-axis represents the brightness temperatures (BT) or BT difference ($BT_{15} - BT_{16}$).

The larger distance $g_{i,i'}(P)$ among different histogram corresponds to the stronger striping effects. We have analyzed the difference among 16 detectors (without separating HAM sides) for each small region.

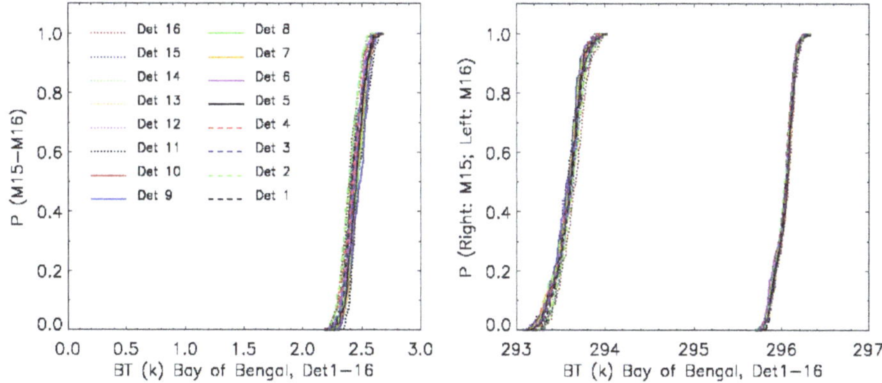

Figure 4. The cumulative histogram for Bay of Bengal over tropical region on 19 June 2013 in M15 − M16 (**left**), as well as M15 and M16 (**right**).

In Figure 4, the Y-axis is the percentage of the pixels with the value less than the value in X-axis. In left panel, X-axis represents the brightness temperatures difference ($BT_{15} - BT_{16}$). In the right panel, the values in X-axis are BT_{15} or BT_{16}. Each line is for one detector. In this figure, we can see that the horizontal distance between histograms is almost a constant. Due to the water vapor absorption difference between M15 and M16, $BT_{15} - BT_{16}$ can also represent the water vapor content. In M15 − M16, the maximum horizontal distance at 50% percentage $g_{50\%}$ is 0.093 K, and the BT difference range is 0.5 K in X-axis. The following relative magnitude, *i.e.*, the ratio of horizontal distance to the X-axis range, is better used to compare among different bands:

$$R = \frac{g\,(P = 50\%)}{Range_{X-axis}} \quad (8)$$

The ratio in M15 − M16, M15 and M16 are 0.187, 0.067 and 0.107, respectively. M15 − M16 has larger ratio than single band. These horizontal distances are larger than those from LBLRTM calculations by 0.027 K, 0.03 K, and 0.037 K in M15 − M16, M15, and M16, respectively. The large variation mainly comes from detector 1.

To better understand the atmospheric effect on striping, we have also analyzed the cases over the polar region. For example, Figure 5 shows the cumulative histogram over Gulf of Alaska on 20 May 2014 in M15 − M16 (left), as well as M15 and M16 (right). In general, the horizontal distance is very small and almost a constant. The ratios are 0.149, 0.044, and 0.015 in M15 − M16, M15 and M16, respectively, which are all smaller than those over topical cases. The histogram divergence in M15 − M16 is very close to the magnitude from LBLRTM, but is larger than LBLRTM magnitude by 0.023 K for M15 and less than LBLRTM by 0.005 K for M16. Polar region has smaller BT difference than tropical region due to smaller water vapor effect.

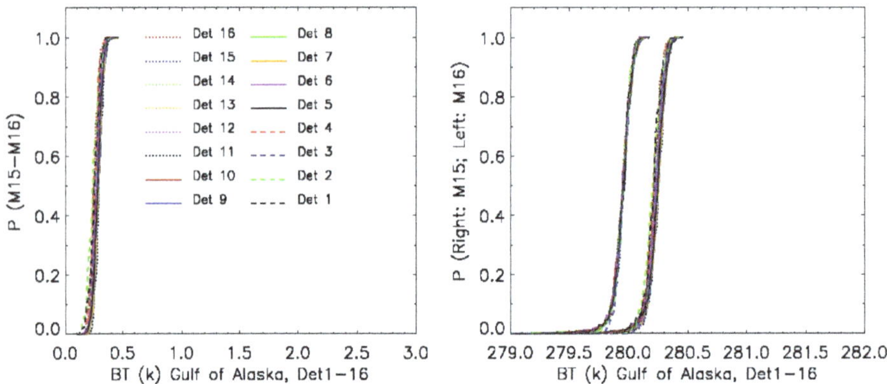

Figure 5. Cumulative histogram over the Gulf of Alaska on 20 May 2014 in M15 − M16 (**left**), as well as M15 and M16 (**right**).

In this study, we have performed analysis on several cases, including three cases over tropical region and three cases over polar region (see Table 2). Figure 6 compares the magnitude of temperature variation among 16 detectors from VIIRS observation data and LBLRTM over tropical and polar region. In M15 − M16 and M16, both LBLRTM and VIIRS observation show larger temperature difference over tropical than over polar region, which means that the tropical region is much more affected by water vapor than polar region because the higher BT (larger than 2 K) difference in tropical region means more water vapor content. In most cases, VIIRS observation has larger magnitude in BT difference among different detectors than LBLRTM except for polar case 3 in M15 − M16. The maximum magnitude differences between VIIRS observation and LBLRTM for tropical cases are 0.028 K, 0.039 K, and 0.06 K in M15 − M16, M16, and M15, respectively. However, the magnitude difference between VIIRS observation and LBLRTM for polar case is less than 0.026 K.

In order to see whether the striping pattern is apparent in an image, the temperature range needs be considered, so the relative magnitude, *i.e.*, the ratio of BT variation magnitude to temperature range, is compared over tropical and polar region (Figure 7). The results indicate that M15 − M16 has smaller temperature variation range than single band (M15 or M16), so the larger relative magnitude represents more obvious striping pattern in M15 − M16 image. The striping pattern in the image of $BT_{15} - BT_{16}$ is more obvious than that in M15 or M16 because the signal to noise ratio (SNR) in the band difference image is larger than that in a single band. In most cases, the ratio in tropical region is larger than that over polar region except for tropical case 1. However, observation data has large variability and the variation cannot be effectively validated. In addition to water vapor, other factors such as instrument stability possibly affect the striping pattern.

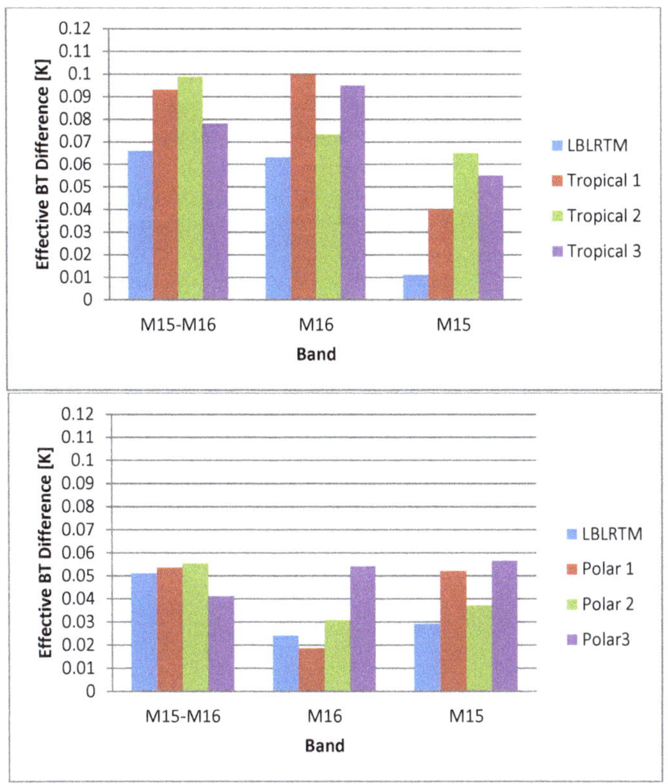

Figure 6. The magnitude of brightness temperature variation among 16 detectors for Visible Infrared Imaging Radiometer Suite (VIIRS) observation data and LBLRTM over tropical (**top**) and polar (**bottom**) region for six cases.

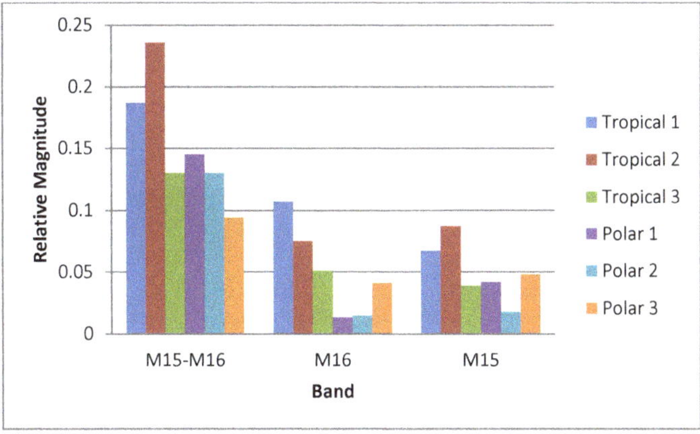

Figure 7. Comparison of the relative magnitude (*i.e.*, ratio of brightness temperature variation magnitude to temperature range) over tropical and polar region.

So far the striping pattern has been analyzed using LBLRTM and VIIRS observation data. Figure 8 compares the averaged effective temperature difference from LBLRTM and VIIRS observation. The results from observation are the averaged over several cases for both tropical and polar atmospheres. In M15 – M16 and M16, both LBLRTM and VIIRS observation show larger temperature difference in tropical than in polar region. In most cases, VIIRS observation has larger magnitude in temperature difference among different detectors than LBLRTM except for polar case in M15 – M16. For tropical atmosphere, VIIRS observation has larger magnitude than LBLRTM by 0.024 K in M15 – M16 and 0.025 K in M16. For polar atmosphere, the difference between observation and LBLRTM is 0.005 K in M15 – M16 and 0.009 K in M16. In general, the magnitude of variation among 16 detectors over tropical region is much more affected by water vapor than that over polar region, *i.e.*, larger for high BT difference $BT_{15} - BT_{16}$ (high water vapor absorption). Therefore, the water vapor has an impact on the striping pattern, but it is not the only factor.

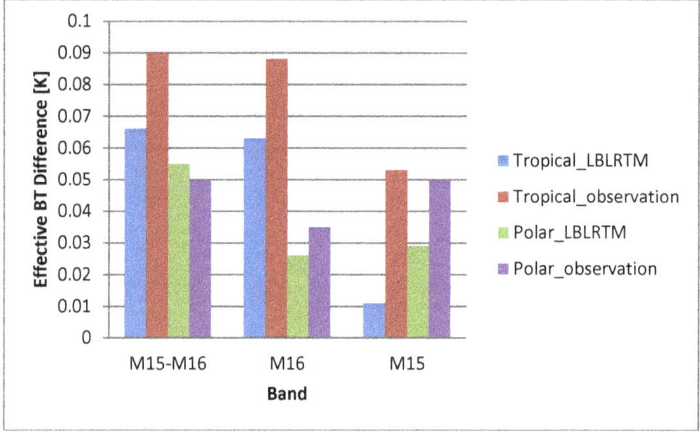

Figure 8. Comparison of effective temperature difference between LBLRTM and the VIIRS observation for tropical and polar cases. Note: VIIRS observation uses the average of several cases to represent the tropical and polar case.

Based on the analysis above, we can see that LBLRTM results are nearly identical to the previous study using MODTRAN [4]. We found that results from MODTRAN are repeatable using LBLRTM. Previous study does not completely solve the problem because there are still some other potential factors related to striping issue, therefore we will look further on detector stability.

Alternatively, we have also used the definition of VIIRS striping index (VSI) as the ratio between the mean along-track variance (V_{along_track}) and the mean along scan variance (V_{along_scan}).

The results show that the VSI is very sensitive. It decreases from 5 to 0.4 as the number of samples increases from 10 to 60, and increases from 4 to 10 when the number of scans increases from 6 to 10. This method is further improved later for onboard data analysis.

When the mean changes with the orbit time, the Allan deviation [18] will be an appropriate method to describe the spread distribution of the measurement essentially. The Allan deviation is used to compute the variance along track and along scan direction, and then use the variance ratio of along track to along scan to represent the VSI. Comparing with the standard deviation method, this Allan deviation is not sensitive to the number of samples, but it is also not an ideal method to separate tropical and polar cases in striping quantification because there is no obvious difference in VSI between tropical and polar cases using this method.

In general, we have used three different methods, including cumulative histogram, standard deviation, and Allan deviation, to quantify VSI in earth view (EV) observation data. Among them, the cumulative histogram method is the best to use without strict limitations. The Allan deviation method is not very sensitive to the sample size of subset but is hard to separate the tropical and polar cases for striping quantification. The standard deviation method is sensitive to the sample size of earth view data, but it will be further improved for onboard data analysis in the following section. The earth view data is complex because the noise is mixed with signal, so we further analyze the onboard calibration data in the next section, which is spatially uniform and eliminates EV effect such as atmospheric effect.

4.2. VIIRS Striping Analysis Using Onboard Calibrated Data

The SDR radiometric calibration algorithms convert raw data record (RDR) from Earth View (EV) observations into SDR radiance, reflectance, and brightness temperature products using onboard calibrators. The VIIRS on board calibrators include Blackbody (BB), Solar Diffuser (SD), and Solar Diffuser Stability Monitor (SDSM). This two-point calibration approach uses two calibration points: the Onboard Calibration BB and SD as the high points for thermal emissive bands (TEB) and reflective solar bands (RSB), respectively, and Space View (SV) for offset subtraction for both bands. The instrument gain can be derived from VIIRS observations at these two calibration points, which provides the basis for the two-point non-linear calibration for TEBs. The BB temperature is accurately measured with six embedded BB thermistors. While majority of the blackbody radiance observed by a given band is emitted and its in-band spectral radiance calculated from the Planck function using BB thermistor temperatures, a small portion is due to thermal emission and reflection from several optical path and surrounding components: including the Rotating Telescope Assembly (RTA), Half Angle Mirror (HAM), scan cavity, and shield. In addition, the response *versus* scan angle effects (RVS) must also be considered [19].

A. Noise Analysis Using Raw Data from BB, SV, and SD

The detector noise based on BB, SV, and SD observations are analyzed. BB is a calibration source for TEB and it is spatially uniform. It can measure the noise of single detector along scan direction, and noise of different detectors along track direction. Each detector has its own gain, the change in gain and offset between scans, the Digital Count Restoration (DCR) or detector instability can lead to striping. DCR is the process to ensure that the lowest level signals never drop below the dynamic range of the Analog to Digital Converter (ADC).

The SD is originally designed as a calibration source for RSB. However, in this study, we use the SD as a calibrated target. The advantage is that it acts as a proxy for SST and can simulate the effect on SST. On the other hand, the SD temperature drifts and changes with time or orbit, so we will use the ratio of SD in HAM side A to B to remove this effect and perform analysis of variance. Now the standard deviation discussed before can be used for the SD radiance ratio. We have analyzed three continuous OBCIP files on 30 June 2015. Figure 9 indicates the variations of BB−SV (*i.e.*, dBB) and SD−SV (*i.e.*, dSD) along track and along scan directions for detector 1 in band M15. There are 48 samples and 72 scans for three granules. The patterns are more consistent along scan. The variation along track is much larger than that along scan, which suggests that it is an important factor causing the striping pattern.

The drifting trend of solar diffuser can also be shown clearly by comparing the pattern between BB−SV (*i.e.*, dBB) and SD−SV (*i.e.*, dSD) along track and along scan (Figure 10). For both dBB and dSD, the pattern within each scan is more consistent. The mean value of dSD increases with the time/orbit, which will be considered in the following analysis.

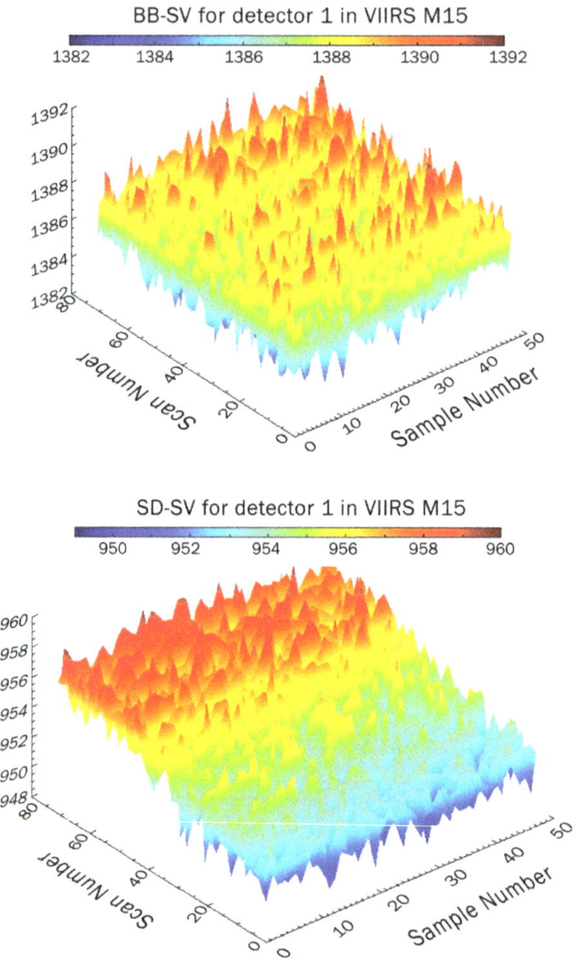

Figure 9. Variations of BB−SV (*i.e.*, dBB; **top**) and SD−SV (*i.e.*, dSD; **bottom**) along track and along scan for detector 1 in M15.

The VIIRS observations of the BB includes the signal from the blackbody emitted radiance at the temperature measured by the platinum resistance thermistors (PRT), as well as instrument radiometric noise. If we assume the BB target and environment to be stable, then the variation of dBB can be mostly due to random noise, detector and DCR instability, (gain is assumed stable short term). The random noise effect can be illustrated by the variation of dBB with 48 samples from each scan, and the effect of DCR and detector change can be observed by the variation of dBB in along track direction. As an example, three continuous granules (t0755_e0756, t0756_e0758, and t0758_e0759) on 30 June 2015 are analyzed. Figure 11 provides the effects of random noise and DCR in M15. The black dashed lines in the figure refer to the averaged value. In the bottom panel, the variation of dBB with scan for detector 1 and 1st sample along track shows larger deviation from the averaged value, and the STD is 1.12612. In the top panel, the STD for dBB of 48 samples along scan direction is about 0.93589, which means the DCR is a dominant factor causing variation in dBB.

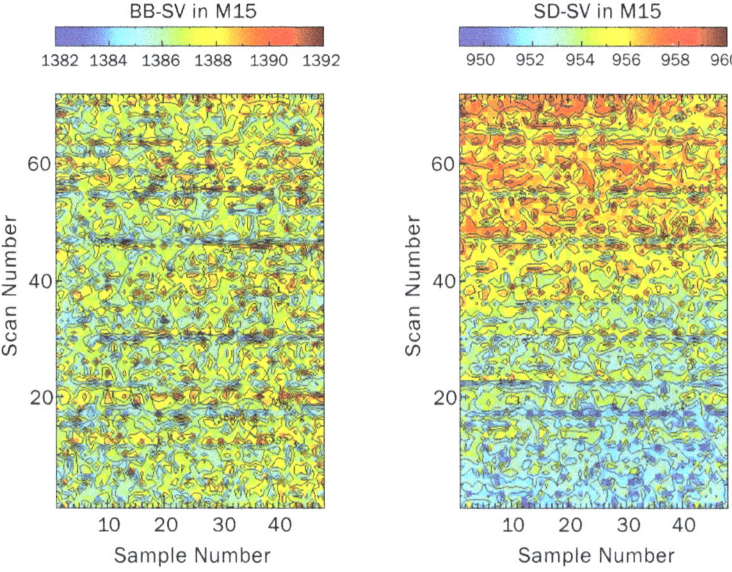

Figure 10. The distribution of BB−SV (*i.e.*, dBB; **left**) and SD−SV (*i.e.*, dSD; **right**) for detector 1 in M15. Color represents the value of data.

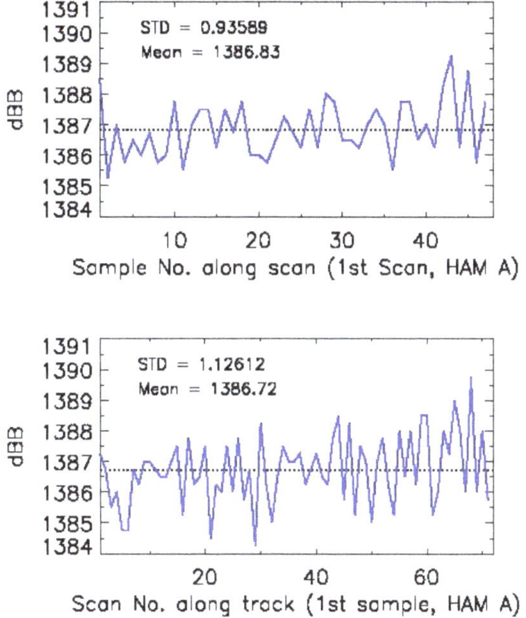

Figure 11. The variation of dBB with 48 samples for one scan (**top**) and the variation of dBB with scan for detector 1 and 1st sample along track (**bottom**) in M15 for three continuous granules (t0755_e0756, t0756_e0758, and t0758_e0759) on 30 June 2015. Dashed line refers to the averaged value.

Analysis of variance (ANOVA) uses the standard deviation method to separate the overall variance into variations among and between groups, and then compare the ratios to determine the dominant statistical effect. Previously we mentioned that the standard deviation method cannot be applied to SD radiance directly because it has trend over the orbital time. However, after removing the drift trend by taking the HAM side ratio, it removes the trend and now we can use the two-way ANOVA method to evaluate the variation in SD radiance ratio along track and along scan directions. This method is not used for EV data analysis because EV data is complicated. When considering the ratio of two HAM sides, each detector has 24 scans and 48 samples in one granule OBCIP file. As an example, we analyzed two continuous granules on 30 June 2015, then each detector totally has 48 scans ($r = nscan = 48$) and 48 samples ($c = nsample = 48$) in these two granules. The following equations are used for the variance analysis of the SD radiance ratio ($SD\_Ratio_{A,B}$):

The mean square between scans is

$$MS_R = \frac{SS_R}{r-1} = \frac{SS_R}{nscan-1} \tag{9}$$

The mean square between samples is

$$MS_C = \frac{SS_C}{c-1} = \frac{SS_C}{nsample-1} \tag{10}$$

The mean square for error is

$$MS_E = \frac{SS_E}{(r-1)\,(c-1)} = \frac{SS_E}{(nscan-1)\,(nsample-1)} \tag{11}$$

The ANOVA test ratio (V) along track and along scan are defined as

$$V_{along\_track} = \frac{MS_R}{MS_E} \text{ and } V_{along\_scan} = \frac{MS_C}{MS_E}$$

$$SS_R = \sum_{i=1}^{nscan} \frac{\left[\sum_{j=1}^{nsample} x_{i,j}\right]^2}{c} - \frac{\left[\sum_{i=1}^{nscan}\left(\sum_{j=1}^{nsample} x_{i,j}\right)\right]^2}{r\,c} \tag{13}$$

$$SS_C = \sum_{j=1}^{nsample} \frac{\left[\sum_{i=1}^{nscan} x_{i,j}\right]^2}{r} - \frac{\left[\sum_{i=1}^{nscan}\left(\sum_{j=1}^{nsample} x_{i,j}\right)\right]^2}{r\,c} \tag{14}$$

$$SS_E = \sum_{i=1}^{nscan}\left(\sum_{j=1}^{nsample} x_{i,j}^2\right) - \frac{\left[\sum_{i=1}^{nscan}\left(\sum_{j=1}^{nsample} x_{i,j}\right)\right]^2}{r\,c} - SS_R - SS_C \tag{15}$$

Where r is the number of scans and c is the number of samples in each scan. SS_R, SS_C, and SS_E are the sum of squares between scans, between samples, and from residual error, respectively. We have performed the ANOVA test for dBB ratio (left panels) and dSD ratio (right panels). In Figure 12, along track variation is larger than along scan variation in M15 (top panels) and M16 (bottom panels). The large variations in the along track direction is one of the major reasons for striping. In M15, detectors 1 and 2 have much larger along track variations and they are more noisy than other detectors, which is likely caused by detector instability. In M16, the detectors 9 and 12 have larger along track variation than other detectors. For along scan variation, the detector noises are at similar level.

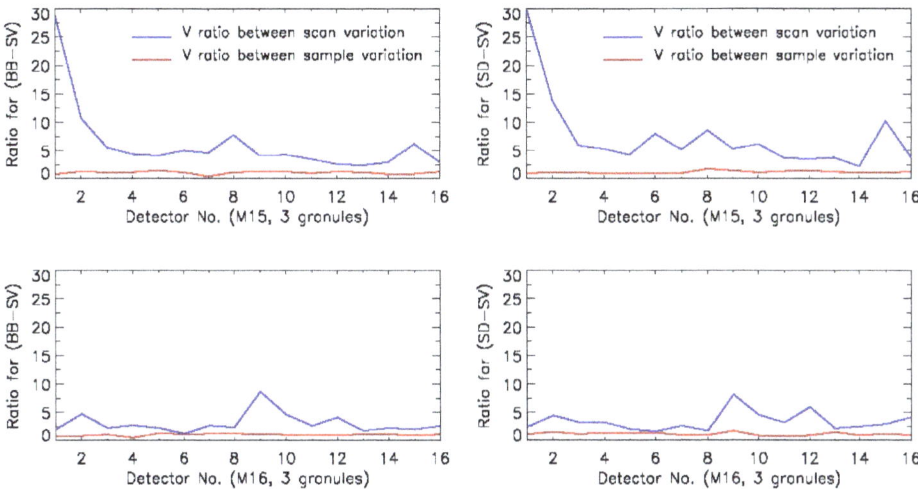

Figure 12. Two way Analysis of Variance (ANOVA) test ratio *versus* 16 detectors for dBB ratio (**left panels**) and dSD ratio (**right panels**) along tracks and along scans in M15 (**top panels**) and M16 (**bottom panels**).

B. Noise Analysis Using Solar Diffuser Thermal Radiance from Operational Calibration Algorithm

The SDR radiometric calibration algorithms convert raw data from Earth View observations into SDR radiance, reflectance, and brightness temperature products. In the operational calibration algorithm, the SD radiance is calculated using the following equation (similar to the equation for earth view) for VIIRS [19,20]:

$$L_{SD}(\theta_{SD}, B) = \frac{(1 - RVS(\theta_{SD}, B)) \cdot \left[\left(\frac{1}{\rho_{rta}(\lambda)} - 1 \right) \cdot L(T_{rta}, \lambda) - \frac{L(T_{ham}, \lambda)}{\rho_{rta}(\lambda)} \right] + F \cdot \sum_{j=0}^{2} c_j (T_{\det}, T_{elec}) \cdot dn_{SD}{}^{j}}{RVS(\theta_{SD}, B)} \tag{16}$$

where $L_{SD}(\theta, B)$is the band-averaged spectral radiance at the solar diffuser for scan angle θ. $RVS(\theta, B)$ refers to Response Versus Scan function at scan angle θ for band B. Here we assume it to be one because there is no RVS Look-Up-Table for the SD (no prelaunch measurement) in thermal infrared band. $\rho_{rta}(\lambda)$ is the spectral reflectance of RTA. $L(T, \lambda)$ is the Blackbody spectral radiance from Planck's function. $c_j(T_{\det}, T_{elec})$ is jth order coefficient of the response function after calibration update.

$$dn_{SD} = DN_{SD} - DN_{SV} \tag{17}$$

F is the factor for radiance coefficient which is computed from:

$$F = RVS(\theta_{obc}) \cdot \frac{\left\{ \begin{array}{c} \left(1 - \frac{1}{RVS(\theta_{obc})}\right) \cdot \left[\left(\frac{1}{\rho_{rta}(\lambda)} - 1 \right) \cdot L(T_{rta}(t), \lambda) - \frac{L(T_{ham}(t), \lambda)}{\rho_{rta}(\lambda)} \right] \\ + \varepsilon_{obc}(\lambda) \cdot L(T_{obc}(t), \lambda) + L_{obc\_rfl}(T_{sh}(t), T_{cav}(t), T_{tele}(t), \lambda) \end{array} \right\}}{\sum_{j=0}^{2} c_j \cdot dn_{obc}(t)^j} \tag{18}$$

where $\varepsilon_{obc}(\lambda)$ is the spectral emissivity of the BB. t is the start time of the acquisition period and the average counts per scan viewing the OBCBB is

$$dn_{obc}(t) = \frac{1}{N_{frame}} \sum_{m=0}^{N_{frame}} dn_{obc}\left(t + \Delta t_{frame}\right) \tag{19}$$

The number of frames per scan N_{frame} is 48 for the "M" band.

In Equation (18); $L_{obc\_rfl}(T_{sh}, T_{cav}, T_{tele}, \lambda)$ is the band-averaged OBCBB reflected radiance:

$$L_{obc\_rfl}(T_{sh}, T_{cav}, T_{tele}, \lambda) = \left[\begin{array}{c} F_{cav} \cdot (1 - \varepsilon_{obc}(\lambda)) \cdot L(T_{cav}, \lambda) + F_{sh} \cdot (1 - \varepsilon_{obc}(\lambda)) \cdot L(T_{sh}, \lambda) \\ +F_{tele} \cdot (1 - \varepsilon_{obc}(\lambda) \cdot L(T_{tele}, \lambda)) \end{array} \right] \tag{20}$$

Where the factors F_{sh}, F_{cav}, and F_{tele} represent the fraction of the reflectance off the OBCBB originating from the sources blackbody shield, cavity, and telescope [19,20].

To compute the SD thermal radiance, several Look Up Tables (LUTs) were used. The BB, SD, and SV observations from VIIRS On-board Calibrator Intermediate Product (OBCIP) and telemetry measurements are retrieved to compute SD radiance in TEB spectral bands (VIIRS M15 and M16).

In Figure 13, top panel shows SD radiance computed from the operational calibration algorithm (Equation (20)) for HAM A *versus* scan number for two continuous granules (t0755175_e0756417 and t0756429_e0758071) on June 30, 2015, and the bottom panel shows the ratio of SD radiance between HAM sides A and B in M15. From top panel, we found that SD radiance increases from detector 1 to detector 16, and there is a trend that SD radiance keep changing with scan number (or time) for different portion of orbit, so a normal standard deviation method cannot be used to analyze SD radiance directly. To remove the impact of Response *versus* Scan angle (RVS), we will analyze the ratio of SD radiance between HAM sides A and B ($SD\_Ratio_{A,B}$). Bottom panel illustrates that the changing trend with time in different portion of the orbit is removed after using the ratio of SD radiance between HAM side A and B. Therefore, we can utilize the normal standard deviation method to evaluate the SD radiance ratio ($SD\_Ratio_{A,B}$) directly.

The V ratios (defined in Equation (12)) computed from SD radiance ratio ($SD\_Ratio_{A,B}$) are compared for 16 detectors in M15 and M16 (Figure 14). We found that V ratio between scans (blue lines) is much larger than that between samples (red lines) for both of M15 and M16, which means that the variation between scans is one of the major reasons for the striping observed in VIIRS imagery. This figure also shows the noise level *versus* detectors along the track direction: detectors 1 and 2 have much higher noise levels than other detectors in M15. In M16, detectors 9 and 12 have about twice and one half higher noise level than other detectors. Within the scan direction, the V ratio values for all detectors are very similar for either M15 or M16. Therefore, the detector noise due to along track variation has higher chance causing the striping observed in the images.

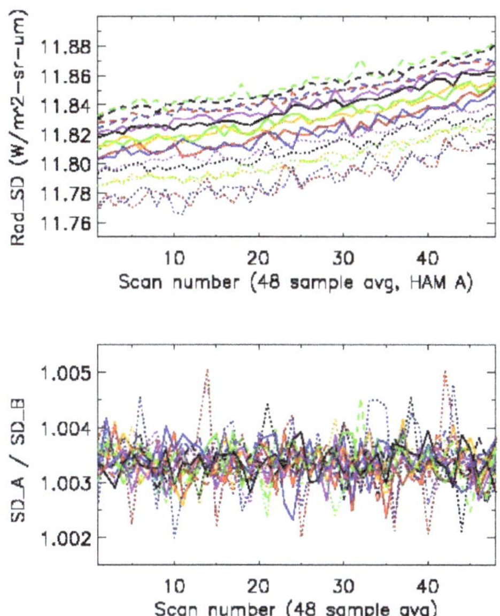

Figure 13. Detector dependent SD radiance for Half Angle Mirror (HAM) side A (**top**) and ratio of SD radiance between HAM sides A and B (**bottom**) in M15 on 30 June 2015. Each line represents a detector. Note: use 48-sample averaged value for each scan.

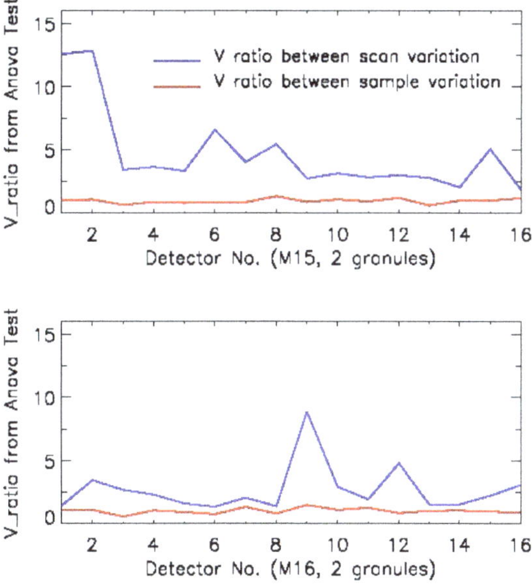

Figure 14. V ratios (defined in Equation (16)) from ANOVA tests for SD radiance HAM side ratio for 16 detectors in M15 (**top**) and M16 (**bottom**) for two granules on 30 June 2015.

The cumulative histogram method discussed in Section 4.1 (Equations (7) and (8) can also be used for SD radiance to show detector-to-detector variation. The cumulative histogram of SD radiance for one granule d20130619_t0746 over the Bay of Bengal in M15 and M16 is provided in Figure 15. During this short period (about 1 min and 24 s), the effect of SD radiance changing over time should be small, so Figure 15 shows the cumulative histogram for SD radiance which is computed using the TEB operational calibration algorithm Equation (16). X-axis represents the SD radiance, and Y-axis is the percentage of the pixels with the value less than X-axis value. Each line is for one detector. The horizontal distance is approximately a constant. The maximum horizontal distance between Det 1 and Det 15 in M15 (left panel) and M16 (right panel) are 0.0655 and 0.050, respectively. Considering the radiance range, the relative magnitude (defined by Equation (8)) in M16 is 0.4036, which is larger than 0.3475 in M15. That means the detectors in M16 are more stable than those in M15, but the striping in M16 is more apparent. Comparing with EV striping analysis which was used for BT, the relative magnitude using onboard SD radiance is much larger and can show the detector-to-detector variation more clearly. On the other hand, since there is no thermistor on the SD, its temperature uniformity is not known so the analysis may be affected by any temperature trend on the SD surface.

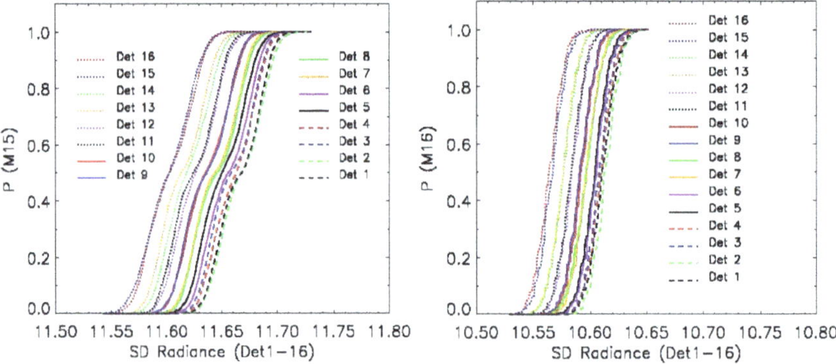

Figure 15. The cumulative histogram for SD radiance ($W/(m^2 \cdot Sr \cdot \mu m)$) over the Bay of Bengal in M15 (**left**) and M16 (**right**) for granule d20130619_t0746 on 19 June 2013.

C. Correlation between dBB and dSD for Detector 1 in M15

One hypothesis is that the DCR and detector instability may have contributed to the striping. As a proof, the correlation between dBB (which is an indicator of DCR and detector noise short term) and dSD (which is proxy for earth view) is analyzed. DCR and detector noise can be propagated to the dSD as well as earth view data. Therefore, in addition to the above analysis, we have also studied the possible relationship between dBB and dSD for same cases as in Table 2. Since detector 1 in M15 is much more noisy than other detectors (see Figures 12 and 14), the following study focuses on detector 1 in M15. dBB has no trend over the orbital time, so dBB variation is mostly due to random noise, detector and DCR instability. If there is correlation between dBB and dSD, then the DCR effect may be related to the striping in dSD. As an example, we use one granule (d20130619_t0748_e0749) OBCIP file in Bengal Case 1, which is about 1 min and 24 s. During this short period, we assume BB temperature does not change. Table 3 summarizes the linear correlation coefficients between dBB and dSD (in the 3rd column), as well as along track and along scan Allan deviation of SD Radiance in M15 and M16. In general, dBB and dSD are more correlated over the tropical region than polar region. We have also analyzed the ratio of along track to along scan Allan Deviation of solar diffuser radiance, which show larger values for tropical cases than polar cases.

Figure 16 uses one granule (d20130619_t0748_e0749) OBCIP file in Bengal Case 1 as an example, left panel shows the correlation between dBB and dSD, and the correlation coefficients is about 0.8345. Right panel shows the horizontal divergence or striping through the cumulative histogram of dBB. The maximum horizontal divergence among detectors for dBB and dSD are 181.985 and

129.953, respectively. Table 4 summarizes the linear correlation coefficients between dBB and dSD and the averaged horizontal divergence of dBB and dSD for six cases. In general, higher correlation corresponds to larger averaged horizontal divergence.

Table 3. Correlation coefficients between dBB and dSD, as well as along track and along scan Allan deviation of solar diffuser radiance in M15 and M16.

Cases	Granule	Coeff.* for det 1 in M15	Allan Deviation for SD_Rad In M15			Allan Deviation for SD_Rad In M16		
			Along Track	Along Scan	Ratio	Along Track	Along Scan	Ratio
Bengal: case1	d20130619_t0746444	0.8345	0.00919	0.00479	1.91827	0.00709	0.00442	1.60429
Bengal: case2	d20140622_t0753301	0.8151	0.00920	0.00486	1.89414	0.00714	0.00449	1.59022
Bengal: case3	d20140703_t0748470	0.7912	0.00901	0.00479	1.88105	0.00704	0.00443	1.58842
Alaska: case 1	d20140520_t2158272	0.5981	0.00827	0.00483	1.71272	0.00652	0.00443	1.47338
Alaska: case 2	d20140603_t2237573	0.4225	0.00842	0.00479	1.75669	0.00647	0.00439	1.47115
Polar: case 3	d20150421_t1802552	0.011	0.00793	0.00484	1.63808	0.00585	0.00441	1.32739

Coeff * refers to the correlation coefficient.

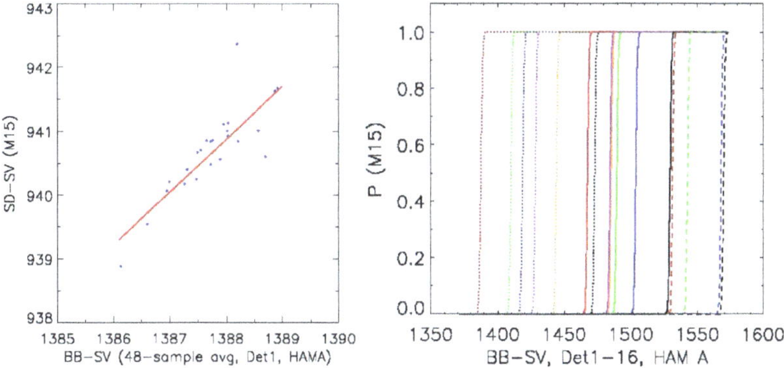

Figure 16. Linear correlation between dBB and dSD (**left**) and the cumulative histogram of 48-sample averaged dBB (**right**) for detector 1 in HAM side A and band M15 over a granule of the Bay of Bengal in case1.

Table 4. Correlation coefficients between dBB and dSD, as well as the maximum horizontal distance from dBB and dSD in M15.

Cases	Granule	Correlation Coefficients for Det 1 in M15	dBB Range	dSD Range
Bengal: case 1	d20130619_t0746444	0.8345	181.985	129.953
Bengal: case 2	d20140622_t0753301	0.8151	179.776	130.252
Bengal: case 3	d20140703_t0748470	0.7912	179.952	130.475
Alaska: case 1	d20140520_t2158272	0.5981	180.691	129.168
Alaska: case 2	d20140603_t2237573	0.4225	181.486	129.872
Polar: case 3	d20150421_t1802552	0.011	179.917	124.966

5. Conclusions

In this study, we have investigated the striping pattern in VIIRS SST imagery using LBLRTM simulation and VIIRS SDR observation data. The analysis indicates that the striping is likely related to the difference in detector level RSR. The results from LBLRTM show that BT difference between the band averaged and detector level RSR has some atmospheric dependency, *i.e.*, the difference over tropical region is larger than that over polar region. The magnitude of the difference between tropical and subarctic summer is 0.017 K in M15 − M16, 0.04 K in M16, and −0.019 K in M15. Results also show that band M16 is more sensitive to the atmospheric conditions due to higher water vapor absorption. M15 − M16 is the critical quantity in SST retrieval, which shows more apparent striping than single bands because of smaller signal-to-noise ratio. Detector # 1–3 shows large atmospheric effect in M15 − M16 and M16. With the band average as reference, detector # 1 has the largest deviation (0.055 K) for tropical case, which is close to 0.05 K in previous study using MODTRAN [4].

In addition to the model simulation analysis, we have also performed case studies to analyze the striping in VIIRS SDR brightness temperature observation data over tropical and polar region. Three different methods including cumulative histogram, standard deviation, and Allan deviation methods, are utilized to quantify the striping in the images of VIIRS earth view data (BT) for bands M15, M16, and M15 − M16. Results show that the histogram divergence difference between tropical and polar region is 0.040 K in, 0.053 K in M16, and 0.003 K in M15 respectively. They are all slightly higher than those from model output. In general, VIIRS SDR brightness temperature observation has larger variability when compared with the model output. However, difference due to atmospheric condition is small in an absolute value, but it also plays a role in causing the striping.

The Earth View data is complex because the noise is mixed with diverse signals. The onboard calibration data excludes the atmospheric effect and is spatially uniform, so it is used to analyze the instrument effect (such as fixed pattern noise) which is a possible factor affecting the striping pattern. The thermal band detector noise characterization has been studied using the raw data of BB, SD, and SV. The SD radiance is also computed from the operational TEB calibration algorithm, which has an advantage because it acts as a proxy of SST and can simulate the impact of SST. ANOVA tests of dBB, dSD, and SD radiance demonstrate that the noise along track direction is the major reason for the striping observed in VIIRS imagery. In along track direction, detectors 1 and 2 in M15 show higher noise levels than other detectors, detectors 9 and 12 in M16 also have much higher noise level than other detectors. Within the scan direction, the ratio (defined by Equation (16)) in ANOVA test for all detectors are very similar for either M15 or M16. Since the dBB has no trend with time, its variation is mostly from random noise, detector instability and DCR short term. The results show that the variation of dBB with scan in along track direction, *i.e.*, detector instability and DCR short term, are the dominant factor for dBB variation. We also found evidence of correlation between dSD and dBB for a detector 1 in M15. Since PRT in dBB is stable during the short period, the DCR instability is likely related to the striping in dSD and earth view data. On the other hand, further analysis is needed with more data samples. These findings will help us better understand the impact of the difference in detector level RSR and noise on VIIRS geophysical retrieval which can also potentially lead to further SDR processing improvements.

Acknowledgments: The authors would like to thank the anonymous reviewers for their constructive comments and suggestions which helped us improve the clarity of the manuscript. We also thank the VIIRS SDR team members in general, and Boryana Efremova and David Moyer, in particular, for fruitful discussions during the team telecons on the root cause of the striping. We also appreciate the help from Zhenping Li for help with the histogram based striping characterization. Thanks are extended to Slawomir Blonski, Wenhui Wang, Bin Zhang, Yan Bai, and Jason Choi for their help in the data processing and analysis. The views, opinions, and findings contained in this paper are those of the authors and should not be construed as official positions, policy, or decisions of the NOAA or the U. S. Government.

Author Contributions: Zhuo Wang contributed to the design of the study, the development of methodology and software code, performed the analysis and wrote the manuscript. Changyong Cao contributed to the design of the study, the development of methodology, and provided technical oversight of the project.

Conflicts of Interest: The authors declare no conflict of interest.

References

1. Miller, S.D.; Lee, T.F.; Fennimore, R.L. Satellite-based imagery techniques for daytime cloud/snow delineation from MODIS. *J. Appl. Meteor.* **2005**, *44*, 987–997. [CrossRef]

2. Bouali, M.; Ignatov, A. Adaptive reduction of striping for improved sea surface temperature imagery from Suomi National Polar Orbiting Partnership (S-NPP) Visible Infrared Imaging Radiometer Suite (VIIRS). *J. Atmos. Ocean. Technol.* **2014**, *31*, 150–163. [CrossRef]
3. Padula, F.; Cao, C. Preliminary study of the Suomi NPP VIIRS detector-level spectral response function effects for the long-wave infrared bands M15 and M16. *Proc. SPIE* **2014**. [CrossRef]
4. Padula, F.; Cao, C. Detector-level spectral characterization of the Suomi NPP VIIRS long-wave infrared bands M15 & M16. *Appl. Opt.* **2015**, *54*, 5109–5116. [PubMed]
5. Godin, R. *VIIRS Sea Surface Temperature Algorithm Theoretical Basis Document (ATBD)*; Goddard Space Flight Center Greenbelt: Greenbelt, MD, USA, 2013.
6. Liou, K.N. *An Introduction to Atmospheric Radiation*, 2nd ed.; International Geophysics Series; Academic Press: San Deigo, CA, USA, 2002; Volume 84.
7. Berk, A.; Anderson, G.P.; Acharya, P.K.; Shettle, E.P. *MODTRAN 5.2.1 User's Manual*; Space Vehicles Directorate: Hanscom AFB, MA, USA, 2011.
8. Clough, S.A.; Iacono, M.J.; Moncet, J.-L. Line-by-line calculation of atmospheric fluxes and cooling rates: Application to water vapor. *J. Geophys. Res.* **1992**, *97*, 15761–15785. [CrossRef]
9. Clough, S.A.; Shephard, M.W.; Mlawer, E.J.; Delamere, J.S.; Iacono, M.J.; Cady-Pereira, K.; Boukabara, S.; Brown, P.D. Atmospheric radiative transfer modeling: A summary of the AER codes. *J. Quant. Spectrosc. Radiat. Transf.* **2005**, *91*, 233–244. [CrossRef]
10. Alvarado, M.J.; Payne, V.H.; Mlawer, E.J.; Uymin, G.; Shephard, M.W.; Cady-Pereira, K.E.; Delamere, J.S.; Moncetm, J.L. Performance of the Line-By-Line Radiative Transfer Model (LBLRTM) for temperature, water vapor, and trace gas retrievals: Recent updates evaluated with IASI case studies. *Atmos. Chem. Phys.* **2013**, *13*, 6687–6711. [CrossRef]
11. Moncet, J.L.; Uymin, G.; Lipton, A.E.; Snell, H.E. Infrared radiance modeling by optimal spectral sampling. *J. Atmos. Sci.* **2008**, *65*, 3917–3934. [CrossRef]
12. Rothman, L.S.; Gordon, I.E.; Barbe, A.; Benner, D.C.; Bernath, P.F.; Birk, M.; Boudon, V.; Brown, L.R.; Campargue, A.; Champion, J.-P.; *et al.* The HITRAN 2008 molecular spectroscopic database. *J. Quant. Spectrosc. Radiat. Transf.* **2009**, *110*, 533–572. [CrossRef]
13. Rosenkranz, P.W. Shape of the 5 mm oxygen band in the atmosphere. *IEEE Trans. Antennas Propag.* **1975**, *23*, 498–506. [CrossRef]
14. Mlawer, E.J.; Payne, V.H.; Moncet, J.-L.; Delamere, J.S.; Alvarado, M.J.; Tobin, D.D. Development and recent evaluation of the MK_CKD model of continuum absorption. *Philos. Trans. R. Soc.* **2012**, *370*, 2520–2556. [CrossRef] [PubMed]
15. Martin, S. *An Introduction to Ocean Remote Sensing*; Cambridge University Press: Cambridge, UK, 2004.
16. Weinreb, M.P.; Xie, R.R.; Lienesch, J.H.; Crosby, D.S. Destriping GOES images by matching empirical distribution functions. *Remote Sens. Environ.* **1989**, *29*, 185–195. [CrossRef]
17. Li, Z.; Yu, F. A Real Time De-Striping Algorithm for Geostationary Operational Environmental Satellite (GOES) 15 Sounder Images. Available online: http://digitalcommons.usu.edu/cgi/viewcontent.cgi?filename=0&article=1197&context=calcon&type=additional (accessed on 6 February 2016).
18. Allan, D.W.; Ashby, N.; Hodge, C.C. Appendix A: Time and frequency measures accuracy, error, precision, predictability, stability, and uncertainty. In *The Science of Timekeeping*; Hewlett Packard Application Note 1289; Hewlett-Packard Company: Englewood, CO, USA, 1997; pp. 56–65.
19. Cao, C.; Xiong, X.; Wolfe, R.; De Luccia, F.; Liu, Q.; Blonski, S.; Lin, G.; Nishihama, M.; Pogorzala, D.; Oudrari, H. *Visible Infrared Imaging Radiometer Suite (VIIRS) Sensor Data Record (SDR) User's Guide*; version 1.2; National Oceanic and Atmospheric Administration, National Environmental Satellite, Data, and Information Service: Washington, DC, USA; September; 2013.
20. Baker, N.; Kilcoyne, H.; NOAA. Joint Polar Satellite System (JPSS) VIIRS Radiometric Calibration Algorithm Theoretical Basis Document. Available online: http://npp.gsfc.nasa.gov/sciencedocs/2015-06/474-00027_ATBD-VIIRS-Radiometric-Calibration_C.pdf (accessed on 6 February 2016).

Article

JPSS-1 VIIRS Radiometric Characterization and Calibration Based on Pre-Launch Testing

Hassan Oudrari [1,*], Jeff McIntire [1], Xiaoxiong Xiong [2], James Butler [2], Qiang Ji [1], Thomas Schwarting [1], Shihyan Lee [1,†] and Boryana Efremova [1,‡]

1 Science Systems and Applications, Inc., Lanham, MD 20706, USA; jeffrey.mcintire@ssaihq.com (J.M.); qiang.ji@ssaihq.com (Q.J.); thomas.schwarting@ssaihq.com (T.S.); shihyan.lee@nasa.gov (S.L.); boryana.efremova@noaa.gov (B.E.)

2 NASA Goddard Space Flight Center, Greenbelt, MD 20771, USA; xiaoxiong.xiong-1@nasa.gov (X.X.); james.j.butler@nasa.gov (J.B.)

* Correspondence: Hassan.oudrari-1@nasa.gov; Tel.: +1-301-867-2094; Fax: +1-301-867-2151

† Current affiliation: SAIC, Beltsville, MD 20705, USA.

‡ Current affiliation: Earth Resources Technology, Inc., Silver Spring, MD 20707, USA.

Academic Editors: Changyong Cao, Richard Müller and Prasad S. Thenkabail
Received: 29 October 2015; Accepted: 28 December 2015; Published: 6 January 2016

Abstract: The Visible Infrared Imaging Radiometer Suite (VIIRS) on-board the first Joint Polar Satellite System (JPSS) completed its sensor level testing on December 2014. The JPSS-1 (J1) mission is scheduled to launch in December 2016, and will be very similar to the Suomi-National Polar-orbiting Partnership (SNPP) mission. VIIRS instrument has 22 spectral bands covering the spectrum between 0.4 and 12.6 μm. It is a cross-track scanning radiometer capable of providing global measurements twice daily, through observations at two spatial resolutions, 375 m and 750 m at nadir for the imaging and moderate bands, respectively. This paper will briefly describe J1 VIIRS characterization and calibration performance and methodologies executed during the pre-launch testing phases by the government independent team to generate the at-launch baseline radiometric performance and the metrics needed to populate the sensor data record (SDR) Look-Up-Tables (LUTs). This paper will also provide an assessment of the sensor pre-launch radiometric performance, such as the sensor signal to noise ratios (SNRs), radiance dynamic range, reflective and emissive bands calibration performance, polarization sensitivity, spectral performance, response-vs-scan (RVS), and scattered light response. A set of performance metrics generated during the pre-launch testing program will be compared to both the VIIRS sensor specification and the SNPP VIIRS pre-launch performance.

Keywords: JPSS; SNPP; VIIRS; pre-launch; radiometric; performance; calibration; spectral

1. Introduction

The purpose of this paper is to provide an overview of the initial assessment of the Visible Infrared Imaging Radiometer Suite (VIIRS) sensor onboard the first Joint Polar Satellite System (JPSS-1 or J1). This VIIRS instrument is the second flight unit of its series, and it has a similar design to the first sensor that was launched on 28 October 2011 on-board the Suomi-National Polar-Orbiting Partnership (SNPP) satellite. SNPP VIIRS has been sending high quality remote sensing data for about four years [1,2], and has not only contributed to the continuity of climate data record initiated by other spaceborne Earth observing instruments, but has also supported many crucial environmental, economic and scientific applications [3–5]. The JPSS program will ensure the launch of multiple VIIRS instruments over the next two decades, starting with J1 in December 2016.

VIIRS is a scanning radiometer that collects visible and infrared imagery and radiometric measurements of the Earth's surface to generate environmental data records (EDRs) in support of data continuity initiated by heritage sensors such as the Advanced Very High Resolution Radiometer

(AVHRR), the Sea-viewing Wide Field-of-view Sensor (SeaWiFS) and the Moderate Resolution Imaging Spectroradiometer (MODIS) [6]. Products derived from VIIRS will support key applications and research studies such as weather forecasting and global measurements of atmospheric, oceanic, and land surface variables [7–9].

J1 VIIRS sensor pre-launch results summarized in this paper are based on the data analyses performed by the VIIRS Characterization Support Team (VCST), with close interaction with other government and contractor teams, including the National Aeronautics and Space Administration (NASA) flight project and science teams, the Aerospace Corporation team, the National Oceanic and Atmospheric Administration (NOAA)-Center for Satellite Applications and Research (STAR) team, and the University of Wisconsin team.

In the following sections, we will provide an overview of the key J1 VIIRS pre-launch radiometric and spectral performance assessments, describing briefly the methodologies used, and comparing the results to SNPP VIIRS pre-launch performance as well as to the sensor requirements. The VIIRS spatial performance assessments are not included in this paper. J1 VIIRS design and testing program will be presented in Section 2. The summary of VIIRS pre-launch testing and sensor performance will be presented in Section 3 for all reflective solar bands (RSB) and thermal emissive bands (TEB), focusing on key performance metrics such as the signal-to-noise (SNR), dynamic range, polarization sensitivity, relative spectral response, and scattered light contamination. A summary and conclusion of J1 VIIRS pre-launch radiometric and spectral performance as well as future performance enhancements will be presented in the last section, Section 4.

2. Sensor Design and Testing Program

2.1. J1 Sensor Design

The VIIRS sensor is designed to operate in a polar, sun-synchronous orbit with a nominal altitude of 828 km at an inclination angle of approximately 98 degrees relative to the equator (1:30 PM local equatorial crossing time, ascending node) [10]. It collects radiometric and imaging data in 22 spectral bands covering the visible and infrared spectral region between 0.4 to 12.6 μm (Table 1). The moderate resolution bands (M-bands) and imaging resolution bands (I-bands) have a spatial resolution of ~750 m and ~375 m respectively, while the ground swath is ~3040 km, permitting daily global coverage.

J1 VIIRS has the same design as SNPP VIIRS with similar on-board calibrators [11]: the Spectralon® solar diffuser (SD) to calibrate the RSB, the solar diffuser stability monitor (SDSM) to track the SD spectral degradation, the V-groove blackbody (BB) to calibrate the TEB, and the space view (SV) to be used for background subtraction (Figure 1).

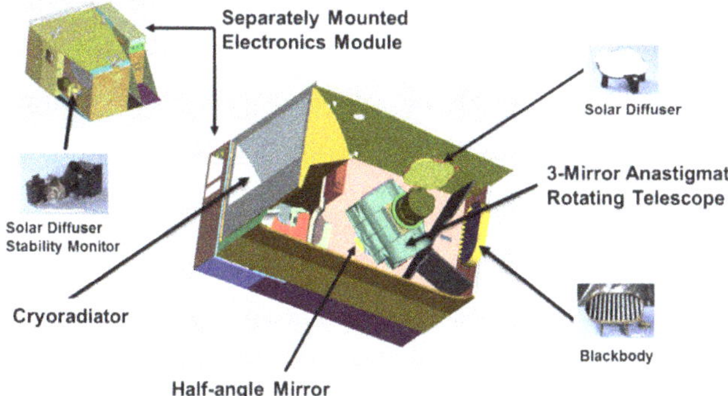

Figure 1. VIIRS instrument design and on-board calibrators: Solar Diffuser (SD), Solar Diffuser Stability Monitor (SDSM), and blackbody (BB).

Table 1. VIIRS 22 bands characteristics and key requirements.

Band Name	Gain	Center Wavelength (nm)	Bandwidth (nm)	Focal Plane Assembly	FOV (km) Nadir/EOS	Polarization Sensitivity (%)	L_{typ}	L_{min}/L_{max}	SNR
Reflective Solar Bands (RSB)									
DNB	VG	700	400	DNB	0.8/1.6	NS	3 e-5 (*)	3e-5/200 (*)	6
M1	HG	412	20	VISNIR	0.8/1.6	3	44.9	30/135	352
	LG						155	135/615	316
M2	HG	445	18	VISNIR	0.8/1.6	2.5	40	26/127	380
	LG						146	127/687	409
M3	HG	488	20	VISNIR	0.8/1.6	2.5	32	22/107	416
	LG						123	107/702	414
M4	HG	555	20	VISNIR	0.8/1.6	2.5	21	12/78	362
	LG						90	78/667	315
M5	HG	672	20	VISNIR	0.8/1.6	2.5	10	8.6/59	242
	LG						68	59/651	360
I1	SG	640	80	VISNIR	0.4/0.8	2.5	22	5/718	119
M6	SG	746	15	VISNIR	0.8/1.6	2.5	9.6	5.3/41	199
M7	HG	865	39	VISNIR	0.8/1.6	3	6.4	3.4/29	215
	LG					3	33.4	29/349	340
I2	SG	865	39	VISNIR	0.4/0.8	3	25	10.3/349	150
M8	SG	1240	20	SMWIR	0.8/1.6	NS	5.4	3.5/164.9	74
M9	SG	1378	15	SMWIR	0.8/1.6	NS	6	0.6/77.1	83
M10	SG	1610	60	SMWIR	0.8/1.6	NS	7.3	1.2/71.2	342
I3	SG	1610	60	SMWIR	0.4/0.8	NS	7.3	1.2/72.5	6
M11	SG	2250	50	SMWIR	0.8/1.6	NS	0.12	0.12/31.8	10
Thermal Emissive Bands (TEB)									
Band Name	**Gain**	**Center Wavelength (nm)**	**Bandwidth (nm)**	**Focal Plane Assembly**	**FOV (km) Nadir/EOS**	**Polarization Sensitivity (%)**	T_{typ} (K)	T_{min}/T_{max} (K)	$NEdT_{(K)}$
I4	SG	3740	380	SMWIR	0.4/0.8	NS	270	210/353	2.5
M12	SG	3700	180	SMWIR	0.8/1.6	NS	270	230/353	0.396
M13	HG	4050	155	SMWIR	0.8/1.6	NS	300	230/343	0.107
	LG					NS	380	343/634	0.423
M14	SG	8550	300	LWIR	0.8/1.6	NS	270	190/336	0.091
M15	SG	10763	1000	LWIR	0.8/1.6	NS	300	190/343	0.07
I5	SG	11450	1900	LWIR	0.4/0.8	NS	210	190/340	1.5
M16	SG	12013	950	LWIR	0.8/1.6	NS	300	190/340	0.072

Notes: * DNB Units are in $Wm^{-2} \cdot sr^{-1}$; The units of spectral radiance for L_{typ}, L_{min}, L_{max} are $Wm^{-2} \cdot sr^{-1} \cdot \mu m^{-1}$; Dual-gain M-bands have two entries, one for high-gain (HG) and one for low-gain (LG); The SNR (at L_{typ}) is the minimum (worse-case) required SNR that applies at the end-of-scan; The NEdT (at T_{typ}) is the maximum (worse-case) required NEdT that applies at the end-of-scan; VG: Variable Gain (DNB has three gains: low-gain, mid-gain and high-gain); SG: Single Gain; NS: Not Specified; EOS: End-of-Scan; FOV: Field of View.

VIIRS design has incorporated multiple enhancements based on lessons learned from heritage sensors such as SNPP VIIRS and MODIS. The light collected by the rotating telescope assembly (RTA) is distributed to three focal planes, the Visible Near-Infrared (VisNIR), the Short- and Mid-Wave Infrared (SMWIR), and the Long-Wave Infrared (LWIR). All bands in the LWIR focal planes are thermally controlled at about 80.5 K, and all bands in the SMWIR are floating very close to the LWIR FPA temperature. Light reaching the focal planes' detectors is converted into analog electric signal then digitized through analog-to-digital conversion. The VIIRS telescope scans the Earth within $\pm$ 56.1 degree of nadir, then the blackbody at about 100 degrees, the SD at about 159 degrees, and the SV at about -65 degrees, in sequence.

Based on lessons learned from SNPP program, a few key changes were made to the J1 instrument to enhance its performance, including: (1) replaced the VIIRS 1394 communication bus with SpaceWire to resolve the anomalies observed by SNPP on-orbit [1,2]; (2) replaced the single board computer with a new design to resolve the computer lock-up issue observed by SNPP [1,2]; (3) the coating on the RTA mirrors was changed from Nickel (Ni) to a proprietary process (VQ) to enhance spatial stability with temperature; (4) the dichroic 2 coating was redesigned to correct the focus between SMWIR and LWIR bands; (5) the proper RTA mirror coating process was used to avoid Tungsten contamination [1,2] and (6) the VisNIR integrated filter assembly was redesigned to significantly reduce filter optical scatter and out-of-band (OOB) features [11].

2.2. J1 Sensor Testing Program

With continuous support from the government independent team, J1 VIIRS instrument was put through a very intensive ground test program, to ensure proper characterization and calibration, to understand the instrument performance, to use reliable and approved data analysis methods, and to investigate, close and document all anomalies identified during testing. The test setup graphs are not shown in this paper since these are considered company property. The VIIRS test program, led by the sensor contractor Raytheon, provided a comprehensive sensor characterization and performance assessment over the full range of instrument operating conditions that will be encountered on orbit. It has also provided the calibration and characterization values needed for each band, detector, gain state, half-angle mirror (HAM) side, electronics side, and instrument temperature. VIIRS general test plan covers three (3) key pre-launch phases: ambient (August 2013–January 2014), sensor level thermal vacuum (TV) (July–October 2014) and spacecraft level TV (expected in the spring of 2016).

By the end of sensor level testing, a large set of performance test procedures were executed to characterize the instrument under various environments (ambient and/or TV), to support the verification of sensor performance requirements [12], and to simulate long-term on-orbit performance. Key sensor performance testing was completed at three (3) instrument plateaus, cold, nominal and hot, during TV testing to cover the range of expected on-orbit conditions. J1 VIIRS key performance testing is listed in Table 2.

Table 2. Key sensor performance testing during ambient and thermal vacuum testing phases.

Ambient Phase Testing August 2013–January 2014	
Polarization sensitivity	Near Field Response (NFR)
Response vs Scan (RVS)	Stray Light Response (SLR)
Radiometric characterization	Electrical and Dynamic Crosstalk
Thermal Vacuum Testing July 2014–November 2014	
RSB and TEB calibration	Relative Spectral Response in-band & OOB [1]
Radiometric stability	Band-to-Band Registration (BBR)
Gain transition determination	Line Spread Function (LSF)

[1] The VisNIR RSR was derived in ambient when the sensor was in the TV chamber (with the chamber door open).

Early in the J1 test program, two (2) important decisions were made that defined how the instrument will be operated: (1) set the cold focal planes temperature at 80.5 K based on the thermal balance testing performed at TV hot plateau; (2) and designated the VIIRS electronics side A as the primary electronics, even though both sides (A and B) showed comparable performance.

3. J1 VIIRS Pre-Launch Performance

An overview of J1 performance assessment will be described in this section based on the NASA VCST team test data analysis. Comparison to sensor specification and SNPP VIIRS will be presented as well. Methodologies will be briefly described for each performance metric, but in general, data analysis methodologies are quite similar to those used for SNPP VIIRS [11].

A series of key VIIRS performance tests were performed at ambient, TV, or both [12]. Because of limited space, this paper will focus on key radiometric performance metrics including the J1 radiometric calibration for the RSB and TEB, dynamic range, signal to noise ratio (SNR), noise equivalent differential temperature (NEdT), polarization sensitivity, relative spectral response (RSR), response *versus* scan-angle (RVS), near field response (NFR) and stray light response (SLR). All of the calibration parameters and SNR were derived in the TV environment at three instrument temperature plateaus: cold, nominal and hot plateau. In this paper, only the performance for electronics side A at the nominal plateau (*i.e.*, closest to on-orbit configuration) will be reported.

3.1. RSB Radiometric Calibration

J1 VIIRS RSB were carefully calibrated in a TV environment using a National Institute of Standards and Technology (NIST) traceable light source, the 100 cm diameter spherical integrating source (SIS100). Another source, the three mirror collimator (TMC) SIS, capable of reaching high radiance values was also used to calibrate M1-M3 low gain. While the SIS100 has a radiance monitor to track the source fluctuations and drifts, the TMC does not, and therefore it is considered less accurate than the SIS100. In addition, the space view source (SVS) was used to collect the dark offset needed to generate the background corrected detector response or digital number (*dn*).

The detector response is related to the light source radiance L ($Wm^{-2} \cdot sr^{-1} \cdot \mu m^{-1}$) through a second degree polynomial:

$$L = c_0 + c_1 \cdot dn + c_2 \cdot dn^2 + U(dn^3) \quad (1)$$

where c_0, c_1, and c_2 are the calibration coefficients and $U(dn^3)$ denotes the truncation error.

The calibration coefficients can be determined from Equation (1) through curve fitting to unsaturated measurements at different radiance levels over the band's dynamic range [L_{min}, L_{max}]; however, the instability of the source over time can lead to high uncertainties for the radiometric calibration algorithm. To mitigate the impact of the source instability, the radiance measurements were performed with and without an attenuator screen inserted into the optical path between the SIS100 and VIIRS RTA entrance aperture. The attenuator screen is an opaque plate with small holes to allow a determined fraction of light through (transmittance of ~56%). The time between attenuator in and out is very short to minimize the effect of source instability on the measurements (~2 min).

Consequently, the ratio of the detected spectral radiances with and without the attenuator is equal to the attenuator transmittance (τ), as:

$$\tau = \frac{\frac{c_0}{c_1} + dn_{in} + \left(\frac{c_2}{c_1}\right) dn_{in}^2}{\frac{c_0}{c_1} + dn_{out} + \left(\frac{c_2}{c_1}\right) dn_{out}^2} \quad (2)$$

where dn_{in} and dn_{out} denote the *dn* with and without the attenuator, respectively. A 3-sigma outlier rejection criteria is used during the calculation.

To facilitate the data regression, Equation (2) is rewritten as:

$$h_0 (\tau - 1) + (\tau\, dn_{out} - dn_{in}) + h_2 \left(\tau\, dn_{out}^2 - dn_{in}^2\right) = 0 \quad (3)$$

where $h_0 = c_0/c_1$, and $h_2 = c_2/c_1$.

The model parameters τ, h_0, and h_2 are determined through a non-linear least-square process. It is worth noting that the RSB on-orbit calibration is performed through the linear coefficient (c_1) updates, and this coefficient or gain is determined pre-launch by inverting Equation (1) and averaging over the selected source levels:

$$c_1 = \left\langle \frac{L}{h_0 + dn_{out} + h_2 dn_{out}^2} \right\rangle \quad (4)$$

The RSB radiometric calibration coefficients are derived in TV at the cold, nominal, and hot temperature plateaus for each detector, HAM side, gain stage, and electronic side, and are then implemented into the SDR LUTs to support on-orbit calibration [13].

In general, the band averaged c_0/c_1 coefficient is very small and is on the order of 10^{-1} (c_0 varies within ± 0.4). Exceptions include M1–M3 low gain and SWIR bands M8–M11, where c_0/c_1 has values up to 3. The high values could be explained by the TMC inaccuracies noted previously, and the SWIR non-linearity issue described in a sub-section below. The c_0/c_1's variation over the three TV temperature plateaus and across detectors as well as its uncertainty is small, and is in general on the order of 10^{-1}, except M1–M3 low gain which has uncertainty values up to 7, probably due to the TMC fluctuations. The quadratic term c_2/c_1 has mean values on the order of 10^{-6} for all RSB bands, and its uncertainty is also on the order of 10^{-6}. The average gain, c_1, shows some sensitivity to instrument temperature, varying by up to 5% between TV plateaus, while its uncertainty is very small, on the order of 10^{-4}.

3.1.1. SNR Calculation and Performance

SNR was computed for each detector using the SIS100 radiance at various levels to cover the dynamic range of each RSB. This SNR is derived for each sample position (*i.e.*, angular position in scan) by dividing the cross-scan average by their standard deviation as:

$$SNR = \frac{1}{M}\sum_{j=1}^{j=M}\left[\frac{\frac{1}{N}\left[\sum_{i=1}^{i=N} dn_{i.j}\right]}{\sigma_j}\right] \tag{5}$$

where M, N are the total number of samples and scans respectively, and *i*, *j* are the scan and sample number respectively. The SNR processing approach is based on the cross-scan standard deviation (σ_j), which avoids the spatial non-uniformity when viewing the SIS aperture across samples.

The SNR estimates derived at each SIS level were fit using a mathematical formula described by Equation (6) in order to compute the SNR at any radiance level, and to smooth out the variability in the SNR over radiance levels:

$$SNR = \frac{L}{\sqrt{k_0 + k_1 L + k_2 L^2}} \tag{6}$$

The band averaged SNR values are shown in Table 3. All J1 RSB meet the SNR specification at L_{typ} with good margins, and overall, these margins are better than those measured for SNPP. It is also worth noting that having a contamination-free RTA mirror coating, J1 SNR performance is expected to be even better than SNPP at the end of mission life. The smallest and largest SNR margins for J1 are 51% and 3172% for M2 high gain and I3, respectively. It is also worth mentioning that the I3 detector 4 is a very noisy detector (SNR = 6.7), and its responsivity is about 50% lower than the other detectors in this band. This detector is being studied to mitigate striping features in I3 images and associated products.

Table 3. J1 VIIRS SNR (at L_{typ}) and L_{sat} values for RSB, and comparison to sensor specification and SNPP. The ratio between measured performance and specification (Spec) is also shown. For dual gain bands, the L_{sat} of high gain stage represents the radiance transition (L_{trans}) to low gain.

Band	Gain Stage	SNR (Spec)	Lmax (Spec)	SNPP SNR	J1 SNR	SNPP SNR/Spec	J1 SNR/Spec	SNPP L_{sat}	J1 L_{sat}	SNPP L_{sat}/Spec	J1 L_{sat}/Spec
M1	High	352	135	613	636	1.74	1.81	172	154	1.27	1.14
M1	Low	316	615	1042	1066	3.30	3.37	696	705	1.13	1.15
M2	High	380	127	554	573	1.46	1.51	138	137	1.09	1.08
M2	Low	409	687	963	986	2.35	2.41	827	880	1.20	1.28
M3	High	416	107	683	706	1.64	1.70	125	113	1.17	1.06
M3	Low	414	702	1008	1063	2.44	2.57	843	838	1.20	1.19
M4	High	362	78	526	559	1.45	1.54	88	87	1.13	1.12
M4	Low	315	667	864	844	2.74	2.68	872	851	1.31	1.28
M5	High	242	59	373	380	1.54	1.57	66	61	1.12	1.04
M5	Low	360	651	776	751	2.16	2.09	726	725	1.12	1.11
M6	High	199	41	409	428	2.06	2.15	48	48	1.16	1.16
M7	High	215	29	524	549	2.44	2.55	31	31	1.07	1.06
M7	Low	340	349	721	760	2.12	2.23	414	409	1.19	1.17
M8	High	74	164.9	358	335	4.84	4.53	126	118	0.77	0.72
M9	High	83	77.1	290	325	3.49	3.91	84	80	1.09	1.04
M10	High	342	71.2	691	765	2.02	2.24	81	77	1.14	1.09
M11	High	10	31.8	105	216	10.49	21.57	35	35	1.09	1.10
I1	High	119	718	261	227	2.19	1.91	771	777	1.07	1.08
I2	High	150	349	273	287	1.82	1.91	413	410	1.18	1.17
I3	High	6	72.5	176	190	29.36	31.72	70	66	0.97	0.91

3.1.2. Dynamic Range

J1 RSB dynamic range was also verified, and RSB saturation was found to be comparable to SNPP pre-launch performance, meeting the specification with acceptable margins, except for bands M8 (72%) and I3 (91%). These two L_{max} non-compliances were expected (similar to SNPP) and have a limited effect on environmental and science products based on SNPP experience. As a note, SNPP M8 and I3 saturation became compliant post-launch because of the RTA mirror degradation on-orbit (decreasing radiometric sensitivity), but J1 will not have that benefit since the mirror coating issue causing this radiometric degradation was corrected in the J1 flight hardware. For dual gain bands, the high gain L_{sat} shown in Table 3 represents the radiance transition (L_{trans}) from high gain to low gain. The requirement for L_{trans} is to be within +50% above L_{max}. As shown in Table 3, all dual gain bands are compliant with this requirement, with M1 having the highest transition, 14% higher than specified high gain L_{max}.

3.1.3. RSB Calibration Uncertainties

VIIRS RSB requirements on the absolute radiometric calibration are against the spectral reflectance accuracy of 2% when viewing a uniform scene of typical radiance. This uncertainty verification was performed using contributors which have been constrained by sensor-level requirements while others are based on the sensor contractor allocations [14]. A key contributor to the radiometric uncertainty includes the detector response characterization uncertainty, which states that the response of a detector to a range of radiance levels from L_{min} to L_{max} must be fit with a second degree polynomial to within 0.3%. This characterization uncertainty requirement was very challenging for many VIIRS RSB to meet, but the spectral reflectance accuracy (2%) was met for all RSB except M11 (2.25%) [14], primarily due to the uncertainty of the SD bi-directional reflectance factor (BRF) at SWIR wavelengths.

3.1.4. SWIR Linearity Issue

J1 SWIR M-bands exhibited more non-linear behavior at low radiance than what was observed on SNPP. Figure 2 shows the ratios of attenuator out signal to attenuator in signal for SWIR bands M8–M11. The non-constant dn ratio reflects the large non-linear response at low radiance levels, while SNPP dn ratios were much flatter. An investigation of this issue discovered that an analog signal processor bias voltage for the SMWIR focal plane was set to a different voltage value (−0.2 V) compared to its setting on SNPP (−0.4 V). The non-linearity present in the SWIR M-bands produces large errors in the attenuator ratios, and therefore large errors in the radiance characterization uncertainty and uniformity. Quantized data are also observed at low radiance values and are harder to address in the calibration algorithm. A mitigation plan is being prepared to enhance the calibration performance of the SWIR bands, based on either a third degree polynomial equation or the adoption of a two-piece calibration (*i.e.*, calibration over two radiance ranges).

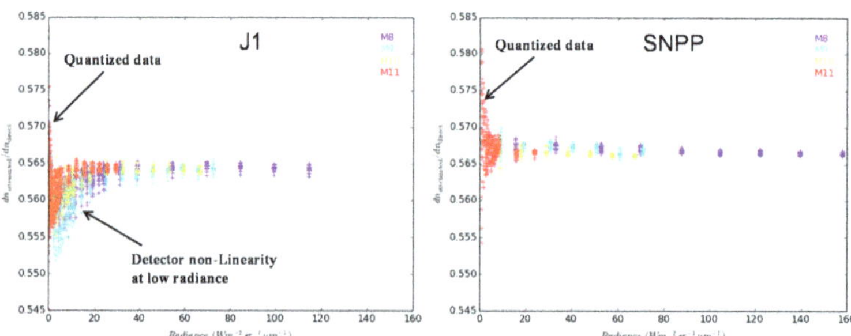

Figure 2. Comparison of J1 and SNPP SWIR (M8–M11) non-linearity as shown by the ratio of sensor response value (dn) for attenuator in and out measurements.

3.1.5. DNB Performance and Linearity Issue

While J1 DNB performance was measured using similar testing as the M-bands and I-bands, additional testing was performed to enhance the DNB characterization (at low light). Data analysis during first baseline testing at cold plateau showed a larger than expected non-linear response for DNB high gain A and B (HGA, HGB). The mid-gain and low gain stages also showed non-linearity features, but to a much lower extent. Upon further analysis, linear fit residuals plotted on a detector basis showed increasing residuals for high gain stages from aggregation modes 21 up to 32. There are 32 aggregation modes that are used by the DNB charge-coupled device (CCD) when scanning from beginning to end of scan. Since the CCD is a matrix of detectors in the scan and track directions, the aggregation consist of averaging a certain number of pixels (averaging is performed in the electronics) in the scan and track directions to keep the spatial resolution almost constant throughout the scan. The root cause of this DNB non-linearity issue was associated to two incorrect biases set in the focal plane interface electronics coupling into the feedthrough effect, and resulting in charge bleeding from one detector to the next (crosstalk) on the signal output line between successive acquisition frames.

Figure 3 shows the DNB response (dn) for HGB (HGA shows similar features) at four aggregation modes (1, 21, 26 and 32), where the non-linearity is clearly increasing from aggregation mode 21 up to aggregation 32. This non-linearity also increases as the radiance decreases. This issue was intensively investigated, and a solution referenced as Option 21 was identified to significantly reduce the non-linearity effect, which consists of fixing the aggregation mode from zone 21 outward, meaning that all aggregation zones from 21 to 32 will be assigned aggregation mode 21. It is also important to note that reducing the non-linearity effect led to better radiometric uncertainty, uniformity (striping) between detectors and between aggregation zones, and SNR performance. The only caveat for this solution is the loss of ability to maintain constant spatial resolution at large scan angles, but this is considered as low risk compared to the non-linearity effect.

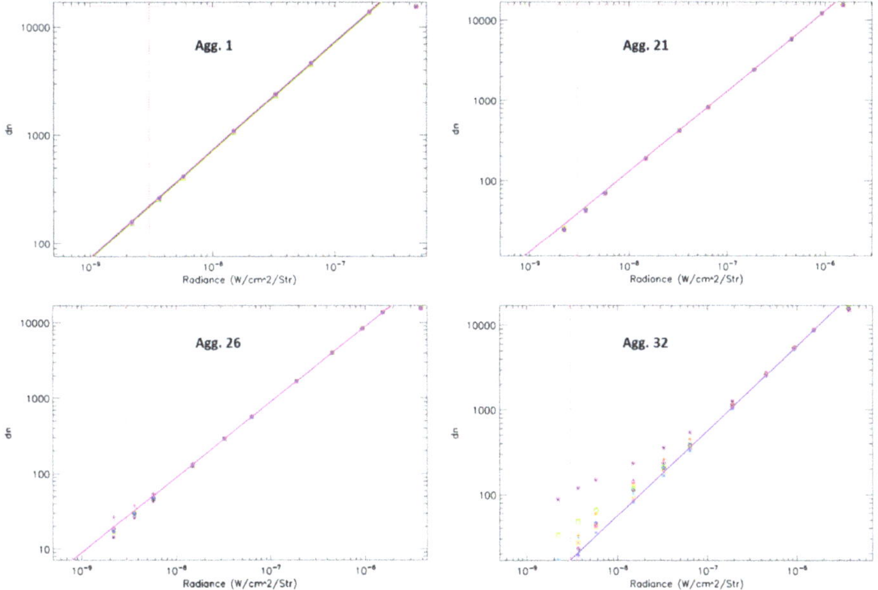

Figure 3. J1 DNB HGB response (dn) as a function of radiance in aggregation zones 1, 21, 26 and 32. Non-linearity feature starts to increase noticeably between aggregation modes 21 and 32 at low radiance. The 16 DNB detectors are represented by different symbols.

Based on Option 21 configuration used to correct for non-linearity, the DNB SNR showed compliance with specification at L_{min} for all aggregation zones, and the DNB dynamic range is compliant with L_{max} specification, 200 $Wm^{-2} \cdot sr^{-1}$, except for aggregation zone 1 (~97% of L_{max}). This is considered as low risk for the DNB since more than 99.9% of the Earth targets seen by SNPP DNB (based on data from 14 orbits) have radiance values less than 150 $Wm^{-2} \cdot sr^{-1}$.

Further concerns associated with on-orbit calibration, and how the non-linearity could impact the radiance quality are being investigated. Additional DNB testing is planned at spacecraft TV level (April 2016) to derive at-launch offset tables based on the new optimized sample-to-zone mapping, and to enhance DNB on-orbit calibration performance.

3.2. Thermal Emissive Bands (TEB) Calibration

J1 VIIRS thermal band calibration is referenced to a NIST traceable external blackbody calibration source (BCS). The radiance reaching the detector is the sum of the source radiance as well as contributors along the optical path (*i.e.*, the RTA, HAM, and aft optics). The reflectance factors represent the total reflectance of the RTA mirrors, HAM and aft-optics. The RVS is the scan angle dependent relative reflectance of the HAM. The temperature of each source is determined from one or more thermistors, and the radiances of the sources are determined via Planck's law convolved over the RSR of each spectral band over the extended band-pass [15].

The path difference radiance between the two sources (BCS and SV) is then:

$$L_{BCS} = RVS_{BCS}\varepsilon_{BCS}L_{BCS} - \frac{(RVS_{BCS} - RVS_{SV})}{\rho_{RTA}}\left[L_{HAM} - (1 - \rho_{RTA})\,L_{RTA}\right] \quad (7)$$

The path difference radiance is modeled as a quadratic polynomial in the offset corrected digital response, or

$$L = c_0 + c_1 dn + c_2 dn^2 \quad (8)$$

The retrieved EV radiance for the BCS is determined by inverting Equation (7), or

$$L_{BCS-ret} = \frac{\left(c_0 + c_1 dn_{BCS} + c_2 dn_{BCS}^2\right)}{RVS_{BCS}} + \frac{(RVS_{BCS} - RVS_{SV})}{RVS_{BCS}\rho_{RTA}}\left[L_{HAM} - (1 - \rho_{RTA})\,L_{RTA}\right] \quad (9)$$

The low gain state of band M13 is calibrated using a second high temperature external blackbody, the TMC blackbody. The calibration of the TMC blackbody was tied to the BCS by cross calibration at scene temperatures where both sources overlap.

NEdT Calculation

The NEdT is the fluctuation in the scene temperature equivalent to the system noise and is computed via the equation

$$NEdT = \frac{NEdL}{\frac{\partial L}{\partial T}} = \frac{L_{BCS}}{SNR\frac{\partial L}{\partial T}} \quad (10)$$

The derivative is of Planck's law with respect to the source temperature. The NEdT was determined at all source levels and compared to the specified value at T_{typ}; this was determined by fitting the SNR as a function of path difference radiance. As shown in Table 4, all TEB band average and per detectors NEdT meet the specification with good margins, and the performance is comparable to SNPP. A mild dependence of NEdT to sensor temperature (TV plateaus) was observed. The smallest and greatest NEdT margins were observed for M14 (39%) and I4 (595%) respectively. Detector noise variability is small in general, with the exception of detector 4 in M15 and detector 5 in M16B which are out-of-family detectors, and could lead to striping in some data products.

Table 4 also shows the J1 maximum temperature and comparison to the specification and SNPP. All bands have saturation values above the specified T_{max}, and comparable to SNPP. For J1, digital saturation occurred first for all bands, while for SNPP two bands, M12 and M14, exhibit analog saturation before digital saturation. The TEB saturation values are consistent among electronics sides and temperature plateaus to within 3 K. M13 low gain saturation was not observed in TV testing, but saturation is expected to be around 670 K based on ambient testing, while SNPP VIIRS has shown saturation around 654 K.

Table 4. J1 VIIRS T_{max} and NEdT performance for TEB derived in TV at nominal plateau, and comparison to Spec and SNPP.

Band	T_{max}			NEdT at T_{typ}		
	Spec	SNPP	J1	Spec	SNPP	J1
I4	353	357	357	2.5	0.41	0.42
I5	340	373	370	1.5	0.42	0.41
M12	353	357	358	0.396	0.13	0.12
M13 HG	343	364	363	0.107	0.044	0.043
M13 LG	634	-	-	0.423	0.340	0.304
M14	336	347	348	0.091	0.061	0.050
M15	343	365	359	0.070	0.030	0.026
M16	340	368	369	0.072	0.038	0.043

The radiometric response uniformity (RRU) represents the detector-to-detector uniformity (or striping) and is quantified by the following equation:

$$RRU = \frac{\left|L_{BCS-ret} - \langle L_{BCS-ret}\rangle_D\right|}{NEdL} \tag{11}$$

where the average EV retrieved radiance is over all detectors in a band and NEdL is derived from Equation (10). The sensor specification is met if the RRU is less than unity within the radiance range from L_{min} to 0.9 L_{max}.

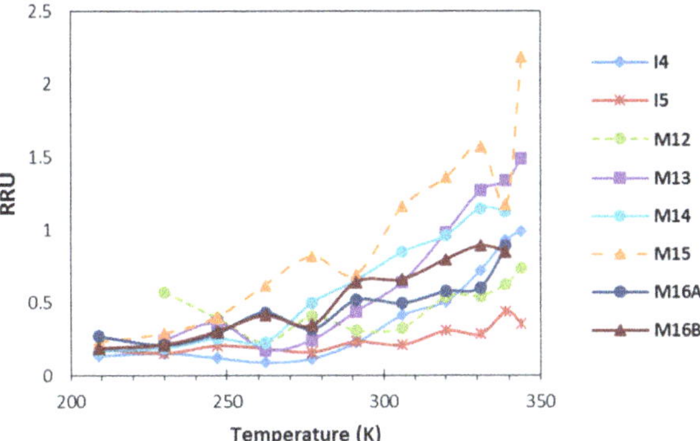

Figure 4. J1 maximum RRU as a function of source (BCS) temperature is shown for all TEB, HAM 0, and nominal plateau (RRU is compliant if less than unity).

Figure 4 shows the detector-to-detector striping performance represented by the RRU metric as a function of scene temperature for all emissive bands. We can easily see that the risk of striping increases with temperature for all TEB bands, because the deviation of the detector retrieved radiance from the band average increases with temperature while the NEdL levels off. Results derived per HAM side, electronics side and temperature plateau have shown some performance variations, with RRU for some bands reaching up to 2 at highest temperatures (larger than 310 K). We should also emphasize the difficulty to meet this requirement for both SNPP and J1 since VIIRS TEB has a very good noise performance (very low NEdL).

The absolute radiance difference (ARD) is the percent difference between the retrieved and calculated BCS radiances, and it is essentially a measure of the fit uncertainty's effect on the accuracy of the retrieved radiance, or

$$ARD = 100\frac{L_{BCS-ret} - L_{BCS}}{L_{BCS}} \tag{12}$$

The band averaged ARD for J1 TEB derived at nominal plateau are shown in Table 5 alongside the specified values in parenthesis. As expected, the ARD results show excellent thermal calibration for all bands at the specified temperatures, and the only exception is M12 at very low temperature (230 K) where the ARD value of 7.6 is slightly higher than the specified value of 7.0. The temperature errors converted from the ARD values are also shown in Table 5 at each specified temperature. The corresponding temperature uncertainty was computed by taking the difference in temperature derived from the radiance with and without the radiance uncertainty. Overall, the ARD performance (and temperature errors) is comparable between J1 and SNPP.

Table 5. J1 Band averaged ARD (%) at specified temperature shown for all TEB, HAM 0, and nominal plateau. ARD specification values are in parenthesis. In addition, shown in the lower part of this table are the corresponding temperature errors (K).

ARD Performance (%)								
Temp (K)	**I4**	**I5**	**M12**	**M13**	**M14**	**M15**	**M16A**	**M16B**
190	~	~	~	~	0.68 (12.30)	0.29 (2.10)	0.17 (1.60)	0.25 (1.60)
230	~	~	7.60 (7.00)	2.95 (5.7)	0.11 (2.40)	0.07 (0.60)	0.08 (0.60)	0.04 (0.60)
267	0.48 (5.00)	0.10 (2.50)	~	~	~	~	~	~
270	~	~	0.24 (0.70)	0.15 (0.70)	0.08 (0.60)	0.05 (0.40)	0.04 (0.40)	0.04 (0.40)
310	~	~	0.25 (0.70)	0.17 (0.70)	0.11 (0.40)	0.06 (0.40)	0.03 (0.40)	0.04 (0.40)
340	~	~	0.27 (0.70)	0.18 (0.70)	0.09 (0.50)	0.05 (0.40)	0.03 (0.40)	0.03 (0.40)
Temperature Error (K)								
Temp (K)	**I4**	**I5**	**M12**	**M13**	**M14**	**M15**	**M16A**	**M16B**
190	~	~	~	~	0.2	0.08	0.07	0.1
230	~	~	1.01	0.71	0.04	0.02	0.03	0.02
267	0.13	0.05	~	~	~	~	~	~
270	~	~	0.05	0.04	0.04	0.03	0.04	0.03
310	~	~	0.06	0.05	0.07	0.04	0.03	0.04
340	~	~	0.08	0.08	0.08	0.04	0.04	0.04

3.3. *Polarization Sensitivity*

During ambient phase testing, J1 polarization sensitivity was characterized using an integrating sphere combined with a sheet polarizer at seven different scan angles in the 400 to 900 nm range covering all VisNIR bands [16,17]. The sheet polarizer was mounted on a rotary stage and was rotated in 15 degree increments from 0 to 360 degrees. Additionally, a long wavelength spectral blocking filter was placed in the optical path when measuring the short wavelength bands, M1 to M3, to eliminate near infrared OOB contributions. The polarization sensitivity was derived for all VisNIR bands, detectors, HAM sides, and seven scan angles (−55, −45, −20, −8, 22, 45, 55), using Fourier series. The quality of

J1 measurements based on the Fourier transform is good, since only the zeroth and second order terms have shown non-negligible results. The polarization sensitivity factors derived based on ambient testing have revealed unexpectedly large non-compliance for four bands, M1–M4. This polarization issue was linked to the redesigned VisNIR spectral filters, and confirmed by the sensor polarization model [16,17]. Based on these findings further performance testing was conducted in the post-TV phase adding four more scan-angles (−37, −30, −15, 4) and limited testing with monochromatic laser source for two bands, M1 and M4, to verify the quality of the sensor polarization modeling [16,17]. The monochromatic polarization testing confirmed the large diattenuation on both edges of the filters bandpass causing the large polarization sensitivity seen in the broadband testing.

The final J1 mean degree of linear polarization (DoLP) values are shown for all VisNIR bands in Table 6 for HAM side 1. The polarization maximum and specification values per band are also shown. The linear polarization sensitivity for bands M1 to M4 was higher than the specified limit, with maximums (both HAM sides) as high as ~6.42% for M1, ~4.36% for M2, ~3.08% for M3, and ~4.35% for M4. Differences in linear polarization sensitivity with HAM side are as high as ~1%, where HAM 1 is generally larger.

Table 6. DoLP (HAM 1) for both SNPP and J1 sensors. The maximum polarization is for both HAM sides. Numbers in bold represent performance non-compliance.

Band	Sensor	Scan Angle											Max Pol.	Spec
		−55	−45	−37	−30	−20	−15	−8	4	22	45	55		
I1	SNPP	0.86	0.76	~	~	0.62	~	0.59	~	0.54	0.58	0.61	1.24	2.5
	J1	0.86	0.9	0.95	0.95	0.94	0.98	0.95	0.98	1	1.03	1.04	1.03	2.5
I2	SNPP	0.49	0.45	~	~	0.47	~	0.51	~	0.56	0.56	0.55	0.56	3
	J1	1.19	0.92	0.75	0.62	0.5	0.51	0.48	0.5	0.53	0.58	0.61	0.92	3
M1	SNPP	3.14	2.73	~	~	2.01	~	1.83	~	1.45	1.23	1.39	2.73	3
	J1	**5.57**	**5.73**	**5.86**	**6.06**	**6.17**	**6.19**	**6.34**	**6.41**	**6.42**	**6.17**	**5.96**	**6.42**	3
M2	SNPP	2.25	2.05	~	~	1.65	~	1.54	~	1.28	1.17	1.3	2.05	2.5
	J1	**4.08**	**4.08**	**4.13**	**4.23**	**4.18**	**4.18**	**4.23**	**4.25**	**4.19**	**4.36**	**4.46**	**4.36**	2.5
M3	SNPP	1.45	1.31	~	~	0.96	~	0.85	~	0.62	0.71	0.81	1.31	2.5
	J1	**2.92**	**2.86**	**2.83**	**2.85**	**2.76**	**2.74**	**2.75**	**2.74**	**2.85**	**3.08**	**3.11**	**3.08**	2.5
M4	SNPP	1.59	1.52	~	~	1.37	~	1.3	~	1.02	0.86	0.82	1.52	2.5
	J1	**4.03**	**4.2**	**4.33**	**4.35**	**4.32**	**4.35**	**4.3**	**4.29**	**4.15**	**3.99**	**3.91**	**4.35**	2.5
M5	SNPP	0.81	0.74	~	~	0.7	~	0.69	~	0.61	0.59	0.57	1.02	2.5
	J1	2.1	2.17	2.23	2.19	2.13	2.14	2.07	2.03	2.02	1.99	1.97	2.23	2.5
M6	SNPP	1.29	1.14	~	~	0.96	~	0.92	~	0.81	0.75	0.7	1.14	2.5
	J1	1.03	0.92	0.89	0.87	0.86	0.91	0.91	0.95	0.96	0.95	0.94	1.32	2.5
M7	SNPP	0.52	0.47	~	~	0.43	~	0.44	~	0.48	0.47	0.45	0.48	3
	J1	1.18	0.92	0.74	0.61	0.48	0.47	0.43	0.46	0.47	0.52	0.56	0.92	3

Large detector-to-detector and scan angle variations were observed within various bands (Figure 5), and these variations could reach up to ~4% (M1). This is likely the result of angle of incidence changes on the VisNIR spectral filter assembly. The final uncertainty analysis has demonstrated compliance with uncertainty specification (0.5%), with an overall polarization factor uncertainty less than 0.22% for all bands except M1 which showed an uncertainty of about 0.37%. While the DNB has no requirement for polarization sensitivity, data analysis has shown this polarization performance to vary between 1.36% and 1.6% over the scan-angle range, detectors and HAM sides.

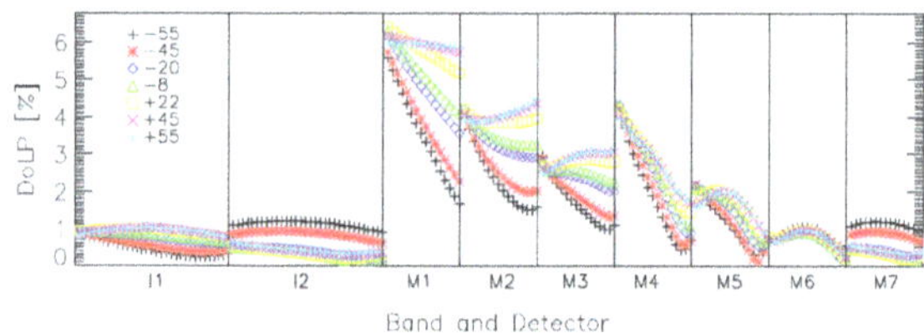

Figure 5. The DoLP (HAM 1) per detector and across scan angles.

Figure 6 shows a comparison of the DoLP results derived for M1 and M4 bands, based on the broadband source and the monochromatic source measurements [16,17]. In general, the results agree well, and the difference is less than 0.5% for the DoLP and less than 14 degrees for the phase. These measurement results were also compared to the outputs of the J1 polarization model. Our preliminary assessment is that the model was not able to represent the detector dependence with the accuracy needed, especially for M4 (maximum difference ~1.5%), and therefore, our recommendation was to revise the model and its component inputs for J1 and future instruments (J2+).

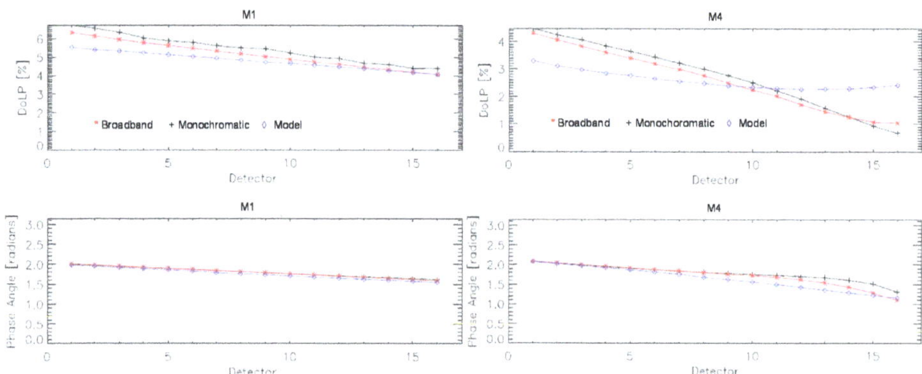

Figure 6. J1 DoLP and phase angle for M1 and M4, using broadband (*), monochromatic (+), and sensor model (◊).

3.4. *Relative Spectral Response (RSR)*

VIIRS sensor level spectral testing was performed using a double monochromator source, the Spectral Measurement Assembly (SpMA) in ambient for the VisNIR bands, and in the TV environment at nominal plateau for the SMWIR and LWIR bands [18]. The monochromator exit slit image was aligned to illuminate one spectral band at a time, and the spectral response was measured over a spectral region covering roughly the acceptance of the dichroic spectral bandpass. The VIIRS spectral data were corrected for the spectral shape of the source using a reference detector data set and normalized to the peak response to determine the VIIRS RSR. The data were analyzed to determine each band's center wavelength, Full Width Half Maximum (FWHM) bandwidth, 1% response points, and the Integrated OOB (IOOB). J1 spectral characterization was completed successfully, and analysis completed by the government team based on the SNR and visual inspection led to the release of high quality J1 VIIRS RSR data set (Version 1, stitched in-band and OOB) in June 2015 [18]. Figure 7 shows the in-band RSR for VisNIR and DNB (top), SMWIR (middle) and LWIR (bottom). As noted previously, the redesign of the VisNIR filters led to major enhancements in terms of IOOB reduction as shown in Figure 8 where we compare the full M1 RSR (in-band and OOB) between J1 and SNPP sensors. Compliance with the band center, bandwidth, extended band-pass, and IOOB requirements were calculated for all bands on a per detector basis. Table 7 provides a band average summary of J1 spectral performance for all bands, showing non-compliances in shaded cells. As a comparison to SNPP VIIRS, performances that were not compliant for SNPP were underlined. We can easily observe the enhancements of J1 IOOB for the VisNIR bands compared to SNPP. Most other non-compliances are either similar to SNPP or not considered a risk to the SDR or EDR performance. Additional RSR testing was performed for VisNIR bands in the post-TV phase using the NIST laser source. Further enhancements to the VisNIR RSRs are expected before J1 instrument launch, which will be based on the combination of best quality measurements from SpMA and laser sources, and this new version is expected to be released in early 2016.

Table 7. J1 Spectral performance (measurement *vs.* specification) summary based on RSR Version 1 data set. Shaded cells represent J1 non-compliances. Underlined cells represent SNPP non-compliances.

Band	Center Wavelength (nm)		Bandwidth (nm)		1% Limits (nm)				IOOB (%)	
	Meas.	Spec.	Meas.	Spec.	Meas.		Spec.		Meas.	Spec.
		Center/ Tolerance		Bandwidth/ Tolerance	Lower Limit	Upper Limit	Lower Limit	Upper Limit		
I1	643.0	640/6	78.6	80/6	594.0	693.5	565	715	0.11	0.5
I2	867.3	865/8	36.4	39/5	841.6	893.5	802	928	0.12	0.7
I3	1603.2	1610/8	60.4	60/5	1544.5	1667.4	1509	1709	0.44	0.7
I4	3747.7	3740/8	386.7	380/5	3474.4	4014.9	3340	4140	0.17	0.5
I5	11483.5	11450/8	1876.5	1900/5	10161.8	13078.7	9900	12900	0.23	0.4
M1	411.2	412/2	17.5	20/2	395.3	425.4	376	444	0.24	1.0
M2	444.8	445/3	17.0	18/2	429.3	457.9	417	473	0.34	1.0
M3	488.6	488/4	19.0	20/3	473.0	504.4	455	521	0.30	0.7
M4	556.3	555/4	18.5	20/3	540.0	573.6	523	589	0.25	0.7
M5	667.1	672/5	19.4	20/3	649.6	684.9	638	706	0.27	0.7
M6	746.0	746/2	13.5	15/2	733.9	758.0	721	771	0.26	0.8
M7	867.5	865/8	36.3	39/5	842.8	892.5	801	929	0.12	0.7
M8	1238.4	1240/5	26.1	20/4	1214.1	1264.9	1205	1275	0.49	0.8
M9	1375.7	1378/4	14.4	15/3	1362.1	1390.0	1351	1405	0.42	1.0
M10	1603.7	1610/14	60.1	60/9	1545.8	1667.5	1509	1709	0.44	0.7
M11	2258.2	2250/13	52.0	50/6	2209.4	2314.3	2167	2333	0.37	1.0
M12	3697.9	3700/32	194.7	180/20	3519.1	3893.7	3410	3990	0.36	1.1
M13	4073.9	4050/34	154.9	155/20	3911.7	4214.0	3790	4310	0.43	1.3
M14	8580.3	8550/70	339.7	300/40	8336.4	8879.1	8050	9050	0.23	0.9
M15	10730.8	10763/113	1000.7	1000/100	9915.8	11638.0	9700	11740	0.41	0.4
M16A	11882.6	12013/88	914.2	950/50	11104.5	12692.3	11060	13050	0.46	0.4
M16B	11883.0	12013/88	933.8	950/50	11101.8	12698.1	11060	13050	0.47	0.4
DNBMGS	693.1	700/14	380.5	400/20	488.2	906.6	470	960	0.03	0.1
DNBLGS	695.6	700/14	380.3	400/20	480.8	904.5	470	960	0.02	0.1

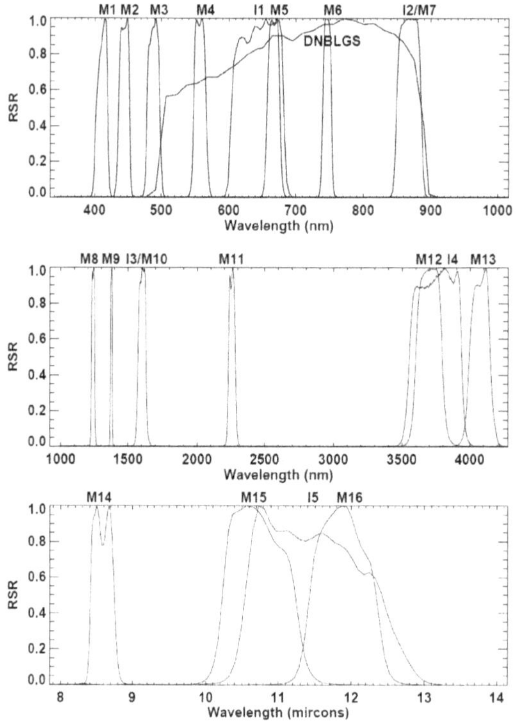

Figure 7. J1 VIIRS in-band RSRs, for VisNIR and DNB (**top**); SMWIR (**middle**); and LWIR (**bottom**).

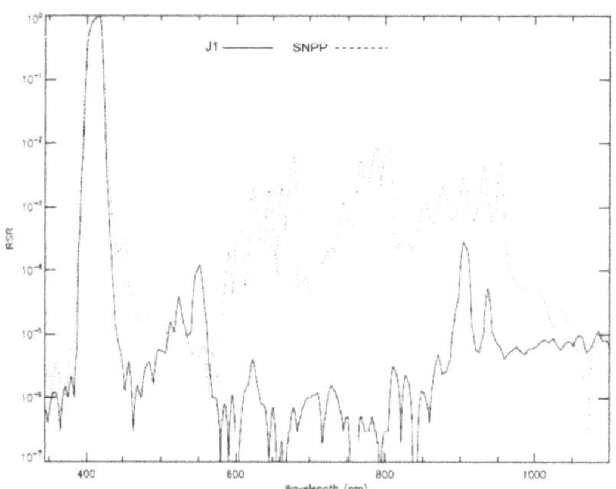

Figure 8. Full M1 RSR comparison between J1 (line) and SNPP (dashes).

3.5. Response Versus Scan-Angle (RVS)

The RVS testing was performed during ambient phase for the RSB using the SIS100 and for the TEB using the laboratory ambient blackbody (LABB) source and the on-board BB. Data taken at eleven angles-of-incidence (AOI) for the RSB and twelve for the TEB were used to fit the RVS function, which is a quadratic polynomial in AOI, after correcting for source drift and background radiances [19].

Figure 9 shows J1 band-average RVS functions for HAM side 0. For all bands, the RVS differences between HAM sides are small with the exception of bands M1, M2, M7 and I2. The variation in the RSB RVS for M1–M11 and I1–I3 is generally small (top and middle plots) and similar to SNPP, varying by less than 1.5% (M1 has largest variation) over the full operational AOI range of 28.6° to 60.2°. The RSB RVS uncertainty target is 0.3% which was determined here as the average fitting residuals. All RSB bands had uncertainties lower than 0.06% with the highest uncertainty observed for DNB, and M9 due to error residuals from water vapor correction [20]. The only exception is I3 detector 4 (very noisy detector) which has shown an uncertainty of about 0.4%.

The J1 band averaged RVS for the MWIR bands, M12–M13 and I4, and the LWIR bands, M14–M16 and I5, are plotted in the middle and bottom plots respectively of Figure 8. The MWIR RVS is generally small varying by less than 1% over the full AOI range. In contrast, the LWIR RVS changes by up to 10% for M14 over the range of AOI. The TEB RVS uncertainty target is 0.2%, which was determined here as the average fitting residuals; all TEB RVS uncertainties are lower than 0.1%, meeting the RVS characterization uncertainty target with good margin. Overall, J1 RVS performance is comparable to SNPP performance.

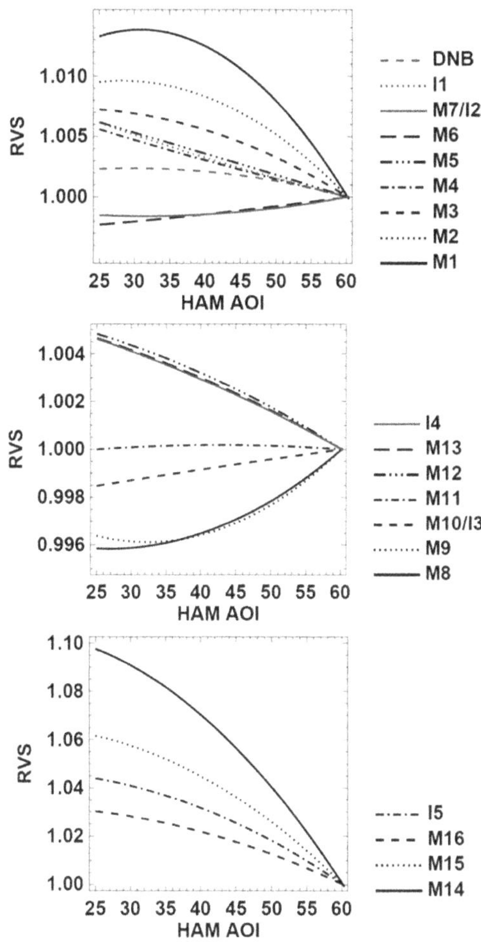

Figure 9. J1 VIIRS band averaged RVS as a function of HAM AOI for HAM 0.

3.6. Near-Field Response (NFR)

NFR is defined as scattered light originating from within 4 degrees of the RTA line of sight. The Scatter Measurement Assembly (ScMA) light source was used with a slit and a band-pass filter to measure the sensor NFR. The VIIRS NFR is limited by the absolute radiometric calibration uncertainty requirement for structured scenes which sets the maximum allowable response at a specified angle limit coming off a 12 by 12 milliradian bright target for each band. To estimate the structured scene response, a Harvey-Shack BRDF scattering model [21,22] was used to fit the measured NFR profile and remove test artifacts and noisy samples.

Figure 10 compares the normalized response for band M5 detector 8 between J1 and SNPP, which represents a typical profile observed in the VIIRS NFR measurements when the sensor is scanning the source through a vertical slit reticle. For both SNPP and J1, the figure shows the NFR falling off rapidly from the peak with additional sharp drops seen around the field baffle locations. In this example, the M5 NFR drops to about 0.1% within 5 samples of the peak (*i.e.*, corresponding to approximately 1.25 km at the Earth surface), and the field baffle reduces the NFR by about one order of magnitude. The NFR is estimated for each detector, and the results show good detector uniformity within the same band. The NFR requirement states that the maximum allowable scattered radiance, as a fraction of typical scene radiance, at the specified angular distance from a bright target shall be less than the specified value. The band averaged VIIRS NFR performance is summarized in Table 8, including the bright target radiance (L_{bright}), the specification (L_{spec}), and the ratio $L_{scat/}/L_{spec}$ for J1 and SNPP VIIRS. The results show all J1 VIIRS bands meet the specification ($L_{scat}/L_{spec}<1$) with margin at the beginning of life, and multiple bands are showing performance enhancements when compared to SNPP, especially M15 and M16.

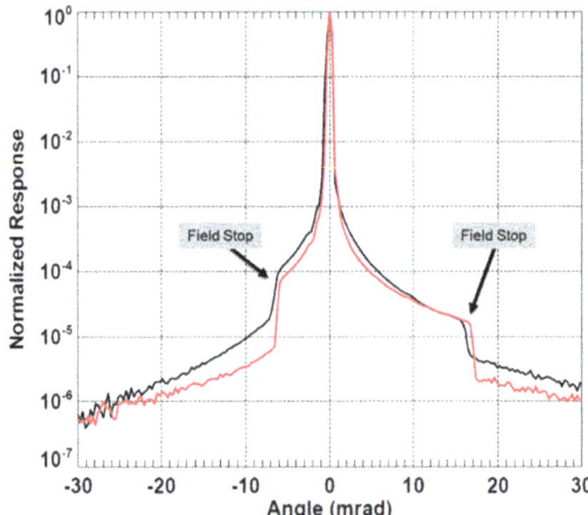

Figure 10. Measured NFR for band M5 (672 nm) detector 8 as a function of scan angle, using J1 data (red line) and SNPP data (black line).

Table 8. The summary of J1 band average NFR performance and comparison to SNPP. The ratio L_{scat}/L_{spec} less than unity means compliance with scatter requirements.

Band	Center Wavelength (nm)	Angular Separation (mrad)	L_{bright}	SNPP		J1	
				L_{spec}	L_{scat}/L_{spec}	L_{spec}	L_{scat}/L_{spec}
M1	412	6	162	2.77×10^{-3}	0.39	4.68×10^{-3}	0.37
M2	445	6	180	2.22×10^{-3}	0.45	3.95×10^{-3}	0.42
M3	488	6	160	2.00×10^{-3}	0.5	3.33×10^{-3}	0.36
M4	555	6	160	1.31×10^{-3}	0.47	2.25×10^{-3}	0.48
M5	672	6	115	8.70×10^{-4}	0.63	9.55×10^{-4}	0.67
M6	746	12	147	1.31×10^{-3}	0.12	1.08×10^{-3}	0.13
M7	865	6	124	5.16×10^{-4}	0.9	6.63×10^{-4}	0.83
M8	1240	6	57	9.47×10^{-4}	0.62	4.92×10^{-4}	0.65
M9	1378	NA	NA	NA	NA	NA	NA
M10	1610	6	86.1	8.48×10^{-4}	0.76	1.03×10^{-3}	0.38
M11	2250	6	1.2	1.00×10^{-3}	0.42	6.03×10^{-6}	0.63
M12	3700	3	0.3	1.67×10^{-3}	0.64	4.78×10^{-4}	0.87
M13	4050	3	1.7	1.86×10^{-3}	0.63	3.15×10^{-3}	0.47
M14	8550	NA	NA	NA	NA	NA	NA
M15	10763	3	12.5	7.75×10^{-4}	1.25	6.69×10^{-3}	0.66
M16	12013	3	11.3	7.92×10^{-4}	1.26	8.95×10^{-3}	0.56
DNB	12013	3	NA	2.00×10^{-3}	NA	2.00×10^{-3}	0.65

3.7. Stray Light Response (SLR)

The far field stray light is defined as the light originating from the region between 4 and 62.5 degrees from the sensor boresight. The far field stray light testing was performed using a radiometrically calibrated 1000 W studio lamp. The VIIRS telescope was staring at a cavity type blackbody while the lamp was moved through 33 equally distanced hemispherical positions to evaluate the angular stray light distribution. The lamp positions covered roughly equally divided annulus out to 62.5 degrees off nadir to simulate the Earth as subtended by VIIRS at the J1 on-orbit operating altitude. The measured instrument response was scaled by the ratio of the studio lamp irradiance and a model of bright cloud irradiance. The VIIRS far-field stray light testing represent a worse case, hole-in-the-cloud scenario. The measured instrument response at each position was weighted by the corresponding annulus, then aggregated to estimate the total stray light contribution from the specified worse-case scenario. The test was conducted with the RTA locked at SV, nadir, and end-of-scan to obtain stray light estimates at different scan angles.

Table 9 shows the estimated J1 VIIRS far field stray light compliance. The sensor test data analysis showed all RSB meet the stray light requirement with large margins at the beginning of life. Margins are between 45% (M5) and 94% (M10, I3). Furthermore, test results indicate that the stray light rejection is comparable between J1 and SNPP. A noticeable difference is that the stray light in SNPP is much stronger along track than along scan direction; whereas in J1, the stray light is more evenly distributed over all angles. Another difference is that band M11 is now compliant for J1 because of the L_{typ} requirement change (from 0.1 to 1.0 $Wm^{-2} \cdot sr^{-1} \cdot \mu m^{-1}$).

Table 9. Summary of J1 VIIRS stray light performance, and the ratio of measured stray light to requirement (1% L_{typ}). The comparison to SNPP stray light performance is also shown.

Band	E_{earth}	L_{typ}	SNPP			J1		
			dn_{spec}	dn_{stray}	dn_{stray}/dn_{spec}	dn_{spec}	dn_{stray}	dn_{stray}/dn_{spec}
M1	1444.1	44.9	17.02	2.58	0.15	7.84	2.42	0.31
M2	1526.1	40	9.39	2.53	0.27	9.03	2.68	0.30
M3	1563.5	32	8.81	2.56	0.29	8.12	2.89	0.36
M4	1510.7	21	7.53	2.39	0.32	7.50	2.96	0.39
M5	1265.7	10	5.06	2.38	0.47	4.69	2.57	0.55
M6	1088.9	9.6	8.00	3.62	0.45	7.22	2.95	0.41
M7	833.2	6.4	6.87	4.06	0.59	6.27	3.29	0.52
M8	353	5.4	1.45	0.77	0.53	1.82	0.44	0.24
M9	262.9	6	2.44	0.92	0.38	2.41	0.36	0.15
M10	165.7	7.3	3.55	1.30	0.36	3.44	0.22	0.06
M11	56.4	0.12	0.11	0.42	3.77	1.15	0.09	0.08
I1	1341.3	22	1.02	0.26	0.26	0.94	0.31	0.33
I2	833.2	25	2.51	0.39	0.15	2.12	0.31	0.15
I3	165.7	7.3	3.19	0.79	0.25	3.92	0.24	0.06

E_{earth} = irradiance of the cloud covered earth scene; dn_{spec} = dn at 1% of typical radiance (L_{typ}); dn_{stray} = dn associated with measured stray light.

4. Conclusions

This paper provides an early assessment of the J1 VIIRS pre-launch radiometric performance, based on an intensive and comprehensive testing program that was designed by the sensor contractor with support from the government calibration and science teams. The instrument testing program was performed to characterize the radiometric, spectral and spatial performance for all 22 spectral bands in various configurations and environments, and simulating the range of on-orbit conditions (ambient and thermal vacuum). The key radiometric and spectral performances were discussed in this paper. The VIIRS spectral bands calibration and characterization have shown very good performance, providing good quality calibration data to populate the SDR LUTs. All performance non-compliances and features were investigated, understood, and characterized using pre-launch test data and/or sensor model simulations. The RSB SNR was compliant for all bands with minimum margin of 51% (M2) and maximum margin of 3172% (I3), while the dynamic range was compliant for all RSB except bands M8 (72%) and I3 (91%). As expected, and similar to SNPP, the TEB have shown very good performance in term of NEdT and dynamic range (full compliance). The spectral testing also provided good quality data to generate the sensor RSRs, and the minor spectral non-compliances are band center for M16, bandpass for M1 and M8, 1% upper limit for I5, and IOOB for M15 and M16. All of these spectral non-compliances are expected to have negligible impact on the data products. The RVS derived for J1 was as expected, very similar to SNPP, and meeting the uncertainty target of 0.3% for RSB and 0.2% for TEB. The scattered light performance represented by the NFR and SLR are meeting the sensor requirements for all bands, and are expected to lead to better or similar performance as SNPP. Overall, J1 performance was comparable to SNPP VIIRS, and it is expected to have high quality J1 data product post-launch. Two minor issues related to high polarization sensitivity (M1–M4) and high non-linearity at low radiance for the DNB and SWIR bands were well characterized, and necessary mitigation plans are being prepared within the ground processing system to generate high quality products on-orbit.

Acknowledgments: The authors of this paper would like to thank the VIIRS team members from Raytheon, NOAA, Aerospace Corporation, NASA Science Team, and University of Wisconsin team for their valuable contributions to VIIRS testing and performance verifications. We also want to thank James McCarthy (Stellar Solutions), David Moyer (Aerospace Corporation), Chris Moeller (University of Wisconsin), and previous VCST members for their valuable support to the VIIRS program and to the pre-launch calibration and characterization effort.

Author Contributions: Hassan Oudrari was the lead author who provided the outline of the paper, contributed to the introduction, the sensor design and the sensor testing, and the compilation of the final paper. Jeff McIntire provided the thermal emissive bands performance assessment, the response *versus* scan-angle performance, and the polarization sensitivity assessment. Xiaoxiong Xiong provided contribution to the sensor design and testing description, and the review of the paper at multiple phases to enhance quality and consistency. James Butler provided the contribution to the sensor design and testing description, and the review of the paper at multiple phases to enhance quality and consistency. Qiang Ji provided the reflective solar bands performance, the stray light performance, and near field performance assessments. Thomas Schwarting provided the relative spectral response assessments, and contributions to the response *versus* scan-angle. Shihyan Lee provided contributions to the DNB performance, the reflective solar bands performance, response versu scan-angle performance assessments. Boryana Efremova provided assessments to the thermal emissive bands performance assessments.

Conflicts of Interest: The authors declare no conflict of interest.

References

1. Cao, C.; de Luccia, F.J.; Xiong, X.; Wolfe, R.; Weng, F. Early On-Orbit Performance of the Visible Infrared Imaging Radiometer Suite Onboard the Suomi National Polar-Orbiting Partnership(S-NPP) Satellite. *IEEE Trans. Geosci. Remote Sens.* **2013**, *52*, 1142–1156. [CrossRef]
2. Xiong, X.; Butler, J.; Chiang, K.; Efremova, B.; Fulbright, J.; Lei, N.; McIntire, J.; Oudrari, H.; Sun, J.; Wang, Z.; *et al.* VIIRS on-orbit calibration methodology and performance. *J. Geophys. Res. Atmos.* **2014**, *119*, 5065–5078. [CrossRef]
3. Jackson, J.; Liu, H.; Laszlo, I.; Kondragunta, S.; Remer, L.A.; Huang, J.; Huang, H.-C. Suomi-NPP VIIRS Aerosol Algorithms and Data Products. *J. Geophys. Res. Atmos.* **2013**, *118*, 12,673–12,689. [CrossRef]
4. Justice, C.O.; Román, M.; Csiszar, I.; Vermote, E.F.; Wolfe, R.E.; Hook, S.J.; Friedl, M.; Wang, Z.; Schaaf, C.B.; Miura, T.; *et al.* Land and cryosphere products from Suomi NPP VIIRS: Overview and status. *J. Geophys. Res. Atmos.* **2013**, *118*, 9753–9765. [CrossRef] [PubMed]
5. Wang, M.; Liu, X.; Tan, L.; Jiang, L.; Son, S.; Shi, W.; Rausch, K.; Voss, K. Impacts of VIIRS SDR performance on ocean color products. *J. Geophys. Res. Atmos.* **2013**, *118*, 10347–10360. [CrossRef]
6. Barnes, W.; Salomonson, V. MODIS: A global image spectroradiometer for the earth observing system. *Crit. Rev. Opt. Sci. Technol.* **1993**, *47*, 285–307.
7. McClain, C.; Hooker, S.; Feldman, G.; Bontempi, P. Satellite data for ocean biology, biogeochemistry, and climate research. *Eos Trans. AGU* **2006**, *87*, 337–339. [CrossRef]
8. King, M.D.; Menzel, W.P.; Kaufman, Y.J.; Tanre, D.; Gao, B.-C.; Platnick, S.; Ackerman, S.A.; Remer, L.A.; Pincus, R.; Hubanks, P.A. Cloud and aerosol and water vapor properties, precipitable water, and profiles of temperature and humidity from MODIS. *IEEE Trans. Geosci. Remote Sen.* **2003**, *41*, 442–458. [CrossRef]
9. Justice, C.O.; Vermote, E.; Townshend, J.R.G.; Defries, R.; Roy, D.P.; Hall, D.K.; Salomonson, V.V.; Privette, J.L.; Riggs, G.; Strahler, A.; *et al.* The moderate resolution imaging spectroradiometer (MODIS): Land remote sensing for global change research. *IEEE Trans. Geosci. Remote Sen.* **1998**, *36*, 1228–1249. [CrossRef]
10. Cao, C.; Xiong, X.; de Luccia, F.; Liu, Q.; Blonski, S.; Pogorzala, D.; Oudrari, H. *VIIRS SDR User's Guide*; Tech. Rep. NESDIS 142; U.S. Dept. Commerce, NOAA: Silver Spring, MD, USA, Febuary 2013.
11. Oudrari, H.; McIntire, J.; Xiong, X.; Butler, J.; Lee, S.; Lei, N.; Schwarting, T.; Sun, J. Prelaunch Radiometric Characterization and Calibration of the S-NPP VIIRS Sensor. *IEEE Trans. Geosci. Remote Sen.* **2015**, *53*, 2195–2210. [CrossRef]
12. Joint Polar Satellite System (JPSS). *VIIRS General Test Plan (GTP)*; Goddard Space Flight Center: Greenbelt, MD, USA, 2011.
13. Joint Polar Satellite System (JPSS). *VIIRS SDR Algorithm Theoretical Basis Document (ATBD)*; Goddard Space Flight Center: Greenbelt, MD, USA, 2011.
14. Joint Polar Satellite System (JPSS). *VIIRS Reflective Solar Bands—Performance Verification Report (PVR)*; Goddard Space Flight Center: Greenbelt, MD, USA, 2011.
15. McIntire, J.; Moyer, D.; Oudrari, H.; Xiaoxiong, X. Pre-launch radiometric characterization of JPSS-1 VIIRS thermal emissive bands. *Remote Sens.* **2015**. in press.
16. McIntire, J.; Young, J.B.; Moyer, D.; Waluschka, E.; Oudrari, H.; Xiong, X. Analysis of JPSS J1 VIIRS Polarization Sensitivity Using the NIST T-SIRCUS. *Proc. SPIE* **2015**, *9607*, 960713.

17. Waluschka, E.; McCorkel, J.; McIntire, J.; Moyer, D.; McAndrew, B.; Brown, S.W.; Lykke, K.; Young, J.B.; Fest, E.; Butler, J.; *et al.* J1 VIIRS polarization narrative: testing and performance. *Proc. SPIE* **2015**, *9607*, 960712.
18. Moeller, C.; Schwarting, T.; McIntire, J.; Moyer, D. JPSS-1 VIIRS pre-launch spectral characterization and performance. *Proc. SPIE* **2015**, *9607*, 960711.
19. Moyer, D.; McIntire, J.; Oudrari, H.; McCarthy, J.; Xiaosiong, X.; de Luccia, F. JPSS-1 VIIRS pre-launch response *versus* scan-angle testing and performance. *Remote Sens.* **2015**, in press.
20. Lee, S.; Cao, C. JPSS-1 VIIRS Prelaunch RSB/DNB RVS characterization and water vapor correction. *Proc. SPIE* **2015**, *9607*, 96071R.
21. Joint Polar Satellite System (JPSS). *VIIRS Product Requirement Document (PRD)*; Goddard Space Flight Center: Greenbelt, MD, USA, 2011.
22. Joint Polar Satellite System (JPSS). *J1 VIIRS Near Field Response (NFR)*; Performance Verification Report (PVR); Goddard Space Flight Center: Greenbelt, MD, USA, 2011.

Article

Pre-Launch Radiometric Characterization of JPSS-1 VIIRS Thermal Emissive Bands

Jeff McIntire [1,*], David Moyer [2], Hassan Oudrari [1] and Xiaoxiong Xiong [3]

[1] Science Systems and Applications, Inc., Lanham, MD 20706, USA; Hassan.oudrari@ssaihq.com
[2] The Aerospace Corporation, El Segundo, CA 90245, USA; david.i.moyer@noaa.gov
[3] NASA Goddard Space Flight Center, Greenbelt, MD 20771, USA; xiaoxiong.xiong-1@nasa.gov
* Correspondence: jeffrey.mcintire@ssaihq.com; Tel.: +1-301-867-2073

Academic Editors: Changyong Cao, Richard Müller and Prasad S. Thenkabail
Received: 28 October 2015; Accepted: 25 December 2015; Published: 7 January 2016

Abstract: Pre-launch characterization and calibration of the thermal emissive spectral bands on the Joint Polar Satellite System (JPSS-1) Visible Infrared Imaging Radiometer Suite (VIIRS) is critical to ensure high quality data products for environmental and climate data records post-launch. A comprehensive test program was conducted at the Raytheon El Segundo facility in 2013–2014, including extensive environmental testing. This work is focused on the thermal band radiometric performance and stability, including evaluation of a number of sensor performance metrics and estimation of uncertainties. Analysis has shown that JPSS-1 VIIRS thermal bands perform very well in relation to their design specifications, and comparisons to the Suomi National Polar-orbiting Partnership (SNPP) VIIRS instrument have shown their performance to be comparable.

Keywords: JPSS; VIIRS; calibration; pre-launch; thermal bands

---

1. Introduction

The Visible Infrared Imaging Radiometer Suite (VIIRS) is a key sensor aboard the Joint Polar Satellite System (JPSS-1) mission, scheduled for launch in late 2016. JPSS-1 VIIRS is the second VIIRS instrument (the first currently flying on the SNPP (Suomi National Polar-orbiting Partnership) satellite [1,2]) and is the follow-on to heritage sensor MODIS [3]. VIIRS is a cross-track scanning radiometer capable of making continuous global observations twice daily. The sensor observes the Earth through 21 spectral bands and one pan-chromatic band covering a spectral range from 0.4 μm to 12.6 μm. These bands are used in support of a number of environmental data records covering land, ocean, and atmospheric science disciplines [4–6]. Seven of these spectral bands are considered thermal emissive (covering 3.7 μm to 12.6 μm) and are listed in Table 1 along with some of their characteristics. The performance and calibration of the VIIRS instrument thermal emissive bands pre-launch is critical to ensure high quality science data records are produced on-orbit, and that the calibration on-orbit is maintained. The pre-launch radiometric calibration and characterization of these thermal bands is the focus of this paper.

1.1. VIIRS Emissive Bands Overview

The VIIRS optical path from the entrance aperture, as viewed by the emissive bands, first passes though an afocal three mirror anastigmat and a fold mirror, known as the rotating telescope assembly (RTA) [1,7]. The RTA is a cross-track scanner that rotates once about every 1.78 s, viewing a ±56 degree swath through the Earth view (EV) port as well as three calibration views, each about 1 degree wide: a view of deep space (SV) used to determine the dark offset at about −66 degrees off nadir; a view of an on-board blackbody (referred to as the OBCBB) at about 100 degrees off nadir, used for on-orbit thermal band calibration; and a view of the solar diffuser at about 159 degrees off nadir, used for reflective band calibration on-orbit. The RTA directs the light onto a two-sided rotating fold mirror, known as the half angle mirror (HAM). The HAM, rotating at half the speed of the RTA, de-rotates the light beam and directs it into the fixed aft-optics. The light entering the aft-optics passes through a fold mirror, a four mirror anastigmat, and two dichroic beamsplitters. The first beamspitter reflects the visible and near-infrared light onto a focal plane assembly (FPA), while longer wavelengths pass through to the second dichroic beamsplitter. The second beamsplitter separates the short and mid-wave infrared (~1–7 μm) from the long-wave infrared (~7–16 μm), and directs the light onto two separate FPAs (via a steering mirror for the short and mid-wave infrared). Both of these FPAs will be cryogenically controlled at about 80.5 K on-orbit by means of a three-stage radiative cooler.

Table 1. VIIRS thermal bands with center wavelengths, bandwidths, gain mode, spatial resolution at nadir, upper and lower dynamic range limits, and typical scene temperature as defined by the sensor specification [8]. SG, HG, and LG refer to single gain, high gain, and low gain, respectively.

Band	Gain Mode	Wavelength [nm]	Bandwidth [nm]	Resolution [m]	T_{MIN} [K]	T_{TYP} [K]	T_{MAX} [K]
I4	SG	3740	380	375	210	270	353
I5	SG	11450	1900	375	190	210	340
M12	SG	3700	180	750	230	270	353
M13	HG	4050	155	750	230	300	343
M13	LG	4050	155	750	343	380	634
M14	SG	8550	300	750	190	270	336
M15	SG	10763	1000	750	190	300	340
M16	SG	12013	950	750	190	300	340

The emissive bands are located on the two temperature controlled FPAs: the mid-wave infrared (MWIR) bands I4, M12, and M13 are located on one FPA (along with five reflective bands covering ~1.2 μm to 2.3 μm) and the long-wave infrared (LWIR) bands I5, M14, M15, and M16 are located on a dedicated FPA. All of the thermal bands use HgCdTe photo-voltaic detector arrays of 16 (or 32 for I4 and I5), staggered perpendicular to the direction of scan on the focal plane. A microlens focuses the light onto each detector after it passes through a spectral bandpass filter, defined by the center wavelength and bandpass listed in Table 1. Two bands, I4 and I5, are imaging bands with nadir resolutions of ~375 m; the remaining thermal bands have resolutions of about 750 m at nadir. One band (M16) has two detector arrays (M16A and M16B) which are combined on-orbit through time-delayed integration. In addition, the 4.05 μm band M13 has two gain states (high and low); the M13 low gain stage is used to access very high temperature scenes (useful on-orbit for fire detection).

On-orbit, the calibration of the thermal bands is referenced to an internal blackbody, the OBCBB. The OBCBB is a V-groove blackbody with an emissivity above 99.6% in the spectral regions accessed by VIIRS thermal bands. The OBCBB is viewed once per scan and on-orbit will be controlled at a fixed temperature (~292 K); it can also be cycled through a series of temperature levels between instrument ambient and 315 K. During pre-launch environmental testing, instrument ambient ranges from ~253–276 K, while on-orbit instrument ambient is expected to be about 267 K (based on SNPP experience [1]).

1.2. Testing Overview

Before launch, a series of comprehensive performance tests were conducted for JPSS-1 VIIRS at the Raytheon El Segundo facility from 2012–2014 [9,10]. This work will focus on the radiometric portions of the test program performed under environmental conditions in the summer/fall of 2014. During the thermal vacuum test program, the bulk of testing was performed at three instrument temperature plateaus (referred to as cold, nominal, and hot) designed to cover the range of possible on-orbit conditions.

Four blackbodies were used as sources during radiometric testing. VIIRS thermal band calibration is referenced pre-launch to an external, NIST traceable blackbody (known as the BCS); the BCS is a cavity type blackbody with an emissivity greater than 99.96% and temperature levels ranging from 190 K to 345 K with a temperature uncertainty of less than 0.06 K (at 10 μm and 300 K). The BCS was located inside the thermal vacuum chamber at approximately 41 degrees scan angle. However, the BCS maximum temperature of 345 K was insufficient to characterize the M13 low gain state; as a result, an additional high temperature blackbody (referred to as the TMCBB), referenced to the BCS, was used for M13 low gain calibration. The TMCBB is an extended area, emissive source that was viewed through collimating optics located outside the thermal vacuum chamber and accessed temperatures ranging from laboratory ambient (about 294 K) up to 763 K. The TMCBB was viewed through a ZnSe window in the thermal vacuum chamber at about −1 degrees scan angle. In addition, an external cold target (known as the SVS) was used to simulate the deep space view. The SVS was located inside the thermal vacuum chamber, viewed at a scan angle of about −65.7 degrees, and controlled at approximately 90 K. The fourth and final blackbody was the OBCBB, which was described above.

The radiometric testing for the thermal bands can be divided into two sections: performance and stability (see Table 2). For the performance testing, the sources (either internal or external) were cycled through a series of discrete temperature levels; for stability testing, the source temperatures remained fixed while the instrument conditions were varied. The performance tests were used to determine the model coefficients relating the detector response to the radiance as well as assess a number of performance metrics, such as response non-linearity and detector-to-detector uniformity. The BCS and TMCBB were cycled through their respective temperature ranges as indicated in Table 2, while the OBCBB was fixed at 292 K. First the BCS was commanded through 12 discrete temperature settings from 190 K to 345 K; while this was happening, the TMCBB temperature also increased from laboratory ambient up to ~375 K. The TMCBB temperature then continued to increase up to 763 K while the BCS temperature remained fixed at 345 K. Note that the TMCBB temperatures listed here refer to blackbody temperature and that the equivalent scene temperature at-aperture is lower. There were additional, special tests in which the BCS temperature was cycled at elevated focal plane temperatures (82 K and 83.5 K). In separate tests, the OBCBB was cycled from instrument ambient to 315 K in discrete temperature levels while the BCS and TMCBB temperatures were fixed (at about 300 K and laboratory ambient, respectively). After reaching 315 K, the OBCBB temperature control was turned off and the OBCBB was allowed to cool down to ~292 K, during which data was also taken.

The stability tests were performed to check the instrument response to variations in operating conditions. For these tests, the BCS and OBCBB were commanded to fixed temperatures (300 K and 292 K, respectively). First, the instrument stability was tested *versus* time for about 3–8 h. The thermal band radiometric stability was also tested with respect to changes in the instrument temperatures while transitioning between plateaus. In particular, the stability was assessed with respect to the electronics temperature and instrument temperature. Stability was also investigated with respect to BUS voltage and FPA temperature changes.

Table 2. Radiometric testing performed for JPSS-1 VIIRS emissive bands during environmental testing [10]. C, N, and H refer to cold, nominal, and hot instrument plateaus; T refers to transitions between instrument plateaus. A and B denote the primary and redundant electronic hardware configuration settings.

Test Type	Instrument Plateaus	Electronics Sides	T_{BCS} [K]	T_{TMCBB} [K]	T_{OBCBB} [K]	T_{FPA} [K]	V_{BUS} [V]
Performance	C, N, H	A, B	190–345	294–375	292	80.5	28
Performance	N, H	A	190–345	294–375	292	82	28
Performance	N	A	190–345	294–375	292	83.5	28
Performance	C, N, H	A, B	345	388–763	292	80.5	28
Performance	C, N, H	A	300	294	Amb - 315	80.5	28
Stability	C, N, H	A, B	270	294	292	80.5	28
Stability	T	A	270	294	292	80.5	28
Stability	C, H	A, B	270	294	292	80.5	27–32
Stability	N	A	270	294	292	80.5–83.5	28

2. Methodology

In this section, the methodology used to determine the radiometric model coefficients is reviewed, as well as the derivation of a number of performance and stability metrics. In addition, an uncertainty assessment is conducted on the radiance and brightness temperature products.

2.1. Radiometric Performance

The at-detector radiance when VIIRS views the BCS is the sum of the source radiance and the radiance from a number of emissive sources along the optical path (RTA, HAM, and aft optics) [7,11], or

$$\begin{aligned} L_{BCS-path} &= \rho_{RTA} \cdot \rho_{HAM} \cdot RVS_{BCS} \cdot \rho_{aft} \cdot \epsilon_{BCS} \cdot L_{BCS} \\ &+ (1-\rho_{RTA})\,\rho_{HAM} \cdot RVS_{BCS} \cdot \rho_{aft} \cdot L_{RTA} \\ &+ (1-\rho_{HAM} \cdot RVS_{BCS})\,\rho_{aft} \cdot L_{HAM} + \left(1-\rho_{aft}\right) L_{aft} \end{aligned} \tag{1}$$

The reflectance factors (ρ_{RTA} and ρ_{aft}) represent the product of individual reflectances of the RTA and aft-optics surfaces respectively and ϵ_{BCS} is the emissivity of the BCS. For VIIRS, the only optic that has a varying angle of incidence (AOI) is the HAM; here, the reflectance is represented by the product $\rho_{HAM}RVS$, where ρ_{HAM} is the reflectance at the SV HAM AOI. RVS is the AOI dependent, relative reflectance of the HAM [12]; in Equation (1), it is at the view angle of the BCS. The radiances of the sources (L) are determined via Planck's law, convolved over the extended bandpass of the spectral transmittance of each band [13]. The temperature of each source is determined from one or more thermistors located on or near each component. Similarly, the at-detector radiance from the OBCBB and SV views are given by [7,11]

$$\begin{aligned} L_{OBCBB-path} &= \rho_{RTA} \cdot \rho_{HAM} \cdot RVS_{OBCBB} \cdot \rho_{aft} \cdot \epsilon_{OBCBB} \cdot L_{OBCBB} \\ &+\rho_{RTA} \cdot \rho_{HAM} \cdot RVS_{OBCBB} \cdot \rho_{aft}\,(1-\epsilon_{OBCBB}) \\ &\times (F_{RTA} \cdot L_{RTA} + F_{SH} \cdot L_{SH} + F_{CAV} \cdot L_{CAV}) \\ &+ (1-\rho_{RTA})\,\rho_{HAM} \cdot RVS_{OBCBB} \cdot \rho_{aft} \cdot L_{RTA} \\ &+ (1-\rho_{HAM} \cdot RVS_{OBCBB})\,\rho_{aft} \cdot L_{HAM} + \left(1-\rho_{aft}\right) L_{aft} \end{aligned} \tag{2}$$

and

$$L_{SV-path} = (1-\rho_{RTA})\,\rho_{HAM}\cdot RVS_{SV}\cdot\rho_{aft}\cdot L_{RTA} + (1-\rho_{HAM}\cdot RVS_{SV})\,\rho_{aft}\cdot L_{HAM} + \left(1-\rho_{aft}\right)L_{aft} \quad (3)$$

The cold reference target (SVS) was at a low enough temperature that the source radiance in the SV was negligible. The relative contributions of light reflected off the OBCBB into the path by the RTA, the blackbody shield (SH), and the scan cavity (CAV) are denoted by F_{RTA}, F_{SH}, and F_{CAV}, respectively. Here ϵ_{OBCBB} is the emissivity of the OBCBB.

The path difference radiance between two sources (the latter of which is the SVS) is [7,11]

$$\begin{aligned}\Delta L_{BCS} &= \frac{L_{BCS-path}-L_{SV-path}}{\rho_{RTA}\cdot\rho_{HAM}\cdot\rho_{aft}} \\ &= RVS_{BCS}\cdot\epsilon_{BCS}\cdot L_{BCS} - \frac{(RVS_{BCS}-RVS_{SV})}{\rho_{RTA}}\left[L_{HAM}-(1-\rho_{RTA})\,L_{RTA}\right]\end{aligned} \quad (4)$$

and

$$\begin{aligned}\Delta L_{OBCBB} &= \frac{L_{OBCBB-path}-L_{SV-path}}{\rho_{RTA}\cdot\rho_{HAM}\cdot\rho_{aft}} \\ &= RVS_{OBCBB}\cdot\epsilon_{OBCBB}\cdot L_{OBCBB} \\ &\quad + RVS_{OBCBB}\,(1-\epsilon_{OBCBB})\,(F_{RTA}\cdot L_{RTA}+F_{SH}\cdot L_{SH}+F_{CAV}\cdot L_{CAV}) \\ &\quad - \frac{(RVS_{OBCBB}-RVS_{SV})}{\rho_{RTA}}\left[L_{HAM}-(1-\rho_{RTA})\,L_{RTA}\right]\end{aligned} \quad (5)$$

Note that the aft optics radiance contribution canceled in the above equations as it is common to both paths and that the reflectance factors (RTA, HAM, and aft-optics) were absorbed into the left hand side of each equation.

The path difference radiance is modeled as a quadratic polynomial in the offset corrected digital response [7,11], or

$$\Delta L = c_0 + c_1\cdot dn + c_2\cdot dn^2 \quad (6)$$

Both the BCS and OBCBB were transitioned through a series of temperature levels; the data acquired was used to determine the coefficients by fitting the path difference radiance *versus* the detector response within the dynamic range (as discussed in Section 3.1). A check on the linearity of Equation (6) was performed by computing

$$NL = \frac{max\left(\Delta_{fit}\right)}{L_{MAX}} \quad (7)$$

where the numerator is the maximum fitting residual for a linear fit and L_{MAX} is the upper limit of the specified dynamic range.

The low gain state of band M13 cannot be calibrated using either the BCS or the OBCBB like the high gain state, so the high temperature external blackbody (TMCBB) was used. The TMCBB was tied to the BCS by cross calibration in the temperature range where both sources overlap (~300–345 K), and then this calibration was extended to the high temperature region, using [7]

$$\begin{aligned}\Delta L_{TMCBB} &= RVS_{TMCBB}\cdot\epsilon_{TMCBB}\cdot L_{TMCBB} + RVS_{TMCBB}\left(1-\tau_{TMC-op}\right)L_{TMC-op} \\ &\quad + RVS_{TMCBB}\left(1-\rho_{window}\right)L_{window} \\ &\quad - \frac{(RVS_{TMCBB}-RVS_{SV})}{\rho_{RTA}}\left[L_{HAM}-(1-\rho_{RTA})\,L_{RTA}\right]\end{aligned} \quad (8)$$

Additional source terms in the TMCBB path were included for the TMC optics (L_{TMC-op}) and the thermal vacuum chamber window (L_{window}), as the TMCBB was outside the thermal vacuum chamber. ρ_{window} is the reflectance of the chamber window. The transmittance of the TMC optics (τ_{TMC-op}) was used to facilitate the calibration transfer from the BCS to the TMCBB. As with Equations (4) and (5), the reflectance factors (RTA, HAM, and aft-optics) were absorbed into the left hand side of the equation.

The radiometric sensitivity was determined by fitting the signal to noise ratio (SNR) to the path difference source radiance [7,11], or

$$SNR = \frac{\Delta L}{NEdL} = \frac{\Delta L}{\sqrt{k_0 + k_1 \cdot \Delta L + k_2 \cdot \Delta L}} \tag{9}$$

As the BCS provided the more stable source, the SNR results derived from that source were closer to the true sensor SNR. The NEdT (the fluctuation in the scene temperature equivalent to the system noise) was computed by the equation

$$NEdT = \frac{\Delta L}{SNR \cdot \frac{\partial L}{\partial T}} \tag{10}$$

The derivative is of Planck's law with respect to the source temperature.

Inverting Equation (4) and substituting in Equation (6), the retrieved EV radiance is [7,11]

$$L_{EV-ret} = \frac{\left(c_0 + c_1 \cdot dn + c_2 \cdot dn^2\right)}{RVS_{EV}} + \frac{\left(RVS_{EV} - RVS_{SV}\right)}{RVS_{EV} \cdot \rho_{RTA}} \left[L_{HAM} - \left(1 - \rho_{RTA}\right) L_{RTA}\right] \tag{11}$$

An additional factor,

$$\frac{\Delta L_{OBCBB}}{c_0 + c_1 \cdot dn_{OBCBB} + c_2 \cdot dn^2_{OBCBB}} \tag{12}$$

is included on-orbit to adjust the calibration coefficients for scan-to-scan variations in the detector responsivity. For the purposes of this work, this factor will also be included when considering the uncertainty estimate in Section 2.3, so as to better estimate the expected on-orbit uncertainty. The detector-to-detector uniformity (or striping) is quantified by the following metric

$$RRU = \frac{\left|L_{EV-ret} - \langle L_{EV-ret}\rangle_D\right|}{NEdL} \tag{13}$$

where the average EV retrieved radiance is over all detectors in a band and the NEdL was derived from Equation (9).

2.2. Radiometric Stability

Limited knowledge of how the instrument behaves with changing instrument conditions was obtained through performance testing. Stability tests were performed to more comprehensively investigate the impact of variations in instrument condition on the thermal band radiometry. The radiometric stability was assessed by trending the linear gain ($1/c_1$), derived from Equation (6) assuming c_0 and c_2 are negligible, or

$$\frac{1}{c_1} \cong \frac{dn}{\Delta L} \tag{14}$$

The linear gain was trended for all the stability cases listed in Table 2: with time, with changes in instrument temperature, with variation of the FPA temperature, and with changes in the BUS voltage. Assessments were made in terms of unit time, temperature, or voltage.

2.3. Uncertainty

The 1-sigma uncertainty of the retrieved EV radiance is estimated by propagating the uncertainty of the terms in Equation (11) following the methodology outlined in [14,15],

$$u^2\left(L_{EV-ret}\right) = \sum_{i=1}^{N}\left(\frac{\partial L_{EV-ret}}{\partial x_i}\right)^2 u^2\left(x_i\right) + 2\sum_{i=1}^{N-1}\sum_{j=i+1}^{N}\left(\frac{\partial L_{EV-ret}}{\partial x_i}\right)\left(\frac{\partial L_{EV-ret}}{\partial x_j}\right) u\left(x_i, x_j\right) \quad (15)$$

Here $u(x_i)$ is the uncertainty of the underlying variable x_i that enters into the calculation of the radiance retrieval and $u(x_i, x_j)$ is the covariance between x_i and x_j. The EV retrieved radiance was defined in Equation (11) and is a function of c_0, c_1, c_2, L_{OBCBB}, L_{HAM}, L_{RTA}, L_{CAV}, L_{SH}, F_{SH}, F_{CAV}, F_{RTA}, RVS_{OBCBB}, RVS_{EV}, RVS_{SV}, ϵ_{OBCBB}, ρ_{RTA}, dn_{EV}, and dn_{OBCBB}. Note that the total uncertainty here is for a single EV pixel; the effects of aggregation and scan angle were also investigated.

In general, the covariance terms were not directly calculated (the exception is the covariance terms between the radiometric coefficients c_0, c_1, and c_2); a direct calculation of these terms is beyond the scope of this work. However, an upper bound on the covariance terms is determined through use of the Schwarz inequality [14], or

$$\left|u\left(x_i, x_j\right)\right| \leqslant u\left(x_i\right) u\left(x_i\right) \quad (16)$$

Results will be presented without covariance terms (excepting those terms between the radiometric coefficients) as a baseline and with covariance terms determined using the Schwarz inequality as a worst case estimate.

What follows is a brief description of the individual uncertainty contributors that enter into the full uncertainty analysis.

The radiance uncertainty for each of the radiances that factor into the present calculation (L_{BCS}, L_{OBCBB}, L_{HAM}, L_{RTA}, L_{SH}, and L_{CAV}) is the RSS of a number of uncertainty contributors. Each of these radiances was converted from a temperature provided by one or more thermistors once per scan (or for the external sources, temperature readings about every 10 s) using the Planck equation, integrated over the spectral response of each band [13]. The error is composed of four components: temperature, spectral, interpolation, and statistical uncertainties. The temperature uncertainties used in this analysis are 0.057 K, 0.03 K, 0.59 K, 9.0 K, 3.0 K, and 6.0 K, respectively [16]. The radiance uncertainty associated with each temperature error was determined by taking the absolute value of the difference between the Planck radiance with and without the temperature uncertainty, or

$$max\left|L\left(T, \lambda\right) - L\left(T \pm \Delta T, \lambda\right)\right| \quad (17)$$

The spectral errors used here are 1.2 nm for the MWIR bands and 4 nm for the LWIR bands [16]. The radiance uncertainties due to spectral errors were determined in the same manner as the temperature uncertainties,

$$max\left|L\left(T, \lambda\right) - L\left(T, \lambda \pm \Delta\lambda\right)\right| \quad (18)$$

On-orbit, the calibration relies on pre-calculated tables to convert source temperature to radiance with temperature intervals of 0.25 K. A linear interpolation is used to determine the radiance at the actual source temperature every scan. The interpolation uncertainty is determined by the following,

$$\frac{1}{2}\left(T - T_0\right)\left(T - T_1\right) f^{(2)}\left(T\right) \quad (19)$$

where T_0 and T_1 are the table values bracketing the measured T and the second derivative is of Planck's radiation law with respect to the measured T. The statistical uncertainties were the standard deviation of the radiances determined within one data collection (or over 100 scans).

The RVS and its associated uncertainties were determined in earlier testing [12]. The derived RVS uncertainty was a combination of fitting error and measurement error; the band averaged values used in this work are 0.05%, 0.12%, 0.07%, 0.06%, 0.08%, 0.07%, 0.06% for bands I4–I5 and M12–M16. No uncertainty due to emission *versus* scan was included in the present work.

The uncertainties for the OBCBB reflectance shape factors are all taken to be 0.01 (the precision error of the calculation). The uncertainty for the reflectance of the RTA is 0.5% at 270 K [16]. This uncertainty was used for all source temperature levels. The OBCBB emissivity was measured at 3.39 μm; the error in the emissivity measurement was determined to be 0.04% for I4, M12, and M13, 0.05% for M14, and 0.06% for I5, M15, and M16 [16].

The uncertainty in the response was the RSS of the precision and accuracy errors for the background subtracted digital response. The precision error was the standard deviation over all analyzed samples and scans. The accuracy error was zero for the purposes of this work; any biases common to all sectors were removed in the background subtraction. The exception is a known bias between fixed and auto gain configurations for M13, which was not included in this work.

The vertical least-squares fitting algorithm used in this work determined the vertical deviations of the set of data points from the fit. The minimum of the vertical deviations was computed by setting the partial derivatives with respect to the coefficients equal to zero. This leads to the a matrix equation, the solution to which determined the radiometric coefficients in Equation (6). This algorithm also produced 1-sigma uncertainties and covariance terms. This approach assumes that the uncertainties in ΔL are roughly constant over the data points used (and uncertainties in dn are negligible). However, this algorithm only included some effects from precision error (random statistical variations), but excluded any bias uncertainties (accuracy error). Furthermore, it should be noted that the uncertainties are valid only insomuch as the model itself is valid. An initial investigation into the quality of the model is discussed below. For the purposes of this work, the radiometric model is considered sufficiently valid to proceed with the uncertainty analysis.

The above procedure was followed at each instrument condition (temperature plateaus and electronics sides) as well as for both internal and external sources. In the present work, only the uncertainty derived using the BCS data is shown. On-orbit, the pre-launch radiometric coefficients will be used as a baseline, to which a scan-by-scan correction is applied (Equation (12)); the coefficients derived from the OBCBB during its warm-up/cool-down cycle will be used as a check on the calibration and replace the BCS coefficients only if a large change in behavior occurs.

2.4. Model Validity

As stated above, the radiometric fitting results are valid only insomuch as the model itself is valid. The functional form of the model was investigated by varying the order of the polynomial used in the fitting. Linear, quadratic, and cubic polynomials were employed. The full error analysis was conducted for each, and the results were compared to determine any relative improvement with increasing polynomial order. In addition, a parametric model was used to vary the source temperatures in the radiometric model in order to minimize the difference between the retrieved radiance and the source radiance determined from Planck's radiation law. This would determine if there was any temperature bias between the BCS and OBCBB sources.

3. Results

3.1. Data Quality and Reduction

This section briefly describes the data selection and reduction performed to determine the radiometric fitting coefficients as well as the performance and stability metrics. The sensor specification defines the dynamic range over which VIIRS thermal bands must be calibrated [8]. However, valid science data may exist for some bands outside the specified range, and in some cases data inside the specified range is of low quality. In this work the fitted range was modified to include all available measured data not contaminated at high temperature by saturation and for which the SNR was greater than 5 at low temperature. The dynamic ranges are shown in Figure 1: the black lines represent the specified dynamic range; the red lines indicate the scene temperatures for which the SNR is equal to 5 (determined from Equation (9)) and the saturation temperatures; and the blue lines indicate the extent of the measurements used in the fitting. Note that the minimum and maximum BCS temperatures were 190 K and 345 K; these were in many cases limiting factors on modifying the fitting range. No band saturated below 345 K. However, the SNR fell below 5 inside the specified dynamic range for I4 (due to high noise, both 210 K and 230 K BCS levels were below the SNR threshold) and M14 (the large difference in RVS from SV to EV angle resulted in a negative offset corrected response at the lowest measured scene temperature). The lowest available measured levels that met the SNR threshold requirement for these two bands were 247 K (I4) and 210 K (M14). M13 low gain is not shown in Figure 1, but all available M13 low gain data was used in the fitting (TMCBB at-aperture scene temperatures from 355 K to 644 K). M13 low gain was not observed to saturate in thermal vacuum testing and the SNR was well above 5 at the lowest measured temperature.

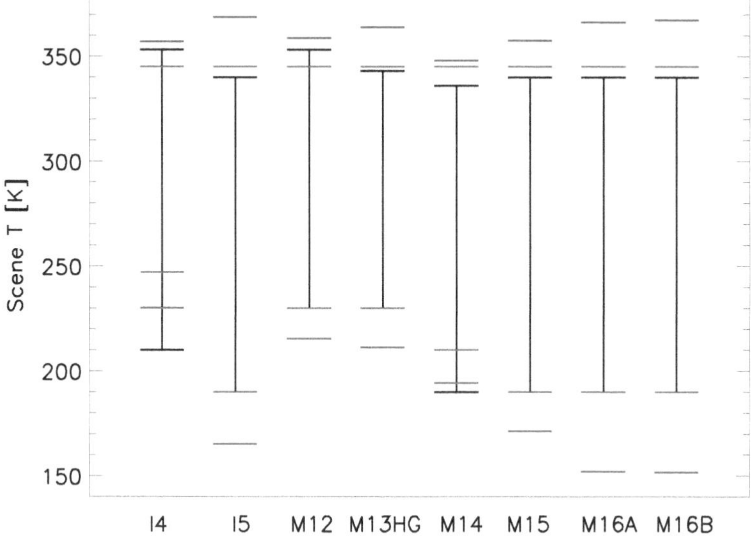

Figure 1. Specified dynamic ranges shown for the VIIRS thermal bands (black lines). Maximum and minimum scene temperatures used in the fitting as shown with blue lines. Saturation and low temperature SNR threshold are also shown (red lines).

Data collections were recorded at various source temperature levels, during each of which the source temperatures were stable (excepting when the OBCBB was cooling down). For each data collection, 100 scans were recorded (50 for each HAM side). For each scan and detector, 48 moderate resolution pixels were recorded for both the OBCBB view and SV (96 pixels for the imaging bands). The BCS and TMCBB views subtended a subset of the EV pixels. Each view was averaged over all valid pixels for a given scan. The SV was used as a dark reference which was subtracted from the signal in all other views to produce the *dn*. The *dn* was then averaged over all valid scans in a given data collection. 1-sigma estimates were also determined for each *dn*, and the corresponding SNR was derived. Outlier rejection was performed at each stage of the calculation. The resulting dn_{BCS}, dn_{OBCBB}, and dn_{TMCBB} were inserted into Equation (6).

VIIRS telemetry data was also extracted and averaged over all scans in each data collection. The telemetry data largely consisted of temperature readings, for both internal and external sources as well as instrument component temperatures. In some cases, more than one thermistor was used for a source in the thermal model; in those cases the average value was used. Data was supplied on the external sources (BCS, TMCBB, and SVS) about every 10 s, from which collection averaged temperature values were determined, and then radiances derived (a Planck function of the temperature reading convolved with the band averaged spectral response functions [13]) and inserted into the equations in Section 2. The telemetry of interest for the thermal model were the RTA, CAV, SH, and HAM temperatures as described in Section 2; these temperatures were also convolved with the band averaged spectral response functions and inserted into the equations in Section 2. Note that there was no direct temperature reading from the RTA, so that the CAV temperature was used with an instrument temperature dependent offset (~4 K). Temperature data from the FPAs, electronics (ELEC), and instrument (OMM) was also collected, and used in the stability assessment.

Some of the input parameters, such as RVS [12], ϵ_{OBCBB}, and ρ_{RTA} [16], were measured in previous testing and their derivation will not be described in this work.

Once the component radiances were determined, the path difference radiances from Equations (4), (5), and (8) were computed. These path difference radiances and their corresponding dn were then inserted into Equation (6), and fits of the radiometric calibration coefficients c_i were performed over all valid data within the fitting range defined in Figure 1. These fit coefficients are next used to derive the retrieved radiance and performance metrics (non-linearity and uniformity). The SNR was also fit over the same range via Equation (9); this SNR fit was used to estimate the SNR and NEdT at T_{TYP}. Lastly, the uncertainty was propagated using Equation (15) and the various uncertainty contributors described in Section 2.3.

To determine the saturation and gain transition scene temperatures, partial views as VIIRS scanned across the BCS and TMCBB sources were used. Here the scan averaged dark offset was subtracted from each pixel of the scan across the sources profiles; the largest dn was averaged over all scans in a data collection and then converted into a scene radiance using Equation (11), from which the scene temperature was determined via Planck's law.

3.2. Radiometric Performance

The offset corrected detector response *versus* path difference radiance using BCS data for the MWIR bands, fit using a quadratic polynomial as described in Equation (6), are shown in Figure 2a for detector 9. The corresponding fitting residuals in % are shown in Figure 2b. For bands I4, M12, and M13 high gain, a significant portion of the 12 bit analog-to-digital converter (ADC) range was used. The radiance residuals for these bands increase at lower temperatures (worst case of up to ~8% at 230 K for M12); the 230 K scene temperature translates to very low radiances in this spectral region, and this places a significant constraint on the fitting. M13 low gain (not shown) also exhibits this behavior at the low end of its measured range (near 355 K). Similarly, Figure 3a,b show the response *versus* radiance curves along with their corresponding radiance residuals in % for the LWIR bands using BCS data. For these bands, most of the 12 bit ADC range was used. The radiance residuals for these bands was generally small (less than 1% for all detectors, and much less when the lowest temperature level was excluded). The exception was detector 5 in M16B, which showed a higher residual than other detectors due to higher than average noise. All measured cases (instrument plateaus, electronics sides, HAM sides, and FPA temperatures as listed in Table 2) showed similar behavior in terms of fitting and residuals.

In general, the c_0 coefficient is on the order of 10^{-2} or less and is roughly consistent over plateaus and electronic sides. I5, M14, and M15 had the largest offsets with absolute values up to ~0.03 $[W/m^2/sr/\mu m]$. The detector dependence is fairly stable *versus* instrument condition for the LWIR bands, while some variation with detector was observed for the MWIR bands. There is a good deal of detector variation in the c_0 trends *versus* instrument temperature. However, it should be noted that the variation of the offset with plateau is on the order of 10^{-2} or less. In the majority of cases, the 2-sigma error bars overlap for the different plateaus (electronics side dependent), indicating that c_0 is generally consistent over instrument temperature conditions. M14 is the exception, in that there appears to be some instrument temperature dependence. In addition, c_0 does not show any noticeable trend for the three FPA temperatures measured at nominal plateau, electronic side A (see Table 2). For M13 low gain, the derived c_0 was non-zero and positive (between 0 and 0.5 $[W/m^2/sr/\mu m]$).

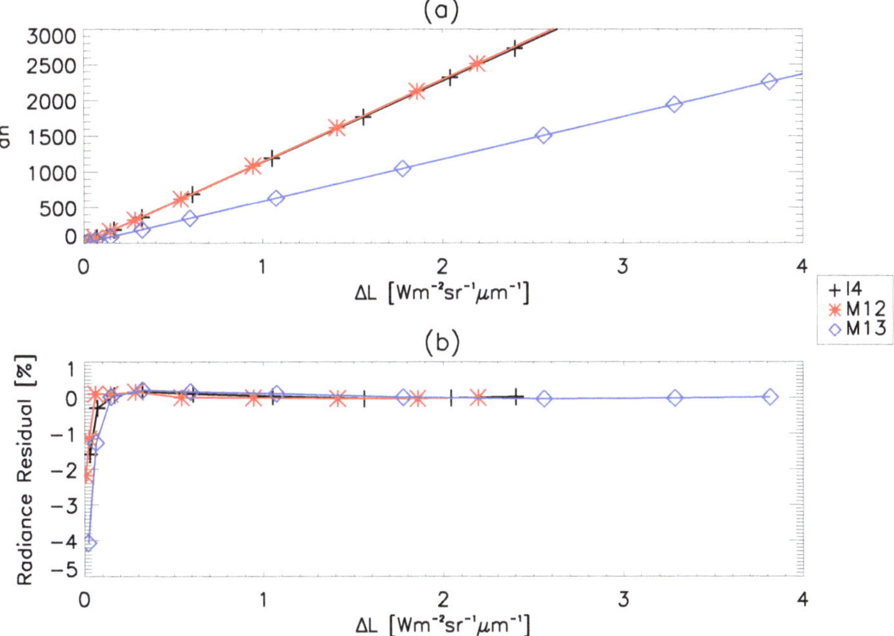

Figure 2. The offset corrected detector response (**a**) and the radiance residual in % (**b**) shown *versus* path difference radiance. The symbols in (**a**) represent measured data and the lines indicate quadratic fits. The measured data was taken from nominal plateau, HAM side A, electronics side A, FPA temperature 80.5 K using detector 9 (BCS data).

The detector-to-detector patterns for the gains ($1/c_1$) are very consistent from instrument temperature plateau to plateau as well as across electronics side. In addition, there is a strong odd – even detector dependence observed in bands I5, M12, M13 (both high and low gain), and M15. The gains trend fairly linearly over instrument temperature plateau (see Table 3), with bands I4, M12, and M13 (both high and low gain) roughly consistent over plateaus and LWIR bands decreasing with instrument temperature. In general, the LWIR bands decrease by about 2% or less and the MWIR bands change by less than 1% over the ~20 K measured instrument temperature range. Here the temperature variation is outside the 2-sigma error bars in most cases for the LWIR bands, but consistent for the MWIR bands. In addition, the electronics sides are not consistent for bands I5 and M14–M16. The gains determined at the three measured FPA temperatures for nominal plateau, electronics side A are consistent for bands I4, M12, and M13 high gain; however, for the LWIR bands, the gains decrease with increasing FPA temperature by between 3%–8% for every 1.5 K (see Table 3).

For the LWIR bands, the detector pattern of the nonlinear term (c_2) is roughly constant over plateau and electronics side; however, for the MWIR bands, the pattern is less well defined. In terms of magnitude, the nonlinear term is consistently on the order of 10^{-7} or less. For most bands and electronics sides, the nonlinear term exhibits a small, roughly linear increase with instrument temperature. There is some noticeable detector variation in the trending for the MWIR bands. In the majority of cases, the 2-sigma error bars overlap. The most glaring exception is M14. c_2 for the

MWIR bands shows almost no change with FPA temperature, while c_2 increases with increasing FPA temperature for the LWIR bands.

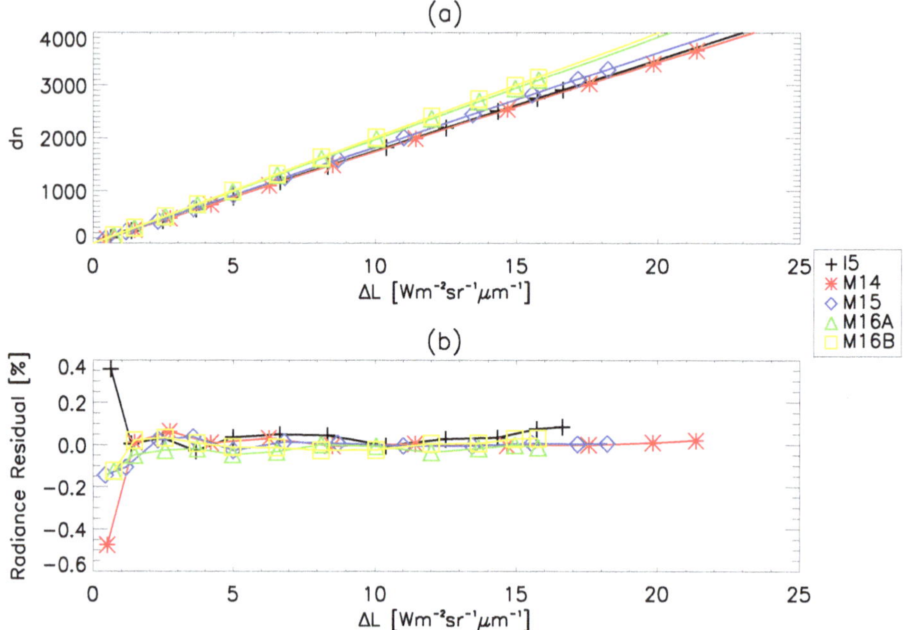

Figure 3. The offset corrected detector response (**a**) and the radiance residual in % (**b**) shown *versus* path difference radiance. The symbols in (**a**) represent measured data and the lines indicate quadratic fits. The measured data was taken from nominal plateau, HAM side A, electronics side A, FPA temperature 80.5 K using detector 9 (BCS data).

The calibration coefficients derived using the OBCBB were compared to those computed from BCS data using a truncated fitting range that approximates the measured OBCBB range from instrument ambient to 315 K. For both cases, 2-sigma uncertainties were also determined for each coefficient. Comparing the two sets of calibration coefficients showed that for the majority of cases all three coefficients were within the 2-sigma uncertainties. Some detectors were exceptions in the following cases: the linear coefficient for M12 at cold and nominal instrument plateaus, the linear coefficient for M13 at hot instrument plateau, and the nonlinear coefficient for M14 at cold instrument plateau.

All of the thermal bands were shown to saturate above their design specifications [8] (see Table 4). All bands digitally saturate first, including both M13 gain stages; two bands (I4 and M12) exhibit analog saturation at some higher radiance within the measured range. For these two bands, the digital response decreases to zero at the highest measured radiance levels. On-orbit, this results in two possible values for a given digital response; fortunately, scenes with temperatures above saturation for these bands are usually fires, and can be correlated to M13 for quality control. The LWIR band saturation temperatures are generally consistent between different instrument plateaus to within ~ 1 K and between electronics sides to within 2–3 K; for the MWIR bands, the separation between electronics sides is less clear, but all instrument conditions are consistent to within 2–3 K. The detector dependence is largely the reflection of the detector gain dependence. Bands I4, M12, and M14 saturate between 4 K and 10 K above the design specification; all other bands saturate more than 10 K above their design specifications.

Table 3. Averaged gains ($1/c_1$) over all detectors and HAM sides measured during radiometric performance testing for all instrument temperature plateaus, electronics sides (A or B), and focal plane temperatures using BCS data.

Band	Cold A 80.5 K	Nominal A 80.5 K	Nominal A 82 K	Nominal A 83.5 K	Hot A 80.5 K	Hot A 82 K	Cold B 80.5 K	Nominal B 80.5 K	Hot B 80.5 K
I4	1134.6	1134.2	1133.8	1133.7	1134.3	1133.5	1134.0	1133.4	1134.3
I5	175.9	174.9	164.3	153.0	174.7	164.1	174.5	173.7	174.0
M12	1137.5	1137.1	1137.7	1138.4	1136.9	1137.5	1135.5	1135.6	1135.4
M13 HG	597.8	597.7	597.2	596.8	597.8	597.4	596.9	596.9	597.2
M13 LG	6.9	6.9	–	–	6.9	–	6.9	6.9	6.9
M14	175.1	174.2	162.5	149.0	173.4	161.6	172.3	171.4	170.5
M15	184.1	183.3	178.0	171.9	183.0	177.7	182.7	182.0	181.9
M16A	201.7	200.0	190.1	179.3	199.4	189.5	200.7	199.3	199.2
M16B	201.0	199.3	189.5	178.7	198.7	188.9	199.9	198.5	198.4

Using Equation (9), the radiance at which the SNR was equal to 5 was calculated. The equivalent scene temperature was estimated, the band average of which is listed in Table 4 using nominal plateau data. For most bands this is below the design specifications [8]; the exceptions were I4 and M14. Estimates of the measured T_{MIN} are consistent over instrument conditions. Note that there is some uncertainty in this estimate for bands I5, M13 low gain, M15, and M16 due to extrapolation well below the measured range.

The gain transition temperatures for M13 were measured to be between 343 and 349 K. There was about a 2 K spread in transition temperature with detector (largely due to detector-to-detector gain differences) and about a 3 K variation over instrument temperature (resulting from increasing background emission at higher instrument temperatures). This is slightly larger than the design specification [8], but has been deemed low risk to the science products.

The NEdT as a function of scene temperature is graphed in Figure 4 for all high gain bands (detector 9). The NEdT increases as the scene temperature decreases for all bands due to the influence of the derivative of the Planck function in Equation (10). For the LWIR bands, the NEdT is below ~0.6 K, even at the lowest scene temperatures; for the MWIR bands, the NEdT increases to between 2.5–3.5 K for M12 and M13 at 210 K and roughly 3.5 K for I4 at 230 K. The NEdT is consistent across instrument conditions, except for the slight increase with instrument temperature due to increasing background emission. M13 low gain is not shown, but behaves similarly to the other bands; the NEdT is below ~0.2 K for scene temperatures above about 400 K and increases to up to ~0.5 K below 400 K. The elevated focal plane measurements had the effect of increasing the NEdT for the LWIR bands at all scene temperatures, with the greatest increases at low temperatures; in contrast, the MWIR bands showed very little difference in NEdT with increasing focal plane temperature.

The NEdT at T_{TYP} for nominal plateau, electronic side A are listed in Table 4 (band maximum over detectors and HAM sides). The specified limit for NEdT at T_{TYP} is also listed; all bands are well below the limit for all conditions. The only exception is detector 5 in band M16B, which is close to, but still below the specified limit. In addition, detector 4 in band M15 was consistently out-of-family. The NEdT at T_{TYP} is very consistent over the range of instrument conditions tested, both in terms of magnitude and detector dependence. The NEdT generally increases with instrument temperature for all bands and both electronics sides (although there is a good deal of detector dependence). This is the result of the increasing background emission in the detectors which occurs at higher instrument temperatures. The NEdT at T_{TYP} increases about 10% over the ~20 K range of measured instrument temperatures (except for M12 which increases roughly 30%).

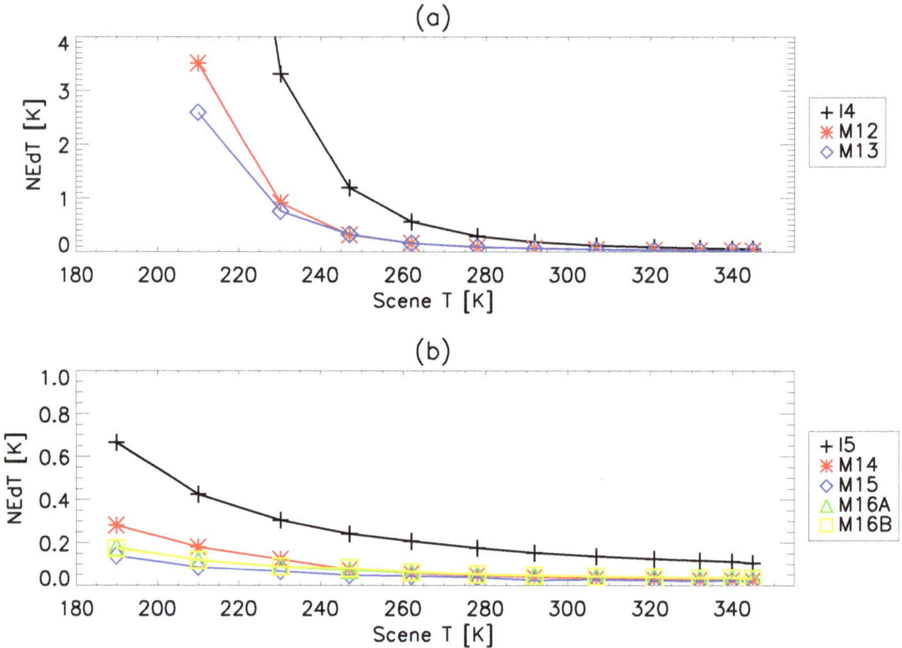

Figure 4. Plots show the measured NEdT as a function of scene temperature for the MWIR bands (**a**) and for the LWIR bands (**b**). Measured data was taken from nominal plateau, HAM side A, electronics side A, FPA temperature 80.5 K using detector 9.

Table 4. Band maximum NL, NEdT at T_{TYP}, T_{SAT}, and T_{MIN} measured during radiometric performance testing at the nominal instrument temperature plateau, electronics sides A, and 80.5 K focal plane temperature compared to their respective sensor specifications [8].

Band	NL [%]		NEdT at T_{TYP} [K]		T_{SAT} [K]		T_{MIN} [K]	
	Meas	**Spec**	**Meas**	**Spec**	**Meas**	**Spec**	**Meas**	**Spec**
I4	0.1	1.0	0.428	2.500	357	353	231	210
I5	0.2	1.0	0.524	1.500	369	340	167	190
M12	0.2	1.0	0.131	0.396	359	353	216	230
M13 HG	0.2	1.0	0.046	0.107	363	343	213	230
M13 LG	0.1	1.0	0.231	0.423	644	634	276	343
M14	0.5	1.0	0.060	0.091	348	336	195	190
M15	0.2	1.0	0.035	0.070	357	340	173	190
M16A	0.2	1.0	0.045	0.072	366	340	154	190
M16B	0.5	1.0	0.072	0.072	367	340	156	190

The non-linearity metric (NL) is generally consistent over instrument conditions and well below the specified limit (see Table 4). Detector 5 in M16B is again an outlier, but is still well below the 1.0% limit. M14 exhibits the largest average non-linearity at around 0.5%.

The potential for detector-to-detector striping was measured by the uniformity metric, graphed in Figure 5 for the worst case detector per band. A value greater then one indicates the potential for striping (dashed horizontal red line in the plots) [8]. The uniformity metric generally increases with increasing scene temperature, indicating increasing potential for striping as the temperatures rises. The potential for striping exists for many bands at the highest scene temperatures (in particular bands M13, M14, and M15). At higher temperatures, the deviation of the retrieved radiance from the band average increases, but the measured NEdL levels off (see Equation (13)); the result is a steadily increasing uniformity metric. On-orbit, the possibility of observing striping with M13 low gain is very limited, given the sparsity of measurements available.

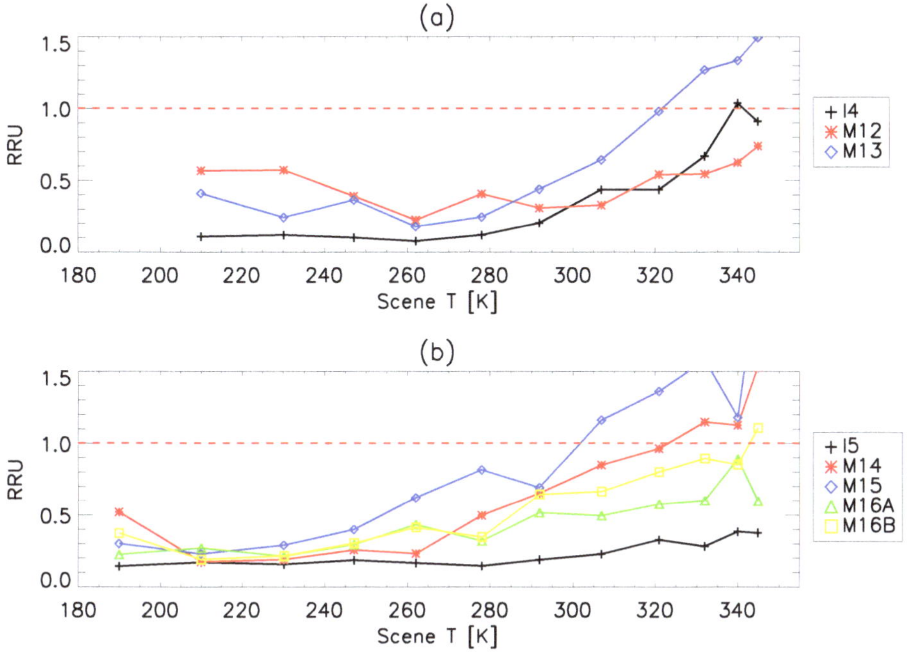

Figure 5. Plots show the measured uniformity metric as a function of scene temperature for the MWIR bands (**a**) and for the LWIR bands (**b**). Measured data was taken from nominal plateau, HAM side A, electronics side A, FPA temperature 80.5 K using the worst case detector.

3.3. Radiometric Stability

The radiometric stability of the JPSS-1 VIIRS instrument was tested for all the cases listed in Table 2. The first stability test listed in Table 2 assessed the instrument variation over time. In this case, the change in the linear gain is determined in the period between successive calibrations. On-orbit, successive calibrations of the linear gain occur every other scan for most thermal bands (M13 is calibrated every fourth scan) by observing the OBCBB. For all of the measured cases, the linear gain varied by under 0.0002% for all thermal bands (the total variation over the course of each test divided by the test time multiplied by the time it takes for two VIIRS scans).

For the stability tests that occurred over transitions between instrument temperature plateaus, the linear gain stability was assessed per degree K in terms of two different instrument temperatures (ELEC and OMM). There were three transition tests: the first from the lowest instrument temperature settings to the cold performance plateau (OMM temperatures ranging from 250–253 K and ELEC temperatures ranging from 262–268 K); the second from cold to nominal performance plateaus (OMM temperatures ranging from 253–262 K and ELEC temperatures ranging from 268–281 K); and the last from nominal to hot performance plateaus (OMM temperatures ranging from 263–273 K and ELEC temperatures ranging from 281–295 K). The linear gains varied by between 0.01%–0.04% (ELEC) and

0.04%–0.11% (OMM) per degree K. The transitions were performed such that the bulk of the ELEC temperture transition occured while the OMM temperature was roughly constant; similarly, most of the OMM temperature variation occured while the ELEC temperature change was small. This allowed the sensitivity to each temperature to be largely isolated and calculated separately. The sensitivity to each of these component temperatures will be included in the on-orbit calibration to remove any residual influences in the thermal model.

The third type of radiometric stability test performed varied the BUS voltage. In each case the BUS voltage was reduced from 32 to 27 Volts, with the sensor taking data at 1 Volt intervals. Results showed that the variation in linear gain was below 0.12% per Volt for all bands.

The last radiometric stability test listed in Table 2 corresponds to variations in the FPA temperatures. For this test, the FPA temperature was set to 83.5 K and then cooled down to 80.5 K (the expected on-orbit conditions). The linear gain for the MWIR bands showed very little change with FPA temperature (less than 0.16% per degree K). In contrast, the linear gain for the LWIR bands changed by between 2.4% (M15) to 6.4% (M14) per degree K. This is consistent with the radiometric performance results shown in Section 3.2.

3.4. Uncertainty

The total uncertainties for the EV retrieved radiance (for a single, un-aggregated pixel) are shown as a function of scene temperature in Figure 6 (worst case detector). For the MWIR bands, the results are between 0.1–0.2 K at high scene temperatures (above ~290 K), but increase rapidly below scene temperatures of about 260–270 K (above about 1 K for I4 below 260 K and for M12 and M13 below 230 K). The LWIR bands uncertainty estimate is between 0.1–0.2 K above 250 K, below which the uncertainty rises slowly to between 0.3–0.5 K at 190 K. I5 shows a little higher uncertainty than the other LWIR bands, with 0.2–0.3 K above 250 K and slowly increasing up to about 0.8 K at 190 K. The calculated total uncertainties (in % radiance) are listed in Table 5 at the scene temperatures required by the design specifications; for most cases, the modeled uncertainty was below the design specifications [8]. The exceptions were bands M12 and M13 at both 230 and 270 K.

The uncertainties for the MWIR bands were dominated by contributions from two terms: L_{OBCBB} and dn_{EV}. The L_{OBCBB} uncertainty was roughly constant with scene temperature at about 0.3% and was dominant above ~320 K for band I4 and about 290 K for the bands M12 and M13. The dn_{EV} uncertainty increased rapidly as the scene temperature decreased and dominated the uncertainty at all other scene temperatures. As this estimate was for a single, un-aggregated pixel and this uncertainty contributor is largely a statistical error, this contribution will decrease when large data sets are used. In those cases, the uncertainties listed in Table 5 for M12 and M13 at both 230 and 270 K will fall below the design specifications. The L_{OBCBB} uncertainty will then dominate above about 260 K for the bands I4, M12, and M13; at lower temperatures the largest contributor is the c_0 error, which increases rapidly as the scene temperature decreases. As a result, the reported uncertainties for these bands at low scene temperatures pose a low risk to the science products.

For most of the LWIR bands, there is no clearly dominant contributor at higher scene temperatures (above about 270 K). The exception is I5, where higher noise causes an increased contribution from the dn_{EV} uncertainty at high scene temperatures. As with the MWIR bands, this contribution will decrease when large data sets are used. At lower scene temperatures, the main contributors are the RVS and dn_{EV} uncertainties, which all increase as the scene temperature decreases.

The impact of pixel aggregation and of scan angle was also investigated. For the bands in which the dn_{EV} was the dominant contributor at some scene temperatures (I4, I5, M12, and M13 especially), pixel aggregation had the effect of reducing the uncertainty. In the bands for which the RVS was a major contributor (bands M14 and M15), the uncertainty tended to be largest at the end of scan and lowest at the beginning of scan. Band M16 showed a mix of the two effects.

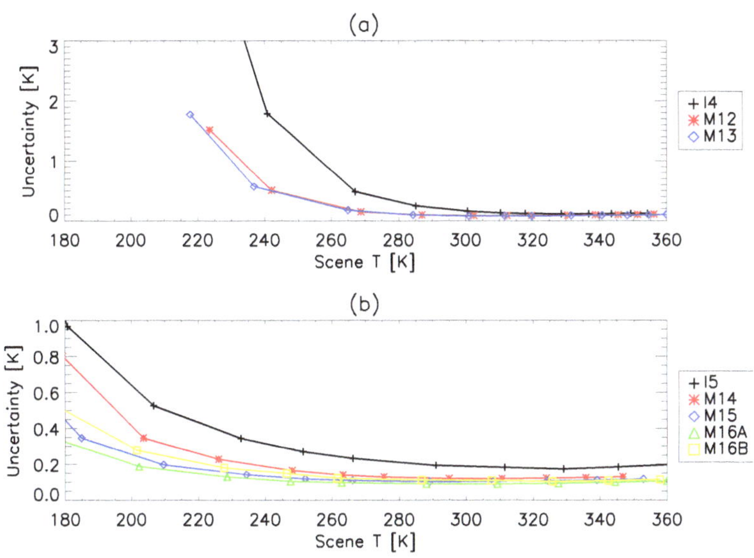

Figure 6. Plots show the modeled total temperature uncertainty as a function of scene temperature for the MWIR bands (**a**) and for the LWIR bands (**b**).

Table 5. Estimated uncertainty [% radiance] compared to the sensor specifications [8].

Band		190 K	230 K	267 K	270 K	310 K	340 K
I4	Spec	–	–	5.00	–	–	–
	Meas	–	–	2.64	–	–	–
I5	Spec	–	–	2.50	–	–	–
	Meas	–	–	0.48	–	–	–
M12	Spec	–	7.00	–	0.70	0.70	0.70
	Meas	–	9.14	–	0.78	0.35	0.33
M13	Spec	–	5.70	–	0.70	0.70	0.70
	Meas	–	7.81	–	0.78	0.29	0.27
M14	Spec	12.30	2.40	–	0.60	0.40	0.50
	Meas	2.86	0.73	–	0.30	0.20	0.18
M15	Spec	2.10	0.60	–	0.40	0.40	0.40
	Meas	1.35	0.41	–	0.20	0.15	0.13
M16A	Spec	1.60	0.60	–	0.40	0.40	0.40
	Meas	0.97	0.30	–	0.17	0.12	0.11
M16B	Spec	1.60	0.60	–	0.40	0.40	0.40
	Meas	1.46	0.41	–	0.20	0.14	0.12

A covariance terms between the radiometric coefficients and the other contributors at the retrieved radiance level were estimated by using the Schwarz inequality as outlined in Equation (16). For all bands, the uncertainty only marginally increased at ~292 K (where the calibration is tied to the OBCBB) and increased as the scene temperature increased or decreased. For the majority of modeled scene temperatures, the increase did not exceed 0.15%; the exceptions were the lowest scene temperatures for bands I4, M12, M13, and M14 as well as high temperature for band I4.

The impact of model validity was investigated by performing linear, quadratic, and cubic fits, and propagating the errors separately for each. In general, only marginal improvement in the uncertainty was derived by increasing the polynomial order from 2 to 3 (the biggest differences were at low temperature in the MWIR bands); in contrast, the quadratic model was a significant improvement over the linear (especially for M14). In addition, the differences between the EV retrieved radiance and BCS source radiance for all bands were minimized at 292 K by removing a bias of ~27 mK from the OBCBB temperature.

3.5. Comparison to SNPP

In general, the performance of the thermal bands from SNPP VIIRS and JPSS-1 VIIRS are comparable. The radiometric fitting coefficients are very similar in magnitude for the thermal bands; there is greater dependence on detector for bands M15 and M16 in SNPP. M16B detector 5 is an outlier for JPSS-1, while M12 detector 1, M16A detector 9, and I5 detector 31 are outliers for SNPP. The band average saturation temperatures and the temperatures at which the SNR falls below 5 are fairly consistent between the two instruments (to within ~5 K). The M13 gain transition temperatures exhibit a comparable spread with instrument conditions. The non-linearity of the radiometric fitting is consistent between the two sensors. Uniformity between detectors shows a similar pattern, with increasing potential for striping as the scene temperature increases. In addition, the uncertainties estimated for both sensors are in general agreement, with the uncertainty generally flat for the LWIR bands and the uncertainty for the MWIR increasing rapidly at low scene temperatures and relatively flat at high scene temperatures. Overall, the comparable pre-launch performance between the two sensors is likely to lead to JPSS-1 VIIRS thermal band science data products of as high a quality as is currently observed for SNPP VIIRS [7].

4. Conclusions

The JPSS-1 VIIRS instrument went through a comprehensive series of performance tests designed to characterize and calibrate the thermal emissive bands before launch at the Raytheon El Segundo facility. Accurate calibration of the VIIRS thermal emissive bands is necessary to ensure that the quality of the science and environmental data products is high. Assessment of the thermal band calibration has found that the instrument has met its design requirements for the majority of cases, and for those few cases where it does not, the risk to science data products is low. JPSS-1 VIIRS thermal band calibration pre-launch indicates that the products expected on-orbit will be comparable to the high quality science data that have been produced by the SNPP VIIRS thermal bands since it was launched in 2011.

Acknowledgments: The authors would like to thank the following: the Raytheon test team including Tung Wang for conducting the performance tests and for developing much of the analysis methodology, and members of the government data analysis working group including James McCarthy and Boryana Efremova for valuable comments. The above mentioned provided valuable information and support to the analysis presented in this work.

Author Contributions: Jeff McIntire wrote the manuscript. Jeff McIntire and David Moyer independently performed the analysis contained in this work. Hassan Oudrari and Xiaoxiong Xiong contributed to the design of this study and to the development of the manuscript.

Conflicts of Interest: The authors declare no conflict of interest.

References

1. Xiong, X.; Butler, J.; Chiang, K.; Efremova, B.; Fulbright, J.; Lei, N.; McIntire, J.; Oudrari, H.; Sun, J.; Wang, Z.; Wu, A. VIIRS on-orbit calibration methodology and performance. *J. Geophys. Res. Atmos.* **2014**, *119*, 5065–5078.
2. Cao, C.; De Luccia, F.; Xiong, X.; Wolfe, R.; Weng, F. Early on-orbit performance of the Visible Infrared Imaging Radiometer Suite (VIIRS) onboard the Suomi National Polar-Orbiting Partnership (S-NPP) satellite. *IEEE Trans. Geosci. Remote Sens.* **2014**, *52*, 1142–1156.

3. Barnes, W.; Salomonson, V. MODIS: A Global Image Spectroradiometer for the Earth Observing System. *Crit. Rev. Opt. Sci. Technol.* **1993**, *CR47*, 285–307.
4. McClain, C.; Hooker, S.; Feldman, G.; Bontempi, P. Satellite Data for Ocean Biology, Biogeochemistry, and Climate Research. *Eos* **2006**, *87*, 337–339.
5. King, M.D.; Menzel, W.P.; Kaufman, Y.J.; Tanre, D.; Gao, B.C.; Platnick, S.; Ackerman, S.A.; Remer, L.A.; Pincus, R.; Hubanks, P.A. Cloud and aerosol and water vapor properties, precipitable water, and profiles of temperature and humidity from MODIS. *IEEE Trans. Geosci. Remote Sen.* **2003**, *41*, 442–458.
6. Justice, C.O.; Vermote, E.; Townshend, J.R.G.; Defries, R.; Roy, D.P.; Hall, D.K.; Salomonson, V.V.; Privette, J.L.; Riggs, G.; Strahler, A.; Lucht, W.; Mynemi, R.B.; Lewis, P.; Barnsley, M.J. The Moderate Resolution Imaging Spectroradiometer (MODIS): land remote sensing for global change research. *IEEE Trans. Geosci. Remote Sen.* **1998**, *36*, 1228–1249.
7. Oudrari, H.; McIntire, J.; Xiong, X.; Butler, J.; Lee, S.; Lei, N.; Schwarting, T.; Sun, J. Prelaunch Radiometric Characterization and Calibration of the S-NPP VIIRS Sensor. *IEEE Trans. Geosci. Remote Sen.* **2015**, *53*, 2195–2210.
8. Joint Polar Satellite System (JPSS) VIIRS Product Requirement Document (PRD). Technical report, Goddard Space Flight Center: Greenbelt, MD, USA, 2014. Revision D.
9. Oudrari, H.; McIntire, J.; Xiong, X.; Butler, J.; Ji, Q.; Schwarting, T.; Lee, S.; Efremova, B. JPSS-1 VIIRS Radiometric Characterization and Calibration Based on Pre-launch Testing. *Remote Sens.* **2015**, doi:10.3390/rs8010041.
10. *(JPSS) General Test Plan (GTP) Visible Infrared Radiometer Suite (VIIRS)*; Technical Report; Raytheon: El Segundo, CA, USA, 2014. Revision B.
11. *Joint Polar Satellite System (JPSS) Visible Infrared Radiometer Suite (VIIRS) Sensor Data Records (SDR) Algorithm Theoretical Basis Document (ATBD)*; Technical Report; Goddard Space Flight Center: Greenbelt, MD, USA, 2013.
12. Moyer, D.; McIntire, J.; Oudrari, H.; McCarthy, J.; Xiong, X.; de Luccia, F. JPSS-1 VIIRS Pre-launch Response *Versus* Scan Angle Testing and Performance. *Remote Sens.* **2015**, under review.
13. Moeller, C.; Schwarting, T.; McIntire, J.; Moyer, D. JPSS J1 VIIRS prelaunch spectral characterization and performance. *Proc. SPIE* **2015**, *9607*, doi:10.1117/12.2188658.
14. Taylor, J.R. *An Introduction to Error Analysis*; University Science Books: Sausalito, CA, USA, 1997.
15. Taylor, B.N.; Kuyatt, C.E. *Guidelines for Evaluating and Expressing the Uncertainty of NIST Measurement Results*; NIST: Gaithersburg, MD, USA, 1994.
16. Monroy, E. *Performance Verification Report—VIIRS J1 Emissive Band Calibration*; Technical Report; Raytheon: El Segundo, CA, USA, 2015.

Article

JPSS-1 VIIRS Pre-Launch Response Versus Scan Angle Testing and Performance

David Moyer [1,*], Jeff McIntire [2], Hassan Oudrari [2], James McCarthy [3], Xiaoxiong Xiong [4] and Frank De Luccia [1]

1 The Aerospace Corporation, El Segundo, CA 90245, USA; Frank.J.DeLuccia@aero.org
2 Science Systems and Applications, Inc., Lanham, MD 20706, USA; jeffrey.mcintire@ssaihq.com (J.M.); hassan.oudrari-1@nasa.gov (H.O.)
3 Stellar Solutions, Palo Alto, CA 90306, USA; jkmccarthy@stellarsolutions.com
4 NASA Goddard Space Flight Center, Greenbelt, MD 20771, USA; xiaoxiong.xiong-1@nasa.gov
* Correspondence: david.i.moyer@aero.org; Tel.: +1-310-336-6170

Academic Editors: Changyong Cao, Dongdong Wang and Prasad S. Thenkabail
Received: 28 November 2015; Accepted: 4 February 2016; Published: 17 February 2016

Abstract: The Visible Infrared Imaging Radiometer Suite (VIIRS) instruments on-board both the Suomi National Polar-orbiting Partnership (S-NPP) and the first Joint Polar Satellite System (JPSS-1) spacecraft, with launch dates of October 2011 and December 2016 respectively, are cross-track scanners with an angular swath of $\pm 56.06^\circ$. A four-mirror Rotating Telescope Assembly (RTA) is used for scanning combined with a Half Angle Mirror (HAM) that directs light exiting from the RTA into the aft-optics. It has 14 Reflective Solar Bands (RSBs), seven Thermal Emissive Bands (TEBs) and a panchromatic Day Night Band (DNB). There are three internal calibration targets, the Solar Diffuser, the BlackBody and the Space View, that have fixed scan angles within the internal cavity of VIIRS. VIIRS has calibration requirements of 2% on RSB reflectance and as tight as 0.4% on TEB radiance that requires the sensor's gain change across the scan or Response Versus Scan angle (RVS) to be well quantified. A flow down of the top level calibration requirements put constraints on the characterization of the RVS to 0.2%–0.3% but there are no specified limitations on the magnitude of response change across scan. The RVS change across scan angle can vary significantly between bands with the RSBs having smaller changes of ~2% and some TEBs having ~10% variation. Within a band, the RVS has both detector and HAM side dependencies that vary across scan. Errors in the RVS characterization will contribute to image banding and striping artifacts if their magnitudes are above the noise level of the detectors. The RVS was characterized pre-launch for both S-NPP and JPSS-1 VIIRS and a comparison of the RVS curves between these two sensors will be discussed.

Keywords: VIIRS; calibration; thermal; emissive; reflective solar; RVS

1. Introduction

The Visible Infrared Imaging Radiometer Suite (VIIRS), aboard both the Suomi National Polar-orbiting Partnership (S-NPP) and the first Joint Polar Satellite System (JPSS-1) spacecraft with launch dates of October 2011 and late 2016 respectively, is a cross-track scanning sensor in a low Earth orbit [1]. VIIRS provides calibrated Top-Of-Atmosphere (TOA) radiance, reflectance and brightness temperature Sensor Data Records (SDRs) for weather and climate applications similar to its heritage sensors Advanced Very High Resolution Radiometer (AVHRR) [2], Operational Linescan System (OLS) [3], and Moderate Resolution Imaging Spectroradiometer (MODIS) [4]. VIIRS has 22 bands on four focal plane assemblies (FPAs). The Visible Near Infrared (VisNIR) FPA (bands I1, I2 and M1–M7) covers a spectral range of 395–900 nm. The Day Night Band (DNB) has its own FPA with multiple detector arrays for each of its three gain stages and is a panchromatic band with a spectral range of

500–900 nm [5]. There are two cold focal planes with the Short- and Mid-wave Wavelength Infrared (SMWIR) FPA (bands I3, I4 and M8–M13) covering a spectral region of 1230–4130 nm while the Long Wavelength Infrared (LWIR) FPA (bands I5 and M14–M16) covers the 8400–12,490 nm wavelength range. The band center wavelength, spatial resolution and gain type information are listed in Table 1. At NADIR, the fourteen moderate resolution bands (M-bands) and the DNB have a ground dynamic field of view (DFOV) of 750 m with 16 detectors in track while the five imaging resolution bands (I-bands) have DFOVs of 375 m with 32 detectors in track.

Table 1. The Visible Infrared Imaging Radiometer Suite (VIIRS) Spectral, Spatial and Radiometric Specifications at Typical Scenes (Ltyp or Typ).

Band Name	Gain	Center Wavelength (nm)	Focal Plane Assembly	DFOV (m)	Calibration Accuracy @Ltyp or Ttyp (%)
DNB	MG	700	DNB	750	5/10/30
M1	DG	412	VNIR	750	2
M2	DG	445	VNIR	750	2
M3	DG	488	VNIR	750	2
M4	DG	555	VNIR	750	2
M5	DG	672	VNIR	750	2
I1	SG	640	VNIR	375	2
M6	SG	746	VNIR	750	2
M7	DG	865	VNIR	750	2
I2	SG	865	VNIR	375	2
M8	SG	1240	SMIR	750	2
M9	SG	1378	SMIR	750	2
M10	SG	1610	SMIR	750	2
I3	SG	1610	SMIR	375	2
M11	SG	2250	SMIR	750	2
I4	SG	3740	SMIR	375	5
M12	SG	3760	SMIR	750	0.7
M13	HG	4050	SMIR	750	0.7
M14	SG	8550	LWIR	750	0.6
M15	SG	10,763	LWIR	750	0.4
I5	SG	11,450	LWIR	375	2.5
M16	SG	12,013	LWIR	750	0.4

VIIRS is in a sun-synchronous orbit with an altitude of ~828 km, an equatorial crossing of 13:30 and swath width of about 3000 km [6]. To achieve this swath, VIIRS covers an Earth View (EV) cross-track scan angle range of $\pm 56.06^\circ$ and uses for the scanner an afocal three mirror anastigmat telescope and fold mirror foreoptic called the Rotating Telescope Assembly (RTA). The RTA rotates 360° to allow the light from the EV as well as from the internal calibration sources to be collected. The light out of the RTA is redirected to the instrument's stationary aft-optics (including the 4 FPAs) using a Half Angle Mirror (HAM) that rotates at half the speed of the RTA. To achieve this, the HAM has a silver mirror coating on both sides that alternates each scan. These are referred to as HAM sides A and B (or sides 0 and 1 with respect to the vendor's nomenclature) and have slightly different reflectance properties since the coating was deposited on the HAM sides at different times. The Angle of Incidence (AOI) on the RTA mirror surfaces and aft-optics are fixed over scan angle but are dependent on detector and band location on the FPAs. The HAM however does have a scan angle dependent AOI variation with 28.60° to 56.47° change over the full EV scan angle. The three calibration targets within the VIIRS cavity, the Space View (SV), On-Board Calibrator BlackBody (OBCBB) and Solar Diffuser (SD) have AOIs around 60.18°, 38.53° and 60.47° respectively. The AOI of the OBCBB is within the AOI range of the EV and matches the EV AOI at a scan angle of -8°, while the SV and SD share almost the same AOI but are located at far different scan angles (-65.7° and 159° respectively), and are outside the range of the EV scan.

Figure 1 illustrates the relationship between the RTA Line of Sight (LOS) and the HAM normal vector. The left most image shows the RTA/HAM geometry at the SV scan angle. The RTA is rotated to -65.7° from the NADIR view and the HAM normal is 9.85° from NADIR. The middle image shows the NADIR scan angle with the RTA LOS at the +Z direction and the HAM normal vector at 23.0°. The last image shows the RTA/HAM combination where the HAM AOI is at a minimum (28.6°) at a scan angle of 46.0° with both the RTA LOS and the HAM normal having the same geometry. However, there is an Out-Of-Plane (OOP) angle of 28.6° on the HAM in the X coordinate direction to fold the light from the RTA towards the aft-optics not shown in the figure. This additional OOP angle along with the angle difference between the RTA LOS and the HAM normal vector are used to compute the AOI of the HAM. The equations to do this conversions will be discussed in the next section. The EV sector uses timing delays to co-register the band imagery so that the SDR products represent the same area on the ground for all bands. This makes the mapping between scan angle and AOI the same for all bands within the EV sector. The calibration sectors however are not co-registered and therefore each band and detector have a unique AOI when viewing these sources that needs to be accounted for when applying their RVS corrections. The calibration sectors record 48 M-band samples or 96 I-band samples during each scan allowing noise suppression through averaging and Signal to Noise Ratio (SNR) determination to be performed on a per scan basis. The EV when in operational mode contains 3200 M-band samples for single gain bands and 6400 for I-bands (see Table 1). These EV samples include aggregation that minimizes the effective pixel growth on the ground due to the Earth's curvature. There are six aggregation zones and their scan angle ranges are listed in Table 2. The dual gain M-bands are packaged unaggregated with 6304 samples due to the need to calibrate each pixel separately depending on gain values (high or low) within an aggregated pixel [7].

Table 2. Sample Aggregation Zones for Non-Day Night Bands (DNBs).

Zone	Start Scan Angle (°)	End Scan Angle (°)	Number of Samples Aggregated
1	−56	−43	1
2	−43	−32	2
3	−32	0	3
4	0	32	3
5	32	43	2
6	43	56	1

The four-mirror RTA, HAM and aft-optics telescope system all use silver mirror coatings to reflect the photons from the EV or calibration sources to the four FPAs. The silver coatings are preferred to other mirror coating options due to its high reflectance over a large spectral range (0.395–13 μm) and very high reflectance (>98%) in the blue spectral region. Similar to its heritage sensor MODIS, VIIRS uses Quantum FSS-99 for its silver mirrors. The process used to deposit the FSS-99 coating onto the mirror substrate is proprietary but consists of a silver mirror with a dielectric overcoat to protect the silver coating from exposure to the environment [8]. The silver mirror and overcoat cause a wavelength dependent reflectance in the mirror surface as a result of spectral variation in the index of refraction. The thickness of the silver and overcoat and the path length variation through these surfaces as a function of AOI on the mirror creates the reflectance variation observed in the RVS measurements. The RTA and aft-optics have fixed AOIs for each band and detector but the HAM AOI does vary throughout the scan. With large AOIs on the HAM of 60.18°, consistency in the coating characteristics (purity of the deposit and coating thickness) are important to maintain reflectance and RVS repeatability between HAM sides and sensor builds. This paper will compare the S-NPP RVS to JPSS-1 RVS performance and will show how mirror coating deposition on the HAM varies both with HAM side (temporally close in deposition time) and sensor builds (very large temporal separation in deposition time).

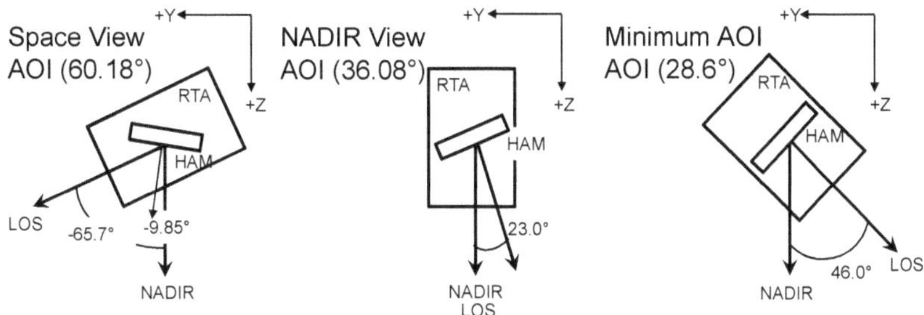

Figure 1. Example of the Rotating Telescope Assembly (RTA) and Half Angle Mirror (HAM) Orientations at three different scan angles. The left most diagram corresponds to the Space View (SV) scan angle, the middle is the NADIR scan angle and the right most is the scan angle where the HAM Angle of Incidence (AOI) is at a minimum.

The radiometric requirements listed in Table 1 for both the RSB and TEB allocate 0.3% and 0.2% (except for band M14 at 0.6%) uncertainty respectively for the RVS characterization. The RVS change over the scan is not restricted in the requirements but must be characterized to within the allotted uncertainty (polarization effects are not included in the RVS and are separately characterized [9] for Environmental Data Record (EDR) processing [10]). The RVS is tested at system level and not at the HAM component level to allow the characterization with the on-orbit view geometry of the detector footprints on the HAM and RTA to be performed. This removes any errors from modeling the view geometry configurations as well as spatial non-uniformity of the HAM coating surface that is not included in the component level measurements.

The RVS requirements are for the full mission of the instrument and thus need to account for not only prelaunch characterization uncertainty but also on-orbit degradation effects. One major on-orbit influence to the silver mirror coating performance is UV exposure [11–13]. Degradation of the reflectance, especially in the blue region, is common for both MODIS and VIIRS on-orbit. The MODIS optical design has the scan mirror as the first surface in optical train which makes it the first surface exposed to UV light and thus degradation is significant over the mission [11]. This is an issue because the AOI on this scan mirror is changing as a function of scan angle and thus the RVS for MODIS has changed shape throughout the mission. It not only changed but each HAM side has changed at different rates [14]. The VIIRS design has the four-mirror RTA before the HAM which absorbs most of the UV light with minimal UV reaching the HAM surface. Since the RTA does not contribute to RVS, only reflectance or gain change, the RVS performance for S-NPP has remained stable. No adjustments to RVS in the SDR ground processing has been performed due to on-orbit degradation. The calibration methodology for both the RSB and TEB will be discussed as well as the RVS characterization results. Comparisons between S-NPP and JPSS-1 will be shown to understand shape differences and their impact to their sensor's calibration.

2. RVS Testing Approach

2.1. Response Versus Scan Angle Test Source Assembly

Pre-launch JPSS-1 VIIRS RVS testing data was collected at Raytheon Space and Airborne Systems in El Segundo, California in late 2013 for the RSB and early 2014 for the TEB. The RVS characterization utilized 3 sources: the Laboratory Ambient BlackBody (LABB) as a target for TEBs, a 100 cm Spherical Integrating Source (SIS-100) for the RSBs and the OBCBB as an ambient blackbody for background subtraction purposes. VIIRS was placed on a Rotating Table to allow the sensor to be rotated with respect to the LABB or SIS100 to facilitate different scan angle measurements.

Table 3 lists the RSB scan angle measurements in the sequence they were acquired. The scan angle positions are purposely not measured in monotonic order and there are several repeats at the −8° scan angle location. The repeats provide stability measurements as well as source drift correction capabilities while the non-sequential scan angles allow for source drift and any other time-dependent Ground Support Equipment (GSE) or VIIRS sensor effects in ambient conditions—to be decoupled from RVS shape characteristics. The SIS-100 source is set to a single radiance level to provide optimum illumination to the blue wavelength and Shortwave Infrared (SWIR) bands at nominal integration time settings. The integration time of the instrument was then adjusted to optimize the remaining RSBs response to the SIS-100 while maintaining testing efficiency. This resulted in three different integration time settings at each scan angle position. Use of fixed sources while varying sensor integration time greatly reduces source drifts during measurements by keeping the lamps within the SIS-100 in a steady state thermal configuration.

Table 3. Response Versus Scan (RVS) Data Collection Configuration with Scan Angles and Diagnostic Window Information for both the Reflective Solar Bands (RSBs) and Thermal Emissive Bands (TEBs).

Collection	RSB Scan Angle (°)	RSB Diagnostic Window	RSB M-Band Sample Offset	TEB Scan Angle (°)	TEB Diagnostic Window	TEB M-Band Sample Offset
1	−65.7	Nominal	0	−8	Third	2128
2	−8	Third	2128	−65.7	First	0
3	−38	First	0	22	Fifth	4254
4	6	Third	2128	-45	First	0
5	−45	First	1064	6	Fourth	3192
6	−8	Third	2128	−8	Third	2128
7	−55.5	First	0	−55.5	First	0
8	22	Fourth	3192	−20	Third	2128
9	−30	First	1064	−38	Second	1064
10	−8	Third	2128	−8	Third	2128
11	−51	First	0	−51	First	0
12	38	Fifth	4254	35	Fifth	4254
13	−20	Second	1064	−30	Second	1064
14	55.5	Fifth	4254	−8	Third	2128
15	−8	Third	2128	−60.6	First	0

Table 3 lists the TEB scan angle measurements in the sequence they were acquired. As with the RSBs, the TEB measurements are purposely non-sequential in scan angle. The scan angle order is slightly different than the RSB measurements but the distribution of scan angles throughout the measurement sequence is very similar. The TEB scan angle measurements add a 35° point near the minimum AOI as well at −60.6° (where the AOI change with scan angle is very steep) to better characterize the RVS for these bands. The TEB RVS dependencies can vary by as much at 10% over scan (M14) and it is important to characterize their shape where the slope of the curve is strongest. The RSB RVS dependences tend to have about 2% variation over scan making them less sensitive to AOI changes at the beginning of scan. The LABB was set to 345 K while the OBCBB was at ambient temperature (~293 K). The TEBs needed only 2 integration time settings to optimize the signal for testing. This allowed the LABB to be at a fixed temperature and provide a more steady state thermal environment during the test. This was important to minimize variation in the background emission within the VIIRS cavity during the test. Both the LABB and SIS-100 are extended sources that allow many VIIRS samples to have their full field of view filled by source illumination. Both the RSB and TEB RVS measurements used an average of 100 M-band or 200 I-band samples across the source and then averaged over 50 scans for each HAM side.

During RVS testing, the VIIRS instrument was configured to be in a special data collection mode called diagnostic mode. This mode turns off on-board sample aggregation in scan, thereby removing any sample averaging complications from the test data analyses. The down side of diagnostic mode is that full scans of data cannot be transmitted with aggregation turned off. As a result only 2048 samples out of the 6304 M-samples within a full scan are reported. To accommodate this lack of full scan range, diagnostic windows of 2048 M-band samples at different scan-angle regions within a sub section of the entire EV scan are used. Columns 3 and 6 in Table 3 list the diagnostic window at each scan angle that was collected. Sample numbers along with the diagnostic window offsets are used to compute the actual AOI of each measurement during the RVS test. For the TEB RVS measurements, a sector rotation was performed to move the SV port into the EV scan data collection range. This did not allow the 55.5° scan angle measurement in the RSB RVS test to be measured but allowed the LABB to be placed at the SV port and still be viewed in the EV sector data window. This allowed 100 M-band samples to be averaged for the SV angle during the TEB test unlike the 48 M-band samples used in the RSB and 15 for the DNB. With the large AOI of the SV and significant sensitivity to RVS for the TEBs, the improvement in the SV RVS measurement accuracy was very important.

2.2. VIIRS RSB RVS Analysis Methodology

The SDRs for the RSBs provide a calibrated TOA radiance and reflectance product. The RVS plays an integral part of calibration algorithm for radiance shown in Equation (1) [15].

$$L = \frac{F\cdot(c_0 + c_1 dn_{EV} + c_2 dn_{EV}^2)}{RVS\,(\theta_{EV})} \tag{1}$$

The c_0, c_1 and c_2 are the calibration coefficients determined pre-launch during Thermal Vacuum (TVAC) testing. The dn_{EV} are the EV Digital Numbers (DNs) minus the SV DNs, the RVS as a function of scan angle (or sample number) is in the denominator and the *F* factor is the gain drift correction determined using the SD as described in Equation (2).

$$F = \frac{RVS\,(\theta_{SD})\cdot E_{SUN}\cdot \tau_{SD}\cdot \cos(\varphi_{SD})\cdot BRF_{SD}\cdot H}{(c_0 + c_1 dn_{SD} + c_2 dn_{SD}^2)} \tag{2}$$

The numerator is the estimated Solar radiance reflected off the SD pin hole screen (τ_{SD}) to attenuate the Solar irradiance (E_{SUN}) that reflects off the lambertian SD (BRF_{SD}). The denominator is the TVAC calibration coefficients scaled by the dn_{SD} (SD DNs) minus the SV DNs. Similar to the RVS in Equation (1), the calibration coefficients are divided by the RVS (θ_{SD}) which moves to the numerator of Equation (2). The RVS (θ_{SD}) is determined from the SD AOI and is band and detector dependent due to a lack of co-registration in the calibration sectors.

The RVS data analysis for the RSBs consists of processing the VIIRS response from the SIS-100 source, mapping the source location in scan angle space to determine the AOI, drift correcting the VIIRS response to account for source instability and fitting the data to a 2nd-order polynomial to allow interpolation of the data to any AOI value. An example of the VIIRS response to the SIS-100 is shown in Figure 2 for band M1. The center of the source profile is identified for each scan angle measurement configuration and 100 M-samples that straddle the center are extracted (only 40 samples for the DNB). These samples have an offset subtraction applied using the mean of the OBCBB samples since the SV port is exposed to allow the SIS-100 to be placed in that location. This is performed for each scan and then the dn_{EV} is averaged over samples and 50 scans for each detector and HAM side. The −65.7° scan angle corresponding to the SV uses the SIS-100 in the SV port and provides only 48 M-samples to be averaged. The mapping of the scan angle to AOI uses the center sample value, diagnostic window offset and Equations (3)–(5).

$$\theta_{scan} = \left(x + \beta - S_{off}\right)\cdot\delta_{scan} - \theta_{start} \tag{3}$$

$$\theta_{HAM} = \frac{\theta_{scan} - \varphi_{ref}}{2} \tag{4}$$

$$AOI = \cos^{-1}\left(\cos(\theta_{HAM})\cdot\cos\left(\varphi_{oop}\right)\right) \tag{5}$$

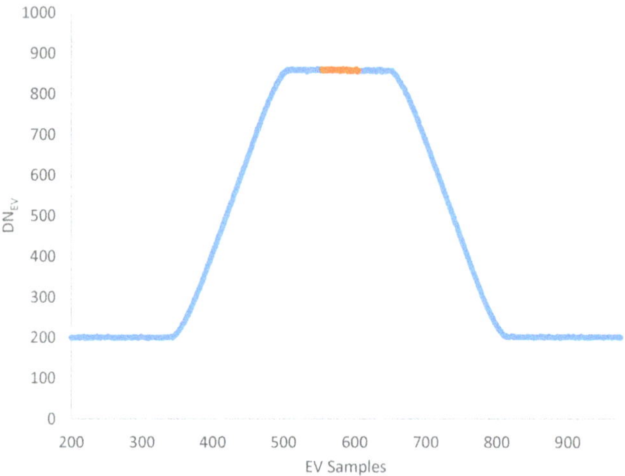

Figure 2. Band M1 DN Response to the SIS-100 Source during the RVS Test (the orange points correspond to the RVS data used during processing).

The angles φ_{oop} and φ_{ref} are fixed angles of 28.6° and 46.0° respectively. The θ_{scan} is the scan angle of VIIRS and is computed for each image sample using Equation (3). The sample number (x) of the image, offset by diagnostic window shift (β), the boresight offset (S_{off}) that integrates the sensor pointing information and scaled by the angular scan rate (δ_{scan} at 0.017785 deg/sample) is combined with the angular offset of the start of scan (θ_{start} at −60.058°) is then subtracted to get the scan angle (3). The scan angle (θ_{scan}) combined with the reference scan angle offset (φ_{ref}), is used to determine the θ_{HAM} angle (4). The θ_{HAM} corresponds to the vector difference between the RTA LOS and the HAM normal vector. The cosine of that θ_{HAM} with the OOP angle (φ_{oop}) corresponding to the angle to fold the light towards the aft optics gives the AOI in (5).

The SIS-100 source drift correction uses the −8° scan angle measurements that are repeated throughout the test to track the source output at a common sensor configuration. These −8° scan angle measurements are assumed to track a linear source drift between repeats. Any non-linear drifts will not be correctly removed from the measurements and will show up as residual error during the RVS fits. Each set of −8° repeat point dn pairs are linear fit and normalized by the first −8° repeat, and used to correct dn values from all scan angle measurement captured between those −8° repeat pairs. The dns of each measurement between repeats is scaled by the drift correction based on the time that measurement was acquired. After the source correction is applied, a 2nd-order polynomial fit of dns normalized to the SV scan angle *versus* AOI is performed to model the RVS characteristics. Since the scan angle measurements are not in a monotonic order, residual source drift correction errors and temporal errors will show up as fit residual errors. This provides confidence that the shape of the RVS curve is capturing true VIIRS gain change *versus* scan angle and not external error sources during the test. The larger the fit residuals, the more uncertainty in the characterization due to these external error sources. One example of this is the band M9 RVS due its spectral response being in a water vapor absorption region. The water vapor varies during the RVS test and will influence the characterization due to these temporal changes. The residuals will be larger but its influence on the RVS shape is reduced since the scan angles are not measured in a sequential fashion. The residuals of the fits are driven by dn noise, GSE uncertainties and other temporal error sources and are used to assess the quality of the RVS characterization.

2.3. VIIRS TEB RVS Analysis Methodology

Unlike the RSBs where the RVS scales the radiance computed using pre-launch calibration coefficients only, the TEBs have a more complicated relationship between RVS and calibrated radiance. Equation 6 shows how the VIIRS response for the TEBs is converted to a calibrated radiance.

$$L_{ap}\left(RVS_{EV}\right) = \frac{\left(RVS_{SV} - \mathrm{RVS}_{EV}\right) \cdot \left[\frac{\left(1-\rho_{rta}\right) \cdot L\left(T_{rta}\right) - L\left(T_{ham}\right)}{\rho_{rta}}\right] + F \cdot \sum_{i=0}^{2} c_i\left(T_{\mathrm{det}}, T_{elec}\right) \cdot \left(DN_{EV} - DN_{SV}\right)^i}{\mathrm{RVS}_{EV}} \quad (6)$$

Similar to the RSBs, the TEBs have c_0, c_1 and c_2 calibration coefficients to convert EV DNs subtracted by the SV DNs (dn_{EV}) into radiance. This is scaled by a gain drift correction (F factor) described in Equation (7). The additional piece to the TEB calibration is the residual background emission terms from the RTA and HAM due to RVS differences between the EV and SV scan angles or OBCBB and SV scan angles. The $L(T_{rta})$ and $L(T_{ham})$ are radiance terms for the RTA and HAM respectively that are estimated using internal thermistors within the VIIRS cavity. The ρ_{rta} is the reflectance of the RTA which combines the reflectance of the four-mirror RTA telescope.

$$F = \mathrm{RVS}_{OBCBB} \frac{\left\{\left(1 - \frac{RVS_{SV}}{\mathrm{RVS}_{OBCBB}}\right) \cdot \left[\frac{\left(1-\rho_{rta}\right) \cdot L\left(T_{rta}\right) - L\left(T_{ham}\right)}{\rho_{rta}}\right]\right\}}{\sum_{j=0}^{2} c_j \cdot \left(DN_{OBCBB} - DN_{SV}\right)^j} + \frac{\varepsilon_{obc} \cdot L\left(T_{OBCBB}\right) + L_{OBCBB\_refl}\left(T_{sh}, T_{cav}, T_{tele}\right)}{\sum_{j=0}^{2} c_j \cdot \left(DN_{OBCBB} - DN_{SV}\right)^j} \quad (7)$$

The L_{OBCBB_refl} is the internal cavity emission reflected off the OBCBB, the ε_{obc} is the emissivity of the OBCBB and the $L(T_{OBCBB})$ is the OBCBB radiance estimated using the 6 potted thermistors. The magnitude in the RVS differences between source and SV views determines how much the RTA and HAM background emission terms contribute to the total radiance. The TEB RVS curves, which can vary by as much as 10% across the scan, must be accurately characterized or brightness temperature biases in the SDR product that are scan angle and scene dependent will occur. Table 4 shows the worst case impact within the EV scan on the SDR brightness temperatures for the TEBs when using an RVS error of 0.2%. The cold scene SDR brightness temperatures are more sensitive to RVS error due to the background emission term being reweighted by the *RVSsv-RVSev* difference in equation 6. Band M14 is the only LWIR band to show large (3 K) change at the low scene temperatures. All of the LWIR bands have > 0.1 K SDR brightness temperature change at warm scenes and would impact the downstream EDR products. These RVS errors would also cause detector-to-detector striping or scan-to-scan banding based on scene temperature and would be difficult to remove with operational code corrections. Therefore, characterizing the RVS to 0.2% uncertainty for the TEB is vital.

Table 4. Maximum Sensor Data Records (SDR Brightness Temperature Error (K) Due to 0.2% RVS Shape Error at Multiple Scene Temperatures (K) for all TEBs.

–	**SDR Brightness Temperature Error (K)**											
Band	190 (K)	210 (K)	230 (K)	247 (K)	262 (K)	278 (K)	292 (K)	307 (K)	321 (K)	332 (K)	340 (K)	345 (K)
I4	3.843	0.946	0.257	0.075	0.015	0.016	0.030	0.041	0.049	0.054	0.057	0.060
I5	0.407	0.207	0.105	0.051	0.014	0.019	0.043	0.068	0.089	0.105	0.116	0.123
M12	4.011	1.086	0.268	0.078	0.015	0.016	0.030	0.041	0.048	0.053	0.057	0.059
M13	5.948	0.950	0.235	0.073	0.015	0.016	0.032	0.043	0.052	0.057	0.061	0.064
M14	3.012	0.361	0.137	0.058	0.015	0.019	0.042	0.064	0.082	0.095	0.104	0.110
M15	0.509	0.231	0.112	0.052	0.014	0.019	0.043	0.067	0.088	0.103	0.114	0.121
M16	0.364	0.195	0.102	0.050	0.014	0.019	0.043	0.068	0.089	0.105	0.117	0.124

The RVS data analysis for the TEBs is slightly different from the RSB methodology. It consists of processing the VIIRS response from the LABB source, mapping the source location in scan angle space to determine the AOI, drift correcting using the LABB and OBCBB temperature information and fitting the data to a 2nd-order polynomial to allow interpolation of the data to any AOI value. The fitting algorithm requires an iteration approach that varies the RVS_{SV} to find the minimum residuals in the polynomial fit. This is needed because there are two RVS values in Equation (6). The purpose of the iteration is to optimize the RVS_{SV}-RVS_{EV} to best correct the background emission term in the radiance retrieval and thus reduce residual errors caused by cavity temperature drifts during the test. The VIIRS response to the LABB is shown in Figure 3 for band M15 and uses 100 M-band samples that straddle the center in a similar fashion as the RSBs. These samples have an offset subtraction applied using the mean of the OBCBB samples since the SV port is used to allow the LABB to be placed in that location. This is performed for each scan and then the dn_{EV} is averaged over samples and 50 scans for each detector and HAM side. The EV sector is rotated for the TEB test to allow the −65.7° scan angle to be viewed in this sector so that 100 sample averaging can be performed for this angle as well. This changes the mapping of the scan angle to AOI compared to the nominal configuration used for the RSB (−60.058°) to −70.056° as the θ_{start} offset in Equation (3). The remaining scan angle to AOI conversion is similar to the RSB portion using Equations (3)–(5).

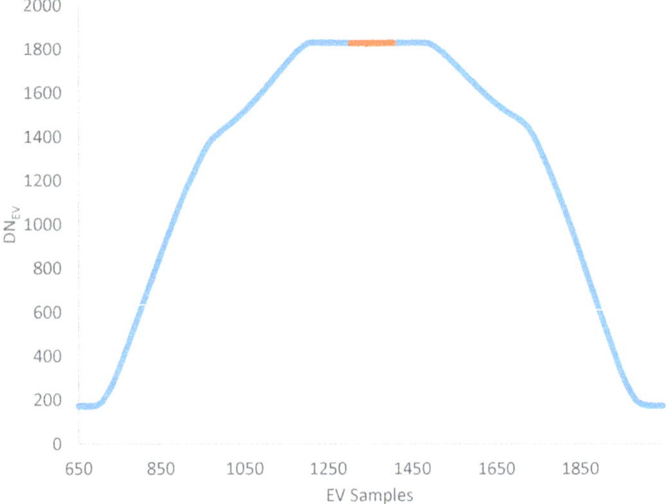

Figure 3. Band M15 DN Response to the LABB Source during the RVS Test (the orange points correspond to the RVS data used during processing).

The temperature of the LABB is set to 345 K, and the OBCBB at an ambient temperature of ~293 K is used to estimate the source radiances. This is then used as the *Lap* value in Equation (6) to solve for RVS_{EV}. The *F* factor is assumed to be 1 and the RVS_{SV} is iterated until the 2nd-order polynomial fit with the minimum fit residuals is determined. The non-sequential measurement of scan angle allows the temporal internal cavity temperature drifts to be decoupled from the RVS shape allowing the RVS_{SV} iteration to mostly impact the residuals and optimize the RVS shape for these bands.

3. S-NPP and J1 VIIRS RVS Results

3.1. Reflective Solar Band Results

The RSB RVS curves were computed for each detector and HAM side and all 3 I-bands and 11 M-bands. These results include the application of the SIS-100 radiance drift correction but does not include water vapor correction for band M9. This correction and its improvements to the M9 RVS has been discussed in [16]. Figure 4 shows the measurement points and 2nd-order polynomial fits for band M1 HAM side A for all detectors. The x-axis is HAM AOI with the left side of the axis at 61° representing the SV angle and beginning of scan while the 28° point on the right side representing the end of scan (the NADIR AOI is 36.08°). The measurement points were not acquired sequentially across the x-axis and show some noise indicating the drift correction error and other sources of temporal errors still remain in the data. The 2nd-order polynomial fit (solid lines) is a least squares fit of these points. The residuals indicate the characterization error in the measurement and are slightly different for each detector due to their noise performance during the test. The points are fairly monotonic and indicate the 2nd-order RVS model is appropriate for characterization purposes. Another example of RVS data is shown in Figure 5 for band I2 HAM side A for all detectors. The peak-to-peak change in RVS for this band is ~0.18% over the entire scan region. Here the temporal errors from the GSE and detector noise are more evident since there is little change in the gain across AOI. A small change in RVS makes detector noise more prominent in the measurements causing the fits look worse for I2 than for M1. However, the 2nd-order fit models the RVS very well (0.05% residuals) even though the variation across scan is very small. Figure 6 shows the RVS for band M9 HAM side A all detectors. The fit residuals in figure 6 for band M9 are much larger due to water vapor variability during the test impacting the radiance reaching the detectors. This water vapor effect is not corrected in this analysis but will be updated to account for water vapor for on-orbit purposes. While the residuals are large, the variation in RVS across the scan is ~0.3% peak-to-peak. This indicates that RVS shape errors for band M9 will not significantly impact the SDR performance on-orbit.

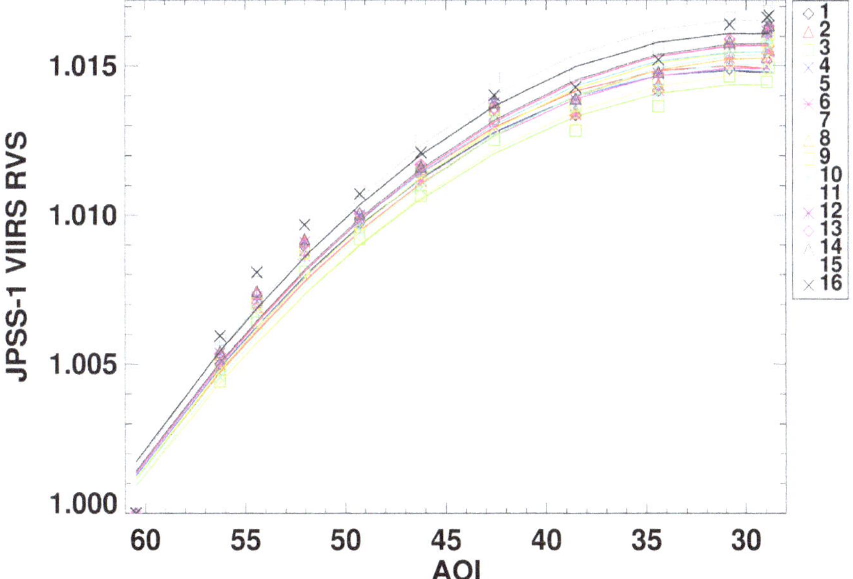

Figure 4. Band M1 HAM side A RVS and Fits for all Detectors.

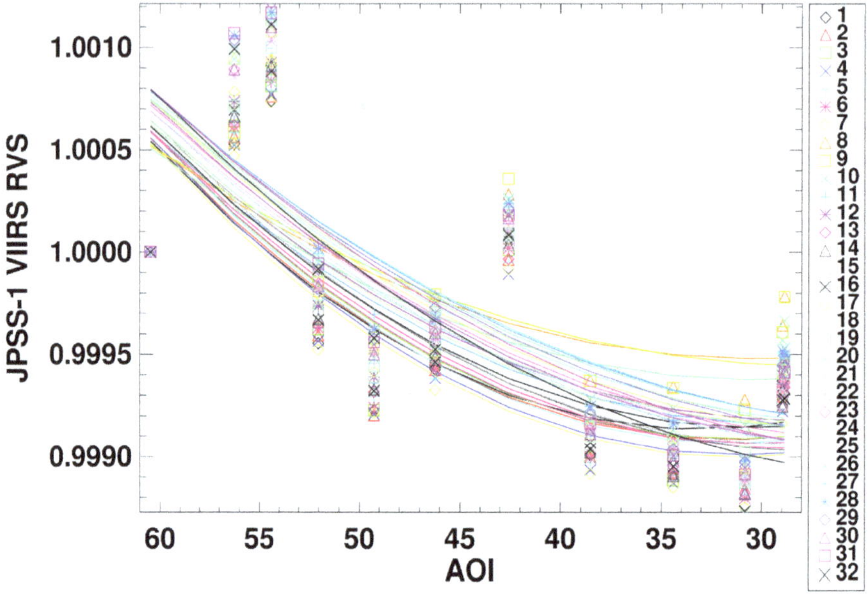

Figure 5. Band I2 HAM side A RVS and Fits for all Detectors.

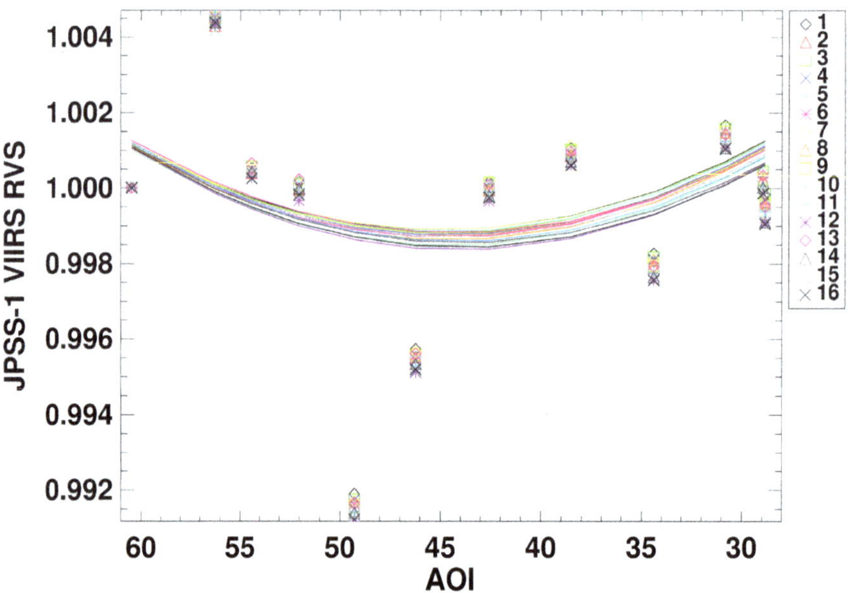

Figure 6. Band M9 HAM side A RVS and Fits for all Detectors.

Figure 7 shows the RVS across scan for the DNB HAM side A all detectors. The variation in RVS over the scan is very small on the order of ~0.3% change peak-to-peak. This change across scan will have minimal impact on the SDR calibration. The edge detectors 1 and 16 do have unique RVS shapes but are most likely due to DN noise in the SV AOI since there were only 14 samples of data averaged for that point. Overall the DNB RVS matches well with the VisNIR RSBs with wavelengths.

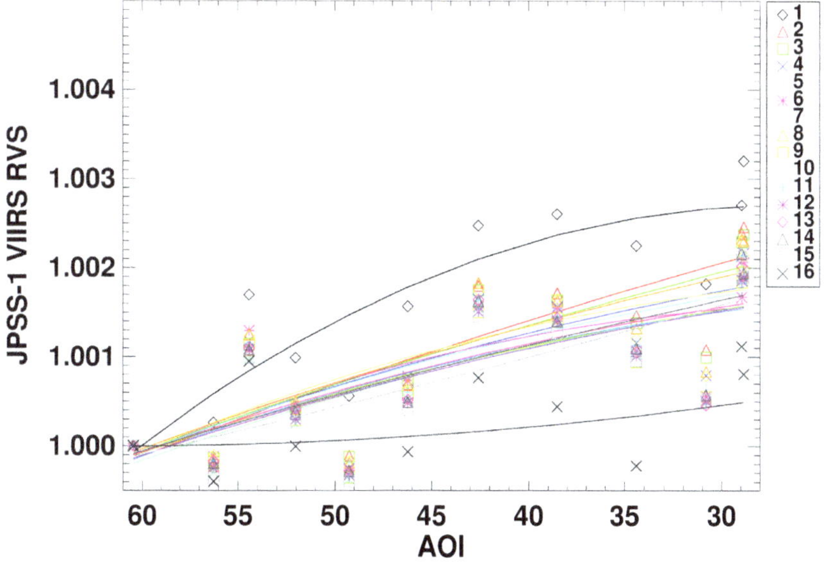

Figure 7. DNB HAM side A RVS and Fits for all Detectors.

Table 5. Maximum RVS Variation across Scan and Characterization Uncertainty for all RSBs.

Band	HAM A Maximum Peak-to-Peak Change in RVS Over Scan (%)	HAM A RMS Residuals (%)	HAM B Maximum Peak-to-Peak Change in RVS Over Scan (%)	HAM B RMS Residuals (%)
DNB	0.28	0.06	0.38	0.06
M1	1.49	0.08	0.93	0.07
M2	1.00	0.05	1.00	0.05
M3	0.74	0.04	0.70	0.05
M4	0.52	0.04	0.54	0.04
M5	0.59	0.04	0.35	0.03
I1	0.57	0.05	0.40	0.04
M6	0.67	0.04	0.50	0.03
M7	0.18	0.04	0.53	0.05
I2	0.18	0.05	0.57	0.06
M8	0.42	0.02	0.53	0.02
M9	0.27	0.28	0.34	0.28
M10	0.15	0.04	0.24	0.03
I3	0.11	0.04	0.20	0.04
M11	0.11	0.03	0.09	0.03

Figures 8–10 shows the detector-averaged RVS curves for each RSB and HAM side to show the general behavior for each band. The SDR products uses band, detector and HAM side dependent RVS corrections. The solid lines are HAM A and the dashed lines are HAM B. The largest change in RVS for the RSBs is band M1 (blue line) with ~1.015 change from the SV AOI on the left side of the plot to the end of EV scan on the right side. The dashed line for HAM B shows considerable change with respect to HAM side A (solid blue line) across the scan. This difference in HAM side RVS is due to temporal separation between when the two mirror sides were coated. HAM side A was coated by the vendor in 2009 while HAM B was coated in early 2010. This meant that the chamber used to coat these mirrors was not consistent and therefore the deposition of the silver mirror onto the HAM was slightly

different between the two sides. Figures 11–13 shows the band-averaged RVS curves for S-NPP. Unlike JPSS-1 VIIRS, S-NPP VIIRS HAM sides were coated very close temporally and have very similar silver mirror reflectance between sides. The overall change in RVS is different between S-NPP HAM sides but the shape across scan angle is very consistent. The magnitude of the RVS between S-NPP and JPSS-1 VIIRS is very similar with the VNIR bands showing an increase in RVS across scan while the SWIR bands mostly decrease with scan. Table 5 lists the maximum peak-to-peak variation in RVS over the scan for each band and HAM side. The RMS of the fit residuals are also listed to show the characterization uncertainty in the fits for each band.

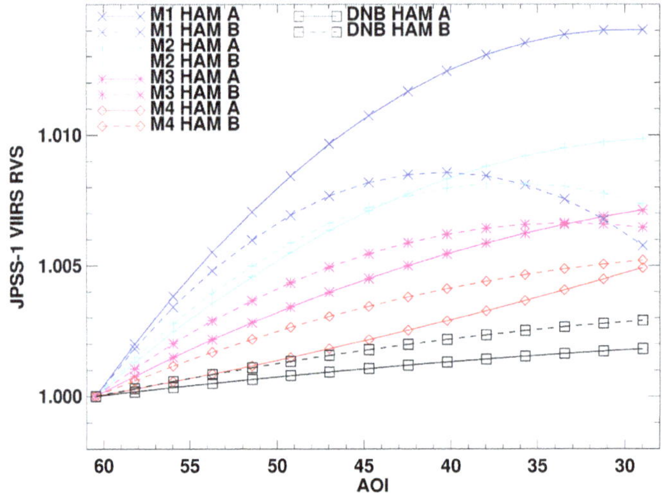

Figure 8. Band-averaged RVS for JPSS-1 VIIRS from the SV AOI to the end of scan AOI for bands M1–M4 and DNB.

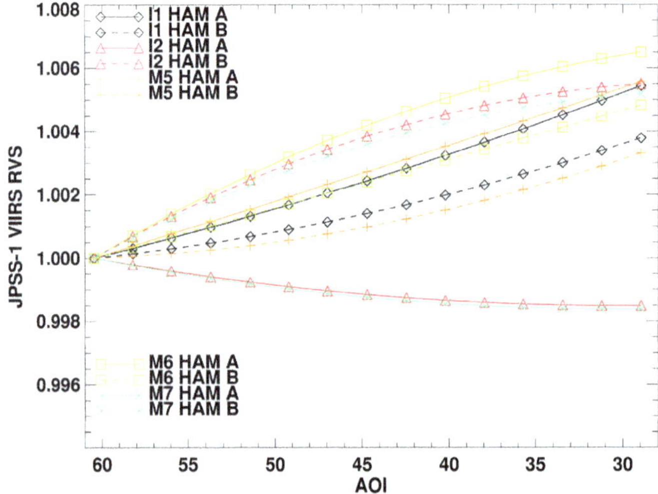

Figure 9. Band-averaged RVS for JPSS-1 VIIRS from the SV AOI to the end of scan AOI for bands I1, I2 and M5–M7.

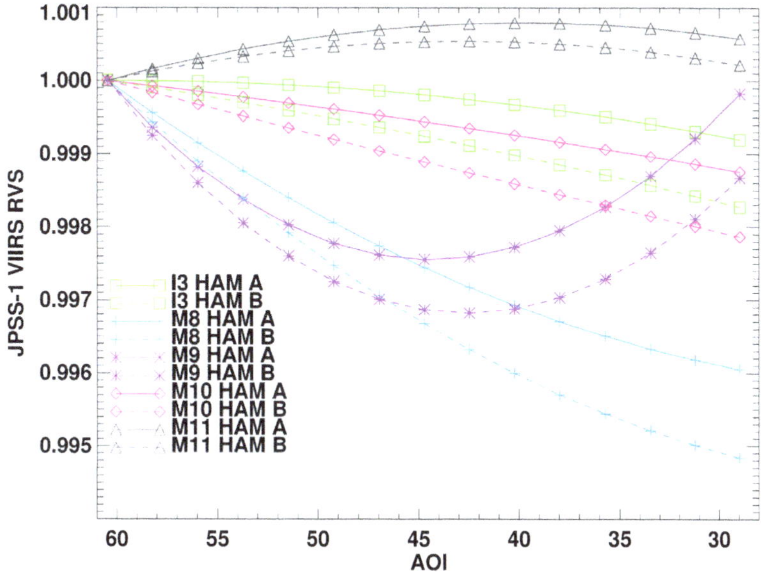

Figure 10. Band-averaged RVS for JPSS-1 VIIRS from the SV AOI to the end of scan AOI for bands I3 and M8-M11.

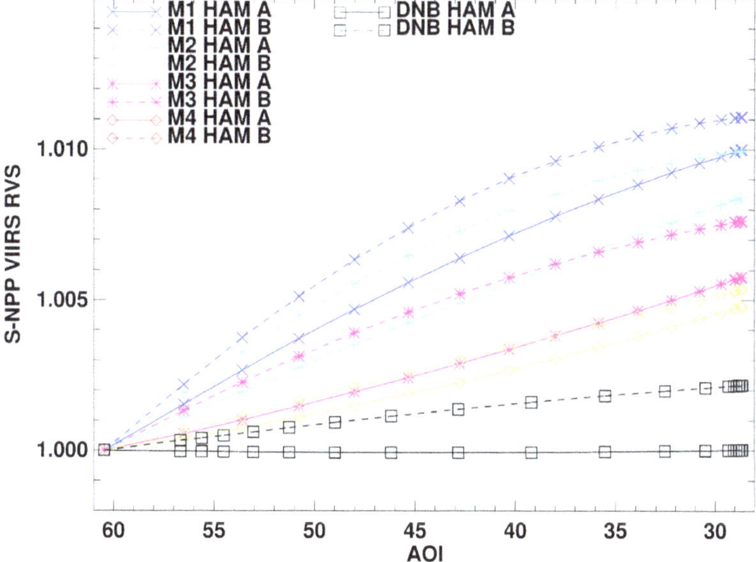

Figure 11. Band-averaged RVS for the Suomi National Polar-orbiting Partnership (S-NPP) VIIRS RSBs from the SV AOI to the end of scan AOI for bands M1–M4.

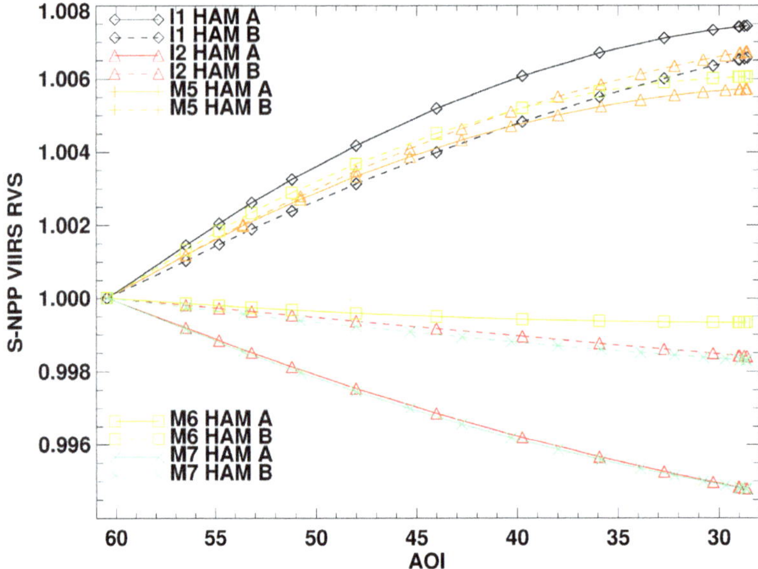

Figure 12. Band-averaged RVS for S-NPP VIIRS RSBs from the SV AOI to the end of scan AOI for bands I1, I2 and M5–M7.

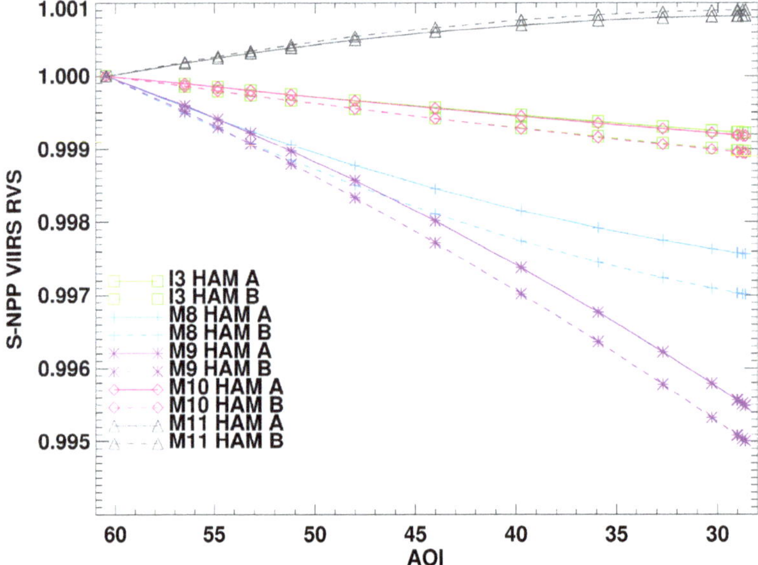

Figure 13. Band-averaged RVS for S-NPP VIIRS RSBs from the SV AOI to the end of scan AOI for bands I3 and M8–M11.

3.2. Thermal Emissive Band Results

The TEB RVS curves are essential to the SDR calibration. Each TEB (2 I-bands and 5 M-bands) detector and HAM side has an individually characterized RVS to reduce detector striping and scan-to-scan banding effects in the SDR. Figure 14 shows the RVS for band M14 HAM side A for all detectors. The variation across scan in Figure 14 is a little less than 10% with a detector dependence of ~0.25% at the end of scan. The points do show some noise as a result of LABB and OBCBB temperature drifting during the test that was not fully accounted for, some emission *versus* scan angle influence and noise in the DNs. Figure 15 shows an RVS curve for band M12 HAM side A all detectors. Similar to the other Mid-Wave Infrared (MWIR) bands, the RVS variation over the scan is relatively small compared to the LWIR bands. The peak-to-peak change is less than 0.6% for M12 which limits the impact of background emission effects in the MWIR. The RVS is much larger in the TEBs than in the RSBs and therefore most likely have larger polarization sensitivity in these mirrors than the RSBs (only the VNIR bands are characterized for polarization sensitivity). However, unlike the VNIR bands which have significant polarization in the scene radiance from the Earth, the TEBs observe unpolarized emission from the Earth and do not have polarization sensitivity requirements. The Emission Versus Scan angle (EVS) was measured prelaunch for S-NPP and computed on-orbit using a deep space maneuver to look at cold space. Significant polarization effects from emitted light off the HAM as a function of HAM AOI would cause the EVS and RVS to have different shapes. However, the on-orbit deep space maneuver EVS matched the RVS shape very well and indicated polarization effects are not a significant contributor to the RVS performance in the SDR algorithm [17].

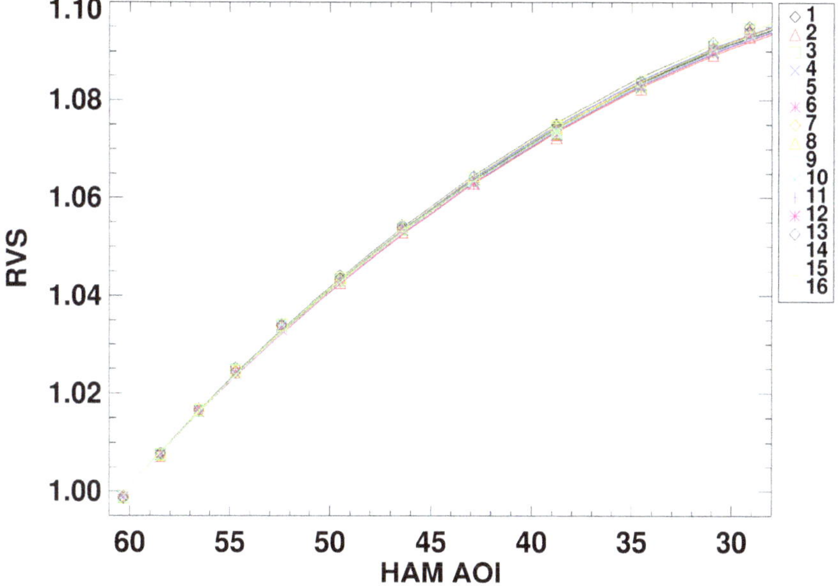

Figure 14. Band M14 HAM side A RVS Response and Fits for all Detectors.

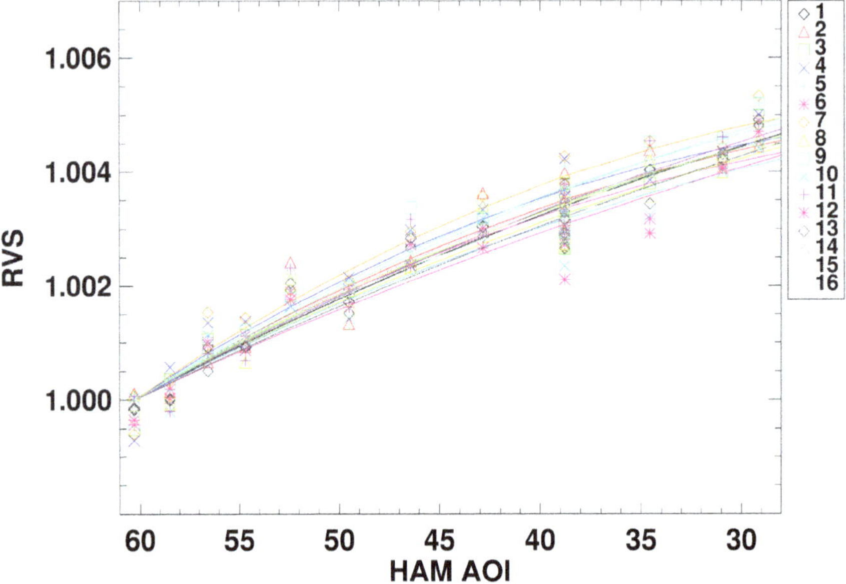

Figure 15. Band M12 HAM side A RVS Response and Fits for all Detectors.

Figures 16 and 17 show the detector averaged JPSS-1 VIIRS TEB RVS curves for both HAM side A (solid lines) and B (dotted lines) across the scan of VIIRS. Unlike the RSBs which showed significant shape differences between HAM sides (especially for blue wavelength bands), the TEBs have consistent shapes. There is some separation in the curves between the HAM sides within a band that is not consistent with S-NPP VIIRS performance but the overall magnitude of the RVS curves for each band are similar to S-NPP. Figures 18 and 19 show the detector-averaged S-NPP VIIRS RVS for all the TEBs and HAM sides. The magnitudes for both S-NPP and JPSS-1 VIIRS are very similar but HAM side difference were larger on the S-NPP compared to JPSS-1. Band M14 shows much better HAM side consistency on JPSS-1 compared to S-NPP. Table 6 lists the worst case peak-to-peak variation over scan for both HAM sides for each TEB as well as the worst case RMS fit residuals.

Table 6. Maximum RVS Variation across Scan and Characterization Uncertainty for all TEBs.

Band	HAM A Maximum Peak-to-Peak Change in RVS Over Scan (%)	HAM A RMS Residuals (%)	HAM B Maximum Peak-to-Peak Change in RVS Over Scan (%)	HAM B RMS Residuals (%)
I4	0.56	0.05	0.60	0.05
M12	0.59	0.07	0.60	0.05
M13	0.56	0.06	0.56	0.05
M14	9.64	0.08	9.60	0.08
M15	6.13	0.07	6.53	0.07
I5	4.55	0.12	4.93	0.11
M16A	3.03	0.05	3.35	0.06
M16B	2.93	0.05	3.28	0.06

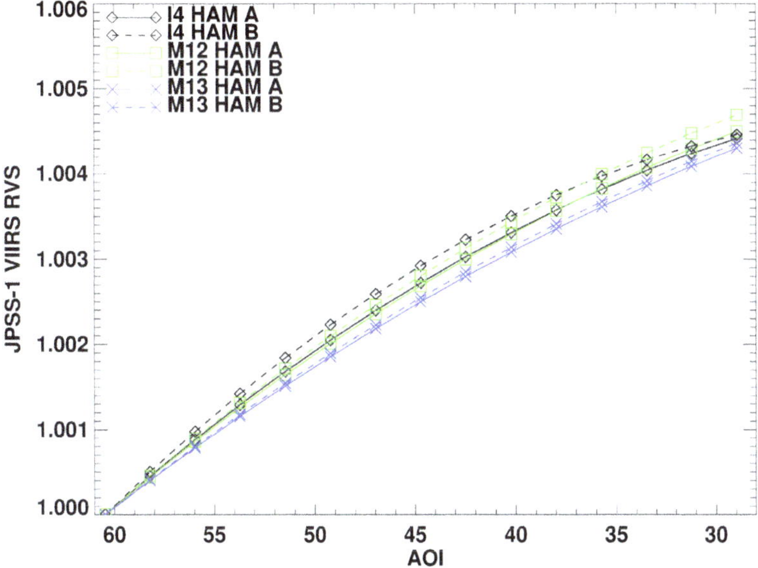

Figure 16. Band-averaged RVS for JPSS-1 VIIRS all TEBs from the SV AOI through to the end of scan AOI for the Mid-wave Wavelength Infrared (MWIR) bands.

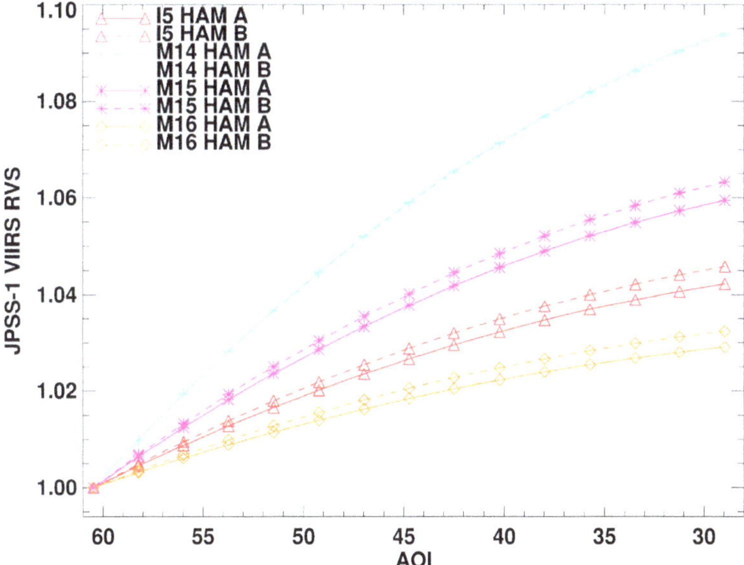

Figure 17. Band-averaged RVS for JPSS-1 VIIRS all TEBs from the SV AOI through to the end of scan AOI for the Long Wavelength Infrared (LWIR) bands.

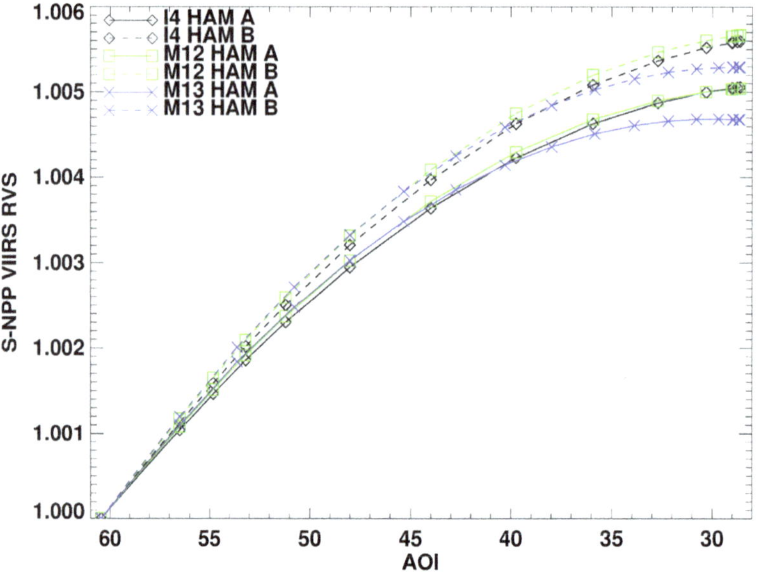

Figure 18. Band-averaged RVS for S-NPP VIIRS all TEBs from the SV AOI through to the end of scan AOI for the MWIR bands.

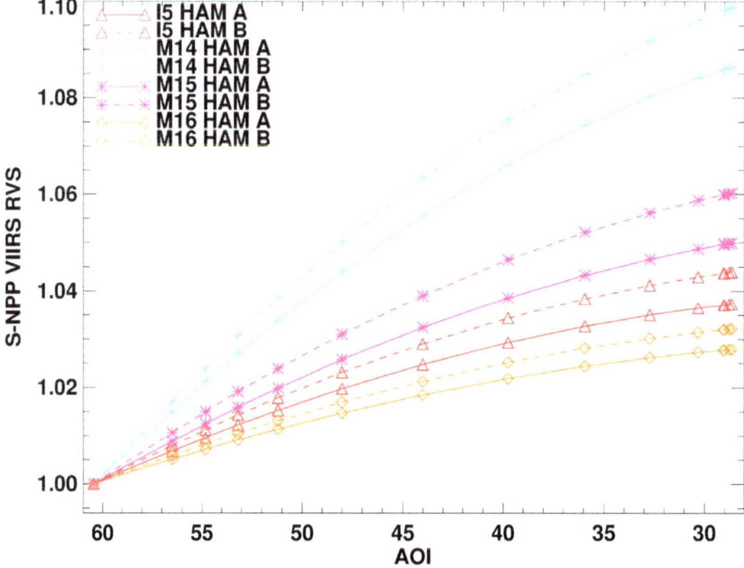

Figure 19. Band-averaged RVS for S-NPP VIIRS all TEBs from the SV AOI through to the end of scan AOI for the LWIR bands.

4. Conclusions

The Response Versus Scan angle for JPSS-1 VIIRS was measured using stable wide-field sources to provide accurate characterization. The results showed that the JPSS-1 VIIRS RVS shapes are consistent with its predecessor on the S-NPP spacecraft with the exception of band M1 being larger for JPSS-1 (Figure 8). Some HAM side differences are present on the JPSS-1 VIIRS that were not present on S-NPP, especially for blue wavelength bands that can be traced back to HAM side coating differences. These differences were caused by a long time period between when the two sides of the JPSS-1 VIIRS HAM coating were deposited. The RSBs show less than 2% variation in RVS across the scan and the MWIR TEBs also show very small (~1%) change across scan as well (similar between JPSS-1 and S-NPP). The LWIR TEBs do show significant (~10%) RVS change across scan making their curves very important to the SDR calibration for these bands. These large RVS variations in the LWIR for JPSS-1 were also present in S-NPP VIIRS. The quality of the S-NPP RVS characterization allows for very good SDR performance on-orbit. It is expected that the JPSS-1 RVS will perform at a similar level. The characterization accuracy is very good (~0.12%) and well below the requirement of 0.3% (RSB) or 0.2% (TEB). The only band with large uncertainties (0.28%) is band M9, but is impacted by water vapor drifting during the test.

Acknowledgments: The authors would like to thank the Raytheon test team, especially Eric Johnson, for their acquisition of the test data, test plans and algorithm development efforts as well as the government data analysis working group for their contributions to this analysis.

Author Contributions: David Moyer wrote the manuscript. David Moyer and Jeff McIntire independently performed the analysis contained in this work. Hassan Oudrari, James McCarthy, Xiaoxiong Xiong and Frank De Luccia contributed to the design of testing used to acquire the RVS measurements, methodologies for characterizing the RVS and the development of the manuscript.

Conflicts of Interest: The authors declare no conflict of interest.

References

1. Cao, C.; Xiong, X.; Blonski, S.; Liu, Q.; Uprety, S.; Shao, X.; Bai, Y.; Weng, F. Suomi NPP VIIRS sensor data record verification, validation, and long-term performance monitoring. *J. Geophys. Res.* **2013**, *188*. [CrossRef]
2. Holben, B.N.; Kaufman, Y.J.; Kendall, J.D. NOAA-11 AVHRR visible and near-IR inflight calibration. *Int. J. Remote Sens.* **1990**, *11*, 1511–1519. [CrossRef]
3. Kramer, H.J. *Observation of the Earth and Its Environment: Survey of Missions and Sensors*, 4th ed.; Springer-Verlag: Berlin, Germany, 2002.
4. Guenther, B.; Xiong, X.; Salomonson, V.V.; Barnes, W.L.; Young, J. On-orbit performance of the Earth Observing System Moderate Resolution Imaging Spectroradiometer; first year of data. *Remote Sens. Environ.* **2002**, *83*, 16–30. [CrossRef]
5. Moeller, C.; Schwarting, T.; McIntire, J.; Moyer, D. JPSS-1 VIIRS pre-launch spectral characterization and performance. *Proc. SPIE* **2015**, *9607*. [CrossRef]
6. Cao, C.; Xiong, X.; de Luccia, F.; Liu, Q.; Blonski, S.; Pogorzala, D.; Oudrari, H. *VIIRS SDR User's Guide*; Technical Report NESDIS 142; U.S. Department Commerce. NOAA: Silver Spring, MD, USA, 2013.
7. Joint Polar Satellite System (JPSS) Program. *VIIRS Radiometric Calibration Algorithm Theoretical Basis Document (ATBD)*; Technical Report D43777; Joint Polar Satellite System: Lanham, MD, USA, 2010.
8. Quantum Coating Inc., 1259 N. Church Street, Moorestown, NJ 08057. Available online: http://www.quantumcoating.com/fss99/ (accessed on 28 November 2015).
9. Waluschka, E. VIIRS polarization testing. *Proc. SPIE* **2009**, *7452*. [CrossRef]
10. Kulkarny, V.; Hauss, B.; Jackson, J.; Ip, J.; Pratt, P.; Snodgrass, C.; Tsugawa, R.; Bendow, B.; Mineart, G. The impact of VIIRS polarization sensitivity on ocean color. In Proceedings of the 2010 IEEE International Geoscience and Remote Sensing Symposium (IGARSS), Honolulu, HI, USA, 25–30 July 2010; pp. 220–223.
11. Xiaoxiong, X.; Choi, T.; Che, N.; Wang, Z.; Dodd, J.; Xie, Y.; Barnes, W. Results and lessons from a decade of Terra MODIS on-orbit spectral characterization. *Proc. SPIE* **2010**, *7862*. [CrossRef]
12. Xiong, X.; Wenny, B.N.; Sun, J.; Angal, A.; Wu, A.; Chen, H.; Choi, T.; Madhavan, S.; Geng, X.; Link, D.; *et al.* Overview of Aqua MODIS 10-year on-orbit calibration and performance. *Proc. SPIE* **2012**, *8533*. [CrossRef]
13. Xiong, X.; Chiang, K.; McIntire, J.; Oudrari, H.; Wu, A.; Schwaller, M.; Butler, J. Early assessment of VIIRS on-orbit calibration and support activities. In Proceedings of the 2012 IEEE International Geoscience and Remote Sensing Symposium (IGARSS), Munich, Germany, 22–27 July 2012; pp. 7189–7192.

14. Sun, J.; Xiong, X.; Angal, A.; Chen, H.; Wu, A.; Geng, X. Time-dependent response versus scan angle for MODIS reflective solar bands. *IEEE Trans. Geosci. Remote Sens.* **2014**, *52*, 3159–3174. [CrossRef]
15. Oudrari, H.; McIntire, J.; Xiong, X.; Butler, J.; Lee, S.; Lei, N.; Schwarting, T.; Sun, J. Prelaunch radiometric characterization and calibration of the S-NPP VIIRS Sensor. *IEEE TGRS* **2015**, *53*, 2195–2210. [CrossRef]
16. Lee, S.; Cao, C. JPSS-1 VIIRS Prelaunch RSB/DNB RVS Characterization and Water Vapor Correction. *Proc. SPIE* **2015**, *9607*. [CrossRef]
17. Wu, A.; Xiong, X.; Chiang, K.; Sun, C. Assessment of the NPP VIIRS RVS for the thermal emissive bands using the first pitch maneuver observations. *Proc. SPIE* **2012**, *8510*. [CrossRef]

Chapter 3: Sensor Data Records Intercomparison and Monitoring

Article

Inter-Comparison of S-NPP VIIRS and Aqua MODIS Thermal Emissive Bands Using Hyperspectral Infrared Sounder Measurements as a Transfer Reference

Yonghong Li [1,*], Aisheng Wu [1] and Xiaoxiong Xiong [2]

[1] Science Systems and Applications, Inc., 10210 Greenbelt Rd., Lanham, MD 20706, USA; aisheng.wu@ssaihq.com
[2] Sciences and Exploration Directorate, NASA/GSFC, Greenbelt, MD 20771, USA; xiaoxiong.xiong-1@nasa.gov
* Correspondence: yonghong.li@ssaihq.com; Tel.: +1-301-867-2067; Fax: +1-301-867-2151

Academic Editors: Changyong Cao, Magaly Koch and Prasad S. Thenkabail
Received: 26 October 2015; Accepted: 8 January 2016; Published: 19 January 2016

Abstract: This paper compares the calibration consistency of the spectrally-matched thermal emissive bands (TEB) between the Suomi National Polar-orbiting Partnership (S-NPP) Visible Infrared Imaging Radiometer Suite (VIIRS) and the Aqua Moderate Resolution Imaging Spectroradiometer (MODIS), using observations from their simultaneous nadir overpasses (SNO). Nearly-simultaneous hyperspectral measurements from the Aqua Atmospheric Infrared Sounder(AIRS) and the S-NPP Cross-track Infrared Sounder (CrIS) are used to account for existing spectral response differences between MODIS and VIIRS TEB. The comparison uses VIIRS Sensor Data Records (SDR) in MODIS five-minute granule format provided by the NASA Land Product and Evaluation and Test Element (PEATE) and Aqua MODIS Collection 6 Level 1 B (L1B) products. Each AIRS footprint of 13.5 km (or CrIS field of view of 14 km) is co-located with multiple MODIS (or VIIRS) pixels. The corresponding AIRS- and CrIS-simulated MODIS and VIIRS radiances are derived by convolutions based on sensor-dependent relative spectral response (RSR) functions. The VIIRS and MODIS TEB calibration consistency is evaluated and the two sensors agreed within 0.2 K in brightness temperature. Additional factors affecting the comparison such as geolocation and atmospheric water vapor content are also discussed in this paper.

Keywords: VIIRS; MODIS; AIRS; CrIS; thermal emissive bands; calibration

---

1. Introduction

Information from space-borne instruments has provided long term observational science data by acquiring imagery of the Earth. Various sensors collect data with different spatial resolutions and their spectra span wavelengths from the visible through the infrared. Satellite thermal infrared remote sensing data have many applications, from surface materials and features, such as soil moisture, to surface temperature mapping [1]. Calibration is crucial before these Earth observations are used.

The Visible Infrared Imaging Radiometer Suite (VIIRS), a major Earth observing instrument aboard the Suomi National Polar-orbiting Partnership (S-NPP) satellite, has operated on-orbit for over three years [2]. The VIIRS instrument is designed to extend the measurements from the Moderate Resolution Imaging Spectroradiometer (MODIS), another key instrument of NASA Earth observation missions. MODIS is onboard the Terra and Aqua satellites and has operated for over fifteen and thirteen years, respectively [3]. It is a multispectral sensor that collects data for studying land, atmosphere, and ocean features [4,5]. Nearly 40 scientific products have been derived from MODIS calibrated data [5].

Collection 6 is the latest released L1B product version, which includes several major improvements to the calibration to handle various issues with the aging MODIS sensors [6].

Toller *et al.* [3] and Xiong *et al.* [7] have provided details on MODIS thermal emissive bands (TEB), on-board calibration algorithms, characteristics, performance, challenging issues, and lessons learned, such as a long wavelength infrared focal plane assembly optical leak and accurately tracking the scan mirror response *versus* scan angle changes over time. A number of approaches are used to independently track the performance of Aqua MODIS (hereafter MODIS) Level 1B TEB, which includes reference to near-surface temperature observations over Dome Concordia, Antarctica [8], comparison of the simultaneous nadir overpasses (SNO) measurements with the Infrared Atmospheric Sounding Interferometer [9] and the Atmospheric Infrared Sounder (AIRS) [10,11]. VIIRS on-orbit calibration methodology and early performance has been reported in several studies [12–14]. In addition to the evaluation based solely on VIIRS calibration data, other investigators have conducted inter-comparisons with different sensors. Example MODIS-VIIRS inter-comparisons include (1) scene-based cross-comparison [15]; (2) SNO data for tracking and evaluation of the RSB stability and performance [16]; and (3) SNO data for assessing the calibration consistency of TEB bands [17]. In the study by Efremova *et al.* [17], the SNO data of S-NPP VIIRS (hereafter VIIRS), MODIS, and S-NPP CrIS were collected in the CrIS footprints for VIIRS-MODIS inter-comparison, in which CrIS measurements were used as a transfer reference to derive the bias correction caused by different relative spectral response (RSR) functions between VIIRS and MODIS. Their results showed that the RSR-corrected brightness temperature (BT) differences between VIIRS and MODIS are exceptionally small and are generally within ±0.2 K over the entire scene temperature range using ten SNO datasets between August 2013 and July 2014.

AIRS and CrIS are hyperspectral infrared sounders on board the Aqua and S-NPP satellites, respectively. AIRS is a grating spectrometer with multiple detector arrays for the corresponding spectral channels while CrIS is an interferometer. Both of the two hardware approaches achieve nearly the same spectral characteristics. Details on their calibration and performance are found in [18–20].

Since MODIS is a heritage sensor for VIIRS, their consistency in measurements is very important for extending the current MODIS scientific products. This study extends the previous work [17] comparing VIIRS and MODIS TEB using CrIS in two respects: (1) AIRS hyperspectral measurements are included. This allows for a RSR correction to achieve a better spectrally-matched VIIRS and MODIS TEB than the use of CrIS hyperspectral data alone; (2) this study extends the previous results from limited SNO events to an extensive dataset which covers a complete latitude range from 82°S to 82°N.

Section 2 introduces data features of the four sensors. Section 3 provides general information on the TEB on-orbit calibration. Inter-comparison methodology and data processing are introduced in Section 4; Section 5 shows inter-comparison results and provides a discussion of the results; and Section 6 is a summary of our analysis.

2. Sensor Data Features

VIIRS has 16 moderate-resolution bands (750 m at nadir) and five image-resolution bands (375 m at nadir). Among them, there are 14 reflectance solar bands (RSBs, imaging bands I1-3, moderate-resolution bands M1-11) with wavelengths ranging from 0.4 to 2.3 μm and seven TEBs (imaging bands I4-5, moderate-resolution bands M12-16) covering a spectral range from 3.6 to 12.5 μm. VIIRS Earth view scenes in the moderate-resolution bands contain 3200 samples in the scan direction within a scan angle range of $\pm56°$ off nadir and 16 detectors in the track direction.

MODIS has 20 RSBs (bands 1–19, 26) with wavelengths ranging from 0.4 to 2.3 μm and 16 TEBs (bands 20–25, 27–36) covering a spectra range from 3.6 to 14.4 μm. The RSB sensors have three ground resolutions (250, 500, and 1000 m) at nadir, while all TEB sensors have 1000 m ground resolution at nadir. MODIS Earth view scenes in the 1-km resolution bands contain 1354 samples in the scan direction within a scan angle range of $\pm55°$ off nadir and 10 detectors in the track direction.

AIRS provides atmospheric emission spectra to derive temperature and humidity profiles with high precision. Its thermal infrared spectra span 3.7 to 15.4 μm with 2378 spectral channels. AIRS is an across track scanning system with scan range of ±49.5°, centered at nadir. A nominal scan line covers 90 infrared footprints, which corresponds to a ground resolution of 13.5 km at nadir at a satellite altitude of 705.3 km. In contrast to AIRS, CrIS is a Fourier transform spectrometer, which measures radiance for retrieving profiles of temperature, pressure, and moisture. It has 1305 spectral channels covering three wavelength ranges: shortwave infrared (3.92–4.64 μm), mid-wave infrared (5.71–8.26 μm), and long-wave infrared (9.14–15.38 μm). Its scan range covers ±50° with 30 Earth-scene views (also called field of regard). Each view position includes a 3 × 3 field-of-view (FOV) array. The spatial resolution of each FOV is 14 km at nadir for a satellite altitude of 824 km. Similar to VIIRS and MODIS, AIRS and CrIS also use an internal blackbody (BB) and a deep space view (SV) to maintain their calibration on-orbit.

Both VIIRS and MODIS TEB wavelengths are within the AIRS and CrIS hyperspectral coverage. Since there are regular SNO occurrences between S-NPP and Aqua, both AIRS and CrIS hyperspectral measurements can be used to provide a simultaneous RSR correction between VIIRS and MODIS spectrally-matched TEB, enabling a precise evaluation of the calibration difference between the two sensors. Furthermore, since MODIS and AIRS are on board of the Aqua satellite and VIIRS and CrIS are on board of the S-NPP satellite, more frequent SNO data could be collected due to the fact that their orbits are nearly parallel.

3. VIIRS and MODIS TEB On-Orbit Calibration

The TEB calibration describes the relationship between the digital response of a detector and the sensor's at-aperture radiance. Both VIIRS and MODIS TEB use the onboard BB and SV observations to perform their onboard calibration. BB and SV are viewed every scan to provide scan-by-scan calibration coefficients. The BB is designed to ensure a high effective emissivity and temperature uniformity. The emissivity are greater than 0.992 for MODIS and 0.997 for VIIRS [21,22]. The temperature of the onboard blackbody are controlled at 292.5 K during normal operations on VIIRS and 285 K on MODIS. Quarterly warm-up cool-down (WUCD) activities are scheduled so the measurements can be used to derive the calibration offset and nonlinear coefficient. During each WUCD event, blackbody temperatures vary from 267 K to 315 K for VIIRS and from 270 K to 315 K for MODIS.

A quadratic algorithm is currently used in the TEB calibration. The expression for VIIRS TEB radiance calculation is [23]:

$$L_{EV} = \frac{1}{RVS_{EV}} \left(F \cdot \sum_{i=0}^{2} c_i \cdot dn_{EV}^{i} - (RVS_{EV} - RVS_{SV}) \cdot \frac{(1-\rho_{rta}) \cdot L(T_{rta}) - L(T_{ham})}{\rho_{rta}} \right) \tag{1}$$

where RVS and T represent response *versus* scan-angle and temperature; L and ρ are temperature-dependent spectral radiance and spectral reflectance; dn is the background corrected digital count; EV, SV, rta, and ham represent Earth view, space view, rotating telescope assembly, and half angle mirror (HAM), respectively. c_0, c_1, and c_2 are calibration coefficients for each band, detector, and HAM side. Currently, prelaunch calibration coefficients (c_0, c_1, and c_2) are used in VIIRS TEB calibration. The on-orbit calibration scaling factor, F, is calculated on a scan-by-scan basis from BB measurements [23].

For MODIS, the TEB radiance is computed by:

$$L_{EV} = \frac{1}{RVS_{EV}} \left(a_0 + b_1 \cdot dn_{EV} + a_2 \cdot dn_{EV}^2 - (RVS_{SV} - RVS_{EV}) \cdot L(T_{sm}) \right) \tag{2}$$

where L is spectral band averaged radiance and SM represents scan mirror. a_0, b_1, and a_2 are calibration coefficients for each band, detector, and mirror side. a_0 and a_2 are based on prelaunch tests and are adjusted using an iterative approach to account for the on-orbit drifts [24], and b_1 is calculated from the BB radiance on a scan-by-scan basis (except for band 21).

4. VIIRS-MODIS Inter-Comparison Methodology

In this paper, the VIIRS-MODIS inter-comparison is conducted using SNO observations. The orbital parameters of the Aqua and S-NPP satellites were obtained from the two-line element sets [25]. The SNO criterion is set to within 30 s in determining the crossover periods between Aqua and S-NPP orbits. A set of L1B 5-min granules of VIIRS, MODIS, as well as corresponding AIRS L1B granules and CrIS SDR granules, are first selected based on predicted SNO time periods. Then the radiances of each infrared sounder FOV from the SNOs are extracted according to overpass time and geolocation.

For the comparison of measurements from different sensors, all data were converted to the same spectral and spatial grid. Table 1 lists information on VIIRS, MODIS, AIRS, and CrIS spectral bands. In our analysis, the high spatial resolutions of MODIS L1B data and VIIRS Level 1 5-min data were aggregated into each AIRS and CrIS FOV in order to match the footprint. The radiance difference between VIIRS and MODIS TEB due to their RSR differences was derived by the comparison to the integrated AIRS and CrIS hyperspectral data.

Table 1. VIIRS, MODIS, AIRS, and CrIS spectral bands.

	Onboard Satellite	Spectral Channels		Spectral Range (μm)	Nadir Spatial Resolution	# of Samples in Scan	Scan Angle Range
VIIRS	S-NPP: 3+years	TEB	I-bands I4–I5 M-bands M12–M16	3.6~12.5	375 m 750 m	3200	±56°
MODIS	Aqua: 13+years	TEB	band 20–25, 27–30 band 31–36	3.6~9.8 11~14.4	1 km	1354	±55°
CrIS	S-NPP: 3+years	1305 spectral channels		3.92~4.64 5.71~8.26 9.14~15.38	14 km	30; 3 × 3 FOV	±50°
AIRS	Aqua: 13+years	2378 spectral channels		3.74~4.61 6.2~8.22 8.8~15.4	13.5 km	90	±49.5°

4.1. *Aggregation of Higher Spatial Data in Infrared Sounder FOV*

AIRS data at nadir were taken from six footprints (~±3.3°) around the center. VIIRS and MODIS radiance at nadir were defined as the data within a range of ±10 degrees from nadir to get a stable aggregated value. For each AIRS footprint, all pixels of VIIRS and MODIS radiance measurements within a 6.75 km radius of the AIRS pixel center during each SNO were averaged, respectively. Here we assume the FOV spatial response is evenly distributed. Typically, one AIRS footprint includes 140 MODIS pixels and more than 240 VIIRS pixels (750 m resolution). In our analysis, only footprints that are covered by more than 100 MODIS pixels and 180 VIIRS pixels (750 m) were accepted in the SNO data collection.

The same method was applied to CrIS SDR nadir pixels (two center field of regards, around ±3.3° scan angle), where all MODIS and VIIRS measurements within a 7 km radius of a CrIS pixel center were averaged. In the inter-comparison below, only CrIS FOVs that include more than 110 MODIS pixels and 190 VIIRS pixels (750 m resolution) were extracted in the SNO data collection since each CrIS FOV typically contains ~150 MODIS pixels and more than 260 VIIRS pixels (750 m resolution).

4.2. *Sensor Spectral band RSR Matching*

The spectral profiles of VIIRS and MODIS TEBs as well as spectra coverage of AIRS and CrIS are illustrated in Figure 1. The RSR in Figure 1 describes the system transmission as well as the detector sensitivity. To compare the radiance measurements of a spectrally-matched VIIRS and MODIS band, the bias caused by the RSR difference between the two sensors needs to be removed. In this VIIRS-MODIS inter-comparison, AIRS and CrIS hyperspectral measurements were used to derive the bias. Data from four pairs of spectrally-overlapping VIIRS-MODIS bands, M13-B22 (3.9~4.05 μm), M13-B23 (4.0~4.2 μm), M15-B31 (10.1~11.5 μm), and M16-B32 (11.5~12.5 μm), were analyzed in the following inter-comparisons. Both AIRS L1B data and CrIS SDR data were integrated over MODIS bands 22, 23, 31–32, and VIIRS bands M13, M15-16 separately. For each band, the sensor-dependent RSR function of VIIRS and MODIS was first applied to each infrared sounder (AIRS and CrIS) channel using linear interpolation. Then, for each VIIRS and MODIS TEB band, the simulated radiances from the infrared sounder (AIRS and CrIS) measurements were calculated by:

$$L_{simulated} = \frac{\int_{\lambda_1}^{\lambda_2} L_{hi}(\lambda) \cdot RSR(\lambda) \cdot d\lambda}{\int_{\lambda_1}^{\lambda_2} RSR(\lambda) \cdot d\lambda} \tag{3}$$

where "L_{hi}" is the band-dependent interpolated radiance and $[\lambda_1, \lambda_2]$ is the wavelength range of a band. The difference in simulated radiance between VIIRS RSR and MODIS RSR was derived and used as a bias correction in the VIIRS-MODIS inter-comparison. The RSR correction was implemented for each SNO pixel.

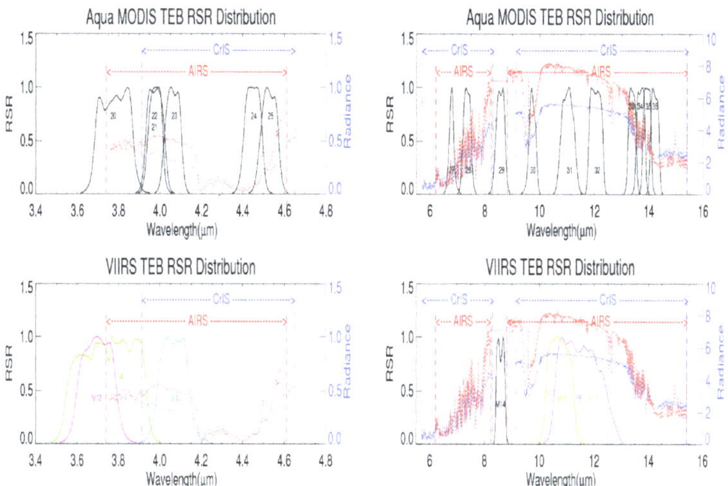

Figure 1. Spectral distribution of VIIRS and MODIS TEBs, as well as AIRS and CrIS spectra coverage. (radiance unit: W/m^2/μm/sr).

5. Results and Discussions

In the following inter-comparisons, we extracted SNO data from 24 S-NPP/Aqua orbits during April 2014–March 2015, which includes ~57,500 AIRS footprints along with ~13.3 million VIIRS pixels and ~7.7 million MODIS pixels, as well as from ~72,000 CrIS FOVs matching with ~17 million VIIRS pixels and ~9.7 million MODIS pixels. The SNOs span a latitude ranging from −82 degrees to +82 degrees. To get good quality AIRS data, 467 bad channels and noisy channels were removed based on the channel property list given in [26] and QA parameter given in each AIRS granule. Only channels with QA less than three were used.

The inter-comparison results in this section were averaged over a radiance interval of 0.02 W/m^2/μm/sr for VIIRS band M13 and 0.2 W/m^2/μm/sr for VIIRS bands M15-16. The latitude bin size in the spatial distribution figures below is five degrees. To determine the VIIRS-MODIS inter-comparison in brightness temperature (BT), the radiance of each SNO was converted into BT using the Planck function. The inter-comparison results below were averaged at BT intervals of 1 K for all bands. ΔL and ΔBT represent the difference between VIIRS and MODIS measurements in radiance and in BT as calculated by:

$$\Delta L = L_{VIIRS} - L_{MODIS}, \; \Delta BT = BT_{VIIRS} - BT_{MODIS} \tag{4}$$

The RSR corrected MODIS measurements were derived by:

$$L_{MODIS_{corr}} = L_{MODIS} \cdot factor, \text{ where } factor = \frac{L_{simulated}(RSR_{VIIRS})}{L_{simulated}(RSR_{MODIS})} \tag{5}$$

ΔL_{corr} and ΔBT_{corr} represent the difference between VIIRS measurements and RSR corrected MODIS measurements in radiance and BT:

$$\Delta L_{corr} = L_{VIIRS} - L_{MODIS\_corr}, \; \Delta BT_{corr} = BT_{VIIRS} - BT_{MODIS\_corr} \tag{6}$$

5.1. VIIRS-MODIS Inter-Comparison With RSR Correction from AIRS Measurements

In order to derive the VIIRS-MODIS RSR correction using AIRS measurements, MODIS-AIRS SNOs were first selected and then VIIRS-MODIS-AIRS SNOs were collected based on the geolocation and overpass time for each MODIS-AIRS SNO pixel. Figure 2 shows the distributions of VIIRS-MODIS-AIRS SNO pixels for VIIRS(M13)-MODIS(B23) and VIIRS(M15)-MODIS(B31).

For each aggregation area (1 K BT by 5° Latitude) shown in Figure 2, the differences between VIIRS and MODIS measurements from all AIRS pixels within the area were extracted first. A 3-sigma filtering was applied to remove outlier data. Then the mean difference was determined by the average of the filtered data. Figure 3a,c,e,g gives the distribution of VIIRS-MODIS inter-comparison directly from VIIRS and MODIS measurements. Since VIIRS-MODIS-AIRS SNOs were collected based on AIRS pixels, the simulated VIIRS radiance and simulated MODIS radiance were calculated for each AIRS pixel. Hence, the ratio between the two simulated radiances was used to correct the radiance caused by the existing difference in RSR for a pair of VIIRS-MODIS bands. Figure 3b,d,f,h illustrates the distribution of VIIRS-MODIS inter-comparison data after applying the RSR correction.

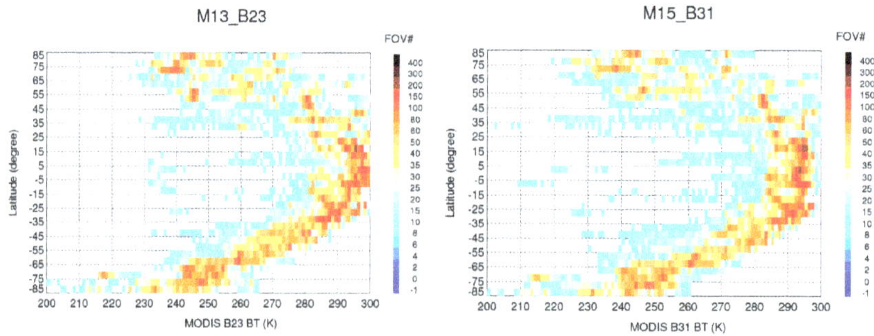

Figure 2. VIIRS-MODIS-AIRS SNO FOV number distribution.

As shown in Figure 3, the relatively large differences in the M13-B22, M13-B23, and M16-B32 comparisons are mainly caused by differences in their spectral coverage, since the RSR correction significantly improves the agreement. The differences after the RSR correction are well within 0.2 K and nearly constant across the entire temperature and latitude range. There is a slight latitude dependence for M16-B32, which is expected since the bandwidth of M16 is nearly double that of B32. For M15-B31, the RSR correction only produces a small improvement for the differences because they have a smaller bandwidth difference and they are already significantly smaller than the differences for the other band pairs. Moreover, some high differences, shown in the tropical area for BT lower than 270 K, are caused by limited sampling and scene variation (see Figure 2).

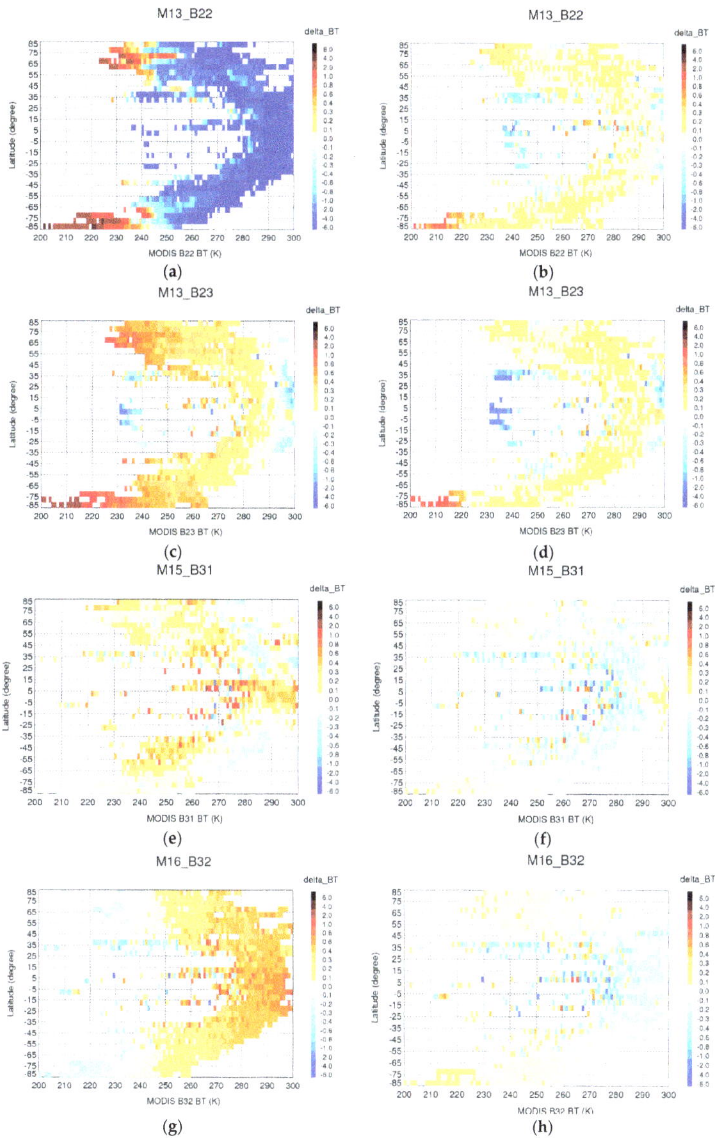

Figure 3. Distribution of VIIRS-MODIS inter-comparison without/with RSR correction from AIRS measurements. (**a**) M13_B22 without RSR correction; (**b**) M13_B22 with RSR correction; (**c**) M13_B23 without RSR correction; (**d**) M13_B23 with RSR correction; (**e**) M15_B31 without RSR correction; (**f**) M15_B31 with RSR correction; (**g**) M16_B32 without RSR correction; (**h**) M16_B32 with RSR correction.

The VIIRS-MODIS differences are plotted as a function of BT before and after the RSR correction (see Figure 4). The error bars represent the standard deviation for the corresponding BT bin. Results show that the strong dependence of the differences on temperature for M13-B22, M13-B23, and M16-B32 is almost completely removed and the remaining differences are well within 0.2 K. Relatively larger differences are observed at lower BT (less than 240 K) in the M13-B22 and M13-B23 comparisons. Table 2 lists a quantitative comparison of the results for the four pairs of the bands over a range of BT. VIIRS M14 and MODIS B29 have very close spectral features (see Figure 1) and their measurements have less than 0.2 K differences (see Table 2). Since AIRS spectra does not cover M14-B29 range, no RSR correction can be applied to the comparison of the two bands.

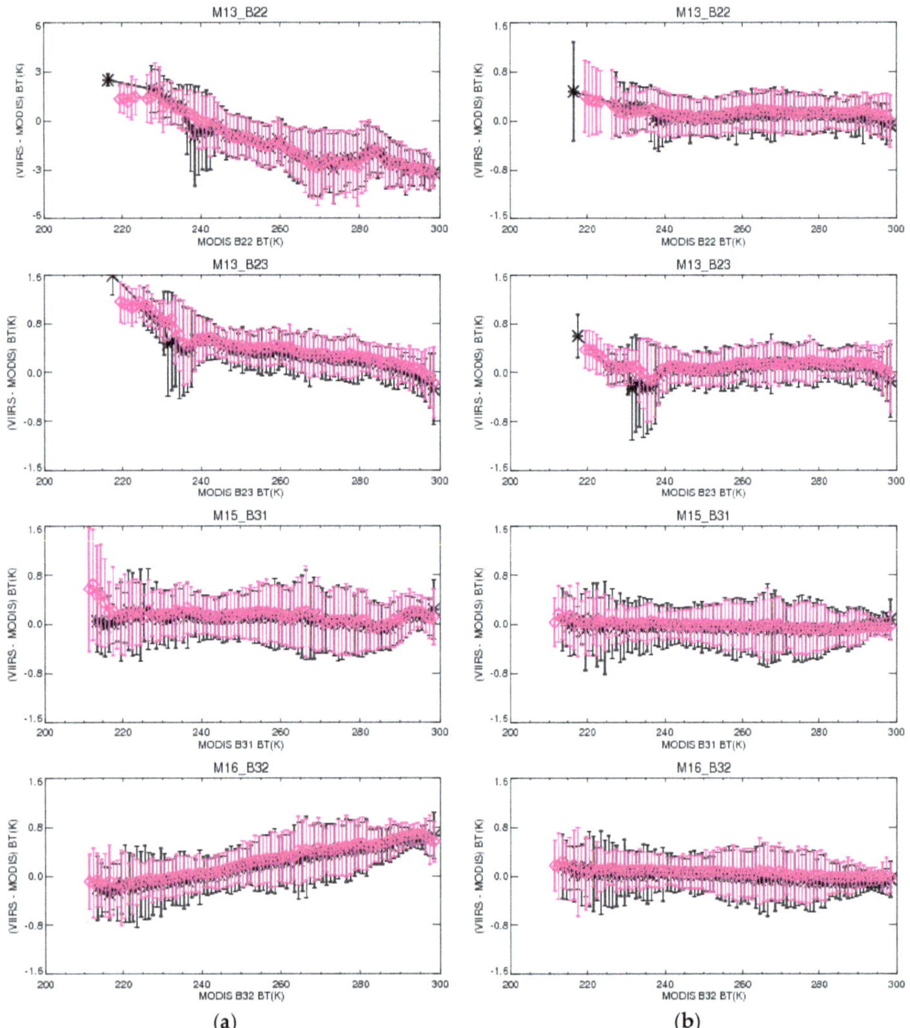

Figure 4. VIIRS-MODIS inter-comparison without/with RSR correction from AIRS (black stars) and CrIS (blue diamonds) measurements, (**a**) without RSR correction; and (**b**) with RSR correction.

Table 2. Difference in brightness temperature without/with RSR correction from AIRS measurements (unit: K; RMS: root-mean-square of the bias).

	M13-B22		M13-B23		M15-B31		M16-B32		M14-B29	
BT_{MODIS}	ΔBT	ΔBT_{corr}	ΔBT	ΔBT_{corr}	ΔBT	ΔBT_{corr}	ΔBT	ΔBT_{corr}	ΔBT	ΔBT_{corr}
220	-	-	-	-	0.11	0.00	−0.12	0.12	-	-
230	1.47	0.16	0.80	0.04	0.11	−0.01	−0.08	0.04	−0.02	-
240	−0.61	0.05	0.53	0.04	0.11	−0.02	0.01	0.07	−0.02	-
250	−0.96	0.06	0.36	0.04	0.11	−0.04	0.13	0.02	−0.04	-
260	−1.31	0.10	0.37	0.14	0.14	−0.02	0.25	−0.00	−0.08	-
270	−2.54	0.10	0.25	0.16	0.06	−0.05	0.36	−0.03	−0.11	-
280	−2.30	0.13	0.19	0.14	−0.00	−0.08	0.45	−0.08	−0.15	-
290	−2.80	0.07	0.04	0.12	0.11	−0.04	0.58	−0.10	−0.19	-
RMS	1.88	0.10	0.43	0.11	0.10	0.04	0.31	0.07	0.11	-

5.2. VIIRS-MODIS Inter-Comparison With RSR Correction from CrIS Measurements

As with the VIIRS-MODIS-AIRS SNO data extraction, VIIRS-CrIS SNOs were collected first because both are onboard the S-NPP satellite. Then VIIRS-MODIS-CrIS SNOs were collected based on the geolocation and overpass time for each VIIRS-CrIS SNO pixel. The spatial distributions of VIIRS-MODIS-CrIS SNO pixel number for M13-B23 and M15-B31 are shown in Figure 5.

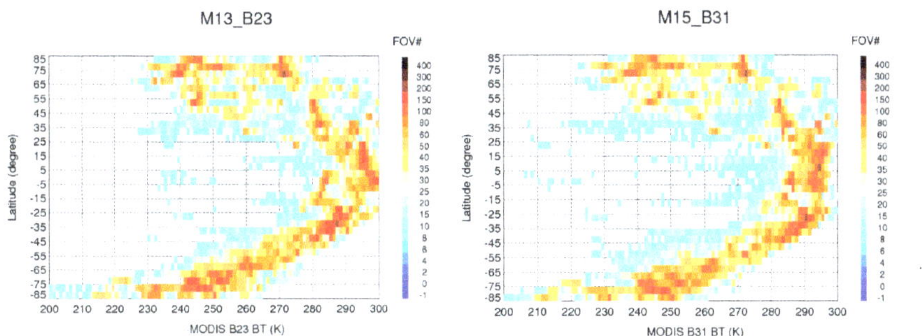

Figure 5. VIIRS-MODIS-CrIS SNO FOV number distribution.

Like the data processing method discussed in Section 5.1, the differences between VIIRS and MODIS measurements from all VIIRS-MODIS-CrIS pixels were extracted. 3-sigma filtering was then applied to remove outliers. The VIIRS-MODIS difference was then determined by the average of the filtered data. The ratio of the simulated VIIRS radiance to the simulated MODIS radiance was derived from CrIS spectral data for each SNO pixel and was used to correct for the RSR difference between the spectrally-matched band pairs. Figure 6 shows the distributions of VIIRS-MODIS BT differences before and after the RSR correction.

Compared with the distributions resulting from AIRS, Figure 6 illustrates that the distributions with CrIS SNOs have a much larger spatial coverage. This could be the reason that the BT differences before the RSR correction are slightly larger than those observed in the comparison using AIRS. After the RSR correction, the VIIRS-MODIS differences are much smaller and they are comparable with those after the RSR correction using AIRS. It is very significant that the two independent hyperspectral measurements from AIRS and CrIS, which could have differences due to existing calibration issues with each of the two instruments, can produce nearly the identical results in terms of the RSR correction for the spectrally matched bands between VIIRS and MODIS. The impact of existing measurement offsets in either CrIS and AIRS on the comparison of MODIS and VIIRS should be small because the

RSR correction is relative according to Equation (5) and the effective MODIS and VIIRS band widths are narrower than 0.5 μm.

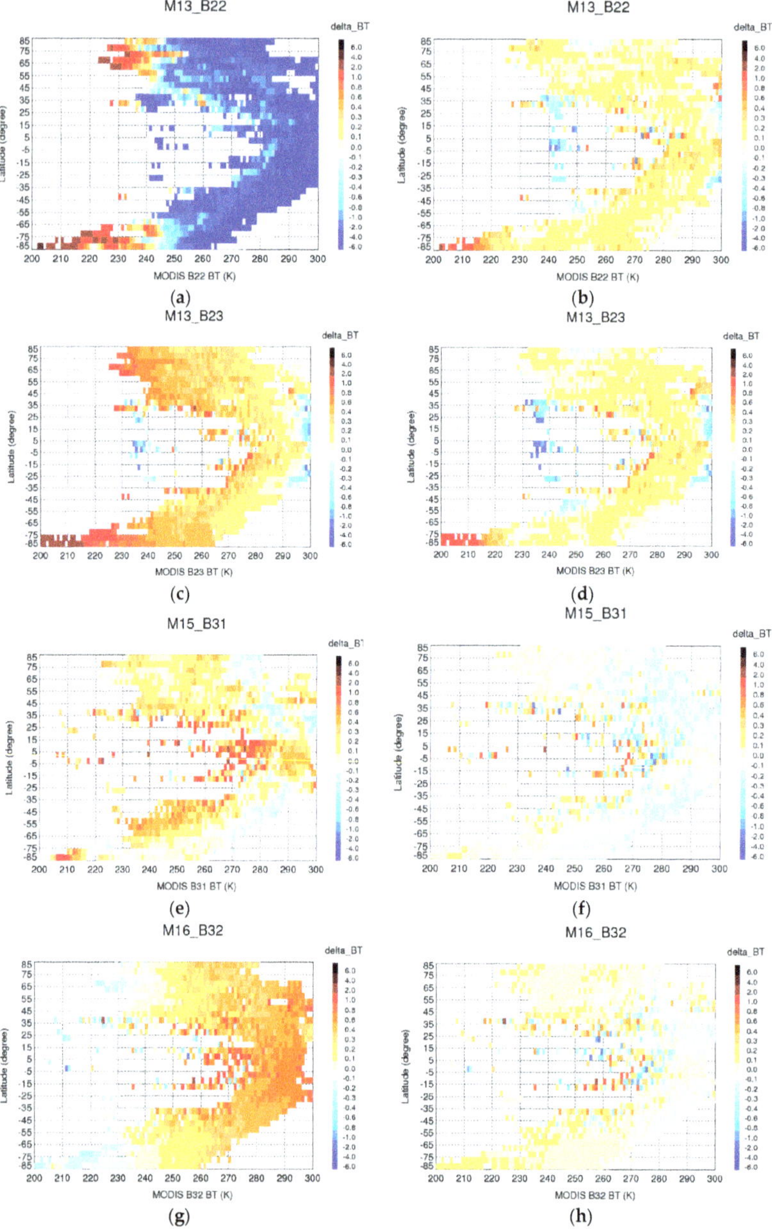

Figure 6. Distribution of VIIRS-MODIS inter-comparison without/with RSR correction from CrIS measurements. (**a**) M13_B22 without RSR correction; (**b**) M13_B22 with RSR correction; (**c**) M13_B23 without RSR correction; (**d**) M13_B23 with RSR correction; (**e**) M15_B31 without RSR correction; (**f**) M15_B31 with RSR correction; (**g**) M16_B32 without RSR correction; (**h**) M16_B32 with RSR correction.

Similarly, the VIIRS-MODIS differences are plotted *versus* BT, as shown in Figure 4. As expected, the results are consistent with those derived using AIRS. It is also noted that the CrIS results cover a wider range of BT than the results using AIRS because more SNO pixels and larger spatial coverage are used. Table 3 lists the quantitative comparison results over a range of BT and the results indicate that the differences after the RSR correction are well within 0.20 K for all the comparison bands. As seen from Figure 4 and Tables 2 and 3 and [27], the differences between AIRS and CrIS measurements could be caused by AIRS and CrIS radiometric differences.

Table 3. Difference of brightness temperature without/with RSR correction from CrIS measurements (unit: K).

	M13-B22		M13-B23		M15-B31		M16-B32		M14-B29	
BT_{MODIS}	ΔBT	ΔBT_{corr}	ΔBT	ΔBT_{corr}	ΔBT	ΔBT_{corr}	ΔBT	ΔBT_{corr}	ΔBT	ΔBT_{corr}
220	1.36	0.35	1.12	0.36	0.15	0.01	−0.15	0.08	−0.09	-
230	0.82	0.11	0.84	0.10	0.16	0.01	−0.04	0.09	−0.04	-
240	−0.12	0.10	0.54	0.06	0.15	0.00	0.04	0.08	−0.04	-
250	−0.96	0.08	0.44	0.05	0.18	0.00	0.20	0.07	−0.05	-
260	−1.40	0.15	0.41	0.16	0.13	−0.05	0.30	0.04	−0.11	-
270	−2.64	0.15	0.30	0.15	0.05	−0.06	0.42	0.04	−0.13	-
280	−2.26	0.14	0.25	0.15	0.06	−0.07	0.53	0.01	−0.18	-
290	−2.65	0.16	0.11	0.15	0.10	−0.06	0.65	0.00	−0.25	-
RMS	1.75	0.17	0.59	0.17	0.13	0.04	0.36	0.06	0.13	-

The RSR correction factor is defined as the ratio of simulated radiances using VIIRS RSR and MODIS RSR in Equation (5). Figure 7 plots of RSR correction factors *versus* BT. For M13-B22, M13-B23, M15-B31, and M16-B32, the RSR correction factors match well from AIRS and CrIS measurements. This also indicates that AIRS and CrIS instruments are radiometrically consistent in the proposed SNO inter-comparison approach.

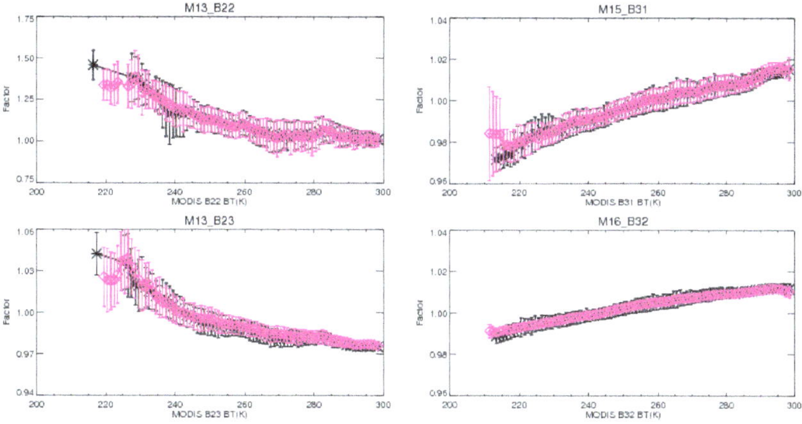

Figure 7. RSR correction factors from AIRS (black stars) and CrIS (blue diamonds) measurements.

The SNO data was also analyzed in radiance. Instead of the absolute radiance difference, it is expressed as a percentage. Table 4 summarizes the percent differences of radiance before and after the RSR correction. For longer wavelength bands, after the RSR correction, the differences of VIIRS(M15)-MODIS(B31) and VIIRS(M16)-MODIS(B32) are within 0.1% and 0.2%, respectively, except at very low radiance of 1.0 W/m ^{2}/μm/sr.

Table 4. Percent difference in radiance without/with RSR correction (L_{MODIS} unit: W/m^2/µm/sr; Δ unit: %).

L_{MODIS}	M13-B22				M13-B23				L_{MODIS}	M15-B31				M16-B32			
	AIRS		CrIS		AIRS		CrIS			AIRS		CrIS		AIRS		CrIS	
	ΔL	ΔL_{corr}	ΔL	ΔL_{corr}	ΔL	ΔL_{corr}	ΔL	ΔL_{corr}		ΔL	ΔL_{corr}	ΔL	ΔL_{corr}	ΔL	ΔL_{corr}	ΔL	ΔL_{corr}
0.1	10.42	0.65	10.02	0.88	−0.74	0.36	−0.29	0.62	1.0	−3.28	0.28	−3.69	0.24	−1.07	0.53	-	-
0.2	2.27	0.62	3.79	0.63	−1.08	0.71	−0.84	0.72	2.0	−1.77	0.01	−1.78	0.10	−0.48	0.22	−0.47	0.22
0.3	6.40	0.53	5.78	0.61	−1.24	0.64	−0.84	0.88	3.0	−1.01	−0.07	−0.87	0.03	−0.10	0.09	0.04	0.18
0.4	3.75	0.32	2.99	0.50	−1.41	0.50	−1.19	0.61	4.0	−0.41	−0.08	−0.30	−0.01	0.33	0.05	0.45	0.14
0.5	2.34	0.19	2.19	0.52	−1.75	0.49	−1.41	0.72	5.0	0.01	−0.11	0.07	−0.09	0.58	−0.05	0.70	0.08
0.6	0.24	−0.19	1.62	0.37	−2.12	0.33	−1.72	0.61	6.0	0.28	−0.14	0.31	−0.10	0.79	−0.07	0.85	0.01
0.7	−1.10	−0.21	1.66	−0.49	−2.41	0.01	−2.21	0.09	7.0	0.61	−0.14	0.73	−0.10	0.90	−0.11	0.96	−0.03
0.8	−3.71	−0.46	−4.51	−1.99	−3.12	-0.57	−3.96	−1.5	8.0	1.00	−0.10	1.01	−0.10	1.07	−0.14	1.16	−0.02
0.9	-	-	−9.70	−6.53	-	-	-	-	9.0	1.40	−0.01	1.4	−0.08	1.11	−0.11	1.27	0.02

5.3. Discussions

For the shorter wavelength bands, VIIRS M13 spectra covers part of MODIS B22 spectra, while the MODIS B23 spectra range is within M13 coverage. This is why larger radiance differences appear in the M13-B22 comparison than between M13 and B23 before the RSR correction, and the differences are still radiance-dependent after the RSR correction. Sparsity of data points causes some fluctuations at the edges of data range.

The BT uncertainties are roughly unchanged after the RSR correction. This is because they are mainly caused by the existing footprint mismatch between the individual hyperspectral pixel and integrated VIIRS and MODIS pixels. Since the AIRS and CrIS footprint sizes are similar, the uncertainty results obtained from using the two hyperspectral sensors are consistent. In this study, the uncertainties are based on 1 sigma and they are within 0.50 K over the entire scene temperature range from 220 K to 300 K. For the atmospheric window bands at 11 and 12 μm, the uncertainty varies with scene temperature. Larger uncertainties occur between BT at 250 and 280 K, where the scene variation is dominant.

The calculation of the RSR correction in this study is based on the in-band RSR for VIIRS and MODIS, respectively. The impact of out-of-band RSR is considered to be negligible. This is based on pre-launch measurements of the in-band and out-of-band RSR. For the thermal emissive bands, the magnitudes of the out-of-band RSR are well within 10^{-4}.

6. Conclusions

VIIRS TEB performance was evaluated by comparing to MODIS using AIRS and CrIS hyperspectral measurements as a transfer reference to correct the existing spectral band RSR differences between VIIRS and MODIS. All the sensor data are chosen from SNOs so they are completely matched in both the time and spatial domains. Results show that the two sensors agree within 0.2 K in brightness temperature after the RSR correction between the following spectrally-matched band pairs at the wavelength range of 3.9~4.2 μm (VIIRS(M13)-MODIS(B22) and VIIRS(M13)-MODIS(B23)), 10.1~11.5 μm (VIIRS(M15)-MODIS(B31)), and 11.5~12.5 μm (VIIRS(M16)-MODIS(B32)). Relatively larger differences are observed at lower BT (less than 240 K) in M13-B22 and M13-B23 comparisons. The results of this study demonstrate that the real-time hyperspectral measurements can facilitate high quality sensor calibration inter-comparisons.

Acknowledgments: The authors would like to thank Jeffrey McIntire's suggestions/comments in preparing the manuscript ion as well as Andrew Wald's help in the revision.

Author Contributions: Yonghong Li designed and implemented the algorithm, processed the data, and conducted the analysis of the results. She drafted the manuscript and provided the major responses to the reviewers. Aisheng Wu was involved in the refinement of the algorithm, data analysis and interpretation. He also helped to revise the manuscript to address some of the major concerns. Xiaoxiong Xiong initiated this study and revised the final version of the manuscript.

Conflicts of Interest: The authors declare no conflict of interest.

References

1. Lillesand, T.; Kiefer, R.W.; Chipman, J. *Remote Sensing and Image Interpretation*, 7th ed.; John Wiley & Sons, Inc.: Hoboken, NJ, USA, 2015; pp. 609–630.
2. Cao, C.; Blonski, S.; Wang, W.; Shao, X.; Choi, T.; Bai, Y.; Xiong, X. Overview of Suomi NPP VIIRS performance in the last 2.5 years. *Proc. SPIE* **2014**, *9264*. [CrossRef]
3. Toller, G.; Xiong, X.; Chiang, V.; Kuyper, J.; Sun, J.; Tan, L.; Barnes, W. Status of earth observing system Terra and Aqua moderate-resolution imaging spectroradiometer level 1B algorithm. *J. Appl. Remote Sens.* **2008**, *2*, 023505.
4. Franz, B.A.; Werdell, P.J.; Meister, G.; Bailey, S.W.; Eplee, R.E., Jr.; Feldman, G.C.; Kwiatkowskaa, E.; McClain, C.R.; Patt, F.S.; Thomas, D. The continuity of ocean color measurements from SeaWiFS to MODIS. *Proc. SPIE* **2005**, *5882*, 304–316.

5. Data Products. Available online: http://modis.gsfc.nasa.gov/data/dataprod/ (accessed on 13 January 2016).
6. Wenny, B.; Wu, A.; Madhavan, S.; Wang, Z.; Li, Y.; Chen, N.; Chiang, V.; Xiong, X. MODIS TEB calibration approach in Collection 6. *Proc. SPIE* **2012**, *8533*, 11.
7. Xiong, X.; Wenny, B.N.; Wu, A.; Barnes, W.; Salomonson, V. Aqua MODIS thermal emissive bands on-orbit calibration, characterization, and performance. *IEEE Trans. Geosci. Remote Sens.* **2009**, *47*, 803–814. [CrossRef]
8. Wenny, B.N.; Xiong, X. Using a cold earth surface target to characterize long-term stability of the MODIS thermal emissive bands. *IEEE Geosci. Remote Sens. Lett.* **2008**, *5*, 162–165. [CrossRef]
9. Li, Y.; Wu, A.; Xiong, X. Evaluating calibration of MODIS thermal emissive bands using infrared atmospheric sounding interferometer measurements. *Proc. SPIE* **2013**, *8724*. [CrossRef]
10. Tobin, D.C.; Revercomb, H.E.; Moeller, C.C.; Pagano, T.S. Use of atmospheric infrared sounder high-spectral resolution spectra to assess the calibration of Moderate resolution imaging spectroradiometer on EOS Aqua. *J. Geophys. Res.* **2006**, *111*, D09S05. [CrossRef]
11. Xie, Y.; Wu, A.; Xiong, X. Tracking long-term stability of Aqua MODIS and AIRS at different scan angles. In Proceedings of the Earth Observing Systems XV, San Diego, CA, USA, 1 August 2010.
12. Cao, C.; de Luccia, F.; Xiong, X.; Wolfe, R.; Weng, F. Early on-orbit performance of the visible infrared imaging radiometer suite onboard the Suomi National Polar-Orbiting Partnership (S-NPP) satellite. *IEEE Trans. Geosci. Remote Sens.* **2014**, *52*, 1142–1156. [CrossRef]
13. Efremova, B.; McIntire, J.; Moyer, D.; Xiong, X.; Wu, A. SNPP VIIRS thermal emissive bands on-orbit calibration and performance. *J. Geophys. Res.* **2014**, *119*, 10859–10875.
14. Moeller, C.; Tobin, D.; Quinn, G. S-NPP VIIRS thermal band spectral radiance performance through 18 months of operation on-orbit. *Proc. SPIE* **2013**, *8866*. [CrossRef]
15. Pahlevan, N.; Lee, Z.; Lawson, A.; Arnone, R. Scene-based cross-comparison of SNPP-VIIRS and Aqua-MODIS over oceanic waters. *Proc. SPIE* **2013**, *8866*. [CrossRef]
16. Wu, A.; Xiong, X. NPP VIIRS and Aqua MODIS RSB comparison using observations from simultaneous nadir overpasses (SNO). *Proc. SPIE* **2012**, *8510*. [CrossRef]
17. Efremova, B.; Wu, A.; Xiong, X. Relative spectral response corrected calibration inter-comparison of S-NPP VIIRS and Aqua MODIS thermal emissive bands. *Proc. SPIE* **2014**, *9218*. [CrossRef]
18. Elliott, D.A.; Aumann, H.H. Comparison of AIRS and IASI surface observations of DomeC in Antarctica with surface temperatures reported by AWS8989. In Proceedings of the American Geophysical Union, Fall Meeting, San Francisco, CA, USA, 15–19 Decemeber 2008.
19. Wang, L.; Goldberg, M.; Wu, X.; Cao, C.; Iacovazzi, R.A., Jr.; Yu, F.; Li, Y. Consistency assessment of Atmospheric Infrared Sounder and Infrared Atmospheric Sounding Interferometer radiances: Double differences *versus* simultaneous nadir overpasses. *J. Geophys. Res.* **2011**, *116*, D11111. [CrossRef]
20. Tobin, D.; Revercomb, H.; Knuteson, B.; Best, F.; Taylor, J.; Deslover, D.; Borg, L.; Moeller, C.; Martin, G.; Kuehn, R.; *et al.* Suomi NPP/JPSS cross-track infrared sounder (CRIS): Intercalibration with AIRS, IASI, and VIIRS. In Proceedings of the AMS Annual Meeting, Austin, TX, USA, 6–10 January 2013.
21. Blackbody Assembly. Available online: Modis.gsfc.nasa.gov/about/blackbody.php (accessed on 13 January 2016).
22. Kloepfer, J.; Toylor, C.; Murgai, V. Characterization of the VIIRS blackbody emittance. In Proceedings of the Conference on CALCON, Logan, UT, USA, 18 August 2013.
23. Joint Polar Satellite System (JPSS) VIIRS Radiometric Calibration Algorithm Theoretical Basis Document (ATBD). Available online: npp.gsfc.nasa.gov/sciencedocs/2015-06/474-00027_ATBD-VIIRS-Radiometric-Calibration_C.pdf (accessed on 13 January 2016).
24. Wu, A.; Wang, Z.; Li, Y.; Madhavan, S.; Wenny, B.N.; Chen, N.; Xiong, X. Adjusting Aqua MODIS TEB Nonlinear Calibration Coefficients Using Iterative Solution. *Proc. SPIE* **2014**, *9264*. [CrossRef]
25. CelesTrack. Available online: http://www.celestrak.com/ (accessed on 13 January 2016).
26. AIRS Version 5 Documentation. Available online: http://disc.sci.gsfc.nasa.gov/AIRS/documentation/v5_docs (accessed on 13 January 2016).
27. Wang, L.; Han, Y.; Jin, X.; Chen, Y.; Tremblay, D.A. Radiometric consistency assessment of hyperspectral infrared sounders. *Atmos. Meas. Tech.* **2015**, *8*, 4831–4844. [CrossRef]

Article

Preliminary Inter-Comparison between AHI, VIIRS and MODIS Clear-Sky Ocean Radiances for Accurate SST Retrievals

Xingming Liang [1,2,*], Alexander Ignatov [1], Maxim Kramar [1,3] and Fangfang Yu [1,4]

1 NOAA Center for Satellite Application and Research (STAR), College Park, MD 20740, USA; Alex.Ignatov@noaa.gov (A.I.); Maxim.Kramar@noaa.gov (M.K.); Fangfang.Yu@noaa.gov (F.Y.)
2 CSU, Cooperative Institute for Research in the Atmospheres (CIRA), Fort Collins, CO 80523, USA
3 GST Inc., Greenbelt, MD 20740, USA
4 ERT Inc., Laurel, MD 20707, USA
* Correspondence: Xingming.Liang@noaa.gov; Tel.: +1-301-683-3362; Fax: +1-301-683-3301

Academic Editors: Changyong Cao, Magaly Koch and Prasad S. Thenkabail
Received: 19 November 2015; Accepted: 18 February 2016; Published: 1 March 2016

Abstract: Clear-sky brightness temperatures (BT) in five bands of the Advanced Himawari Imager (AHI; flown onboard Himawari-8 satellite) centered at 3.9, 8.6, 10.4, 11.2, and 12.3 μm (denoted by IR37, IR86, IR10, IR11, and IR12, respectively) are used in the NOAA Advanced Clear-Sky Processor for Oceans (ACSPO) sea surface temperature (SST) retrieval system. Here, AHI BTs are preliminarily evaluated for stability and consistency with the corresponding VIIRS and MODIS BTs, using the sensor observation minus model simulation (O-M) biases and corresponding double differences. The objective is to ensure accurate and consistent SST products from the polar and geo sensors, and to prepare for the launch of the GOES-R satellite in 2016. All five AHI SST bands are found to be largely in-family with their polar counterparts, but biased low relative to the VIIRS and MODIS (which, in turn, were found to be stable and consistent, except for Terra IR86, which is biased high by 1.5 K). The negative biases are larger in IR37 and IR12 (up to ~−0.5 K), followed by the three remaining longwave IR bands IR86, IR10, and IR11 (from −0.3 to −0.4 K). These negative biases may be in part due to the uncertainties in AHI calibration and characterization, although uncertainties in the coefficients of the Community Radiative Transfer Model (CRTM, used to generate the "M" term) may also contribute. Work is underway to add AHI analyses in the NOAA Monitoring of IR Clear-Sky Radiances over Oceans for SST (MICROS) system and improve AHI BTs by collaborating with the sensor calibration and CRTM teams. The Advanced Baseline Imager (ABI) analyses will be also added in MICROS when GOES-R is launched in late 2016 and the ABI IR data become available.

Keywords: SST; AHI; VIIRS; MODIS; Himawari-8; S-NPP; Terra; Aqua; MICROS; ACSPO

1. Introduction

With the launch of the first Visible Infrared Imager Radiometer Suite (VIIRS) onboard the Suomi National Polar Partnership (S-NPP) satellite in October 2011, NOAA has entered a new era of the polar-orbiting environmental satellite operations. Four more VIIRS sensors will fly onboard the follow-on US Joint Polar Satellite System (JPSS) satellites, J1–J4, planned for launch from 2017 to 2026. With its improved spatial and spectral resolution, and radiometric accuracy and stability, the VIIRS instrument is in many ways superior to its operational and research predecessors, the Advanced Very High Resolution Radiometers (AVHRR) flown onboard multiple NOAA and Meteorological Operational (Metop) satellites, and the NASA Moderate-resolution Imaging Spectroradiometers (MODIS) flown onboard Terra and Aqua satellites [1–3].

The NOAA geostationary satellite fleet is also due for a major upgrade, with the launch of the first Advanced Baseline Imager (ABI) onboard the Geostationary Operational Environmental Satellite-R (GOES-R) in October 2016 [4]. The Himawari-8 (H8) geostationary satellite, launched by the Japan Aerospace Exploration Agency (JAXA) in October 2014 and declared operational by the Japan Meteorological Agency (JMA) in July 2015, carries the Advanced Himawari Imager (AHI) onboard, a very close proxy of the ABI. The ABI/AHI sensors have significantly improved upon their predecessors, the imagers onboard the US GOES and Japanese Multi-functional Transport Satellites (MTSAT) [5,6]. In July 2015, the JMA switched over the satellite operations from MTSAT-2 (also known as Himawari-7) to H8, and designated the MTSAT-2 as a standby satellite [7].

Both VIIRS and ABI/AHI provide enhanced capabilities for weather and climate monitoring. Sea surface temperature (SST) is one of the key environmental variables, derived from the clear-sky brightness temperatures (BTs) over ocean in the window bands. Assessing the radiometric accuracy, stability, and cross-platform consistency of the VIIRS and ABI/AHI BTs is critically important to ensure accurate, stable, and consistent SST records from the current polar and geostationary sensors. In this study, we preliminarily cross-evaluate the AHI and VIIRS BTs, in support of the H8 and S-NPP SST production at NOAA, and in preparation for the future GOES-R and JPSS SST operations.

VIIRS and AHI SST retrievals are performed at NOAA using the Advanced Clear Sky Processor for Ocean (ACSPO) system. In addition to the primary SST product, ACSPO also reports the top-of-atmosphere (TOA) BTs over the ocean in the window bands [8]. The clear-sky ocean pixels are identified using the ACSPO clear-sky mask [9]. The clear-sky sensor BTs are another NOAA product, which can be directly assimilated in the Numerical Weather Prediction (NWP) models. To facilitate their validation, the simulated BTs are also calculated in ACSPO using the Community Radiative Transfer Model (CRTM) [10], with the National Centers for Environmental Prediction Global Forecast System (NCEP GFS) [11] upper air fields and the Canadian Meteorological Centre daily L4 SST analysis (CMC 0.2°) [12] as inputs. The model BTs can be used to validate the sensor BTs, among many other applications [13,14].

ACSPO system is operational at NOAA with the S-NPP VIIRS and with several Advanced Very High Resolution Radiometers (AVHRR) Global Area Coverage (GAC; 4 km resolution) onboard NOAA satellites, and with AVHRR Full Resolution Area Coverage (FRAC; 1 km) onboard Metop satellites. Data of two MODIS, onboard Terra and Aqua, and AHI, onboard H8, are also routinely processed in an experimental mode. The SST and model minus observation (M-O, which in this study are inverted to O-M, to facilitate analyses of sensor BTs) biases are monitored in near real-time (NRT) using the Monitoring of IR Clear-Sky Radiances over Oceans for SST (MICROS) web-based system [13,15]. The previous evaluation of the AVHRR, MODIS, and VIIRS radiometric stability and cross-platform consistency in the three SST bands historically used in ACSPO (IR37, IR11, and IR12) confirmed that the VIIRS is a stable and well-calibrated sensor [14]. MODIS BTs were also found to be generally consistent with VIIRS BTs (with the exception of the IR37 on Aqua, which was ~0.2 K out of the VIIRS and Terra family).

With the launch of H8, the ACSPO system was adapted to process AHI data. Note that AHI has five IR window bands, centered at 3.9 (band 7), 8.6 (11), 10.4 (13), 11.2 (14), and 12.4 µm (15). Some of these bands are unique to the AHI/ABI (e.g., *cf.* the three-band configuration in the longwave atmospheric window, IR10/IR11/IR12, with the customary IR11/IR12 two-band configuration on the polar sensors). Additionally, AHI and ABI have an IR86 band. A similar band is also available on MODIS and VIIRS, but not on AVHRR, and so far has not been explored in the ACSPO processing. These new bands are instrumental for SST and clear-sky mask (in particular, to ensure their day/night continuity) and they are currently being actively explored in ACSPO. The MICROS system is also being updated to include the AHI data and to prepare for monitoring of the GOES-R ABI, and to include the new SST bands.

This paper documents preliminary analyses of the AHI O-M biases in all five SST bands, with the emphasis on comparisons with the corresponding VIIRS and MODIS O-M biases. The double differences are also analyzed to evaluate the SST radiances from the three sensors for stability and cross-platform consistency, and to get ready for the GOES-R ABI planned for launch in 2016.

2. SST Bands and ACSPO Data Selection

2.1. AHI, VIIRS, and MODIS SST Bands

Sensor bands analyzed in this study are summarized in Table 1, and the corresponding relative sensor response functions (SRFs) are shown in Figure 1.

Table 1. AHI, VIIRS, and MODIS SST bands (B: band; CW: center wavelength; SR: spectral range).

Band Name	AHI			VIIRS			MODIS		
	B	CW (μm)	SR (μm)	B	CW (μm)	SR (μm)	B	CW (μm)	SR (μm)
IR37	7	3.85	3.59–4.11	M12	3.70	3.66–3.84	20	3.75	3.66–3.84
IR86	11	8.60	8.12–9.07	M14	8.58	8.40–8.70	29	8.55	8.40–8.70
IR10	13	10.45	9.90–10.96						
IR11	14	11.20	10.31–12.18	M15	10.73	10.26–11.26	31	11.03	10.78–11.28
IR12	15	12.35	11.17–13.66	M16	11.85	11.54–12.49	32	12.02	11.77–12.27

The MODIS and VIIRS SRFs were provided by the NASA MODIS calibration support team (MCST) and by the VIIRS calibration team at the University of Wisconsin, respectively. The AHI SRFs were obtained from the JMA AHI imager website [7]. Here, the bands centered at 3.7/3.9, 8.6, 10.5, 10.7/11.0/11.2, and 11.9/12.0/12.4 μm are denoted IR37, IR86, IR10, IR11, and IR12, respectively, consistently for all three sensors. Note, however, that their SRFs, center wavelengths, and bandwidths do differ. In the atmospheric windows away from the absorption lines, these spectral variations are expected to only minimally affect the TOA BTs. Recall that the previous comparisons between different AVHRR, MODIS, and VIIRS sensors in [14] have shown that the corresponding O-M biases are typically within several hundredths of a Kelvin in the three bands—IR37, IR11, and IR12. The AHI IR37 SRF is shifted to the smaller wavenumbers (longer wavelengths) and covers two N_2O absorption lines around 3.9 μm [16]. Similar SRF behavior is found in the spectrally wide AVHRR band IR37, which however results in only minimal differences between the AVHRR, MODIS, and VIIRS [14]. Note that in addition to the band 20 centered at 3.7 μm, MODIS also has band 22 centered at 3.9 μm and band 23 centered at 4.0 μm, which can be used in concert with band 20 for more accurate nighttime SST retrievals [17]. However, similar bands are not available on VIIRS and AHI, and are beyond the scope of this study, which is focused on the comparison between these three sensors.

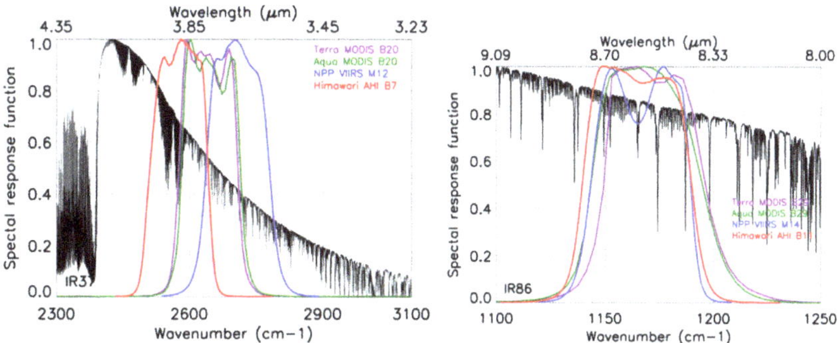

Figure 1. *Cont.*

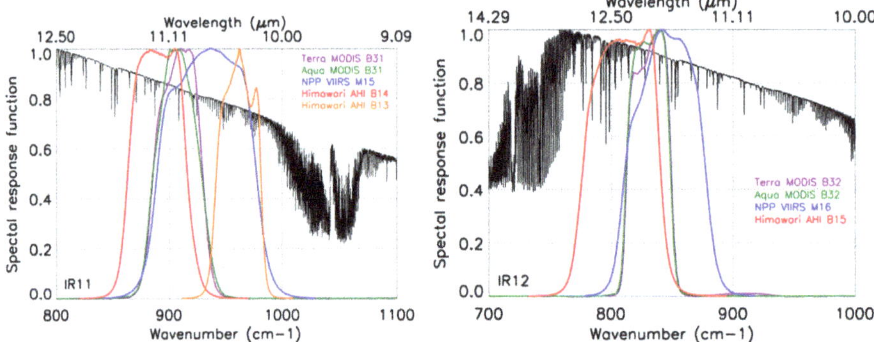

Figure 1. Relative Sensor spectral response functions for IR37 (MODIS B20, VIIRS M12, AHI B7), IR86 (MODIS B29, VIIRS M14 and AHI B11), IR11 (MODIS B31, VIIRS M15, AHI B13 and B14), and IR12 (MODIS B32, VIIRS M16, AHI B15). The scaled TOA radiances (black, scaled at maximum) are superimposed, calculated with Line-By-Line Radiative Transfer Model (LBLRTM, http://rtweb.aer.com/lblrtm_frame.html) using an average atmospheric profile over the 81 ECMWF profiles commonly used for training CRTM coefficients.

2.2. Data

The VIIRS and MODIS provide near-global coverage twice daily, during both day and night, in a swath of ~3040 km (2330 km for MODIS), at 0.75 km resolution at nadir and ~1.5 km at swath edge (1 and 5 km for MODIS, respectively). The ACSPO clear-sky mask identifies from 100–130 M (38–45 M for MODIS) clear-sky pixels on both day and night sides of the Earth ("M" stands for million.)

The H8 is positioned over the equator at 140.7°E. The AHI generates global full disk (FD) images every 10 min, resulting in a total of 142 FD images per day (two slots at 02:40 and 14:40 UTC are reserved for the housekeeping operations). With 2 km resolution at nadir (which degrades to ~6.5 km at view zenith angle, VZA ~ 70°), each AHI FD image comprises a total of 5500 × 5500 pixels, of which from 2.5–5.0 M are identified as clear SST pixels by the current ACSPO clear-sky mask.

Cross-sensor comparisons in MICROS employ double differences (DD) defined as follows [11,12]:

$$\text{SAT} - \text{H8} = \mathit{mean}\,(\text{SAT}\,[\text{O-M}]) - \mathit{mean}\,(\text{H8}\,[\text{O-M}]) \tag{1}$$

Here, the H8 is used as a reference, and SAT may denote the S-NPP, Terra, or Aqua, which are all evaluated against the H8. To approximately represent the AHI domain, VIIRS and MODIS data were sub-sampled within ±90° longitude around the ~140.7°E H8 sub-satellite point (the latitude range in the sub-sampling was retained from −90° to +90°). The mean is calculated over the full domain for H8, and over the corresponding subsamples for the polar satellite data. Additionally, the VZA for AHI was restricted to <70°, consistently with the maximum VZAs for VIIRS and MODIS. Note that the "M" terms can be accurately calculated for large VZAs with the CRTM which takes into account the sphericity of the Earth [13,14].

2.3. Observation Time and Matchup

Polar sun-synchronous satellites observe the same target on Earth twice daily, during the day and at night, at approximately same local times (LT). Recall that for the afternoon ("PM") S-NPP and Aqua, the Equatorial Crossing Time (ECT) is ~1:30 pm/am LT whereas for the mid-morning ("AM") Terra, the ECT is ~10:30 am/pm LT. The actual LT "in-pixel" is approximately centered at these ECTs but may vary within several hours, depending on the latitude (due to orbital inclination) and pixel position in the scan. Between the AM and PM platforms, the LT differs by 3:00 h on average. At night, the effect of the diurnal cycle is minimal, and the corresponding global mean O-M biases differ by no more than several hundredths of a Kelvin [14]. During the daytime, diurnal variability, and the corresponding sensitivity of the O-M biases to LT differences, is larger. In MICROS, all polar analyses are stratified by day and night, and using nighttime data (*i.e.*, data with the solar zenith angle, SZA > 90°) is preferred for cross-platform comparisons [8,13,14].

The 10-min AHI FDs provide continuous coverage of the full diurnal cycle, with each FD image covering a wide range of different time zones, often including day and night. Special provision should be made when polar and geo data are compared. In this study, the LT was calculated for each VIIRS, MODIS, and AHI pixel, given its Coordinated Universal Time (UTC) and longitude, and stratified into hourly LT bins. Figure 2 shows the number of clear-sky ocean pixels (NCSOP) in each LT bin, for one full day of 2 June 2015. For AHI, the LT distribution is nearly uniform with ~18 M pixels per hour, whereas the VIIRS and MODIS data are clustered in 3–4 daytime, and 3–4 nighttime bins, centered approximately at the satellite ECTs (~1:30 a.m./p.m. for VIIRS and Aqua, and ~10:30 am/pm for Terra). The NCSOP in the most populated (central) polar bins are ~15 M, ~12 M, and ~27 M for Terra, Aqua, and S-NPP, respectively, and these numbers are reduced ~3~5 times in the adjacent neighboring bins. The sparsely populated polar bins (with NCSOP < 1% of the total sample) were removed, to minimize the outliers.

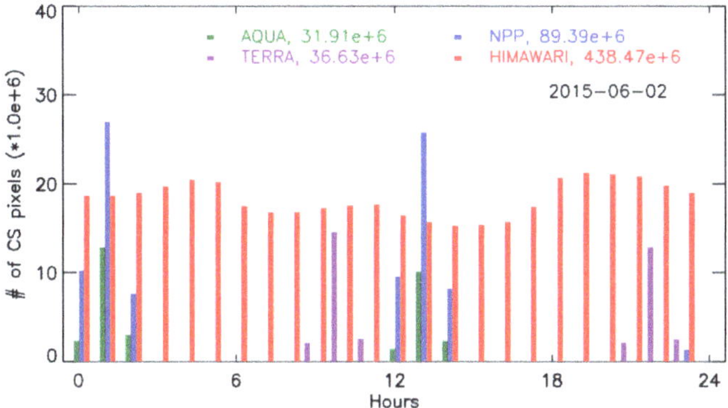

Figure 2. The number of clear-sky pixels in every hour of local time for the day of 2 June 2015 (polar data are sub-sampled to represent the H8 domain).

In what follows, only nighttime data are used to eliminate and minimize the effect of the diurnal cycle and solar refection. The exception is only made when the diurnal cycle is analyzed. In all cases, the specifics of the data sample used in each particular case are explicitly defined.

3. Results and Discussion

3.1. O-M Biases on 2 June 2015

Evaluation of the AHI O-M biases by the NOAA SST team commenced in February 2015, when the first AHI IR data were made available to NOAA by the JMA. The results have been continuously reported to the NOAA AHI/ABI calibration team and used to improve the sensor calibration. While the H8 was declared operational by the JMA in early July 2015, this study uses one month of AHI data from 28 May–27 June 2015. Though still in a commissioning phase, the NOAA AHI/ABI calibration team has confirmed that the calibration in the five IR window bands has not changed.

Figure 3 shows geographical distribution of the nighttime O-M biases in IR37 from S-NPP VIIRS and H8 AHI on 2 June 2015. Note that the H8 has no data around the FD Earth limb, due to the VZA < 70° cut off. The cloud patterns and coverage are similar between the polar and geo sensors. However, the AHI BTs are biased negative relative to the VIIRS.

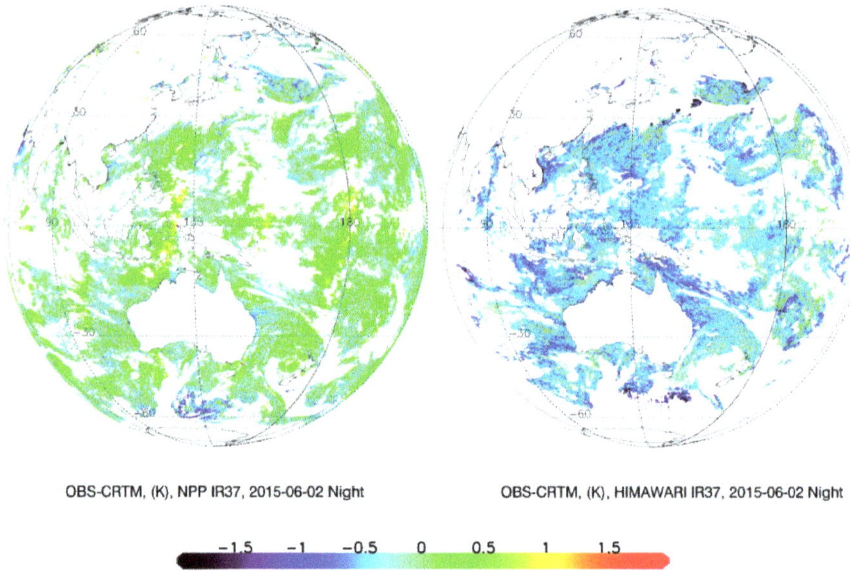

Figure 3. Nighttime O-M biases in IR37 for S-NPP VIIRS and H8 AHI. Data are gridded to 1° × 1°. For VIIRS, nighttime data are defined as those with solar zenith angle, SZA > 90°. For AHI, only data from 1–2 a.m. LT were used (which come from different FD images taken at different UTC times). If more than one VIIRS overpass or AHI FD image satisfying these conditions were available, they were all averaged within a 1° box, for mapping purposes.

Figure 4 shows global histograms of the O-M biases in the domain of Figure 3, but in the IR86 band. The shape is close to Gaussian, for all sensors. Terra shows a large anomaly, due to the electronic cross-talk in IR86 [18]. Other than this known and documented problem, the Aqua and H8 bracket the S-NPP within ~±0.25 K, with the AHI found on the cold end of the family. Note that the IR86 band was not analyzed before [14] and work is currently underway to add it in MICROS.

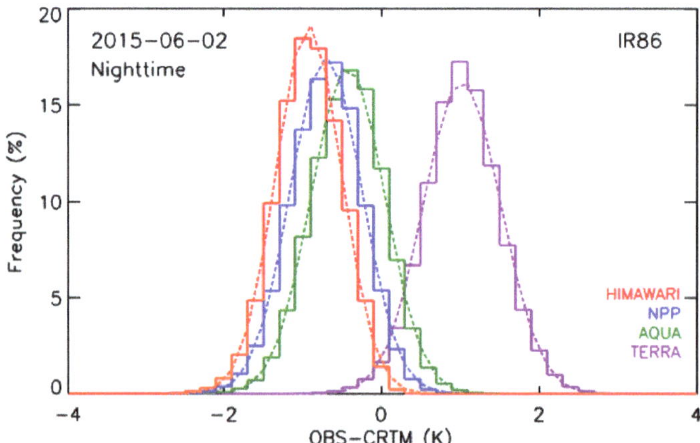

Figure 4. Histograms of the nighttime O-M biases in IR86. Corresponding statistics are listed in Table 2. The dotted lines are Gaussian fits corresponding to the median and robust standard deviation.

Table 2 shows the first and second order O-M statistics for IR86 from Figure 4, as well as for the other four SST bands. In all cases, AHI is in family, although on its low end. In the IR37, the AHI bias with respect to VIIRS is ~−0.5 K. A smaller negative bias of ~−0.3 K is seen in Aqua IR37. This bias was previously observed and attributed to the suboptimal characterization of the Aqua IR37 [14]. The S-NPP and Terra agree to within 0.06 K. In the IR11 and IR12, all three polar sensors closely agree to within several hundredths of a Kelvin, whereas AHI is again biased cold, from ~−0.3 to −0.5 K. The AHI IR10 has no polar counterpart, and its closest polar proxy is IR11. If anything, the AHI IR10 also appears biased several tenths of a Kelvin cold.

Table 2. The nighttime O-M mean biases, standard deviations, and number of clear-sky ocean pixels (NCSOP) for 2 June 2015.

Band Name	Mean, K				SD, K			
	H-8	S-NPP	Aqua	Terra	H-8	S-NPP	Aqua	Terra
IR37	−0.51	−0.04	−0.26	0.02	0.39	0.33	0.33	0.32
IR86	−0.82	−0.60	−0.33	1.13	0.42	0.46	0.47	0.49
IR10	−0.79				0.50			
IR11	−0.86	−0.54	−0.51	−0.54	0.60	0.53	0.54	0.52
IR12	−1.09	−0.64	−0.67	−0.70	0.69	0.63	0.62	0.59
	NCSOP (million)				18.6	46.9	18.2	17.8

By and large, all AHI BTs are in-family with their polar counterparts which is, by itself, an encouraging result. Biases range from 0.2–0.3 K in the longwave bands, IR86-IR11. The way how the CRTM coefficients are defined and calculated, can contribute up to several tenths of a Kelvin which may explain, at least a part of these biases [14]. In IR37 and IR12, biases are larger, from 0.4–0.5 K, and are more likely to be caused by the SRF differences or possible uncertainties in the AHI calibration. To independently verify and complement the RTM-based radiance monitoring, hyperspectral sensors such as the Atmospheric Infrared Sounder (AIRS) onboard Aqua, the Infrared Atmospheric Sounding Interferometer (IASI) onboard Metop satellites, and the cross-track infrared sunder (CrIS) onboard S-NPP may be explored (e.g., [19,20]).

The fact that all AHI bands are biased cold may also suggest a larger residual cloud contamination in the ACSPO AHI data compared to the corresponding polar products. Note, however, that the initial implementation of the ACSPO clear-sky mask for AHI was aimed to be conservative. Visual inspection of the AHI SST imagery indeed confirms that the cloud leakages do not exceed those in the polar products (*cf.* e.g., Figure 3), and are unlikely to be the only (or the main) cause of the observed negative AHI biases. Work continues to fine-tune the ACSPO clear-sky mask and re-evaluate the AHI O-M biases.

Figure 5 shows the O-M biases in IR10 and IR11 on 2 June 2015 as a function of the LT (*cf.* Figure 2). (Note that the diurnal cycle in the O-M is mostly attributed to the "O" term because the "M" term is produced from the daily first-guess "foundation" CMC L4 (which most closely represents the diurnal minimum, typically achieved at night before the sunrise), whereas the six-hourly GFS profiles have only secondary effect in the window bands. The polar curves were obtained by a parallel shift of the AHI IR11 curve and anchoring it to the most populated nighttime polar bin. (Note that this fit may not go through all remaining polar bins, including daytime, due to noise in the data which is expected to increase in the less populated data bins.) The diurnal cycle goes through the minimum at ~4–5 a.m., then warms up reaching its peak at 1–2 p.m., and gradually cools off again. The amplitude of the diurnal cycle in the IR11 is ~0.42 K. For comparison, the diurnal cycle in GOES BTs in a comparable IR11 band was estimated to be ~0.55 K [21].

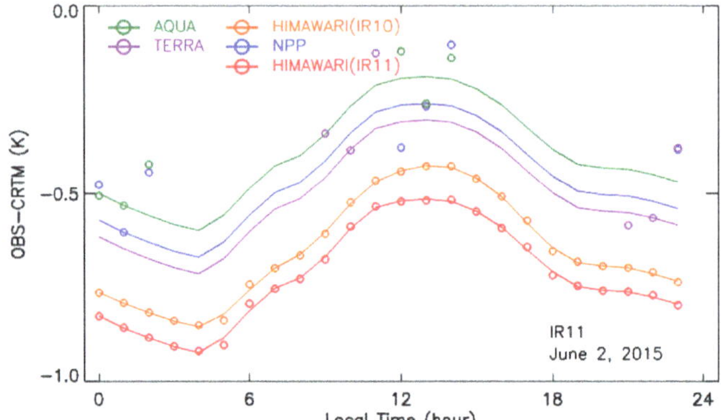

Figure 5. Mean O-M biases over the H8 domain as a function of LT on 2 June 2015 for S-NPP VIIRS, Terra and Aqua MODIS (IR11) and AHI (IR10 and IR11). The polar curves are obtained by a parallel shift of the AHI IR11 curve, and anchoring to the respective most populated nighttime polar bins.

3.2. Stability of the O-M Biases and Double Differences (DDs)

In this section, one month of data from 28 May to 27 June 2015 is analyzed to verify the stability and representativeness of the one day analyses in the previous section.

Figure 6 shows the time-series of the hourly O-M biases in the five SST bands. It is produced as Figure 5 but for a one month time period. Two observations are worth noting. First, the diurnal cycle is much larger in IR37 than in all other bands. This is due to the effect of solar reflectance in this band during the daytime, which is not modeled fully accurately in the current CRTM [22]. The other observation is that the O-M biases are relatively stable in time and do not show obvious systematic changes. However, The day-to-day changes in the geo and polar O-M data (by several tenths of a Kelvin) appear to occur "in-phase", likely due to the unstable "M" term (*i.e.*, unstable CMC SST and the GFS atmospheric profiles first guess fields used as input into CRTM), which will cancel out when the corresponding DDs are calculated.

The nighttime DDs are plotted in Figure 7. For each day, only one data point out of 24 h is plotted, representing the most populated nighttime polar bin. Although the LTs are consistent within each bin, the BTs may nevertheless slightly differ, because the corresponding samples are not spatially collocated point-by-point (as it is done, for instance, in the simultaneous nadir overpasses analyses [23]). Using the most populated nighttime bin does minimize the effect of the diurnal cycle on cross-platform comparisons. Also, using the most statistically significant sample (bin) helps to suppress the residual random noise in the data. All observations made earlier for only one day of data, 2 June 2015 (summarized in Table 2 and discussed in previous section), continue to hold, including in particular the relative signs and magnitudes of the cross platform biases. However, Figure 7 now additionally confirms that the statistics in Table 2 are representative and stable in time.

The AHI IR37 is biased ~−0.47 K cold relative to the S-NPP VIIRS. Terra and Aqua disagree by ~0.25 K, with Terra being closer to the VIIRS. This observation was already made earlier [14] and is now additionally confirmed. In the newly analyzed IR86, AHI BTs are biased cold by ~−0.23 K, whereas both Aqua and Terra are biased warm, by +0.24 K and +1.72 K, respectively. The large Terra bias has been documented elsewhere [18] and is independently confirmed here. In the two longwave bands, AHI BTs are again biased low, from ~−0.3 K in IR11 to ~−0.4 K in IR12.

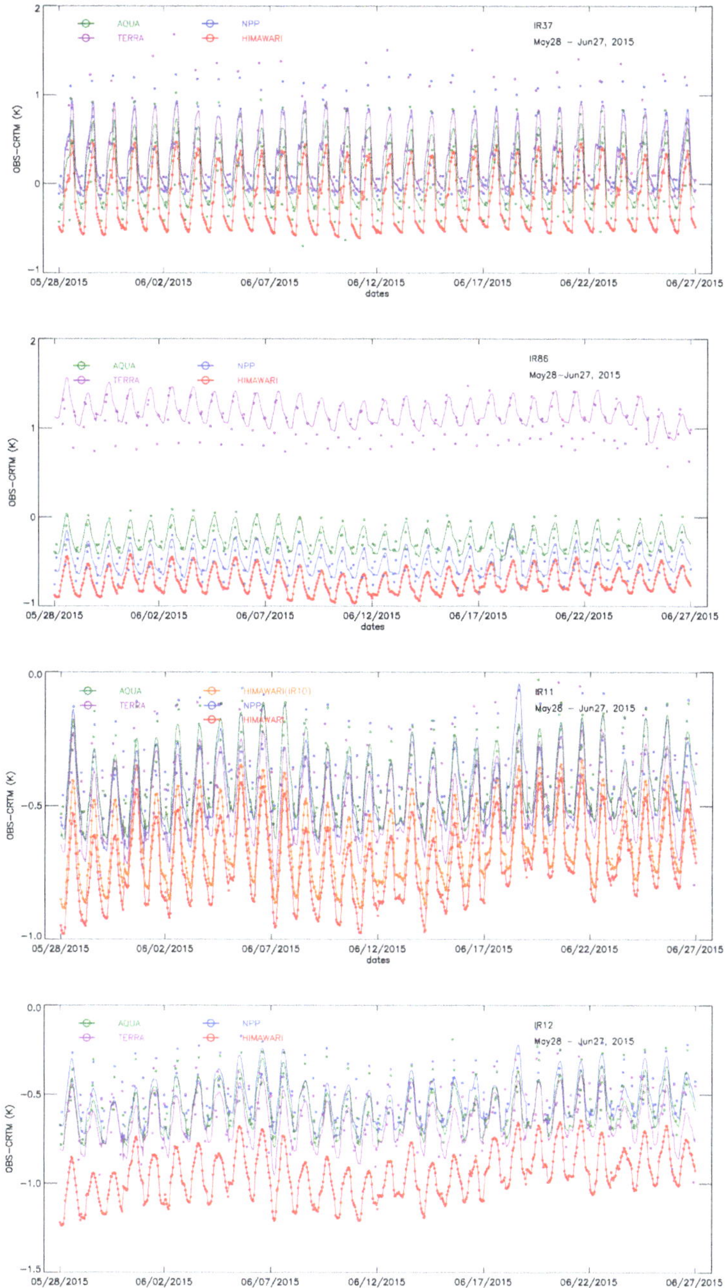

Figure 6. Time-series of the O-M biases for AHI, VIIRS, and MODISs for all five SST bands. The diurnal curves for the VIIRS and MODIS are generated using the same method as in Figure 5.

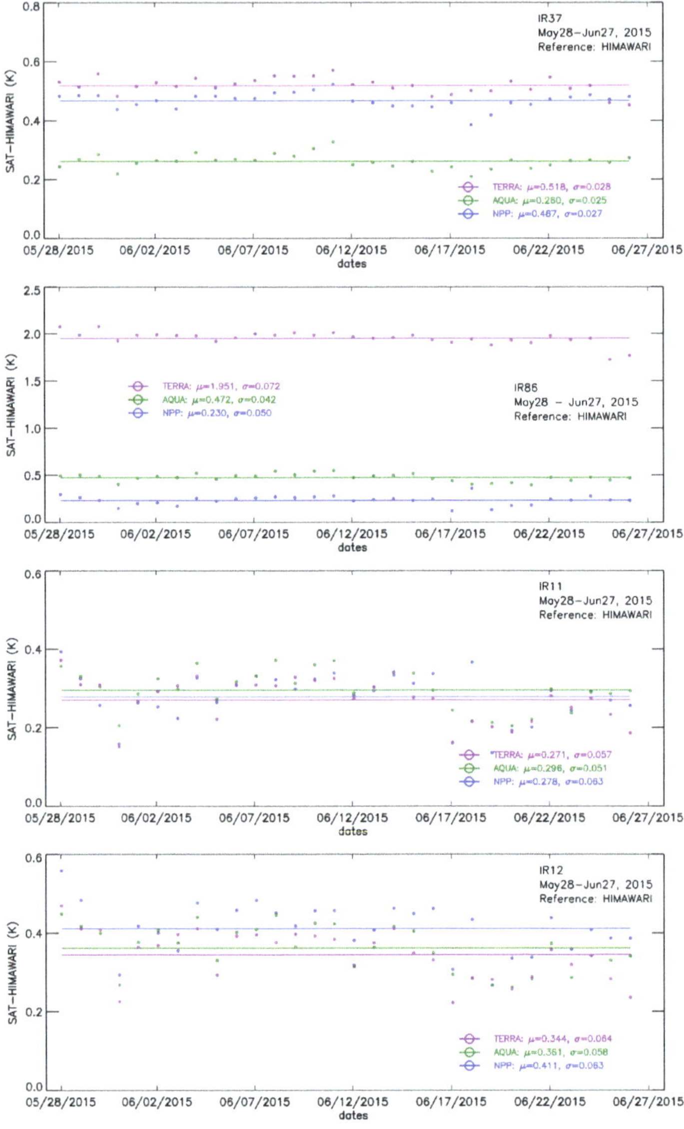

Figure 7. Double differences (DDs) calculated from Figure 6 using Equation (1). For each day, only one data point (out of 24 in Figure 6) is saved corresponding to the most populated polar nighttime bin (which is always 1:30 a.m. for the S-NPP and Aqua and 10:30 p.m. for the Terra—*cf.* Figure 2). Horizontal lines represent mean values over the 30 days. Corresponding mean and standard deviation statistics are also superimposed.

4. Conclusions and Future Work

NOAA is generating an experimental ACSPO SST product from the newly-launched AHI onboard Himawari-8. To support this product, O-M biases in five AHI IR SST window bands at IR37, IR86, IR10, IR11, and IR12 have been calculated and compared against the better understood and characterized VIIRS and MODIS. The objectives are to understand the accuracy and stability of the AHI BTs and check them for consistency with polar SST sensors and, thus, enhance readiness for the launch of the next-generation US geostationary satellite, GOES-R in 2016, which will carry the ABI sensor similar to the AHI. These analyses are critically important to ensure accuracy, stability, and consistency of the different NOAA geostationary and polar SST products. For the first time, two additional SST bands, IR86 and IR10, were analyzed, to facilitate their use in the ACSPO clear-sky mask and improved SST algorithms.

The major conclusion from our analyses is that all AHI bands are largely in-family with their polar counterparts, but biased cold by −0.25 to −0.5 K, depending upon the band. Additionally, Terra IR86 is biased high by 1.5 K due to electronic crosstalk. The diurnal variability (DV) of the O-M biases for AHI was applied to polar data by anchoring the AHI DV curves to the most populated corresponding nighttime polar bin. The DV in IR37 is much larger than in the other IR bands, due to the effect of solar reflectance in this band during the daytime which is modeled not fully accurately by the current CRTM. One month time series of O-M biases and double differences indicate that the AHI data are relatively stable and the temporal day-to-day variations are due to the unstable "M" term (*i.e.*, the CMC SST and GFS profiles used as input into CRTM).

Overall, AHI BTs are generally stable in time. Although the negative O-M biases may be, in principle, due to the residual cloud in ACSPO, the initial clear-sky mask in ACSPO AHI product was set conservatively and our experience suggests that these factors cannot fully explain the observed magnitudes of the biases. The other possible cause is the way the CRTM coefficients are calculated. The accuracy of the CRTM simulations in MICROS may be independently verified using the hyperspectral infrared sensors, such as the AIRS, IASI, and CrIS as references, similarly to the current JMA implementation [24]. Work is underway to attribute, minimize, and reconcile the O-M biases, and improve their utility for the evaluation of sensor performance for both geo and polar data. However, our experience suggests that these factors are unlikely to fully explain all of these cold biases and, at least a part of them, may be due to the uncertainties in AHI calibration and characterization.

The future work will be adding AHI SST bands in MICROS, including two new bands, IR86 and IR10. The work with the AHI sensor calibration team will continue towards understanding of the root cause of the cold AHI sensor BT biases, and their minimization. The ABI will be added in MICROS when the GOES-R is launched, and compared with the geo AHI and polar VIIRS and MODIS sensors.

Acknowledgments: This work is supported by the JPSS and GOES-R Program Offices, and by the NOAA Product System Development and Implementation (PSDI) and Ocean Remote Sensing Programs. Xingming Liang acknowledges the CSU CIRA research scientist fellowship. Thanks go to Boris Petrenko, Irina Gladkova for their Himawari-8 ACSPO analyses and productive discussions. Yury Kihai, John Stroup, Yaoxian Huang, and John Sapper of NOAA also contributed to the ACSPO development, data collection and analyses, and provided critical feedback and advice. Thanks also go to Fred Wu, co-lead of the NOAA geo calibration team, and to the JAXA and JMA for providing timely and highly accurate AHI L1b data. Thanks to Yong Chen of the NOAA Joint Center for Satellite Data Assimilation (JCSDA) for providing the spectral radiance data plotted in Figure 1. Thanks to Xiaoxiong Xiong of NASA, Junqiang Sun and Changyong Cao of NOAA STAR for helpful discussions and to Quanhua Liu, Paul Van Delst, David Groff and Fuzhong Weng of the JCSDA for providing the CRTM. We also thank three anonymous reviewers for constructive suggestions. The views, opinions, and findings contained in this report are those of the authors and should not be construed as an official NOAA or U.S. Government position, policy, or decision.

Author Contributions: Xingming Liang performed analyses of the data, drafted an initial version of the manuscript, and worked with co-authors on its improvements. Alexander Ignatov significantly contributed to discussions, data analyzes and manuscript preparation. Maxim Kramar set up the ACSPO system to process AHI L1b data, optimized its performance, worked with the calibration team to improve and better handle the AHI data, and produced a high quality Himawari-8 L2 SST product. Fangfang Yu helped with spectral response functions for different sensors and discussions.

Conflicts of Interest: The authors declare no conflict of interest.

References

1. Cao, C.; Xiong, X.; Blonski, S.; Liu, Q.; Uprety, S.; Shao, X.; Bai, Y.; Weng, F. Suomi NPP VIIRS sensor data record verification, validation, and long-term performance monitoring. *J. Geophys. Res.* **2013**, *118*, 11664–11678. [CrossRef]
2. Cao, C.; DeLuccia, F.J.; Xiong, X.; Wolfe, R.; Weng, F. Early on-orbit performance of the Visible Infrared Imaging Radiometer Suite (VIIRS) onboard the Suomi National Polar-orbiting Partnership (S-NPP) Satellite. *IEEE Trans. Geosci. Remote Sens.* **2014**, *52*, 1142–1156. [CrossRef]
3. Xiong, X.; Butler, J.; Chiang, K.; Efremova, B.; Fulbright, J.; Lei, N.; Mclntire, J.; Oudrari, H.; Sun, J.; Wang, Z.; *et al.* VIIRS on-orbit calibration methodology and performance. *J. Geophys. Res.* **2014**, *119*, 5065–5078. [CrossRef]
4. Schmit, T.J.; Gunshor, M.M.; Menzel, W.P.; Gurka, J.J.; Li, J.; Bachmeier, S. Introducing the next-generation advanced baseline imager on GOES-R. *Bull. Am. Meteor. Soc.* **2005**, *86*, 1079–1096. [CrossRef]
5. Murata, H.; Takahashi, M.; Kosaka, Y. VIS and IR Bands of Himawari-8/AHI Compatible with Those of MTSAT-2/Imager. Available online: http://www.data.jma.go.jp/mscweb/technotes/msctechrep60.pdf (accessed on 16 December 2015).
6. Okuyama, A.; Andou, A.; Date, K.; Hoasaka, K.; Mori, N.; Murata, H.; Tabata, T.; Takahashi, M.; Yoshino, R.; Bessho, K. Preliminary validation of Himawari-8/AHI navigation and calibration. *Proc. SPIE 9607* **2015**. [CrossRef]
7. Japan Meteorological Agency Himawari-8 Advanced Himawari Imaer (JMA AHI) Webpage. Available online: http://www.data.jma.go.jp/mscweb/en/himawari89/himawari89plan.html (accessed on 16 December 2015).
8. Liang, X.; Ignatov, A.; Kihai, Y. Implementation of the Community Radiative Transfer Model (CRTM) in Advanced Clear-Sky Processor for Oceans (ACSPO) and validation against nighttime AVHRR radiances. *J. Geophys. Res.* **2009**, *114*. [CrossRef]
9. Petrenko, B.; Ignatov, A.; Kihai, Y.; Heidinger, A. Clear-sky mask for the advanced clear-sky processor for oceans. *J. Atmos. Oceanic Technol.* **2010**, *27*, 1609–1623. [CrossRef]
10. Han, Y.; van Delst, P.; Liu, Q.; Weng, F.; Yan, B.; Treadon, R.; Derber, J. *Community Radiative Transfer Model (CRTM)—Version 1*; NOAA Technicl Report NESDIS 122. NOAA: Silver Spring, MD, USA, 2006.
11. National Centers for Environmental Prediction Global Forecast System (NCEP GFS) webpage. Available online: http://www.emc.ncep.noaa.gov/index.php?branch=GFS (accessed on 6 November 2015).
12. Brasnett, B. The impact of satellite retrievals in a global sea-surface-temperature analysis. *Q. J. R. Meteorol. Soc.* **2008**, *134*. [CrossRef]
13. Liang, X.; Ignatov, A. Monitoring of IR clear-sky radiances over oceans for SST (MICROS). *J. Atmos. Oceanic Technol.* **2011**, *28*. [CrossRef]
14. Liang, X.; Ignatov, A. AVHRR, MODIS, and VIIRS radiometric stability and consistency in SST bands. *J. Geophys. Res.* **2013**, *118*. [CrossRef]
15. Monitoring of IR Clear-Sky Radiances over Oceans for SST (MICROS) Webpage. Availabe online: http://www.star.nesdis.noaa.gov/sod/sst/micros (accessed on 24 February 2016).
16. Griffin, M.K.; Burke, H.K.; Kerekes, J.P. Understanding radiative transfer in the midwave infrared, a precursor to full spectrum atmospheric compensation. *Proc. SPIE 5425* **2004**. [CrossRef]
17. Kilpatrick, K.A.; Podestá, G.; Walsh, S.; Williams, E.; Halliwell, V.; Szczodrak, M.; Brown, O.B.; Minnett, P.J.; Evans, R. A decade of sea surface temperature from MODIS. *Remote Sens. Environ.* **2015**, *165*, 27–41. [CrossRef]
18. Sun, J.; Madhavan, S.; Xiong, X.; Wang, M. Long-term drift induced by the electronic crosstalk in Terra MODIS Band 29. *J. Geophys. Res.* **2015**, *120*. [CrossRef]
19. Wang, L.; Han, Y.; Jin, X.; Chen, Y.; Tremblay, D. Radiometric consistency assessment of hyperspectral infrared sounders. *Atmos. Meas. Tech.* **2015**, *8*. [CrossRef]
20. Wang, L.; Cao, C. On-orbit calibration assessment of AVHRR longwave channels on Metop-A using IASI. *IEEE Trans. Geosci. Remote Sens.* **2008**, *46*, 4005–4013. [CrossRef]
21. Garand, L. Toward an integrated land–ocean surface skin temperature analysis from the variational assimilation of infrared radiances. *J. Appl. Meteor.* **2003**, *42*, 570–583. [CrossRef]

22. Liang, X.; Ignatov, A.; Han, Y.; Zhang, H. Validation and Improvements of Daytime CRTM Performance Using AVHRR IR 3.7 um Band. Available online: https://ams.confex.com/ams/pdfpapers/170593.pdf (accessed on 24 February 2016).
23. Cao, C.; Weinreb, M.; Xu, H. Predicting simultaneous nadir overpasses among polar-orbiting meteorological satellites for the inter-satellite calibration of radiometers. *J. Atmos. Ocean. Technol.* **2004**, *21*, 537–542. [CrossRef]
24. JMA Himawari-8 AHI IR Inter-calibration with AIRS, IASI-A/B and CrIS webpage. Available online: http://www.data.jma.go.jp/mscweb/data/monitoring/gsics/ir/monit_geoleoir.html (accessed on 18 January 2016).

Article

Radiometric Inter-Calibration between Himawari-8 AHI and S-NPP VIIRS for the Solar Reflective Bands

Fangfang Yu [1,*] and Xiangqian Wu [2]

1 Earth Resources Technology Inc., 14401 Sweitzer Lane Suite 300, Laurel, MD 20707, USA
2 National Oceanic and Atmospheric Administration (NOAA)/National Environmental Satellite, Data, and Information Service (NESDIS)/Center for Satellite Applications and Research (STAR), 5830 University Research Court, College Park, MD 20740, USA; Xiangqian.Wu@noaa.gov
* Correspondence: Fangfang.Yu@noaa.gov; Tel.: +1-301-683-2555; Fax: +1-301-683-3526

Academic Editors: Changyong Cao, Richard Müller and Prasad S. Thenkabail
Received: 19 November 2015; Accepted: 12 February 2016; Published: 23 February 2016

Abstract: The Advanced Himawari Imager (AHI) on-board Himawari-8, which was launched on 7 October 2014, is the first geostationary instrument housed with a solar diffuser to provide accurate onboard calibrated data for the visible and near-infrared (VNIR) bands. In this study, the Ray-matching and collocated Deep Convective Cloud (DCC) methods, both of which are based on incidently collocated homogeneous pairs between AHI and Suomi NPP (S-NPP) Visible Infrared Imaging Radiometer Suite (VIIRS), are used to evaluate the calibration difference between these two instruments. While the Ray-matching method is used to examine the reflectance difference over the all-sky collocations with similar viewing and illumination geometries, the near lambertian collocated DCC pxiels are used to examine the difference for the median or high reflectance scenes. Strong linear relationships between AHI and VIIRS can be found at all the paired AHI and VIIRS bands. Results of both methods indicate that AHI radiometric calibration accuracy agrees well with VIIRS data within 5% for B1-4 and B6 at mid and high reflectance scenes, while AHI B5 is generally brighter than VIIRS by ~6%–8%. No apparent East-West viewing angle dependent calibration difference can be found at all the VNIR bands. Compared to the Ray-matching method, the collocated DCC method provides less uncertainty of inter-calibration results at near-infrared (NIR) bands. As AHI has similar optics and calibration designs to the GOES-R Advanced Baseline Imager (ABI), which is currently scheduled to launch in fall 2016, the on-orbit AHI data provides a unique opportunity to develop, test and examine the cal/val tools developed for ABI.

Keywords: Himawari AHI; GOES-R ABI; S-NPP VIIRS; inter-calibration; collocation; ray-matching; solar reflective bands; Deep Convective Cloud (DCC)

---

1. Introduction

As the first in a series of next-generation geostationary (GEO) weather imagers, the Advanced Himawari Imager (AHI) was successfully launched on-board of Himawari-8 by the Japan Meteorological Agency (JMA) on 7 October 2014. It has 16 multispectral bands, including six visible and near infrared (VNIR) and 10 IR bands. The first AHI images of the Earth captured on 18 December 2014 demonstrate a significant increase in high spatial and spectral resolutions, compared to its predecessor MTSAT-series satellite images, which only have one visible and four infrared (IR) bands. For the first time, a solar diffuser is equipped on a geostationary weather instrument to provide bright reference for the on-orbit radiometric calibration of VNIR bands to reduce the calibration uncertainty. A new design, including double scan mirrors and controlled calibration target temperature, is used to improve the radiometric calibration accuracy of the IR bands. The instrument produces full-disk imagery every 10 min and rapid scanning of Japan and target areas at 2.5-min intervals to sense the reflective and emitted energy from the environment of the Asia–Pacific region. With the significant improvements in spectral, spatial and temporal resolutions and calibration techniques, AHI greatly improves the capacity of weather forecasting, environmental monitoring, and weather prediction accuracy, and provides valuable data for climate and weather research studies.

AHI has a very similar optical design as the Advanced Baseline Imager (ABI) onboard the American GOES-R satellite, which is currently planned to be launched in fall 2016. Both AHI and ABI have similar spectral and spatial characteristics, except that the AHI 1-km green band (0.51 μm) is replaced with the ABI 2-km 1.38 μm band and the normal spatial resolution of 1.61 μm band is 2-km for AHI but 1-km for ABI. For the calibration of collected data, both instruments use internal calibration target of blackbody and deep space for infrared band calibration, while the Solar Diffuser (SD) and deep space observations are used for the calibration of VNIR bands. On 8 June 2015, JMA, for the first time, updated the solar calibration Look-Up-Table (LUT) derived from the on-orbit SD calibration target measurements [1]. Apparent striping and banding issues, which were previously observed in the AHI VNIR images calibrated with ground measurement have been significantly reduced with the LUT update. JMA officially announced the operation of AHI data on 7 July 2015.

Like ABI, the SD on-board AHI is sub-aperture, making it challenging for accurate pre-launch and on-board calibration [2,3]. Sensor-to-sensor inter-calibration provides one practical way to examine the calibration difference between two instruments and, thus, is often applied to validate the calibration accuracy of a newly launched instruments. The AHI VNIR spectral responses roughly match those of the Visible Infrared Imaging Radiometer Suite (VIIRS) (Figure 1), one of the key payloads on Suomi-NPP (S-NPP), which was launched as a Low Earth Orbit (LEO) satellite on 28 October 2011. Both AHI and VIIRS use a Spectralon® SD for solar reflective calibration. However, the VIIRS SD is covered with a fixed attenuation screen and accompanies a Solar Diffuser Stability Monitor (SDSM) to track SD degradation. Special spacecraft maneuvers were performed during the S-NPP post-launch period to characterize the SD solar attenuation screen and SDSM screen transmission functions, and also the SD Bidirectional Reflectance Distribution Function (BRDF) [4]. Results of vicarious calibration indicate that the VIIRS VNIR radiometric uncertainty is comparable to that of the Moderate-Resolution Imaging Spectroradiometer (MODIS) Collection 6 data within 2% at typical scenes [5]. Long-term monitoring of the VIIRS measurements at well-characterized desert calibration sites shows that the VIIRS moderate bands are very stable [6].

However, unlike the AHI solar calibration, in which the SD data are collected with a special integration time to achieve about 100% albedo [7], VIIRS has a Solar Attenuation Screen (SAS), which results in radiances equivalent to ~10% albedo for the calibration of all VNIR detectors [8,9]. The objective of this study is, therefore, to validate the AHI VNIR radiometric calibration accuracy using GEO-LEO inter-calibration techniques over the VIIRS collocated observations. Two methods are applied: (1) the Ray-matching method to examine the calibration difference at all-sky collocations ranging from dark open ocean pixels to high reflectance cloud ones; and (2) the collocated Deep Convective Cloud (DCC) method to assess the difference at median or high reflectance scenes. As the Ray-matching method can also provide matched scenes across the East–West (E–W) field of regard along the Equator area, the E–W viewing angle dependent calibration difference is also investigated for the VNIR bands.

2. AHI and VIIRS Collocations

2.1. AHI and VIIRS VNIR Data

Located at about 35,800 km above the Equator, at 140.7°E longitude, AHI provides a full-disk scan of the Earth at every 10 min and target area scans every 2.5 min. The six AHI VNIR bands have three nadir spatial resolutions: 0.5-km for B3 (0.64 μm), 1-km for B1 (0.47 μm), B2 (0.51 μm) and B4 (0.86 μm), and 2-km for B5 (1.6 μm) and B6 (2.3 μm) (Table 1). Raw data are re-sampled to fixed grids as pixels in each AHI Earth image. The fixed grids are a set of static pixel locations relative to an ideal geostationary satellite viewpoint and are used to aid users by providing continuity in locations of geographic, features throughout the satellite mission's life. Since April, 2015, the mean Image Navigation Registration (INR) error at 2-km bands is less than 0.5-pixel [1,10]. Apparent stripes were observed in each VNIR band data before 8 June 2015, but they were substationally reduced with the update of on-orbit derived calibration coefficients. In this study, only the full disk AHI data from the HimawariCloud [10,11], a JMA service to disseminate Himawari AHI L1B data, are used for the analyses. While the potential residual stripes may cause relative radiance differences among the detectors, we focus on the mean reflectance differences over the collocated scenes between these two instruments. Here, reflectance is converted from calibrated radiance with spectral solar irradiance, normalized with the solar zenith angle and the sun-Earth distance.

Table 1. Spectra of AHI VNIR bands and their spectrally matched VIIRS bands.

Himawari-8 AHI			S-NPP VIIRS		
Central Wavelength (μm)	Band Name	Nadir Spatial Resolution (km)	Central Wavelength (μm)	Band Name	Nadir Spatial Resolution (km)
0.47	B1	1.0	0.486	M3	0.750
0.51	B2	1.0	0.486	M3	0.750
0.64	B3	0.5	0.639	I1	0.375
0.86	B4	1.0	0.862	M7	0.750
			0.862	I2	0.375
1.6	B5	2.0	1.602	M10	0.750
			1.602	I3	0.375
2.3	B6	2.0	2.257	M11	0.750

S-NPP is an afternoon LEO orbit satellite with a 16-day repeat cycle of data collection. The local Equator crossing time at ascending node is ~1:30 p.m. VIIRS has 22 spectral bands, including 14 VNIR bands, severn IR bands, and one Day-Night Band (DNB), and views the Earth at scan angle ranging from -56.28° to $+56.28^\circ$ for a wide-swath of 3000-km with no gap between orbits. The scan angle dependent reflectance is greatly reduced with the optical design of the rotating telescope assembly, a rotating half-angle mirror, and the careful pre-launch calibrations [5,8,12]. There are two types of spatial resolutions, 750-m nadir spatial resolution for the 16 moderate (M) resolution bands and the DNB and the 375-m nadir spatial resolution for the 5 imaging (I) bands (Table 1). The VIIRS M-bands have a better Signal-to-Noise Ratio (SNR) and radiometric accuracy, while the I-bands have a high spatial resolution with similar spectral response [5]. The geolocation accuracy of VIIRS data is better than 0.1 km [13].

Figure 1 shows the Spectral Response Function (SRF) of the six AHI VNIR bands [12], together with the spectrally-matched VIIRS bands [5]. Both AHI B4 and B5 have two spectrally matched VIIRS bands, one M-band and one I-band, respectively (Table 1). Unfortunately, there is no such well-matched VIIRS SRF for AHI B1 and B2. VIIRS M3, which is spectrally the most similar to these two bands, is used as the reference. The VIIRS analog-to-digital (A/D) conversion is 14-bit quantization, but is truncated to 12 bits for Earth data, while all the AHI VNIR data are truncated at 11 bits. The NOAA operational VIIRS Senser Data Record (SDR) data [12] are used in this study. These VIIRS data are available at NOAA's Comprehensive Large Array-data Stewardship (CLASS).

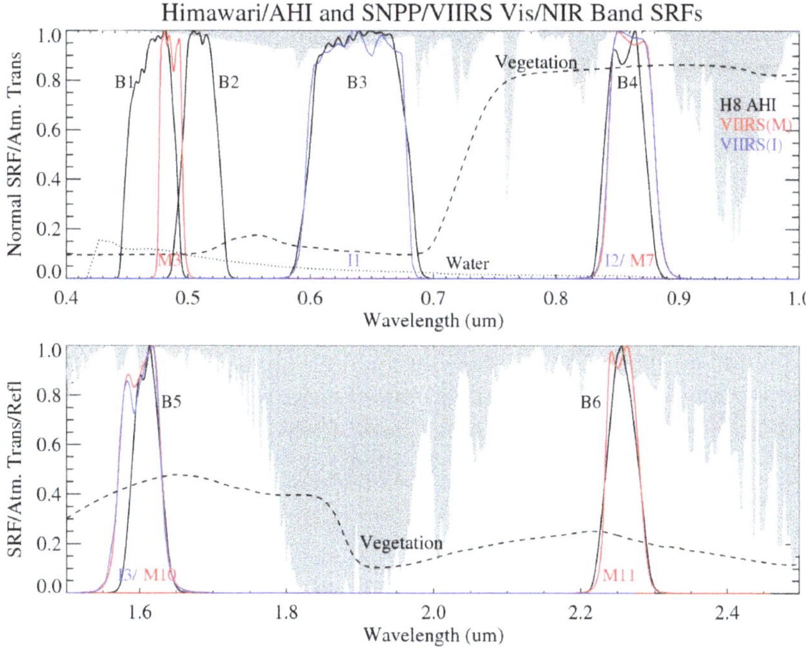

Figure 1. Spectral response function (SRF) of AHI (black) and VIIRS (red for M bands and in blue for I bands), as well as the atmospheric transmission (gray shadow), and vegetation and water spectra (dark, dashed lines). The VIIRS SRFs plotted are of Non-Government (NG) version [6].

2.2. GEO-LEO Collocations

Satellite inter-comparison/inter-calibration is usually based on the comparison of measurements of the same targets. In order to reduce the uncertainty caused by impacts of atmospheric component variations and target surface BRDF, both instruments should view the geo-spatially collocated targets at similar viewing geometries within a short time interval. The following criteria are used to identify the collocated pairs without the match of viewing azimuth angles, similar to the collocation criteria applied to the GEO-LEO IR inter-calibration by the Global Satellite Inter-Calibration System (GSICS) community for the GEO-LEO IR inter-calibration [14,15].

- The time difference between GEO and LEO observations is less than 5 min. This criterion is used to reduce the impact of atmospheric variations on the Top Of Atmosphere (TOA) reflectance, as well as to ensure similar solar illumination angles to reduce the BRDF impact.
- The cosine of viewing zenith angle difference between the GEO and LEO instrument is less than 1%. As the optical path is proportional to the cosine of the viewing zenith angle, this criterion is to ensure similar optical paths for the atmospheric absorption and scattering effects, as well as similar viewing zenith angles.
- To ensure the same targets observed by the two instruments, the spatial distance between the centers of each GEO and LEO collocated pairs is less than the nominal spatial resolution of the corresponding VIIRS band, that is, 375 m for the VIIRS I bands and 750 m for the VIIRS M bands.
- To reduce the computing time, all the AHI B1, 2, 3 and 4 images are degraded to 2-km spatial resolution by averaging the radiane and reflectance at every 2 pixels x 2 pixels (for AHI B1, 2 and 4) or 4 pixels x 4 pixels (for AHI B3). To match the AHI pixel spatial size, the arrays of 3 VIIRS pixels × 3 VIIRS pixels and 7 VIIRS pixels × 7 VIIRS pixels centered at the collocated pixel are

considered as the spatially collocated VIIRS scenes for the M-band and I-band data, respectively. The mean values of 3 VIRS pixels × 3 and 7 VIIRS pixels × 7 pixels are used to simulate the AHI measurements. As the clouds may be moving within the time interval, the statistical information (mean and standard deviation) of the environmental (ENV) pixels, which are three times of the AHI pixel in size, and centered at the collocated pixels, are also archived for both AHI and VIIRS data.

Again, note that no matching criterion on viewing azimuth angles is applied to these collocations yet. As shown in Figure 2, two sets of collocation data are archived from different spatial and temporal domains: (1) the collocations within 20°S < latitude < 20°N and 120.7°E < longitude < 160.7°E and during 12:00 p.m.–3:00 p.m. Satellite Local Time (SLT) from 20 July to 20 August 2015. This collocation dataset is collected near the AHI sub-satellite region around the VIIRS ascending time. This collocation data is also used for the general calibration accuracy inter-comparison described in Section 3.1, and also used to identify the collocated Deep Convective Clouds (DCC) pixels as described in Section 3.2; and (2) day-time collocations, from 9:00 a.m. SLT to 5:00 p.m. SLT, within 10°S < latitude < 10°N and 60.7°E < longitude < 220.7°E from 20 June to 20 August 2015. The latitude threshold is used because the collocations usually occur near the Equator when matching with line of sight is required. The spatial domain is within ±80° longitude from the GEO sub-satellite point used, because the maximum angle of the Earth observed from the GEO satellite is 81.3° [16]. This set of collocations is used to assess the E–W scan mirror uniformity with the similar algorithm described in Yu *et al.* [17], and reported in Section 3.2.

The spatial and temporal information of the collocated data of AHI and Cross-track Infrared Sounder (CrIS), a hyperspectral radiometer also onboard the S-NPP satellite, are used to reduce the pair searching time from AHI and VIIRS images. More than ten thousand collocation scenes near the AHI sub-satellite region can be obtained every day with these criteria.

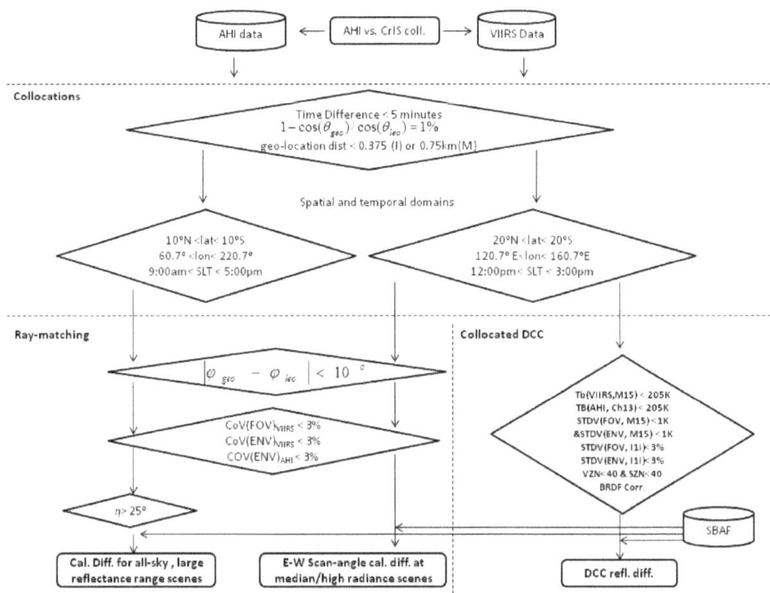

Figure 2. Flow-chart of AHI *vs.* VIIRS inter-calibration processes. The time and spatial location information of AHI and CrIS collocations are used to restrict the temporal and spatial searching domains. θ_{geo} and θ_{leo} are the GEO and LEO viewing zenith angles, respectively. φ_{geo} and φ_{leo} are the viewing azimuth angles for GEO and LEO instruments, respectively. η is the sun-glint angle.

3. Ray-Matching and Collocated DCC Methods

3.1. Ray-Matching Method

The Ray-matching method provides a straight-forward tool for the direct inter-comparison between GEO and LEO instruments over all-sky scenes [18]. The method is based on the comparison of the observations over the same Earth surface targets, at which point both GEO and LEO instruments scan with similar viewing and illumination angles within similar atmospheric conditions. Matching in the viewing azimuth angle between these two instruments is also applied to reduce the BRDF impact. To ensure that the same targets observed by the two instruments, the homogeneous scenes are used by filtering with the Coefficients of Variance (*CoV* = standard deviation/mean) of the reflectance. The homogeneous scenes are also used to compensate for the potential navigation difference between these two satellites.

- $CoV(ENV)_{VIIRS}$ < 3%, $CoV(ENV)_{AHI}$ < 3% and $CoV(FOV)_{VIIRS}$ < 3%. FOV is the nominal spatial size (2 km) of AHI pixel used in this study, corresponding to the field-of-view (FOV) for B5 and B6.
- $|\varphi g_{eo} - \varphi_{leo}| < 10^\circ$

where φg_{eo} and φ_{leo} are the GEO and LEO viewing azimuth angles, respectively.

3.2. Collocated DCC Method

DCC is highly convective clouds that overshoots the tropical tropopause layer. With a distance from the ground of more than 10 km above the tropopause, DCC strongly reduces the influences of the surface and atmosphere on the TOA visible reflectance, particularly for the contributions from the molecular scattering and atmospheric water vapor absorption. At the visible spectra (e.g., 0.4 μm–1.0 μm), DCC reflectance is nearly invariant (e.g., spectral white), as the size of the ice particles of the clouds are large enough for Mie scattering and the absorption impact is negligible [19] and is considered near Lambertian when the viewing and solar zenith angles are less than 40° [20]. However, as the wavelength increases, the impact of absorption of cloud particles increases, resulting in certain variations in the reflectance. However, as DCC is selected by the GSICS community as a common calibration target to inter-calibrate visible bands, it would be also interesting to study the performance of DCC reflectance at near infrared (NIR) bands (>1.0 μm), for example AHI B5 (1.6 μm) and B6 (2.3 μm).

The DCC calibration method is a statistical method that requires sufficient DCC pixels for robust results [21,22]. As DCC is mainly distributed within the Inter-Tropical Convergence Zone (ITCZ), the spatial domain to identify DCC pixels is defined as ±20° latitude and longitude region from the GEO sub-satellite point [20]. However, although DCC reflectance is found to be stable in the long-term, some slight intra-annual variations can be observed, probably associated with the intra-annual variations of cloud physical properties [21,22]. This seasonal variation in the DCC reflectance may result in uncertainty in the absolute calibration if not properly corrected, especially at the early stage in a satellite mission's life.

The GEO-LEO inter-calibration over the collocated DCC pairs can overcome the problem of the seasonal variation of DCC reflectance. Assessment of different satellite global data indicated that the West Pacific Ocean area has the highest frequency of DCC pixels [23]. Therefore, it is possible to identify sufficient collocated DCC pixels within the short study period. In addition to the spatial domain, further criteria, following Doelling *et al.* [21], are also applied to identify the DCC pixels over the collocated pairs, as described in Section 2.2. Note that no viewing azimuth matching criterion was applied to DCC selections. Each DCC pixel reflectance is corrected with the Hu *et al.* [24] model for BRDF correction.

- Brightness temperature (Tb) of AHI B13 (10.4 μm) and VIIRS M15 (10.7 μm) are less than 205 K. Selection of DCC pixels is sensitive to the Tb threshold [20]. Although both AHI B13 and VIIRS

M15 are, in general, well-calibrated [1,12], in this study, the Tb threshold value of 205 K is applied to both instruments to reduce the possible impact of radiometric calibration difference at extremely cold DCC pixels.

- Standard deviation of Tb for VIIRS M15 FOV and ENV arrays are less than 1 K
- Standard deviation of Tb for AHI B13 ENV arrays are less than 1 K
- CoV of reflectance for VIIRS I1 FOV and ENV arrays are less than 3%
- Both the GEO and LEO viewing zenith angle (θ_v) and solar zenith angle (θ_s) should be less than 40°, that is, $\theta_v < 40°$ and $\theta_s < 40°$

3.3. Spectral Band Adjustment Factor (SBAF)

The Scanning Imaging Absorption spectrometer for Atmospheric CHartographY (SCIAMACHY)-derived SBAFs, provided by NASA Langley, are applied to correct the SRF differences between AHI and VIIRS [25]. As NOAA employs the VIIRS Non-Government (NG) version of SRFs for the operational calibration, the linear fit SBAF, derived with NPP-VIIRS-NG SRF, are selected from the SBAF web-tools. As the spatial domain of the Ray-matching method is dominated by ocean, the all-sky tropical ocean SBAF values are chosen in this study. Table 2 lists the SBAF coefficients and their uncertainties from the web-tools. Unfortunately, there is no SCIAMACHY-based SBAFs for AHI B6 (2.3 μm). We assume that the SBAF for this band is 1.0 due to the similar SRFs between AHI B6 and VIIRS M11, and there is no strong absorption within the spectral range (Figure 1). In this study, all the inter-calibration analyses are based on reflectance. All the AHI data used for inter-comparison with VIIRS are corrected with SBAF coefficients as follows:

$$R_{VIIRS,AHI} = (R_{AHI} - SBAF\_Offset)/SBAF\_Slope \tag{1}$$

where R_{AHI} is the AHI reflectance and $R_{VIIRS,AHI}$ is the corresponding AHI reflectance corrected with VIIRS SRF. SBAF_Slope and SBAF_Offset are the slope and offset of the SCIAMACHY-based linear regression coefficients, respectively.

Table 2. SBAF (AHI/VIIRS) coefficients (slope and offset) for AHI VNIR bands and the uncertainty values.

SBAF	AHI	B1	B2	B3	B4		B5		B6
	VIIRS	M3	M3	I1	M7	I2	M10	I3	M11
Ray-matching (all-sky tropical Ocean)	SBAF_Slope	0.991	1.005	1.000	0.998	0.998	1.019	1.022	1.0
	SBAF_Offset	9.5e−3	−1.341e−2	-2.07e−4	−4.18e−4	−3.67e−4	−2.216e−4	−2.465e−4	0.0
	Uncertainty (%)	0.820	1.172	0.187	0.448	0.422	1.701	1.839	-
Coll. DCC	SBAF_Slope	0.992	1.014	1.000	1.003	1.003	1.035	1.038	1.0
	SBAF_Offset	9.989e−3	−2.124e−2	1.594e−5	−1.545e−3	−1.459e−3	2.472e−3	2.875e−3	0.0
	Uncertainty (%)	0.238	0.596	0.033	0.106	0.100	0.736	0.753	-

4. Results and Discussions

4.1. Ray-Matching Inter-Calibration: Large Measured Radiance/Reflectance Range

Figure 3 are the scatter-plots of AHI and VIIRS reflectance from the homogeneous scenes selected with the Ray-matching method. To avoid directional reflectance from the sun-glint area caused by the specular reflection, only collocations beyond the sun-glint angle of 25° ($\eta > 25$), which is selected to balance the requirements of sufficient collocation for analysis and, meanwhile, to reduce the sun-glint impact, are used for this analysis. The sun-glint angle is calculated with the following equation [26]:

$$\cos(\eta) = \cos(\theta_s)\cos(\theta_v) + \sin(\theta_s)\sin(\theta_v)\cos(180 - \varphi)$$

where η is the sun glint angle, θ_s and θ_v are the solar zenith angle and satellite viewing zenith angle, respectively, and φ is the relative azimuth angle between solar and viewing azimuths.

As shown in Figure 3, there are strong linear relationships between the AHI reflectance and the corresponding VIIRS M and I data. Two of VIIRS M bands used in this study, M3 and M7, have dual gain design. Although the homogenous criteria removes most scenes at a median-radiance level when they are partially covered with cloud, the selected scenes still cover large radiance ranges from the low radiance of clear ocean scenes to the high reflectance of cloud covered or bright scenes. The offsets of the linear fitting functions are very small and positive, less than 0.6% for AHI B3-6. Due to the Rayleigh scattering of which effect on the TOA radiance or reflectance decreases with an increase in wavelength, the minimum reflectance values are about 7.8% and 2.5% for VIIRS M3 and I1, respectively, and about 9.1%, 6.4% and 2.4% for AHI B1, B2 and B3, respectively. It is also expected that AHI B2 reflectance of clear ocean scenes should be smaller than VIIRS M3 as the AHI B2 central wavelength is larger than that of VIIRS M3. However, the fitting slope of the collocated scenes at M3 high-gain data is 1.2 (AHI is generally bright than VIIRS), while the slope of scenes at the low-gain part is 0.97 (AHI is generally darker than VIIRS). The discrepancy of the fitting slopes at high-gain and low-gain data may suggest that the SCIAMACHY-based Ray-matching SBAF has some uncertainty for the clear-sky open ocean data for this pairs of instrument bands. Two possible reasons can be attributed to the SBAF uncertainty: (1) the large SRF difference between AHI B2 and VIIRS M3, and (2) the large foot-print of SCIAMACHY made it hard to find clear sky ocean pixels in the low radiance end pixels.

The collocated data in the figure are classified with VIIRS specification values: the high gain and low gain switchover radiace and the minimum values of dynamic ranges. Due to the stringent requirements from the user community, the minimums of dynamic ranges are specified at VIIRS VNIR bands and used to generate meaningful physical products. Some extremely low VIIRS radiance scenes with a mean FOV radiance below the minimum of the VIIRS dynamic range values can be observed at the bands of M7, M10, M11, I2 and I3. These are most likely very clean ocean surface scenes, from which the signals reaching the instrument's aperture are very small at these bands. For the two short-wave bands (M3 and I1), the effect of Rayleigh scattering increases the TOA reflectance and, thus, all the radiance at the two bands are larger than the minimum specifications. There is no minimum dynamic range specification for AHI data [27].

The calibration difference between these two instruments can be displayed with the scene dependent reflectance ratios (ratio = AHI/VIIRS) (Figure 4). All the paired bands display relatively large variation in the reflectance ratio at the low radiance scenes, while the ratio is relatively consistent at median and high radiance ones. The most likely reason is that impact of noises is most apparent at low radiance scenes. Thus the outliers, which have the VIIRS radiance less than the minimum of VIIRS dynamic range, display the largest variations in the ratios. Several reasons may cause the relatively large variation in the reflectance ratio at low reflectance scenes. One possible source could be the small AHI instrument coherent noise, as the magnitude of ratios increases with the central wavelength from visible to NIR bands, while the reflected solar energy decreases at these bands (Figure 4). Like ABI, AHI also applies a resampling algorithm in generating the fixed grid L1B data [28]. According to Pgorzala *et al.* [29], the resampling process has the potential to both smooth and amplify the radiance variations within the neighouring samples [29]. In this study, although the CoV values of AHI and VIIRS ENV arrays' reflectance are applied to ensure the homogeneous collocations, the resampling process may cause deviation of the collocated AHI reflectance and, thus, be attributed to the ratio variation. Slight difference in detector-to-detector responsivity may also contribute to the ratio variations. Another possible noise source is the different directional reflectance resulting from the movement of sun-glint mask residual, if any, within the collocation time interval. Additionally, interestingly, at the VIIRS dual-gain bands (M3 and M7), there is no homogeneous collocation available at the low-gain side near the switchover point between high and low gains [5,8]. Further effort is needed to understand the causes of these ratio variations at the VIIRS high gain or low radiance scenes.

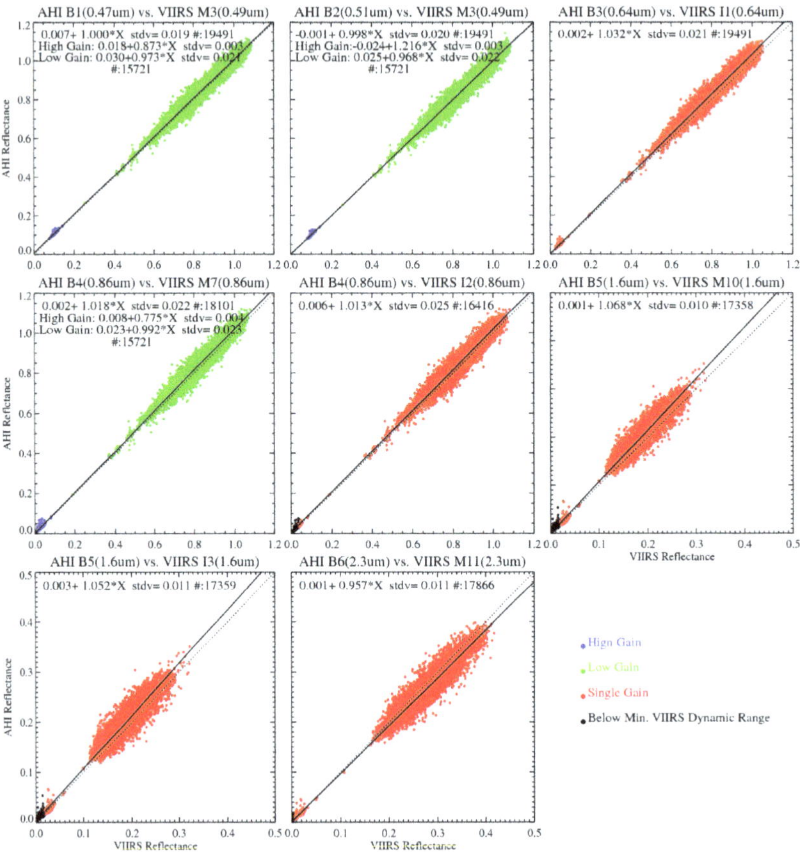

Figure 3. Relationship between homogeneous ray-matched AHI and VIIRS data from 20 July to 20 August 2015. The x-axis is VIIRS reflectance and y-axis is AHI reflectance.

As shown in Figure 4, the median/high reflectance scenes have relatively consistent reflectance ratios. A set of radiance and reflectance thresholds are then used to separate the median/high radiance scenes from the low radiance scenes to further analyze the AHI and VIIRS calibration differences. For the VIIRS dual-gain bands (M3 and M7), the radiance values of the switchover points are used as the threshold to define the median/high radiance scenes. A reflectance value of 0.2 for AHI B3 and B4 and reflectance of 0.1 for AHI B5 and B6 are used as thresholds. The mean and standard deviation of the reflectance ratios from the median/high collocated scenes are reported in Table 3. As shown in this table, AHI radiometric calibration accuracy agrees well with that of VIIRS, within a 5% difference, except for B5: AHI reflectance is generally higher than VIIRS data at B1, B3, and B4 by 1.0%, 3.7%, and 2.1%, respectively, and lower than VIIRS data at B2 by 0.1% and B6 by 3.7%. The largest reflectance difference occurs at AHI B5, which is, in general, larger than VIIRS M10 and I3 data by 7%–8%. Similar results between the AHI and VIIRS inter-calibration were also reported by JMA with the Ray-matching method [1].

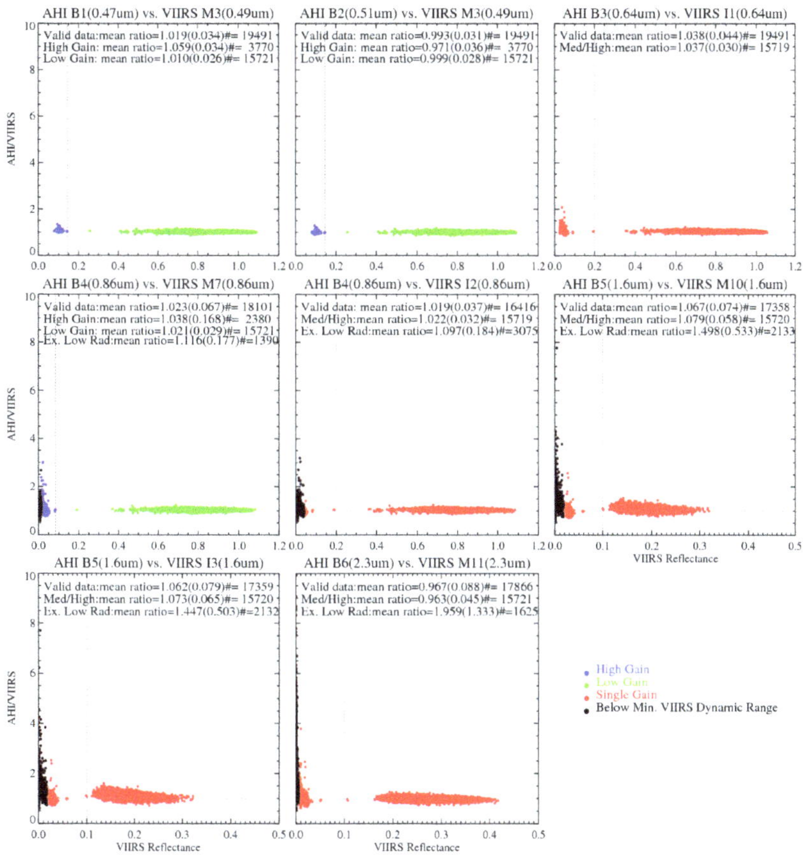

Figure 4. Scene dependent reflectance ratio for the homogeneous ray-matched AHI and VIIRS collocations from 20 July to 20 August 2016. The x-axis is the VIIRS reflectance and y-axis is AHI and VIIRS reflectance ratio. The vertical dashed lines are the thresholds used to separate the median/high radiance scenes from the low radiance ones. The horizontal dashed line is the ratio of 1.0. The outliers are the collocations with VIIRS radiance less than the VIIRS minimum dynamic ranges.

Table 3. Reflectance ratio between AHI and VIIRS M and I bands for the ray-matched median/high reflectance scenes and collocated DCC measurements.

AHI		B1	B2	B3	B4		B5		B6
VIIRS		M3	M3	I1	M7	I2	M10	I3	M11
Ray-matching		1.010 (±0.026)	0.999 (±0.028)	1.037 (±0.030)	1.021 (±0.029)	1.022 (±0.032)	1.079 (±0.058)	1.073 (±0.065)	0.963 (±0.045)
DCC	Median	1.002	0.992	1.031	1.014	1.015	1.067	1.061	0.955
	Mode	0.992	0.985	1.030	1.024	1.014	1.102	1.084	0.977
	Mean	1.003	0.994	1.031	1.014	1.015	1.064	1.058	0.958
	Statistics *	1.003 (±0.024)	0.995 (±0.026)	1.032 (±0.028)	1.015 (±0.024)	1.015 (0.025)	1.065 (±0.030)	1.059 (±0.032)	0.959 (±0.026)

*: mean and standard deviation of the reflectance ratios for all the collocated DCC scenes at each paired bands.

4.2. Ray-Matching Inter-Calibration: E–W Viewing Angle Dependent Calibration Difference

Figure 5 shows the spatial distribution of the Ray-matched collocations within 10°S to 10°N from 20 June to 20 August 2015. Due to the inclination of S-NPP (98.69°), the collocations east of the GEO nadir are mainly distributed in the Northern hemisphere, while those west of the sub-satellite points are mainly in the Southern hemisphere. Such a wide distribution of the Ray-matched collocations along the Equator, thus, provides an opportunity to examine the viewing angle dependent calibration difference between these two satellites. This assessment is applied to the homogenous median/high reflectance scenes. No mask of sun-glint area is applied because the sun-glint effects on the median/high reflectance scenes are statistically very small. This is because clouds usually do not have strong directional forward or backward reflectances as those of clear ocean surfaces [30].

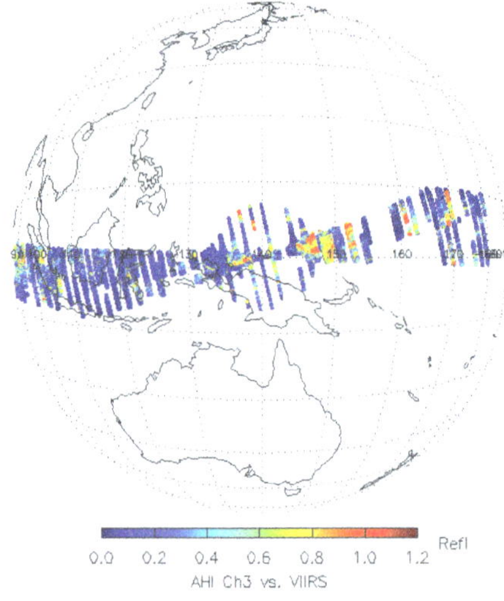

Figure 5. Spatial distribution of Ray-matched collocations within ±10° latitude for the collocation data collocated from 20 June to 20 August 2015.

According to the GEO satellite and Earth geometry, the maximum viewing angle from the GEO satellite to the Earth surface is 8.7° [14]. The viewing angle is defined as the angle between the sub-satellite point, GEO satellite and the viewing target on the Earth. The collocation data are equally split into 10 bins between −8.7° and +8.7°. The AHI pixel coordinates are used to calculate the viewing angle. For the AHI pixels at 2 km normal spatial resolution, the Instantaneous Geometric Field of View (IGFOV) is 56 μrad. A minimum of 100 homogeneous scenes is required to calculate the mean and standard variation values of reflectance ratio at each valid bin. The angular dependent reflectance ratios for each AHI and corresponding VIIRS bands are shown in Figure 6. No apparent trending can be observed at these bands with a GEO viewing angle of <7°, which is less than about 50° of GEO and LEO viewing zenith angles. Relatively large ratio variations can be observed at a few viewing angle bins of AHI B5, but they are associated with relatively larger uncertainty. Slight positive deviation can be observed at AHI B1 *vs.* M3, B3 *vs.* I1, B4 *vs.* M7, B4 *vs.* I2, and B6 *vs.* M11 for the data with AHI viewing zenith angle less than −7° (about 50° for the GEO and LEO viewing zenith angles, west of the nadir point). However, the deviations are within the measurement uncertainty level. Unfortunately, no collocation data are selected for viewing angles larger than 7° in this study. Extension of AHI and VIIRS collocation time threshold earlier than 9:00 a.m. SLT should be applied to cover the viewing angles in future studies. Unfortunately, a large portion of the collocations had low reflectance during this study period (Figure 5), a longer period or a different study period will be needed to ensure robust results with sufficient median/high reflectance numbers for all the viewing angle bins.

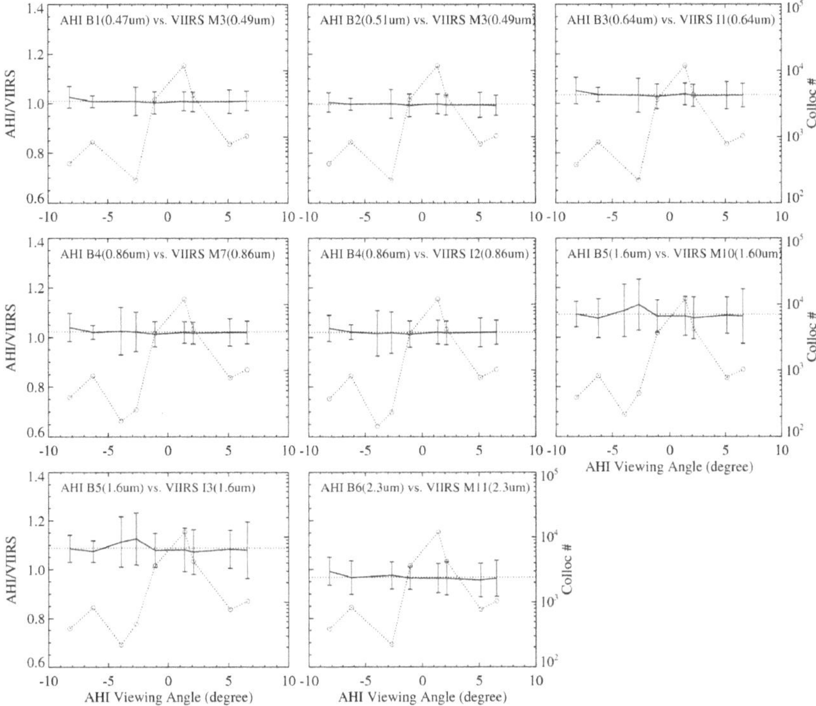

Figure 6. The E–W viewing angle dependent reflectance ratio between AHI and VIIRS for the data collected from 20 June to 20 August 2015. The red segments represent the standard deviation of the reflectance ratio at each valid bin. The open circles connected with dashed lines are the number of selected scenes and are referred with the second y-axis which ranges from 10^2 to 10^5. The x-axis is AHI viewing angle in degree.

4.3. Collocated DCC Results

The DCC pixels are selected with AHI B3 and B13, and VIIRS I1 and M15 measurements (Figure 2). About 2500 collocated DCC pixels are identified during the study period, from 20 July to 20 August 2015. Figure 7 shows the histograms of the DCC reflectances. Each pair of AHI and VIIRS data exhibit very similar but shifted histogram shapes. While the DCC shows single mode and high reflectance at the bands with wavelength less than 1.0 µm (AHI B1-4 and VIIRS M3, M7, I1 and I2), the reflectance is relatively low with bi-model histogram patterns at 1.6 µm (AHI B5 and VIIRS M10 and I3) and 2.3 µm (AHI B6 and VIIRS M11). The VIIRS DCC data at all the short-wavelength bands display sharper-shaped histograms than the corresponding AHI data. This is because AHI has wider dynamic ranges than VIIRS at these visible bands with less quantization bits [8,26]. While the VIIRS data are truncated to 12 bits, AHI data are truncated to 11 bits. The narrower dynamic range and more quantization bits make VIIRS data more sensitive to the slight variations of DCC reflectance, which are believed to be statistically very stable at the visible wavelength.

Three types of DCC measurements are used to represent the DCC reflectance: median, mode and mean values of the DCC data. Therefore, in addition to the statistics of each DCC pair, there are four methods to calculate the reflectance ratios of DCC of these two instruments. As reported in

Table 3, the results of the four types of reflectance ratios agree very well with those derived from the Ray-matching method: the calibration differences between AHI and VIIRS are less than 5% for AHI B1-4 and B6, however, AHI B5 is brighter than VIIRS by about 6%–8%. Although DCC is not among the brightest Earth targets at the NIR bands, the collocated DCC method provides less uncertainty for the GEO-LEO inter-calibration than the Ray-matching method. For the visible wavelength bands (<1.0 μm), collocated DCC have very comparable inter-calibration uncertainty with the Ray-matching method. Due to the existence of bi-model reflectance histograms, the median or mean value of DCC reflectance may be considered for long-term trending of calibration accuracy for these two NIR bands. Further study is needed to understand the causes to the bi-mode pattern.

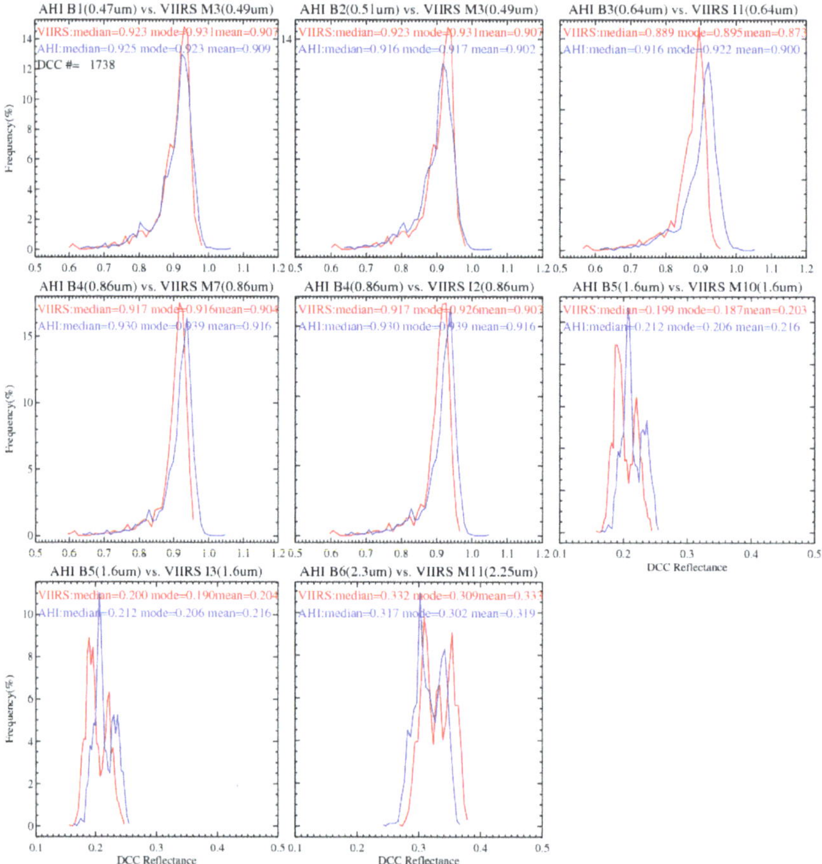

Figure 7. Histograms of collocated DCC reflectance for the data collected from 20 July to 20 August 2015, All the bands have the same DCC pixels. The y-axis is the DCC frequency (%) and the x-axis is the DCC reflectance.

5. Conclusions

Both AHI and VIIRS use a solar diffuser as a bright target to provide on-orbit calibration for the VNIR bands. In this study, two inter-calibration methods are used to compare the AHI VNIR radiometric calibration accuracy with the collocated VIIRS measurements for the six AHI VNIR bands. These two methods are based on spatially collocated pairs, which the two instruments view with the same viewing zenith angles within a 5 min difference. The Ray-matching method uses collocated homogeneous scenes with similar viewing azimuth angles, while the collocated DCC method uses the near-Lambertion effect of the DCC pixels. Therefore, the Ray-matching method can provide inter-calibration results for all sky scenes, while the collocated DCC method focuses on the calibration differences at median or high reflectance pixels. Viewing angle dependent calibration difference at homogenous median/high reflectance scenes is also examined with the Ray-matching method.

The Ray-matching method shows there is a strong linear relationship between AHI and VIIRS at all the paired bands. Results of both the Ray-matching and collocated DCC methods indicate that AHI radiance quality agrees well with that of VIIRS, within a 5% difference, except for AHI B5, which is brighter than VIIRS M10 and I3 by 6%–8%. Relatively larger reflectance ratio variation, which increases with wavelength, can be observed at low radiance scenes, probably associated with certain AHI instrument noises. No trending in the E–W viewing angular dependent calibration difference between these two instruments can be observed within the uncertainty of the Ray-matching method.

The good agreement of the inter-calibration results between the Ray-matching and collocated DCC methods for all the paired bands indicate that the DCC method can also be used for the calibration of NIR bands. Compared to the Ray-matching method, the collocated DCC method has less uncertainty on the inter-calibration results for the NIR bands. Further study may be needed to understand the bi-mode patterns of the NIR DCC reflectance histograms. Overall, the earlier launch of AHI provides a unique opportunity to develop, test and examine the ABI cal/val tools.

Acknowledgments: This work was funded by GOES-R program. We would like to thank Japan Meteorological Agency for sharing the Himawari AHI data and many valuable discussions with NOAA colleagues: Boryana Efremova and Changyong Cao on the VIIRS calibration, Aaron Pearlman on ABI pre-launch calibration, Pubu Ciren on sun glint effect, and Wenhui Wang on the VIIRS DCC reflectance. We also appreciate the time and valuable internal review comments from Bob Iacovazzi and Boryana Efremova. This manuscript content is solely the opinions of the authors and do not constitute a statement of policy, decision, or position on behalf of NOAA or the U.S. government. The authors thank anonymous reviewers for valuable comments that greatly helped in improving the quality of this manuscript.

Author Contributions: Fangfang Yu designed the study, collected the collocation data, developed the processing codes, performed the analyses, and wrote the manuscript. Xiangqian Wu contributed to the design of this study and provided in-depth discussions.

Conflicts of Interest: The authors declare no conflict of interest.

References

1. Takahashi, M.; Andou, A.; Date, K.; Hosaka, K.; Mori, N.; Murata, H.; Okuyama, A.; Tabata, T.; Yoshino, R.; Bessho, K. Himawari-8 post-launch radiometric calibration and navigation analysis. In Proceedings of the 2015 EUMETSAT Conference, Toulouse, France, 21–25 September 2015.
2. Cao, C. Overview and progress update on GOES-R calibration/validation. In Proceedings of the 2012 AMS 92nd Annual Meeting/8th Annual Symposium on Future Operational Environment Satellite System, New Orleans, LA, USA, 22–26 January 2012.
3. Pearlman, A.; Pogorzala, D.; Cao, C. GOES-R Advanced Baseline Imager: Spectral functions and radiometric biases with the NPP visible infrared imaging radiometer suite evaluated for desert calibration sites. *Appl. Opt.* **2013**, *31*, 7660–7668. [CrossRef] [PubMed]
4. Xiong, J.; Butler, J.; Chiang, K.; Efremova, B.; Fulbright, J.; Lei, N.; McIntire, J.; Oudrari, H.; Sun, J.; Wang, Z.; *et al.* VIIRS on-orbit calibration methodology and performance. *J. Geophys. Res.: Atmos.* **2013**, *118*, 5065–5078. [CrossRef]
5. Cao, C.; Xiong, X.; Blonski, S.; Liu, Q.; Uprety, S.; Shao, X.; Bai, Y.; Weng, F. Suomi NPP VIIRS sensor data record verification, validation, and long-term performance monitoring. *J. Geophys. Res.: Atmos.* **2013**, *118*, 1–15. [CrossRef]
6. Uprety, S.; Cao, C. Suomi NPP VIIRS reflective solar band on-orbit radiometric stability and accuracy assessment using desert and Antarctica Dome C sites. *Remote Sens. Environ.* **2015**, *166*, 106–115. [CrossRef]
7. Griffith, C. ABI's unique calibration and validation capabilities. In Proceedings of the 2015 EUMETSAT Conference, Toulouse, France, 21–25 September 2015.

8. NOAA/STAR VIIRS SDR Team and the science team members. Joint Polar Satellite System (JPSS) Visible Infrared Imaging Radiometer Suite (VIIRS) Sensor Data Records (SDR) Radiometric Calibration Algorithm Theoretical Basis Document (ATBD) 2013. Available online: http://www.star.nesdis.noaa.gov/smcd/spb/nsun/snpp/VIIRS/ATBD-VIIRS-RadiometricCal_20131212.pdf (accessed on 19 November 2015).
9. McIntire, J.; Moyer, D.; Efremova, B.; Oudrari, H.; Xiong, X. On-orbit characterization of S-NPP VIIRS transmission functions. *IEEE Trans. Geosci. Remote Sens.* **2015**, *53*. [CrossRef]
10. Bessho, K.; Date, K.; Hayashi, M.; Ikeda, A.; Imai, T.; Inoue, H.; Kumagai, Y.; Miyakawa, T.; Murata, H.; Ohno, T.; *et al.* An introduction to Himawari-8/9—Japan's new-generation geostationary meteorological satellites. *J. Meteorol. Soc. Jpn.* **2016**. [CrossRef]
11. Japan Meteorological Agency (JMA) Himawari-8 Advanced Himawari Imaer (AHI) Webpage. Available online: http://www.data.jma.go.jp/mscweb/en/himawari89/space_segment/spsg_ahi.html (accessed on 19 November 2015).
12. Cao, C.; de Luccia, F.; Xiong, X.; Wolfe, R.; Weng, F. Early on-orbit performance of the visible infrared imaging radiometer suite onboard the Suomi National Polar-Orbiting Partnership (S-NPP) satellite. *IEEE Trans. Geosci. Remote Sens.* **2014**, *52*, 1142–1156. [CrossRef]
13. Wolfe, R.; Lin, G.; Nishihama, M.; Tewari, K.; Tilton, J.; Isaacman, A. Suomi NPP VIIRS prelaunch and on-orbit geometric calibration and characterization. *J. Geophys. Res.* **2013**, *118*, 11508–11521.
14. Wu, X.; Hewison, T.; Tahara, Y. GSICS GEO-LEO inter-calibration: Baseline algorithm and early results. *Proc. SPIE* **2009**, *7456*. [CrossRef]
15. Hewison, T.; Wu., X.; Yu, F.; Tahara, Y.; Hu, X.; Kim, D.; Koenig, M. GSICS inter-calibration of infrared channels of geostationary imagers using Metop/IASI. *IEEE Trans. Geosci. Remote Sens.* **2013**, *51*, 1160–1170. [CrossRef]
16. Gardashov, R.; Eminov, M. Determination of sunglint location and its characteristics on observation from a METEOSAT 9 satellite. *Int. J. Remote Sens.* **2015**, *36*, 2584–2598. [CrossRef]
17. Yu, F.; Wu, X.; Varma Raja, M.; Li, Y.; Wang, L.; Goldberg, M. Diurnal and scan angle variations in the calibration of GOES Imager infrared channels. *IEEE Trans. Geosci. Remote Sens.* **2013**, *51*, 671–683. [CrossRef]
18. Doelling, D.; Minnis, P.; Nguyen, L. Calibration comparison between SEVERI, MODIS and GOES data. In Proceedings of the 2004 MSG RAO Workshop, Salzburg, Austria, 6–10 September 2004.
19. Sohn, B.-J.; Ham, S.; Yang, P. Possibility of the visible-channel calibration using deep convective clouds overshooting the TTL. *J. Appl. Meteorol. Climatol.* **2009**, *48*, 2271–2283.
20. Doelling, D.; Nguyen, L.; Minnis, P. On the use of deep convective clouds to calibrate AVHRR data. *SPIE Proc.* **2004**, *5542*. [CrossRef]
21. Doelling, D.; Morstad, D.; Scarino, B.; Bhatt, R.; Gopalan, A. The characterization of deep convective clouds as an invariant calibration target and as a visible calibration technique. *IEEE Trans. Geosci. Remote Sens.* **2013**, *51*, 1147–1159. [CrossRef]
22. Yu, F.; Wu, X. An integrated method to improve the GOES Imager visible radiometric calibration accuracy. *Remote Sens. Environ.* **2015**, *164*, 103–113. [CrossRef]
23. Beuchler, D.; Koshak, W.; Christian, H.; Goodman, S. Assessing the performance of the Lightning Imaging Sensor (LIS) using deep convective clouds. *Atmos. Res.* **2014**, *135–136*, 397–403. [CrossRef]
24. Hu, Y.; Wielicki, B.; Yang, P.; Stackhouse, P.; Lin, B.; Young, D. Application of deep convective cloud albedo observation to satellite-based study of terrestrial atmosphere: Monitoring the stability of spaceborn measurement and assessing absorption anomaly. *IEEE Trans. Geosci. Remote Sens.* **2004**, *42*, 2594–2599.
25. Scarino, B.; Doelling, D.; Minnis, P.; Gopalan, A.; Chee, T.; Bhatt, R.; Lukashin, C.; Haney, C. A web-based tool for calculating spectral band difference adjustment factors derived from SCIAMACHY hyper-spectral data. *IEEE Trans. Geosci. Remote Sens.* **2016**. [CrossRef]
26. NOAA/NESDIS/STAR, Algorithm Theoretical Basis Document, ABI Aerosol Detection Product, May 2010. Available online: http://www.star.nesdis.noaa.gov/goesr/docs/ATBD/ADP.pdf (accessed on 19 November 2015).
27. Japan Meteorological Agency. AHI-8 Performance Test Results, 2015. Available online: http://www.data.jma.go.jp/mscweb/en/himawari89/space_segment/doc/AHI8_performance_test_en.pdf (accessed on 19 November 2015).
28. Okuyama, A.; (Japan Meterological Agency, Tokoyo, Japan). Personal communication, 2014.

29. Pogorzala, D.; Padula, F.; Cao, C.; Wu, X. The GOES-R ABI fixed format: Overview and case sudy. Available online: http://adsabs.harvard.edu/abs/2014AGUFMIN43C3702P (accessed on 19 November 2015).
30. Zuidema, P.; Davies, R.; Moroney, C. On the angular radiance closure of tropical cumulus cognestus clouds observed by the Multiangle Imaging Spectroradiometr. *J. Geophys. Res.* **2003**, *108*, 4626–4637. [CrossRef]

Article

Fast and Accurate Collocation of the Visible Infrared Imaging Radiometer Suite Measurements with Cross-Track Infrared Sounder

Likun Wang [1,*], Denis Tremblay [2], Bin Zhang [3] and Yong Han [4]

1 Earth System Science Interdisciplinary Center, University of Maryland, 5825 University Research Court, Suite 4001, College Park, MD 20740, USA
2 Science Data Processing Inc., Laurel, MD 20723, USA; Denis.Tremblay@noaa.gov
3 Earth Resource Technology Inc., 14401 Sweitzer Lane, Laurel, MD 20707, USA; Bin.Zhang@noaa.gov
4 Center for Satellite Applications and Research, NESDIS/NOAA, 5830 University Research Court, College Park, MD 20740, USA; Yong.Han@noaa.gov
* Correspondence: wlikun@umd.edu; Tel.: +1-301-683-3551; Fax: +1-301-405-8468

Academic Editors: Changyong Cao, Dongdong Wang and Prasad S. Thenkabail
Received: 29 November 2015; Accepted: 18 January 2016; Published: 21 January 2016

Abstract: Given the fact that Cross-track Infrared Sounder (CrIS) and the Visible Infrared Imaging Radiometer Suite (VIIRS) are currently onboard the Suomi National Polar-orbiting Partnership (Suomi NPP) satellite and will continue to be carried on the same platform as future Joint Polar Satellite System (JPSS) satellites for the next decade, it is desirable to develop a fast and accurate collocation scheme to collocate VIIRS products and measurements with CrIS for applications that rely on combining measurements from two sensors such as inter-calibration, geolocation assessment, and cloud detection. In this study, an accurate and fast collocation method to collocate VIIRS measurements within CrIS instantaneous field of view (IFOV) directly based on line-of-sight (LOS) pointing vectors is developed and discussed in detail. We demonstrate that this method is not only accurate and precise from a mathematical perspective, but also easy to implement computationally. More importantly, with optimization, this method is very fast and efficient and thus can meet operational requirements. Finally, this collocation method can be extended to a wide variety of sensors on different satellite platforms.

Keywords: VIIRS; CrIS; inter-calibration; collocation; cloud detection

---

1. Introduction

The Suomi National Polar-orbiting Partnership (Suomi NPP) satellite, successfully launched in October 2011, is a weather satellite to serve as a gap-filler between NOAA's heritage Polar Operational Environmental Satellites (POES) and the new generation Joint Polar Satellite Systems (JPSS). Five key instruments are carried on Suomi NPP, that is, the Advanced Technology Microwave Sounder (ATMS), the Cross-track Infrared Sounder (CrIS), the Ozone Mapping and Profiler Suite (OMPS), the Visible Infrared Imaging Radiometer Suite (VIIRS), and Clouds and the Earth's Radiant Energy System (CERES). Among them, VIIRS is a whiskbroom scanning imaging radiometer, collecting visible and infrared imagery of the Earth through 22 spectral bands between 0.412 μm and 12.01 μm with a resolution of 375 m or 750 m at nadir [1]. CrIS is a Michelson interferometer with 1305 spectral channels over three wavelength ranges: long-wave infrared (LWIR) (9.14–15.38 μm), middle-wave IR (MWIR) (5.71–8.26 μm), and short-wave IR (SWIR) (3.92–4.64 μm) [2]. In contrast to a state-of-the-art high-spatial-resolution imager instrument of VIIRS, the sounder instrument CrIS provides information on the vertical profiles of temperature, water vapor, and critical trace gases of the atmosphere, albeit with coarse spatial resolution (14.0 km at nadir). The combination of high spatial resolution measurements from an imager and high spectral resolution measurements from an infrared (IR) sounder can take advantage of both spectral and spatial capabilities; hence, it can further improve atmospheric and surface geophysical parameter retrievals [3,4] and data utilization for numerical weather prediction models [5]. Furthermore, owing to its hyperspectral nature and accurate radiometric and spectral calibration, radiance spectra from the IR sounder can be integrated through the spectral response functions (SRF) to simulate imager radiance measurements and thus independently assess spectral and radiometric calibration accuracy of IR channels of the imager [6–10]. Finally, recent study demonstrated that, by taking advantage of high spatial resolution and accurate geolocation of VIIRS measurements, spatially collocated measurements from the VIIRS image bands can effectively evaluate the geolocation accuracy of CrIS that has a coarse spatial resolution [11]. All these applications are dependent on accurate and fast collocation of sounder measurements with imager measurements.

Collocation of the measurements from two satellite sensors (either on the same satellite platform or not) involves pairing measurements from two sensors that observe the same location on the Earth but with different spatial resolutions. In other words, it involves finding overlapped measurements from two sensors. After this step, the observational time and atmospheric path (e.g., satellite zenith and azimuth angle differences) of paired and collocated measurements can be further checked by constraining the observational time and view geometry differences dependent on applications. For the two sensors on the same platform, time and angle difference is not an issue. For two sensors on different platforms, once the first step is accomplished, it is relatively easy to further filter the data by choosing suitable thresholds of time and angle differences. For both cases, spatial collocation is the key step. Therefore, the collocation in this study is mainly referred to as spatial collocation that associates overlapped measurements from two sensors.

To the best of our knowledge, not many studies have been published on spatial collocation methods in existing literature. A traditional but simple method is to check ground pixel distance to find overlapped measurements [12,13]. This method cannot precisely and accurately deal with off-nadir field of view (FOV) distortion because the FOV footprint projected on the ground becomes egg-shaped, or oviform as the scan moves away from nadir. As a pioneer study, Aoki (1980) [14,15] describes how to match AVHRR and HIRS/2 on the same satellite. The first peer-reviewed publication on spatial collocation is the work by Nagle and Holz (2009) [16], which provides a guidance for a general methodology that can be applied to a wide range of satellite, aircraft, and surface measurements and allow for efficient collocation with measurements having varying spatial and temporal sampling. Specifically, two methods are discussed for spatial collocation (referred to as "overlap detection" in the paper), including (1) the quasi-elliptical approach and (2) the quasi-conical approach. The quasi-elliptical approach basically assumes that the projected footprints on the Earth Surface from the sounder's FOVs are approximately elliptical. A coordinate system based on the major and minor axes of the quasi ellipse are created on the terrestrial surface. The collocation task is then simplified to determine whether a given point on the Earth's surface falls within the quasi ellipse by examining the distance to the major and minor axes. By contrast, the quasi-conical approach avoids any exercise in analytic geometry and is much more straightforward. Basically, any sounder instrument views the underlying Earth as if through a cone whose angular opening is determined by the size of the FOV angle. Then, the imager observation viewed within the solid angle of the cone overlaps the sounder FOV on the ground. Compared to the quasi-elliptical approach method, one need not be concerned with the size, shape, or orientation of the FOV projected on the surface. In both studies by Aoki (1980) [14] and Nagle and Holz (2009) [16], the quasi-elliptical approach has been described in detail and thus recommended for use, while the second method—the conical approach—is not suggested by Nagle and Holz (2009) because of its computational inaccuracy for small angles [16]. However, for hyperspectral IR instruments like CrIS, which contains 3×3 detectors in one field of regards (FOR), the corresponding FOV footprints are rotated along with scan positions due to the 45°-mounted scan mirror, resulting in complicated geometric calculations. In addition, the quasi-elliptical approximation inevitably causes uncertainties owing to a perceptible curvature of the Earth when the scan angle increases. For applications that require accurate spatial collocation (e.g., geolocation assessment), the first approach is inadequate.

In this study, we argue that the conical approach is accurate and precise in essence and much easier to apply in computation algorithms. With appropriate optimization, this method is not only fast and efficient but also can meet accuracy requirements. Finally, given the fact that CrIS and VIIRS will continue to be onboard the same platform of future Joint Polar Satellite System (JPSS) satellites for the next decade, it is desirable to discuss the implementation details of spatial collocation between CrIS and VIIRS to assist applications that rely on the combination of measurements from two sensors. More importantly, as we demonstrate later, without loss of generality, this collocation method can also be applied to a wide variety of sensors on different satellites.

The paper is organized as follows: Section 2 summarizes instruments of characteristics, Section 3 describes the methodology, Section 4 presents the results and applications and Section 5 concludes the paper.

2. Instrument Characteristics

Both CrIS and VIIRS are onboard Suomi NPP spacecraft at a nominal altitude of ~829 km in a Sun-synchronous orbit with local equatorial crossing times of ~13:30 (ascending) and ~01:30 A.M (descending). As a step-scan Fourier transform spectrometer, CrIS takes 8 s for each scan sweep, each collecting 34 Fields of Regard (FORs). Among them, 30 are the Earth scenes and four are the embedded space and blackbody calibration views. The scan mirror stepwise "stares" at the Earth step by step in the cross-track direction from $-48.3°$ to +48.3 with a 3.3° step angle, equaling a 2200 km swath width on the Earth. Nine field stops define the 3×3 detector array for each IR wavelength band, which are arrayed as 3×3 0.963° circles and separated by 1.1°. CrIS radiance spectrum (without apodization) covers three IR bands from 650 to 1095 cm^{-1}, 1210 to 1750 cm^{-1}, and 2155 to 2550 cm^{-1} with spectral resolutions of 0.625 cm^{-1}, 1.25 cm^{-1}, and 2.5 cm^{-1} at the normal operational mode (a total of 1305 spectral channels). After apodization by using a Hamming function, the effective spectral resolution decreases by a factor of 1.82.

The VIIRS instrument is a whiskbroom scanning radiometer with a field of regard of ±56.3° in the cross-track direction. The swath width is 3060 km. VIIRS has 22 spectral bands covering the spectrum between 0.412 μm and 12.01 μm, including 16 moderate resolution bands (M-bands) with a spatial resolution of 750 m at nadir, 5 imaging resolution bands (I-bands)–375 m at nadir, and 1 panchromatic Day-Night-Band (DNB) with a 750 m spatial resolution throughout the scan. The M-bands include 11 Reflective Solar Bands (RSB) and 5 Thermal Emissive Bands (TEBs). The I-bands include three RSBs and two TEBs. VIIRS uses a unique approach of pixel aggregation which controls the pixel growth towards the end of the scan. As a result, the VIIRS spatial resolutions for nadir and edge-of-scan data are comparable. For example, for typical I-bands, the resolution changes from 375 m at nadir to ~800 m at the end of a scan. On the other hand, in order to save transmission bandwidth, VIIRS also uses a "bow-tie removal" approach that removes duplicated pixels in the off-nadir areas where there is an overlap of several pixels between adjacent scans [1].

There are four VIIRS IR channels that are fully covered by CrIS spectra at the longwave IR (LWIR) and shortwave IR (SWIR) bands as shown in Figure 1, including three M-bands and one I-band, *i.e.*, M13 (4.05 μm) , M15 (10.8 μm) , M16 (12.0 μm), and I5 (11.5 μm). In this study, we only use high spatial resolution VIIRS I5 data. However, the method proposed in this study can also be applied to other bands and VIIRS product. In order to compare CrIS hyperspectral radiances with VIIRS band radiances, we need to perform spectral convolution to reduce the high resolution CrIS spectrum to match the band radiances from VIIRS. Specifically, given the CrIS hyperspectral radiance R at each wavenumber, it can be convolved with the VIIRS SRF S to generate the CrIS-convolved VIIRS band radiance L as

$$L = \frac{\int_{\nu_1}^{\nu_2} R(\nu)S(v)d\nu}{\int_{\nu_1}^{\nu_2} S(\nu)d\nu} \quad (1)$$

where v_1 and v_2 are band pass limits. Henceforward, except for those specially noted below, the CrIS data in this study are referred to as the simulated VIIRS band radiances for the I5 band.

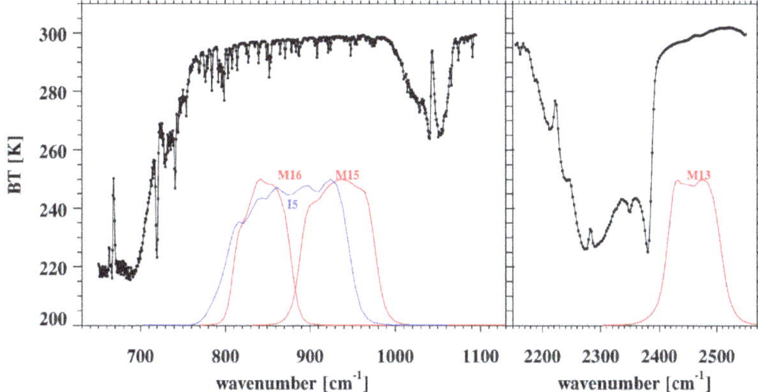

Figure 1. CrIS spectra from the LWIR and SWIR bands (bands 1 and 3) and the VIIRS spectral response functions of I5, M13, M15, and M16 bands.

Shown in Figure 2 are examples of CrIS and VIIRS I5 images from 1024UTC to 1032UTC on 5 September 2015 when the SNPP satellites passed over the Red Sea region. The scan characteristics between the imager and sounder are clearly revealed. Basically, with high spatial resolution and continuous scan (6400 pixel per each scan line), VIIRS has the advantage to resolve the details of clouds and surface features. Given the relatively large footprint and "step-and-stare" scan mechanism (30 FORs per each scan and 9 FOVs in each FOR), there are spatial gaps among the FOVs. In addition, due to the different maximum scan angles of CrIS and VIIRS, that is 48.3° *versus* 58.3°, the VIIRS swath is larger than CrIS. As a result, a part of the Earth surface cannot be detected by CrIS (but can be detected by VIIRS). Figure 3 gives the enlarged four scans of CrIS FOV footprints projected on the Earth overlapped with VIIRS I5 image, which are computed using the method by Wang *et al.* [11]. These spatial gaps are clearly shown. In addition, the FOV footprints close to nadir are projected as a circle on the Earth and steadily changed to an ellipse. Specifically, the center FOV (FOV 5) is changed from a 14.0 km circle at nadir into an ellipse with major and minor axes of 43.6 km and 23.2 km at the end of the scan. Finally, the FOV footprints are also rotated with FORs because of the 45-degree mounted scan mirror. Based on Figure 3, the key of collocation of CrIS and VIIRS images is to find the VIIRS pixels within each large CrIS FOVs footprint.

When we compute the FOV footprint, the CrIS detector size is treated as a 0.963° circle. This value is from CrIS engineering packets and also used for CrIS spectral calibration [17]. However, in reality, the CrIS detectors have their own spatial response functions, namely, the spatial distribution of the contributions to the total radiance. The detector spatial response was measured during prelaunch testing. The response function is approximately a Gaussian distribution and is normalized by the peak value. The FOV size of 0.963° actually corresponds to ~41% of the peak response but already collects ~98% of total radiation falling on the detector. In other words, for a typical sounder instrument like CrIS, its spatial response function quickly reaches its peak and then becomes flat. Theoretically, when spatially averaging collocated VIIRS pixels, the spatial response function should be used as weights [4]. However, the sensitivity test indicated that it is accurate enough to assume spatial response function as a box shape with 0.963° as a cut-off value and the BT differences caused by this assumption are ~0.002 K.

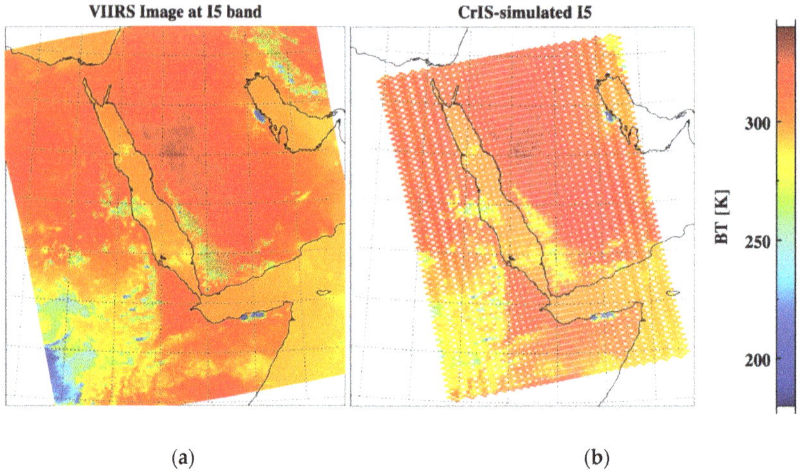

Figure 2. Examples of VIIRS I5 (**a**) and CrIS (**b**) images from 1024UTC to 1032UTC on 5 September 2015. The CrIS spectra have been convolved with VIIRS spectral response function to match VIIRS I5 band radiances.

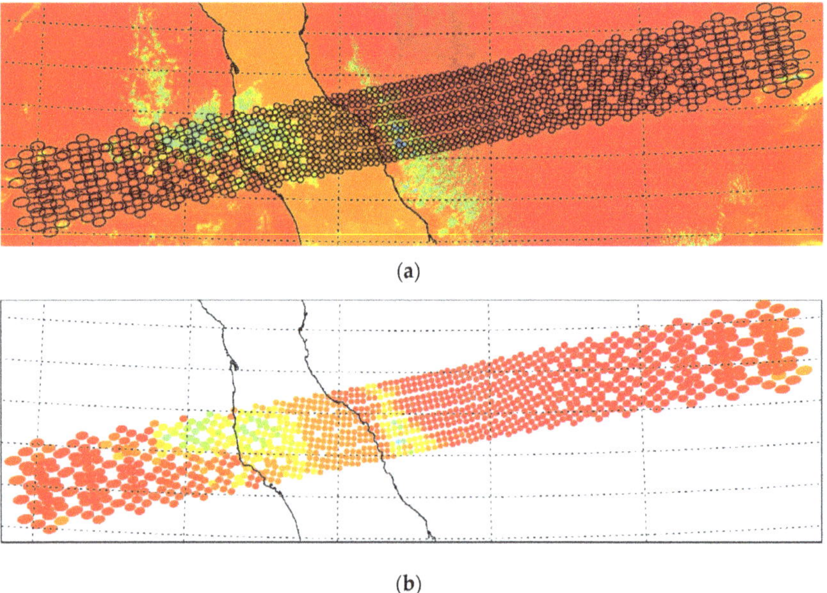

Figure 3. Enlarged plots four CrIS scans in Figure 2, including (**a**) projected CrIS FOV footprints overlapped with VIIRS image and (**b**) CrIS FOV images.

3. Method

The uniqueness of this study is to collocate the VIIRS and CrIS measurements based on VIIRS and CrIS Line-of-Sight (LOS) pointing vectors, defined as the vector from the satellite position to the Earth surface pixel location. As a first step, CrIS and VIIRS geolocation datasets, which contain latitude, longitude, satellite range, satellite azimuth and zenith angles of CrIS and VIIRS measurements, are used to compute the LOS Pointing Vectors. In the following step, VIIRS and CrIS LOS vectors are matched by examining the angle between them. The corresponding indexes of VIIRS pixels are retrieved through the matched VIIRS LOS vectors. The details of the method as well as the optimization and implementation of the algorithm are described in this section.

3.1. Computation of Line-of-Sight (LOS) Pointing Vector

3.1.1. Coordinate Systems

Table 1 and Figure 4 summarize the coordinate systems used in this study. Geolocation datasets of CrIS and VIIRS contain geodetic latitude and longitude of each satellite measurement, characterizing the intersection location of the LOS pointing vector with the Earth. In a typical geographic coordinate system, geodetic latitude and longitude usually represent horizontal position. The third variable represents the vertical position above the Earth's ellipsoid. Hence, this coordinate system can be simply called the geodetic latitude, longitude, and altitude coordinate system (LLA). The reference Earth ellipsoid can be characterized by the so-called geodetic datum. Both CrIS and VIIRS geolocation algorithms use the World Geodetic System 1984 (WGS84) as a geodetic reference. Finally, we should point out that there are actually two geolocation datasets for VIIRS SDR datasets, that is, the one with terrain correction and the other without terrain correction. The geolocation data without terrain correction are used in order to be consistent with CrIS because CrIS only has a geolocation dataset on the Earth ellipsoid without terrain correction.

Table 1. Summary of coordinate systems used in this study.

Coordinates	Type	Origin	Variables
Local Spherical Coordinate	Spherical	Measurement location	(R, Θ, Φ) R: Range (meter) Θ: Zenith Angle (degree) Φ: Azimuth Angle (degree)
Local East, North, Up (ENU) Coordinate	Cartesian	Measurement Location	(East, North, Up) in meter
Geodetic Latitude, Longitude, and Altitude (LLA) Coordinate	Spherical	Earth Center	(ψ, λ, h) ψ: Geodetic Latitude (degree) λ: Longitude (degree) h: Altitude (meter)
Earth-centered, earth-fixed (ECEF) Coordinate	Cartesian	Earth Center	(X, Y, Z) in meter

Given the location of satellite measurements, the satellite position at a given location can be described in a local spherical coordinate system. Specifically, three variables—including satellite azimuth and zenith angle as well as satellite range (the distance from the pixel location to satellite position)—can accurately determine the satellite position at a given point on the Earth. All of these values are contained in CrIS and VIIRS geolocation datasets. Correspondingly, the local East, North, Up (ENU) Cartesian coordinate system (ENU) is formed from a plane tangent to the Earth's surface fixed to the same specific location, and the east axis is labeled E, the north N, and the up U, by convention. It consists of three numbers: one represents the position along the northern axis, one along the eastern axis, and one represents the vertical position pointing up from a local tangent plane. This ENU coordinate system is far more intuitive and practical when computing the LOS pointing vector, which is simplified as the inverse satellite position vector—the vector that points from the satellite position to the origin (0, 0, 0) in ENU.

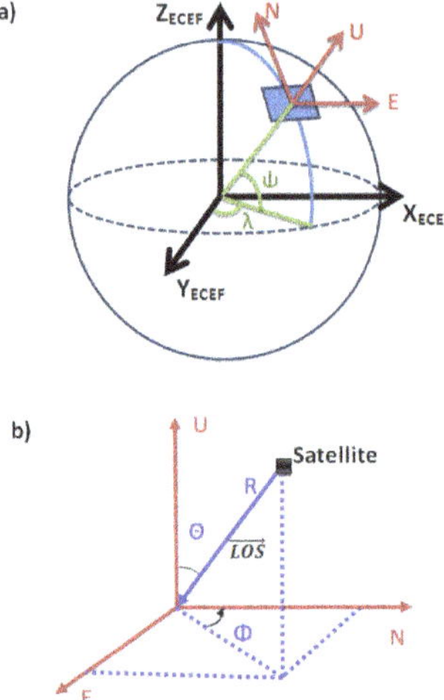

Figure 4. Illustration of the coordinate systems listed in Table 1, including (**a**) ENU (red color), LLA (green color), ECEF (black color) coordinate systems; and (**b**) local spherical coordinate (blue color).

For the local ENU coordinate, however, the origin varies with different satellite measurement locations. It is desirable to choose a common coordinate to show the LOS vector at different locations. Hence, the Earth-Centered, Earth-Fixed (ECEF) coordinate, also known as the Earth Centered Rotational (ECR) coordinate, can be utilized to facilitate the computation. It represents positions as X, Y, and Z coordinates. The point (0, 0, 0) is defined as the center of mass of the Earth, hence the name Earth-Centered. Its axes are aligned with the International Reference Pole (IRP) and International Reference Meridian (IRM) that are fixed with respect to the surface of the Earth. The third axis is formed by the cross product of another two axes. Shown in Figure 5a are three vectors. The satellite position vector **P** is defined as the one pointing towards the satellite from the Earth Center—the ECEF origin. The satellite measurement location vector **G** is the vector pointing towards the location of satellite measurements from the Earth Center. The satellite LOS vector **LOS** is the one pointing towards the satellite measurement location on the Earth Ellipsoid surface from the satellite position. Mathematically, a relationship among these three vectors can be shown as,

$$P = G - LOS \tag{2}$$

Henceforth, except for when the vectors are explicitly denoted in a specific coordinate, they are referred to as those expressed in ECEF. Basically, the collocation CrIS and VIIRS measurements are carried out in ECEF by matching CrIS and VIIRS **LOS** vectors.

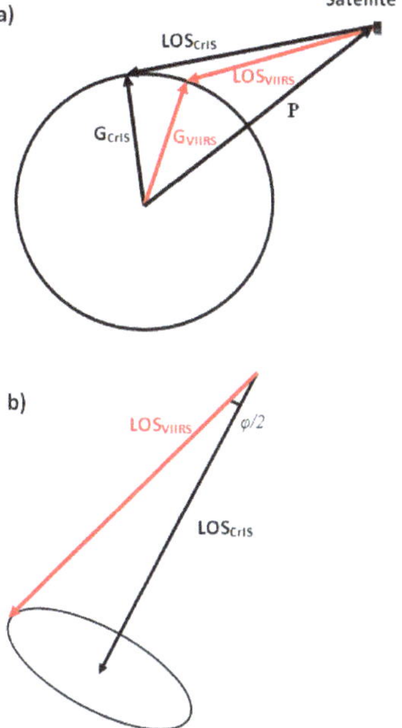

Figure 5. Schematic diagrams showing how to collocate VIIRS pixels CrIS FOV through VIIRS and CrIS LOS pointing vector, including (**a**) computation of the VIIRS and CrIS LOS vectors in ECEF and (**b**) examination of the angle between CrIS and VIIRS LOS vectors. Note that φ is the CrIS detector FOV angle of 0.963°.

3.1.2. Computation of LOS *Vector*

First, given the satellite range, zenith angle, and azimuth angles as (R, Θ, Φ), the $\mathbf{LOS_{ENU}}$ vector in a ENU coordinate can be determined as (E, N, U) as,

$$\boldsymbol{LOS_{ENU}} = \begin{pmatrix} E \\ N \\ U \end{pmatrix} = \begin{pmatrix} R\ \sin\Theta\sin\Phi \\ R\ \sin\Theta\cos\Phi \\ R\ \cos\Theta \end{pmatrix} \tag{3}$$

Second, any vector in a local ENU coordinate can be converted into ECEF through two rotations if we know the geodetic latitude (ψ) and longitude (λ) where the local ENU coordinate is formed, including Equation (1) a clockwise rotation over east-axis by an angle 90-ψ to align the up-axis with the Z-axis and then Equation (2) a clockwise rotation over the Z-axis by and angle 90+λ to align the east-axis with the X-axis. Consequently, these two rotations can be combined to form the following equation as,

$$\boldsymbol{LOS} = \begin{pmatrix} LOS_X \\ LOS_Y \\ LOS_Z \end{pmatrix} = \begin{pmatrix} -\sin\lambda & -\cos\lambda\ \sin\psi & \cos\lambda\cos\psi \\ \cos\lambda & \sin\lambda\sin\psi & \sin\lambda\cos\psi \\ 0 & \cos\psi & \sin\psi \end{pmatrix} \begin{pmatrix} E \\ N \\ U \end{pmatrix} \tag{4}$$

Finally, the satellite measurement location vector **G** can be computed in ECEF using geodetic latitude (ψ) and longitude (λ) of the satellite measurement location on the Earth ellipsoid

$$\mathbf{G} = \begin{pmatrix} G_X \\ G_Y \\ G_Z \end{pmatrix} = \begin{pmatrix} [N(\psi) + h]\cos\psi\cos\lambda \\ [N(\psi) + h]\cos\psi\sin\lambda \\ \left[\frac{b^2}{a^2}N(\psi) + h\right]\sin\psi \end{pmatrix} \quad (5)$$

whereh h is height above the Earth ellipsoid, N(ψ) is Prime Vertical radius of Curvature at a given location that can be computed using geodetic latitude (ψ) as an input, and a and b are semi-major and semi-minor axes for the defined the Earth ellipsoid. For any location on the Earth ellipsoid surface, h simplifies to zero.

Based on the above two steps, the $\boldsymbol{LOS}$ and $\boldsymbol{G}$ vectors for CrIS and VIIRS are calculated in ECEF respectively, symbolized as $\boldsymbol{LOS_{VIIRS}}$, $\boldsymbol{LOS_{CrIS}}$, $\boldsymbol{G_{VIIRS}}$, and $\boldsymbol{G_{CrIS}}$. Using Equation (2), the satellite position vector $\boldsymbol{P_{VIIRS}}$ and $\boldsymbol{P_{CrIS}}$ can be further derived. On the other hand, the satellite position vectors in ECEF in the mid scan are saved in the CrIS and VIIRS geolocation dataset when the geolocation algorithm outputs the latitude, longitude, and other geolocation information. Inter-comparison of these two satellite position vectors—the one derived using the above method and the other saved in datasets—can indirectly evaluate the accuracy of the computation method. Shown in Figure 6 is the magnitude of these two satellite position vectors varying with time. Basically, it clearly shows that they are aligned each other over time. The bottom panel gives the magnitude differences of these two satellite position vectors, which are less than 4.0 m (compared to ~7200 km). It confirms that the above method is very accurate and the uncertainties are at a negligible level.

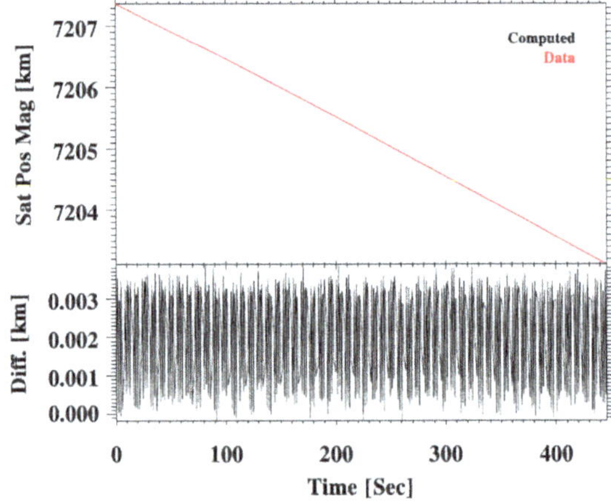

Figure 6. (**Top**) Magnitudes from satellite position vectors in ECEF contained in CrIS geolocation datasets (red color) and derived from Equation (2) (black) as well as their differences (**bottom**).

3.2. Collocation of VIIRS and CrIS LOS Vectors

Once the CrIS and VIIRS LOS vectors are derived, the collocation of VIIRS and CrIS can be simplified as examining the angles between two vectors. As shown in Figure 5b, the circular CrIS detector "sees" the Earth through a solid angle like a cone, whose angular opening is determined by the size of the FOV angle (that is 0.963° for CrIS). Coming from the same satellite position, any VIIRS vectors falling in the cone will overlap with CrIS measurements on the ground. Therefore, we need not be concerned with the size and shape of FOVs on the Earth surface. Generally, the dot product of two Euclidean vectors **A** and **B** is defined by $\boldsymbol{A} \cdot \boldsymbol{B} = ||\boldsymbol{A}||\,||\boldsymbol{B}||\,\boldsymbol{\cos\theta}$, where θ is the angle of the two vector and $||A||$ means the length of the vector **A**. Specifically, the criterion for collocation—which determines whether the angle of two vectors is less than a threshold—can be expressed as

$$\frac{(G_{CrIS} - P_{CrIS}) \cdot ((G_{VIIRS} - P_{CrIS})}{||G_{CrIS} - P_{CrIS}|| \, ||G_{VIIRS} - P_{CrIS}||} > \cos\left(\frac{1}{2}\varphi\right) \quad (6)$$

where $\mathbf{G_{CrIS}}$-$\mathbf{P_{CrIS}}$ equals the CrIS LOS vector, $\mathbf{G_{VIIRS}}$-$\mathbf{P_{CrIS}}$ represents the VIIRS LOS vector, and $\phi$is the FOV size angle of 0.963°. In Equation (6), the numerator is the dot product of the two vectors, while the denominator is the product of the two vectors' lengths. The VIIRS LOS vector has to be re-computed using the same satellite position $\mathbf{P_{CrIS}}$ from CrIS instead of $\mathbf{P_{VIIRS}}$ from VIIRS in order to ensure the two vectors of $\mathbf{G_{VIIRS}}$-$\mathbf{P_{CrIS}}$ and $\mathbf{G_{CrIS}}$-$\mathbf{P_{CrIS}}$ originate from the same position. Since there is still a slight time difference between CrIS and VIIRS measurements, the satellite positions for CrIS and VIIRS are not exactly the same. Furthermore, Equation (6) also avoids using an arccosine routine because it needs high accuracy for a small angle between two nearly parallel vectors.

Equation (6) seems straightforward mathematically and is simple to implement computationally. However, the situation is complicated by the fact that one has to go through time-consuming search loops. For examples, in the granule shown in Figure 2, there are 9 × 30 × 56 CrIS FOVs and 6400 × 9216 VIIRS pixels. For a specific CrIS LOS vector corresponding to a CrIS measurement, one has to check whether the angles of this CrIS LOS vector satisfy the requirement stated in Equation (6) with all 6400 × 9216 VIIRS LOS vectors. It turns out there are a total of 9 × 30 × 56 × 6400 × 9216 loops. In computer sciences, it is called brute-force search, which is a very general problem-solving technique that consists of systematically enumerating all possible candidates for the solution and checking whether each candidate satisfies the problem's statement. While a brute-force search is simple to implement, and will always find a solution if it exists, its cost is proportional to the number of candidate solutions in terms of computing time or memory space.

One way to speed up the brute-force search is to reduce the search space, *i.e.*, the set of searched VIIRS pixels. For example, if the search is only limited to near neighbor VIIRS pixels that are close to a given CrIS measurement, the number of loops will be greatly reduced. The main idea is to pre-process the data set, and selectively obtain a set of pivot pixels. The brute-force search is only applied to these pixels to check if they meet the search requirements. To achieve this, the closest VIIRS pixel that matches a given CrIS FOVs location must be first found as a pivot point to define the search area, which may not be necessarily accurate but has to be close enough to the given CrIS FOV. Once this pixel was found, a rectangle area can be defined composed of 201 × 201 VIIRS pixels centered at this VIIRS pixel position. Consequently, the search area is reduced to 201 × 201 VIIRS pixels by trimming out the vast majority of possible search area from original 6400 × 9216 VIIRS pixels.

Finding the closest matched VIIRS pixels for a specific CrIS measurement in the latitude and longitude coordinate system becomes a nearest-neighbor search (NNS) problem. There are various solutions to the NNS problem that have been proposed in computer science. The quality and usefulness of the algorithms are determined by the time complexity of queries as well as the space complexity of any search data structures that must be maintained. For our purpose, the match_2d function [18] is used to find the closest coordinate match between CrIS and VIIRS based on latitude and longitude datasets within a certain search radius. Specifically, a histogram is created to bin the latitude and longitude of VIIRS pixels. The function then searches for ones which have fallen in that bin as well as the relevant three adjacent bins (depending on location within the bin), and finally finds the closest match position. Through the above two considerations, the performance of the collocation algorithm is dramatically improved. For example, on a linux machine with 16 CPUs of 2400.198 MHz, the collocation algorithms for a typical CrIS granule (composed by 9 × 30 × 4 measurements)—which also includes input and output (IO) time—only takes ~8 s.

4. Results and Application

4.1. Collocation Results

The collocation method is applied to the CrIS and VIIRS images shown in Figure 3. The final collocation results are given in Figure 7a, where only VIIRS pixels falling within CrIS FOVs are shown. By way of illustration, Figure 7b shows enlarged collocation plots of FOR1 and FOR15 that have scan angles of −48.3° and −1.65°, respectively. Using the center FOV (FOV5) as an example, it collects 1068 and 3946 VIIRS pixels at scan angles of −1.65°and −48.3° and resistively (the blow-tie deleted pixels are not counted). Notice that as the CrIS scan angle increases, the projected CrIS FOVs become increasingly more quasi-elliptical. More importantly, the nine FOVs in given FORs at large scan angles (e.g., FOR 1 and 30) are not exactly identical but also slightly vary with position relative to the FOR center.

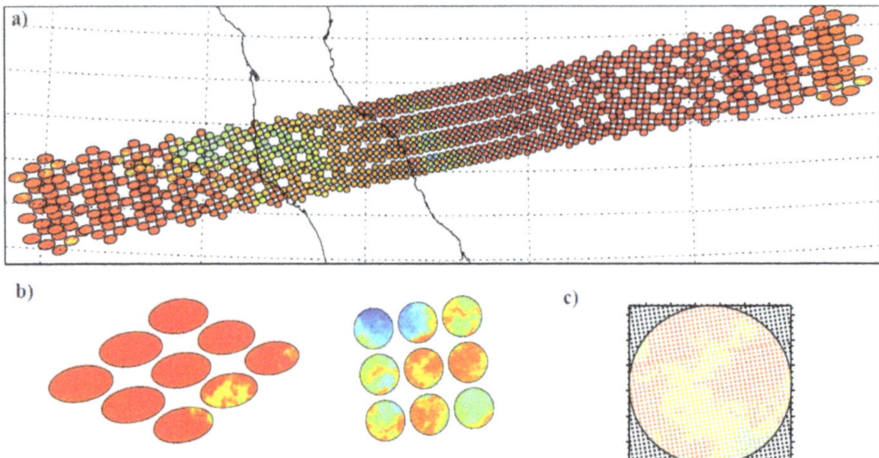

Figure 7. Collocated VIIRS pixels within the CrIS FOVs, including (**a**) the four scans shown in Figure 3; (**b**) the enlarged images for FOR 1 and 15 in the first scan line (from the bottom up) with scan angles of −48.3° and −1.65°; and (**c**) the enlarged plot for the center FOV (FOV 5) in the FOR 15 in the first scan, where the colorful points indicate VIIRS pixels falling within the CrIS FOV and the black ones represent those outside CrIS FOVs. Note that CrIS FOV shapes shown as black lines are independently computed using CrIS geolocation dataset and the FOV angle of 0.963°.

4.2. Collocation Method Evaluation

The accuracy of the collocation is affected by two factors, including Equation (1) the accuracy of collocation method and Equation (2) the accuracy of inputs, e.g., the consistency and accuracy of VIIRS and CrIS geolocation datasets. The latter case is not discussed here because the accuracy of VIIRS and CrIS geolocation data has been discussed in previous studies [11,19]. Therefore, we mainly focus on the former one. In other words, we simply assume that the CrIS and VIIRS geolocation dataset is accurate enough and evaluate the collocation method itself.

Figure 6 indirectly estimates the uncertainties of retrieving CrIS and VIIRS LOS vectors, which are in the order of meters (less than 4.0 m). In the algorithm, we use the double precision approach to perform all the calculations. The final accumulated uncertainties for the collocation method are estimated to be less than 20.0 m. Figure 7c shows the enlarged plots for the center FOV (FOV 5) in the FOR 15 in the first scan, where the colorful points indicate VIIRS pixels falling within the CrIS FOV and the black ones represent those outside CrIS FOVs based on our collocation method. Note that CrIS

FOV shapes shown as black lines are independently computed using CrIS geolocation dataset and the FOV angle of 0.963°. The two independent methods that use the same inputs are consistent with each other, suggesting that the collocation method works well.

The VIIRS radiances values within CrIS FOVs can be spatially averaged together and then compared with CrIS-simulated VIIRS I5 band radiances. It can evaluate the effectiveness and accuracy of the collocation. Shown in Figure 8 are CrIS-VIIRS BT difference maps and a scatter plot of spatial-averaged VIIRS BTs *versus* spectral-convolved CrIS BTs for VIIRS I5 band. If the collocation of CrIS and VIIRS is not accurate enough, the BT differences are expected to have large standard deviations for inhomogeneous scenes because these scenes are very sensitive to spatial match. As a result, these FOVs can be apparently identified from CrIS-VIIRS BT difference map.

Shown in Figure 8, most of the cloudy FOVs and FOVs near coast regions are hard to detect by eyes. Moreover, the scatter plot of CrIS BT *versus* VIIRS BT indicates small standard deviations (spread), suggesting that the collocation method is accurate. On the other hand, the accurate collocation between CrIS and VIIRS allows inter-calibration between the narrow-band imager and hyperspectral sounder to examine each sensor's radiometric calibration [10,20]. The scatter plot in Figure 8 indicates good agreement between CrIS and VIIRS I5 bands, though CrIS is slightly warmer than VIIRS. Demonstrated in Figure 9 is another case that occurred at the North Polar regions on 1 October 2015 and had complicated cloud structures. However, the CrIS-VIIRS BT difference map does not show the corresponding cloud structures, indicating that the collocation method works well.

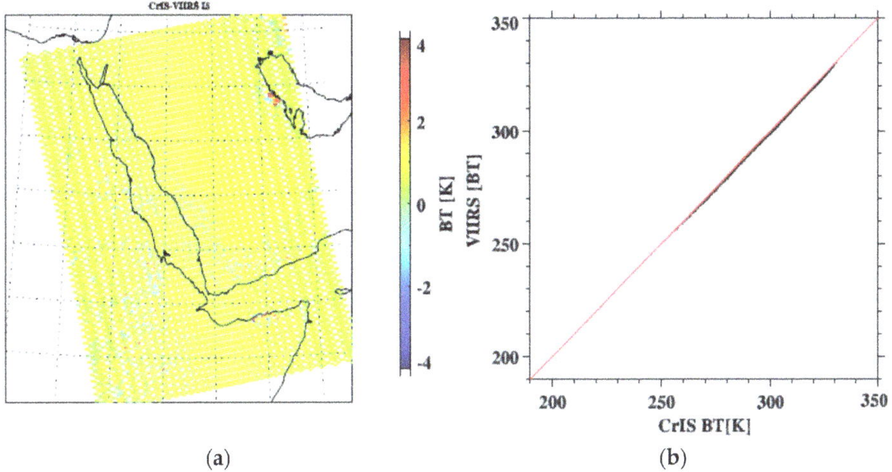

Figure 8. CrIS-VIIRS BT difference image map (**a**) and scatter plot of VIIRS BT *versus* CrIS BT for VIIRS I5 band (**b**). CrIS spectra are convolved with VIIRS SFRs to simulate VIIRS I5 band radiances, while collocated VIIRS radiances are spatially averaged within CrIS FOVs. The original CrIS and VIIRS images can be found in Figure 2.

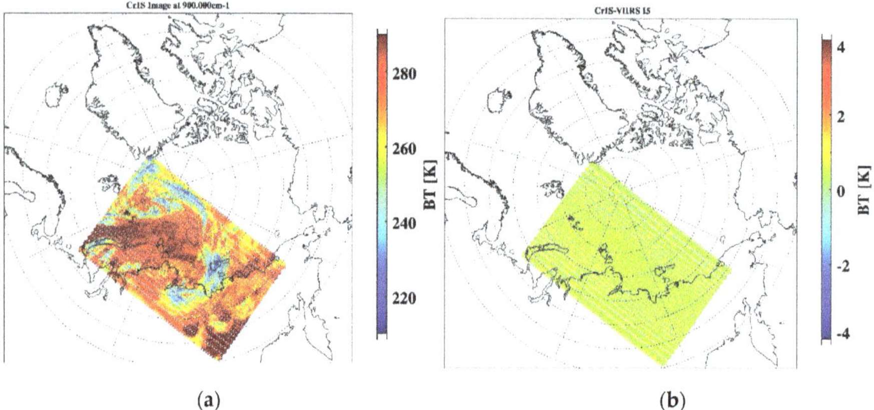

(a) (b)

Figure 9. CrIS image (**a**) at the North Polar Region on 1 October 2015 and CrIS-VIIRS BT difference map (**b**).

4.3. Other Applications

Another application of collocating VIIRS and CrIS measurements is to characterize CrIS FOV scene features by taking advantage of various VIIRS products. For example, the operational data assimilation of hyper-spectral radiances for numerical weather prediction (NWP) models still relies primarily on cloud-free data, even though substantial progress in the widely studied subject of assimilation of cloud-affected radiances has been made [5]. One challenge remaining in the assimilation of cloud-free infrared sounder radiances is to detect clear and cloud contaminated data correctly. With the fast and accurate collocation of CrIS and VIIRS, one way to identify CrIS clear FOVs is to use a collocated VIIRS cloud mask product (VCM). The VCM technique incorporates a number of cloud detection tests and produces a cloud confidence indicator of confidently cloudy, probably cloudy, probably clear, or confidently clear [21]. Specifically, in order to identify clear sky CrIS FOVs, the VCM products are first collocated with CrIS FOVs using the above collocation method. Second, if all the VIIRS pixels within a given CrIS FOV are all flagged as confidently clear, this CrIS FOV is identified as a clear sky FOV. After the CrIS granule in Figure 2 is examined, 7948 of 15,120 CrIS FOVs are identified as clear sky, shown in Figure 10. Compared with the widely-used observation minus background (O-B) method by the data assimilation community [22], we do believe that the imager-collocated cloud-detection scheme has several advantages. First, it does not rely on model background information as well as radiative transfer calculations. Second, the VIIRS measurements are independent of CrIS and thus can provide reliable sub-pixel information. However, a detailed comparison of these two cloud detection schemes is yet to be finalized.

Finally, we would like to demonstrate that this collocation method can be easily extended to two sensors on different satellite platforms. Shown in Figure 11 are the images from the Atmospheric Infrared Sounder (AIRS) on Aqua and CrIS on SNPP at 900 cm^{-1} on 1 May 2015, when two satellites passed over the western Africa with a ~7-min time difference (Aqua followed SNPP). Since Aqua and SNPP satellites have a similar equatorial crossing time (~1330 LST) but different altitudes (*SNPP*'s altitude is 824 km and *Aqua*'s is 705 km), these two satellites follow each other with a wealth of coincident data through full scan swaths. To collocate CrIS measurements with AIRS measurements using the above method, we only need to replace VIIRS LOS vector with AIRS LOS vector in Equation (6). Basically, the CrIS FOVs that are overlapped with AIRS FOVs are paired together because AIRS and CrIS have a spatial resolution of 13.5 km and 14.0 km at nadir, respectively. In this way, the CrIS FOVs have been re-arranged to spatially match AIRS. The collocated image—which is given in the right panel in Figure 11—looks reasonable. However, we should point out that, for a pair of sensors on different satellites, the collocation does suffer observational time differences and angle differences. Particularly, for the collocated image shown in Figure 11, the zenith angle differences for collocated FOVs vary from 1.0° to 27° dependent on scan positions.

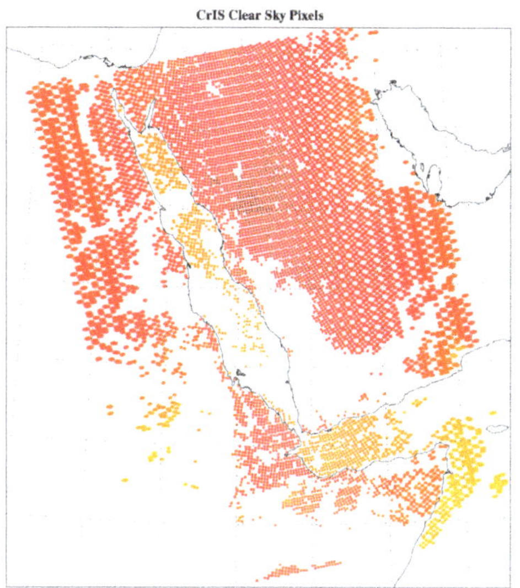

Figure 10. CrIS clear sky FOVs' BTs detected using VIIRS cloud mask product. The original CrIS and VIIRS images as well as the color bar can be found in Figure 2.

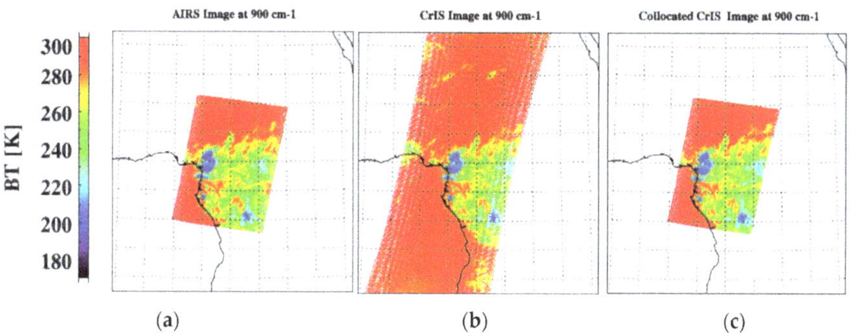

(a) (b) (c)

Figure 11. Examples of CrIS (**a**) and AIRS (**b**) images at 900 cm^{-1} as well as a collocated CrIS image that matches AIRS (**c**).

5. Conclusions

Combining high spatial resolution measurements from an imager and high spectral resolution measurements from an IR sounder can greatly improve atmospheric and surface geophysical parameter retrievals, data utilization for numerical weather prediction models, and inter-calibration capabilities including both radiometric and geometric calibration. All these applications depend on an accurate and fast collocation scheme. In this study, a fast and accurate collocation scheme to collocate VIIRS products and measurements with CrIS is developed and discussed in detail, which can be used in the current SNPP and future JPSS satellite platforms. Directly based on the CrIS and VIIRS LOS pointing vectors, the algorithm is mathematically very accurate. More importantly, with optimization, this method is very fast and efficient and thus can meet operational requirements. We also demonstrate that this collocation method can be extended to a wide variety of sensors on different satellites, for example CrIS with AIRS on different polar orbiting satellites as well as CrIS and geostationary imagers.

As a final note, we would like to point out that the algorithm can be implemented in any language (such as Fortran, Java Python, and C++) that is able to incorporate vector and matrix algebra calculations. It is also feasible to use a nearest-neighbor search in implementing the algorithm in order to respond to operational requirements. Efforts are being made to integrate this algorithm into the operational environments of various applications.

Acknowledgments: The VIIRS and CrIS data for this paper were downloaded at NOAA's Comprehensive Large Array Data Stewardship System. The AIRS data are available at NASA Goddard Earth Sciences (GES) Data and Information Services Center (DISC). The authors thank anonymous reviewers for providing valuable comments for the study. This study is funded by the NOAA JPSS Program Office. Likun Wang is supported by NOAA grant NA14NES4320003 (Cooperative Institute for Climate and Satellites) at the University of Maryland/ESSIC. The manuscript contents are solely the opinions of the authors and do not constitute a statement of policy, decision, or position on behalf of NOAA or the U.S. government.

Author Contributions: Likun Wang is the main author of this research work and wrote the manuscript. Denis Tremblay developed the original version of CrIS footprint shape computation codes in Matlab. Bin Zhang reviewed the IDL collocation codes and identified several bugs. All the authors reviewed and edited the manuscript.

Conflicts of Interest: The authors declare no conflict of interest.

References

1. Cao, C.; Xiong, J.; Blonski, S.; Liu, Q.; Uprety, S.; Shao, X.; Bai, Y.; Weng, F. Suomi NPP VIIRS sensor data record verification, validation, and long-term performance monitoring. *J. Geophys. Res. Atmos.* **2013**, *118*. [CrossRef]
2. Han, Y.; Revercomb, H.; Cromp, M.; Gu, D.; Johnson, D.; Mooney, D.; Scott, D.; Strow, L.; Bingham, G.; Borg, L.; *et al.* Suomi NPP CrIS measurements, sensor data record algorithm, calibration and validation activities, and record data quality. *J. Geophys. Res. Atmos.* **2013**, *118*. [CrossRef]
3. Li, J.; Menzel, W.P.; Sun, F.; Schmit, T.J.; Gurka, J. AIRS subpixel cloud characterization using MODIS cloud products. *J. Appl. Meteorol.* **2004**, *43*, 1083–1094. [CrossRef]
4. Kahn, B.H.; Fishbein, E.; Nasiri, S.L.; Eldering, A.; Fetzer, E.J.; Garay, M.J.; Lee, S.-Y. The radiative consistency of atmospheric infrared sounder and moderate resolution imaging spectroradiometer cloud retrievals. *J. Geophys. Res.* **2007**, *112*, D09201. [CrossRef]
5. Eresmaa, R. Imager-assisted cloud detection for assimilation of infrared atmospheric sounding interferometer radiances. *Q. J. R. Meteorol. Soc.* **2014**, *140*, 2342–2352. [CrossRef]
6. Wang, L.; Wu, X.; Goldberg, M.; Cao, C.; Li, Y.; Sohn, S.-H. Comparison of AIRS and IASI radiances using GOES imagers as transfer radiometers toward climate data records. *J. Appl. Meteor. Climatol.* **2010**, *49*, 478–492. [CrossRef]
7. Wang, L.; Cao, C. On-orbit calibration assessment of AVHRR longwave channels on MetOp-A using IASI. *IEEE Trans. Geosci. Remote Sens.* **2008**, *46*, 4005–4013. [CrossRef]
8. Tobin, D.C.; Revercomb, H.E.; Moeller, C.C.; Pagano, T.S. Use of atmospheric infrared sounder high–spectral resolution spectra to assess the calibration of moderate resolution imaging spectroradiometer on EOS Aqua. *J. Geophys. Res. Atmos.* **2006**, *111*. [CrossRef]
9. Schreier, M.M.; Kahn, B.H.; Eldering, A.; Elliott, D.A.; Fishbein, E.; Irion, F.W.; Pagano, T.S. Radiance comparisons of MODIS and AIRS using spatial response information. *J. Atmos. Oceanic Technol.* **2010**, *27*, 1331–1342. [CrossRef]
10. Wang, L.; Han, Y.; Tremblay, D.; Weng, F.; Goldberg, M. Inter-comparison of NPP/CrIS radiances with VIIRS, AIRS, and IASI: A post-launch calibration assessment. *Proc. SPIE* **2012**, *8528*. [CrossRef]
11. Wang, L.; Tremblay, D.A.; Han, Y.; Esplin, M.; Hagan, D.E.; Predina, J.; Suwinski, L.; Jin, X.; Chen, Y. Geolocation assessment for CrIS sensor data records. *J. Geophys. Res. Atmos.* **2013**, *118*, 690–704. [CrossRef]
12. Cao, C.; Xu, H.; Sullivan, J.; McMillin, L.; Ciren, P.; Hou, Y.-T. Intersatellite radiance biases for the high-resolution infrared radiation sounders (HIRS) on board NOAA-15, -16, and -17 from simultaneous nadir observations. *J. Atmos. Ocean. Technol.* **2005**, *22*, 381–395. [CrossRef]

13. Holl, G.; Buehler, S.A.; Rydberg, B.; Jimenez, C. Collocating satellite-based radar and radiometer measurements—Methodology and usage examples. *Atmos. Meas. Tech.* **2010**, *3*, 693–708. [CrossRef]
14. Aoki, T. *A Method for Matching the HIRS-2 and AVHRR Pictures of TIROS-N Satellites*; Technical Note for Meteorological Satellite Center: Tokyo, Japan; October; 1980; pp. 15–26.
15. Aoki, T. Clear radiance retrieval of HIRS channels with the use of AVHRR data. In Proceedings of the Technical Proceedings of the First TOVS Study Conference, Igls, Austria, 29 August–2 September 1983.
16. Nagle, F.W.; Holz, R.E. Computationally efficient methods of collocating satellite, aircraft, and ground observations. *J. Atmos. Oceanic Technol.* **2009**, *26*, 1585–1595. [CrossRef]
17. Strow, L.; Motteler, H.; Tobin, D.; Hannon, S.; Predina, J.; Suwinski, L. Frequency calibration and validation of CrIS satellite sounder. *J. Geophys. Res. Atmos.* **2013**, *118*, 12486–12496. [CrossRef]
18. Match_2d Function. Available online: http://tir.astro.utoledo.edu/idl/match_2d.pro (accessed on 20 January 2016).
19. Wolfe, R.E.; Lin, G.; Nishihama, M.; Tewari, K.P.; Tilton, J.C.; Isaacman, A.R. Suomi NPP VIIRS prelaunch and on-orbit geometric calibration and characterization. *J. Geophys. Res. Atmos.* **2013**, *118*. [CrossRef]
20. Tobin, D.; Revercomb, H.; Knuteson, B.; Best, F.; Taylor, J.; Deslover, D.; Borg, L.; Moeller, C.; Martin, G.; Kuehn, R.; *et al.* Soumi NPP/JPSS Cross-Track Infrared Sounder (CrIS): Intercalibration with AIRS, IASI, and VIIRS. In Proceedings of the 93rd AMS Annual Meeting, Austin, TX, USA, 6–10 January 2013.
21. Kopp, T.J.; Thomas, W.; Heidinger, A.K.; Botambekov, D.; Frey, R.A.; Hutchison, K.D.; Iisager, B.D.; Brueske, K.; Reed, B. The VIIRS cloud mask: Progress in the first year of S-NPP toward a common cloud detection scheme. *J. Geophys. Res. Atmos.* **2014**, *119*, 2441–2456. [CrossRef]
22. McNally, A.P.; Watts, P.D. A cloud detection algorithm for high-spectral-resolution infrared sounders. *Q. J. R. Meteorol. Soc.* **2003**, *129*, 3411–3423. [CrossRef]

Article

Improved Band-to-Band Registration Characterization for VIIRS Reflective Solar Bands Based on Lunar Observations

Zhipeng Wang [1,*], Xiaoxiong Xiong [2] and Yonghong Li [1]

1 Science Systems and Applications Inc., 10210 Greenbelt Road, Greenbelt, MD 20706, USA; yonghong.li@ssaihq.com
2 Sciences and Exploration Directorate, NASA GSFC, Greenbelt, MD 20771, USA; xiaoxiong.xiong-1@nasa.gov
* Correspondence: zhipeng.wang@ssaihq.com; Tel.: +1-520-309-0559; Fax: +1-301-867-2151

Academic Editors: Changyong Cao, Richard Müller and Prasad S. Thenkabail
Received: 20 October 2015; Accepted: 25 December 2015; Published: 31 December 2015

Abstract: Spectral bands of the Visible Infrared Imaging Radiometer Suite (VIIRS) instrument aboard the Suomi National Polar-orbiting Partnership (S-NPP) satellite are spatially co-registered. The accuracy of the band-to-band registration (BBR) is one of the key spatial parameters that must be characterized. Unlike its predecessor, the Moderate Resolution Imaging Spectroradiometer (MODIS), VIIRS has no on-board calibrator specifically designed to perform on-orbit BBR characterization. To circumvent this problem, a BBR characterization method for VIIRS reflective solar bands (RSB) based on regularly-acquired lunar images has been developed. While its results can satisfactorily demonstrate that the long-term stability of the BBR is well within ± 0.1 moderate resolution band pixels, undesired seasonal oscillations have been observed in the trending. The oscillations are most obvious between the visible/near-infrared bands and short-/middle wave infrared bands. This paper investigates the oscillations and identifies their cause as the band/spectral dependence of the centroid position and the seasonal rotation of the lunar images over calibration events. Accordingly, an improved algorithm is proposed to quantify the rotation and compensate for its impact. After the correction, the seasonal oscillation in the resulting BBR is reduced from up to 0.05 moderate resolution band pixels to around 0.01 moderate resolution band pixels. After removing this spurious seasonal oscillation, the BBR, as well as its long-term drift are well determined.

Keywords: VIIRS; band-to-band registration; Moon; spatial characterization

1. Introduction

The Visible Infrared Imaging Radiometer Suite (VIIRS) is a remote sensing instrument aboard the Suomi National Polar-orbiting Partnership (S-NPP) satellite and provides measurements of large-scale global dynamics in the oceans, on the land and in the lower atmosphere [1]. As a passive whiskbroom imaging spectroradiometer, VIIRS captures data in 15 reflective solar bands (RSB), including a panchromatic day/night band (DNB) and seven thermal emissive bands (TEB), covering a spectral range from 0.4 to 12.5 μm [2]. VIIRS was designed and built by the same instrument vendor as the Moderate resolution Imaging Spectroradiometer (MODIS) on board the NASA EOS Terra and Aqua satellites, for which it is the follow-on instrument [3]. The operation and calibration strategy of VIIRS has been largely inherited from MODIS [4].

VIIRS detectors have two spatial resolutions: the nominal ground pixel sizes at nadir are 375 m for the imaging (I) Bands I1–I5 and 750 m for the moderate (M) resolution Bands M1–M16, respectively. The M and I bands are located on three separate focal plane assemblies (FPA): the visible/near-infrared (VIS/NIR), the short-wave and mid-wave infrared (S/MWIR) and the long-wave infrared (LWIR). The bands are positioned in parallel, slightly separated from each other in the along-scan direction, as shown in Figure 1. The detectors in each band are aligned in the along-track direction. Through a band-dependent time delay, the Earth view images of all of these bands are spatially co-registered. The residual misalignments after co-registration are quantified by the band-to-band registration (BBR). The DNB uses a separate FPA and is not co-registered with other RSB bands. Therefore, it is excluded from the following discussion.

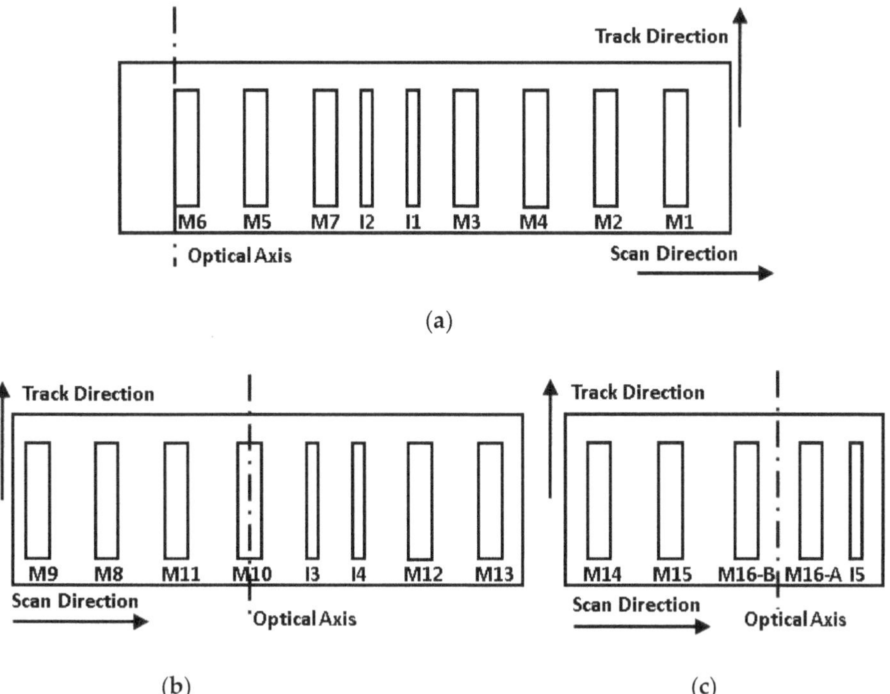

Figure 1. VIIRS band layout on (**a**) VIS/NIR; (**b**) short-wave and mid-wave infrared (S/MWIR) and (**c**) long-wave infrared (LWIR) focal plane assemblies (FPA) [2].

BBR is a key performance parameter in VIIRS spatial characterization. The VIIRS BBR was characterized in a series of pre-launch tests [5,6]. There is no on-board calibrator, such as MODIS's Spectroradiometric Calibration Assembly (SRCA) [7], to monitor the on-orbit BBR for VIIRS. Therefore, alternative approaches using remote targets have been developed, with the general idea of calculating the BBR offset between any two bands by measuring the shift between their images of the same target. Various approaches using Earth view data [8–10] or the lunar images have been proposed [11–13]. Compared to the Earth view data, lunar data are available much less frequently and cover a limited scan-angle range, because the Moon is only viewed at a fixed scan angle corresponding to the space view (SV) port. However, the surface property of the Moon is both spatially and radiometrically stable in the long run. In addition, lunar images are not blurred by atmospheric scattering [14]. Therefore, the shift measured by the lunar images is impacted by less dynamic error sources than Earth view images.

The shift between images can be measured in different ways, such as mutual information [9] or centroid. The lunar centroid approach was developed for MODIS and has been adapted to VIIRS. This paper continues the effort of our group to further improve the algorithm. While previous BBR results from the lunar centroid approach are good enough to demonstrate that the on-orbit BBR for the RSB has been stable and meets the sensor design requirement, there is an apparent band- or wavelength-dependent seasonal oscillation of up to 0.05 M band pixels observed in the BBR trending, which is more like an error instead of actual BBR oscillation [11]. The cause of the oscillation is diagnosed, and a correction based on the study is applied, which significantly reduces the oscillation in the updated BBR trending.

2. VIIRS Lunar Calibration

The Moon has been widely used as a target for on-orbit calibration and characterization of remote sensing instruments for its long-term stability. Regular VIIRS lunar observations are scheduled primarily for the validation of its RSB radiometric calibration stability [15]. The lunar irradiance strongly depends on the illumination and view geometries, especially the lunar phase or the Moon-to-Earth-to-VIIRS angle. To minimize the brightness variation, VIIRS lunar calibrations are scheduled only when the phase is within a small angular range from −51.5 to −50.5 degrees. A spacecraft roll maneuver is usually performed for each lunar calibration to ensure the Moon is viewed through the SV port at an elevation angle of 24.325 degrees below the instrument X-Y plane, which is the plane perpendicular to the nadir direction. The observations can only be scheduled around once a month. Limited by the satellite orbit and allowed roll angle range, the observations cannot be scheduled for three to four months each year. From the launch of VIIRS in late 2011 to September 2015, 33 scheduled lunar observations have been successfully performed.

Figure 1 shows the layout of VIIRS spectral bands on the FPA. Each I band has 32 detectors, and each M band has 16 detectors. In the along-scan direction, the I bands are sampled at twice the frequency of the M bands. An M band image pixel is registered with 2 by 2 I band image pixels. During the scheduled lunar calibration, a data sector rotation is applied, so the Moon image captured through the SV port is actually read out at the center of the Earth view sector. In this zone, the I bands and the single gain M bands (M6, M8–M12, M14–M16) are spatially aggregated onboard every three samples in the along-scan direction by the instrument. For the dual-gain M bands, the same aggregation strategy is applied in ground processing for consistency. After aggregation, an image pixel is of a nearly square shape with a size of 750 m × 750 m. The diameters of the lunar images are approximately 20 pixels for I bands and 10 pixels for M bands, depending on the VIIRS-Moon distances at the time of lunar calibration.

During a lunar calibration event, the Moon moves relative to the field of view (FOV) of the FPA along its track direction. Figure 2 shows the scan-by-scan Band I1 images acquired from the entire SV data sector during the April 2015 lunar calibration. The horizontal direction is the along-scan direction with a width of 96 frames; the vertical direction is the along-track direction with a height of 32 detectors. The illuminated fraction of the Moon at a phase of −51 degrees is about 81.5%. The orientations of the gibbous Moon change with calibration events, which is determined by the position and pointing direction of VIIRS relative to the Moon. The detector response is output as a 12-bit digital number (DN). The lunar images DN must have the dark reference subtracted on a scan-by-scan basis. The subtracted value is then corrected for the detector gain using the instrument temperature-dependent coefficients determined from the pre-launch measurement and the F-factor derived on-orbit using the on-board solar diffuser and the response *versus* scan (RVS) angle [16]. The result is the 2D radiance profile of the lunar surface.

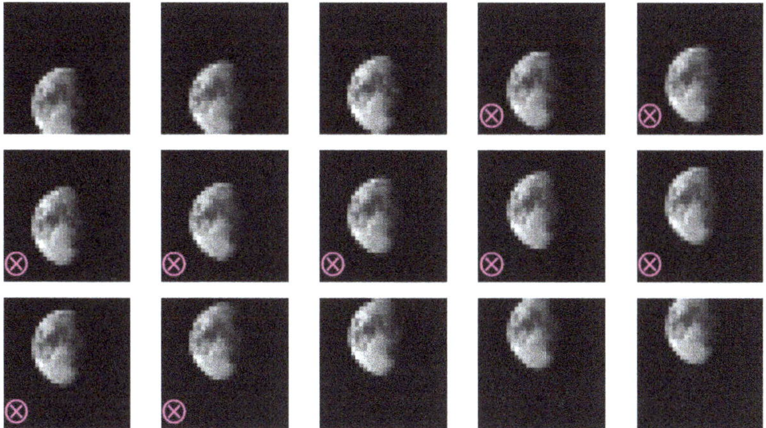

Figure 2. Scan-by-scan Band I1 images during the April 2015 lunar calibration: the horizontal direction is the along-scan direction; the vertical direction is the along-track direction. The symbols in the lower left corner mark those scans with complete lunar images (not being cut-off by the top or bottom edge of the FPA).

3. Lunar BBR Algorithm

3.1. Current Algorithm

The design specification of the VIIRS BBR is based on the overlapping fraction of the matching detectors between two bands. It can also be quantified by the relative offsets between the matching detectors in both the along-scan and along-track directions [5]. These two forms of expression are related by:

$$A \approx (1 - BBR_{scan})\,(1 - BBR_{track}) \tag{1}$$

where BBR_{scan} and BBR_{track} are the offset in units of pixels in the along-scan and along-track directions, respectively. They are calculated for each detector. The results in this paper are presented using their band-averaged values.

In the lunar centroid approach, the centroid of the lunar radiance profiles in the along-scan direction is calculated by:

$$X_{B,D} = \frac{\sum\limits_{f}\left(\sum\limits_{s} L_{B,D}\right)\cdot f}{\sum\limits_{f}\sum\limits_{s} L^{*}_{B,D}} \tag{2}$$

where L is the retrieved radiance of each pixel. f is the frame/sample number, and s is the scan number. B and D are the notations for band and detector, respectively. The centroid in the along-track direction is calculated by:

$$Y_{B,D} = \frac{\sum\limits_{s}\left(\sum\limits_{f} L_{B,D}\right)\cdot s}{\sum\limits_{s}\sum\limits_{f} L_{B,D}} \tag{3}$$

Ignoring the band dependence of the lunar centroid, the BBR offset between two bands is calculated by:

$$BBR_{scan} = \overline{X_{B1,D}} - \overline{X_{B2,D}};\ BBR_{track} = \left(\overline{\Delta Y_{B1,D}} - \overline{\Delta Y_{B2,D}}\right)/\beta \tag{4}$$

where β is the so-called oversampling factor. Because the moving speed of the satellite and the instantaneous field of view (IFOV) of the detector are not synchronized, VIIRS swaths on the lunar surface overlap along the track direction from scan to scan. β is defined as the number of scans taken for a VIIRS swath to move one detector IFOV in the track direction. This factor is calculated from the ephemeris of the VIIRS and the Moon, and the details of the calculation were introduced in a separate work [17]. The factor converts the along-track shift calculated per scan in Equation (3) to per IFOV or image pixel.

For VIIRS, the unit of the BBR result used in this paper is an M-band pixel. For the convenience of the users, it is sometimes converted to meters on the Earth ground by multiplying by the spatial resolution of the pixel on the ground at nadir, which is approximately 750 m for M bands. Figure 3 shows the BBR trending generated using the current approach, with Band I1 as the reference. The plot shows that the BBR for all RSBs have been stable since launch with values of less than 0.05 pixels in both the along-scan and along-track directions; some small drifts were observed.

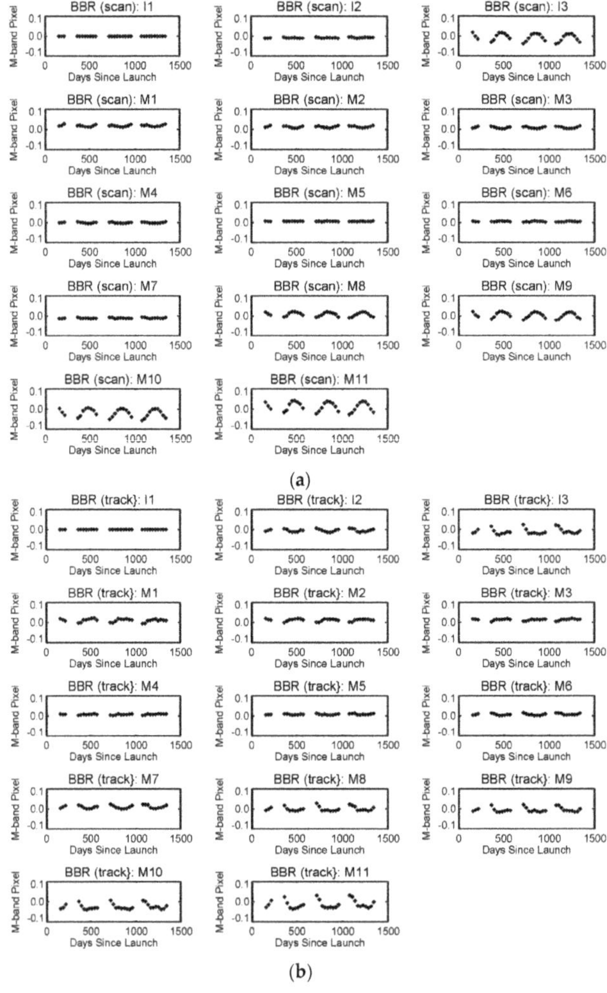

Figure 3. Band-to-band registration (BBR) results using current the centroid approach in both the (**a**) along-scan and (**b**) along-track directions, referenced to Band I1.

3.2. Impact of Lunar Image Rotation on BBR Results

Band- or wavelength-dependent seasonal oscillation patterns are observed in Figure 3. As the center wavelength difference relative to Band I1 increases, the amplitude of the BBR oscillation increases from Band M7 to M11. I2/M7 (865 nm) and I3/M10 (1610 nm) are two band pairs that have almost identical center wavelengths, and their BBR trending is very similar.

Further investigation has related the oscillation to the wavelength-dependence of the image centroids. The current approach assumes that the offset between the centroids of two bands in Equation (4) is solely caused by BBR offset. The actual centroid positions are actually wavelength dependent. Each plot in Figure 4a is a scan of the Band I1 lunar image captured during a calibration event. For nearly identical lunar phases and a small libration effect, we can assume that the centroid positions on the Moon are almost unchanged among events. Because the lunar images rotate from event to event, the centroids of two bands, which are displaced from each other, rotate accordingly. Thus, the centroid displacement, which is denoted by a vector **R**, will also rotate as is illustrated in Figure 4b.

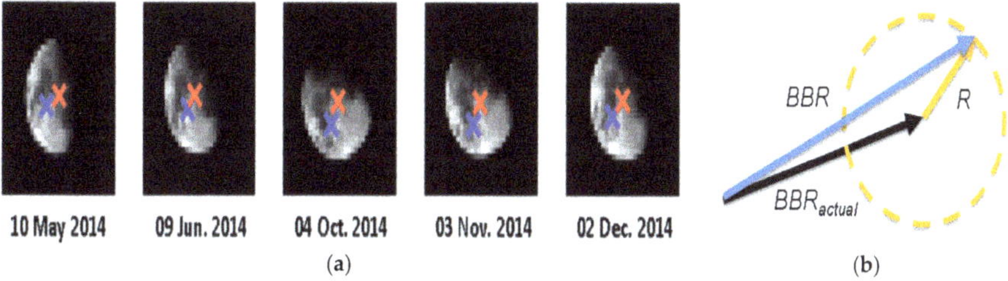

Figure 4. (**a**) VIIRS Band I1 lunar images of the center scans of a few lunar calibration events. Blue and red crosses illustrate the estimated positions of the centroids of two bands; (**b**) a sketch that shows how the measured BBR varies with the rotation of the centroid displacement vector.

Therefore, the centroid differences calculated from Equation (4) are a combination of the actual BBR offset and the components of **R** in the along-scan and along-track direction, respectively:

$$BBR_{scan} = BBR_{scan,actual} + R \cdot \sin(\theta + \theta_0) \tag{5}$$

and:

$$BBR_{track} = BBR_{track,actual} + R \cdot \cos(\theta + \theta_0) \tag{6}$$

where θ is the solar illumination angle, which is newly defined in this paper. It is in reference to FPA coordinates, as is shown in Figure 5. R is the length of the **R**, or the distance between the centroids of the two bands. θ_0 is the fixed angle between **R** and the illumination angle.

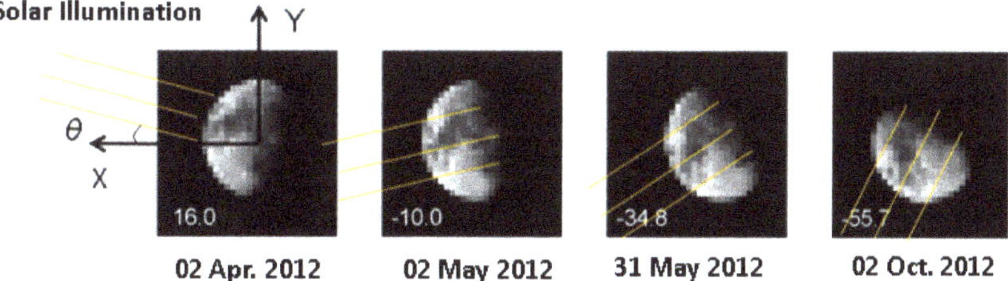

Figure 5. The definition of the solar illumination angle in FPA coordinates. The number at the bottom left corner of each plot is the angle for that event.

The trending of the solar illumination angles for all lunar events is plotted in Figure 6, as well as the trending of the oversampling factor and VIS/NIR FPA temperatures. Compared to the BBR results of the SWIR bands in Figure 3, the correlation between the two sets of data is significant. We conclude that the centroid displacement between bands and the lunar image rotation together contribute to the majority of the seasonal oscillations observed in Figure 3.

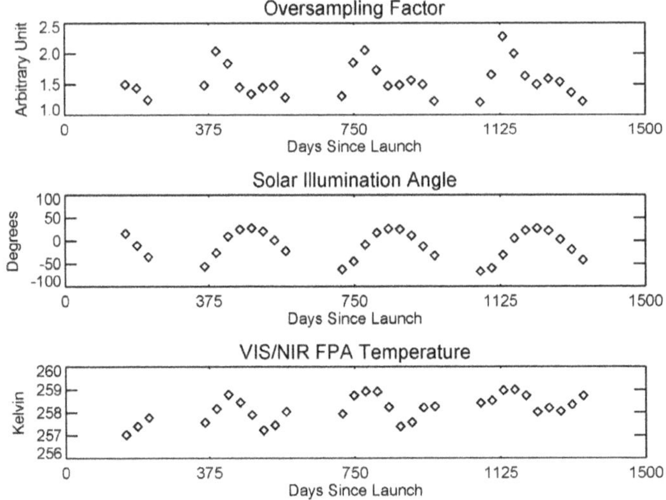

Figure 6. The trending of the instrumental parameters related to BBR characterization.

3.3. Image Rotation Correction

An improvement is proposed for the lunar centroid approach to correct the impact of the image rotation. In this proposed approach, R and θ_0 are considered as unknown parameters and must be derived from lunar data themselves. We assume that the BBR offset remained unchanged in the first year after launch. Then, the lunar calibration events scheduled in this time period are used as the training dataset. Since $BBR_{scan,actual}$ and $BBR_{track,actual}$ are constant, they are fitted as unknown parameters together with R and θ_0. The fitted values are provided in Table 1.

Table 1. The estimated coefficients.

Band	Center Wavelength (μm)	BBR_{actual} *Scan* (m)	BBR_{actual} *Track* (m)	R (m)	θ_0
I1	0.640	–	–	–	–
I2	0.865	0.13	−3.63	7.53	93.8
I3	1.610	14.31	18.94	35.58	22.1
M1	0.412	7.66	4.45	9.46	62.0
M2	0.445	4.70	4.16	9.11	42.7
M3	0.488	3.28	3.75	8.33	26.7
M4	0.555	−1.09	2.33	4.40	22.7
M5	0.672	6.15	3.62	3.15	133.4
M6	0.746	8.45	6.09	5.51	−106.1
M7	0.865	0.09	4.84	10.86	−80.3
M8	1.240	18.34	15.10	20.58	25.9
M9	1.378	17.20	18.12	28.43	21.5
M10	1.610	4.17	19.60	36.85	22.4
M11	2.250	38.27	22.23	46.44	30.9

The centroid displacement is quantified by R. Not surprisingly, its magnitude increases as the difference in center wavelengths between two bands grows. When R increases, the oscillation observed in current BBR_{scan} and BBR_{track} without image rotation correction also increases. In reference to Figure 4b, the magnitude of R defines the maximum BBR variation that is possibly introduced. For Band M11, this value is approximately 46.44/750 = 0.062 pixels. Because the actual rotation of the lunar images does not cover the whole 360 degree range, the observed variation is smaller than this upper limit. This is consistent with the results shown in Figure 3.

With the parameters in Table 1, the actual BBR of the following event is calculated by:

$$BBR_{scan,actual} = BBR_{scan} - R \cdot \sin(\theta + \theta_0) \tag{7}$$

and:

$$BBR_{track,actual} = BBR_{track} - R \cdot \cos(\theta + \theta_0) \tag{8}$$

Again, the amount of correction is limited by $\pm R$. If the actual BBR or its on-orbit change is much larger than this value; then, the effect of this correction will be less significant.

4. Results and Discussion

4.1. VIIRS BBR Results

The lunar BBR results have been updated with lunar image rotation correction for VIIRS RSB in the along-scan and along-track directions. Figure 7 shows the improved BBR trending, still with Band I1 as a reference. Compared to the uncorrected results in Figure 3, the seasonal oscillation in the BBR results is significantly reduced from up to ± 0.05 M band pixels to less than ± 0.01 M band pixels. The long-term drift between the SWIR bands and the VIS/NIR bands has also been identified more accurately.

Figure 8 shows the average (lines) and the standard deviation (error bars) of the BBR during the years 2012 to 2015, before and after the image rotation correction, respectively. The plots also demonstrate how the reduction in the seasonal BBR oscillation reduction helps to determine the actual BBR and its long-term drift more accurately.

The rotation of the lunar image provides an opportunity to separate **R** from the actual BBR offset so that we can calculate its absolute value. If the orientation of the Moon does not rotate so that R and θ are fixed, then the BBR offset derived from Equations (5) and (6) always contains the component of centroid displacement. Therefore, the results can only reflect the on-orbit change of the BBR offset, but not its actual values.

There is still residual seasonal oscillation that is possibly related to the on-orbit instrument temperature change. For VIIRS, while the temperature of S/MWIR FPA is controlled to be stable at around 80 K, the temperature of VIS/NIR FPA has a larger dependence on the instrument temperature that changes on orbit, as is shown in Figure 6. The temperature difference between the two FPAs may cause the BBR offset. A similar relationship between the BBR and instrument temperature has been observed from MODIS, as well [7]. Since the instrument temperature change introduces actual BBR offset instead of an artifact, there is no need to have it corrected.

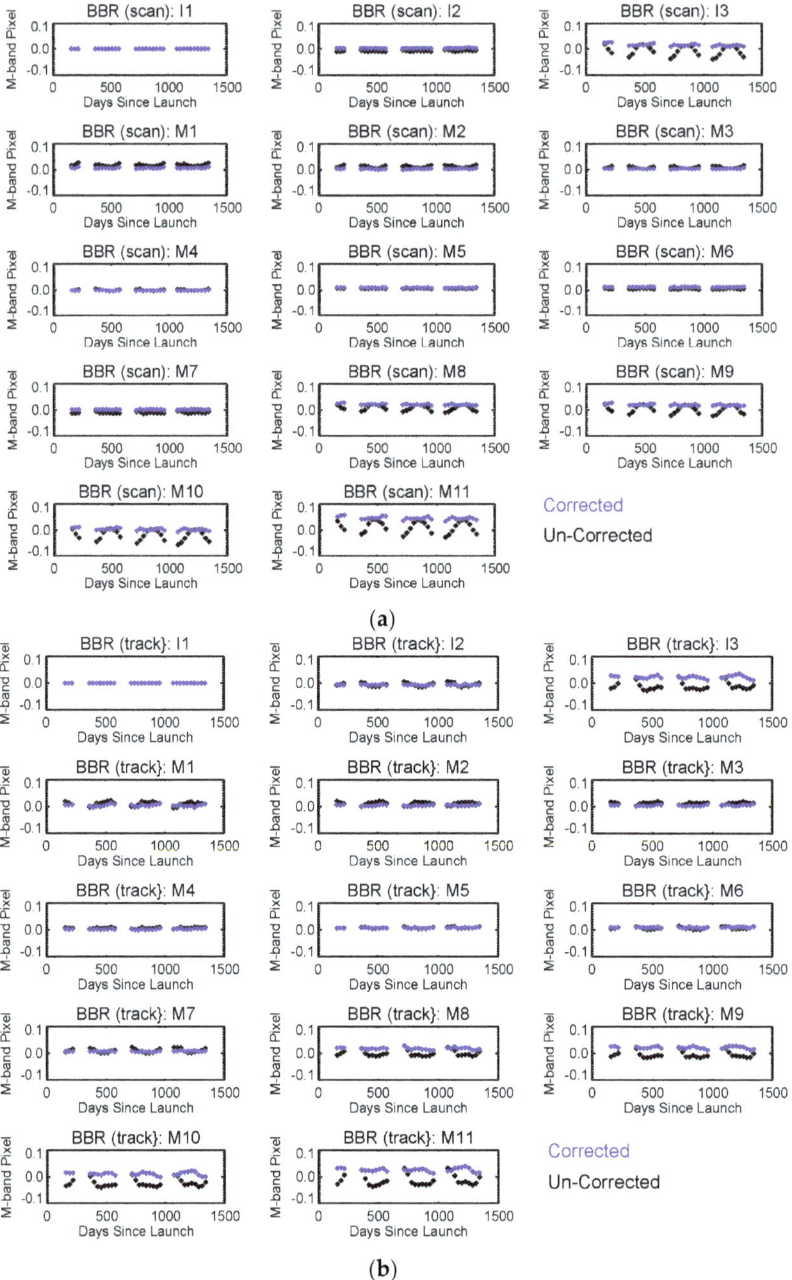

Figure 7. The image rotation-corrected VIIRS BBR results in both (**a**) along-scan and (**b**) along-track directions, referenced to Band I1.

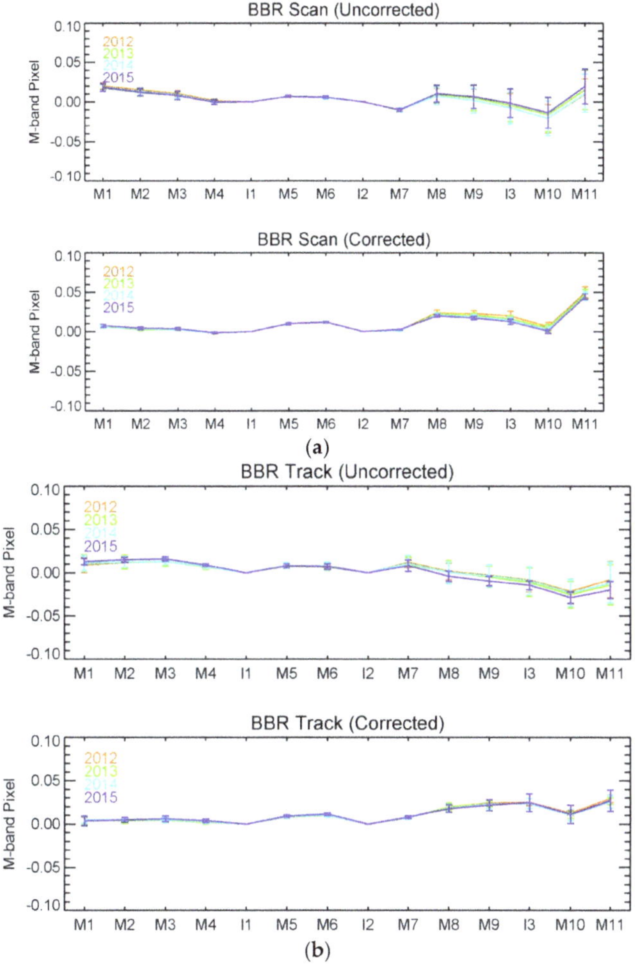

Figure 8. The yearly-averaged BBR results after image rotation correction in both along-scan (**a**) and along-track (**b**) directions before and after image rotation correction, referenced to Band I1.

4.2. Extent of Lunar Centroid Approach for Other Sensors

The centroid displacement on the lunar surface, R_{Moon}, can be estimated by multiplying R in Table 1 by a constant ratio factor 384,400/828, where 384,400 km is the nominal Moon to Earth distance and 828 km is the nominal VIIRS altitude. These values are listed in Table 2. The values in Table 2 are properties of the lunar surface and are thus independent of the sensors.

Table 2. The centroid displacement on the lunar surface (at the phase around −51 degrees).

Band	Center Wavelength (μm)	R_{Moon} (m)	Band	Center Wavelength (μm)	R_{Moon} (m)
I1	0.640	–	M5	0.672	1462
I2	0.865	3495	M6	0.746	2558
I3	1.610	16,518	M7	0.865	5042
M1	0.412	4392	M8	1.240	9554
M2	0.445	4229	M9	1.378	13,199
M3	0.488	3867	M10	1.610	17,108
M4	0.555	2043	M11	2.250	21,560

The values in Table 2 are useful to evaluate the compatibility of the lunar centroid approach for other sensors. It primarily depends on the spatial resolution of the sensors. Because the centroid displacement changes with lunar phase and libration, the value calculated here has certain uncharacterized uncertainty. Therefore, it is better that the centroid displacement, and its uncertainty, is small when compared to the spatial resolution of the sensor.

For sensors with coarser resolutions, such as MODIS and VIIRS, the maximum centroid displacement (0.640 μm *vs.* 2.25 μm) corresponds to about 0.1 pixels, so the centroid approach is likely feasible after the centroid displacement correction. For sensors with higher resolution, such as the Advanced Baseline Imager on the Geostationary Operational Environmental Satellite R-Series (GOES-R) with a resolution of 0.55 microradians (compared to 450 microradians for VIIRS), the centroid displacement between 0.640 μm and 2.25 μm corresponds to 1.15 pixels. The number is likely too large for characterizing sub-pixel level BBR. In this case, using the edge of the Moon instead of the centroid is probably more reasonable.

5. Conclusions

One of the key sensor spatial parameters, BBR, can be monitored on-orbit for VIIRS RSB with data from regularly-scheduled lunar calibration by comparing the centroids of the lunar images of different spectral bands. Seasonal oscillation of up to 0.05 pixels is observed in the BBR trending, and the amplitude of the oscillation increases with the center-wavelength difference between the two bands. It is found that the lunar image rotation among lunar calibration events contributes to the seasonal oscillation of BBR, because the centroid displacement vector between two spectral bands rotates with the lunar image, and the rotation changes its components in both along-scan and along-track directions.

In this paper, the lunar image rotation is quantified, and its impact is compensated. The seasonal oscillation in BBR trending is significantly reduced to a level of 0.01 pixels, especially between VIS/NIR bands and SWIR bands. The results confirm that the BBR in the SV direction, where the lunar image is actually acquired, has been stable and is well within the design specification of 0.1 pixels. The VIIRS BBR will be continuously monitored and evaluated throughout the VIIRS's lifetime, as part of the on-orbit calibration effort. Furthermore, the lunar centroid displacement on the Moon is estimated, and the values can be used to check the applicability of the lunar centroid approach to characterize the BBR of similar sensors.

Acknowledgments: Jeff McIntire kindly reviewed the manuscript and made valuable suggestions to improve it. His effort is very much appreciated. The authors would also like to thank other members of the NASA MODIS Characterization Support Team and the VIIRS Characterization Support Team for their contribution.

Author Contributions: Zhipeng Wang developed and implemented the lunar image rotation algorithm presented in the paper. He drafted the manuscript and the responses to the reviewers' comments. Xiaoxiong Xiong developed the concept of the band-to-band registration characterization using the Moon. He revised the manuscript critically and gave final approval of the version to be published. Yonghong Li was involved in the data acquisition and analysis. She also independently verified the results and revised the manuscript critically.

Conflicts of Interest: The authors declare no conflict of interest.

References

1. Murphy, R.P.; Ardanuy, P.E.; Deluccia, F.; Clement, J.E.; Schueler, C. Earth science satellite remote sensing. In *The Visible Infrared Imaging Radiometer Suite*; Springer-Verlag: New York, NY, USA, 2006; Volume 1, pp. 199–223.
2. Baker, N. *Joint Polar Satellite System VIIRS Geolocation Algorithm Theoretical Basis Document (Section 3.3)*; Doc. No. 474-00053; GSFC JPSS Configuration Management Office: Greenbelt, MD, USA, 2011.
3. Barnes, W.L.; Xiong, X.; Guenther, B.; Salomonson, V. Development, characterization, and performance of the EOS MODIS sensors. *Proc. SPIE* **2003**, *5151*, 337–345.
4. Xiong, X.; Butler, J.; Chiang, K.; Efremova, B.; Fulbright, J.P.; Lei, N.; McIntire, J.; Oudrari, H.; Sun, J.; Wang, Z.; *et al.* VIIRS on-orbit calibration methodology and performance. *J. Geophys. Res.* **2014**, *119*, 5065–5078. [CrossRef]
5. Lin, G.; Wolfe, R.E.; Nishihama, M. VIIRS geometric performance status. *Proc. SPIE* **2011**, *8153*. [CrossRef]
6. Wolfe, R.E.; Lin, G.; Nishihama, M.; Tewari, K.P.; Montano, E. NPP VIIRS early on-orbit geometric performance. *Proc. SPIE* **2012**, *8510*. [CrossRef]
7. Xiong, X.; Che, N.; Barnes, W.L. Terra MODIS on-orbit spatial characterization and performance. *IEEE Trans. Geosci. Remote Sens.* **2005**, *43*, 355–365. [CrossRef]
8. Wolfe, R.E.; Lin, G.; Nishihama, M.; Tewari, K.P.; Tilton, J.C.; Isaacman, A.R. Suomi NPP VIIRS prelaunch and on-orbit geometric calibration and characterization. *J. Geophys. Res. Atmos.* **2013**, *118*, 11508–11521. [CrossRef]
9. Tilton, J.C.; Tan, B.; Lin, G. Measurement of band-to-band registration of the NPP VIIRS instrument from on-orbit data. In Proceedings of the NOAA STAR JPSS Annual Science Team Meeting, College Park, MD, USA, 24–28 August 2015.
10. Liao, L. Band-to-band registration of remotely sensed imagery using mutual information and its application to Suomi NPP VIIRS data. In Proceedings of the AMS 20th Confrence on Satellite Meteorology & Oceanography, Phoenix, AZ, USA, 4–8 January 2015.
11. Xiong, X.; Sun, J.; Xiong, S.; Barnes, W.L. Using the Moon for MODIS on-orbit spatial characterization. *Proc. SPIE* **2003**, *5234*, 480–487.
12. Wang, Z.; Xiong, X. VIIRS on-orbit spatial characterization using the Moon. *IEEE Geosci. Remote Sens. Lett.* **2014**, *11*, 1116–1120. [CrossRef]
13. Wang, Z.; Xiong, X.; Li, Y. Update of VIIRS on-orbit spatial parameters characterized with the Moon. *IEEE Trans. Geosci. Remote Sens.* **2015**, *53*, 5486–5494. [CrossRef]
14. Kieffer, H.H.; Stone, T.C. The spectral irradiance of the Moon. *Astronom. J.* **2005**, *129*, 2887–2901. [CrossRef]
15. Xiong, X.; Sun, J.; Fulbright, J.; Wang, Z. Lunar calibration and performance for S-NPP VIIRS reflective solar bands. *IEEE Trans. Geosci. Remote Sens.* **2015**. [CrossRef]
16. Mills, S. *VIIRS Radiometric Calibration Algorithm Theoretical Basis Document ATBD*; Doc. No.: D43777; Northrop Grumman: Virgina, VA, USA, 2010.
17. Sun, J.; Xiong, X.; Barnes, W.; Guenther, B. MODIS reflective solar bands on-orbit lunar calibration. *IEEE Trans. Geosci. Remote Sens.* **2007**, *45*, 2383–2393. [CrossRef]

Article

Radiometric Stability Monitoring of the Suomi NPP Visible Infrared Imaging Radiometer Suite (VIIRS) Reflective Solar Bands Using the Moon

Taeyoung Choi [1,*], Xi Shao [1,2], Changyong Cao [3] and Fuzhong Weng [3]

1 Earth Resources Technology Inc., 5830 University Research Ct. #2672, College Park, MD 20740, USA; xi.shao@noaa.gov

2 Department of Astronomy, University of Maryland, College Park, MD 20742, USA

3 NOAA/NESDIS/STAR, 5830 University Research Ct., College Park, MD 20740, USA; changyong.cao@noaa.gov (C.C.); fuzhong.weng@noaa.gov (F.W.)

* Correspondence: taeyoung.choi@noaa.gov; Tel.: +1-301-683-3562; Fax: +1-301-683-3526

Academic Editors: Richard Müller and Prasad S. Thenkabail

Received: 20 October 2015; Accepted: 15 December 2015; Published: 25 December 2015

Abstract: The Suomi NPP (S-NPP) Visible Infrared Imaging Radiometer Suite (VIIRS) performs the scheduled lunar roll maneuver on a monthly basis. The lunar calibration coefficients and lunar F-factor are calculated by taking the ratio of the lunar observed radiance to the simulated radiance from the Miller and Turner (MT) lunar model. The lunar F-factor is also validated against that derived from the VIIRS Solar Diffuser (SD). The MT model-based lunar F-factors in general agree with SD F-factors. The Lunar Band Ratio (LBR) is also derived from two channel lunar radiances and is implemented in the National Oceanic and Atmospheric Administration (NOAA) Integrated Calibration and Validation System (ICVS) to monitor the VIIRS long-term radiometric performance. The lunar radiances at pixels are summed for each of the VIIRS Reflective Solar Bands (RSBs) and normalized by the reference band M11 which has the most stable SD-based calibration coefficient. LBRs agree with the SD based F-factor ratios within one percent. Based on analysis with these two independent lunar calibration methods, SD-based and LBR-based calibrations show a lifetime consistency. Thus, it is recommended that LBR be used for both VIIRS radiometric calibration and lifetime stability monitoring.

Keywords: S-NPP; VIIRS; lunar calibration; lunar band ratio; calibration coefficients; F-factor; Solar Diffuser; Miller and Turner; radiometric stability

---

1. Introduction

On 28 October 2011, the Suomi National Polar orbiting Partnership (Suomi NPP) satellite was successfully launched with the Visible Infrared Imaging Radiometer Suite (VIIRS) onboard [1–3]. VIIRS was designed to replace historical sensors such as the Defense Meteorological Satellite Program (DMSP) Operational Line-scanning System (OLS), NASA MODerate-resolution Imaging Spectroradiometer (MODIS), NOAA Polar-orbiting Operational Environment Satellite (POES) Advanced Very High Resolution Radiometer (AVHRR), and Sea-viewing Wide Field-of-view Sensor (SeaWiFS) on GeoEye's SeaStar satellite to fulfill needs of civil, military and science communities [4]. It scans within a large viewing angle of 112.56 degrees at the nominal altitude of 829 km as a whiskbroom scanning radiometer so that the entire Earth can be observed twice daily. VIIRS has 22 spectral bands ranging from 0.412 μm to 12.01 μm and ca2n be further combined into the 14 Reflective Solar Bands (RSB), 7 Thermal Emissive Bands (TEB), and 1 Day Night Band (DNB). The spectral wavelengths from 0.41 μm to 12.01 μm can be observed with moderate resolution and is referred as M-bands whereas those from 0.64 μm to 11.45 μm is used for imaging capabilities, I-bands. DNB is located in the region between 0.5 μm and 0.9 μm.

The nadir spatial resolution of M bands, I bands and DNB are 750 m, 375 m and 750 m, respectively. There are corresponding spectrally similar bands in M bands to those in I bands. The Relative Spectral Responses (RSR) in I2/M7 and I3/M10 band pairs are mostly overlapped, and the I bands provide a higher spatial resolution with similar radiometric responses. In addition, M1, M2, M3, M4, M5, M7, and M13 are designed with low and high gain settings (also known as dual gain) to support a wider dynamic range required for accurate ocean color and land/fire applications.

For monitoring and detection of long-term climate and Earth's environmental changes, a stable radiometric calibration is required for on-orbit imaging sensors. The primary source of the VIIRS RSB band radiometric calibration is based on the Solar Diffuser (SD) and its degradation over time is measured by the Solar Diffuser Stability Monitor (SDSM). The calibration methodology and design of the SD and SDSM are adopted from the two MODIS sensors onboard NASA's Terra and Aqua satellites. The required VIIRS radiometric calibration uncertainty is expected to be less than 2 percent in spectral reflectance unit at the typical radiance [4]. To meet this uncertainty requirement, the SDSM is designed to track the SD reflectance changes over time.

As an alternative light source for calibration, the moon has a long-term stability within the visible and shortwave infrared spectrum ranges. It is widely accepted that the radiometric property of the lunar surface is stable [5,6], however the lunar observations require some further processing for an absolute calibration source. The US Geological Survey (USGS) developed the Robotic Lunar Observatory (ROLO) model used in modeling lunar irradiance for Earth-observing remote sensing instruments covering wavelengths between 0.35 μm and 2.39 μm [6–8]. The ROLO model provides spectral irradiance of the moon considering the variations of the moon phase and libration angles at the sensor's observation point and the Relative Spectral Response (RSR) of the sensor's band pass filter. Recently, Miller and Turner developed a lunar spectral irradiance database covering 0.3 μm to 1.2 μm wavelength range particularly for DNB calibration of the S-NPP VIIRS [5]. Later, the wavelength coverage of this model was extended to up to 2.8 μm. The model is based on convolution of solar irradiance spectra and lunar phase-dependent lunar reflectance, and considers the seasonal changes in Sun/Earth/Moon geometry over seasons using a Look-Up-Table (LUT) approach. In addition, Cao *et al.* developed the Lunar Band Ratio (LBR) method which provides relative radiometric calibration without using a lunar irradiance model [9]. The LBR methodology was successfully implemented for the AVHRR sensor radiometric calibration used in climate change studies [9]. Despite the lunar observation complexity such as incomplete lunar collections, it was demonstrated that the LBR could be used as a scheme to perform lifetime calibration of the Normalized Difference Vegetation Index (NDVI) derived from the AVHRR observation with ± 1 percent (one sigma) level of satiability. The trending results with LBR were also compared to and consistent with the method using high altitude bright cloud developed by Vermote and Kaufman, which was also used in the NASA's Long-Term Data Record (LTDR) project [10].

In this study, two independent methods using a lunar irradiance model and LBR are applied to evaluate radiometric performance of S-NPP VIIRS RSB bands. Using the Miller and Turner (MT) model, the moon-based calibration coefficients (F-factors) are calculated and compared to the primary SD derived F-factors which have been used for generating NOAA's VIIRS Science Data Record (SDR) products. The MT model provides source irradiances after correcting geometric parameters and it acts as a known illumination reference. The lunar F-factor plays an important role in validating the SD based trending results. The LBRs are calculated as the ratio between the sum of the valid pixel digital numbers (DN) and the reference band. The ratio approach factors out the geometric dependencies of the Moon-Earth and Sun-Moon distances significantly reducing seasonal variations. Finally, the radiometric calibration differences among the two lunar based calibration results and SD-based F-factor responses are compared and discussed over the VIIRS' lifetime.

2. SD-Based Calibration and Selection of Reference Band

The VIIRS onboard calibrators include the SD, blackbody (BB) and space view (SV).The RSB calibration depends on SD and SV observations. When S-NPP satellite moves from the night side toward the day side of the Earth near the South Pole, the Sun illuminates the SD panel and the Sun view port of the SDSM through attenuation screens. Since the geometric orientation of the light source, *i.e.*, azimuth and elevation of the Sun, changes for each VIIRS scan, an accurate estimation of the bidirectional reflectance distribution function (BRDF) of SD is necessary. The SD BRDF measurements were performed during the pre-launch testing and updated with the on-orbit yaw maneuver measurements. The BRDF is modified by the SD degradation due to the exposure of SD to sunlight over time. The SD degradation (H-factor) has been monitored by SDSM. The SDSM is a ratio radiometer that measures the rate of SD degradation by taking the ratio between Sun and SD view digital counts (DCs) with proper corrections of the SD and Sun attenuation screens at the time of observation. All the SD and SDSM related trending plots are routinely monitored and available at the NOAA Integrated Calibration Validation System (ICVS) webpage [11]. The details of H-factor are well documented in the VIIRS radiometric calibration algorithm theoretical basis document (ATBD) [4]. Figure 1 shows the ICVS version of the lifetime trending plot of SD degradation.

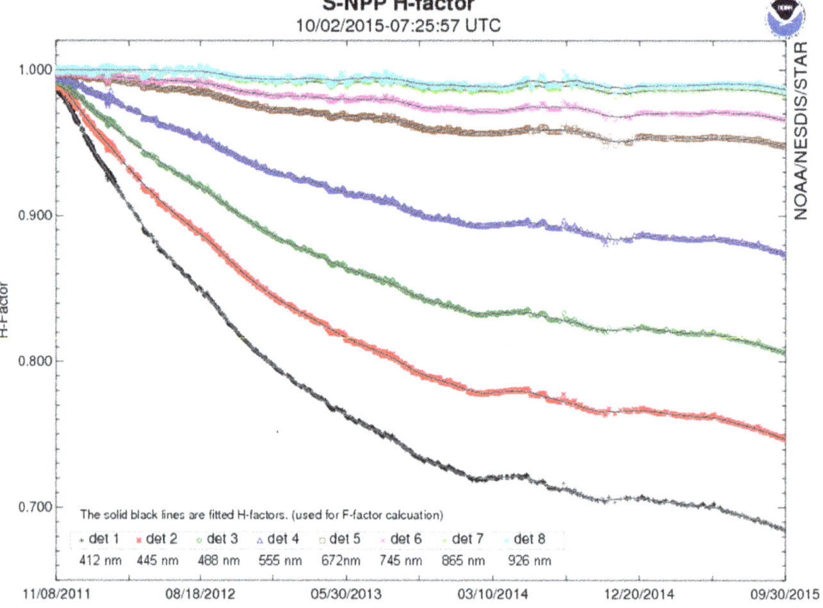

Figure 1. Visible Infrared Imaging Radiometer Suite Solar Diffuser (VIIRS SD) degradation over time as revealed by H factor monitored by Solar Diffuser Stability Monitor (SDSM). In the legend, "det" means SDSM detectors and the detector center wavelength are listed below the detector number. (from ICVS webpage on 30 September 2015 [11]).

Once the H-factor is determined, the RSB calibration coefficients (F-factors) can be calculated using the ratios between the estimated and measured radiance at the time of the SD observations as shown in Equation (1).

$$F = \frac{\cos(\theta_{inc}) \cdot \left[\overline{E_{sun} \cdot \tau_{sds} \cdot H(t) \cdot BRDF}\right] \cdot RVS_{SD}}{c_0 + c_1 \cdot dn_{SD} + c_2 \cdot dn_{SD}^2} \quad (1)$$

The "dn_{SD}" is the offset corrected SD DN, and "RVS_{SD}" is response *versus* scan function at the angle of SD observation, "$c_{0,1,2}$" are the detectors and electronics temperature dependent calibration coefficients, "θ_{inc}" is solar incident angle to the SD screen, "E_{sun}" is solar irradiance, "τ_{sds}" is screen transmittance function, "*BRDF*" is the BRDF function out of on-orbit yaw maneuvers, and "*H(t)*" is SD degradation over time. The H-factors are applied to corresponding VIIRS bands according to the SDSM detector center wavelengths at the SD observation angle. Figure 2 shows the ICVS version of F-factors and they are averaged over the Half Angle Mirror (HAM) sides, gain states, and all the detectors in each band. The F-factors are quite stable over the lifetime of the VIIRS in the visible bands of M1 to M4. On the other hand, the NIR bands M5 to M10 show a gradual increase over time due to detector gain changes along with the S-NPP mirror contamination [2,3]. The SD degradation applied bands, M1~M7, I1 and I2, have a small drop in F factor near the end of each year due to the reduced number of data points (or mis-alignment). The oscillations in the H-factor are caused by the limitations in the definition of the SDSM and SD "sweet spot" [12]. The "sweet spot" is the predefined SDSM and SD geometric ranges that are defined in the Operational Algorithm Description (OAD) at Joint Polar Satellite System (JPSS) webpage [13]. Among all the RSB bands, the band M11 in Figure 2 shows the most stable F-factor response over the VIIRS lifetime. Band M11 has its center wavelength at 2.25 μm and is not affected by SD degradation (H-factor) effects. Therefore, the band M11 is selected as a reference band for calculating LBR in this study.

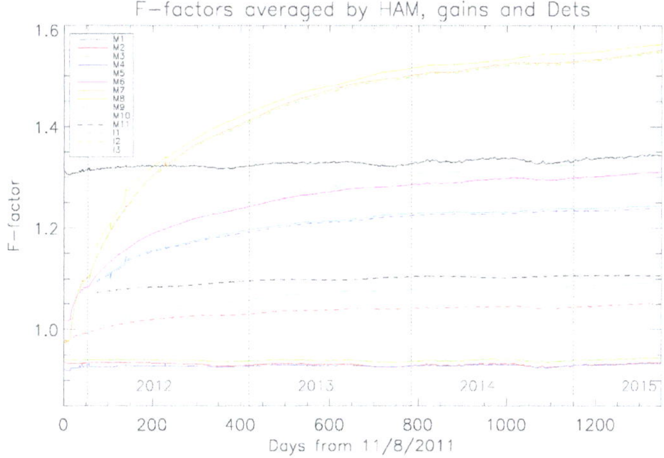

Figure 2. NOAA ICVS VIIRS F-factors averaged in Half Angle Mirror (HAM) sides, gain states and detectors for all the RSB bands.

3. Lunar Calibration and Methodology

3.1. VIIRS Scheduled Lunar Collections

Similar to MODIS, VIIRS views the moon through the SV port from −66.10 to −65.25 degrees from the nadir before the start of the Earth View (EV) reading as shown in Figure 3 [2]. Since the Earth limb angle is located −62 degrees from the nadir for a sensor with 828 km altitude, the location of SV port is approximately 3 degrees above it [14]. The SV port provides 48 frames (or samples) in the scan direction for the M bands and 96 frames for the I bands. The *frame* means the sample in the scan direction. For the scheduled lunar collections, two major constraints of the VIIRS roll maneuver are the roll and lunar phase angles. During the maneuver, the roll angle along with the moon location was determined to place the moon at the center of SV frame. The sector rotation shifts the SV frame to the center of EV. The range of roll angle is limited between 0 and −14 degrees by the operational limits [15]. Because of the narrow SV viewing angle of 0.85 degrees, a spacecraft roll maneuver is required to cover the entire disk for the desired lunar phase angle of −51 degrees.

During the scheduled lunar data collection, a "sector rotation" is performed so that the VIIRS EV angle range is shifted to place the SV data at the center of the EV frame. By doing so, the data coverage of the 0.85 degrees of the SV port is expanded to the 112.56 degrees of the EV range. The other purpose of the sector rotation is that the EV data is corrected for the band-to-band sampling location differences in each band. This means that the sampling time difference of each band has been corrected so that the locations of the moon are aligned in all bands. Figure 4 shows the EV image of band M11 together with the start and end of the sector rotation of the scheduled lunar data collection on 29 May 2015. As expected, the multiple scans of the moon are located at the center of EV frame. Since the DN level of the moon is much lower than the bright EV scene and the fill values, a color map was chosen to visualize the moon acquisition in Figure 4. After the sector rotation, the SV DN values are not valid for the background offset determination and the offset levels need to be calculated using the dark space DN values on either sides of the moon. All of the scheduled lunar data collections of VIIRS used in this study are listed in Table 1. The actual lunar phase angles range between −52 and −50 degree, which greatly limits the uncertainty due to variations in phase angles.

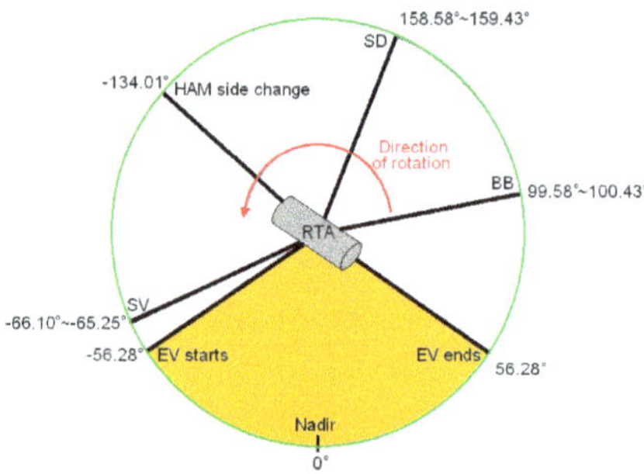

Figure 3. A simplified schematic of VIIRS view *versus* scan angle [2].

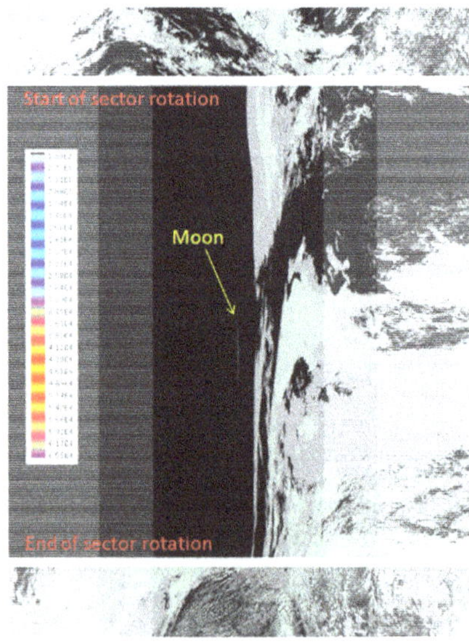

Figure 4. A scheduled lunar data collection for band M11 on 29 May 2015 with the sector rotation.

Table 1. The scheduled moon data collection information for VIIRS.

Date	Time	Phase Angle	Date	Time	Phase Angle
2 April 2012	23:05:32	−51.24	12 March 2014	01:12:08	−51.05
2 May 2012	10:20:25	−50.92	10 April 2014	20:53:40	−50.60
25 October 2012	06:58:38	−51.01	10 May 2014	13:13:21	−50.91
23 November 2012	21:18:43	−50.73	9 June 2014 *	03:49:02	−51.04
23 December 2012	15:01:16	−50.90	4 October 2014 *	17:29:33	−50.80
21 February 2013	09:31:50	−50.71	3 November 2014	01:08:00	−50.52
23 March 2013	03:29:24	−51.15	31 December 2014	19:38:32	−50.73
21 April 2013	19:48:16	−50.82	30 January 2015	08:22:39	−51.16
14 October 2013	21:39:42	−50.94	30 March 2015	16:49:30	−51.29
13 November 2013	06:58:03	−50.66	29 April 2015	12:29:48	−50.43
12 December 2013	19:36:11	−50.39	29 March 2015	04:47:30	−51.07
11 January 2014	10:00:10	−51.30	27 June 2015 *	14:17:10	−54.42
10 February 2014	05:34:37	−51.03	-	-	-

* The asterisks above the date denote that no roll maneuvers were required for the corresponding collections.

3.2. Lunar Irradiance Models

The amount of moonlight observed by satellite radiometers depends on Sun-Moon-Earth geometry, lunar phase (ranging from a new to full moon) and libration of the moon. In this paper, we use a top-of-atmosphere spectral lunar irradiance model developed by Miller and Turner (MT) to evaluate the spectral irradiance of moon derived from VIIRS observation [7]. This model was originally developed in preparation of calibrating nighttime low-light measurements from the Day-Night-Band (DNB) sensors of S-NPP VIIRS to enable quantitative nighttime multispectral applications [16]. The MT model uses solar source observations called Solar Radiation and Climate Experiment (SORCE), lunar spectral albedo data, and accounts for the time varying Sun/Earth/Moon geometry and lunar phase. It produces 1-nm resolution irradiance spectra over the interval from 0.3 μm to 2.8 μm for any given lunar phase. In this model, the scattering property of lunar surface is assumed to be Lambertian. The model has been benchmarked against lunar observations such as with SeaWiFs and MODIS-Aqua satellites and the ROLO lunar irradiance model [5].

As mentioned earlier, ROLO has been extensively used for as an alternative on-orbit radiometric calibration source. Recently, European Organization for the Exploitation of Meteorological Satellites (EUMETSAT) implemented a ROLO based lunar calibration tool for the Global Space-based Inter-Calibration System (GSICS) community [17]. The name of the tool is called GSICS Implementation of the ROLO model (GIRO) and it was successfully developed and validated among different agencies. We have also successfully implemented GIRO as well as MT model. Unfortunately detailed results from the GIRO model cannot be included this paper because the GIRO usage policy has not yet been finalized.

3.3. VIIRS Observed Lunar Irradiance and Lunar F-Factor

As shown in Figure 4, VIIRS lunar irradiance calculation starts with identifying the granules based on the collection time from Table 1. For all of the VIIRS RSB, the selected lunar scans are around the center of the raw EV images. The lunar data for irradiance calculation are selected to include lunar images with a sufficient number of dark pixels from either sides of the moon for dark offset calculation. The center frame number along the scan direction of the lunar scans used in this study is #3152 for dual gain M bands, #1600 for single gain M bands, and #3195 for I bands. Based on the moon frame number, offset values are calculated from dark pixels within two windows to the left and right of the moon along the scan direction. For the dual gain M bands, the chosen window width is 50 frames and the center of the two windows is located at frame #3077 and #3227, respectively. For the single gain M bands and I band, the same window locations are chosen corresponding to the different sampling frequencies. The scan-averaged DN offset-value calculated from the two windows is applied to derive offset-corrected lunar DN for each band (B), detector (D), HAM sides (H), Frame (F) and Scan (N) as shown in Equation (2). This offset-removed DN value is denoted as "*dn*". For all of the scheduled lunar data collections, the gain state is at high gain because of the low radiance level of the moon.

$$\text{dn}(\text{B}, \text{D}, \text{H}, \text{F}, \text{N}) = \text{lunar DN}(\text{B}, \text{D}, \text{H}, \text{F}, \text{N}) - \left\langle \text{DN}(\text{B}, \text{D}, \text{H}, \text{N}) \text{ offset} \right\rangle_{\text{scan averaged}} \quad (2)$$

At aperture-lunar radiance is calculated based on Equation (3), once the background offset is removed from lunar data. The on-orbit "*c*" coefficients are calculated with simultaneous detector and focal plane temperature measurements (T_{det} and T_{ele}) for each *B*, *D*, and *H*. The "*c*" coefficients correct the combined thermal effects caused by the Analog Signal Processor (ASP) and Analog to Digital Converter (ADC) [4]. They are represented as the second order polynomial function of detector and focal plane temperatures considering the non-linear effect. The F-factor is interpolated and applied to Equation (3) at the time of lunar scan for each *B*, *D*, and *H*. Finally, the response *versus* scan angle (RVS) is applied which is very close to unity for the view angle near SV.

$$\text{L}_{\text{pixel}}(B, D, H, F, N) = \frac{F(B, D, H, N) \cdot \sum_{i=0}^{2} c_i(B, D, H, F, N, T_{\text{det}}, T_{ele}) \cdot dn(B, D, H, F, N)^i}{RVS(\theta_{sv}, B, D)} \quad (3)$$

Starting on 9 May 2014, the c_0 was set to zero and c_2 values were recalculated using the prelaunch test results [18]. The updated *c* coefficients reduced errors in data at low radiance level especially for the spectrally similar band pairs of I2/M7 and I3/M10. An in-depth study on the c coefficient update was reported and showed improved accuracy in radiance data measured over a Hawaiian ocean scene [19].

The observed lunar irradiance, E, is calculated by the summation of radiances of all pixels as shown in Equation (4).

$$\text{E(B,D)} = \sum_{Pixel} \frac{\Omega_{\text{Pixel}}}{\text{F}_{\text{Pixel}}} \text{L}_{\text{pixel}} \quad (4)$$

where B denotes band, D denotes detector, Ω_{Pixel} is the solid angle and F_{pixel} is the oversampling factor of the corresponding pixel. The oversampling factor is defined as the number of times that the same lunar surface is viewed by the detector. By the nature of the lunar collection and sensor scanning process, the same detector may see the same lunar surface more than once. For lunar observation of MODIS, the oversampling factor ranged from 1.39 to 5.56 depending on the spatial overlaps between the scans and resolution of the band [20].

To avoid pixel-based solid angle calculation, the lunar irradiance calculation can be simplified with Equation (5) by introducing a new measurable parameter: the effective number of lunar pixels (N_{pixel}).

$$\text{E(B,D)} = \sum_{Pixel} \frac{\Omega_{\text{Pixel}}}{\text{F}_{\text{Pixel}}} \text{L}_{\text{pixel}} = \frac{\Omega_{\text{Pixel}}}{\text{F}_{\text{Pixel}}} \sum_{Pixel} \text{L}_{\text{pixel}} = \frac{\sum_{Pixel} \text{L}_{\text{pixel}}}{N_{pixel}} \cdot \frac{N_{pixel} \cdot \Omega_{\text{Pixel}}}{\text{F}_{\text{Pixel}}} \quad (5)$$

In this study, only complete lunar scans are chosen when lunar F-factor and LBR are calculated. To ensure that a full moon disk is chosen, 5 complete lunar scans are selected for each collection with a 2-pixel margins (750 m resolution) at the start and end of the scans. Since we are using the complete lunar scans only, the pixel solid angle and oversampling factors are constant at the time of the complete lunar scans and these two parameters can be moved out of the summation as common factors. Then, by introducing the factor of N_{pixel} into the equation, the right side of Equation (5) can rearranged to become the multiplication of two factors. The first part becomes mean radiance of the moon and the second part of the equation becomes moon solid angle as shown in Equation (6).

$$\text{E(B,D)} = \overline{Rad_{\text{moon}}} \cdot \Omega_{\text{moon}} \quad (6)$$

Here, $\overline{Rad}_{\text{moon}}$ is the mean radiance of the moon, and Ω_{moon} is the effective moon solid angle at the time of lunar observation.

Since the effective lunar solid angle Ω_{moon} can be expressed as a function of the lunar phase angle (φ) and distance between satellite and Moon ($Dist_{Sat\_Moon}$), Equation (6) is expanded in Equation (7) to provide details of the lunar irradiance calculation.

$$\text{E(B,D)} = \overline{Rad}_{\text{moon}} \cdot \frac{\pi \cdot R_{moon}^2}{Dist_{Sat\_Moon}^2} \cdot \frac{1+\cos(\varphi)}{2} \tag{7}$$

The onboard calibration of the reflective M and I bands of VIIRS uses SD with known BRDF properties. Using a predetermined solar irradiance model, the F-factor is defined as the ratio between the model-based SD radiance and VIIRS-measured radiance and can be used to trend radiometric performance of VIIRS sensors. Similarly, the lunar calibration coefficient (lunar F-factor) can be calculated with Equation (8) based on the ratio between lunar irradiance derived from MT model-and measured lunar irradiance by VIIRS. Since the scheduled lunar data collections are for high gain state only in the case of dual gain M bands, the derived lunar F-factors only trend the radiometric performance of high gain state for these bands.

$$\text{F(B,D)} = \frac{I_{MT\_model}(B)}{E(B,D)} \tag{8}$$

3.4. Lunar Band Ratio (LBR)

Previously, most of the lunar calibrations relied on using the lunar irradiance model such as RObotic Lunar Observatory (ROLO) and recently MT model [6–8] as the reference. The Lunar Band Ratio (LBR) approach is developed and applied as a relative radiometric calibration method and eliminates the need of a lunar irradiance model. As mentioned in the introduction section, the LBR was implemented using lunar observations from the AVHRR and the results were successfully used for climate change trending [9]. In this study, the LBR method is applied to monitor the lifetime radiometric calibration accuracy of the VIIRS RSBs. The LBR used in this paper is defined by Equation (9), which is calculated as a ratio between the sum of the offset-removed "*dn*" for the band of interest and sum of the "*dn*" of the reference band M11. Because the LBR is calculated as a ratio between the band of interest and a reference band for lunar data acquired at the same time, it factors out residual geo-related variations such as Sun-Moon and Moon-Earth distance effects and lunar phase dependence.

$$\text{LBR(B)} = \frac{\sum dn_{Pixel}(B)}{\sum dn_{Pixel}(Band\ M11)} \tag{9}$$

4. Results

4.1. VIIRS Observed Lunar Irradiance Comparisons

The lunar surface observed by the VIIRS sensor is partially illuminated with lunar phase angle of −51.07 degree as shown in Figure 5. Near the nadir frame in band M1~M5, M7 and I bands provides 3 extra samples in scan direction which results in 3 times wide moon shapes in these bands. The phase angle (angle between Moon-Sun and Moon-VIIRS vector) determines the illuminated portion on the moon surface facing the sensor. In addition to the phase angle, the Sun-Moon and Moon-VIIRS distances also vary and depend on the time of data collection. The solid angle of Moon viewed by VIIRS is inversely proportional to the square of the moon and sensor distance and lunar phase as described in Equation (7). Taking all accounts of these lunar view geometry factors at the time of data collection, the lunar irradiances observed by VIIRS are shown in Figure 6 for all VIIRS RSBs since 2012. For each band, lunar irradiance is calculated over all detectors by combining lunar data from multiple lunar scans. Figure 6 shows lunar irradiance has an annual oscillation. The oscillation, has larger impact than the lunar phase because the phase angles of scheduled lunar observations are mostly around −51 degrees as listed in Table 1. During the months of June to September, there are no lunar data collections, because the moon goes below the Earth limb [14]. This is the reason that some of the lunar collections do not require lunar roll maneuver to place the moon in the SV port. In these roll maneuver-free lunar collections, manual inspections of lunar images for dark offset selection are required to ensure correct offset removal for lunar radiance calculation.

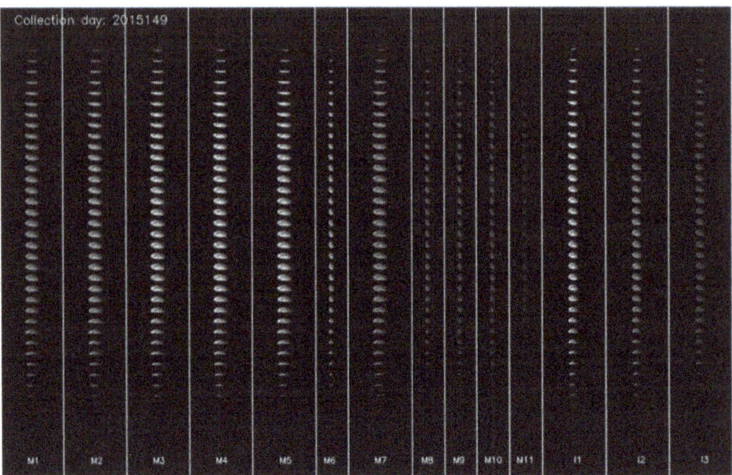

Figure 5. Scheduled lunar data collection for VIIRS on 29 May 2015.

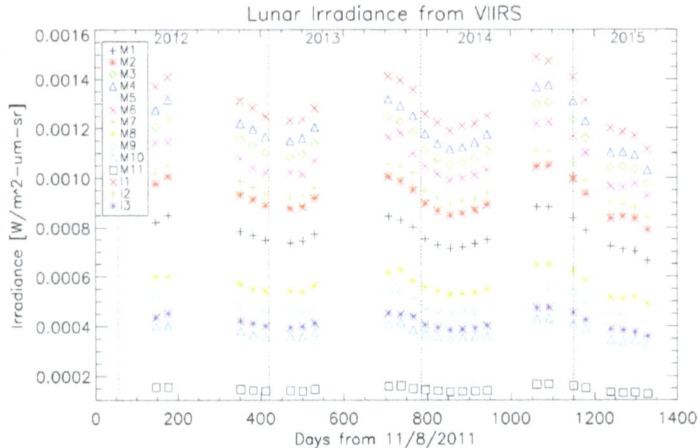

Figure 6. Lunar irradiance from VIIRS lunar observations for all RSBs of VIIRS.

Lunar irradiance models such as the MT model also provide predictions of lunar irradiance for corresponding Sun-Moon-Satellite geometry. Figure 7 shows derived MT model-based lunar irradiance profiles for all VIIRS RSBs. The pattern is very similar with a slight difference when compared to the lunar irradiance derived from VIIRS lunar observations. To visualize the difference, Figure 8 shows the mean and standard deviation of lunar irradiances derived from the MT model and VIIRS observation with regard to the center wavelengths of VIIRS RSBs. Basic shape of irradiance profile is similar to the solar spectrum, since moon reflects solar illumination modulated by the lunar surface reflectance properties. In most cases, MT model has slightly larger values than VIIRS observed values except for the I1 and I2 bands. These differences are mostly caused by the degradation of VIIRS RSB sensors with additional contribution from uncertainty in the lunar model. The S-NPP VIIRS is based on the Solar Spectrum (SOLSPEC) instrument flown on the Atmospheric Laboratory for Applications and Science (ATLAS) spacecraft [4,21], whereas MT model is based on SORCE. Consequently, the band based static biases are expected between the VIIRS observed radiance and the MT model results. The spectrally matched band pairs of I2/M7 responses are very consistent with VIIRS observation and MT model cases in Figures 6 and 7. The symbols are mostly overlapped and the differences are hardly recognizable. However, I3/M10 has notable differences between them, which is cause by the calibration coefficient $c_0 = 0$ update [22].

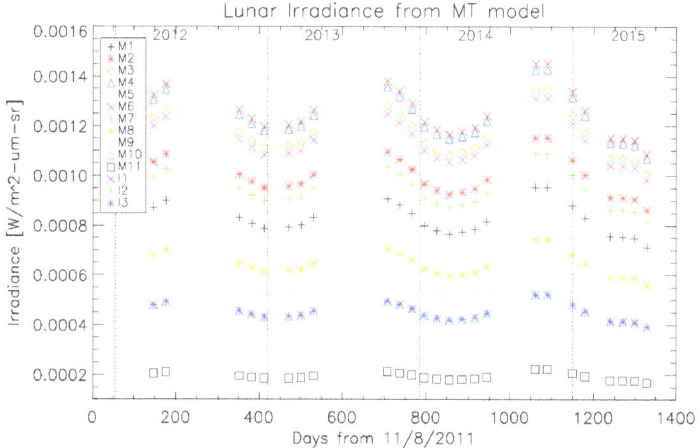

Figure 7. Lunar irradiance calculated with Miller and Turner (MT) model at the center time of VIIRS scheduled lunar collections.

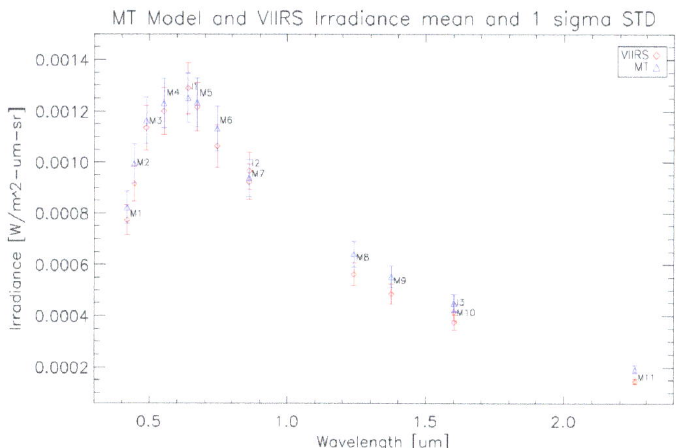

Figure 8. Lifetime mean and 1 sigma range of the Miller and Turner (MT) model and observed lunar irradiance.

4.2. Lunar F-Factor versus SD F-Factor

The model predictions are often used to assist relative trending of sensor performance, because there are uncertainties in the absolute lunar irradiances [5,6]. Therefore, to compare the long term trending of the SD- and lunar F-factors, proper scaling factors need to be applied to the lunar F-factors. For all the VIIRS RSB bands, a best fitting scaling factor is calculated by minimizing the differences between lunar and SD based F-factors. The scaling factors are then applied to the lunar F-factors. The scaled lunar F-factors are over plotted in Figure 9 along with the SD F-factors. There are apparent annual oscillations in the lunar F-factor trending which is similar to the oscillations in the SD F-factor trending. Despite these oscillations, the lunar F-factors in general follow the lifetime trends of the SD F-factors. We further calculate the differences between the two F-factors by interpolating SD-F factors at the time of lunar data collection. The standard deviations of the difference range from 1.20 percent for band M4 to 2.37 percent for band M8. Figure 9 also provides all the standard deviation values on the right side of the figure. Based on these standard deviation values, the lifetime lunar F-factors track SD F-factors are within 2 percent for most bands and within 2.5 percent for bands M8 and M9.

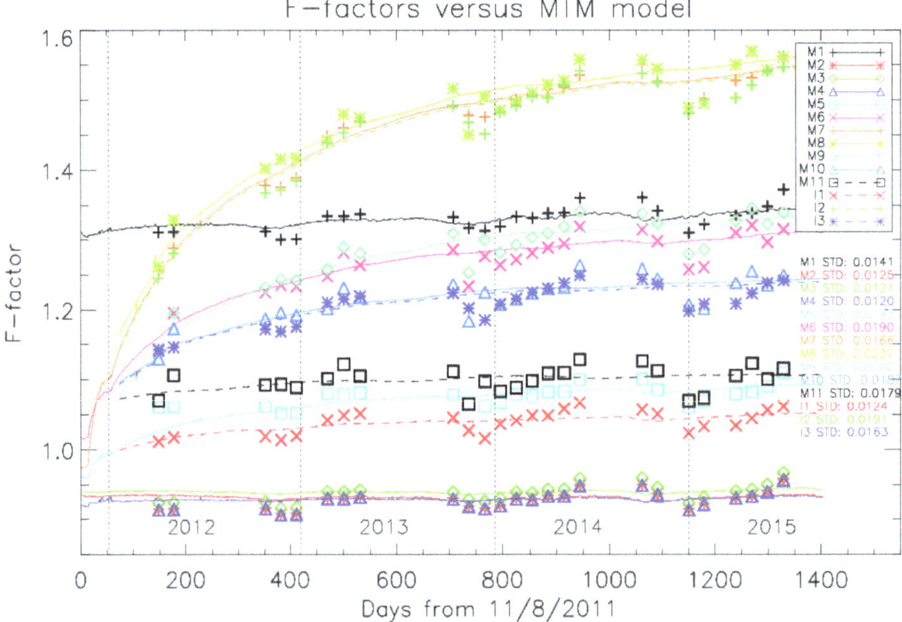

Figure 9. Evolution of VIIRS SD and lunar F-factors for all VIIRS RSBs. The lunar F-factors have been applied with the best fitting scaling factors.

The same F-factor method is applied to the ROLO mode based on EUMETSAT developed GIRO lunar irradiance tool. The ROLO based F-factors provide better agreements to the SD F-factors when they are compared to the MT model based results. Unfortunately, the detailed results cannot be included in this paper because the GIRO data policy has not yet been settled.

4.3. Lunar Band Ratio (LBR) and SD F-Factor Ratio Comparisons

Using Equation (9), LBRs are calculated and normalized with respect to the reference band M11. SD-based F-factors are similarly normalized by M11 in order to compare and validate the lifetime trends between the two independent calibration sources. Figure 10 shows the lifetime trend comparison between the SD-based F-factor ratio and LBR values. We note that the ratio of band M11 response is on the unity line since SD F-factors and LBR are referenced with this band. One advantage of using a ratio approach that can be seen is that the annual oscillations in the lunar model (Figure 9) mostly disappear. The SD F-factor ratios (lines in the figure) are normalized to the value on 18 January 2012 when the cryocooler door was opened and SWIR bands M8 and M11 data became available. A best fitting scaling factor to the LBRs is calculated and applied. The standard deviation values for the differences between LBRs and SD F-factor ratios are also shown next to the band legend on the left side in Figure 10. Compared to the differences between MT model-based and SD-based lunar F-factor, the standard deviations between band ratios are significantly reduced and they are all less than 1 percent level. These improvements in standard deviations are summarized in Table 2. Table 2 also shows the standard deviation differences (Δ%, third row) between the MT-based lunar F-factor and LBR in comparison to SD-based F factors. Significant improvements in comparison are observed with the LBR approach. From Table 2, the percent reduction in standard deviation can be seen to grow larger from 47 percent in the short wavelength M1 band to 92 percent in the longer wavelength M10.

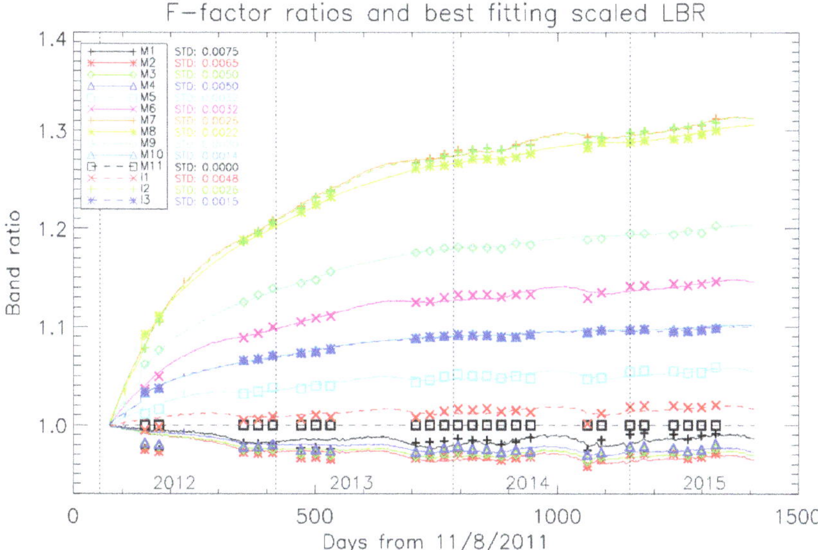

Figure 10. F-factor ratios and lunar LBRs for all VIIRS RSBs. The SD based F-factors and LBR are both normalized to reference band M11.

Table 2. Percent standard deviation improvements when comparing MT model-based lunar F-factor to LBR approaches. Both use SD-F factor as reference in calculating standard deviations.

Band	M1	M2	M3	M4	M5	M6	M7	M8	M9	M10	M11	I1	I2	I3
MT Model STD (%)	1.41	1.25	1.21	1.20	1.43	1.90	1.66	2.37	2.02	1.87	1.79	1.24	1.91	1.63
LBR STD (%)	0.75	0.65	0.50	0.59	0.32	0.32	0.26	0.22	0.20	0.14	-	0.48	0.26	0.15
STD Imp. Δ %	47	48	59	51	78	83	84	90	90	92	-	61	86	90

4.4. Lifetime Trends in Short Wavelength M1-M4 Bands

Recently, a hybrid method was developed combining advantages of ROLO based lunar calibration and SD based calibration [23]. Lunar calibration provides long-term baseline because it has monthly updates whereas SD calibration provides smooth and daily updates. In the hybrid method, the SD F-factors are modulated to match the lifetime lunar calibration trends in M1 to M4 only. Since SD degraded non-uniformly with respect to the incident angle for SDSM and RTA view direction, there was 0.1 percent per degree for detector 1 in band M1 wavelength [23]. With their hybrid approach, F-factor ratios were fitted to quadratic polynomials over time to correct the SD F-factors to the lifetime lunar calibration. To further analyze corresponding lifetime trends in band M1 to M4, Figure 11 shows MT model-based lunar and SD F-factors aligned with the best fitting scaling factor for each band. The lunar F-factors (denoted as symbols) have lifetime trend with positive slopes. They also have relatively larger annual oscillations compared to the SD F-factors. This lifetime drift can be caused by uncertainties in the MT lunar model with respect to geometry parameters.

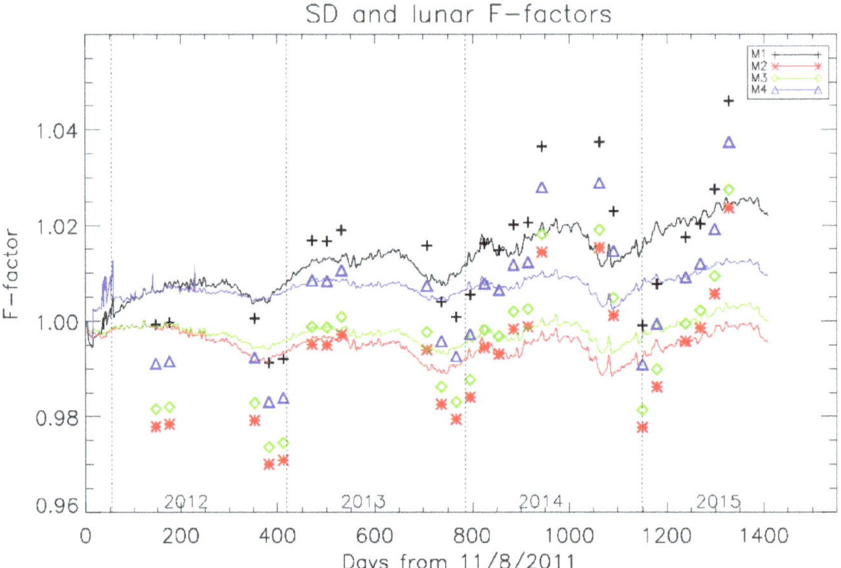

Figure 11. Normalized SD and MT-based lunar F-factors in bands M1 to M4. The lunar F-factors have been scaled to the SD F-factors with the best fitting scaling factors.

When F-factors are compared with to ROLO model *versus* SD, no lifetime time dependent discrepancies are observable. However there are similar annual oscillations to the MT model, the amplitudes of the oscillation are smaller and constant over the years. In addition, the ROLO based lunar F-factors and SD F-factors show excellent agreement between them especially in bands M1 to M4. To validate the lifetime drifts in the VIIRS short wavelengths, LBR results are further investigated in Figure 12.

Figure 12 shows LBR trending results without using a lunar irradiance model. Since the F-factor changes of the reference band M11 over time was faster in 2012, the normalized SD F-factor ratios in M1 to M4 show rapid drop. Once the reference band is stabilized after 2013, the SD F-factor ratios show stable trends with similar annual oscillations as in Figure 11. Comparing the LBR and F-factor ratios show that they are highly consistent with each other without any significant lifetime deviation in the short wavelength bands. The LBR and F-factor ratios agree to within 0.75 percent level of standard deviation for all M1–M4 spectral bands except for the first two LBR values in year 2012. Considering all the inter-comparison results from lunar irradiance model-based and model-free LBR methods, there are no direct evidences of lifetime calibration discrepancies (especially in short wavelength bands) between the SD and lunar based calibration.

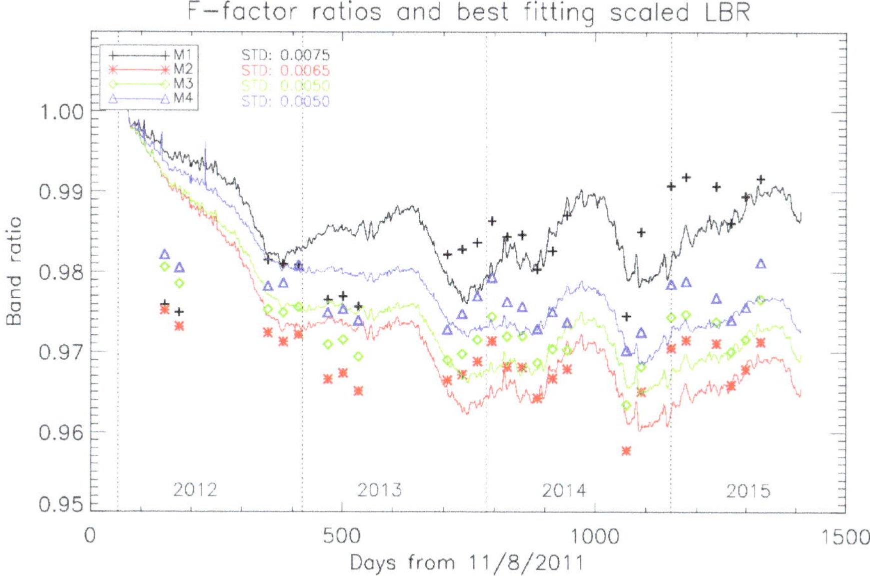

Figure 12. Normalized SD and lunar F-factor ratios in all RSB bands zoomed in M1 to M4 range.

5. Conclusions

The moon serves as a robust and independent on-orbit calibration source for the S-NPP VIIRS sensor without complications of atmospheric effects. In this paper, we presented two independent lunar data processing techniques for sensor radiometric performance trending. One is lunar irradiance model (MT model)-based lunar F-factor method and the other is the reference band-based LBR method. In the MT mode-based lunar F-factor method, we presented details of lunar irradiance calculation steps, including offset estimations, pixel radiance calculation, and effective moon solid angle calculation. The effective lunar solid angle formulation enables us to avoid calculation of complex and unclear over-sampling factors, which is the key factor for lunar calibrations. The MT model-based F-factors are compared to the SD F-factors. The lunar F-factors are within 2 percent standard deviation range for most spectral bands except for M8 and M9 bands, which is within 2.5 percent.

The Lunar Band Ratio (LBR), a simple band ratio method, is applied without using a lunar irradiance model. After the proper offset-removal and effective pixel aggregation; all lunar responses of RSBs are normalized to the reference band M11. The LBR comparison analysis reveals that all the bands consistently agree with the SD F-factor ratio within 1 percent level. The advantage of applying LBR for radiometric instrument performance trending is that it is simple to implement and does not depend on lunar irradiance model. This band ratio approach is not subject to the uncertainties included in the lunar model and factors out geometric dependences. One recognized limitation of the LBR method is that a stable reference band needs to be selected and it can only trend the relative radiometric stability.

This paper presented validations of the SD-based VIIRS lifetime radiometric performances using MT model-based lunar calibration and LBR. The two methods did not show evidences of lifetime differences between SD and lunar based calibrations. The MT model based results show that most of the RSB bands are within 2 percent standard deviation whereas LBR results provide sub percent accuracy especially in bands M1 to M4. The LBR can potentially be very useful to monitor and validate lifetime radiometric trends of on-orbit hyper- or multi-spectral imaging sensors, because it is conceptually simple to apply.

The two lunar calibration methods, MT model and LBR, will be added to the VIIRS ICVS website as independent validations of the current SD based calibration results. These two methods are also going to be applied to the next JPSS1 (J1) VIIRS as a part of radiometric calibration and validation plan.

Acknowledgments: Authors would like to thank Steven Miller for providing the MT lunar irradiance model, which is used for F-factor calculation. Thanks are extended to all NOAA ICVS and VIIRS SDR team members for the invaluable supports. Authors thank to Mitch Schull for his detailed review of the manuscript. The manuscript contents are solely the opinions of the authors and do not constitute a statement of policy, decision, or position on behalf of NOAA or the U.S. government. This work is funded by the NOAA JPSS program.

Author Contributions: Taeyoung Choi collected all the VIIRS lunar data, developed the processing code, performed the analysis, and wrote the manuscript. Xi Shao provided MT lunar irradiance model results and developed the effective lunar solid angle method. Changyong Cao suggested the LBR methodology and provided directions of the study. Fuzhong Weng supported the study and provided technical guidance.

Conflicts of Interest: The authors declare no conflict of interest.

References

1. Schueler, C.; Clement, J.E.; Ardanuy, P.; Welsh, C.; DeLuccia, F.; Swenson, H. NPOESS VIIRS sensor design overview. *Proc. SPIE* **2013**, *4483*. [CrossRef]
2. Cao, C.; Xiong, X.; Blonski, S.; Liu, Q.; Uprety, S.; Shao, X.; Bai, Y.; Weng, F. Suomi NPP VIIRS sensor data record verification, validation, and long-term performance monitoring. *J. Geophys. Res. Atmos.* **2013**, *118*. [CrossRef]
3. Cao, C.; DeLuccia, F.; Xiong, X.; Wolfe, R.; Weng, F. Early on-orbit performance of the visible infrared imaging radiometer suite (VIIRS) onboard the suomi national polar-orbiting partnership (S-NPP) Satellite. *IEEE Trans. Geosci. Remote Sens.* **2014**, *52*, 1142–1156. [CrossRef]
4. Baker, N.; Kilcoyne, H. Joint Polar Satellite System (JPSS) VIIRS Radiometric Calibration Algorithm Theoretical Basis Document (ATBD). Available online: http://jointmission.gsfc.nasa.gov/sciencedocs/2015-06/474-00027_ATBD-VIIRS-Radiometric-Calibration_C.pdf (accessed on 9 October 2015).
5. Kieffer, H.H.; Stone, T.C. The spectral irradiance of the Moon. *Astron. J.* **2005**, *129*, 2887–2901. [CrossRef]
6. Stone, T.C.; Kieffer, H.H. An absolute irradiance of the Moon for on-orbit calibration. *Proc. SPIE* **2002**, *4814*. [CrossRef]
7. Miller, S.D.; Turner, R.E. A dynamic lunar spectral irradiance data set for NPOESS/VIIRS day/night band nighttime environmental applications. *IEEE Trans. Geosci. Remote Sens.* **2009**, *47*, 2316. [CrossRef]
8. Stone, T.C.; Kieffer, H.H. Use of the Moon to support onorbit sensor calibration for climate change measurements. *Proc. SPIE* **2006**, *6296*. [CrossRef]
9. Cao, C.; Vermote, E.; Xiong, X. Using AVHRR lunar observations for NDVI long-term climate change detection. *J. Geophys. Res.* **2009**, *114*. [CrossRef]
10. The Land Long Term Data Record (LTDR) Webpage. Available online: http://ltdr.nascom.nasa.gov/ (accessed on 17 November 2015).
11. The NOAA Integrated Calibration Validation System (ICVS) Webpage. Available online: http://www.star.nesdis.noaa.gov/icvs/status_NPP_VIIRS.php (accessed on 9 October 2015).
12. Fulbright, J.P.; Lei, N.; McIntire, J.; Efremova, B.; Chen, X.; Xiong, X. Improving the characterization and performance of the Suomi-NPP VIIRS solar diffuser stability monitor. *Proc. SPIE* **2013**, *8866*. [CrossRef]
13. Joint Polar Satellite System (JPSS) Webpage. Available online: http://jointmission.gsfc.nasa.gov/documents.html (accessed on 9 October 2015).
14. Patt, F.S.; Eplee, R.E.; Barnes, R.A.; Meister, G.; Butler, J.J. Use of the moon as a calibration reference for NPP VIIRS. *Proc. SPIE* **2005**, *5882*. [CrossRef]
15. Fulbright, J.P.; Wang, Z.; Xiong, X. Suomi-NPP VIIRS lunar radiometric calibration observations. *Proc. SPIE* **2014**, *9218*. [CrossRef]
16. Shao, X.; Cao, C.; Uprety, S. Vicarious calibration of S-NPP/VIIRS day-night band. *Proc. SPIE* **2013**, *8866*. [CrossRef]
17. Lunar Calibration Algorithm Work Area. Available online: https://gsics.nesdis.noaa.gov/wiki/Development/LunarWorkArea (accessed on 9 October 2015).

18. DeLuccia, F.J. VIIRS reflective solar band (RSB) performance and uncertainty estimates. In Proceedings of Suomi NPP SDR Science and Products Review, College Park, MD, USA, 18–20 December 2013.
19. Blonski, S.; Cao, C.; Shao, X.; Uprety, S. VIIRS reflective solar bands calibration changes and potential impacts on ocean color applications. *Proc. SPIE* **2014**, *9111*. [CrossRef]
20. Xiong, X.; Sun, J.; Barnes, W. Intercomparison of On-Orbit Calibration Consistency between Terra and Aqua MODIS Reflective Solar Bands Using the Moon. *IEEE Trans. Geosci. Remote Sens.* **2008**, *5*, 778–782. [CrossRef]
21. Thuillier, G.; Herse, M.; Labs, D.; Foujolos, T.; Peetermans, W.; Gillotay, D.; Simon, P.C.; Mandel, H. The Solar Spectral Irradiance from 200 to 2400 nm as Measured by the SOLSPEC Spectrometer from the ATLAS and EURECA Missions. *Solar Phys.* **2013**, *214*, 1–22. [CrossRef]
22. Choi, T.; Sun, N.; Chen, W.; Cao, C.; Weng, F. S-NPP VIIRS significant events in 2014 monitored by NOAA integrated calibration validation system. In Proceedings of the NOAA STAR JPSS 2015 Annual Science Team Meeting, College Park, MD, USA, 24–28 August 2015.
23. Sun, J.; Wang, M. VIIRS reflective solar bands calibration improvements with hybrid approach. In Proceedings of the NOAA STAR JPSS 2015 Annual Science Team Meeting, College Park, MD, USA, 24–28 August 2015.

Article

Monitoring the NOAA Operational VIIRS RSB and DNB Calibration Stability Using Monthly and Semi-Monthly Deep Convective Clouds Time Series

Wenhui Wang [1,*] and Changyong Cao [2]

1 Earth Resource Technology, Inc., NCWCP, E/RA21, 5830 University Research Ct., Suite 2664, College Park, MD 20740, USA

2 NOAA (National Oceanic and Atmospheric Administration)/NESDIS (National Environmental Satellite, Data, and Information Service)/STAR (Center for Satellite Applications and Research), NCWCP, E/RA2, 5830 University Research Ct., Suite 2730, College Park, MD 20740, USA; changyong.cao@noaa.gov

* Correspondence: wenhui.wang@noaa.gov; Tel.: +1-301-683-3531; Fax: +1-301-683-3526

Academic Editors: Richard Müller and Prasad S. Thenkabail

Received: 30 October 2015; Accepted: 25 December 2015; Published: 4 January 2016

Abstract: The Visible and Infrared Imaging Radiometer Suite (VIIRS) onboard the Joint Polar Satellite System (JPSS)/Suomi National Polar-Orbiting Partnership (SNPP) satellite provide sensor data records for the retrievals of many environment data records. It is critical to monitor the VIIRS long-term calibration stability to ensure quality EDR retrieval. This study investigates the radiometric calibration stability of the NOAA operational SNPP VIIRS Reflective Solar Bands (RSB) and Day-Night-Band (DNB) using Deep Convective Clouds (DCC). Monthly and semi-monthly DCC time series for 10 moderate resolution bands (M-bands, M1–M5 and M7–M11, March 2013–September 2015), DNB (March 2013–September 2015, low gain stage), and three imagery resolution bands (I-bands, I1–I3, January 2014–September 2015) were developed and analyzed for long-term radiometric calibration stability monitoring. Monthly DCC time series show that M5 and M7 are generally stable, with a stability of 0.4%. DNB has also been stable since May 2013, after its relative response function update, with a stability of 0.5%. The stabilities of M1–M4 are 0.6%–0.8%. Large fluctuations in M1–M4 DCC reflectance were observed since early 2014, correlated with F-factor (calibration coefficients) trend changes during the same period. The stabilities of M8-M11 are from 1.0% to 3.1%, comparable to the natural DCC variability at the shortwave infrared spectrum. DCC mean band ratio time series show that the calibration stabilities of I1–I3 follow closely with M5, M7, and M10. Relative calibration changes were observed in M1/M4 and M5/M7 DCC mean band ratio time series. The DCC time series are generally consistent with results from the VIIRS validation sites and VIIRS/MODIS (the Moderate-resolution Imaging Spectroradiometer) simultaneous nadir overpass time series. Semi-monthly DCC time series for RSB M-bands and DNB were compared with monthly DCC time series. The results indicate that semi-monthly DCC time series are useful for stability monitoring at higher temporal resolution.

Keywords: Suomi NPP; VIIRS; calibration stability monitoring; deep convective clouds; reflective solar bands; visible near infrared bands; shortwave infrared bands; day-night-band

---

1. Introduction

The Visible and Infrared Imaging Radiometer Suite (VIIRS) onboard the Joint Polar Satellite System (JPSS)/Suomi National Polar-Orbiting Partnership (SNPP) satellite has 22 spectral bands, with 14 Reflective Solar Bands (RSB), 7 Thermal Emissive Bands (TEB) and 1 Day-Night-Band (DNB) [1,2]. VIIRS RSBs are calibrated using a full-aperture Solar Diffuser (SD) and the degradation of SD is monitored by a Solar Diffuser Stability Monitor (SDSM). Significant SD, SDSM detectors, and Rotating

Telescope Assembly (RTA) mirror degradations were observed, especially early after launch. The rate of RTA mirror and SD degradations has decreased since mid-2013 and early 2014, respectively. Their impacts on the instrument performance are negligible (at 0.1% level) due to weekly updates of RSB calibration coefficients (F-factor) Look-Up Table (LUT) [3,4]. However, larger F-factor fluctuations have been observed since early 2014 in some shorter wavelength bands, with band M1 fluctuating the most (~3%). The underlying cause for the F-factor anomalies has not been identified so far. VIIRS RSB Sensor Data Records (SDR) provide input data for the retrieval of many Environment Data Records (EDR) products, such as ocean color, vegetation, cloud, and surface albedo. It is therefore critical to monitor the long-term stability of VIIRS RSBs to ensure quality EDR products.

VIIRS DNB is a panchromatic imagery band (0.5–0.9 μm) with dynamic range of approximately seven orders of magnitude [5]. It has three gain stages: the Low Gain Stage (LGS, daytime), the Medium Gain Stage (MGS, twilight) and the High Gain Stage (HGS, nighttime). The onboard calibration of DNB LGS is similar to that for RSBs, with LGS gain derived using SD/SDSM data. DNB MGS and HGS are calibrated using LGS gain and MGS/LGS and HGS/LGS gain ratios estimated using Earth View (EV) data at the terminator orbit where EV observations at the three gain stages co-exist. VIIRS DNB is originally designed as an imagery band to continue the night observation heritage of the Operational Linescan System (OLS) onboard the Defense Meteorological Satellite Program (DMSP). DNB nighttime observations have been widely used in many areas such as detecting power outrage, monitoring city lights, urban expansion and fishing boats, as well as studying air glow, aurora, and lightning. Due to its superior spatial and radiometric performance, quantitative applications of the VIIRS DNB band have been developed or are under development, such as estimating population and economic output [6,7] and nighttime aerosol retrieval [8].

The JPSS Interface Data Processing Segment (IDPS) ground processing team has produced nearly four years of NOAA operational VIIRS SDRs to date, with several major calibration changes applied since launch. The JPSS program is currently planning to reprocess the entire VIIRS SDR products. To better facilitate quantitative applications and the future reprocessing of the VIIRS RSB and DNB SDRs, it is important to use independent validation time series to evaluate their long-term post-launch calibration stability. Many efforts have been devoted to characterize the long-term stability of the NOAA operational VIIRS RSBs and DNB (HGS). Uprety *et al.* [9] and Uprety and Cao [10] investigated the VIIRS onboard RSB radiometric performance using the extended Simultaneous Nadir Overpass (SNO-x) approach over desert sites, and the Antarctica Dome C site. Wang and Cao [11] provided a preliminary assessment of the VIIRS bands M1–M5 and M7 calibration stability using monthly Deep Convective Clouds (DCC) time series. Liao *et al.* [5] evaluated DNB HGS radiometric calibration accuracy using vicarious calibration under lunar illumination. Ma *et al.* [12] investigated the DNB HGS calibration accuracy by comparing the observed VIIRS DCC radiance with nighttime DCC radiances simulated using a radiative transfer model. Cao and Bai [13] investigated the feasibility of using point light sources for monitoring DNB HGS calibration stability.

The purpose of this study is to investigate the NOAA operational VIIRS RSB and DNB (LGS) long-term radiometric calibration stabilities using the DCC technique. It extends the previous study [11], which provides preliminary results for the moderate resolution visible and near infrared (VIS/NIR) bands (M-bands) only, by using longer time series and analyzing more bands, including DNB, shortwave infrared bands (SWIR) and imagery resolution bands (I-bands). Moreover, semi-monthly DCC time series were also developed and analyzed for stability monitoring at higher temporal resolution. In this study, the DCC technique was applied to all RSB bands, including 10 M-bands and 3 I-bands. Stable DNB LGS radiometric calibration is a prerequisite for quality DNB nighttime calibration to support quantitative applications of DNB data. Previous studies have focused on characterizing HGS data directly, but the radiometric calibration stability of DNB LGS has not been well studied so far. This paper is organized as follows. Section 2 introduces VIIRS SDR products used in this study. Section 3 presents the DCC technique for RSBs and DNB. Section 4 presents and discusses monthly and semi-monthly DCC time series for calibration change and long-term stability monitoring. Finally Section 5 summarizes the findings of the study.

2. VIIRS SDR Products Used

VIIRS has 11 RSB M-bands (M1–M11, 750 m) and 3 RSB I-bands (I1–I3, 375 m). M6 saturates over DCCs, therefore it was excluded from this study. TEB band M15 (10.763 μm) provides brightness temperature measurements that are required to identify DCC pixels for RSB M-bands and DNB. I1–I3 DCCs are identified using I5 (11.469 μm) brightness temperature. The center wavelength, spatial, and radiometric characteristics of all bands used in this study are summarized in Table 1. Figure 1 presents the Relative Spectral Response (RSR) functions for these bands.

Table 1. Spectral, spatial, and radiometric characteristics of Visible and Infrared Imaging Radiometer Suite (VIIRS) spectral bands used in this study. M1–M5 and M7 SNRs at low gain stage and typical scene radiances are listed because Deep Convective Clouds (DCCs) are highly reflective; band M15 and I5 NEdTs is estimated at 205 K scene temperature.

	Band	Center Wavelength (μm)	Spatial Resolution at Nadir (m)	SNR/NEdT (Spec)	SNR/NEdT (On-Orbit)
VIS/NIR	M1	0.411	750	316	1045
	M2	0.444	750	409	1010
	M3	0.486	750	414	988
	M4	0.551	750	315	856
	M5	0.672	750	360	631
	I1	0.639	375	119	214
	M7	0.862	750	340	631
	I2	0.862	375	150	264
SWIR	M8	1.238	750	74	221
	M9	1.375	750	83	227
	M10	1.602	750	342	586
	I3	1.602	375	6	149
	M11	2.257	750	10	22
DNB	DNB	0.700	750	⩾6@Lmin	>9 across scan after degradation
TEB	M15	10.729	750	0.26 K	0.10 K
	I5	11.469	375	1.7 K	0.43 K

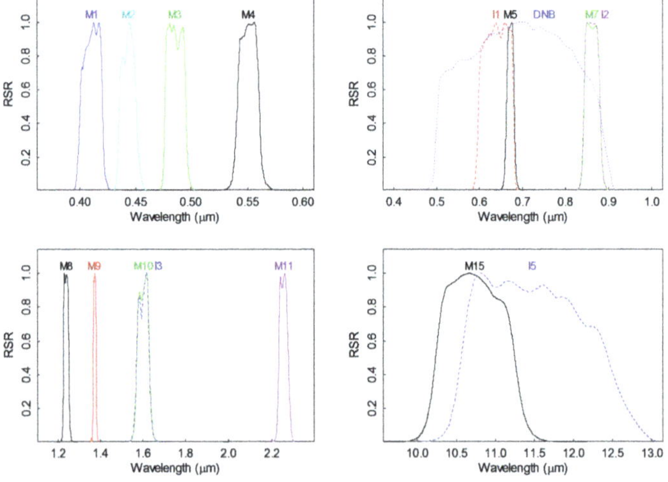

Figure 1. Relative spectral response functions (RSR) of all VIIRS bands used in this study.

The area of interests of this study is a region defined as 25°S to 25°N and 150°W to 60°W. This area covers a portion of the InterTropical Convergence Zone (ITCZ) over the western end of tropical Pacific Ocean and its adjacent South America Continent. DCCs over the same region were also used in previous studies [11,14]. VIIRS daytime Top of Atmosphere (TOA) reflectance/brightness temperature, and radiance were downloaded from the NOAA's Comprehensive Large Array-data Stewardship System (CLASS), the NOAA/NESDIS/STAR Central Data Repository (a 4 months revolving mirror site of CLASS for SNPP data), and the NASA Atmosphere Science Investigator-led Processing Systems (SIPS). All VIIRS SDR products used in this study are generated by the NOAA operational ground processing unit at the JPSS IDPS. It is worth noting that no vicarious calibration is applied to the IDPS version of VIIRS RSB and DNB SDR products.

3. The DCC Technique for VIIRS

DCCs are extremely cold clouds above which the absorptions due to water vapor and other gases are minimal in the VIS/NIR spectrum. DCCs are abundant over the ITCZ and can be simply identified using a single longwave infrared (LWIR) channel centered at ~11 μm brightness temperature (TB11 hereafter). Hu *et al.* [15] first demonstrated that DCCs have a constant mean albedo over the lifetime of the Clouds and the Earth's Radiant Energy System (CERES) on board the Tropical Rainfall Measuring Mission (TRMM) satellite. The DCC technique outlined by Hu *et al.* [15] was further improved through various studies such as Doelling *et al.* [16,17], Minnis *et al.* [18], and Fougnie and Bach [19]. It has been widely used for post-launch calibration and stability monitoring in the solar reflective spectrum during the past decade [11,16–21]. The DCC technique generally consists of the following steps: (1) collecting satellite data over an area of interest; (2) identifying DCC pixels; (3) correcting for the anisotropic effect in DCC reflectance; (4) calculating monthly DCC probability distribution functions (PDF); (5) generating and analyzing monthly DCC mean and mode time series. It is a statistical-based vicarious calibration method; therefore sufficient DCC samples need to be collected to ensure robust statistical analysis results.

In this study, VIIRS DCC pixels are identified using the Wang and Cao [11,14] method, which is similar to those described in Minnis *et al.* [18] and Doelling *et al.* [17]. The criteria for identifying VIIRS M-bands DCC pixels are summarized as follow: (1) M15 brightness temperature is less than 205 K; (2) standard deviation of TB11 of the subject pixel and eight adjacent pixels is less than 1 K; (3) standard deviation of TOA reflectance of the subject pixel and eight adjacent pixels is less than 3% relative to the mean reflectance of the nine pixels; (4) solar zenith angle is less than 40°; (5) sensor view zenith angle is less than 35°.

The identifications of DNB and I-bands DCCs are similar to that for M-bands. DNB radiances were mapped to M15 lat/lons and converted to TOA reflectance before DCC pixels were identified using band M15 brightness temperatures. Band I5 brightness temperatures were used for the I-band DCC identifications. Wang and Cao [14] found that the mode and mean of DCC reflectance is a function of spatial resolution but not sensitive to brightness temperature difference on the order of 0.5 K. In this study, I1–I3 TOA reflectance and I5 brightness temperatures were down-sampled to M-band resolution before DCCs are identified to facilitate inter-comparison of DCC time series between M-bands and I-bands. Though different TB11 bands were used for M-bands/DNB and I-bands DCC identifications, the mean M15 brightness temperature is only ~0.2 K higher than the mean I5 brightness temperatures for the down-sampled I-bands DCCs, therefore its impact can be ignored.

RSB M-bands and DNB DCC TOA reflectance as well as TB11 brightness temperature datasets over the area of interest were generated using the DCC identification criteria described above from March 2012 to September 2015. The study period for RSB I-bands is from January 2014 to September 2015. Though DCCs have nearly Lambertian behavior, the anisotropic effect still exists in the DCC TOA reflectance and Angular Distribution Models (ADM) were developed to account for the effect [15,17,22]. For each DCC pixel, the ADM-adjusted DCC reflectance datasets were also generated using an ADM developed by Hu *et al.* [15]. Monthly PDFs, as well as their means and modes, were calculated for the DCC TOA reflectance and the ADM-adjusted DCC reflectance (DCC reflectance hereafter) with a 0.003 increment (in reflectance) [14]. Figure 2 shows examples of monthly PDFs for bands M1, M7, M9, and DNB ADM-adjusted DCC reflectance. Standard deviations (sd), minimum (Min), maximum (Max), and range (Max−Min) were calculated using DCC datasets for the selected months. Monthly DCC statistics for all bands during the entire study period are summarized in Table 2 (before the ADM adjustment) and Table 3 (after the ADM adjustment).

The monthly DCC statistics derived from this study are generally consistent with previous studies in the VIS/NIR spectrum. The averaged monthly DCC TOA reflectance is larger than 0.94 for all VIS/NIR bands and DNB, confirming that DCCs are highly reflective [15,17,19]. The range of band M5 monthly DCC mean TOA reflectance is ~2.8% (see Table 2), smaller then the ±2% range reported by Fougnie *et al.* [19]. The ADM-adjusted DCC reflectance is more invariant compared to the DCC TOA reflectance (see Tables 2 and 3). The range of band M5 monthly DCC reflectance is further reduced after the ADM-adjustment, agreeing with the results from Doeling *et al.* [17]. Therefore, the ADM-adjusted monthly DCC reflectance (DCC reflectance hereafter) was used to analyze VIIRS VIS/NIR and DNB radiometric calibration stabilities. Detailed analysis of VIIRA VIS/NIR bands and DNB DCC time series are presented in Sections 4.1, 4.2 and 4.4.

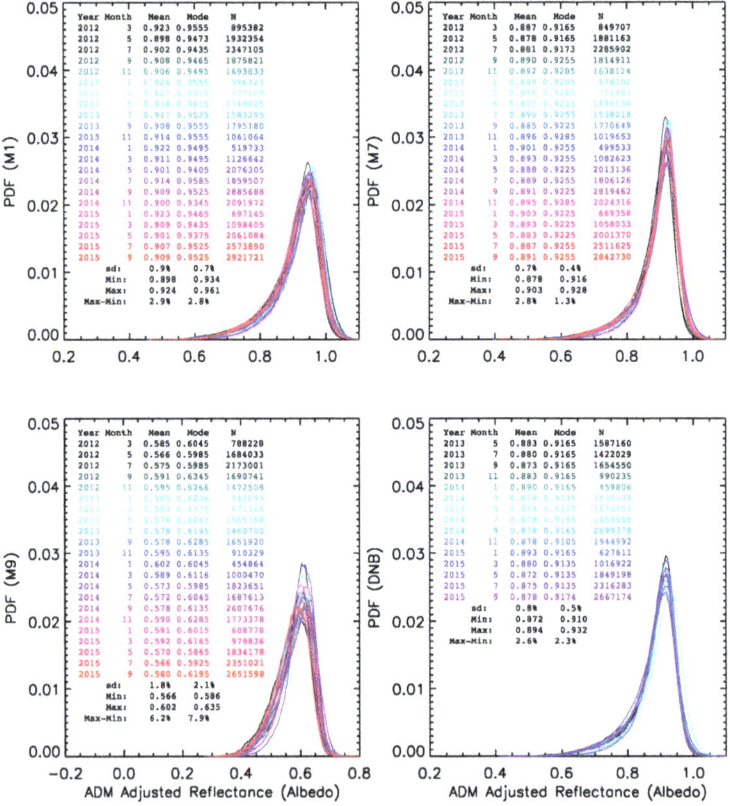

Figure 2. Probability distribution functions of VIIRS bands M1, M7, M9, and Day-Night-Band (DNB) Angular Distribution Models (ADM)-adjusted monthly DCC reflectance. Standard deviation (sd), minimum (Min), maximum (Max), and range (Max−Min) were calculated using DCC data from selected months.

Table 2. Statistics of monthly DCC Top of Atmosphere (TOA) reflectance (before the ADM adjustment) mean and mode time series for Reflective Solar Bands (RSB) M-bands (M1–M5, M7–M11, March 2012–September 2015), DNB (May 2013–September 2015), and RSB I-bands (I1–I3, January 2014–September 2015).

Band		DCC Mode			DCC Mean		
		Avg	sd (%)	Max−Min (%)	Avg	sd (%)	Max−Min (%)
VIS/NIR	M1	1.018	1.0	4.124	0.977	1.0	3.825
	M2	1.009	1.1	4.461	0.967	1.1	5.107
	M3	1.005	1.0	4.178	0.963	1.0	4.187
	M4	0.973	0.9	4.009	0.931	0.9	3.534
	M5	1.002	0.8	3.294	0.959	0.7	2.804
	I1	0.963	0.8	3.740	0.919	0.9	3.544
	M7	0.991	0.6	2.725	0.955	0.7	2.637
	I2	0.991	0.6	1.816	0.954	0.8	3.355
SWIR	M8	0.704	1.0	3.834	0.698	1.0	3.250
	M9	0.655	1.7	6.387	0.625	1.7	6.777
	M10	0.230	3.1	14.332	0.232	3.1	10.342
	I3	0.230	3.2	13.930	0.232	3.1	10.050
	M11	0.371	2.2	8.398	0.371	2.2	7.199
DNB	DNB	0.982	0.8	3.361	0.944	0.9	3.110

Table 3. Statistics of monthly DCC reflectance (after the ADM adjustment) mean and mode time series for all RSB M-bands (M1–M5, M7–M11, March 2012–September 2015), DNB (May 2013–September 2015), and RSB I-bands (I1–I3, January 2014–September 2015).

Band		DCC Mode			DCC Mean		
		Avg	sd (%)	Max−Min (%)	Avg	sd (%)	Max−Min (%)
VIS/NIR	M1	0.950	0.8	3.157	0.911	0.8	2.961
	M2	0.942	0.8	4.461	0.902	1.0	4.782
	M3	0.938	0.8	4.158	0.897	0.9	4.232
	M4	0.908	0.6	2.972	0.868	0.7	3.606
	M5	0.936	0.4	1.924	0.894	0.6	2.756
	I1	0.898	0.5	2.004	0.856	0.8	3.407
	M7	0.924	0.4	1.299	0.891	0.7	2.803
	I2	0.924	0.4	1.624	0.889	0.8	3.227
SWIR	M8	0.656	1.1	3.938	0.650	1.1	3.378
	M9	0.611	1.9	7.860	0.582	1.8	6.628
	M10	0.214	3.5	14.478	0.216	3.3	10.862
	I3	0.214	3.3	14.037	0.216	3.3	11.024
	M11	0.345	2.6	10.427	0.346	2.5	8.065
DNB	DNB	0.917	0.5	2.291	0.881	0.8	3.701

The monthly DCC statistics derived from this study also generally agree with previous studies in the SWIR spectrum [17,20]. DCCs are less reflective in the SWIR spectrum, with the averaged monthly DCC mean reflectance ranging from ~0.2 to ~0.7 for VIIRS band M8–M11 and I3. DCCs are also less stable in the SWIR spectrum. The standard deviations of ranges of SWIR bands monthly mean and mode of DCC TOA reflectance are much larger than those for the VIS/NIR bands in majority of cases (see Table 2). Our results indicate that the Hu *et al.* [15] ADM used in this study is ineffective for the SWIR bands. The ADM-adjusted DCC reflectance has larger variance than the DCC TOA reflectance before the ADM adjustment (see Tables 2 and 3). Therefore, the DCC TOA reflectance without ADM adjustment was used to characterize the VIIRS SWIR bands calibration stability. More effort is needed to develop ADM for the SWIR bands in the future. In-depth analysis of VIIRS SWIR bands DCC time series are presented in Sections 4.3 and 4.4.

4. Results and Discussions

4.1. Calibration Stability for VIS/NIR M-Bands

4.1.1. Monthly VIS/NIR DCC Time Series

The DCC technique uses the mean and mode of the monthly DCC PDFs as the two important indices for calibration stability monitoring. Doelling *et al.* [17] shows that the mode of monthly DCC reflectance in the VIS/NIR spectrum is more stable than the mean in terms of TB11 threshold, spatial uniformity thresholds, and the choice of ADMs, as well as regional and temporal variations. Wang and Cao's [14] DCC radiometric sensitivity studies based on VIIRS bands M5, M7, and I2 DCCs indicated that the mode of DCC reflectance is also more stable in terms of spatial resolution, TB11 threshold and calibration bias, and DCC cluster size. Therefore we used the mode method for monitoring the calibration stability of individual VIS/NIR bands. It is worth noting the accuracy of the mode method is restricted by the the 0.003 increment used in this study.

Figure 3 shows monthly DCC mode time series for M1–M5 and M7 from March 2012 to September 2015. DCC mean time series are illustrated in Figure 4 for comparison purposes. The DCC mean time series exhibit similar patterns, however, with noticeable seasonal variations compared to the mode time series. Table 3 summarizes the statistics of monthly DCC time series for all individual VIS/NIR bands discussed in this subsection and the following subsections. The DCC mode time series reveal that the stabilities (calculated as the standard deviations of time series) are ~0.8% or less, with ranges less than 4.5% for all VIS/NIR M-bands investigated. Similar to our previous study for the time period from March 2013 to August 2014 [11], the calibration of bands M5 and M7 continue to be more stable than the other four VIS/NIR bands, with a stability of 0.4% and ranges of 1.9% and 1.3%, respectively. However, band M7 shows an obvious calibration jump (~1%) around September 2012; an M5 calibration anomaly was observed in February of 2015, similar to bands M1–M4. The calibrations of bands M1–M4 are less stable. The standard deviations for these four bands are 0.6% to 0.8% and ranges are from ~3.0% to 4.5%. Moreover, noticeable degradations can be observed in M2–M4 DCC mode time series, with M2 showing the most degradation (0.45%/year from May 2012 to September 2015). 95% confidence intervals shown in Figure 3 indicate the downward trends are robust for these 3 bands. Uprety and Cao [10] investigated the radiometric calibration stability of the NOAA operational calibrated VIIRS RSB bands. The bands M1 and M5 trends derived from this study are generally consistent with their findings based on normalized Bidirectional Reflectance Distribution Function (BRDF) time series over desert and Dome C sites. However, the results for bands M2–M4 and M7 from this study disagree with the results reported by Uprety and Cao's [10].

Several major VIIRS calibration change events [23] are clearly shown in these time series. The April 2012 IDPS code and LUT changes and the October 2013 calibration changes due to the misalignment of SDSM with VIIRS instrument [24] are obvious in M2–M4 DCC mode time series. The F-factor trend changes since February 2014 and the switching of VIIRS operational H-Factor in the RSB calibration to the high fidelity version on 22 May 2014, have the most significant impacts on bands M1–M4. M1–M4 monthly DCC reflectance for June 2014 jumped by ~2.3%, 1.8%, 1.8%, and 0.9%, respectively, compared to the previous two months (see Figure 3).

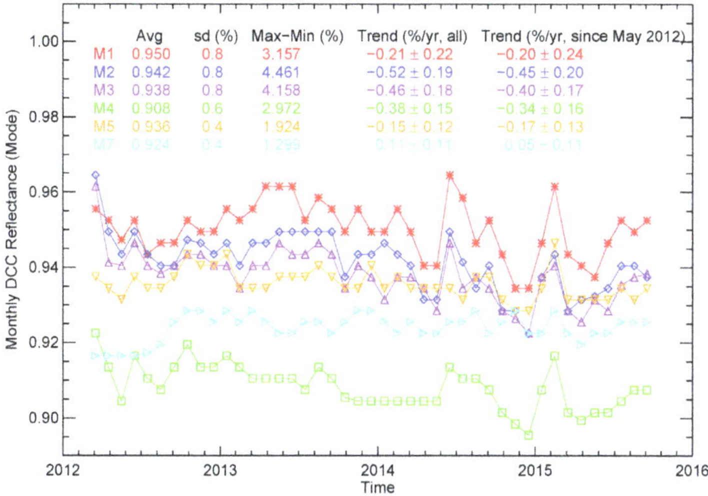

Figure 3. Monthly DCC mode time series for bands M1–M5 and M7 (March 2012–September 2015).

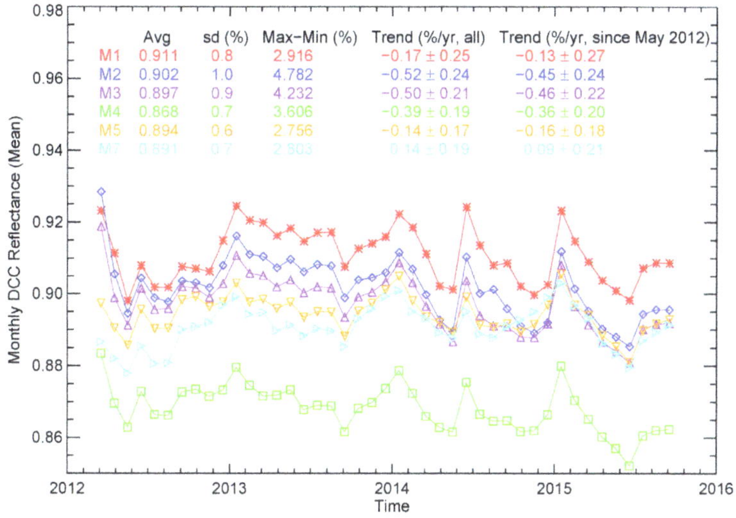

Figure 4. Monthly DCC mean time series for bands M1–M5 and M7 (March 2012–September 2015).

Figure 5 compares M1–M4 DCC mode time series and the corresponding IDPS F-factor time series. Our results indicate that the VIS/NIR DCC reflectance variations are correlated with IDPS F-factor changes, with correlation coefficients (R) ranging from 0.35 to 0.54. Larger fluctuations in the band M1–M4 monthly DCC reflectance observed since early 2014 are generally consistent with IDPS F-factor trend changes during this period. The coincidence of changes in monthly DCC reflectance and F-factor time series further supports that the DCC technique is valuable for VIIRS RSB calibration monitoring.

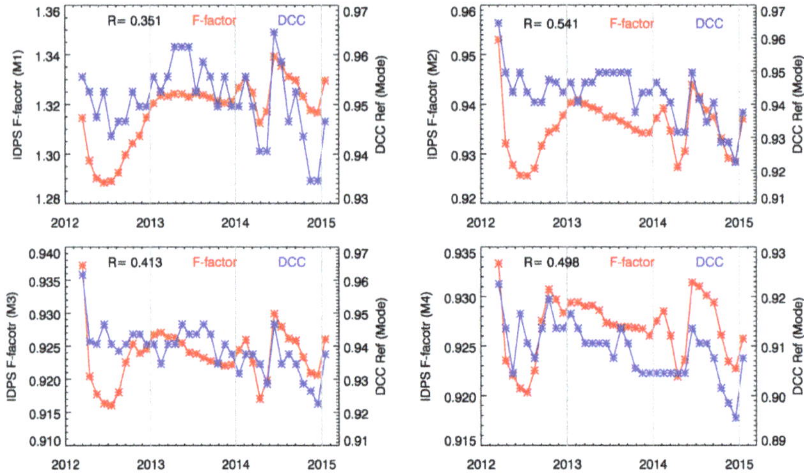

Figure 5. Comparison of DCC mode and IDPS F-factor time series for bands M1–M4 (R represents correlation coefficient).

4.1.2. Semi-Monthly VIS/NIR DCC Time Series

The number of monthly DCC pixels over the study area varies from ~0.5 to ~3 million. DCCs are more abundant during late spring, summer, and fall (see Figure 2). We investigated the feasibility of splitting monthly DCC datasets to monitor the VIIRS calibration stability at a higher temporal resolution from April to November, while monthly DCCs are continued to be used for the remaining months. The mean and mode were calculated independently for each semi-monthly DCC dataset using the same method as the monthly DCCs. Figure 6 compares monthly and semi-monthly DCC mode time series for M1, M4, and M7. Semi-monthly DCC time series show similar stability statistics as those from monthly DCC time series. The major calibration changes (see Section 4.1.1) occurred during the DCC abundant months can be observed clearly in semi-monthly DCC time series. However, the semi-monthly time series are noisier, especially for M1 and M4, which experience larger calibration changes. On the other hand, for M7 for which the calibration is relatively stable, the semi-monthly and monthly time series are very close to each other. Semi-monthly DCC may better capture calibration fluctuations because IDPS F-factors are updated weekly.

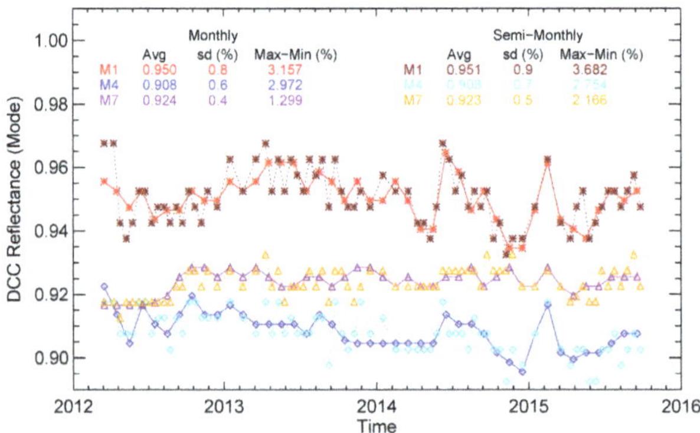

Figure 6. Comparison of monthly and semi-monthly DCC mode time series for M1, M4, and M7.

4.1.3. DCC Mean Band Ratio Times Series

VIS/NIR DCC spectral ratio is insensitive to solar and viewing geometries, cloud optical thickness, and cloud particle type [19] in the VIS/NIR spectrum. Previous studies found that DCC mean ratio time series is generally more stable than the mode time series for inter-channel relative calibration using the DCC technique [14,19]. Fougnie *et al.* [19] indicated that the accuracy of DCC mean band ratio time series is ~0.2% if the reference band is perfectly characterized. Figure 7 shows the DCC band ratio time series for M1/M4 and M5/M7, with the mean ratio time series plotted in blue and the mode ratio time series plotted in gray. The statistics of DCC band ratio time series, including average, standard deviation, range, and trend with 95% confidence interval, are also shown for each time series. It can be observed that the mode and mean band ratio time series are generally consistent with each other. However, the mean ratio time series are smoother than the mode ratio time series. As a result, the mean band ratio time series were used for relative calibration stability monitoring in this study.

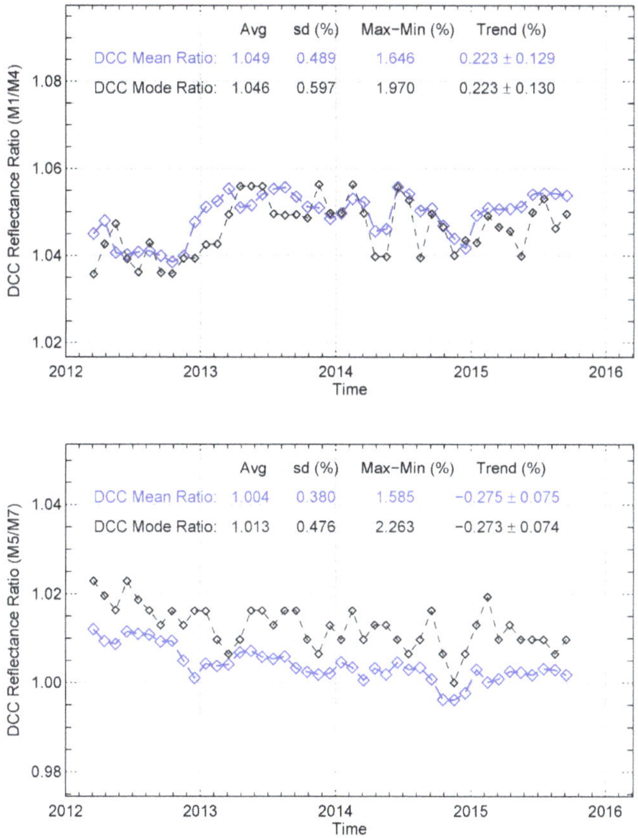

Figure 7. Monthly DCC band ratio time series for M1/M4 and M5/M7.

DCC mean band ratio time series reveals relative calibration changes that is not obvious in the time series for individual bands. Band M4 has been used as reference by the ocean color EDR retrievals. M1/M4 DCC band ratio time series reveal a ~1.5% relative calibration change between 2012 and 2013 due to the November 2012 SD processing parameter change [23]. It generally agrees with vicarious calibration results reported by the NOAA STAR ocean color group based on the MOBY Hawaii *in situ* observations [25]. M5/M7 band ratio time series decreases by ~1.0% from March 2012 to September 2015, which may negatively impact the green vegetation and ocean color EDR long-term trends studies if these calibration issues are not properly addressed. 95% confidence intervals indicate that the M1/M4 and M5/M7 band ratio changes are significant. Figure 8 compares IDPS F-factor ratio time series and DCC mean band ratio time series. DCC band ratio time series are correlated with F-factor band ratio time series, especially for M1/M4. The DCC technique also has uncertainty associated with it. However, our results indicate that the DCC band ratio changes are at least partially caused by the imperfect IDPS F-factor calculations.

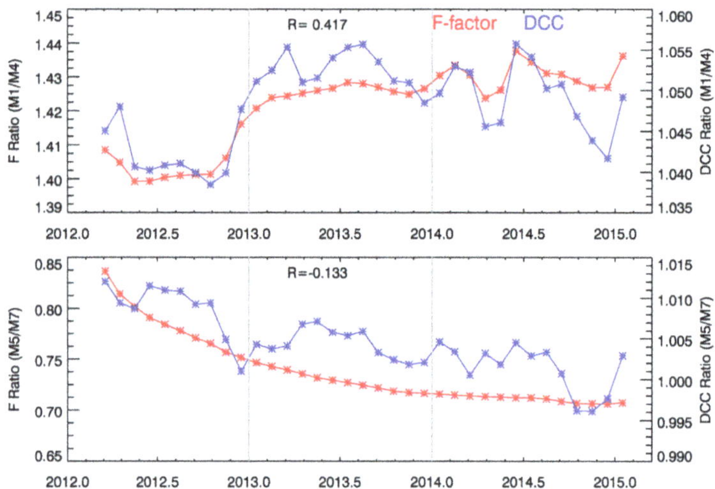

Figure 8. Comparison of IDPS F-factor and DCC mean band ratio time series for M1/M4 and M5/M7.

4.1.4. Comparing DCC Time Series with MODIS/VIIRS SNO-x Time Series

The radiometric calibration stability was previously studied using VIIRS and MODIS (the Moderate-resolution Imaging Spectroradiometer) Aqua SNO-x time series [9]. We compared the VIS/NIR VIIRS DCC mean and mode time series with the VIIRS-MODIS SNO-x time series using two years of data (March 2012 to February 2014). Three time series are plotted for each band using different colors in Figure 9: DCC mode (blue), DCC mean (dark gray), and SNO-x (red) time series. To facilitate visual comparison, the DCC mean time series were adjusted using constant offsets to match the DCC mode time series. The correlation coefficients between the SNO-x and the DCC mean time series were calculated using the un-adjusted values. Assuming the radiometric calibration of MODIS Aqua bands is stable, SNO-x time series indicate that M5 and M7 are more stable than bands M1–M4, same as the results from DCC time series (see Section 4.1.1). Interestingly, DCC mean time series have stronger correlations with the SNO-x time series for bands M1–M4, the calibration of which are less stable. However, DCC mode time series are closer to the SNO-x time series for bands M5 and M7, the radiometric calibration of which are relatively stable.

Another feature that can be observed is that DCC time series are more invariant compared to the SNO-x time series for all VIS/NIR bands. Previous studies indicated that the accuracy of the SNO method is affected by difference in solar geometry between SNO match-ups and imperfect cloud screening [9]. The DCC technique is not affected by cloud screening. Moreover, it is also less sensitive to solar geometry due to the fact that DCCs over tropics were used and the anisotropic effect is mostly accounted for. In addition, the statistics from the DCC technique are more robust because large number of DCC samples is available each month, while the number of SNO match ups is limited by satellite orbits, clouds, and solar geometry.

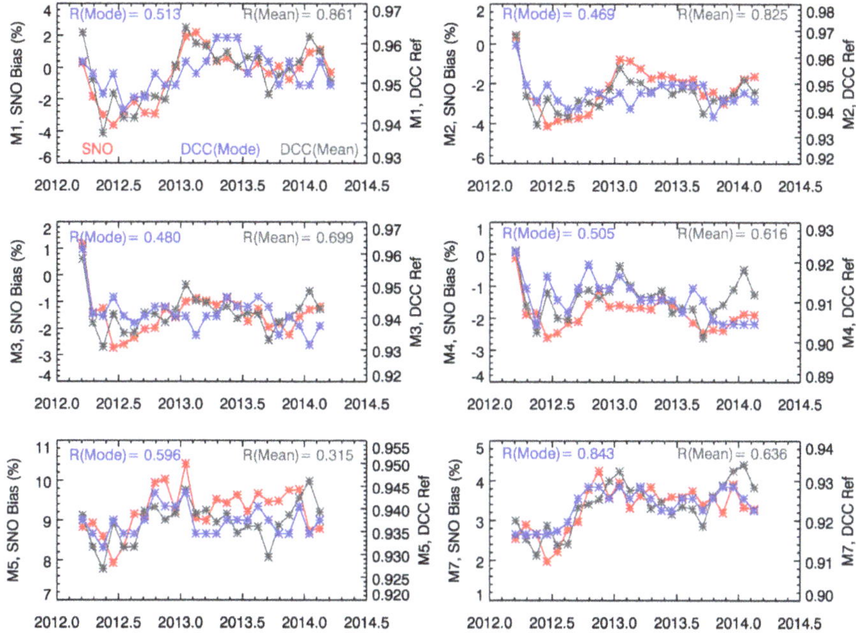

Figure 9. Comparison of DCC mode and mean time series and SNO-x bias time series (the DCC mean time series were adjusted using constant offsets to match the DCC mode time series to facilitate visual comparison).

4.2. Calibration Stability for DNB LGS

DNB LGS gain characterization is indispensable for the calibration of DNB HGS (nighttime) observations that have attracted much attention from the DNB user community. While the DNB RSR was carefully characterized prelaunch, it was found the DNB RSR shifted toward the blue spectra after launch due to the VIIRS RTA mirror degradation with UV exposure. As a result of the degradation, the on-orbit spectral bandpass shifted slightly towards the blue and the full-width half maximum decreased slightly [5]. In this study, DNB TOA radiances were converted to TOA reflectance using the model-predicted RSR at orbit 6557 (February 2013) [5,26] that was implemented in operations in April 2014.

Figure 10 shows monthly and semi-monthly DCC mode time series for DNB daytime observations from April 2012 to September 2015. March 2012 DNB data was not used because of bad calibration. Though the RSR used in this study should be effective for DNB observations since February 2013, the monthly DNB DCC reflectance is not stable until May 2013. The mode of monthly DNB DCC reflectance increases from ~0.83 to ~0.92 from April 2012 to May 2015, at least partially due to LGS gain estimation errors caused by RSR shifts and the mismatch between DNB TOA radiance and RSR used to convert TOA radiance to TOA reflectance. This issue will be studied in the future after more RSRs for observations prior to May 2013 become available. DNB monthly DCC reflectance becomes generally stable since May 2013, with standard deviation of 0.5% and variation of 2.3%. Since the DNB RSR update, the DNB radiometric calibration performance is consistent with those for bands M5 and M7, the RSRs of which overlap partially with that for DNB (see Figure 1).

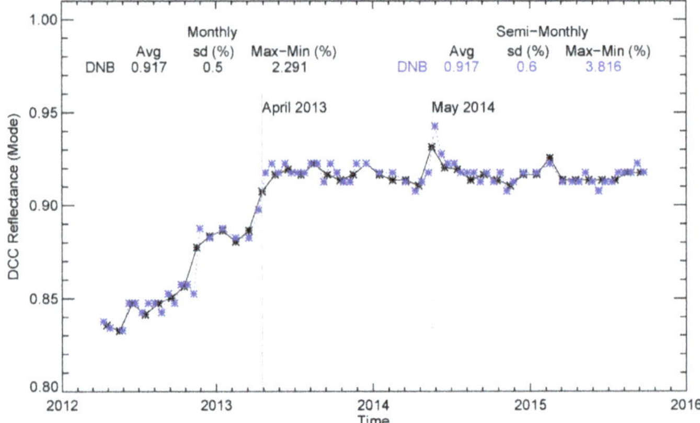

Figure 10. Monthly (black solid line) and semi-monthly (blue dash line) DNB DCC mode time series (April 2012–September 2015).

While similar patterns were observed in the semi-monthly DNB DCC mean time series, the May 2014 DNB calibration anomaly is better captured by the semi-monthly time series. Because of a rare solar eclipse event on 29 April 2014 over the Antarctica region, the 9 May 2014 IDPS operational DNB LGS LUT was updated using the contaminated SD data. This error affected the DNB calibration during 9–15 May 2014 before the next LGS LUT update. As a result, monthly DNB DCC reflectance increased by ~2% in May 2014. This calibration normally is more clearly seen in the semi-monthly DCC time series, with the DNB DCC reflectance during the first half of May significantly higher than the average. Moreover, the semi-monthly time series show that the calibration of DNB becomes stable since the second half of April 2013, instead of May 2013 indicated by the monthly time series.

4.3. Calibration Stability for SWIR M-Bands

Figure 11 shows monthly DCC mode and mean time series for VIIRS SWIR M-bands from March 2012 to September 2015. DCC mean time series are represented by black color and DCC mode time series are represented by blue color. DCC mean and mode time series statistics, including averaged DCC TOA reflectance, standard deviation, and range, also are compared in Table 2. It is worth noting that the DCC TOA reflectance (without ADM adjustment) was used for SWIR bands stability monitoring due to the fact that the ADM adjustment is ineffective for the SWIR bands (see Section 3). Moreover, compared to VIS/NIR M-bands and DNB, DCC mean time series are slightly more stable than mode time series in the SWIR spectrum in most cases. Similar results was reported by Doelling *et al.* [17] based on MODIS Aqua DCCs and by Bhatt *et al.* [20] using the NASA Land Product Evaluation and Analysis Tool Elements (PEATE) version AS3100 VIIRS DCCs. Therefore, DCC mean time series are used for calibration stability monitoring for individual SWIR bands in this study. However, the difference in standard deviation between the mode and mean time series are small in the SWIR spectrum.

Monthly and semi-monthly DCC mean time series for SWIR M-bands are illustrated in Figure 12. M8–M11 calibration stability derived from monthly and semi-monthly DCC mean time series are very close to each other in all cases, similar to those for the stable VIS/NIR bands M5 and M7. However, DCCs are less invariant in the SWIR spectrum. Nearly constant annual cycles can be observed in the M8–M11 DCC mode and mean time series, which is different from the fact that no obvious annual cycles exist in the VIS/NIR DCC mode time series. DCC reflectance is mainly influenced by cloud particle type/size in SWIR spectrum [19,27]. For the area of interest of this study, Wang and Cao [14] found that DCCs mostly occur over the land during winter and over ocean during summer. The annual cycles may be due to the differences in cloud particle type/size over land and ocean. Similar annual cycles were observed in the MODIS Terra and Aqua DCC time series in the SWIR spectrum [20].

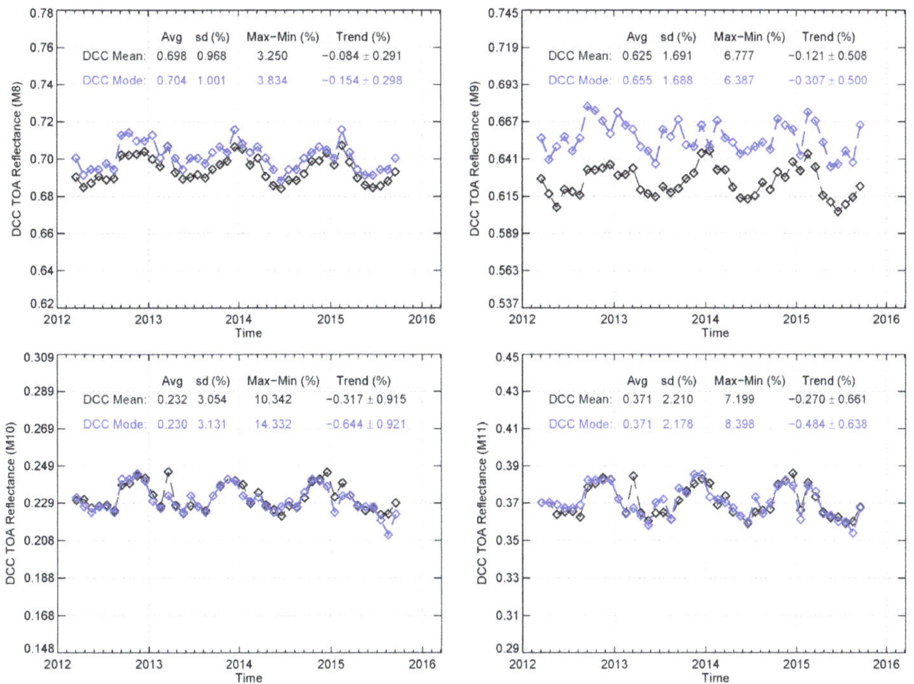

Figure 11. Monthly DCC mode and mean time series for SWIR bands (M8–M11).

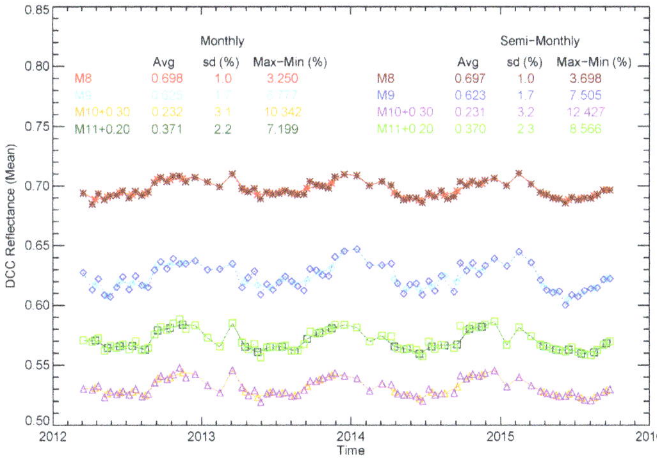

Figure 12. Comparison of monthly and semi-monthly DCC mean time series for SWIR bands M8-M11 (monthly DCC TOA reflectance mean time series were used).

The stabilities of bands M8 to M11 DCC mean time series (see Figure 11) are ~1.0%, 1.7%, 3.1%, and 2.2%, respectively, much larger than those for VIS/NIR M-bands (see Figure 3). The ranges of

SWIR bands DCC mean time series are also larger, ranging from 3.3% to 10.3%. However, no obvious degradation (trend) can be observed in all SWIR bands DCC time series. Uprety and Cao [10] also indicated that there are no significant trends in bands M8 and M11 normalized BRDF time series over deserts; however, a trend is observed in their band M10 time series. The DCC statistics for M8 (1.238 μm) are consistent with the natural DCC temporal variability derived from MODIS band B5 (1.24 μm) [17], indicating the calibration of M8 is likely to be stable. VIIRS band M9 (1.375 μm) has slightly larger variations than band M8, but still comparable with the natural variability derived from its MODIS counterpart (B26, 1.37 μm). The RSR of band M9 lies in a water absorption region. Though atmosphere absorption is generally negligible over DCCs, water vapor absorption may have a larger impact on M9 DCC reflectance. The standard deviation of band M11 is ~2.2%, slightly higher than the 1.8% natural DCC temporal visibility observed by MODIS band B6 (2.12 μm). The standard deviation of VIIRS M10 (1.602 μm) is the highest (3.1%) among all SWIR bands. The stabilities of the NOAA operational version M10 and M11 derived from this study are slightly worse than those for the NASA Land PEATE version [20], however, the later covers a shorter time period compared to this study.

4.4. Calibration Stability for RSB I-Bands

Monthly DCC time series for RSB I-bands (I1–I3) are presented in Figure 13 (January 2014–September 2015). The central wavelengths of I1–I3 are comparable to those for M5, M7, and M10, respectively (see Figure 1). DCC time series for the equivalent M-bands are also plotted in the background for comparison purpose. Moreover, DCC mode time series (with ADM-adjustment) were used for the VIS/NIR bands I1 (M5) and I2 (M7); while DCC mean time series (without ADM-adjustment) were used for the SWIR band I3 (M10). Statistics of RSB I-bands DCC mode and mean time series are available in Tables 2 and 3. It can be observed that I-bands DCC time series follow closely their equivalent M-bands. The nearly constant DCC reflectance difference between M5 and I1 is due to RSR differences between the two bands.

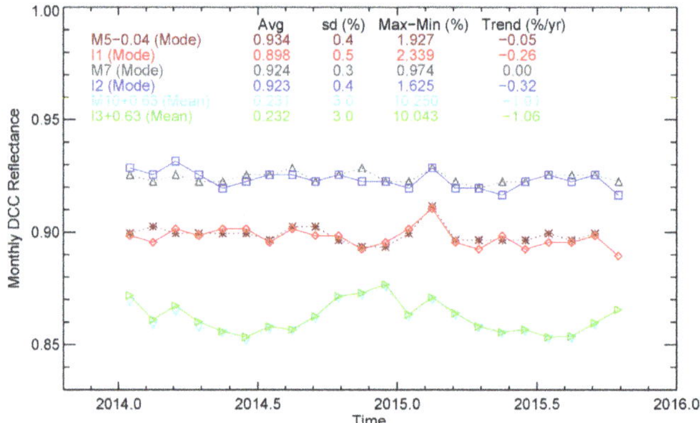

Figure 13. Monthly DCC time series for bands I1–I3 and their equivalent M-bands (ADM-adjusted monthly DCC reflectance mode time series were used for I1,I2, M5, and M7; monthly DCC TOA reflectance mean time series were used for I3 and M10).

Figure 14 compares the I1/M5, I2/M7, and I3/M10 DCC mean ratio time series and the band ratio time series over 4 desert sites, including Libyan4, Egypt3, Sonoran Desert, and Mali1 [28,29]. The vertical line in each panel marks the 1 May 2014 C0 = 0 calibration change. This calibration change has little impact on I1 and M5. I1/M5 band ratio time series show the two bands are relatively stable to each other, though the absolute TOA reflectance over deserts and DCCs are different. I2/M7 also generally agree with each other based on DCC mean ratio time series, consistent with desert sites band ratio time series and results from granule level bands I2 and M7 radiance measurements comparisons [23]. The C0 = 0 calibration change is clearly observable in band I2/M7 DCC and desert band ratio time series. The calibration change can also be observed in I3/M10 validation sites and DCC band ratio time series. Compared to I2 and M7, I3 and M10 have a slightly larger RSR difference. Before the calibration change, I3 and M10 differ by ~2% over deserts and ~1% over DCCs. The difference between the two bands is significantly reduced after the 1 May 2014, calibration change. Results from this study are potentially useful for JPSS program, which is currently investigating the feasibility of replacing band M10 with a water vapor band for VIIRS onboard the future JPSS satellites to support column water vapor retrieval.

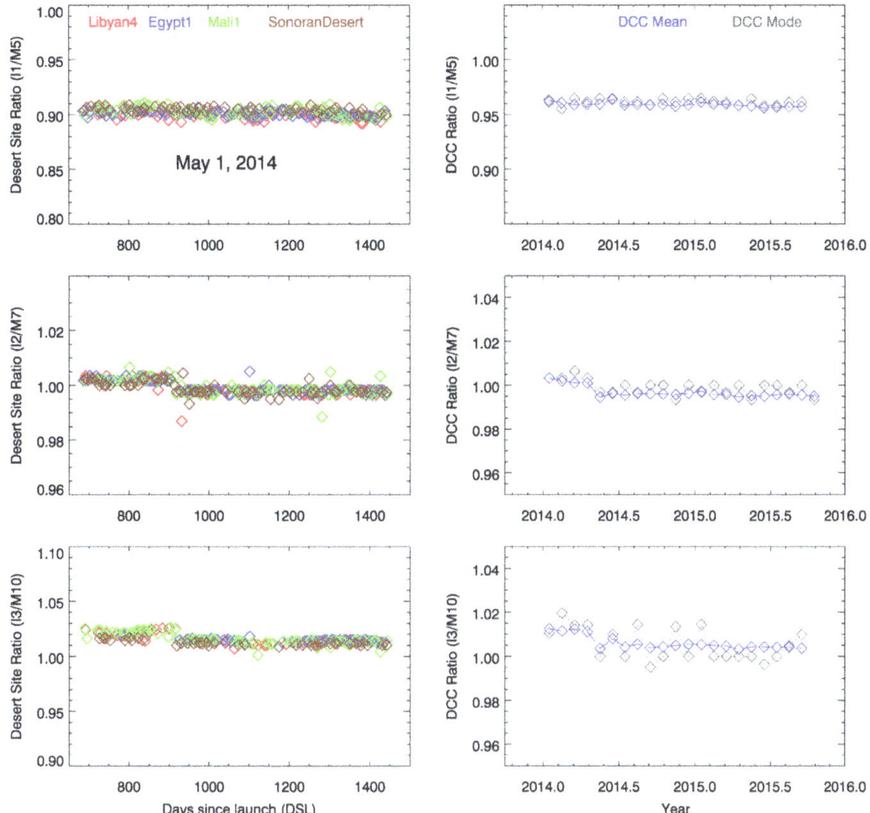

Figure 14. Comparison of DCC mean band ratio and validation site band ratio time series for I1/M5, I2/M7 and M10/I3. The vertical lines in the plots mark 1 May 2015.

5. Summary and Conclusions

The NOAA operational VIIRS RSBs and DNB calibration stability using the DCC technique is demonstrated in this study. Monthly DCC time series for bands M1–M5, M7–M11, DNB, I1-I3 were developed and extensively analyzed. The mode of monthly DCC reflectance (after ADM-adjustment) was used for calibration stability monitoring for the VIS/NIR bands and DNB. Our results show that the NOAA operational radiometric calibration for bands M5 and M7 are generally stable, with stabilities of 0.4% and ranges less than 1.9% during the entire time period. The stabilities of bands M1–M4 are 0.6%–0.8%, with ranges of 3.0%–4.5%. Large fluctuations in bands M1-M4 monthly DCC reflectance were observed since early 2014, correlated with F-factor trend changes during this period. DNB is stable since May 2013 (after the RSR update), similar to bands M5 and M7. The calibration stability of DNB is 0.5% from May 2014 to September 2015. The mean of monthly DCC TOA reflectance (without ADM adjustment) was used for the SWIR bands calibration stabiwality monitoring. The calibration stabilities for M8–M11 are from 1.0% to 3.1%, similar to the natural DCC temporal variability at the SWIR spectrum. The calibration stabilities of I1-I3 are very close to those of their spectral equivalent M-bands (M5, M7, and M10).

VIIRS RSBs inter-channel relative calibration stability was analyzed using DCC mean band ratio time series. I1/M5, I2/M7, and I3/M10 bands ratio time series show consistent calibrations for the 3 band pairs from January 2014 to September 2015. The May 2014 C0 = 0 calibration change is clearly observable in the I2/M7 and I3/M10 band ratio time series. Band ratio time series reveal relative calibration changes for M1/M4 and M5/M7, which may have negative impacts on long-term trend analyses for ocean color and vegetation EDRs if the calibration changes are not accounted for.

The DCC time series were also compared with the IDPS F-factor time series, VIIRS validation site time series, and the VIIRS-MODIS SNO-x time series. Comparison results further support that the DCC time series are capable of detecting sub-percent calibration changes in VIIRS VIS/NIR RSB and DNB bands. Semi-monthly DCC time series were also developed and our results show the monthly and semi-monthly DCC time series generally agree with each other. The semi-monthly time series are slightly noisier than the monthly time series, especially for bands M1–M4, in which large F-factor and DCC reflectance fluctuations exist but the DCC reflectance are known to be more invariant. Our results indicate that semi-monthly DCC time series are useful for stability monitoring at higher temporal resolution. The VIIRS DCC time series developed in this study may contribute to the EDR long-term trend studies and the future SNPP VIIRS SDR reprocessing. The monthly DCC time series used in this study are updated each month and available online [29].

Acknowledgments: This work is funded by the JPSS program office. The authors thank Slawomir Blonski for providing IDPS F-factor time series and Sirish Uprety for providing VIIRS-MODIS SNO-x time series. The authors would also like to thank GSICS vis/nir sub-group for kindly providing the Hu *et al*, (2004) Angular Distribution Model. The manuscript contents are solely the opinions of the authors and do not constitute a statement of policy, decision, or position on behalf of NOAA or the U.S. government.

Author Contributions: Wenhui Wang designed the study, developed methodology and software code, performed the analysis and wrote the manuscript. Changyong Cao contributed to the design of the study and provided technical oversight of the project.

Conflicts of Interest: The authors declare no conflict of interest.

References

1. Cao, C.; de Luccia, F.J.; Xiong, X.; Wolfe, R.; Weng, F. Early on-orbit performance of the Visible Infrared Imaging Radiometer Suite onboard the Suomi National Polar-Orbiting Partnership (S-NPP) satellite. *IEEE Trans. Geosci. Remote Sens.* **2014**, *52*, 1142–1156. [CrossRef]
2. Cao, C.; Xiong, J.; Blonski, S.; Liu, Q.; Uprety, S.; Shao, X.; Bai, Y.; Weng, F. Suomi NPP VIIRS sensor data record verification, validation, and long-term performance monitoring. *J. Geophys. Res. Atmos.* **2013**. [CrossRef]
3. Cardema, J.C.; Rausch, K.W.; Lei, N.; Moyer, D.I.; de Luccia, F.J. Operational calibration of VIIRS reflective solar band sensor data records. *Proc. SPIE* **2012**, 8510.
4. Moy, G.; Rausch, K.; Haas, E.; Wilkinson, T.; Cardema, J.; de Luccia, F. Mission history of reflective solar band calibration performance of VIIRS. *Proc. SPIE* **2015**, 9607.
5. Liao, L.B.; Weiss, S.; Mills, S.; Hauss, B. Suomi NPP VIIRS day-night band on-orbit performance. *J. Geophys. Res. Atmos.* **2013**, *118*. [CrossRef]
6. Chen, X.; Nordhaus, W. A test of the new VIIRS lights data set: Population and economic output in Africa. *Remote Sens.* **2015**, *7*, 4937. [CrossRef]

7. Shi, K.; Yu, B.; Huang, Y.; Hu, Y.; Yin, B.; Chen, Z.; Chen, L.; Wu, J. Evaluating the ability of NPP-VIIRS nighttime light data to estimate the gross domestic product and the electric power consumption of China at multiple scales: A comparison with DMSP-OLS data. *Remote Sens.* **2014**, *6*, 1705. [CrossRef]
8. Johnson, R.S.; Zhang, J.; Hyer, E.J.; Miller, S.D.; Reid, J.S. Preliminary investigations toward nighttime aerosol optical depth retrievals from the VIIRS Day/Night Band. *Atmos. Meas. Technol.* **2013**, *6*, 1245–1255. [CrossRef]
9. Uprety, S.; Cao, C.; Xiong, X.; Blonski, S.; Wu, A.; Shao, X. Radiometric inter-comparison between Suomi NPP VIIRS and Aqua MODIS reflective solar bands using simultaneous nadir overpass in the low latitudes. *J. Atmos. Ocean. Technol.* **2013**, *30*, 2720–2736. [CrossRef]
10. Uprety, S.; Cao, C. Suomi NPP VIIRS reflective solar band on-orbit radiometric stability and accuracy assessment using desert and Antarctica Dome C sites. *Remote Sens. Environ.* **2015**, *166*, 106–115. [CrossRef]
11. Wang, W.; Cao, C. Assessing the VIIRS RSB calibration stability using deep convective clouds. *Proc. SPIE* **2014**, 9264.
12. Ma, S.; Yan, W.; Huang, Y.-X.; Ai, W.-H.; Zhao, X. Vicarious calibration of S-NPP/VIIRS day-night band using deep convective clouds. *Remote Sens. Environ.* **2015**, *158*, 42–55. [CrossRef]
13. Cao, C.; Bai, Y. Quantitative analysis of VIIRS DNB nightlight point source for light power estimation and stability monitoring. *Remote Sens.* **2014**, *6*, 11915–11935. [CrossRef]
14. Wang, W.; Cao, C. DCC radiometric sensitivity to spatial resolution, cluster size, and LWIR calibration bias sased on VIIRS observations. *J. Atmos. Ocean. Technol.* **2015**, *32*, 48–60. [CrossRef]
15. Hu, Y.; Wielicki, B.A.; Ping, Y.; Stackhouse, P.W., Jr.; Lin, B.; Young, D.F. Application of deep convective cloud albedo observation to satellite-based study of the terrestrial atmosphere: Monitoring the stability of spaceborne measurements and assessing absorption anomaly. *IEEE Trans. Geosci. Remote Sens.* **2004**, *42*, 2594–2599.
16. Doelling, D.R.; Minnis, P.; Nguyen, L. On the use of deep convective clouds to calibrate AVHRR data. In Proceedings of the International Symposium on Optical Science and Technology SPIE 49th Annual Meeting, Denver, CO, USA, 2–6 August 2004.
17. Doelling, D.R.; Morstad, D.; Scarino, B.R.; Bhatt, R.; Gopalan, A. The characterization of deep convective clouds as an invariant calibration target and as a visible calibration technique. *IEEE Trans. Geosci. Remote Sens.* **2013**, *51*, 1147–1159. [CrossRef]
18. Minnis, P.; Doelling, D.R.; Nguyen, L.; Miller, W.F.; Chakrapani, V. Assessment of the visible channel calibrations of the VIRS on TRMM and MODIS on Aqua and Terra. *J. Atmos. Ocean. Technol.* **2008**, *25*, 385–400. [CrossRef]
19. Fougnie, B.; Bach, R. Monitoring of radiometric sensitivity changes of space sensors using deep convective clouds: Operational application to PARASOL. *IEEE Trans. Geosci. Remote Sens.* **2009**, *47*, 851–861. [CrossRef]
20. Bhatt, R.; Doelling, D.; Wu, A.; Xiong, X.; Scarino, B.; Haney, C.; Gopalan, A. Initial stability assessment of S-NPP VIIRS reflective solar band calibration using invariant desert and deep convective cloud targets. *Remote Sens.* **2014**, *6*, 2809. [CrossRef]
21. Aumann, H.H.; Pagano, T.; Hofstadter, M. Observations of deep convective clouds as stable reflected light standard for climate research: AIRS evaluation. *SPIE* **2007**, 6684.
22. Loeb, N.G.; Kato, S.; Loukachine, K.; Manalo-Smith, N. Angular distribution models for top-of-atmosphere radiative flux estimation from the Clouds and the Earth's Radiant Energy System instrument on the Terra satellite part I: Methodology. *J. Atmos. Ocean. Technol.* **2005**, *22*, 338–351. [CrossRef]
23. Blonski, S.; Cao, C.; Shao, X.; Uprety, S. VIIRS reflective solar bands calibration changes and potential impacts on ocean color applications. *Proc. SPIE* **2014**, 9111.
24. Fulbright, J.P.; Lei, N.; McIntire, J.; Efremova, B.; Chen, X.; Xiong, X. Improving the characterization and performance of the Suomi-NPP VIIRS solar diffuser stability monitor. *Proc. SPIE* **2013**, 8866.
25. Wang, M.; Liu, X.; Tan, L.; Jiang, L.; Son, S.; Shi, W.; Rausch, K.; Voss, K. Impacts of VIIRS SDR performance on ocean color products. *J. Geophys. Res. Atmos.* **2013**, *118*, 10347–10360. [CrossRef]
26. Lei, N.; Wang, Z.; Guenther, B.; Xiong, X.; Gleason, J. Modeling the detector radiometric response gains of the Suomi NPP VIIRS reflective solar bands. *Proc. SPIE* **2012**, 8533.
27. Minnis, P.; Hong, G.; Ayers, J.K.; Smith, W.L.; Yost, C.R.; Heymsfield, A.J.; Heymsfield, G.M.; Hlavka, D.L.; King, M.D.; Korn, E.; *et al.* Simulations of infrared radiances over a deep convective cloud system observed during TC4: Potential for enhancing nocturnal ice cloud retrievals. *Remote Sens.* **2012**, *4*, 3022. [CrossRef]

28. Wang, W.; Cao, C.; Uprety, S.; Bai, Y.; Padula, F.; Shao, X. Developing an automated global validation site time series system for VIIRS. *Proc. SPIE* **2014**, 9218.
29. SNPP VIIRS Validation Site Time Series. Available online: http://ncc.nesdis.noaa.gov/VIIRS/VSTS.php (accessed on 30 October 2015).

Article

Evaluation of VIIRS and MODIS Thermal Emissive Band Calibration Stability Using Ground Target

Sriharsha Madhavan [1,*], Jake Brinkmann [1], Brian N. Wenny [1], Aisheng Wu [1] and Xiaoxiong Xiong [2]

1 Science Systems and Applications, Inc., 10210 Greenbelt Rd., Lanham, MD 20706, USA; jake.brinkmann@ssaihq.com (J.B.); brian.wenny@ssaihq.com (B.N.W.); aisheng.wu@ssaihq.com (A.W.)
2 Sciences and Exploration Directorate, NASA/GSFC, Greenbelt, MD 20771, USA; xiaoxiong.xiong-1@nasa.gov
* Correspondence: sriharsha.madhavan@ssaihq.com; Tel.: +1-301-867-2071; Fax: +1-301-867-2151

Academic Editors: Changyong Cao, Richard Muller and Prasad S. Thenkabail
Received: 2 November 2015; Accepted: 4 February 2016; Published: 19 February 2016

Abstract: The S-NPP Visible Infrared Imaging Radiometer Suite (VIIRS) instrument, a polar orbiting Earth remote sensing instrument built using a strong MODIS background, employs a similarly designed on-board calibrating source—a V-grooved blackbody for the Thermal Emissive Bands (TEB). The central wavelengths of most VIIRS TEBs are very close to those of MODIS with the exception of the 10.7 μm channel. To ensure the long term continuity of climate data records derived using VIIRS and MODIS TEB, it is necessary to assess any systematic differences between the two instruments, including scenes with temperatures significantly lower than blackbody operating temperatures at approximately 290 K. Previous work performed by the MODIS Characterization Support Team (MCST) at NASA/GSFC used the frequent observations of the Dome Concordia site located in Antarctica to evaluate the calibration stability and consistency of Terra and Aqua MODIS over the mission lifetime. The near-surface temperature measurements from an automatic weather station (AWS) provide a direct reference useful for tracking the stability and determining the relative bias between the two MODIS instruments. In this study, the same technique is applied to the VIIRS TEB and the results are compared with those from the matched MODIS TEB. The results of this study show a small negative bias when comparing the matching VIIRS and Aqua MODIS TEB, implying a higher brightness temperature for S-VIIRS at the cold end. Statistically no significant drift is observed for VIIRS TEB performance over the first 3.5 years of the mission.

Keywords: VIIRS; MODIS; thermal emissive bands; Dome Concordia; calibration

1. Introduction

The Suomi Visible Infrared Imaging Radiometer Suite (S-VIIRS) and the MODerate resolution Imaging Spectroradiometer (MODIS) are cross track scanning radiometers orbiting the Earth on a sun synchronous polar orbit with the corresponding altitudes of approximately 824 km and 705 km, respectively [1,2]. The S-VIIRS is on board the S-National Polar-orbiting Partnership platform whereas the MODIS is on board the Terra (T) and Aqua (A) platforms. As of 28 October 2015, the S-VIIRS has completed four years of on orbit flight. The T- and A-MODIS have completed 15 and 13 years of successful on-orbit operation recording the geophysical changes of the Earth in a wide range of spectral channels. Together, the three sensors provide high quality radiometric measurements of the Earth; the S-VIIRS mission is to ensure the continuity of the valuable data records from MODIS. Examples of science products derived from S-VIIRS and compared with MODIS sensor are available in [3–5], which cover the ocean, land, and atmospheric science discipline areas. In order to ensure the long term continuity of these data products, the calibrations of the VIIRS and MODIS are extremely critical. In order to achieve the traceability to ground based references, both the VIIRS and MODIS instruments are bestowed with robust on-board calibrators (see Figure 1a,b). In this paper, we focus on the S-VIIRS and A-MODIS bands with wavelengths longer than 3.7 μm. These bands are referred to as Thermal Emissive Bands. The calibration of the S-VIIRS and A-MODIS is based on a similar v-grooved BlackBody (BB) as shown in Figure 2a,b, whose temperature measurements were traceable to the National Institute of Standards and Technology temperature scales [6]. Further, the MODIS BB is monitored using 12 thermistors whereas the VIIRS BB is monitored via 6 uniformly spaced thermistors. The individual thermistor locations are roughly shown in Figure 2a,b.

S-VIIRS has two types of bands providing ground observations at different spatial resolutions. The moderate-resolution (M-) bands have a spatial resolution of 750 m, which is similar to the 1 km bands of MODIS. Additionally, to achieve a wide dynamic range some of the M-bands are dual gain, with high gain intended for detection at the low end of the dynamic range and the low gain for the higher end. M13 is the only dual gain TEB band which is primarily used for fire detection and is similar to MODIS bands 21 and 22. The two imaging (I-) bands are fine resolution that have a spatial resolution of 375 m. The I4 band is very similar to the MODIS band 20 while the I5 band covers MODIS bands 31 and 32, respectively. Figure 3 shows the Relative Spectral Response (RSR) of all the VIIRS TEB overlaid with the response curve of the BB radiance at a set temperature of 290 K [7]. It is important to note the RSRs for 3 μm–4 μm band pairs cover parts of the spectrum influenced by scattered solar irradiance. Table 1 gives the spectral center wavelengths of the VIIRS TEBs along with the matching MODIS TEB. In the rest of the paper we use the following match up of M-bands as mentioned in Table 1 for cross comparison of the two instruments.

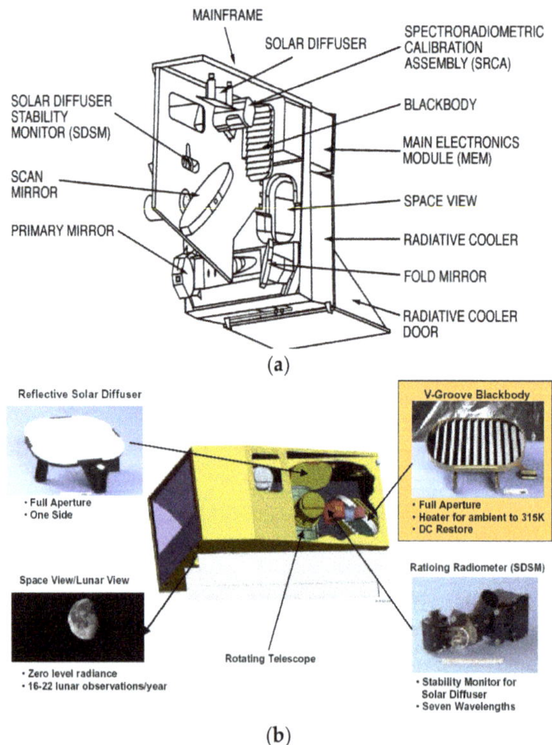

Figure 1. Instrument setup with on-board calibrators (**a**) MODIS; (**b**) VIIRS [1,2].

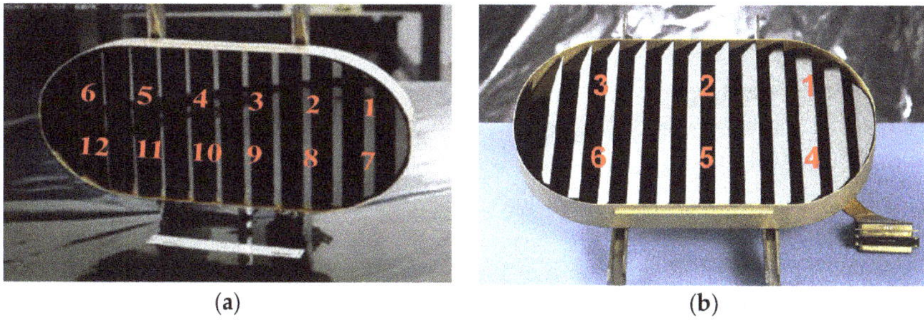

Figure 2. V-grooved BlackBody controlled using various thermistors (**a**) MODIS; (**b**) VIIRS [6].

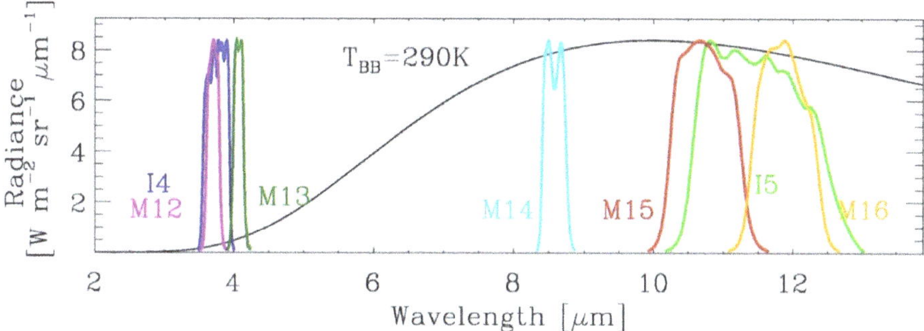

Figure 3. Relative Spectral Response of the VIIRS TEBs overlaid with spectral radiance of the BlackBody at 290 K [7].

Table 1. VIIRS and MODIS matching TEBs [7].

VIIRS Band (C.W. (μm))	**I4** (3.74)	**I5** (11.45)	**M12** (3.70)	**M13** (4.05)	**M14** (8.55)	**M15** (10.76)	**M16** (12.01)
MODIS Band (C.W. (μm))	**B20** (3.78)	**B31** (11.03) **B32** (12.04)	**B20** (3.78)	**B22** (3.96)	**B29** (8.56)	**B31** (11.03)	**B32** (12.04)

In previously reported works [8–10], cold Earth View (EV) targets, such as Dome Concordia (C), served as a reference to evaluate the sensor calibration deficiencies at the low end of the dynamic range. In this paper, we extend the methodology developed in [8] to assess the calibration stability and consistency of S-VIIRS and A-MODIS. Since both S-VIIRS and A-MODIS are on afternoon orbits, with near-simultaneous scene acquisition times, the two instrument responses can be cross verified using coincident automatic weather station (AWS) ground measurements. The objective of the work is to track the S-VIIRS TEBs on-orbit performance for 3.5 years since launch, for identifying potential instrument based dependencies, useful in improving future reprocessing of the Level 1B datasets.

With the defined objective as stated above, the rest of the paper is prepared as follows. The next section briefly reviews the TEB calibration algorithm for both VIIRS and MODIS, laying down the foundation for the various calibration terms. The third section describes the evaluation methodology of the TEB of both instruments using the proxy reference. The fourth section provides the results and discussions. Finally, the paper is tied with a summary of the work.

2. TEB On-Orbit Calibration

For MODIS TEB, a quadratic model is applied to describe the relationship between the "at" sensor aperture radiance L and background subtracted instrument response dn [11]. For the BB calibration, the "at" sensor radiance models the thermal environment as:

$$L_{CAL} = RVS_{BB}\varepsilon_{BB}L_{BB} + (RVS_{SV} - RVS_{BB})L_{SM} + RVS_{BB}(1 - \varepsilon_{BB})\varepsilon_{CAV}L_{CAV} \quad (1)$$

where L_{CAL} is the at sensor aperture radiance for the BB view, RVS_{BB} (RVS_{SV}) is the response *versus* scan angle (RVS) at the sensor's BB (SV) view angle. ε_{BB} is the BB emissivity and ε_{CAV} is the effective scan cavity emissivity. The radiance contains contributions primarily from the BB, in addition has minor contributions from the scan mirror and instrument cavity. The terms L_{BB} (L_{SM}, L_{CAV}) are computed using Planck's equation at a measured T_{BB} (T_{SM}, and T_{CAV}). Using Equation (1), the linear calibration term b_1 for BB observation can be related to the at sensor aperture radiance L_{CAL} by

$$b_1 = (L_{CAL} - a_0 - a_2 dn_{BB}^2)/dn_{BB} \quad (2)$$

For both T- and A- MODIS, a BB Warm-up/Cool-down (WUCD) process on a quarterly basis derives the minor temporal change in calibration coefficients of the quadratic model, especially the offset (a_0) and quadratic (a_2) terms. In MODIS Collection 5 (C5) or earlier collections, Equation (2) is fitted to the WUCD measurements without constraint. In T-MODIS Collection 6, the offset term a_0 is constrained to be zero and only the linear and quadratic terms are fitted to the measured data. The approach of setting a_0 to zero was suggested based on the results reported in [9,10].

The calibration equation for S-VIIRS TEB is similar with slight modifications from MODIS, wherein the background removed BB response is related to the spectral radiance as seen by the instrument aperture (L_{ap}) by Equation (3) and is as follows [12]:

$$F(B) = \frac{RVS_{BB}(B)L_{ap}(B) + \Delta L_{bg}(B, \theta_{BB})}{\sum_{i=0}^{2} c_i dn_{BB}{}^i} \quad (3)$$

where the terms c_0, c_1, and c_2 are pre launch coefficients, ΔL_{bg} is the residual self emission background term , determined for each band B and at the angle of incidence θ and finally the term F-Factor that relates to the linear change calculated scan-by-scan. The slight modifications from Equations (2) and (3) stem from the fact of using the quadratic terms in VIIRS as a lumped sum of products, the linear calibration change captured from the scan by scan measurements of the BB. Apart from the afore mentioned differences the calibration methodology are based on similar looking equations for the TEB of the two instruments.

3. Evaluation Methodology

Vicarious calibration using well characterized EV targets have proven to be great sources for evaluating and validating the calibration accuracy. Works reported in [8,13–17] have shown and identified stable EV targets that serve the above mentioned purpose. The Dome C (75.102°S, 123.395°E) is one such EV target that has been characterized for Long Wave InfraRed (LWIR) satellite retrievals [8]. Some of the key highlights of this site are: minimum spatial variability, the most homogenous surface with a slope of approximately 0.004%; High Infrared emissivity and relatively uniform surface temperatures; High surface elevation (3233 meters above sea level); Extremely dry and rarified atmospheric conditions allowing the outgoing surface radiation to be very close to the Top Of Atmosphere (TOA) radiance in the spectral window of 11 to 12 μm. The Dome C site is well characterized as being very stable and uniform. Previous works have demonstrated that spatial variability of MODIS Band 31 BTs was typically less than 0.3 K in the Dome C region [18]. From the stand point of polar orbiting EV based remote sensing satellites this site provides a unique advantage of several overpasses per day over the region. Thus, both

A-MODIS and S-VIIRS possess thousands of high quality Dome C scene acquisitions in the Level 1 B repository. Figure 4 gives one such S-VIIRS image acquired over Antarctica that comprises the Dome C site (identified by the small black square). The image shown is an orthographic map projection, depicting the retrieved Brightness Temperature (B.T.) from M15 band of S-VIIRS. In general, the scene as expected is a cool target with mean B.T. of approximately 235 K.

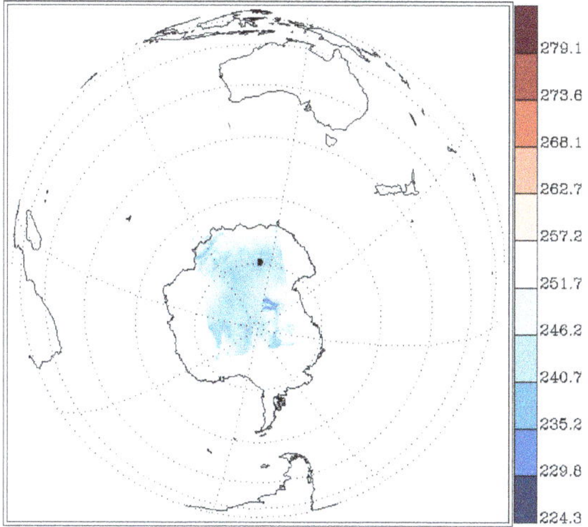

Figure 4. S-VIIRS Band M15 image illustrating the Antarctic EV location comprising the Dome C site.

Additionally, the Dome C site is characterized by the Concordia research stations jointly operated by research scientists from France and Italy. An AWS, operated by Univ. Wisconsin, has been fully functional since installed in 1995. The AWS is physically installed approximately three meters above the ground base at Dome C. The AWS measurements are calibrated with a radiometric accuracy of $\pm$0.5 K over a range of approximately 233 K–293 K respectively. Additionally, the typical errors were less than 1.0 K for measurements varying from approximately 198 K–298 K [19]. As of May 2015 a twenty year record of various meteorological measurements have been archived that serve as a great ground reference to assess the calibration stability and consistency of MODIS and VIIRS sensors. Figure 5 shows the lifetime temperature observations from both AWS and the MODIS measurements over Dome C since 2002. Based on the trends it can be seen that the lifetime trends are very stable with mean temperatures at approximately 220 K, with seasonal oscillations of approximately $\pm$25 K. Thus the Dome C site provides a useful reference to assess the lower order deficiencies in calibration models for systems such as MODIS and VIIRS.

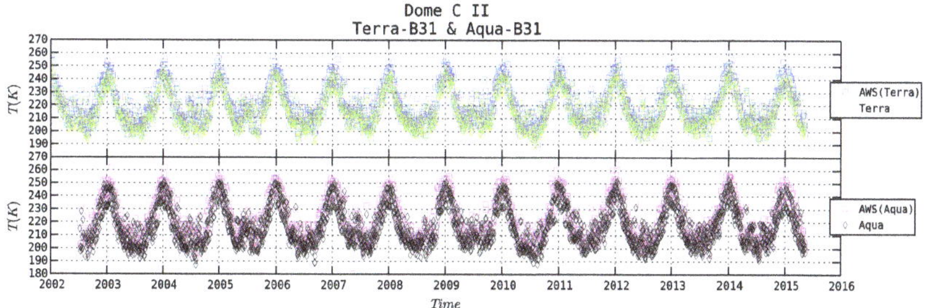

Figure 5. Lifetime temperature observations over Dome C using AWS, T- and A- MODIS band 31.

In the current work, the AWS is the proxy that is employed to assess the calibration consistency and stability of A-MODIS and S-VIIRS since January 2012. The Dome C measurement extractions from the two instruments are described as follows. About 250 nadir viewing Dome C scene acquisitions were identified for both instruments. These scenes were screened for clouds using two filtering steps. The first set of cloud screening was performed using the level 2 cloud mask products. Further, a clear sky restore was performed for MODIS using the algorithm mentioned in [20]. Antarctica presents a significant challenge to discriminate clouds from surface, where the surface is often colder than cloud. In order to obtain the maximum number clear-sky the MODIS clear-sky-restore algorithm was used to identify clear-sky pixels incorrectly flagged as cloudy. Unfortunately, the inversions required for this test can not be attained with the limited number VIIRS bands. Next, the spatially equivalent pixels were aggregated up for both the instruments. The total Dome C area covered was a small square region of 20 km $\times$ 20 km with the center pixels corresponding to the above mentioned latitude and longitudes of the Dome C site. After acquiring the matching co-located data, the L1B radiances from both the sensors is converted to a retrieved B.T. using the Planck equation and the spectral center wavelengths for the matching wavelengths. The 10-min average AWS data closest to the scene acquisitions is separately recorded. The data source used for this research work is as follows: For A-MODIS the latest collection C6 L1 B radiances were used, the latest version V3.1 L1B radiances for VIIRS from the Land PEATE distribution. The data are acquired from the site provided in [21]. Further, the AWS measurements are made available by the Antarctic Meteorology Research Center at the University of Wisconsin. The data can be downloaded from reference shown by [22].

4. Results and Discussion

This section presents the results of the calibration stability and consistency of the two sensors. Thus the results and subsequent discussions for each of the above mentioned analyses is broken into two subsections.

4.1. Calibration Stability

The first aspect of the study incorporated the temperature differences of the afore-mentioned 5 TEB of A-MODIS and S-VIIRS compared against the AWS measurements. The trends are provided in Figure 6a–e. The trends show the temperature differences of the two sensors from the AWS measurements for three and a half years from January 2012. The red diamond points are temperature differences with S-VIIRS while the blue diamond points are for differences with A-MODIS. Additionally shown is a black solid line that references a zero difference. The trends are expected to remain consistent between both sensors, while not necessarily approaching zero. A seasonal oscillation is seen in the difference trends, the difference oscillations between the various bands are smaller with increasing wavelengths. This is particularly true for wavelengths upwards of 8 μm (Figure 6c–e); the oscillations are approximately within ± 10 K. It is also noted from the temperature trends shown in Figure 5 that Dome C being located in Southern Hemisphere experiences temperature highs in December (summer) and lows in July (winter), respectively. As a result the coldest scenes are acquired during the May–August time span. In the colder months, retrieved temperatures from S-VIIRS have generally tended to be higher when compared to the AWS reference. A plausible reasoning for this is given later. For the same time frame the A-MODIS is in general closer to the AWS measurements. In the warmer scene retrievals, both S-VIIRS and A-MODIS recorded slightly lower temperatures in comparison to the AWS measurements. This can be expected as the TOA radiance reaching the sensor is attenuated by the atmospheric absorption due to higher water vapor content. Overall, over the three year period from 2012 there are no observable drifts in temperature trends for both the instruments. This indicates the 3.5 years fidelity of the TEBs of both A-MODIS and S-VIIRS.

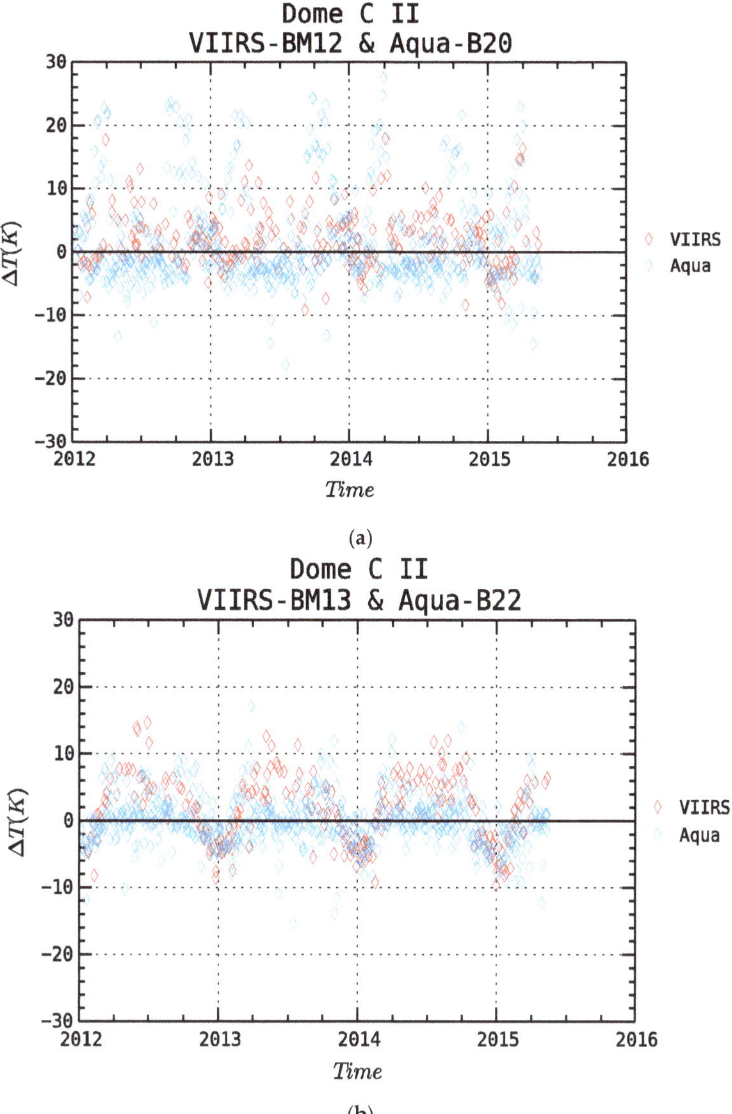

Figure 6. *Cont.*

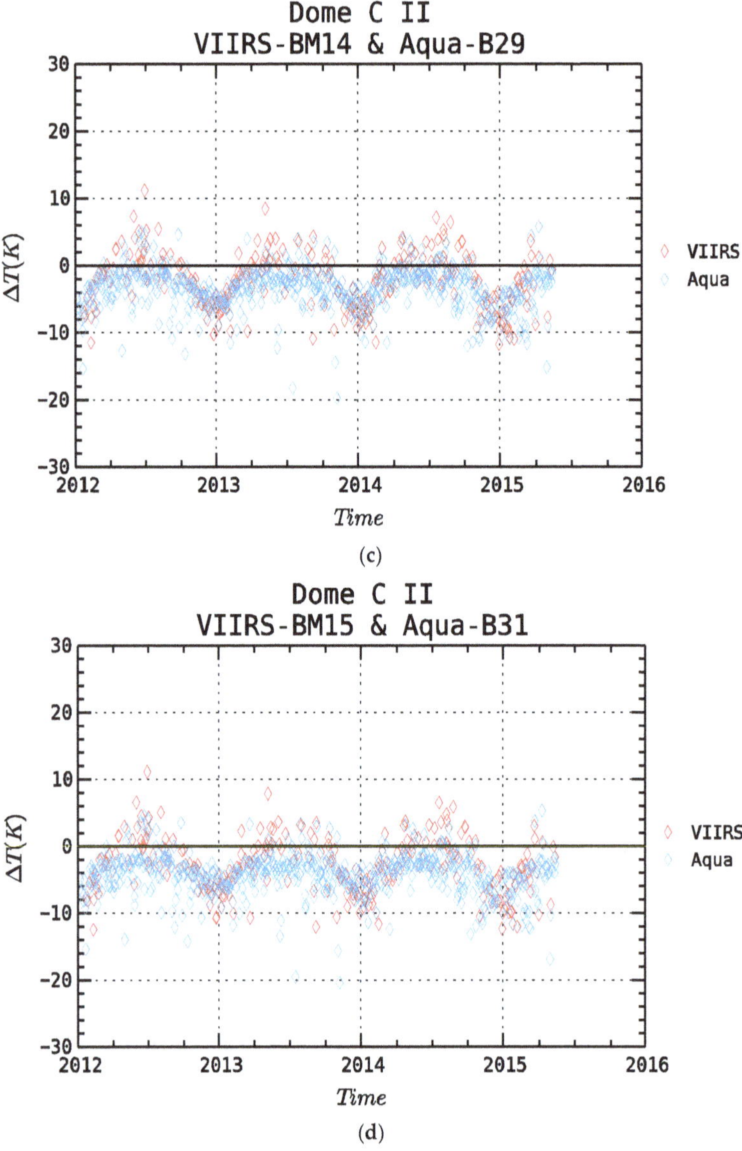

(c)

(d)

Figure 6. *Cont.*

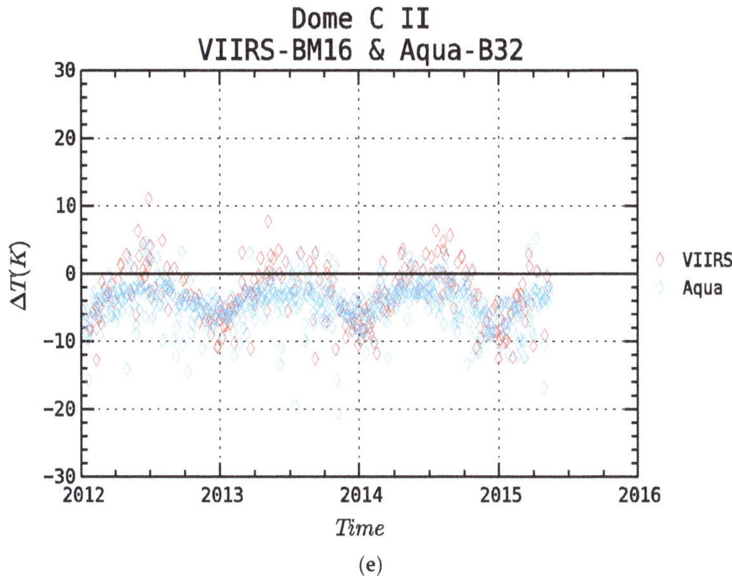

Figure 6. Long term Brightness Temperature difference trends (Sensor-AWS) (**a**) M12/B20; (**b**) M13/B22; (**c**) M14/B29; (**d**) M15/B31; (**e**) M16/B32.

4.2. Calibration Consistency

In order to assess the calibration consistency between the two instruments which is extremely important for long term continuity of climate data records, a double difference of the Brightness Temperatures between the two sensors is done to remove the differences caused by using the AWS as the proxy. This measurement is known as "relative bias" and has been reported for T- and A-MODIS in [8]. The AWS air-temperature measurement is different than the 20 km $\times$ 20 km skin surface measured by the sensor. The difference is assumed to be random due to the fluid dynamics of air, and stable within the small satellite acquisition time differences. Figure 7a–e presents the weekly-averaged relative bias trends between S-VIIRS and A-MODIS for the 5 TEB. For each of the bands a linear fit is performed to assess any significant trend. Additionally, shown in each subplot are various statistical quantities such as the mean (μ), standard deviation (σ), delta change in relative bias over 3.5 years (ΔT), and the p-value. Due to the differences between AWS and satellite measurements in the SMIR spectrum that positively correlate with non-negative solar elevation, only the night time measurements were used for computing the relative bias in M12/B20 and M13/B22 band pairs, removing approximately 10 weeks per year during Antarctic Summer.

Overall, based on the trends and fit statistics shown there is no statistical evidence for any drift in the relative bias between the two instruments. This is indicative by the high p-value suggesting the linear predictor to be not a representative of the response, the null hypothesis of the slope being not statistically significant be accepted. However, a warm relative bias is assessed between S-VIIRS and A-MODIS. In other words, at colder scenes the S-VIIRS is found to retrieve slightly warmer temperatures in comparison to A-MODIS. The average warm bias (BT [K]/3.5 years) for the various TEB are as follows: approximately 4.94 K for the M12, 4.90 K for M13, 1.48 K for M14, 1.53 K for M15 and 1.41 K for M16, respectively. Some of the differences reported above would have to be absorbed by the uncertainties in the calibration itself. The expected differences are typically based on the design requirements of the calibration sources used for both the instruments. The Blackbody calibration sources are expected to be within 50 mK and 30 mK for MODIS and VIIRS respectively. The differences between the two instruments are expected to be between 0.2 K and 0.5 K [6]. Further, a small portion of this relative warm bias can be attributed to the RSR differences between the two sensors and results to the effect have been previously reported in [7]. The RSR curves for the matching TEB for MODIS and VIIRS are shown in Figure 8. It is observed that indeed the RSR differences are bound to affect the various bands, probably minimal in the case of M14 and band 29. The RSR curves are very similar and hence expected to show the least impact for this band. Despite the fact that the RSRs are close for the M14 band the mean relative bias between the two instruments is quite significant. This implies that only a small factor would be accounted for if the RSR shifts were corrected. The remaining bias does point to the inadequacy of the calibration model of S-VIIRS. The impacts of spectral correction for the RSR differences using measured simultaneous hyper-spectral data from A-Atmospheric Infrared Sounder (AIRS) for M15/B31 and M16 B32 are described in [23]. The RSR correction factor is derived as a radiance ratio between S-VIIRS and A-MODIS theoretical values, derived from RSR integration using AIRS hyper-spectral data. Results from the work reported in [23] indicate that the RSR correction reduces the differences by approximately 0.10 K and 0.29 K for M15/B31 and M16/B32 respectively. For M14/B29, since there is not enough spectral coverage for both A-AIRS and S-Cross track Infrared Sounder (CrIS), there is no RSR factor derived for this band. However, it should be noted that Infrared Atmospheric Sounding Interferometer (IASI) would provide contiguous spectral coverage over the difficult M14/B29.

In terms of the drift in the relative bias over 3.5 years, it is assessed to be significantly small for most bands with the exception of band M13/B22. The largest drift is observed to be about 0.163 K for band M13. Results for all bands have been summarized in Table 2. Further, it is noted that drift in all the matching TEB are well within the noise requirements given in terms of Noise Equivalent Temperature difference (NEdT) of both instruments [6].

Based on the results shown in Figures 6 and 7 it is suggestive a small cold scene retrieval bias in S-VIIRS TEB. In the current Collection 6 for T-MODIS the offset calibration term is set to zero whereas A-MODIS uses the pre-launch/on-orbit based non-zero coefficients [24]. Similar to the current study, T-MODIS suffered from the warm bias in the cold scene retrievals in comparison to A-MODIS. Though on different orbits, and the Dome C acquisition times not being as close to A-MODIS and S-VIIRS, the relative bias study was extended, comparing S-VIIRS and T-MODIS. Since the path radiance may be considerably different for the two sensor acquisitions, only the surface sensing bands are used in the discussion. Figure 9 shows the relative bias trends between S-VIIRS and T-MODIS for M15/Band 31, M16/Band 32, and M14/Band 29, respectively. Figure 9 also provides similar fit statistics as illustrated in the results shown in Figure 7. From these two figures the following observations are made. First, the mean relative biases between S-VIIRS and T-MODIS for the three surface bands are smaller in comparison to the mean relative bias between S-VIIRS and A-MODIS, but still positive in sign. The differences in the mean relative biases between (A-MODIS~S-VIIRS) and (T-MODIS~S-VIIRS) instruments are approximately 0.63 K, 0.29 K, and 0.23 K, respectively, for the M14, M15 and M16 bands.

A similar cold scene bias in S-VIIRS TEB has been reported in a different study that comprised of early mission inter comparison of S-VIIRS with Cross-track Infrared Sounder (CrIS) and S-VIIRS with the IASI [25]. The RSR differences between MODIS and VIIRS are also expected to impact the differences seen in the relative bias estimates though at a very small magnitude. Overall, over the first 3.5 years of the mission both A-MODIS and S-VIIRS TEB performance are very stable and consistent. Improvements suggested here are very useful in future reprocessing of the science data records of S-VIIRS should tie the two sensors on an even keel.

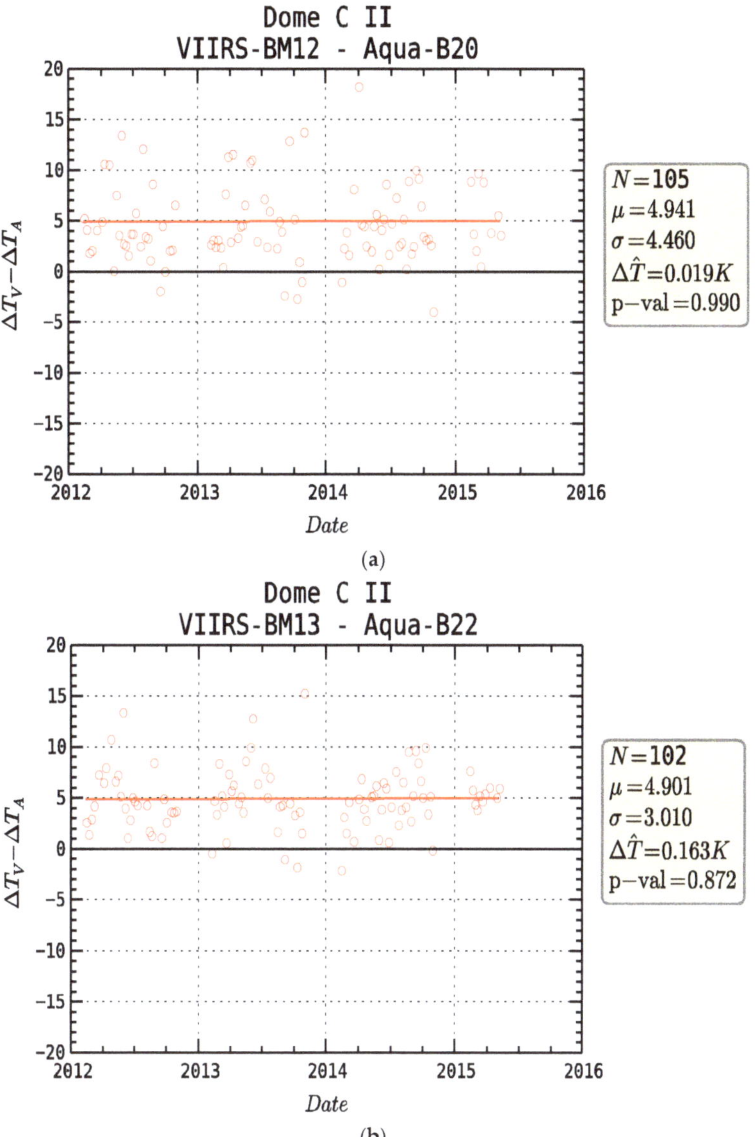

Figure 7. *Cont.*

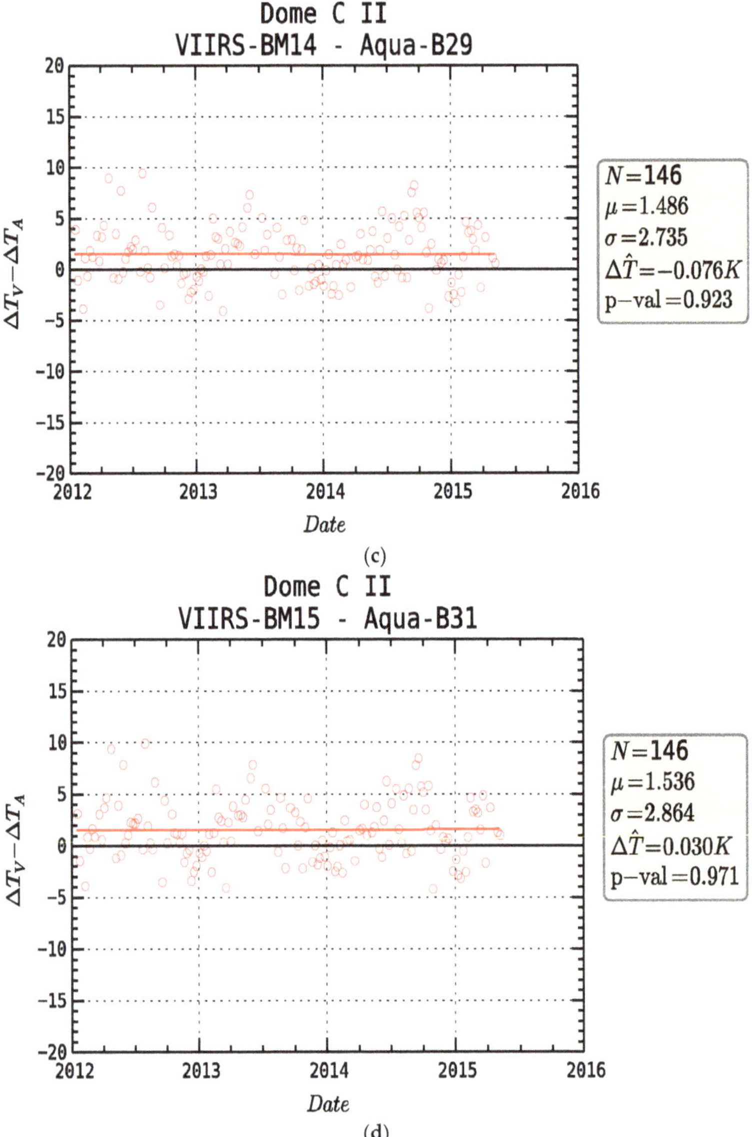

(c)

(d)

Figure 7. *Cont.*

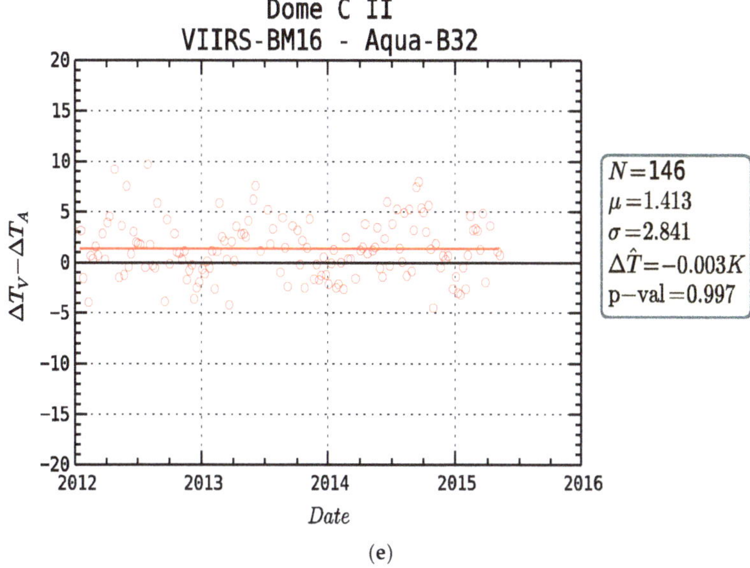

(e)

Figure 7. The 3.5-year relative bias trend between S-VIIRS and A-MODIS. (**a**) M12/B20; (**b**) M13/B22; (**c**) M14/B29; (**d**) M15/B31; (**e**) M16/B32.

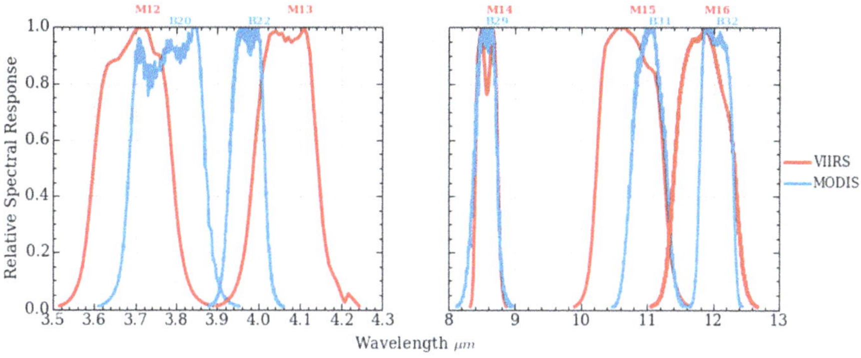

Figure 8. RSR Curves for matching TEB for VIIRS and MODIS.

Table 2. Relative Bias drift [K/3.5 years] between S-VIIRS and A-MODIS.

VIIRS/MODIS Matching Band	M12/B20	M13/B22	M14/B29	M15/B31	M16/B32
ΔT Relative Bias (K) *	0.019	0.163	−0.076	0.030	−0.003

* *p*-value (shown previously) indicates statistically insignificant for the relative bias drift.

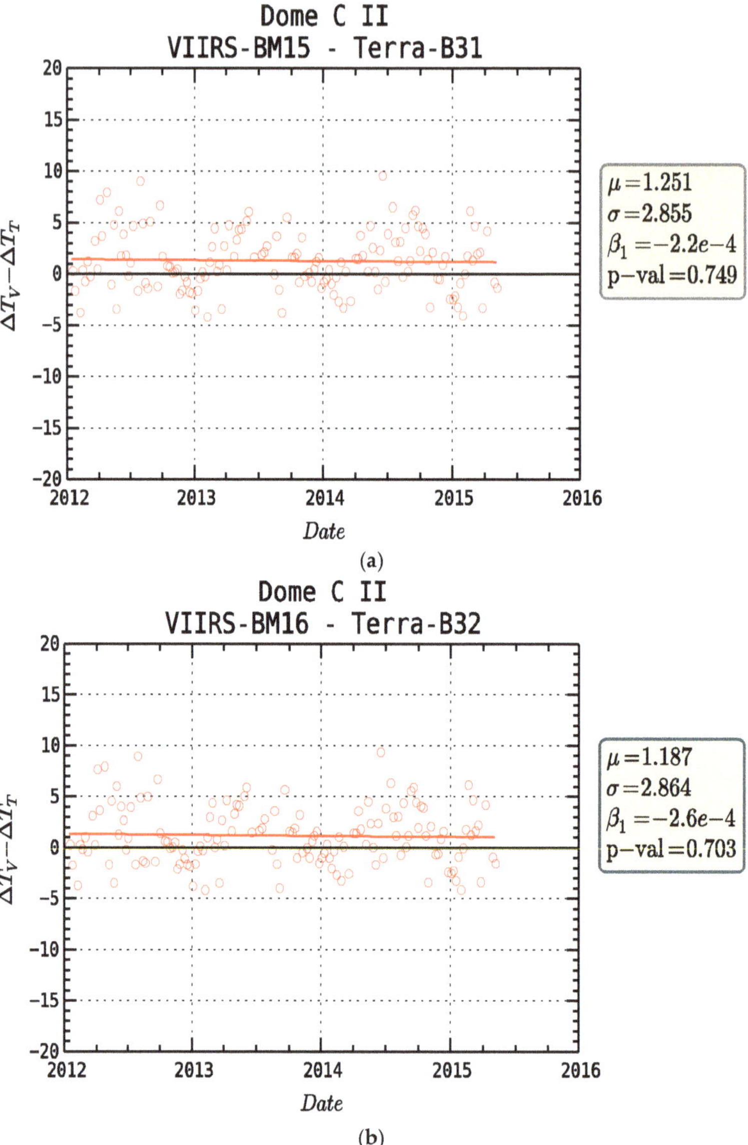

Figure 9. *Cont.*

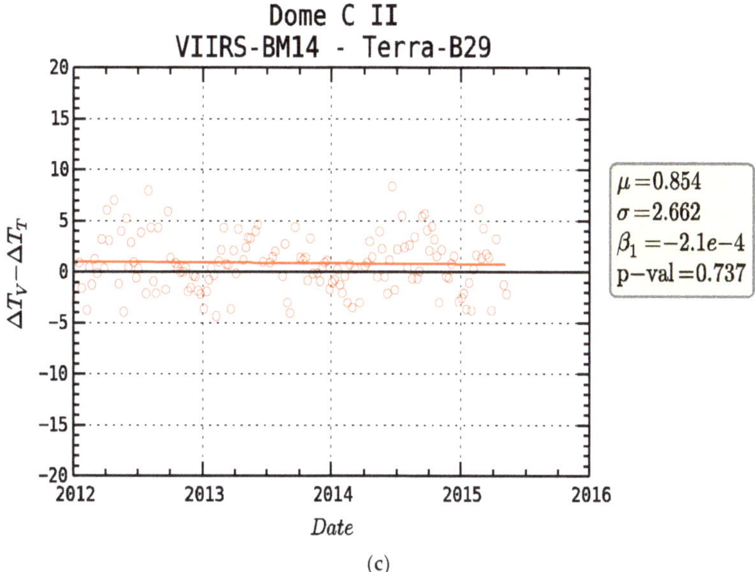

(c)

Figure 9. Relative bias trends between S-VIIRS and T-MODIS. (**a**) M15/B31; (**b**) M16/B32; (**c**) M14/B29.

5. Conclusions

Earth-observing remote sensing instruments such as MODIS and VIIRS are critical for global monitoring of the various geophysical retrievals that shape the Earth's Climate. Further, given the legacy of the MODIS instruments, S-VIIRS becomes a key bridge to ensuring long term continuity of the climate data records. With such high requirements on the radiometric fidelity for both instruments, we have tried to assess the radiometric consistency and stability using well characterized EV target "Dome C". An AWS is used as a stable proxy to assess the 3.5 year on-orbit performance of S-VIIRS, and perform an inter comparison of S-VIIRS and A-MODIS using a previously established methodology for the 2 MODIS instruments [8]. In general, S-VIIRS TEBs performance is very similar to A-MODIS. Statistically, no significant temporal drift in the relative bias measurements for the two instruments is observed. The temporal drift in relative bias over 3.5 years of performance is assessed to be within ± 0.08 K with the exception of the M13 band of S-VIIRS. A small warm bias is noted in the cold scene retrieval for all the 5M- bands of S-VIIRS suggesting an offset impact in the S-VIIRS TEB calibration model. The relative bias comparisons between S-VIIRS and T-MODIS provided evidence the deficiency could reside in the offset calibration term in the VIIRS TEB model. In summary, the TEB performance for both S-VIIRS and A-MODIS based on the BB as a calibrator source has been found to be within design requirements.

Acknowledgments: The authors would like to thank all current (in particular Andrew Wald for proof reading the article) and past members of the MCST and VIIRS Characterization Support Team (VCST) for their many and varied contributions over the years.

Author Contributions: Sriharsha Madhavan was responsible in writing the manuscript. Jake Brinkmann provided the dataset and comphrensive analysis for this work. Brian Wenny was chief investigator for the relative bias efforts in the previous work using MODIS instruments. His insights were quite useful for this work. Aisheng Wu provided support in extending the work to VIIRS. Funding for this research was provided through NASA contract NNG15HQ01C, as represented by Xiaoxiong Xiong.

Conflicts of Interest: The authors declare no conflict of interest.

References

1. Xiong, X.; Butler, J.; Chiang, K.; Efremova, B.; Fulbright, J.; Lei, N.; McIntire, J.; Oudrari, H.; Sun, J.; Wang, Z.; Wu, A. VIIRS on-orbit calibration methodology and performance. *J. Geophys. Res. Atmos.* **2014**, *119*, 5065–5078. [CrossRef]
2. Xiong, X.; Chiang, K.; Esposito, J.; Guenther, B.; Barnes, W. MODIS on-orbit calibration and characterization. *Metrologia* **2003**, *40*, 89–92. [CrossRef]
3. Turpie, K.R.; Robinson, W.D.; Franz, B.A.; Eplee, R.E.; Meister, G.; Fireman, G.F.; Patt, F.S.; Barnes, R.A.; McClain, C.R. Suomi NPP VIIRS ocean color data product early mission assessment. *SPIE Proc.* **2012**, 8510. [CrossRef]
4. Justice, C.O.; Romàn, M.O.; Csiszar, I.; Vermote, E.F.; Wolfe, R.E.; Hook, S.J.; Friedl, M.; Wang, Z.; Schaaf, C.B.; Miura, T.; *et al.* Land and cryosphere products from Suomi VIIRS: Overview and status. *J. Geophys. Res. Atoms.* **2013**, *118*, 9753–9765. [CrossRef] [PubMed]
5. Liu, H.; Remer, L.A.; Huang, J.; Huang, H.-C.; Kondragunta, S.; Laszlo, I.; Oo, M.; Jackson, J.M. Preliminary evaluation of S-NPP VIIRS aerosol optical thickness. *J. Geophys. Res. Atoms.* **2013**, *119*, 3942–3962. [CrossRef]
6. Xiong, X.; Butler, J.; Wu, A.; Chiang, V.; Efremova, B.; Madhavan, S.; Mcintire, J.; Oudrari, H. Comparison of MODIS and VIIRS onboard blackbody performance. *SPIE Proc.* **2012**, *8533*. [CrossRef]
7. Efremova, B.; Wu, A.; Xiong, X. Relative spectral response corrected calibration inter-comparison of S-NPP VIIRS and Aqua MODIS thermal emissive bands. *SPIE Proc.* **2014**, *9218*. [CrossRef]
8. Wenny, B.N.; Xiong, X. Using a Cold Earth Surface Target to Characterize Long-term Stability of the MODIS Thermal Emissive Bands. *IEEE Geosci. Remote Sens. Lett.* **2008**, *5*, 162–165. [CrossRef]
9. Wenny, B.N.; Xiong, X.; Madhavan, S.; Wu, A.; Li, Y. Long-term band-to-band calibration stability of MODIS thermal emissive bands. *SPIE Proc.* **2013**, *8724*. [CrossRef]
10. Wenny, B.N.; Xiong, X.; Madhavan, S. Evaluation of Terra and Aqua MODIS thermal emissive band calibration consistency. *SPIE Proc.* **2012**, *8533*. [CrossRef]
11. Xiong, X.; Wu, A.; Wenny, B.N.; Madhavan, S.; Wang, Z.; Li, Y.; Chen, N.; Barnes, W.; Salomonson, V. Terra and Aqua MODIS thermal emissive bands on-orbit calibration and performance. *IEEE Trans. Geosci. Remote Sens.* **2015**, *53*, 5709–5721. [CrossRef]
12. Efremova, B.; McIntire, J.; Moyer, D.; Wu, A.; Xiong, X. S-NPP VIIRS thermal emissive bands on-orbit calibration and performance. *J. Geophys. Res. Atmos.* **2014**, *119*, 10859–10875. [CrossRef]
13. Hook, S.J.; Clodius, W.B.; Balick, L.; Alley, R.E.; Abtahi, A.; Richards, R.C.; Schladow, S.G. In-Flight validation of mid- and thermal infrared data from the Multispectral Thermal Imager (MTI) using an automated high-altitude validation site at Lake Tahoe CA/NV, USA. *IEEE Trans. Geosci. Remote Sens.* **2005**, *43*, 1991–1999. [CrossRef]
14. Barsi, J.A.; Schott, J.R.; Hook, S.J.; Raqueno, N.G.; Markham, B.L.; Radocinski, R.G. Landsat-8 Thermal Infrared Sensor (TIRS) vicarious radiometric calibration. *Remote Sens.* **2014**, *6*, 11607–11626. [CrossRef]
15. Cao, C.; Uprety, S.; Xiong, X.; Wu, A.; Jing, P.; Smith, D.; Chander, G.; Fox, N.; Ungar, S. Establishing the antarctic Dome C community reference standard site towards consistent measurements from earth observation satellites. *Can. J. Remote Sens.* **2010**, *36*, 498–513. [CrossRef]
16. Xiong, X.; Wu, A.; Wenny, B. Using Dome C for MODIS Calibration Stability and Consistency. *J. Appl. Remote Sens.* **2009**, *3*, 033520.
17. Wu, A.; Xiong, X.; Cao, C. Tracking the calibration stability and consistency of the 3.7, 11.0 and 12.0 micron channels of the NOAA-KLM AVHRR with MODIS. *Int. J. Remote Sens.* **2009**, *30*, 5901–5917. [CrossRef]
18. Walden, V.P.; Roth, W.L.; Stone, R.S.; Halter, B. Radiometric validation of the atmospheric infrared sounder over the Antarctic plateau. *J. Geophys. Res. Atoms.* **2006**, *111*. D09 S03.
19. Weidner, G.; Thom, J.; Lazzara, M. Legacy calibration of the Automatic Weather Station Model 2 of the United States Antarctic Program (A primer). In Proceedings of the 8th Antarctic Meteorological Observation, Modeling, and Forecasting Workshop, Madison, WI, USA, 10–12 June 2013.
20. Ackerman, S.; Strabala, K.; Menzel, W.; Frey, R.; Moeller, C.; Gumley, L. Discriminating clear sky from clouds with MODIS. *J. Geophys. Res. Atoms.* **1998**, *103*, 32141–32157. [CrossRef]
21. Level 1 and Atmosphere Archive and Distribution System (LAADS). Available online: https://ladsweb.nascom.nasa.gov/data/search.html (accessed on 6 August 2015).

22. AMRC/AWS FTP Data Server. Available online: ftp://amrc.ssec.wisc.edu/pub/aws/ (accessed on 29 June 2015).
23. Li, Y.; Wu, A.; Xiong, X. Inter-comparison of S-NPP VIIRS and Aqua MODIS thermal emissive bands using hyperspectral infrared sounder measurements as a transfer reference. *Remote Sens.* **2016**, *8*. [CrossRef]
24. Wenny, B.N.; Wu, A.; Madhavan, S.; Wang, Z.; Li, Y.; Chen, N.; Chiang, V.; Xiong, X. MODIS TEB calibration approach in collection 6. *SPIE Proc.* **2012**, *8533*. [CrossRef]
25. Moeller, C.; Tobin, D.; Quinn, G. S-NPP VIIRS thermal band spectral radiance performance through 18 months of operation on-orbit. *SPIE Proc.* **2013**, *8866*. [CrossRef]

Article

Using Ground Targets to Validate S-NPP VIIRS Day-Night Band Calibration

Xuexia Chen [1,*], Aisheng Wu [1], Xiaoxiong Xiong [2], Ning Lei [1], Zhipeng Wang [1] and Kwofu Chiang [1]

[1] Science Systems and Applications Incorporation, Lanham, MD 20706, USA; aisheng.wu@ssaihq.com (A.W.); ning.lei@ssaihq.com (N.L.); zhipeng.wang@ssaihq.com (Z.W.); kwofu.chiang@ssaihq.com (K.C.)

[2] NASA Goddard Space Flight Center, Greenbelt, MD 20661, USA; xiaoxiong.xiong-1@nasa.gov

* Correspondence: xuexia.chen@ssaihq.com; Tel.: +1-301-867-2047

Academic Editors: Changyong Cao, Jose Moreno, Xiaofeng Li and Prasad S. Thenkabail
Received: 29 June 2016; Accepted: 22 November 2016; Published: 30 November 2016

Abstract: In this study, the observations from S-NPP VIIRS Day-Night band (DNB) and Moderate resolution bands (M bands) of Libya 4 and Dome C over the first four years of the mission are used to assess the DNB low gain calibration stability. The Sensor Data Records produced by NASA Land Product Evaluation and Algorithm Testing Element (PEATE) are acquired from nearly nadir overpasses for Libya 4 desert and Dome C snow surfaces. A kernel-driven bidirectional reflectance distribution function (BRDF) correction model is used for both Libya 4 and Dome C sites to correct the surface BRDF influence. At both sites, the simulated top-of-atmosphere (TOA) DNB reflectances based on SCIAMACHY spectral data are compared with Land PEATE TOA reflectances based on modulated Relative Spectral Response (RSR). In the Libya 4 site, the results indicate a decrease of 1.03% in Land PEATE TOA reflectance and a decrease of 1.01% in SCIAMACHY derived TOA reflectance over the period from April 2012 to January 2016. In the Dome C site, the decreases are 0.29% and 0.14%, respectively. The consistency between SCIAMACHY and Land PEATE data trends is good. The small difference between SCIAMACHY and Land PEATE derived TOA reflectances could be caused by changes in the surface targets, atmosphere status, and on-orbit calibration. The reflectances and radiances of Land PEATE DNB are also compared with matching M bands and the integral M bands based on M4, M5, and M7. The fitting trends of the DNB to integral M bands ratios indicate a 0.75% decrease at the Libya 4 site and a 1.89% decrease at the Dome C site. Part of the difference is due to an insufficient number of sampled bands available within the DNB wavelength range. The above results indicate that the Land PEATE VIIRS DNB product is accurate and stable. The methods used in this study can be used on other satellite instruments to provide quantitative assessments for calibration stability.

Keywords: VIIRS; SCIAMACHY; Day-Night Band; M bands; calibration; radiance; reflectance; RSR; BRDF; Libya 4; Dome C

---

1. Introduction

1.1. NPP VIIRS Sensor Overview

The Suomi National Polar-orbiting Partnership (S-NPP) is one of the modern Earth-orbiting Earth-observing weather satellites. Since S-NPP was launched successfully on 28 October 2011 from Vandenberg Air Force Base in California, it has provided observations of the entire Earth's surface twice daily, including the land, ocean, and atmosphere. S-NPP represents a bridge providing consistent and continuous observation data between NASA's Earth Observing System (EOS) satellites and the Joint Polar Satellite System satellites [1–3]. The five instruments onboard S-NPP are the Visible Infrared Imaging Radiometer Suite (VIIRS), the Cross-Track Infrared Sounder, the Ozone Mapping and Profiler Suite, the Advanced Technology Microwave Sounder, and the Cloud and Earth Radiance Energy System.

VIIRS is designed based on the Moderate Resolution Imaging Spectroradiometer (MODIS) heritage, and is a wide-swath (3034 km), cross-track scanning radiometer which observes the Earth's surface in 22 visible and infrared spectral bands (0.4–12.5 μm). Among the 22 spectral bands, there are 15 reflective solar bands (RSB) and 7 thermal emissive bands (TEB). The VIIRS observation altitude is 828 km and its Earth view observation scan angle range is ±56.28° off nadir. The VIIRS instrument was powered on 8 November 2011 and the instrument nadir door was opened on 21 November 2011. Among the RSB, the DNB on VIIRS is a visible/near-infrared panchromatic band, which can observe the Earth during both daytime and nighttime. The spatial resolution of the DNB is 750 m at nadir and there are 32 aggregation zones through each half of the instrument swath on either side of nadir. These aggregation zones are designed to maintain a near 750 m spatial resolution for pixels away from the nadir. In order to provide imagery of clouds and Earth targets from full sunlight to quarter moon illumination status, the DNB comprises three gain stages: the low gain stage (LGS), the medium gain stage (MGS), and the high gain stage (HGS). The LGS is used for observing bright earth scenes at day time, the MGS is used for scenes near the terminator or at twilight, and the HGS is used for night scenes. In our study, we focus on the DNB LGS performance. The M4 to M7 wavelength ranges are the matching M bands within the DNB wavelength range. Of these bands, M4, M5, and M7 have Low Gain and High Gain stages. Table 1 lists characteristics for the DNB for the narrower bandpass bands M4 through M7.

Table 1. List of characteristics of DNB and M bands within DNB wavelength range.

VIIRS Band	Wavelength Range (nm)	Bandpass (nm)	Gain Stages	Band Explanation	Spatial Resolution at Nadir (m)
DNB	500–900	400	Low, Middle, High	Visible/Reflective	750
M4 M5	545–565 662–682	20 20	Low, High Low, High	Visible/Reflective	750
M6 M7	739–754 846–885	15 39	Single Low, High	Near IR	750

1.2. Reference Sites

Several Earth surface sites with stable radiometric characteristics are widely used as references to monitor satellite sensor long-term stability [4–7]. In this study, we chose the Libya 4 desert site and Dome C in Antarctica as reference sites, as they have been used extensively in previous studies to monitor satellite performance.

The Libya 4 desert site (Latitude: 28.55°N, Longitude: 23.39°E) is close to horizontally uniform and is a relatively homogeneous area covered by sands in Africa [4,5]. The Libya 4 site is often selected as a calibration study site based on its high spatial uniformity and temporally invariant surface cover properties for stable reflectance and bidirectional reflectance distribution function (BRDF) [4,5,8–10]. This site is located in an arid region and thus has a low probability of cloudy weather and precipitation.

The Dome C site (Latitude: 74.50°S, Longitude: 123.00°E) is one of the several plateaus on the Antarctic ice sheet and it is considered to be a nearly homogeneous surface [5–7]. Located in the east Antarctic, Dome C is covered by snow and has an extremely small terrain surface slope. The altitude of the site is 3200 m. Its uniformity properties produce temporally invariant reflectance and BRDF spectral response [11,12]. It has a clear atmosphere with minimized influence of aerosol and atmospheric water vapor when compared with other reference sites [12]. Since the Dome C site is close to the South Pole of the Earth, VIIRS can make several overpasses in any given day and it is easy to find a good daytime image near nadir observation during the southern hemisphere summer period, from October to March [9].

1.3. Relative Spectral Response (RSR)

Prelaunch RSRs of all VIIRS bands were measured during sensor thermal vacuum testing. A double monochromator operating in subtractive mode was used to collect spectral test data in the lab environment for generating VIIRS RSRs [13,14]. Shortly after S-NPP launch, it was observed that several VIIRS reflective bands had decreased optical throughput due to the contamination of the mirror coating on the rotating telescope assembly (RTA) [1,15,16], which also caused the RSRs to change over time. Without updating the RSRs, the sensor radiometric calibration and the computed top-of-the-atmosphere (TOA) spectral reflectance delivered in the Sensor Data Records (SDRs) could be biased [17].

The modulated RSR is derived from a wavelength-dependent optical throughput degradation model developed by NASA VIIRS Characterization Support Team (VCST) [13]. These modulated band-averaged RSRs are generated at different times during SNPP mission, and are used to update post-launch SDR Look-Up-Tables (LUT) to correct the RTA mirror contamination effect [15,18–20]. Figure 1 shows DNB and M4–M7 RSRs at orbit 154 (8 November, 2011, same as prelaunch RSRs) and at orbit 20875 (8 November 2015, four years after launch). The near infrared bands have been more affected by RTA contamination than those in the visible wavelength range. In the past four years, the DNB RSR peak wavelength has shifted to shorter wavelengths with the peak RSR wavelength changing from 784 nm to 680 nm. Based on the degradation observation and analysis in the past, the RSRs have been more stable now than in the past, and may change less in the future.

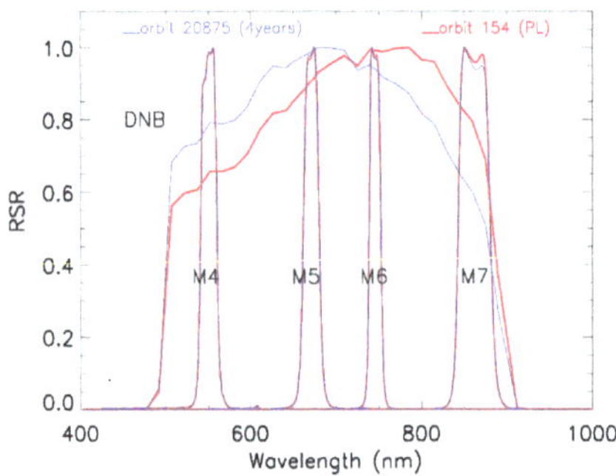

Figure 1. DNB and M4–M7 bands prelaunch RSRs at orbit 154 and modulated RSRs at orbit 20875.

1.4. VIIRS Radiometric Calibration Stability

Radiometric calibration is the process of characterizing the instrument response to signal inputs, which usually are known, controlled, and stable [21]. Through the calibration process, systematic errors may be discovered and evaluated and can also help to determine the relative biases between different instruments. Calibration stability aims to measure the relative on-orbit calibration change and bias with time [22]. When a sensor's calibration is stable and consistent, the response of that sensor to the same inputs is repeatable, traceable, and expectable over time [22]. Calibration stability studies usually look at long-term records, which are built upon current calibration methodologies, and consider all of the currently known system errors. These long-term records could be stable if the instrument and the inputs are stable whereas the records could experience jumps and drifts when new events occur over the sensor mission period. New systematic errors can be identified when specific causes are found to produce the specific effect that causes a drift of the observations. Long-term calibration stability can be achieved by frequently monitoring the instrument performance and recalibration with all the discovered system errors. Calibration stability can also evaluate whether the sensor meets mission requirements with an accurate interpretation of the data in order to make mission decisions.

The fine spatial resolution of VIIRS RSB and DNB products allow science community users to develop global surface type classification products [23], to estimate light power [24], to study nighttime lighting patterns [25], to monitor nighttime surface air quality [26], to monitor and forecast land surface phenology [27], and to research other topics. These growing interests bring a greater need for better understanding the calibration stability and absolute accuracy of the VIIRS RSB and DNB products.

The VIIRS RSB and DNB on-orbit calibration methodology used by the VCST has been described in several studies [1,13,18–20,28,29]. Some studies in the past have also demonstrated independent techniques for validation of VIIRS on-orbit calibration using Earth view data. Uprety et al. [10,30,31] have analyzed the stability of VIIRS RSB in the past few years through inter-comparison with Aqua MODIS and Landsat 8 OLI at Libya 4, Sudan 1, Dome C, and Ocean sites. Wu et al. [9] have done an assessment of the radiometric calibration stability of the first three years of VIIRS visible and near infrared spectral bands using measurements from Suomi-NPP VIIRS and Aqua MODIS simultaneous nadir overpasses over the Libya 4 and Dome C sites. Bhatt et al. [8] and Wang and Cao [32] have also assessed the VIIRS RSB using Libya 4 and deep convective clouds. Chen et al. [33] have validated the VIIRS DNB performance at Libya 4 and Dome C using both NOAA Interface Data Processing Segment (IDPS) and the NASA Land Product Evaluation and Algorithm Testing Element (PEATE) data. Before 5 April 2013, IDPS did not have accurate data due to the prelaunch RSR LUT. After excluding data before 5 April 2013, IDPS DNB radiance data are consistent with Land PEATE data with 0.6% or less difference for Libya 4 and 2% or less difference for Dome C [33]. Time series of the bridge light data are also used to validate the stability of the light measurements and the calibration of VIIRS DNB radiance [24]. These results indicate that the VIIRS variability is less than 0.6% per year for RSB and 2.5% for shortwave infrared (SWIR) bands [8,9]. The difference between the absolution calibration of VIIRS and MODIS for most bands are within 2.0% [9,10,30,32]. The global surface classification type products generated using S-NPP VIIRS data maintain (78.64% $\pm$ 0.57%) overall accuracy [23].

Some research used Simultaneous Nadir Overpass (SNO) data with similar geometry angles from two or more sensors which allowed for inter-comparison [9,30,32]. The SNO method is relatively simple and robust. Since it is possible that a pair of polar-orbiting satellites can observe the same point on the Earth's surface at nearly the same time close to nadir, there can be very little ground surface and atmospheric path difference for observed collocated pixels [22]. However, SNO calibration alone is not sufficient to produce a long-term time series for absolute calibration. The difference in the spectral response functions between different sensors and the change of surface targets for different SNO observations can introduce uncertainties in long-term data records and makes the inter-calibration results difficult to understand. The SNO method is more efficient if one satellite has an accurate absolute calibration and can be used as a stable standard for other sensors.

Other research uses pseudo-invariant ground targets, such as desert, snow, bridge light power, and deep convective clouds, for sensor calibration and validation [4,8,10,24,32,33]. There are several advantages of using pseudo-invariant ground targets for validation. The ground targets are usually selected based on their spectral and spatial uniformity and radiometric stability, thus they have minimal effects from the environment and are suitable for long-term monitoring. The ground sites are usually accessible for ground-based measurements of the radiometric calibration and the measurements can be done if it is necessary. In addition, monitoring the ground targets long-term trends allows for investigation of the influence from RSRs changes, F-factor calculation, and other onboard calibration events [32,33].

Besides the above research, our study is the first one to validate the VIIRS DNB most recent radiometric calibration quality at two Earth view targets using the NASA Land PEATE products. The main objective in this research is to track the first four years of on-orbit calibration stability of the VIIRS DNB and M bands (M4, M5, and M7) using two stable ground reference sites, Libya 4 and Dome C. In the above paragraphs, we have presented an overview of the instrument, study sites,

RSRs, and previous study results. In the Section 2, we describe the derivation of the data and the methodology. In Section 3, we provide analysis and the results. Finally, in Section 4, we summarize our findings and conclusions. The results shown in this paper are useful for validating the current VIIRS instrument calibration stability and can help guide future calibration algorithm improvements and data reprocessing. The methods can be adopted for validating other instruments on board satellites, such as Moderate Resolution Imaging Spectroradiometer (MODIS), Satellites Pour l'Observation de la Terre or Earth-observing Satellites (SPOT), China Brazil Earth Resources Satellite (CBERS), and future Joint Polar Satellite System (JPSS).

2. Data and Methodology

2.1. Data

The data products used in this study are provided by the NASA Land PEATE which can be downloaded from Level 1 and Atmosphere Archive and Distribution System (LAADS) C1.1 Reprocessing version (http://ladsweb.nascom.nasa.gov/). Three products were downloaded including NPP_VDNE_L1-VIIRS/NPP Day/Night Band 5-Min L1 Swath SDRs 750 m, NPP_VMAE_L1-VIIRS/NPP Moderate Resolution 5-Min L1 Swath SDRs, and GEO 750 m. We selected the data to be near nadir overpasses of Libya 4 and Dome C to cover the period between S-NPP launch and January 2016. The Land PEATE data has been reprocessed with improved calibration algorithms and LUTs [9,20,34]. In this study, the solar irradiance LUT is provided from the Aerospace Corporation and the modulated RSR LUTs are provided by the VCST. Among the data we used, the on-orbit SDSM screen transmission functions combined with the solar diffuser BRDF functions were generated from both yaw maneuver and regular on-orbit data conducted early in the mission [20] and the RSR functions have been updated on 31 March 2012, 15 July 2012, and 1 February 2013 and were implemented in data processing.

Images are visually examined and we exclude those with visible clouds or shadow over the Libya 4 and Dome C sites. In addition, the distances from the study site center pixel to nadir are less than 60 pixels, which is about 45 km on ground distance, and the satellite zenith angles are less than 3.5 degrees. The solar zenith angle range is from 14 to 55 degrees for Libya 4 and from 58 to 70 degrees for Dome C. We extract 32×32 pixels to calculate the DNB mean radiance and reflectance and use co-located 32×32 M bands pixels to calculate the M bands mean radiance and reflectance. All observations are at daytime, thus the DNB data are all observed in low gain stage at both the Libya 4 site and the Dome C site. At the Libya 4 site, M1 and M2 bands are in the high gain stage; M3 band has both high gain and low gain stage observation due to different seasons; and M4, M5, and M7 bands are all in the low gain stage. At the Dome C site, all dual gain M bands observations are in the low gain stage. More details on the algorithm are provided in Section 2.2. We also exclude the data that have shown less homogeneity, where the standard deviation of the DNB radiances is higher than 0.0001 ($\text{W}\cdot\text{cm}^{-2}\cdot\text{Sr}^{-1}$). Based on the above requirements, Libya 4 and Dome C site images are selected from 16-day repeatable orbits so all data have approximately the same viewing geometry relative to the site.

2.2. Calculate Radiance and Reflectance Data

DNB SDRs include calibrated and geolocated radiance data, while the M band SDRs contain both radiance and reflectance data. L_m is provided by the DNB SDRs product directly, which is calibrated considering the response versus scan angle of a rotating half angle mirror and the solar diffuser BRDF degradation using the time-dependent modulated RSR [1,15]. The DNB radiance with the Earth-Sun distance, solar zenith angle, and RSR normalization are calculated using the following Equations (1)–(4).

$$L_{normalized} = \frac{L_m \times d^2}{\cos(\theta)} \tag{1}$$

$$L_{rsr\_normalized} = \frac{L_m \times d^2}{\cos(\theta) \times F_{ESUN}} \tag{2}$$

$$ESUN_t = \frac{\sum(RSR_{\lambda,t} \times SolarIrradiance_\lambda \times \Delta\lambda)}{\sum(RSR_{\lambda,t} \times \Delta\lambda)} \tag{3}$$

$$F_{ESUN} = \frac{ESUN_t}{ESUN_{t_0}} \tag{4}$$

$L_{normalized}$ = Measured solar radiance after Earth-Sun distance and solar zenith angle normalization ($W \cdot cm^{-2} \cdot Sr^{-1}$).

$L_{rsr\_normalized}$ = Measured solar radiance after Earth-Sun distance, solar zenith angle, and RSR normalization ($W \cdot cm^{-2} \cdot Sr^{-1}$).

F_{ESUN} = Ratio of solar irradiance (ESUN) between the data collection time t and the reference time t_0 (prelaunch at orbit 154).

L_m = Measured solar radiance imported from the SDRs DNB product ($W \cdot cm^{-2} \cdot Sr^{-1}$).

d = Earth-Sun distance (astronomical units).

θ = Solar zenith angle (degree).

$ESUN_t$ = Band dependent and time dependent mean solar irradiance ($W \cdot m^{-2} \cdot nm^{-1}$). The summation is taken over lambda ranging from 200 nm to 200,000 nm.

$SolarIrradiance_\lambda$ = Solar spectral irradiance ($W \cdot m^{-2} \cdot nm^{-1}$).

$RSR_{\lambda,t}$ = Wavelength dependent and time dependent relative spectral response of band λ at time t.

$\Delta\lambda$ = Wavelength spectral interval (nm).

To retrieve the DNB TOA reflectance data, we use the Equation (5).

$$\rho_{normalized} = \frac{\pi \times L_m \times d^2}{\cos(\theta) \times ESUN_t} \tag{5}$$

$\rho_{normalized}$ = Top of atmosphere reflectance with Earth-Sun distance, solar zenith angle, and RSR normalization.

M bands radiance (L_{m_Mband}) and reflectance ($\rho_{normalized\_Mband}$) data are calculated by using the scaled integer (SI), scale, and offset values provided by the SDRs products following Equations (6)–(9).

$$L_{m\_Mband} = SI \times scale + offset \tag{6}$$

$$L_{normalized\_Mband} = \frac{L_{m\_Mband} \times d^2}{\cos(\theta)} \tag{7}$$

$$L_{rsr\_normalized\_Mband} = \frac{L_{m\_Mband} \times d^2}{\cos(\theta) \times F_{ESUN}} \tag{8}$$

$$\rho_{normalized\_Mband} = \frac{\pi \times L_{m\_Mband} \times d^2}{\cos(\theta) \times ESUN_t} \tag{9}$$

2.3. A Kernel-Driven BRDF Correction Model

When we use satellite data from inter-comparison sensors, as well as from adjoining paths or long term time series data of the same sensor, a BRDF model is required to remove the anisotropy influence from the surface and different geometry angles [9,30,32,35]. In this study, we use a semi-empirical BRDF correction model, consisting of two kernel-driven components [36]. This model has been found to be functional when applied to reflectances observed by MODIS over the Libya desert and Antarctic surface sites and by Landsat 7 Enhanced Thematic Mapper Plus over Sonoran and Libya desert sites [4,5].

This BRDF model assumes that the ground target subpixel surface contains a large number of identical protrusions with rectangular and vertical wall shapes. These protrusions are randomly distributed on a horizontal surface. The calculation of the BRDF relies on the empirical weighted sum of two functions (or kernels) of view and illumination geometry and an isotropic parameter [4,5,36].

The kernel-driven BRDF corrected radiance or reflectance R is expressed as Equation (10) [36]:

$$R^i(\theta,\ \varphi,\ \psi) = K_0^i + K_1^i f_1^i(\theta,\ \varphi,\ \psi) + K_2^i f_2^i(\theta,\ \varphi,\ \psi) \tag{10}$$

where θ, φ are the solar zenith, satellite view zenith angles; ψ is the relative azimuth angle between the solar and the satellite sensor directions with value from 0 to π; i is the index for the band wavelength; f_1 is estimated from the bidirectional radiance/reflectance associated with the geometric scattering component of the geometrical structure of opaque reflectors on the subsurface and shadowing effects, while f_2 is derived from the volume scattering contribution by a collection of dispersed facets which simulates the volume scattering properties of surface features such as canopies, bare soils, sands, and other factors. Coefficients K_0, K_1, and K_2 are parameters related to the surface's subpixel-scale physical structure, optical properties, and the background and protrusion reflectance. The parameter K_0 also represents the bidirectional reflectance at 0 solar zenith angle and view angle, which can provide a basis for the inter-comparison of sensor data acquired with different viewing or solar angles [36].

K_0, K_1, and K_2 are surface type-specific parameters, thus the coefficients are determined empirically for each site in this study. The values of these coefficients are calculated by minimizing the differences between the observed and modeled reflectance using an optimization algorithm, given the measured values of θ, φ and ψ from the training data sets for each site. In this study, we use the first three full years available RSR corrected VIIRS DNB and M bands radiance/reflectance data (derived from Equations (2), (5), (8) and (9)) as the training data set in the kernel-driven BRDF correction model to cover a complete seasonal oscillation cycle. Please note that these training data are good for validating a relative trend but are not good for calculating the absolute calibration. In our study, Equation (11) is used to calculate the BRDF corrected Land PEATE DNB radiance (L_{brdf}) and TOA reflectance (ρ_{brdf}), which can be used to validate long-term sensor calibration stability. The DNB $R_{radiance}/R_{reflectance}$ and $K_{0\_radiance}/K_{0\_reflectance}$ are derived from the Equation (10) by using radiance data from Equation (2) or reflectance data from Equation (5) as BRDF model inputs.

$$L_{brdf} = L_{normalized} \times \frac{K_{0\_radiance}}{R_{radiance}};\ \rho_{brdf} = \rho_{normalized} \times \frac{K_{0\_reflectance}}{R_{reflectance}} \tag{11}$$

Equation (12) is used to calculate the BRDF corrected Land PEATE M bands radiance (L_{brdf_Mband}) and TOA reflectance (ρ_{brdf_Mband}), which can be used in the comparison with DNB long-term calibration stability. The $R_{radiance}/R_{reflectance}$ and $K_{0\_radiance}/K_{0\_reflectance}$ are derived from Equation (10) for each M band using radiance data from Equation (8) or reflectance data from Equation (9).

$$\begin{aligned} L_{brdf\_Mband} &= L_{normalized\_Mband} \times \frac{K_{0\_radiance}}{R_{radiance}}; \\ \rho_{brdf\_Mband} &= \rho_{normalized\_Mband} \times \frac{K_{0\_reflectance}}{R_{reflectance}} \end{aligned} \tag{12}$$

2.4. SCIAMACHY Spectral Data Simulated TOA Reflectance

In this study, we also use the Libya 4 and Dome C spectral data recorded by SCIAMACHY to retrieve the TOA reflectance observed by VIIRS. The SCIAMACHY data used in this study is provided by the European Space Agency. The data is the averaged reflectance spectra collected over the study sites from 2002 to 2010. We assume the ground targets do not change significantly and their reflectance spectra are stable over the period of study. The SCIAMACHY data has a very fine spectral resolution (<1 nm) [16]. Thus, SCIAMACHY data can provide high spectral precision measurements of our study sites by sampling several absorption features (Figure 2). These features are within common atmospheric absorption wavelengths and are affected by atmospheric status and ground materials status. SCIAMACHY provides the measured hyperspectral data with enough spectral and spatial resolutions and spectral coverage over the surface sites selected in this study since the SCIAMACHY data are stable and their average data, our input of the hyperspectral data, is constant over time. Thus, any impacts due to the DNB RSR changes can be easily simulated and identified.

We use Equation (13) to calculate the modulated TOA reflectance considering the RSR influence with time. $\rho_{SCIAMACHY}$ is the SCIAMACHY spectral reflectance indicated in Figure 2. The summation is taken over the wavelength parameter, λ, ranging from 200 nm to 200,000 nm. We assume that the surface spectral and solar irradiance did not change, and the time-dependent RSR is the major source for the change of TOA reflectance.

$$\rho_{TOA\_SCIAMACHY} = \frac{\sum(\rho_{SCIAMACHY} \times SolarIrradiance_{\lambda} \times RSR_{\lambda} \times \Delta\lambda)}{\sum(RSR_{\lambda} \times SolarIrradiance_{\lambda} \times \Delta\lambda)} \quad (13)$$

The SCIAMACHY derived TOA reflectance ($\rho_{TOA\_SCIAMACHY}$) is compared with the BRDF corrected Land PEATE DNB TOA reflectance (ρ_{brdf}) for each study site. First, we use the first three full years of available Land PEATE DNB TOA reflectance data (ρ_{brdf}) to generate a linear fit model. Second, we use this linear model to derive fitted reflectance values for all of the available dates used in this study. Third, we normalize ρ_{brdf} and all of the fitted reflectance data from the second step to the first data point of the fitted reflectance value. We also normalize $\rho_{TOA\_SCIAMACHY}$ to its first data value. After the above normalization process, the changes of all three data trends can be easily compared.

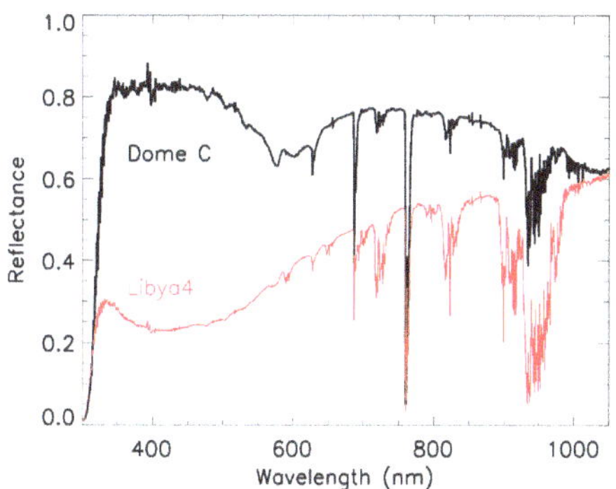

Figure 2. SCIAMACHY spectra of Libya 4 and Dome C sites.

2.5. DNB Comparison with M bands

In this study, we compare the DNB with the M bands within the DNB wavelength range, except M6. The M6 band lacks sufficient data due to its band radiance saturation and 'fold-over' [20].

An integrated Mbands radiance using M4, M5, and M7 bands is compared with the DNB for long term trend monitoring following Equations (14)–(17).

$$R_{mi} = \frac{\int RSR(Miband)_{\lambda} \times RSR(DNB)_{\lambda} \times \rho_{SCIAMACHY} \times d\lambda}{\int RSR(DNB)_{\lambda} \times \rho_{SCIAMACHY} \times d\lambda} \quad (14)$$

$$W_{mi} = \frac{R_{mi}}{R_{m4} + R_{m5} + R_{m7}} \quad (15)$$

$$DNB_{bandwide} = \int RSR(DNB)_{\lambda} \times d\lambda \quad (16)$$

$$Integral_{Mbands} = (W_{m4} \times L_{m4} + W_{m5} \times L_{m5} + W_{m7} \times L_{m7}) \times DNB_{bandwide} \quad (17)$$

R_{mi} (i = 4, 5, 7) are the weight parameters of M4, M5, and M7 bands, which are computed from the RSR of these bands [*RSR(Miband)*]. W_{mi} are the adjusted weights that contain the scale factor for making the sum of all the weights equal to 1. The integral M bands radiance ($Integral_{Mbands}$) is calculated in Equation (16) using the M bands radiance normalized by solar zenith angle, Earth-Sun distance, and BRDF correction (L_{mi}), and the DNB integral RSR ($DNB_{bandwide}$).

For the Dome C site, atmospheric Rayleigh scattering and the change of solar zenith angle causes variations in the reflected radiance (isotropic) [6]. Field experiments demonstrate that when the snow surface is smooth, there is very little effect from varying azimuth angle between the sun and the surface roughness features' dominant orientation direction [6,37]. The methodology used to characterize the long-term trend of the Dome C site is similar to the Libya 4 site reported in the above sections.

3. Results and Discussion

3.1. DNB RSR Influence on ESUN Stability

To accurately compute the DNB and M bands reflectance, a time-dependent, modulated RSR is used to model the change of incident solar radiance expected during solar calibration events. The band dependent observed solar irradiance (ESUN) changes due to the change in RSR. Figure 3 indicates the DNB and M4-M7 ESUN difference derived from the RSR at orbit 154 (same as prelaunch RSR) and from the modulated RSRs after launch. There is an approximately 3.5% increase of ESUN for the DNB and about 0.1% increase for M7 during the past four years (1473 days since launch). ESUN increases by approximately 0.1%, 0.095%, and 0.065% for M4, M5, and M6 bands, respectively.

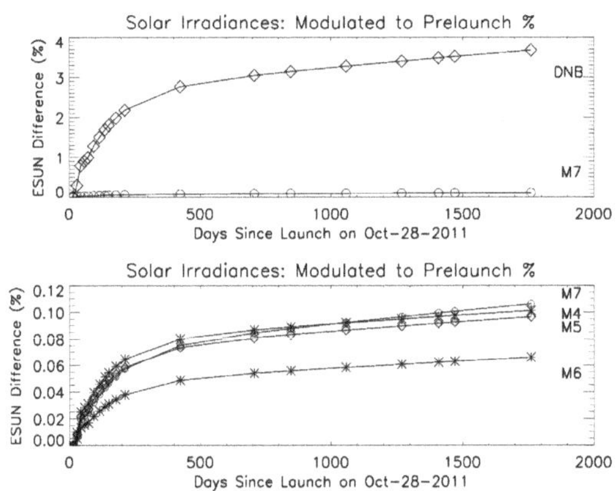

Figure 3. DNB and M4–M7 ESUN difference derived from using the modulated RSRs and using the prelaunch RSRs.

3.2. Libya 4 Site Radiance and Reflectance Long Term Stability

Among all of the selected data, the Libya 4 site view zenith angles are within a narrow range from 1 to 3.5 degrees. The variation of reflectance and radiance are relative to the solar zenith angle. Although the fluctuation patterns are roughly the same, the reflectance with the kernel driven BRDF correction shows less seasonally related fluctuations than a previous study that uses only the Lambertian model correction for DNB and M bands [33] (Figure 4). Our results also find that the longer the wavelength, the better the kernel-driven BRDF correction of TOA reflectance (Figure 4). The solar irradiance has more interaction with the material at shorter wavelengths than at longer wavelengths, thus there is reduced scattering and increased direct reflection at longer wavelengths [38]. This causes an increase in the BRDF at the longer wavelengths. These results confirm that the reflectance for desert sites are slightly non-Lambertian and thus the kernel-driven BRDF model performs better than the Lambertian correction at these sites, especially for longer wavelengths bands.

To investigate the VIIRS DNB data calibration stability, a comparison of the Libya 4 DNB normalized reflectance of Land PEATE and SCIAMACHY data is performed (Figure 5). In this figure, the Land PEATE reflectance data are normalized to the first fitted value of the Land PEATE linear fit model. The SCIAMACHY reflectance data are normalized to their first data point. We assume the surface SCIAMACHY reflectance spectra and Solar Irradiance LUT do not change in the past, when the DNB modulated RSR change. The SCIAMACHY derived TOA reflectance has a 1.01% decrease in the past four years. The trend in the TOA reflectance derived from Land PEATE data has some oscillation and the linear fit of the Land PEATE data indicates a 1.03% decrease in the past. The difference between the two trends is relatively small, which demonstrates that the Land PEATE reflectance changes are mostly caused by the RSR changes. Comparing with another study result that used IDPS data on the deep convective cloud reflectance time series [32], our Land PEATE reflectance data at Libya 4 are more stable in earlier days before April 2013. This is because the Land PEATE data have been reprocessed based on the latest version of RSR LUTs and the data are more accurate. After April 2013, both IDPS and Land PEATE data trends are stable.

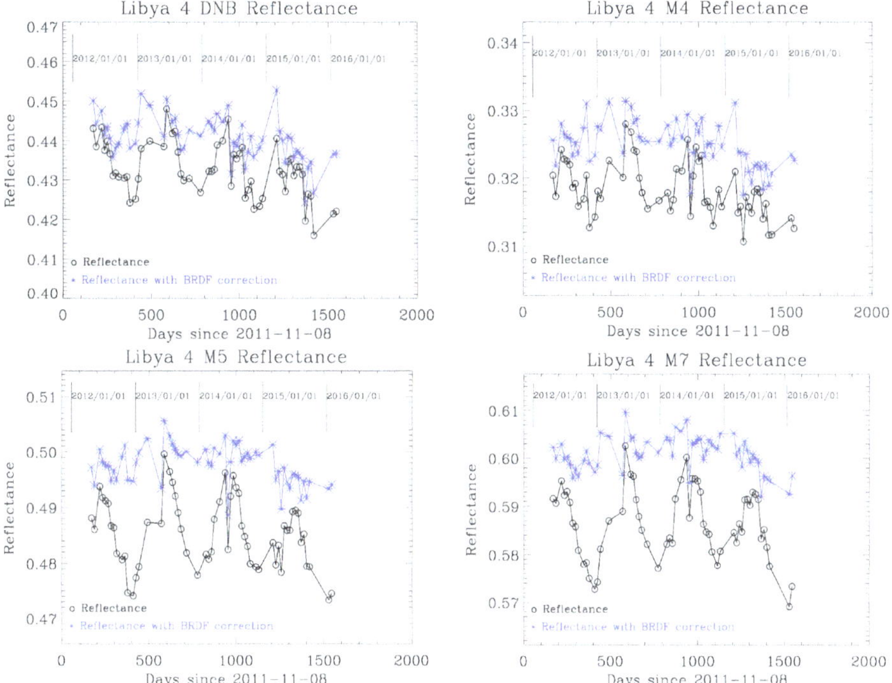

Figure 4. Comparison of Libya 4 TOA reflectance before and after kernel driven BRDF correction for DNB, M4, M5, and M7 bands.

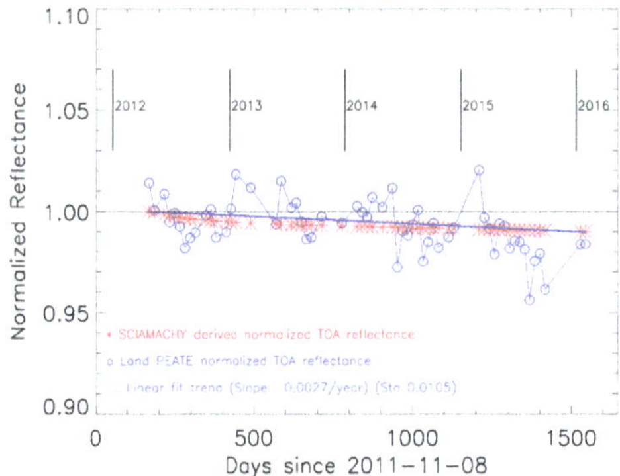

Figure 5. Libya 4 Land PEATE and SCIAMACHY DNB normalized reflectance data.

The comparison of Land PEATE DNB and M bands normalized reflectance is shown in Figure 6. In this figure, each band's reflectance data are normalized to its first data point. The Libya 4 reflectance spectra increase between 400 nm and 900 nm (Figure 2) and the M bands' RSRs have very little change (Figure 3), thus the normalized DNB reflectance has a larger decrease compared to the M bands when the DNB RSR peak wavelength moves to the shorter wavelength side at later times. For data after 2015, both DNB and M bands have some notable decrease which may be mostly due to surface changes or atmosphere status changes since the RSR change is much smaller than earlier years.

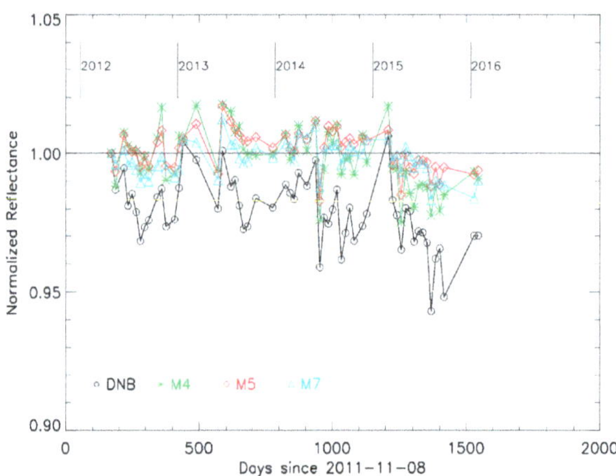

Figure 6. Comparison of Land PEATE DNB and M bands normalized reflectance at Libya 4 site.

The integral Mbands radiances from M4, M5, and M7 bands are compared with the measured DNB radiance after Earth-Sun distance, solar zenith angle, and BRDF correction (Figure 7). Their seasonal oscillations are very similar and the radiance values are close. In the Figure 7, the mean values and 1 standard deviation error bars of DNB 32 × 32 pixels' are indicated and the percentages of standard deviation range from 0.96% to 2.08%. The integral Mbands radiance values are also indicated in Figure 7. For the M4, M5, and M7 individual bands, their percentage of standard deviation ranges are between 0.94% and 2.14%.

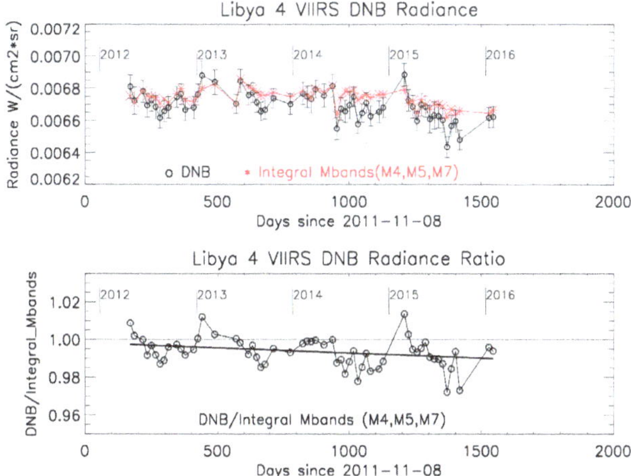

Figure 7. Comparison of Libya 4 DNB radiance and M bands integral radiance.

The use of the M band integrated radiance is to simulate the DNB, thus, ideally the ratio of the two is 1. Currently, the weights for M bands are changed with time due to the RSRs change and the ratio between DNB radiance to integral M bands radiance ranges from 1.01 to 0.97 (Figure 7). The fitting trend indicates that the DNB to integral M band ratio has a 0.75% decrease at Libya 4 site, which is partly due to the fact that the M bands used for integration are insufficient to simulate the whole DNB bandpass.

3.3. Dome C site Radiance and Reflectance Long Term Stability

The long-term trends of Dome C reflectance are indicated in Figure 8. Though the DNB modulated RSR has significant changes in the past, the Dome C surface spectral reflectance is relatively flat in the visible band range from 400 nm to 1000 nm (Figure 2), the SCIAMACHY derived TOA reflectance has only a 0.14% decrease in the past four years. The Land PEATE data derived TOA reflectance trend has large oscillations and the linear fit of Land PEATE data indicates a 0.29% decrease. These two trends are in good agreement.

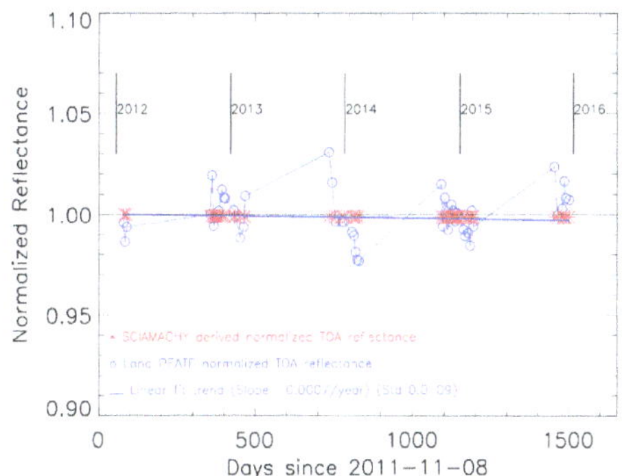

Figure 8. Dome C DNB normalized reflectance of Land PEATE and SCIAMACHY data.

The reflectance data of the Dome C site has significantly larger solar zenith angles (58–70 degrees) than those of the Libya 4 site (15–55 degrees). After the kernel-driven BRDF model correction, the reflectance data still show large fluctuations. This indicates that there is still some uncertainty due to surface and atmosphere status variation at the Dome C site.

The DNB and M bands normalized reflectance over the Dome C site are shown in Figure 9. Both M bands and DNB have little impact from changing RSRs, thus the trends of DNB and M bands reflectance are flat. However, they do show large seasonal oscillations due to the BRDF impact.

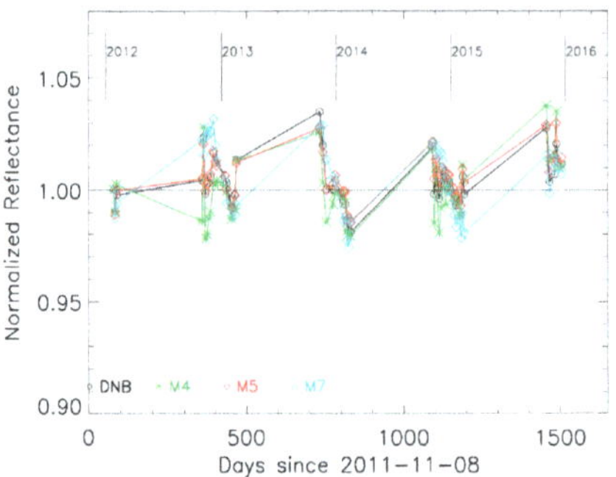

Figure 9. Comparison of Dome C DNB reflectance and M bands reflectance.

The Dome C integral M bands radiance from M4, M5, and M7 bands are compared with the DNB radiance in Figure 10. Their seasonal oscillations are very similar and the values are close. In Figure 10, the mean values and 1 standard deviation error bars of the DNB 32 × 32 pixels are indicated and the percentages of standard deviation range from 0.29% to 1.45%. For M4, M5, and M7 bands, their individual percentages of standard deviation range from 0.19% to 1.73%. The ratio between DNB to integral M bands radiance ranges from 1.02 to 0.99, with about 3% difference. The ratio at the Dome C site is slightly different from the Libya 4 site because the surface spectral characteristics are different, though the modulated RSR changes are the same. The linear fitting trend indicates the DNB to Integral M bands ratio has decreased about 1.9% in the past four years.

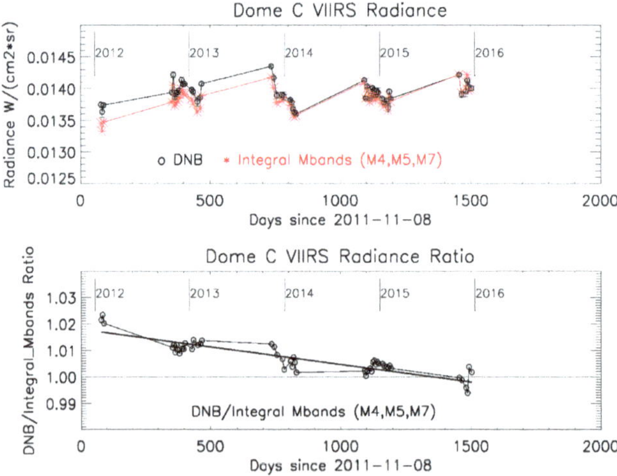

Figure 10. DNB radiance comparison with integral M bands radiance for Dome C site.

4. Conclusions

S-NPP VIIRS has been operating normally collecting global data on a daily basis for more than four years since opening the nadir door on 21 November 2011. In this study, we use two independent sites, Libya 4 and Dome C, to validate the VIIRS DNB and M bands calibration performance using nadir overpass images. Because of the DNB wide wavelength range, the DNB observed solar irradiance is quite sensitive to the incoming radiative spectrum as well as the RSR. In our investigation, the DNB modulated RSR introduces an approximately 3.5% increase on the DNB observed solar irradiance, while the M4-M7 bands RSR influences are 0.1%, 0.095%, 0.065%, and 0.1%, respectively. Long-term trend monitoring of the Earth's surface using satellite observations always has certain challenges because the ground surface is usually viewed, at most, once a day with different viewing angles and solar zenith angles. Even after having performed atmospheric corrections and Earth-Sun distance correction, a bidirectional reflectance model is also required to reduce the oscillation. The surface reflectance bidirectional effects in many types of surfaces can add a significant component of noise-like fluctuations to the time series. The magnitude of these effects can introduce some uncertainties in the calibration stability assessments when using ground targets.

The kernel-driven BRDF model has been adopted in our study for the correction of surface bidirectional reflectance effects in a time series of satellite observations. When both the sun and viewing angles vary during a long-term observation period on the same invariant target, the model can follow a semi-empirical approach to derive BRDF coefficients specific for the heterogeneous surfaces. This method is suitable for long-term data series, which can have sufficient training data for the model. The BRDF kernel model used here contains only three adjustable parameters: the solar zenith angle, the satellite zenith angle, and the relative azimuth angle between the sun and the satellite, to describe the surface BRDF effects. Based on our previous experience, this model can also be tested in more heterogeneous areas if the model input parameters can be derived from the observed time series. The regression of observed data should be made in a chosen time period that is short enough to consider the surface as time invariant and long enough to contain all seasonal oscillation data for a sufficient regression. The minimum data inputs for this regression are at least four training data sets [36].

In our study, the kernel-driven BRDF model can remove most of the solar zenith angle and atmospheric Rayleigh scattering influence due to different seasons and atmosphere status in both the Libya 4 and Dome C site. The linear BRDF correction also works well for the Dome C site when applied to data with a solar zenith angle of less than 70 degrees. Long-term trending results from the Libya 4 desert and Dome C sites are based on the assumptions that the sites are stable in surface reflectance. Short-term weather events, such as dust storms or rains in the desert site, winds, clouds, and moisture changes can cause some variations of surface reflectance changes in the sites. These variations may not be removed based on our procedures because those are true observations. However, since they are not frequently observed events, their presence should have very little impact on the trending results if our observation time is long enough and covers seasonal oscillation cycles. As illustrated, the methods used in this study are suitable for tracking VIIRS calibration stability.

The theoretical TOA reflectance long term trend calculated from SCIAMACHY spectra and DNB RSR data indicate a decreasing trend of approximately 1.01% for the Libya 4 site and 0.14% for the Dome C site during the past four years while the linear fit of Land PEATE DNB reflectance data show decreases of 1.03% and 0.29% for these two sites, respectively. The differences of the DNB reflectance trends between the Land PEATE and SCIAMACHY are partially due to the Land PEATE early data processed RSRs and SDSM transmittance screens. The current trends also demonstrate that the Land PEATE data long-term trend changes are mostly caused by RSR influence instead of ground target changes.

Though we will need longer VIIRS observation times and more ground data to better understand and verify the S-NPP VIIRS Land PEATE data quality, results of this study provide useful information on VIIRS post-launch calibration assessment and preliminary analysis of its calibration stability for the past four years of operations. Based on the above results, VIIRS Land PEATE data is suitable for long-term monitoring tasks due to its stable radiometric calibration and data reprocessing procedure with improved calibration algorithm. Scientists need to consider the RSR degradation influence before using DNB for long-term land cover change detection or monitoring.

Acknowledgments: All the SCIAMACHY data is distributed by the European Space Agency. The authors would like to thank Truman Wilson and Kevin Twedt for providing comments on this manuscript. We also thank the reviewers and editors for their useful comments.

Author Contributions: X.C. designed the study, developed the methodology, performed the analysis and wrote the manuscript. A.W., X.X., N.L., Z.W. and K.C. contributed to the design of the study, improvement of the methodology, and manuscript reviews. A.W. also developed the program tool for Kernel-driven BRDF Correction Model and processed the SCIAMACHY data. N.L. also developed the DNB and M bands RSR degradation model and provided the program to interpolate RSR for different time period.

Conflicts of Interest: The authors declare no conflict of interest.

Abbreviations

The following abbreviations are used in this manuscript:

S-NPP	Suomi National Polar-orbiting Partnership
VIIRS	Visible Infrared Imaging Radiometer Suite (VIIRS)
VCST	VIIRS Characterization Support Team
Land PEATE	Land Product Evaluation and Algorithm Testing Element
IDPS	Interface Data Processing Segment
SDRs	Sensor Data Records
RSB	Reflective solar bands
DNB	Day-Night band
M bands	Moderate resolution bands
BRDF	Bidirectional reflectance distribution function
TOA	Top-of-atmosphere
RSR	Relative Spectral Response
SDSM	Solar Diffuser Stability Monitor
LUT	Look-Up-Table
RTA	Rotating telescope assembly
DCC	Deep convective clouds
SNO	Simultaneous Nadir Overpass
ESUN	Solar irradiance

References

1. Xiong, X.; Butler, J.; Chiang, K.; Efremova, B.; Fulbright, J.; Lei, N.; McIntire, J.; Oudrari, H.; Sun, J.; Wang, Z.; Wu, A. VIIRS on-orbit calibration methodology and performance. *J. Geophys. Res. Atm.* **2014**, *119*, 1–11. [CrossRef]
2. Butler, J.; Xiong, X.; Oudrari, H.; Pan, C.; Gleason, J. NASA calibration and characterization in the NOESS Preparatory Project (NPP). In Proceedings of the 2009 IEEE International Geoscience and Remote Sensing Symposium, Cape Town, South Africa, 12–17 July 2009.
3. Cao, C.; De Luccia, F.J.; Xiong, X.; Wolfe, R.; Weng, F. Early on-orbit performance of the Visible Infrared Imaging Radiometer Suite onboard the Suomi National Polar-Orbiting Partnership (S-NPP) Satellite. *IEEE Trans. Geosci. Remote Sens.* **2013**, *52*, 1142–1156. [CrossRef]
4. Angal, A.; Xiong, X.; Choi, T.; Chander, G.; Wu, A. Using the Sonoran and Libyan desert test sites to monitor the temporal stability of reflective solar bands for Landsat 7 ETM+ and Terra MODIS sensors. *J. Appl. Remote Sens.* **2010**, *4*. [CrossRef]
5. Wu, A.; Xiong, X.; Cao, C.; Angal, A. Monitoring MODIS calibration stability of visible and near-IR bands from observed top-of-atmosphere BRDF-normalized reflectances over Libyan desert and Antarctic surfaces. *Proc. SPIE* **2008**. [CrossRef]

6. Hudson, S.R.; Wang, S.G.; Brandt, R.E.; Grenfell, T.C.; Six, D. Spectral bidirectional reflectance of Antarctic snow: Measurements and parameterization. *J. Geophys. Res.* **2006**. [CrossRef]
7. Madhavan, S.; Wu, A.; Brinkmann, J.; Wenny, B.; Xiong, X. Evaluation of VIIRS and MODIS thermal emissive band calibration consistency using Dome C. *Proc. SPIE* **2015**. [CrossRef]
8. Bhatt, R.; Doelling, D.; Wu, A.; Xiong, X.; Scarino, B.; Haney, C.O.; Gopalan, A. Initial stability assessment of S-NPP VIIRS reflective solar band calibration using invariant desert and deep convective cloud targets. *Remote Sens.* **2014**, *6*, 2809–2826. [CrossRef]
9. Wu, A.; Xiong, X.; Cao, C.; Chiang, K. Assessment of SNPP VIIRS VIS/NIR radiometric calibration stability using Aqua MODIS and invariant surface targets. *IEEE Trans. Geosci. Remote Sens.* **2016**, *54*, 2918–2924. [CrossRef]
10. Uprety, S.; Cao, C. Suomi NPP VIIRS reflective solar band on-orbit radiometric stability and accuracy assessment using desert and Antarctica Dome C sites. *Remote Sens. Environ.* **2015**, *166*, 106–115. [CrossRef]
11. Chander, G. Questionnaire for Information Regarding the CEOS WGCV IVOS subgroup Cal/Val Test Sites for Land Imager Radiometric Gain. Available online: http://calval.cr.usgs.gov/PDF/QA4EO-WGCV-IVO-CSP-001.pdf (accessed on 15 June 2016).
12. Moeller, C.; Mcintire, J.; Schwarting, T.; Moyer, D. VIIRS F1 "best" relative spectral response characterization by the government team. *Proc. SPIE* **2011**. [CrossRef]
13. Lei, N.; Xiong, X.; Guenther, B. Modeling the detector radiometric gains of the Suomi NPP VIIRS reflective solar bands. *IEEE Trans. Geosci. Remote Sens.* **2015**, *53*, 1565–1573. [CrossRef]
14. Oudrari, H.; McIntire, J.; Xiong, X.; Butler, J.; Lee, S.; Lei, N.; Schwarting, T.; Sun, J. Prelaunch radiometric characterization and calibration of the S-NPP VIIRS Sensor. *IEEE Trans. Geosci. Remote Sens.* **2015**. [CrossRef]
15. Barrie, J.D.; Fuqua, P.D.; Meshishnek, M.J.; Ciofalo, M.R.; Chu, C.T.; Chaney, J.A.; Moision, R.M.; Graziani, L. Root cause determination of on-orbit degradation of the VIIRS rotating telescope assembly. *Proc. SPIE* **2012**. [CrossRef]
16. Gyanesh, C.; Mishra, N.; Helder, D.L.; Aaron, D.B.; Angal, A.; Choi, T.; Xiong, X.; Doelling, D.R. Applications of spectral band adjustment factors (SBAF) for cross-calibration. *IEEE Trans. Geosci. Remote Sens.* **2013**, *51*, 1267–1281.
17. Joint Polar Satellite System (JPSS) VIIRS Radiometric Calibration Algorithm Theoretical Basis Document (ATBD). Available online: http://www.star.nesdis.noaa.gov/jpss/documents/ATBD/D0001-M01-S01-003_JPSS_ATBD_VIIRS-SDR_C.pdf (accessed on 15 June 2016).
18. Chen, H.; Sun, C.; Chen, X.; Chiang, K.; Xiong, X. On-orbit calibration and performance of S-NPP VIIRS DNB. *Proc. SPIE* **2016**. [CrossRef]
19. Lee, S.; Chiang, K.; Xiong, X.; Sun, C.; Anderson, S. The S-NPP VIIRS Day-Night Band on-orbit calibration/characterization and current state of SDR products. *Remote Sens.* **2014**, *6*, 12427–12446. [CrossRef]
20. Lei, N.; Chen, X.; Xiong, X. Determination of the SNPP VIIRS SDSM screen relative transmittance from both yaw maneuver and regular on-orbit data. *IEEE Trans. Geosci. Remote Sens.* **2016**, *54*, 1390–1398. [CrossRef]
21. Guide to Meteorological Instruments and Methods of Observation (WMO-No. 8). Provisional 2014 edition for CIMO-16 Approval. World Meteorological Organization. Available online: https://www.wmo.int/pages/prog/www/IMOP/publications/CIMO-Guide/Provisional2014Edition.html (accessed on 15 June 2016).
22. Ohring, G. *Achieving Satellite Instrument Calibration for Climate Change (ASIC^3)*; National Oceanic and Atmospheric Administration and Other Organizations at the National Conference Center: Lansdowne, VA, USA, 2006.
23. Zhang, R.; Huang, C.; Zhan, X.; Dai, Q.; Song, K. Development and validation of the global surface type data product from S-NPP VIIRS. *Remote Sens. Lett.* **2016**, *7*, 51–60. [CrossRef]
24. Cao, C.; Bai, Y. Quantitative analysis of VIIRS DNB nightlight point source for light power estimation and stability monitoring. *Remote Sens.* **2014**. [CrossRef]
25. Baugh, K.; Hsu, F.C.; Elvidge, C.; Zhizhin, M. Nighttime lights compositing using the VIIRS Day-Night Band: preliminary results. *Proc. Asia-Pac. Adv. Netw.* **2013**, *35*, 70–86. [CrossRef]
26. Wang, J.; Aegerter, C.; Xu, X.; Szykman, J.J. Potential application of VIIRS Day/Night Band for monitoring nighttime surface PM2.5 air quality from space. *Atm. Environ.* **2016**, *124*, 55–63. [CrossRef]
27. Zhang, X.; Friedl, M.; Henebry, G.; Jayavelu, S.; Ray, J.; Liu, Y.; Schaaf, C.; Wang, J. Development and validation of a land surface phenology product from VIIRS. In Proceedings of the MODIS/VIIRS 2016 Science Team Meeting, Silver Spring, MD, USA, 6–10 June 2016.

28. Lee, S.; McIntire, J.; Oudrari, H.; Schwarting, T.; Xiong, X. A new method for Suomi-NPP Day-Night Band on-orbit radiometric calibration. *IEEE Trans. Geosci. Remote Sens.* **2015**, *53*, 324–334.
29. Lei, N.; Xiong, X. Functional form of the radiometric equation for the SNPP VIIRS reflective solar bands. *Proc. SPIE* **2016**. [CrossRef]
30. Uprety, S.; Cao, C.; Xiong, X.; Blonski, S.; Wu, A.; Shao, X. Radiometric intercomparison between Suomi-NPP VIIRS and Aqua MODIS reflective solar bands using simultaneous nadir overpass in the low latitudes. *J. Atm. Ocean. Technol.* **2013**, *30*, 2720–2736. [CrossRef]
31. Uprety, S.; Blonski, S.; Cao, C. On-orbit radiometric performance characterization of S-NPP VIIRS reflective solar bands. *Proc. SPIE* **2016**. [CrossRef]
32. Wang, W.; Cao, C. Monitoring the NOAA operational VIIRS RSB and DNB calibration stability using monthly and semi-monthly deep convective clouds time series. *Remote Sens.* **2016**. [CrossRef]
33. Chen, X.; Wu, A.; Xiong, X.; Lei, N.; Wang, Z.; Chiang, K. Validation of S-NPP VIIRS Day-Night band and M bands performance using ground reference targets of Libya 4 and Dome C. *Proc. SPIE* **2015**. [CrossRef]
34. McIntire, J.; Moyer, D.; Efremova, B.; Oudrari, H.; Xiong, X. On-Orbit characterization of S-NPP VIIRS transmission functions. *IEEE Trans. Geosci. Remote Sens.* **2015**. [CrossRef]
35. Roy, D.P.; Zhang, H.K.; Ju, J.; Gomez-Dans, J.L.; Lewis, P.E.; Schaaf, C.B.; Sun, Q.; Li, J.; Huang, H.; Kovalskyy, V. A general method to normalize Landsat reflectance data to nadir BRDF adjusted reflectance. *Remote Sens. Environ.* **2016**, *176*, 255–271. [CrossRef]
36. Roujean, J.L.; Leroy, M.; Deschamps, P.Y. A bidirectional reflectance model of the Earth's surface for the correction of remote sensing data. *J. Geophys. Res.* **1992**, *97*, 20455–20468. [CrossRef]
37. Aoki, T.; Aoki, T.; Fukabori, M. Effects of snow physical parameters on spectral albedo and bidirectional reflectance of snow surface. *J. Geophs. Res.* **2000**, *105*, 10219–10236. [CrossRef]
38. Ientilucci, E.; Gartley, M. Impact of BRDF on physics based modeling as applied to target detection in hyperspectral imagery. *Proc. SPIE* **2009**. [CrossRef]

Chapter 4: Environmental Data Records Product Calibration/Validation

Article

Spectral Cross-Calibration of VIIRS Enhanced Vegetation Index with MODIS: A Case Study Using Year-Long Global Data

Kenta Obata [1,2,*], Tomoaki Miura [2], Hiroki Yoshioka [3], Alfredo R. Huete [4] and Marco Vargas [5]

1. National Institute of Advanced Industrial Science and Technology (AIST), Geological Survey of Japan, the Research Institute of Geology and Geoinformation, Central 7, 1-1-1 Higashi, Tsukuba, Ibaraki 305-8567, Japan
2. Department of Natural Resources and Environmental Management, University of Hawaii at Manoa, 1910 East West Road, Sherman 101, Honolulu, HI 96822, USA; tomoakim@hawaii.edu
3. Department of Information Science and Technology, Aichi Prefectural University, 1522-3 Ibara, Nagakute, Aichi 480-1198, Japan; yoshioka@ist.aichi-pu.ac.jp
4. The Plant Functional Biology and Climate Change Cluster, University of Technology Sydney, P.O. Box 123, Broadway NSW 2007, Australia; Alfredo.Huete@uts.edu.au
5. Center for Satellite Applications and Research, National Oceanic and Atmospheric Administration, College Park, MD 20740, USA; marco.vargas@noaa.gov

* Correspondence: kenta.obata@aist.go.jp; Tel.: +81-29-861-3623

Academic Editors: Changyong Cao, Dongdong Wang and Prasad S. Thenkabail
Received: 6 November 2015; Accepted: 29 December 2015; Published: 5 January 2016

Abstract: In this study, the Visible Infrared Imaging Radiometer Suite (VIIRS) Enhanced Vegetation Index (EVI) was spectrally cross-calibrated with the Moderate Resolution Imaging Spectroradiometer (MODIS) EVI using a year-long, global VIIRS-MODIS dataset at the climate modeling grid (CMG) resolution of 0.05°-by-0.05°. Our cross-calibration approach was to utilize a MODIS-compatible VIIRS EVI equation derived in a previous study [Obata *et al.*, *J. Appl. Remote Sens.*, vol.7, 2013] and optimize the coefficients contained in this EVI equation for global conditions. The calibrated/optimized MODIS-compatible VIIRS EVI was evaluated using another global VIIRS-MODIS CMG dataset of which acquisition dates did not overlap with those used in the calibration. The calibrated VIIRS EVI showed much higher compatibility with the MODIS EVI than the original VIIRS EVI, where the mean error (MODIS minus VIIRS) and the root mean square error decreased from −0.021 to −0.003 EVI units and from 0.029 to 0.020 EVI units, respectively. Error reductions on the calibrated VIIRS EVI were observed across nearly all view zenith and relative azimuth angle ranges, EVI dynamic range, and land cover types. The performance of the MODIS-compatible VIIRS EVI calibration appeared limited for high EVI values (*i.e.*, EVI > 0.5) due likely to the maturity of the VIIRS dataset used in calibration/optimization. The cross-calibration methodology introduced in this study is expected to be useful for other spectral indices such as the normalized difference vegetation index and two-band EVI.

Keywords: EVI; VIIRS; MODIS; spectral compatibility; cross-calibration; CMG data

---

1. Introduction

Biophysical parameters retrieved from Earth observation data are crucial for improving our understanding of biosphere–atmosphere interactions (e.g., [1]). Spectral vegetation indices (VIs) derived from remotely sensed data have been used successfully to estimate biophysical parameters, for example, the fraction of photosynthetically active radiation (FAPAR), leaf area index (LAI) [2], and green vegetation fraction [3]. The normalized difference vegetation index (NDVI) has been the most widely used index. The NDVI has been found to be highly correlated with the biophysical properties of vegetation canopies and able to reduce the effects of topographic shading and shadowing for seasonal trend analyses of terrestrial vegetation [4] and global carbon cycle modeling [5]. However, the NDVI is affected by other factors such as soil background brightness and aerosol contamination [6]. The enhanced vegetation index (EVI), developed for Moderate Resolution Imaging Spectroradiometer (MODIS) of the National Aeronautics and Space Administration's (NASA) Earth Observing System (EOS), was designed to optimize the vegetation signal with improved sensitivity in high biomass regions and improved vegetation monitoring through a decoupling of the canopy background signal and a reduction in atmospheric aerosol influences that affect the NDVI [7]:

$$\text{EVI} = G\frac{\rho_n - \rho_r}{\rho_n + C_1\rho_r - C_2\rho_b + L} \quad (1)$$

where ρ are the total- or partial (uncorrected for aerosols)-atmosphere corrected reflectances, subscripts "n", "r", and "b" represent the near-infrared (NIR), red, and blue bands, respectively, L is the canopy background brightness adjustment factor, and C_1 and C_2 are the coefficients of the aerosol resistance term. The coefficients adopted in the MODIS EVI algorithm are: $L = 1.0$, $C_1 = 6.0$, $C_2 = 7.5$, and G (gain factor) = 2.5 [8]. MODIS was designed to be highly calibrated and to have explicit atmospheric corrections and, thereby, EVI has been used in a wide range of applications including ecosystem resilience studies [9], an estimation of gross primary production (GPP) [10], and evapotranspiration estimates [11].

The Visible Infrared Imaging Radiometer Suite (VIIRS) sensor onboard Suomi-National Polar-orbiting Partnership (S-NPP) has begun to collect data, which is slated to replace the Advanced Very-High Resolution Radiometer (AVHRR) onboard the National Oceanic and Atmospheric Administration (NOAA) polar-orbiting satellite series with afternoon overpass, and to continue the MODIS highly calibrated data stream. The VIIRS geophysical product, termed environmental data records (EDRs), includes the top-of-atmosphere (TOA) NDVI and top-of-canopy (TOC) EVI [12]. VI continuity/compatibility across AVHRR, MODIS, and VIIRS is of great importance for understanding spatial and temporal dynamics of global vegetation over several decades.

Differences in sensor and platform characteristics, and product generation algorithms, however, cause systematic errors in VI time series across sensors. Spectral bandpasses are one major issue in using multi-sensor data [13] and, thus, this study was focused on inter-sensor spectral compatibility and calibration of the EVI between MODIS and VIIRS. Numerical experiments using Earth Observing-1 (EO-1) Hyperion data showed that VIIRS EVI was higher than MODIS EVI with the maximum differences reaching 0.040 EVI units over a tropical forest-savanna eco-gradient in Brazil [14] and the same trend was observed in a Hyperion-based bandpass simulation analysis conducted over AERONET sites in the conterminous United States [15]. An initial assessment of actual VIIRS EVI data was reported in [16]; an average (bias) of VIIRS EVI minus MODIS EVI (a gain factor G = 2.5) was zero when EVI was zero, but always positive for the rest of EVI dynamic range, indicating that VIIRS EVI was always higher than the MODIS counterpart. Another EVI compatibility analysis using Aqua MODIS (L2G daily 500 m, Collection 5) and VIIRS (L2G daily 500 m) showed that an average of MODIS EVI minus VIIRS EVI was −0.022 for North America for August 2013 [17], indicating that VIIRS EVI was generally higher than MODIS EVI. The positive errors in VIIRS EVI over MODIS EVI were consistently observed in both Hyperion-simulated and actual sensor data. It should be noted that the errors might contain not only the spectral effects but the effects of other factors including differences in the product generation algorithm such as atmospheric correction and quality flags, geolocation errors, spatial resolution differences, and radiometric calibration uncertainties, to name a few.

Numerous cross-sensor VI translation equations, especially for the NDVI, have been proposed and the techniques can basically be applied to the EVI. These techniques for cross-sensor VI translations can be categorized into three approaches as summarized in [17]: (1) polynomial based approach [18–23]; (2) band-averaging approach [24–26]; and (3) vegetation isoline-based approach [27,28]. In our previous study, the vegetation isoline-based approach, that translates a reflectance of an arbitrary wavelength to that of another wavelength in the solar-reflective region based on the radiative transfer theory [29], was employed and a MODIS-compatible EVI with VIIRS spectral bandpasses was derived [17]. The derived equation had four coefficients that were a function of soil, canopy, and atmosphere, e.g., soil line slope, leaf area index (LAI), and aerosol optical thickness (AOT). The MODIS-compatible EVI resulted in a reasonable level of accuracy when the coefficients were fixed at values found via optimization for model-simulated and actual sensor data (the North American continent in August 2013), demonstrating the potential practical utility of the derived equation [17].

The primary objective of this study was to calibrate the MODIS-compatible VIIRS EVI equation for global conditions. "Optimized" coefficients were sought using a year-long, global VIIRS-MODIS dataset at the climate modeling grid (CMG) resolution of 0.05°-by-0.05°. A secondary objective of this study was to develop a cross-calibration protocol that can be used to revise the optimum coefficients when a new dataset becomes available. We evaluated the extent to which errors decreased by applying the obtained VIIRS EVI equation with the optimized coefficients on global data and the degree to which errors varied as a function of sun-target-sensor viewing geometry, EVI values, and land cover type.

2. MODIS-Compatible VIIRS EVI

The MODIS-compatible VIIRS EVI was derived using the vegetation isoline equations [17]. The equations analytically approximate and describe the vegetation isoline, which is defined as the line formed between the reflectances at two different wavelengths for an optically and structurally constant canopy and a constant atmospheric condition over varying canopy background brightness [29]. A horizontally infinite homogeneous atmospheric layer was assumed and a portion of the target area was assumed to be covered with a homogeneous canopy. It was further assumed that the radiative transfer problem in both the covered and uncovered areas could be simulated independently by modeling a horizontally homogeneous canopy and Lambertian soil surface, respectively. The first-order photon interactions between soil and canopy and between canopy and atmosphere were considered (higher order interaction terms were truncated) for deriving the vegetation isoline equations.

In deriving the MODIS-compatible VIIRS EVI, we first obtained equations that related the VIIRS blue, red, and NIR (M3, I1, and I2 bands) to the MODIS respective counterparts (band 3, band 1, and band 2) [17] using the vegetation isoline equations

$$\rho_{b,m} = A_b \rho_{b,v} + D_b \tag{2a}$$

$$\rho_{r,m} = A_r \rho_{r,v} + D_r \tag{2b}$$

$$\rho_{n,m} = A_n \rho_{n,v} + D_n \tag{2c}$$

where $\rho_{b,m}$, $\rho_{r,m}$, and $\rho_{\mathrm{n,m}}$ are MODIS blue, red, and NIR band reflectances which are TOC or partial (uncorrected for aerosol)-atmosphere corrected reflectances that are modeled by adding aerosol layer over the canopy; $\rho_{b,v}$, $\rho_{r,v}$, and $\rho_{n,v}$ are VIIRS counterparts. A_b, A_r, and A_n (slopes of the isoline equation for blue, red, and NIR bands) and D_b, D_r, and D_n (offsets of the isoline equation for blue, red, and NIR bands) are dependent on the reflectance and transmittance of the canopy and atmospheric layers, and the soil line slope and offset for MODIS blue and VIIRS blue, MODIS red and VIIRS red, and MODIS NIR and VIIRS NIR band pairs, respectively.

We then substituted Equations (2a–c) for the MODIS reflectances in Equation (1) to express the MODIS EVI equation as a function of VIIRS reflectances [17],

$$\hat{v}_m = G \frac{\rho_{v,n} - K_1 \rho_{v,r} + K_2}{\rho_{v,n} + K_1 C_1 \rho_{v,r} - K_3 C_2 \rho_{v,b} + K_4} \tag{3}$$

and

$$K_1 = \frac{A_r}{A_n} \tag{4a}$$

$$K_2 = \frac{D_n - D_r}{A_n} \tag{4b}$$

$$K_3 = \frac{A_b}{A_n} \tag{4c}$$

$$K_4 = \frac{C_1 D_r + D_n - C_2 D_b + L}{A_n} \tag{4d}$$

where $\hat{v}_m$ is the MODIS-compatible VIIRS EVI. The coefficients in Equation (3), K_1, K_2, K_3, and K_4 vary with the soil, vegetation, and aerosol conditions [17] and that the exact translation is possible only when the exact conditions of soil, vegetation, and atmosphere are known.

3. Calibration of MODIS-Compatible VIIRS EVI Using Global Data

3.1. Cross-Calibration (Optimization) Algorithm

Our approach for calibrating MODIS-compatible VIIRS EVI was to obtain a single set of K_i (i = 1, 2, 3, 4) that minimizes differences between the MODIS-compatible VIIRS EVI ($\hat{v}_m$) and the MODIS EVI (v_m) via non-linear regression. As in our previous study [17], the mean absolute difference (MAD) between $\hat{v}_m$ and v_m was used as the merit function for the non-linear regression:

$$\min_{K_i \in \mathbb{R}} MAD\left(K_1, K_2, K_3, K_4\right) \tag{5}$$

and

$$MAD\left(K_1, K_2, K_3, K_4\right) = \frac{1}{N}\sum_{i=1}^{N} \left|v_{m,i} - \hat{v}_{m,i}\left(K_1, K_2, K_3, K_4\right)\right| \tag{6}$$

This merit function could include multiple local minima and thereby the optimum, best solution was determined by searching for the global minimum using the Nelder-Mead simplex method in Optimization Toolbox of MATLAB® R2015a (MathWorks Inc., Natick, MA, USA). The algorithm was run without derivatives, but with 100 initial guesses. In the reminder, the best set of coefficients obtained by the algorithm are denoted by K_i^* (i = 1, 2, 3, 4).

3.2. Data Extraction

Aqua MODIS (Collection 5) and S-NPP VIIRS (Archive Set (AS) 3001) global daily 0.05-degree (climate modeling grid, CMG) resolution surface reflectance data were obtained for a one-year period of 1 August 2012 through 31 July 2013 for the calibration/optimization exercise. This VIIRS surface reflectance dataset (AS 3001) was produced at the NASA Science Investigator-led Processing System (SIPS), formerly known as the VIIRS Land Product Evaluation and Analysis Tool Element (Land PEATE), using the Joint Polar Satellite System (JPSS) near-real-time Interface Data Processing Segment (IDPS) software. The products of AS 3001 from SIPS should match the corresponding product from IDPS [30]. The VIIRS CMG surface reflectance algorithm was based on the MODIS Collection 5 CMG surface reflectance product generation algorithm [31]. The MODIS and VIIRS data were screened for cloud, cloud shadow, high aerosol loading, and snow/ice using quality assessment (QA) information and land quality flags (QF), respectively.

The spectral data employed in the calibration/optimization consisted of pairs of MODIS and VIIRS reflectances that were obtained on the same dates at the same CMG locations for the same specific angle bins. Seven view zenith angle bins were defined as $\theta_j \leqslant \theta_v < \theta_j + 8.0$ where θ_v is the view zenith angle in degrees and θ_j's for j = 1, 2, $\cdots$, 7 were 0.0, 8.0, 16.0, 24.0, 32.0, 40.0, 48.0, respectively. The two relative azimuth angle (θ_a) bins in degree were defined, corresponding to the backward scattering ($-90.0 < \theta_a < 90.0$) and forward scattering ($-180.0 < \theta_a -< 90.0$, $90.0 < \theta_a < 180.0$) directions. The mean and standard deviation (STD) of solar zenith angle differences between MODIS and VIIRS (MODIS minus VIIRS) were −1.2 and 1.8, respectively, and differences between relative azimuth angles were 1.9 and 5.7, respectively.

More than 6 million pairs were found in and extracted from the one-year data. The calibration/optimization exercise of the MODIS-compatible VIIRS EVI would have been too computationally intensive if all the pairs had been used. Thus, the minimum necessary number of sample pairs was determined by a numerical experiment in which the number of sample pairs for the optimization was increased from 10^2 to 10^6 with a logarithmic increment in the sample size. MAD between MODIS-compatible EVI and MODIS EVI, represented by p_n (n = 1, 2, $\cdots$, 30), was computed for the n-th dataset. Differences between adjacent MADs (d_n), or between p_n and p_{n+1}, were computed to provide information regarding the rate of change in MAD for n = 1, 2, $\cdots$, 29,

$$d_n = p_{n+1} - p_n \tag{7}$$

A random sample of the size of 10^2 reflectance pairs were selected from the extracted pairs in the above step, from which p_1 was computed. An additional number of randomly selected reflectance pairs were added to the first random sample to create a sample for p_2 from which first p_2 and then d_1 were computed. Likewise, d_n were computed until n = 29. This procedure was repeated 10 times to provide 10 sets of d_n. To evaluate variability in d_n, d_n was plotted as a function of the sample size (Figure 1). d_n changed by $\pm$0.0003 when the sample size was equal to or greater than 41,754 highlighted by the vertical dashed line in Figure 1. This converges to zero when the number of samples is further increased. The threshold of 0.0003 was qualitatively determined after computing d_n, but the change in MAD was small enough or nearly independent of the sample size when it was equal to or greater than 41,754. The minimum necessary number of samples in this study, therefore, was assumed to be equal to or greater than 41,754.

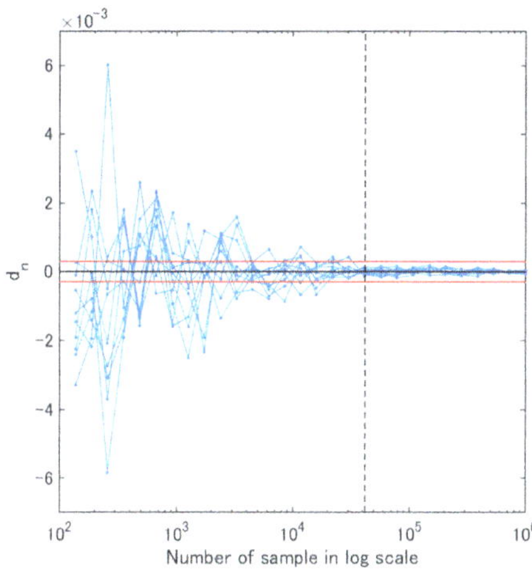

Figure 1. Variability in rate of change in the mean absolute difference (MAD) between MODIS-compatible EVI and MODIS EVI (d_n) as a function of the number of sample data. A vertical dashed line corresponds to the number of sample 41,754.

Prior to data extraction for the optimization, we eliminated anomalies in the sample data. Spectral data pairs were rejected if EVIs were lower than −0.05 or greater than 1.0. This is because, some VIIRS EVI exceeded the upper value of the valid range of MODIS EVI used in MOD13 series [32], *i.e.*, 1.0, due probably to cloud leakage and/or residual snow/ice contamination. In addition, when VIIRS M3 surface reflectance exceeded 0.3 (possible cloud leakage or residual snow/ice contamination), the pair was removed. If $\delta_1 <$ median $(\delta_1) - \sigma$ or $\delta_1 >$ median $(\delta_1) + \sigma$ (δ_1 is MODIS EVI minus VIIRS EVI (see Equation (8)); σ is 0.09, the STD of δ_1), the pair was removed as an outlier. The value of σ, 0.09 was considered large enough since the maximum difference between MODIS and VIIRS EVI calculated from model-simulated (noise free) data was less than 0.02 for nadir view [17] and less than 0.03 for off-nadir (60 degrees) view (results not shown here). The possible causes of these outliers as identified are discussed in next subsection.

In this study, sample data were selected in order for the EVI frequencies to be as uniform as possible. Also, the data were selected to be more uniform with respect to sun-target-sensor viewing geometry. First, 200,000 pairs of MODIS and VIIRS spectra, which satisfied the conditions for the data extraction described above, were randomly selected and extracted from each of the 14 angular bins (seven variations in view zenith angle and two variations in relative azimuth angle), and then 2,800,000 pairs of data were created by combining the 14 data sets. The sample data were then selected randomly from the 2,800,000 pairs but to be uniform with respect to the EVI value. The relative frequency of EVI of the 2,800,000 data is shown by the blue line in Figure 2 whereas the frequency of sample data that were used in the optimization is shown by the red line in Figure 2. The total number of reflectance pairs was 137,278, which is greater than the minimum necessary number of sample (41,754) and computationally-acceptable for the optimization.

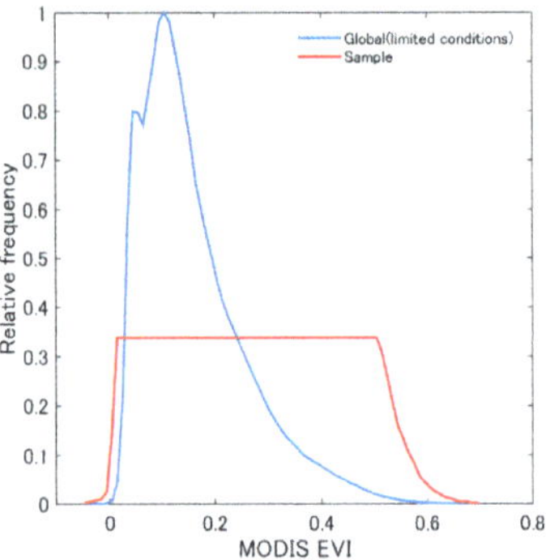

Figure 2. Relative frequency of MODIS EVI extracted from the climate modeling grid (CMG) data with the limited conditions regarding view zenith and azimuth angles and time period, denoted by blue line and that of data used for optimization (sample data), denoted by red line.

The percentage of the pair of reflectances in respective angular ranges is shown in Table 1. Since the angular distribution in each EVI bin (e.g., 0.01 in EVI unit) would not be identical, the resultant values of the percentages were not uniform. The number of samples for backward scattering tended to increase with increasing view angle. On the contrary, it decreased as view angle increased in the forward scattering direction when the view angle was larger than 32.0.

Table 1. Percentage for the number of pairs of MODIS and VIIRS reflectance spectra in each angular condition.

		Relative Azimuth Angle	
		$-90.0 < \theta_a < 90.0$ (Backward Scattering)	**$-180.0 < \theta_a < -90.0$, $90 < \theta_a < 180$ (Forward Scattering)**
View zenith angle	$0.0 < \theta_v < 8.0$	6.7	6.3
	$8.0 < \theta_v < 16.0$	6.9	6.3
	$16.0 < \theta_v < 24.0$	7.7	6.5
	$24.0 < \theta_v < 32.0$	9.3	6.4
	$32.0 < \theta_v < 40.0$	10.4	5.1
	$40.0 < \theta_v < 48.0$	9.6	4.3
	$48.0 < \theta_v < 56.0$	10.7	3.8

3.3. Optimization Results

The errors between MODIS EVI and VIIRS EVI are written as δ_1, and errors between MODIS EVI and MODIS-compatible EVI with the optimized coefficients (K_i^*, i = 1, 2, 3, 4) are written as δ_2^*,

$$\delta_1 = v_m - v_v \tag{8}$$

$$\delta_2^* = v_m - \hat{v}_m^* \left(K_1^*, K_2^*, K_3^*, K_4^*\right) \tag{9}$$

where v_m, v_v, and $\hat{v}_m^*$ are MODIS EVI, VIIRS EVI, and MODIS-compatible VIIRS EVI with the optimized set of coefficients.

In Figure 3a, MODIS EVI minus VIIRS EVI (δ_1) are plotted as a function of MODIS EVI. On average, δ_1 was the smallest at EVI = 0, decreased until approximately EVI = 0.2, and remained nearly the same for higher EVI values (Figure 3a). The mean, STD, and root mean square error (RMSE) of δ_1 were −0.019, 0.028, and 0.034, respectively (Table 2). The mean values of MODIS minus VIIRS reflectances were −0.009, −0.003, and −0.011 for the blue, red, and NIR bands, respectively. In general, higher VIIRS blue and NIR reflectances contributed to higher VIIRS EVI values than the MODIS counterparts. The effect of spectral bandpasses and additional uncertainties caused by radiometric calibration could add systematic errors to surface reflectances. While δ_1 had less scattering about the trend at lower (0 to 0.2) EVI values, large positive values in δ_1 were observed with increasing EVI values (Figure 3a). This indicates that VIIRS EVI shows relatively smaller values than MODIS EVI, which can likely be attributed to cloud leakage in VIIRS Cloud Mask (VCM). Other sources of systematic and random errors arise from differences in the maturity of the atmospheric correction algorithms, including aerosol optical properties, relative geolocation errors between the sensors, and influences associated with differences in spatial compositing for generating CMG data. Although the input reflectance pairs were carefully selected to have the same sun-target-view geometric conditions, subtle differences in the geometric conditions remained in many pairs, being an additional source of random/systematic errors that are inevitable for this cross-sensor comparison.

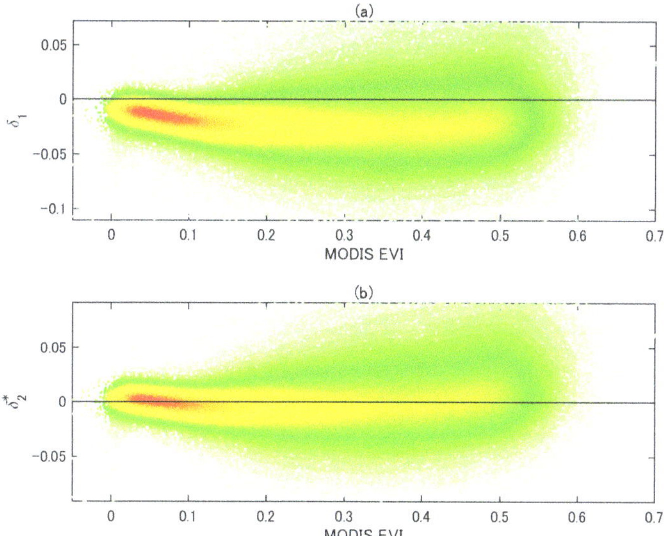

Figure 3. Density plots of EVI differences using the data for the calibration. (**a**) Density plot of MODIS EVI minus VIIRS EVI (δ_1) *vs.* MODIS EVI; (**b**) Density plot of MODIS EVI minus MODIS-compatible VIIRS EVI (δ_2^*) *vs.* MODIS EVI.

Table 2. Statistics of MODIS EVI minus VIIRS EVI (δ_1) and MODIS EVI minus MODIS-compatible VIIRS EVI obtained in the optimization (δ_2^*).

	Mean	STD	RMSE
δ_1	−0.019	0.028	0.034
δ_2^*	0.002	0.028	0.028

The optimization was performed using the extracted data and resultant coefficients are summarized in Table 3. K_1^* and K_4^* were close to unity and K_2^* was close to zero, whereas K_3^* were 0.874. The validity of these values can be evaluated based on the slope and offset calculated from the linear regression between MODIS and VIIRS reflectances. The slope and offset of the linear regression between the VIIRS blue reflectance (x-axis) and MODIS counterpart (y-axis) were 0.813 and 0.0032, respectively, and similarly, 0.934 and 0.0039 for the VIIRS red reflectance and MODIS counterpart, and 0.915 and 0.013 for the VIIRS NIR reflectance and MODIS counterpart. We consider the slopes of the regressions as the averages of the slope of the vegetation isoline equations in Equations (2a–c) for the MODIS-VIIRS CMG data, $\overline{A}_b$, $\overline{A}_r$, and $\overline{A}_n$ (= 0.813, 0.915, and 0.939) and consider the offsets of the regressions as the average values of offsets of the isoline equations, $\overline{D}_b$, $\overline{D}_r$, and $\overline{D}_n$ (= 0.0032, 0.0039, and 0.013). The averages of K_i ($\overline{K}_i$) using these averages can be computed as: $\overline{K}_1 = \overline{A}_r/\overline{A}_n = 1.026$, $\overline{K}_2 = \left(\overline{D}_n - \overline{D}_r\right)/\overline{A}_n = 0.010$, $\overline{K}_3 = \overline{A}_b/\overline{A}_n = 0.888$, and $\overline{K}_4 = (C_1\overline{D}_r + \overline{D}_n - C_2\overline{D}_b + L)/\overline{A}_n = 1.107$. These values are very similar to K_i^*; especially they are the same for K_1. Therefore, the optimized, best coefficients K_i^* can be reasonable.

Table 3. Values of optimized coefficients obtained using the sample data.

K_1^*	K_2^*	K_3^*	K_4^*
1.026	−0.001	0.874	1.022

Thus, the calibrated MODIS-compatible EVI ($\hat{v}_m^*$) is written as

$$\hat{v}_m^* = G\frac{\rho_{v,n} - K_1^*\rho_{v,r} + K_2^*}{\rho_{v,n} + K_1^*C_1\rho_{v,r} - K_3^*C_2\rho_{v,b} + K_4^*} \tag{10}$$

Figure 3b shows MODIS-compatible VIIRS EVI minus MODIS EVI (δ_2^*) and that the MODIS-compatible VIIRS EVI with optimized coefficients resulted in a lower error magnitude than VIIRS EVI, *i.e.*, a reduction of the mean difference from 0.019 (δ_1) to 0.002 (δ_2^*). The STD of δ_2^* (0.028) was the same as that of δ_1 (0.028) because the errors were included in both sensors, especially in VIIRS. The RMSE of δ_2^* (0.028) was lower than that of δ_1 (0.034) because of the reduction in the systematic differences.

Figure 4a,b show the spatial distributions of δ_1 and δ_2^*, respectively. In plotting δ_1 and δ_2^* in map format, when multiple error values were found for a single grid, their average value is plotted for the grid. Errors were not measured over all of the grids because the MODIS and VIIRS sun-target-view geometry matchup were found over limited geographic areas. Note that the size of the colored dots is larger than the actual grid size (0.05°-by-0.05°) for improved visualization. Fewer samples seen in tropical forests and high northern latitude regions were due to frequent cloud cover and snow/ice cover. Generally, large negative values in δ_1 were found everywhere, especially in the southeast area of the South American continent, India, West Europe, the western and southern part of Africa, and the northern part of the Eurasian continent. Positive errors were found in the regions that also showed large negative errors and where vegetation are found. Such samples might have been impacted by factors other than spectral bandpasses as mentioned before. Positive errors were also observed in the regions showing lower magnitudes of negative and positive values in δ_1 (−0.03~0.01) such as sparsely/less vegetated areas and desert including Australia, northern Africa, the central part of Asia, and the southern part of South American continent.

Figure 4b shows the spatial distribution of δ_2^*. The negative values in δ_1 were corrected over the globe by the MODIS-compatible EVI. Areas that showed negative errors in Figure 4a resulted in smaller magnitude of errors. However, positive errors were not properly corrected and remained or increased in their magnitude, in particular, over the southeast of South America and Europe. Clusters of negative errors (approximately −0.01) were observed in Africa, Australia, east South America, and the northern Eurasian continent (the areas heavily impacted by either persistent cloud cover or winter snow cover).

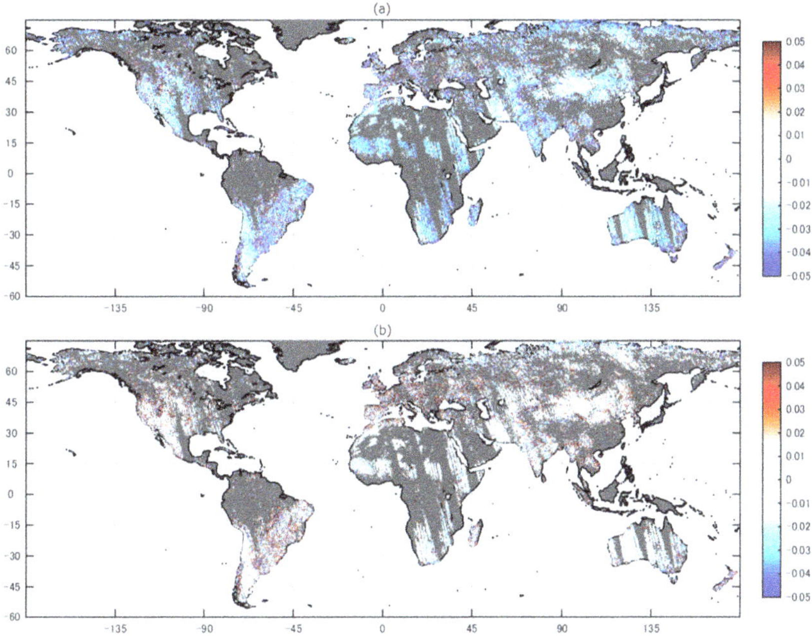

Figure 4. (**a**) Spatial distribution of MODIS EVI minus VIIRS EVI (δ_1) over geographic coordinate for the calibration; (**b**) Spatial distribution of MODIS EVI minus MODIS-compatible EVI (δ_2^*).

4. Evaluation of MODIS-Compatible VIIRS EVI

The MODIS-compatible VIIRS EVI with the optimized coefficients (in Table 3) was evaluated using another dataset, *i.e.*, MODIS and VIIRS global, daily CMG data of which acquisition dates did not overlap with those used in the calibration/optimization. The period of data for the evaluation spanned between 8/1/2013 and 11/30/2013 (four months). The total number of MODIS and VIIRS reflectance pairs, which fell in any of the angular bins described in Section 3.2, was more than 2.8 million from which an evaluation dataset was randomly extracted. The size of this subsample was 200,000. This evaluation dataset had different frequencies over angular and EVI bins from the one used in the previous section.

A density plot of MODIS EVI minus VIIRS EVI (δ_1) as a function of MODIS EVI is shown in Figure 5a. Trends of negative biases and high density in lower EVI values were similar to Figure 3a but the appearance of the plot (distribution of dots) was different because of different scenarios in extracting sample data. The mean of the error (−0.021) was similar to that obtained from the calibration dataset in Figure 3a (−0.019), whereas STD and RMSE (0.021 and 0.029) of this evaluation dataset were smaller than those in Figure 3a (0.028 and 0.034) as summarized in Table 4. Figure 5b shows a density plot of MODIS EVI minus MODIS-compatible EVI (δ_2^*) as a function of MODIS EVI. The absolute average of δ_2^* (0.003) was reduced from that of δ_1 (0.021). STD of δ_2^* (0.020) was slightly lower than that of δ_1 (0.021). RMSE of δ_2^* (0.020) was reduced from that of δ_1 (0.029). These results indicate that the coefficients obtained in the previous section were reasonable.

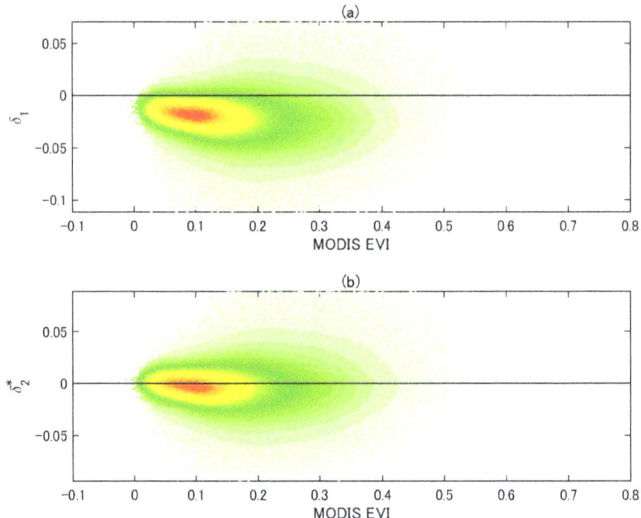

Figure 5. Density plots of EVI differences using the data for the evaluation. (**a**) Density plot of MODIS EVI minus VIIRS EVI (δ_1) *vs.* MODIS EVI; (**b**) Density plot of MODIS EVI minus MODIS-compatible VIIRS EVI (δ_2^*) *vs.* MODIS EVI.

Table 4. Statistics of MODIS EVI minus VIIRS EVI (δ_1) and MODIS EVI minus MODIS-compatible VIIRS EVI (δ_2^*) for the evaluation.

	Mean	**STD**	**RMSE**
δ_1	−0.021	0.021	0.029
δ_2^*	−0.003	0.020	0.020

Figure 6a,b show the spatial distribution of MODIS EVI minus VIIRS EVI (δ_1) and MODIS EVI minus MODIS-compatible EVI (δ_2^*), respectively, for evaluation data. Spatial trends of errors in Figure 6 are very similar to those in Figure 4. The spatial coverage of Figure 6 is slightly wider than Figure 4 since the sample size of the evaluation dataset (200,000) was larger than that of the calibration/optimization dataset (137,278). The number of positive error occurrences in Figure 6b appears slightly smaller than that in Figure 4b because of the two different scenarios used in obtaining the evaluation and calibration datasets. For example, the southern part of the South American continent shows fewer dots of positive errors (red dots) compared to Figure 4b.

The MODIS-compatible VIIRS EVI was further evaluated by examining its performance with respect to sun-target-sensor viewing angle, EVI dynamic range, and land cover type. Mean, STD, and RMSE of δ_2^* and δ_1 were computed and those of δ_2^* were divided by the respective counterparts of δ_1, referred to as the ratio of mean (RM), ratio of STD (RS), and ratio of RMSE (RR), for each angle bin, each EVI bin, and each land cover type (Figures 7–9 respectively). The MODIS Land Cover Type Yearly CMG (MCD12C1 [33], resolution of 0.05°-by-0.05°) for 2012 was used to identify the land cover type of each CMG pixel (the 17-class International Geosphere-Biosphere Programme (IGBP) classification) [34].

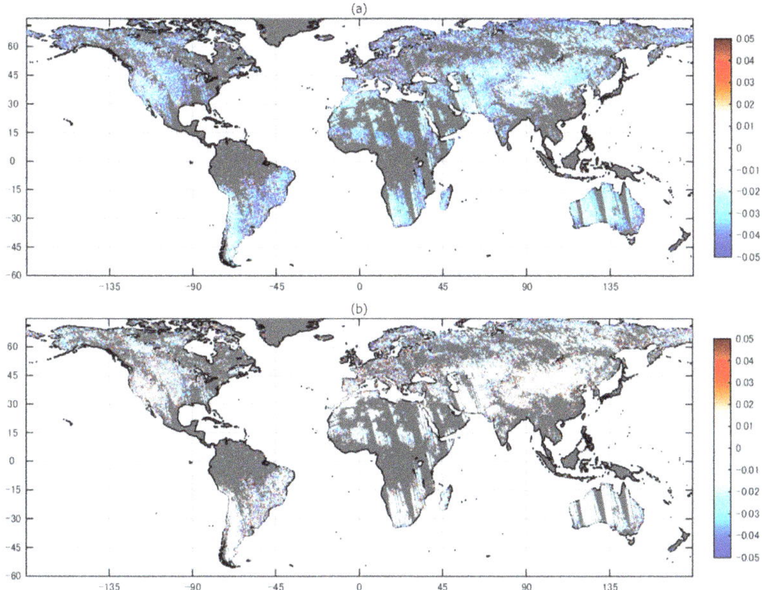

Figure 6. (**a**) Spatial distribution of MODIS EVI minus VIIRS EVI (δ_1) over geographic coordinate for the evaluation; (**b**) Spatial distribution of MODIS EVI minus MODIS-compatible EVI (δ_2^*).

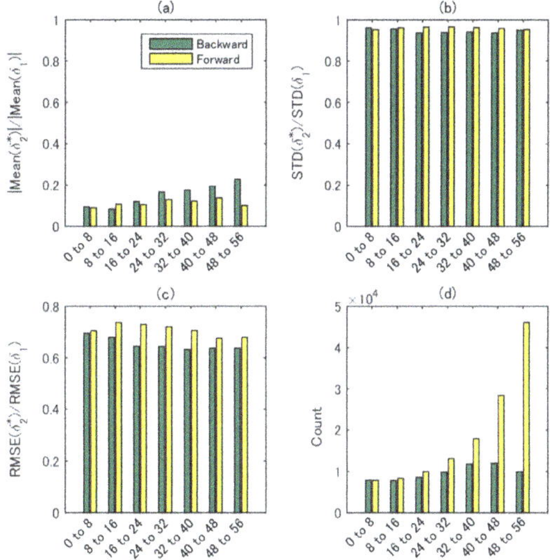

Figure 7. (**a**) Ratio of absolute mean of difference between MODIS EVI minus MODIS-compatible VIIRS EVI (δ_2^*) to that between MODIS EVI minus VIIRS EVI (δ_1) for angular-dependent data (characterized by backward or forward scatterings and view zenith angle); (**b**) Ratio of STD for δ_2^* to that for δ_1; (**c**) Ratio of RMSE for δ_2^* to that for δ_1; (**d**) Frequency of angular distribution in data for evaluation.

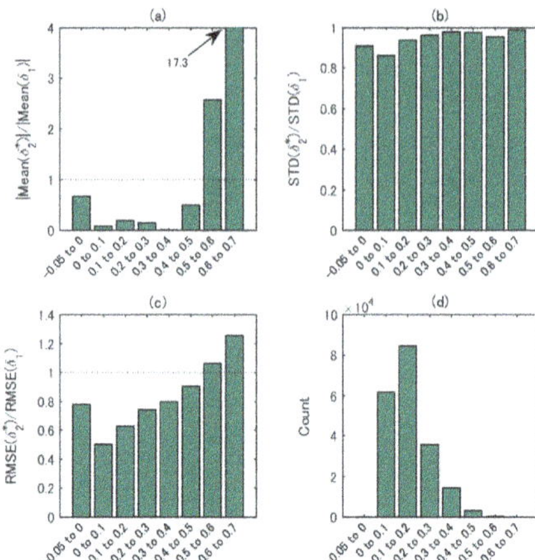

Figure 8. (**a**) Ratio of absolute mean of difference between MODIS EVI minus MODIS-compatible VIIRS EVI (δ_2^*) to that between MODIS EVI minus VIIRS EVI (δ_1) for EVI-dependent data; (**b**) Ratio of STD for δ_2^* to that for δ_1; (**c**) Ratio of RMSE for δ_2^* to that for δ_1; (**d**) Frequency of EVI distribution in data for evaluation.

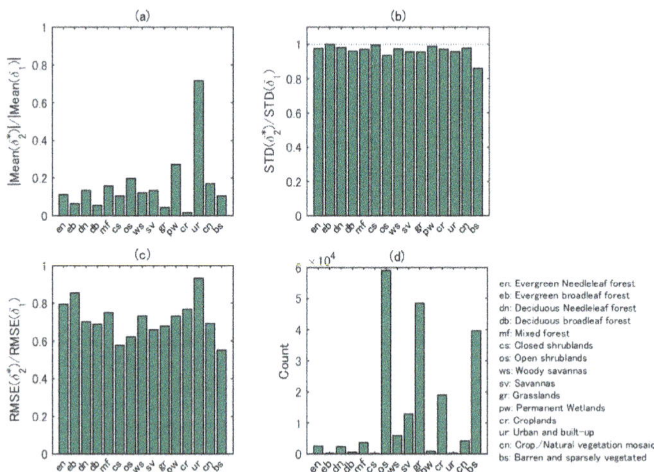

Figure 9. (**a**) Ratio of absolute mean of difference between MODIS EVI minus MODIS-compatible VIIRS EVI (δ_2^*) to that between MODIS EVI minus VIIRS EVI (δ_1) for land cover-dependent data; (**b**) Ratio of STD for δ_2^* to that for δ_1; (**c**) Ratio of RMSE for δ_2^* to that for δ_1; (**d**) Frequency of land cover type distribution in data for evaluation.

The RM increased with view zenith angle for the backward scattering geometry, while it remained nearly the same over the view zenith angle range examined here for the forward scattering direction (Figure 7a). The RS was close to one for both the backward and forward scattering directions, but was slightly smaller for the former than the latter (Figure 7b). The RR was less than 0.8 and smaller for the backward scattering (Figure 7c). It tended to decrease with increasing view angles. The MODIS-compatible EVI performed better for the forward scattering direction in terms of bias but better for the backward scattering direction from the perspective of variability. Figure 7d shows the frequency distribution of each angular condition for this evaluation dataset. The frequency of backward scattering does not show a strong angular dependency, whereas that of forward scattering showed a monotonic increasing trend.

Dependencies of the improvement of RM, RS, and RR on EVI values from MODIS are shown in Figure 8. Note that the maximum MODIS EVI in our evaluation was 0.688. The RM differed largely across EVI bins (Figure 8a). The bias was the smallest for the EVI range of 0.3–0.4 and was the largest for higher EVI values (from 0.5 to 0.7). Whereas the RS was less than unity and varied a little across EVI bins (Figure 8b), the RR varied greatly from 0.5 to 1.25 (Figure 8c). According to the histogram of this dataset (Figure 8d), most of samples were in the EVI range between 0 and 0.5. Since the coefficients of the MODIS-compatible VIIRS EVI are sensitive to canopy greenness [17], a smaller number of samples in higher EVI values would have been a source of the large RM and RR for that EVI value range. It should be noted that cloud leakage in VCM could degrade the performance of the atmospheric correction algorithm, which may likely be associated with densely vegetated areas due to persistent cloud coverage.

Land cover dependencies of RM, RS, and RR are shown in Figure 9a–c, respectively. Note that data corresponding to "Water" and "Permanent snow and ice" were excluded from this evaluation. The RM was less than 0.4 except for the urban and built-up (ur) cover type (Figure 9a). The RS was close to unity (Figure 9b). The RR was smaller than unity, and showed large improvements for barren and sparsely vegetated area (bs), closed shrublands (cs), and open shrublands (os) followed by savanna (sv) and grass land (gr) (Figure 9c). The urban and built-up area, in general, shows totally different spectral characteristics from vegetated areas, which might have caused the largest RM and RR. Figure 9d shows the land cover frequency of this evaluation dataset. The sample sizes of os, gr, and bs were large, and a general negative correlation between the sample size and the improvements in statistics was observed. The forest cover types where high EVI values are expected (first five cover types in Figure 9a–d) showed some improvements in the statistics, although samples with large EVI values resulted in larger errors. The MODIS/VIIRS EVI tended to fall approximately between 0.1 and 0.5 (0.25 in average) for forest types, in which the MODIS-compatible VIIRS EVI showed higher performance than in the higher EVI value range.

5. Discussions

The coefficients obtained through the calibration/optimization using the one-year global dataset were (K_1^*, K_2^*, K_3^*, K_4^*) = (1.026, −0.001, 0.874, 1.022). The optimum coefficients obtained for North America in August 2013 in our previous study [17], represented here by K'_i (i = 1, 2, 3, 4), were (K'_1, K'_2, K'_3, K'_4) = (0.947, 0.010, 0.265, 0.995). These differences in the optimum coefficient values resulted from the spatiotemporal coverages of the calibration datasets used in the present and previous studies. It should be noted that K_i^* obtained in this study fell in or approached the range of coefficients simulated by radiative transfer models (PROSAIL and 6S code) [17]. The maximum and minimum of K_i (i = 1, 2, 3, 4) obtained in the model simulation were (0.980, 1.024), (−0.002, 0.004), (0.695, 0.935), (0.879, 1.020), respectively. K_2^* (−0.001) and K_3^* (0.874) fell within their corresponding ranges, and K_1^* (1.026) and K_4^* (1.022) slightly greater than the ranges. Since the simulation of coefficients was performed by varying optical thickness of vegetation canopy and aerosol layer but by fixing angular condition, considerations of angular variations in the present study could widen the ranges of coefficients characterized by the maximum and minimum to include K^*_i obtained in this study. Further investigation is required to fully understand the relationship between the optimum K^*_i values from actual sensor data and ranges of coefficients simulated using radiative transfer models.

The derived equation of MODIS-compatible VIIRS EVI was a non-linear function of K_i, and therefore the optimization of K_i corresponds to the non-linear regression. The small non-linear biases, however, still remained in the δ_2^* (MODIS EVI minus MODIS-compatible VIIRS EVI) in Figure 3b. The comparison of our algorithm with alternative models such as polynomials is certainly important and worthy to evaluate, which is a focus of future work.

The calibrated MODIS-compatible VIIRS EVI showed less compatibility with MODIS EVI over higher EVI values (>0.5). MODIS-VIIRS reflectance pairs that had large positive values in δ_2^* (MODIS EVI minus MODIS-compatible EVI) also had positive values in δ_1 (MODIS EVI minus VIIRSE EVI). However, a Hyperion-based bandpass simulation analysis conducted over AERONET sites only found negative values in MODIS EVI minus VIIRS EVI [15]. This discrepancy implies that these large positive δ_1 values were likely caused by other potential factors than the bandpass differences, such as cloud leakage in VCM which decreases VIIRS EVI and increases δ_1 and δ_2^*, and the maturity status (quality level) of the input surface reflectance data.

The VIIRS surface reflectance product was on a "beta" maturity status for the data period used in this study. Likewise, the VIIRS sensor data record (SDR), the input to the VIIRS surface reflectance product, reached a provisional status in March 2013 [31]. Quality issues associated with the VIIRS surface reflectance data that we noticed included cloud leakage [17,35], aerosol optical thickness [36,37], and VIIRS algorithmic differences from MODIS. The leakage of small clouds resulted in large biases in the surface reflectance intermediate product (IP) which were not documented in QFs [16]. Therefore, spatial averaging for generating CMG pixels could have operated differently between MODIS and VIIRS due primarily to the cloud leakage in VCM. The MODIS/VIIRS pixels involved in area-averaging for generating CMG pixels heavily depend on algorithmic accuracy for cloud detection over the cloudy area. Such dependency caused the situation that the number of pixels, *i.e.*, the areas used to compute CMG reflectances, was not identical between sensors even if the area for the averaging (0.05°-by-0.05°) was the same location. This could subsequently impose substantial differences between CMG reflectances of sensors, in addition to cloud contamination. Significant improvements have, however, been made to these input products since the commencement of this study and additional improvements are planned (e.g., [38,39]). The same and additional analyses using the VIIRS surface reflectance product with a higher maturity status (e.g., provisional or validated stage-1) should allow to examine and identify the factor(s) that caused this apparent poor calibration results for high EVI values and to obtain more refined calibration results of the equation/coefficients.

Coefficients in the MODIS-compatible EVI to be optimized are also influenced by characteristics in sample data and an optimization algorithm. The sample data are characterized by several factors including maturity (accuracy) of the product, the number of samples, frequencies in sun-target-viewing geometry, observation time period, and land cover types. The optimization of coefficients depends on the merit function and the algorithm to search for the optimum solution. The algorithm comprised by the merit function of MAD and the search algorithm of the Nelder-Mead simplex method starting from multiple initial guesses performed properly for the problem of the EVI optimization, which involves multiple local minima.

6. Conclusions

The MODIS-compatible VIIRS EVI was obtained via cross-calibration using a year-long global dataset (MODIS and VIIRS CMG) and was subsequently evaluated with an additional dataset not used for the calibration/optimization. The dataset for obtaining the optimum set of coefficients was selected in order for EVI (and angular) frequencies to be as uniform as possible across all view zenith angles, relative azimuthal angles, and EVI values. The evaluation results showed a significant decrease in the difference between the VIIRS EVI and MODIS EVI upon cross-calibration, the mean error (absolute value) and RMSE having decreased from 0.021 to 0.003 EVI units, and from 0.029 to 0.020 EVI units, respectively. The random and systematic errors in VIIRS data due to the VIIRS product generation algorithms are expected to decrease, which should further improve the performance of the MODIS-compatible VIIRS EVI upon re-calibration with a new, higher maturity level dataset.

The cross-calibration methodology introduced in this study should be useful not only for the EVI cross-calibration, but also for cross-calibrating other vegetation indices, such as the NDVI and two-band EVI (EVI2) [40] and, thus, has the potential to contribute to addressing continuity of spectral vegetation indices across optical sensors for long term global vegetation monitoring.

Acknowledgments: This work was supported by a NOAA JPSS contract (T.M.) and was partially supported by a JSPS KAKENHI Grant Number 15H02856 (H.Y.).

Author Contributions: Kenta Obata and Tomoaki Miura designed the concept of this study, developed the calibration protocol, and wrote the manuscript; Kenta Obata performed the numerical experiment; Hiroki Yoshioka and Alfredo Huete participated in the discussions on the interpretation of results and contributed to editing and revision of the manuscript; Marco Vargas assisted in the interpretation and discussion of results with his knowledge on the VIIRS algorithm and product characteristics.

Conflicts of Interest: The authors declare no conflict of interest.

References

1. Los, S.O. Analysis of trends in fused AVHRR and MODIS NDVI data for 1982–2006: Indication for a CO_2 fertilization effect in global vegetation. *Glob. Biogeochem. Cycles* **2013**, *27*, 318–330. [CrossRef]
2. Nemani, R.R.; Keeling, C.D.; Hashimoto, H.; Jolly, W.M.; Piper, S.C.; Tucker, C.J.; Myneni, R.B.; Running, S.W. Climate-driven increases in global terrestrial net primary production from 1982 to 1999. *Science* **2003**, *300*, 1560–1563. [CrossRef] [PubMed]
3. Jiang, L.; Kogan, F.N.; Guo, W.; Tarpley, J.D.; Mitchell, K.E.; Ek, M.B.; Tian, Y.; Zheng, W.; Zou, C.-Z.; Ramsay, B.H. Real-time weekly global green vegetation fraction derived from advanced very high resolution radiometer-based NOAA operational global vegetation index (GVI) system. *J. Geophys. Res. Atmos.* **2010**, *115*. [CrossRef]
4. Eastman, J.; Sangermano, F.; Machado, E.; Rogan, J.; Anyamba, A. Global Trends in Seasonality of Normalized Difference Vegetation Index (NDVI), 1982–2011. *Remote Sens.* **2013**, *5*, 4799–4818. [CrossRef]
5. Hall, F.; Masek, J.G.; Collatz, G.J. Evaluation of ISLSCP Initiative II FASIR and GIMMS NDVI products and implications for carbon cycle science. *J. Geophys. Res.* **2006**, *111*, D22S08. [CrossRef]
6. Matsushita, B.; Yang, W.; Chen, J.; Onda, Y.; Qiu, G. Sensitivity of the Enhanced Vegetation Index (EVI) and Normalized Difference Vegetation Index (NDVI) to Topographic Effects: A Case Study in High-density Cypress Forest. *Sensors* **2007**, *7*, 2636–2651. [CrossRef]
7. Huete, A.; Didan, K.; Miura, T.; Rodriguez, E.P.; Gao, X.; Ferreira, L.G. Overview of the radiometric and biophysical performance of the MODIS vegetation indices. *Remote Sens. Environ.* **2002**, *83*, 195–213. [CrossRef]
8. Huete, A.; Justice, C.; Liu, H. Development of vegetation and soil indices for MODIS-EOS. *Remote Sens. Environ.* **1994**, *49*, 224–234. [CrossRef]
9. Ponce Campos, G.E.; Moran, M.S.; Huete, A.; Zhang, Y.; Bresloff, C.; Huxman, T.E.; Eamus, D.; Bosch, D.D.; Buda, A.R.; Gunter, S.A.; *et al.* Ecosystem resilience despite large-scale altered hydroclimatic conditions. *Nature* **2013**, *494*, 349–352. [CrossRef] [PubMed]
10. Xiao, X.; Hollinger, D.; Aber, J.; Goltz, M.; Davidson, E.A.; Zhang, Q.; Moore, B. Satellite-based modeling of gross primary production in an evergreen needleleaf forest. *Remote Sens. Environ.* **2004**, *89*, 519–534. [CrossRef]
11. Nagler, P.; Scott, R.; Westenburg, C.; Cleverly, J.; Glenn, E.; Huete, A. Evapotranspiration on western U.S. rivers estimated using the Enhanced Vegetation Index from MODIS and data from eddy covariance and Bowen ratio flux towers. *Remote Sens. Environ.* **2005**, *97*, 337–351. [CrossRef]
12. Shabanov, N.; Vargas, M.; Miura, T.; Sei, A.; Danial, A. Evaluation of the performance of Suomi NPP VIIRS top of canopy vegetation indices over AERONET sites. *Remote Sens. Environ.* **2015**, *162*, 29–44. [CrossRef]
13. Miura, T.; Turner, J.P.; Huete, A.R. Spectral Compatibility of the NDVI Across VIIRS, MODIS, and AVHRR: An Analysis of Atmospheric Effects Using EO-1 Hyperion. *IEEE Trans. Geosci. Remote Sens.* **2013**, *51*, 1349–1359. [CrossRef]
14. Kim, Y.; Huete, A.R.; Miura, T.; Jiang, Z. Spectral compatibility of vegetation indices across sensors: Band decomposition analysis with Hyperion data. *J. Appl. Remote Sens.* **2010**, *4*, 43520. [CrossRef]
15. Miura, T.; Yoshioka, H. Hyperspectral data in long-term, cross-sensor continuity studies. In *Hyperspectral Remote Sensing of Vegetation*; Thenkabail, P.S., Huete, A.R., Eds.; CRC Press: Boca Raton, FL, USA, 2011; pp. 607–627.
16. Vargas, M.; Miura, T.; Shabanov, N.; Kato, A. An initial assessment of Suomi NPP VIIRS vegetation index EDR. *J. Geophys. Res. Atmos.* **2013**, *118*, 12301–12316. [CrossRef]

17. Obata, K.; Miura, T.; Yoshioka, H.; Huete, A.R. Derivation of a MODIS-compatible enhanced vegetation index from visible infrared imaging radiometer suite spectral reflectances using vegetation isoline equations. *J. Appl. Remote Sens.* **2013**, *7*, 073467. [CrossRef]
18. Steven, M.D.; Malthus, T.J.; Baret, F.; Xu, H.; Chopping, M.J. Intercalibration of vegetation indices from different sensor systems. *Remote Sens. Environ.* **2003**, *88*, 412–422. [CrossRef]
19. Trishchenko, A.P.; Cihlar, J.; Li, Z. Effects of spectral response function on surface reflectance and NDVI measured with moderate resolution satellite sensors. *Remote Sens. Environ.* **2002**, *81*, 1–18. [CrossRef]
20. Trishchenko, A.P. Effects of spectral response function on surface reflectance and NDVI measured with moderate resolution satellite sensors: Extension to AVHRR NOAA-17, 18 and METOP-A. *Remote Sens. Environ.* **2009**, *113*, 335–341. [CrossRef]
21. Van Leeuwen, W.J.D.; Orr, B.J.; Marsh, S.E.; Herrmann, S.M. Multi-sensor NDVI data continuity: Uncertainties and implications for vegetation monitoring applications. *Remote Sens. Environ.* **2006**, *100*, 67–81. [CrossRef]
22. D'Odorico, P.; Gonsamo, A.; Damm, A.; Schaepman, M.E. Experimental Evaluation of Sentinel-2 Spectral Response Functions for NDVI Time-Series Continuity. *IEEE Trans. Geosci. Remote Sens.* **2013**, *51*, 1336–1348. [CrossRef]
23. Gonsamo, A.; Chen, J.M. Spectral Response Function Compatibility Among 21 Satellite Sensors for Vegetation Monitoring. *IEEE Trans. Geosci. Remote Sens.* **2013**, *51*, 1319–1335. [CrossRef]
24. Gitelson, A.A.; Kaufman, Y.J. MODIS NDVI Optimization To Fit the AVHRR Data Series-spectral Considerations. *Remote Sens. Environ.* **1998**, *66*, 343–350. [CrossRef]
25. Gao, B.-C. A practical method for simulating AVHRR-consistent NDVI data series using narrow MODIS channels in the 0.5–1.0 μm spectral range. *IEEE Trans. Geosci. Remote Sens.* **2000**, *38*, 1969–1975.
26. Gunther, K.P.; Maier, S.W. AVHRR compatible vegetation index derived from MERIS data. *Int. J. Remote Sens.* **2007**, *28*, 693–708. [CrossRef]
27. Yoshioka, H.; Miura, T.; Huete, A.R. An isoline-based translation technique of spectral vegetation index using EO-1 Hyperion data. *IEEE Trans. Geosci. Remote Sens.* **2003**, *41*, 1363–1372. [CrossRef]
28. Yoshioka, H.; Miura, T.; Obata, K. Derivation of Relationships between Spectral Vegetation Indices from Multiple Sensors Based on Vegetation Isolines. *Remote Sens.* **2012**, *4*, 583–597. [CrossRef]
29. Yoshioka, H. Vegetation isoline equations for an atmosphere-canopy-soil system. *IEEE Trans. Geosci. Remote Sens.* **2004**, *42*, 166–175. [CrossRef]
30. NASA Goddard Space Flight Center. NASA NPP VIIRS LAND PEATE QA. Available online: http://landweb.nascom.nasa.gov/cgi-bin/NPP_QA/NPPpage.cgi?fileName=dataOrder&subdir=forPage (accessed on 25 December 2015).
31. Justice, C.O.; Román, M.O.; Csiszar, I.; Vermote, E.F.; Wolfe, R.E.; Hook, S.J.; Friedl, M.; Wang, Z.; Schaaf, C.B.; Miura, T.; *et al.* Land and cryosphere products from Suomi NPP VIIRS: Overview and status. *J. Geophys. Res. Atmos.* **2013**, *118*, 9753–9765. [CrossRef] [PubMed]
32. Solano, R.; Didan, K.; Jacobson, A.; Huete, A. *MODIS Vegetation Index User's Guide*; Terrestrial Biophysics and Remote Sensing Lab, the University of Arizona: Tucson, AZ, USA, 2010; Volume 2010.
33. Friedl, M.A.; Sulla-Menashe, D.; Tan, B.; Schneider, A.; Ramankutty, N.; Sibley, A.; Huang, X. MODIS Collection 5 global land cover: Algorithm refinements and characterization of new datasets. *Remote Sens. Environ.* **2010**, *114*, 168–182. [CrossRef]
34. Loveland, T.R.; Belward, A.S. The IGBP-DIS global 1 km land cover data set, DISCover: First results. *Int. J. Remote Sens.* **1997**, *18*, 3289–3295. [CrossRef]
35. Vermote, E.; Justice, C.; Csiszar, I. Early evaluation of the VIIRS calibration, cloud mask and surface reflectance Earth data records. *Remote Sens. Environ.* **2014**, *148*, 134–145. [CrossRef]
36. Jackson, J.M.; Liu, H.; Laszlo, I.; Kondragunta, S.; Remer, L.A.; Huang, J.; Huang, H.-C. Suomi-NPP VIIRS aerosol algorithms and data products. *J. Geophys. Res. Atmos.* **2013**, *118*, 12673–12689. [CrossRef]
37. Liu, H.; Remer, L.A.; Huang, J.; Huang, H.-C.; Kondragunta, S.; Laszlo, I.; Oo, M.; Jackson, J.M. Preliminary evaluation of S-NPP VIIRS aerosol optical thickness. *J. Geophys. Res. Atmos.* **2014**, *119*, 3942–3962. [CrossRef]
38. Levy, R.C.; Remer, L.A.; Kleidman, R.G.; Mattoo, S.; Ichoku, C.; Kahn, R.; Eck, T.F. Global evaluation of the Collection 5 MODIS dark-target aerosol products over land. *Atmos. Chem. Phys.* **2010**, *10*, 10399–10420. [CrossRef]

39. Kopp, T.J.; Thomas, W.; Heidinger, A.K.; Botambekov, D.; Frey, R.A.; Hutchison, K.D.; Iisager, B.D.; Brueske, K.; Reed, B. The VIIRS Cloud Mask: Progress in the first year of S-NPP toward a common cloud detection scheme. *J. Geophys. Res. Atmos.* **2014**, *119*, 2441–2456. [CrossRef]
40. Jiang, Z.; Huete, A.R.; Didan, K.; Miura, T. Development of a two-band enhanced vegetation index without a blue band. *Remote Sens. Environ.* **2008**, *112*, 3833–3845. [CrossRef]

Article

Retrieval of Leaf Area Index (LAI) and Fraction of Absorbed Photosynthetically Active Radiation (FAPAR) from VIIRS Time-Series Data

Zhiqiang Xiao [1,*], Shunlin Liang [1,2], Tongtong Wang [1] and Bo Jiang [1]

1 State Key Laboratory of Remote Sensing Science, School of Geography, Beijing Normal University, Beijing 100875, China; sliang@umd.edu (S.L.); ttwang@mail.bnu.edu.cn (T.W.); bojiang@bnu.edu.cn (B.J.)
2 Department of Geographical Sciences, University of Maryland, College Park, MD 20742, USA
* Correspondence: zhqxiao@bnu.edu.cn; Tel.: +86-10-5880-7698

Academic Editors: Changyong Cao, Parth Sarathi Roy, Jose Moreno and Prasad S. Thenkabail
Received: 15 January 2016; Accepted: 14 April 2016; Published: 21 April 2016

Abstract: Long-term high-quality global leaf area index (LAI) and fraction of absorbed photosynthetically active radiation (FAPAR) products are urgently needed for the study of global change, climate modeling, and many other problems. As the successor of the Moderate Resolution Imaging Spectroradiometer (MODIS) sensor, the Visible Infrared Imaging Radiometer Suite (VIIRS) will continue to provide global environmental measurements. This paper aims to generate longer time series Global LAnd Surface Satellite (GLASS) LAI and FAPAR products after the era of the MODIS sensor. To ensure spatial and temporal consistencies between GLASS LAI/FAPAR values retrieved from different satellite observations, the GLASS LAI/FAPAR retrieval algorithms were adapted in this study to retrieve LAI and FAPAR values from VIIRS surface reflectance time-series data. After reprocessing of the VIIRS surface reflectance to remove remaining effects of cloud contamination and other factors, a database generated from the GLASS LAI product and the reprocessed VIIRS surface reflectance for all Benchmark Land Multisite Analysis and Intercomparison of Products (BELMANIP) sites was used to train general regression neural networks (GRNNs). The reprocessed VIIRS surface reflectance data from an entire year were entered into the trained GRNNs to estimate the one-year LAI values, which were then used to calculate FAPAR values. A cross-comparison indicates that the LAI and FAPAR values retrieved from VIIRS surface reflectance were generally consistent with the GLASS, MODIS and Geoland2/BioPar version 1 (GEOV1) LAI/FAPAR values in their spatial patterns. The LAI/FAPAR values retrieved from VIIRS surface reflectance achieved good agreement with the GLASS LAI/FAPAR values (R^2 = 0.8972 and RMSE = 0.3054; and R^2 = 0.9067 and RMSE = 0.0529, respectively). However, validation of the LAI and FAPAR values derived from VIIRS reflectance data is now limited by the scarcity of LAI/FAPAR ground measurements.

Keywords: retrieval; VIIRS; LAI; FAPAR; GLASS

---

1. Introduction

Leaf area index (LAI) is defined as the one-sided green leaf area per unit ground area in broadleaf canopies and as one-half the total needle surface area per unit ground area in coniferous canopies [1]. The fraction of absorbed photosynthetically active radiation (FAPAR) is defined as the fraction of incident photosynthetically active radiation (400–700 nm) absorbed by the green elements of a vegetation canopy [2]. LAI and FAPAR are widely used in agriculture and ecology and are two of the essential variables identified by the Global Climate Observing System (GCOS) [3]. Estimating LAI and FAPAR from satellite observations is the most useful way to generate LAI/FAPAR products at regional and global scales.

Many algorithms have been developed to retrieve LAI or FAPAR from satellite remote-sensing data [4]. In general, two types of methods can be distinguished: empirical and physical. The empirical methods are based on statistical relationships between LAI/FAPAR and spectral vegetation indices (VIs), which are calibrated using field measurements of LAI/FAPAR and reflectance data recorded by a remote sensor or simulated using canopy radiation models (e.g., Wang *et al.* [5]). These statistical relationships depend not only on vegetation types but also on the points in time and the geographical area the field measurements are collected at. The physical methods are based on inversion of canopy radiative transfer models [6–8]. Inversion techniques based on iterative minimization of a cost function require hundreds of runs of the canopy radiative-transfer model for each pixel and are therefore computationally too demanding. For practical applications, lookup table (LUT) methods [9] and machine learning methods [10] are two popular inversion techniques that are based on a pre-computed reflectance database.

Currently, multiple global LAI and FAPAR products have been produced from data acquired by the Advanced Very High Resolution Radiometer (AVHRR) [11], the Multiangle Imaging SpectroRadiometer (MISR) [12], the Moderate Resolution Imaging Spectroradiometer (MODIS) [9,13], VEGETATION [10], the Sea-Viewing Wide Field-of-View Sensor (SeaWiFS) [14], and the Medium-Resolution Imaging Spectrometer (MERIS) [15]. However, these LAI/FAPAR products are generated using different methods developed independently by different investigators under different assumptions, resulting in inconsistencies between LAI/FAPAR products [16,17]. Furthermore, currently available LAI products have limited length of the time series. In fact, a consistent long-term data set is urgently needed for the study of global change, climate modeling, and many other problems. As part of the efforts for generating the Global LAnd Surface Satellite (GLASS) products for the long-term environmental change studies [18,19], Xiao *et al.* [20] developed a method to retrieve LAI values from satellite observations based on general regression neural networks (GRNNs) and Xiao *et al.* [21] proposed a method to calculate FAPAR values from the LAI values. These methods were used to generate a long time series of GLASS LAI/FAPAR products (1981–2014) from AVHRR and MODIS reflectance data.

It is well known that the Terra and Aqua satellites equipped with the MODIS sensor had a life expectancy of six years and have far exceeded their design life, and the Visible Infrared Imaging Radiometer Suite (VIIRS) sensor on board the Suomi National Polar-Orbiting Partnership (SNPP) satellite will continue to provide global environmental measurements [22,23]. This paper aims to generate longer time series GLASS LAI and FAPAR products after the era of the MODIS sensor. To ensure spatial and temporal consistencies between GLASS LAI/FAPAR values retrieved from different satellite observations, the GLASS LAI/FAPAR retrieval algorithms were adapted in this study to retrieve LAI and FAPAR values from the VIIRS surface reflectance data. The VIIRS surface reflectance was reprocessed to remove the remaining effects of cloud contamination and other factors, and a database generated from the GLASS LAI product and the reprocessed VIIRS surface reflectance from the Benchmark Land Multisite Analysis and Intercomparison of Products (BELMANIP) sites for 2013 and 2014 were used to train GRNNs. Reprocessed VIIRS surface reflectance data for an entire year were entered into the trained GRNNs to estimate the one-year LAI values, which were then used to calculate FAPAR values. The LAI and FAPAR values retrieved from VIIRS reflectance data in this study were compared with the GLASS, MODIS, and Geoland2/BioPar version 1 (GEOV1) LAI/FAPAR products.

This paper is organized as follows. Section 2 describes the VIIRS reflectance product and the existing global LAI and FAPAR products, including GLASS, MODIS, and GEOV1. Several details of the algorithm implementation, including reprocessing of the VIIRS surface reflectance, retrieval of LAI values using GRNNs, and calculation of FAPAR values from the retrieved LAI values, are described in Section 3. Comparisons of the LAI/FAPAR values retrieved from time-series VIIRS surface reflectance data with the MODIS and GEOV1 LAI/FAPAR products are presented in Section 4. The paper concludes in Section 5.

2. Data

2.1. Visible Infrared Imaging Radiometer Suite (VIIRS) Surface Reflectance

As with MODIS, surface reflectance is one of the key products from VIIRS. The VIIRS surface reflectance intermediate product, including three bands (Table 1) with a 500 m spatial resolution and an eight-day temporal sampling period, is based on the heritage MODIS Collection 5 product [24]. The VIIRS surface reflectance intermediate product, in a sinusoidal projection system, is available from the LAADS (Level 1 and Atmosphere Archive and Distribution System) Web [25].

Table 1. VIIRS surface reflectance bands.

Band No.	Band Name	Bandwidth (nm)
B1	Red	603–677
B2	NIR	846–884
B3	SWIR	1580–1640

2.2. Global Leaf Area Index (LAI) and Fraction of Absorbed Photosynthetically Active Radiation (FAPAR) Products

The GLASS LAI and FAPAR products are one of the longest-duration LAI and FAPAR products in the world and were generated and released by the Center for Global Change Data Processing and Analysis of Beijing Normal University [26]. They are also available from the Global Land Cover Facility [27]. The GLASS LAI and FAPAR products have a temporal resolution of eight days, and spans 1981–2014. For 2000–2014, the GLASS LAI product was provided in a sinusoidal projection at a spatial resolution of 1 km and derived from MODIS collection 5 surface reflectance data using GRNNs which were trained using fused time-series LAI values from the CYCLOPES and MODIS collection 5 LAI products and reprocessed time-series MODIS surface reflectance for 2001–2003 over the BELMANIP sites [20]. For 1981–1999, the GLASS LAI product was provided in a geographic latitude/longitude projection at a spatial resolution of 0.05° (~5 km at the Equator) and derived from AVHRR surface reflectance data from NASA's Land Long-Term Data Record (LTDR) project [28] using GRNNs. The fused LAI values from the CYCLOPES and MODIS collection 5 LAI products were aggregated to 0.05° spatial resolution using a spatial-average method. Then, the aggregated LAI time-series values and the corresponding reprocessed AVHRR reflectance values over the BELMANIP sites for 2003 and 2004 were used to train GRNNs [20]. Unlike existing neural network methods that use remote-sensing data acquired only at a specific time to retrieve LAI [10,29], the reprocessed MODIS/AVHRR reflectance values from an entire year were inputted to the GRNNs to estimate the one-year LAI profiles. To ensure physical consistency between LAI and FAPAR retrievals, the GLASS FAPAR product was generated from the GLASS LAI product and has the same properties as the GLASS LAI product. The GLASS FAPAR values correspond to total FAPAR at 10:30 A.M. local time, which is a close approximation of daily average FAPAR [21]. The GLASS LAI and FAPAR products are spatially complete and have continuous trajectories. Direct comparison with ground-based estimates from the Validation of Land European Remote sensing Instrument (VALERI) project [30] and the BigFoot project [31] demonstrated that the GLASS LAI values were closer to the ground-based LAI estimates (RMSE = 0.7848 and R^2 = 0.8095) than the GEOV1 LAI values (RMSE = 0.9084 and R^2 = 0.7939) and the MODIS LAI values (RMSE = 1.1173 and R^2 = 0.6705) and the GLASS FAPAR values were also more accurate (RMSE = 0.0716 and R^2 = 0.9292) than the GEOV1 FAPAR values (RMSE = 0.1085 and R^2 = 0.8681) and the MODIS FAPAR values (RMSE = 0.1276 and R^2 = 0.8048) [20,21].

The MODIS LAI and FAPAR products have been available since 2000 and are provided in a sinusoidal projection. The MODIS LAI/FAPAR retrieval algorithm includes a main algorithm and a backup algorithm. The main algorithm is based on LUTs simulated using a three-dimensional radiative-transfer model. When the main algorithm fails, the backup algorithm is used to estimate

LAI and FAPAR from biome-specific relationships between LAI/FAPAR and normalized difference vegetation index (NDVI) [13]. Generally, LAI and FAPAR estimates using the backup algorithm are of lower quality, mainly because of residual clouds and poor atmospheric correction [32]. The MODIS FAPAR product is defined as the instantaneous black sky FAPAR (*i.e.*, under direct illumination) at the time of the Terra overpass (10:30 A.M.). Until now, MODIS LAI and FAPAR products have been released in a total of six versions. In collection 5, parameters of the main and backup algorithms are defined for eight main biome classes according to MODIS land cover, and a new stochastic radiative-transfer model was used to provide a better representation of canopy structure and the spatial heterogeneity intrinsic to woody biomes. Collection 5 MODIS LAI/FAPAR products are generated at 1 km spatial resolution and an eight-day time step. Unlike MODAGAGG used in Collection 5, the collection 6 retrieval algorithm uses daily L2G−lite surface reflectance (MOD09GA) as input and uses an improved multi-year land-cover product [33]. Collection 6 MODIS LAI/FAPAR products are generated at a spatial resolution of 500 m. The temporal compositing periods are eight and four days [33]. In this study, both collection 5 and 6 LAI/FAPAR products are compared with the LAI and FAPAR retrieved from VIIRS surface reflectance data. For clarification, the collection 5 LAI and FAPAR products are denoted by MODC5, and the collection 6 LAI and FAPAR products are denoted by MODC6.

The GEOV1 LAI and FAPAR products have been available since 1998 [34]. The products are provided in a Plate Carrée projection at 1/112° spatial resolution and 10-day frequency. The GEOV1 LAI and FAPAR products were derived from SPOT/VEGETATION sensor data using back-propagation neural networks. The MODIS and CYCLOPES LAI/FAPAR products were fused and scaled to generate best estimates of LAI and FAPAR, which were used to train the back-propagation neural networks using the SPOT/VEGETATION top-of-canopy nadir reflectance values over the BELMANIP sites [29]. The calibrated neural networks were used to generate the GEOV1 LAI and FAPAR products from SPOT/VEGETATION top-of-canopy nadir reflectance data. The GEOV1 FAPAR product corresponds to the instantaneous black sky FAPAR by green parts at 10:15 A.M. local time.

2.3. Experimental Area

Tile h09v05 was selected to investigate spatial patterns specific to the LAI/FAPAR values retrieved from the VIIRS surface reflectance data. It covers Texas and New Mexico, USA. The main biome types of this tile include grasses/cereal crops, shrubs, savannah, needleleaf forests, and broadleaf forests according to the MODIS land-cover map (MCD12Q1).

In addition, six sites with different biome types were selected to compare the time series of the reconstructed VIIRS and MODIS surface reflectance and the time series of LAI and FAPAR values. The six sites we chose were: Konza, Argo, Counami, Larose, Laprida, Donga, and their attributes are shown in Table 2.

Table 2. Selected site information.

Site Name	Country	Latitude (°)	Longitude (°)	Land-Cover Types
Konza	USA	39.08907	−96.57140	Grasses and cereal crops
Argo	USA	40.00666	−88.29154	Broadleaf crops
Counami	French Guyana	5.34363	−53.23691	Broadleaf forests
Larose	Canada	45.38057	−75.21714	Needleleaf forests
Laprida	Argentina	−36.99026	−60.55291	Savannah
Donga	Benin	9.76970	1.74528	Shrub

3. Methodology

3.1. Reprocessing of the VIIRS Surface Reflectance Data

VIIRS surface reflectance was estimated from satellite observations using atmospheric correction algorithm which consists of aerosol LUTs that are populated by the Second Simulation of the Satellite Signal in the Solar Spectrum (6S) radiative transfer model [35]. Although great efforts have been made, the VIIRS surface reflectance product still contains considerable noise caused, for instance, by cloud or mixed-cloud pixels, resulting in temporal and spatial inconsistencies in subsequent downstream products.

To remove residual clouds from the land-surface reflectance product, Xiao *et al.* [36] developed a method (referred to hereafter as VIRR algorithm for clarification) which incorporates upper envelopes of time series VIs as constraint conditions to reconstruct time series of surface reflectance for the red, near-infrared (NIR), and shortwave infrared (SWIR) bands. Satellite-retrieved surface reflectance data are used to calculate VIs, and a penalized least square regression based on three-dimensional discrete cosine transform (DCT-PLS) [37] is used to calculate continuous and smooth upper envelopes of VIs. Cloud-contaminated surface reflectance values were detected using the time-series VIs and the upper envelopes of the time-series VIs. Surface reflectance data with good quality in a given time window, along with the continuous and smooth upper envelopes of VIs, were used to estimate the optimal values of coefficients of quadratic polynomial functions fitted to the surface reflectance data. Then, time series of surface reflectance can be reconstructed using the quadratic polynomial functions according to the optimal values of the coefficients. If there are large gaps because of either cloud cover or missing observations, the reflectance values were filled using those in preceding or succeeding years according to better quality. Detailed information about the VIRR method can be found in Xiao *et al.* [36]. The method was used to reconstruct time series of surface reflectance from the MODIS/TERRA surface reflectance product (MOD09A1). The cloud-free MODIS/AQUA surface reflectance product (MYD09A1) was used as reference data for evaluation of the reconstructed time series of surface reflectance. Comparisons of the collocated surface reflectance values from MYD09A1 and the reconstructed surface reflectance over the BELMANIP sites in 2003 demonstrate that the VIRR method provides good performance for the red band (R^2 = 0.8606 and RMSE = 0.0366) and NIR band (R^2 = 0.6934 and RMSE = 0.0519) [36].

As the successor of the MODIS sensor, the VIIRS sensor has similar spectral bands with the MODIS sensor. In this study, the VIRR method was used to identify cloud-contaminated pixels in the VIIRS surface reflectance data and to generate continuous surface reflectances from VIIRS. The reconstructed surface reflectance values were then used to retrieve LAI and FAPAR.

3.2. LAI Retrieval Using General Regression Neural Networks (GRNNs)

In this study, the GLASS LAI retrieval algorithm was adapted to retrieve LAI values consistent with the GLASS LAI product from VIIRS reflectance time-series data using GRNNs. In order to retrieve global LAI from VIIRS surface reflectance using the GRNNs, the training database should be broadly representative. The BELMANIP sites provide a good sampling of biome types and conditions throughout the world [38]. There are a total of 402 BELMANIP sites [39]. A database for the GRNNs was generated over all BELMANIP sites. For each BELMANIP site, 7×7 pixels of the GLASS LAI product and the reprocessed VIIRS reflectance data from 2013–2014 were extracted. The database was randomly split into a training dataset (85% of the data) to train the GRNNs and a testing dataset (15% of the data) to evaluate their performance. To achieve better performance and convergence, the range of variation of input and output values were normalized between -1 and 1 according to the minimum and maximum values of the input and output values respectively.

The GRNN, developed by Specht [40], is a generalization of radial basis function networks and probabilistic neural networks. The advantage of this type of neural network is that it can approximate the map inherent in any sample data set. In addition, the GRNN does not require iterative training; the functional estimate is computed directly from the training data. If the kernel function of a GRNN is Gaussian, the fundamental formulation of the GRNN can be expressed as follows:

$$\mathbf{Y}'(\mathbf{X}) = \frac{\sum_{i=1}^{n} \mathbf{Y}^i \exp\left(-\frac{D_i^2}{2\sigma^2}\right)}{\sum_{i=1}^{n} \exp\left(-\frac{D_i^2}{2\sigma^2}\right)}, \tag{1}$$

where $D_i^2 = \left(\mathbf{X} - \mathbf{X}^i\right)^T \left(\mathbf{X} - \mathbf{X}^i\right)$ represents the squared Euclidean distance between the input vector $\mathbf{X}$ and the ith training input vector $\mathbf{X}^i$, $\mathbf{Y}^i$ is the output vector corresponding to the vector $\mathbf{X}^i$, $\mathbf{Y}'(\mathbf{X})$ is the estimate corresponding to the vector $\mathbf{X}$, n is the number of samples, and σ is a smoothing parameter that controls the size of the receptor region. Equation (1) shows that the estimate $\mathbf{Y}'(\mathbf{X})$, given an input vector $\mathbf{X}$, is the weighted average of all the sample observations $\mathbf{Y}^i$, where the weight for each observation is proportional to the Euclidean distance between the vector $\mathbf{X}$ and the training input vector $\mathbf{X}^i$. When the input to the GRNN is given, the architecture and weights of the GRNN are determined.

To retrieve LAI profiles from VIIRS surface reflectance time-series data using a GRNN, the input vector $\mathbf{X}$ of the GRNN is the reprocessed VIIRS time-series reflectance values (for a one-year period); that is, $\mathbf{X} = \left(R_1^1, R_2^1, \cdots, R_{46}^1, R_1^2, R_2^2, \cdots, R_{46}^2, \cdots, R_1^m, R_2^m, \cdots, R_{46}^m\right)^T$ and contains $46 \times m$ components, where m is the number of bands. The output vector $\mathbf{Y}' = (LAI_1, LAI_2, \cdots, LAI_{46})^T$ is the corresponding LAI time series for the year and contains 46 components.

The smoothing parameter σ in Equation (1) is the only free parameter in the GRNN formulation. Therefore, GRNN training essentially involves optimizing the smoothing parameter. In this study, the holdout method was used to find the optimal smoothing parameter at which the following cost function reaches its minimum:

$$f(\sigma) = \frac{1}{n}\sum_{i=1}^{n} \left(\hat{\mathbf{Y}}_i(\mathbf{X}_i) - \mathbf{Y}_i\right)^2, \tag{2}$$

where $\hat{\mathbf{Y}}_i(\mathbf{X}_i)$ is the estimate corresponding to $\mathbf{X}_i$ using the GRNN trained over all the training samples except the ith sample.

The performance of the trained GRNNs was evaluated over the testing dataset. Figure 1 shows the density scatterplots between the testing LAI values and the LAI values retrieved from VIIRS surface reflectance in the red, NIR, and SWIR bands using the GRNNs. It shows that the training was very efficient for the LAI retrieval. Most points in Figure 1 are around the 1:1 line, which indicates that the retrieved LAI values using the GRNNs achieve excellent agreement with the testing LAI values across the LAI range (R^2 = 0.9478, RMSE = 0.3891 and Bias = −0.0184).

The trained GRNNs were then used to retrieve LAI from the reprocessed VIIRS reflectance data. The reprocessed VIIRS reflectance data for a one-year period were entered into the trained GRNNs, and the output of the trained GRNNs represented the one-year LAI profile for each pixel.

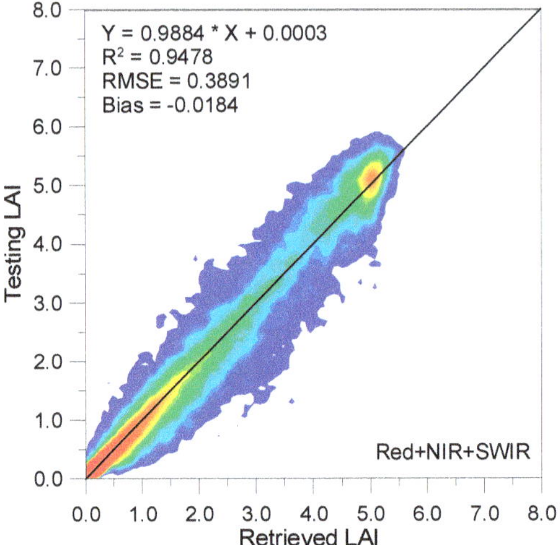

Figure 1. Comparison between the testing GLASS LAI values and the LAI values retrieved from VIIRS surface reflectance in the red, NIR and SWIR bands using the GRNNs. The size of the testing samples is 1758.

3.3. Calculation of FAPAR

In this study, the GLASS FAPAR retrieval algorithm was adapted to calculate FAPAR values consistent with the GLASS FAPAR product from the LAI values retrieved from VIIRS surface reflectance. Let τ_{PAR} be the transmittance of PAR down to the soil; FAPAR can then be approximately calculated as follows [34]:

$$FAPAR = 1 - \tau_{\text{PAR}}. \tag{3}$$

The radiation into the vegetation canopy includes direct and diffuse PAR. Therefore, the transmittance of PAR down to the soil can also be expressed as:

$$\tau_{\text{PAR}} = \tau_{\text{PAR}}^{\text{dir}} - \left(\tau_{\text{PAR}}^{\text{dir}} - \tau_{\text{PAR}}^{\text{dif}}\right) \times f_{\text{skyl}}, \tag{4}$$

where $\tau_{\text{PAR}}^{\text{dir}}$ and $\tau_{\text{PAR}}^{\text{dif}}$ are the fractions of the radiative flux originating from the direct illumination source and the transmitted fraction of the incident diffuse illumination source respectively, and f_{skyl} is the fraction of diffuse skylight. If the leaf area index of a canopy is lai and the absorptivity of leaves for radiation is a, the fraction of the total beam radiation transmitted through a canopy can be approximated using an exponential model [41]:

$$\tau_{\text{PAR}}^{\text{dir}} = e^{-\sqrt{a} \times k_c(\varphi) \times \Omega \times lai}, \tag{5}$$

where Ω is the clumping index, φ is the solar zenith angle, and $k_c\,(\varphi)$ is the canopy extinction coefficient for PAR. For an ellipsoidal leaf angle distribution, $k_c\,(\varphi)$ is calculated as follows [41].

$$k_c\,(\varphi) = \frac{\sqrt{x^2 + \tan^2(\varphi)}}{x + 1.774 \times (x + 1.182)^{-0.733}}, \tag{6}$$

where x is the ratio of average projected areas of canopy elements on horizontal and vertical surfaces. Diffuse radiation comes from all directions. Therefore, the diffuse transmission coefficient can be calculated by integrating the direct transmission coefficient over all illumination directions:

$$\tau_{\mathrm{PAR}}^{\mathrm{dif}} = 2\int_{0}^{\frac{\pi}{2}} \tau_{\mathrm{PAR}}^{\mathrm{dir}} \sin\varphi \cos\varphi d\varphi. \quad (7)$$

In the method just described, LAI is an important input parameter to estimate FAPAR. The LAI values retrieved from VIIRS surface reflectance were used to calculate FAPAR values at 10:30 A.M. local time, which are close approximations of daily average FAPAR. The clumping index is another input parameter. He *et al.* [42] used the MODIS bidirectional reflectance distribution function (BRDF) parameter to derive a global clumping index map at 500 m resolution. In the present study, the MODIS-derived clumping index map was used to calculate canopy transmittance.

3.4. Comparison and Analysis

The reconstructed VIIRS surface reflectance data at the AGRO site are presented to illustrate the performance of the VIRR algorithm. The similarities between the time series of the reconstructed VIIRS and MODIS surface reflectance over the selected sites were quantitatively evaluated. In this study, the following agreement index (S_{AI}) [43,44] was adopted to measure similarity.

$$S_{\mathrm{AI}} = 100 - \frac{\sum\limits_{i=1}^{M} (y_i - x_i)^2}{\sum\limits_{i=1}^{M} (|y_i - \overline{x}| + |x_i - \overline{x}|)^2} \times 100\%, \quad (8)$$

where x_i and y_i are the time series values at time i, respectively, $\overline{x}$ is the mean value of x_i, M is the length of the time series. The similarity ranges from 0.0% to 100.0%, where 0.0% represents complete disagreement between the time series x_i and y_i while 100.0% indicates that the time series x_i and y_i are identical.

The VIIRS surface reflectance data, including three bands (red, NIR, and SWIR), were used to retrieve LAI values. For clarification, the LAI and FAPAR products derived in this study are denoted as VIIRS products. The VIIRS LAI and FAPAR products have eight-day temporal and 1 km spatial resolution and are provided in a sinusoidal projection. A cross-comparison was performed to evaluate the spatial and temporal consistencies between the VIIRS LAI/FAPAR products and other existing global LAI/FAPAR products, including GLASS, MODC5, MODC6, and GEOV1 LAI/FAPAR products. For comparisons of spatial consistency, the maps of these LAI and FAPAR products over MODIS tile h09v05 were compared to investigate spatial patterns as well to check the distribution in space of the missing data, and density scatterplots among these LAI/FAPAR products over the tile were also analyzed. To compare temporal consistency, temporal profiles of the differences between GLASS, MODC5, MODC6 LAI/FAPAR products and VIIRS LAI/FAPAR products were checked over the selected sites. The similarities between the series of VIIRS LAI/FAPAR values and GLASS, MODC5, and MODC6 LAI/FAPAR values were calculated to evaluate the agreement between these LAI/FAPAR time series.

For the evaluations of spatial and temporal consistencies, the GEOV1 LAI and FAPAR products were re-projected onto the sinusoidal projection used in the MODIS and GLASS LAI/FAPAR products using the General Cartographic Transformation Package (GCTP) map projection library [45] and resampled to 1 km spatial resolution using the nearest-neighbor resampling method in order not to alter the product values.

In this study, only valid LAI and FAPAR values of these products were used for comparison. For the VIIRS, GLASS and GEOV1 LAI/FAPAR products, all LAI and FAPAR values were considered to be valid. For the MODC5 and MODC6 LAI/FAPAR products, the LAI and FAPAR values retrieved from the backup algorithm were not used for comparison of the LAI and FAPAR products because of their overall lower quality originating from residual clouds and poor atmospheric correction [32]. In other words, only the LAI and FAPAR values retrieved from the main algorithm (QC < 64) were considered to be valid.

4. Results and Discussion

4.1. Evaluation of the Reconstructed VIIRS Surface Reflectance

Figure 2 depicts time series of reconstructed surface reflectance using the VIRR method at the AGRO site in 2013. For the purpose of comparison, the time series of VIIRS surface reflectance at this site are also shown in Figure 2. The VIIRS surface reflectance, especially in the red and NIR bands, has several abnormal values at this site. The VIRR algorithm identified those observations with high reflectance values as clouds, and the residual clouds were removed in the reconstructed surface reflectance. It is apparent that the reconstructed surface reflectance from the VIRR method is in good agreement with the cloud-free VIIRS surface reflectance.

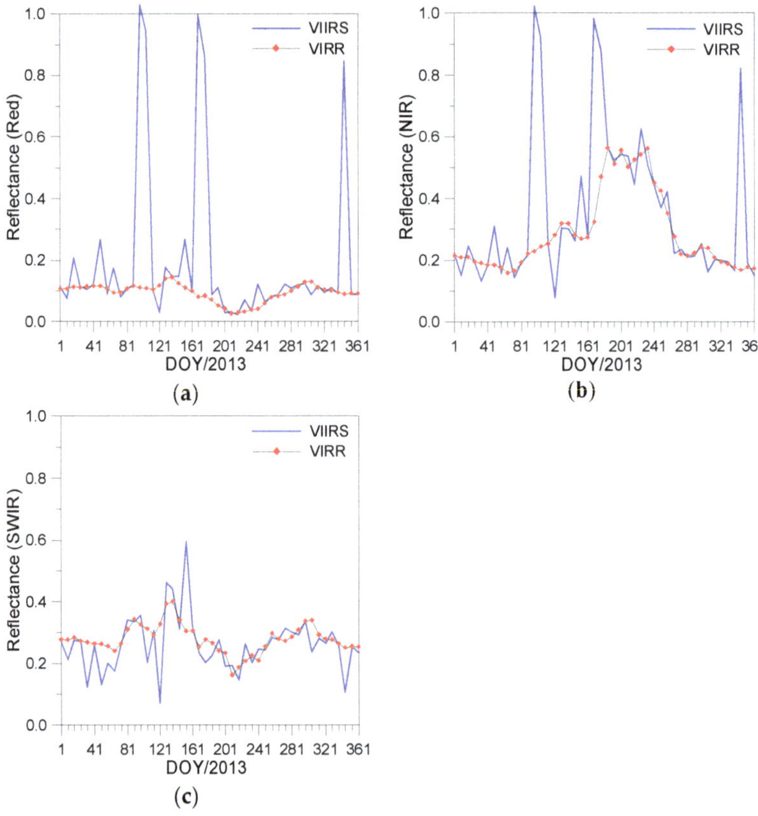

Figure 2. Temporal profiles of VIIRS surface reflectance and reconstructed surface reflectance using the VIRR method for the center pixel of the AGRO site in 2013. (**a**) Surface reflectance in the red band; (**b**) surface reflectance in the NIR band; (**c**) surface reflectance in the SWIR band.

To further evaluate the reconstructed VIIRS surface reflectance, the similarities between the series of the reconstructed VIIRS and MODIS surface reflectance data were calculated (Table 3).

The time-series similarities are high, especially at the Argo and Konza sites where the similarities for the red, NIR and SWIR reflectance are larger than 88%. However, the similarities between the series of the reconstructed VIIRS and MODIS surface reflectance especially in the red and SWIR bands are relatively poor at the Counami site. The similarity for the red reflectance is less than 30%. The low similarity for the Counami site is due to the fact that the number of high-quality satellite observations was very limited in tropical rain forests.

Table 3. Similarities (%) between the series of the VIIRS and GLASS, MODC5 and MODC6 LAI/FAPAR values and the series of the reconstructed VIIRS and MODIS surface reflectance.

Site Name	VIIRS and GLASS		VIIRS and MODC5		VIIRS and MODC6		Reconstructed VIIRS and MODIS Reflectance		
	LAI	FAPAR	LAI	FAPAR	LAI	FAPAR	Red	NIR	SWIR
Laprida	99.0802	95.7917	83.5879	55.7401	66.4576	60.6284	80.9976	55.0921	67.1797
Counami	49.9244	45.9245	17.7743	38.4644	13.1316	41.2487	29.8642	40.9023	35.0129
Konza	99.1311	98.2412	93.3907	96.2746	96.0984	94.7275	91.7717	89.2229	88.4589
Argo	99.5227	99.0775	90.4616	96.3072	93.6396	96.6767	95.4829	96.6589	92.6440
Larose	99.6944	98.5516	87.3624	77.2915	97.2239	95.7807	72.0348	72.0870	39.9923
Donga	98.3687	98.5982	80.3088	78.9755	87.3224	79.2447	86.2486	74.4560	58.2819

4.2. *Evaluation of the Retrieved LAI and FAPAR*

4.2.1. Spatial Comparison

An example of VIIRS, MODC5, MODC6, GLASS, and GEAV1 LAI products for tile h09v05 on day 185, 2013 is shown in Figure 3. Areas masked in grey correspond to pixels where these LAI products did not provide valid LAI values. As in the GLASS LAI product, no missing values exist in the VIIRS LAI product. It is apparent that the VIIRS, MODC5, MODC6, GLASS, and GEAV1 LAI products are generally consistent in their spatial patterns, depicting a clear correlation with the MODIS biome map (Figure 3f). These LAI products took on their highest values over broadleaf forest regions. LAI values were intermediate over grass/cereal crop, needleleaf forest and savannah regions and were very low over shrub regions. However, discrepancies existed in the magnitudes of these LAI products. Over the broadleaf forest regions, the MODC5 LAI values could reach 6.9 LAI units and were larger than the corresponding VIIRS, GLASS and GEOV1 LAI values, which was partly due to overestimation of the MODC5 LAI values associated with broadleaf forests [16]. Obviously, the MODC6 LAI values have significant improvements. Over needleleaf forest regions, the VIIRS, GLASS and GEOV1 LAI products ranged between 3.0 and 5.0 LAI units and were clearly higher than the MODC5 and MODC6 LAI values (around 1.8 LAI units).

Figure 4 shows density scatterplots among the VIIRS, GLASS, MODC5, MODC6, and GEOV1 LAI values for tile h09v05 on day 185, 2013. Only the collocated LAI values among these LAI products are included in Figure 4. Figure 4c,d illustrates that the VIIRS LAI values were slightly higher than those of MODC5 and MODC6 for lower LAI values, but clearly lower than those of MODC5, MODC6, and GEOV1 when the LAI values were greater than 3.5 LAI units. Compared with MODC5, MODC6, and GEOV1 LAI values, those of GLASS are distributed more closely around the 1:1 line against the VIIRS LAI values, which showed that the VIIRS LAI values achieved better agreement with the GLASS LAI values (R^2 = 0.8972 and RMSE = 0.3054) across the LAI range than did the other products. This can be attributed to the same GRNN structure for the GLASS and VIIRS LAI retrieval, and the LAI training data from the GLASS LAI product.

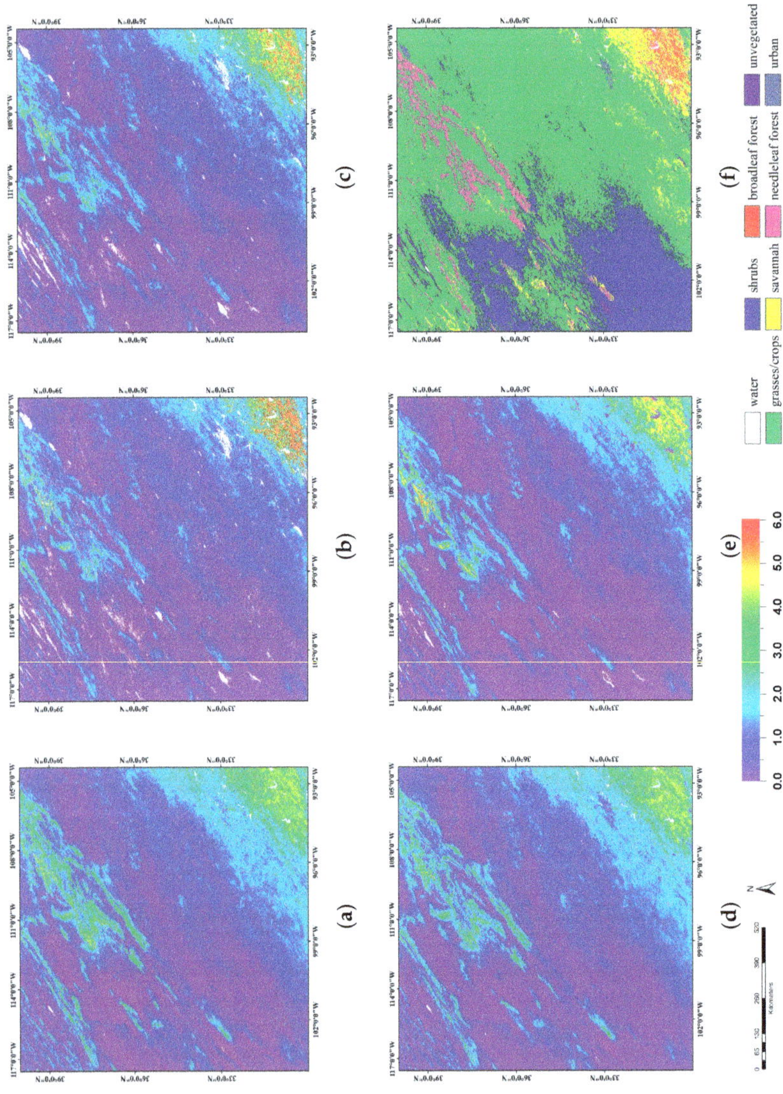

Figure 3. Maps of **(a)** VIIRS; **(b)** MODC5; **(c)** MODC6; **(d)** GLASS; and **(e)** GEOV1 LAI products for tile h09v05 on day 185, 2013. White color denotes water, and areas masked in grey correspond to pixels where these LAI products did not provide valid LAI values; **(f)** shows the MODIS 8-biomes land cover map.

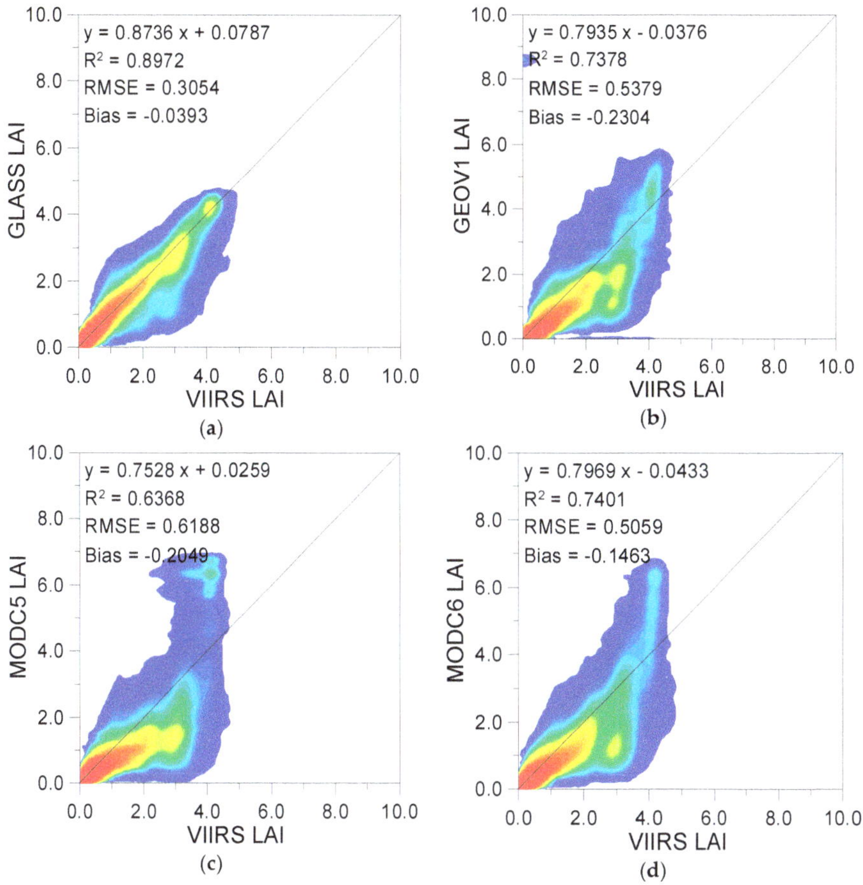

Figure 4. Density scatterplots between the VIIRS LAI values and the GLASS (**a**); GEOV1 (**b**); MODC5 (**c**); MODC6 (**d**) LAI values for tile h09v05 on day 185, 2013. Only the collocated LAI values among these LAI products are included.

Figure 5 shows maps of the VIIRS, MODC5, MODC6, GLASS, and GEAV1 FAPAR products for tile h09v05 on day 185, 2013. Similar to the LAI products, the VIIRS, MODC5, MODC6, GLASS, and GEAV1 FAPAR products are generally consistent in their spatial patterns. The MODC5, MODC6, and GEOV1 FAPAR values were between 0.1 and 0.2 higher than the VIIRS and GLASS FAPAR values over the broadleaf forest regions and slightly larger than the VIIRS and GLASS FAPAR values over the needleleaf forest and grass/cereal crop regions.

The density scatterplots among the VIIRS, GLASS, MODC5, MODC6, and GEOV1 FAPAR values for tile h09v05 on day 185, 2013 are shown in Figure 6. Like the VIIRS LAI product, the VIIRS FAPAR values achieved good agreement with the GLASS FAPAR values (R^2 = 0.9067 and RMSE = 0.0529) across the FAPAR range. However, most points in Figure 6b–d is above the 1:1 line, which indicates that the VIIRS FAPAR values are lower than those from MODC5, MODC6, and GEOV1. Previous observations indicate that the MODC5 and GEOV1 FAPAR values are also slightly larger than those of GLASS [21]. These discrepancies can be partly explained by overestimation in the MODC5, MODC6, and GEOV1 values [17,21].

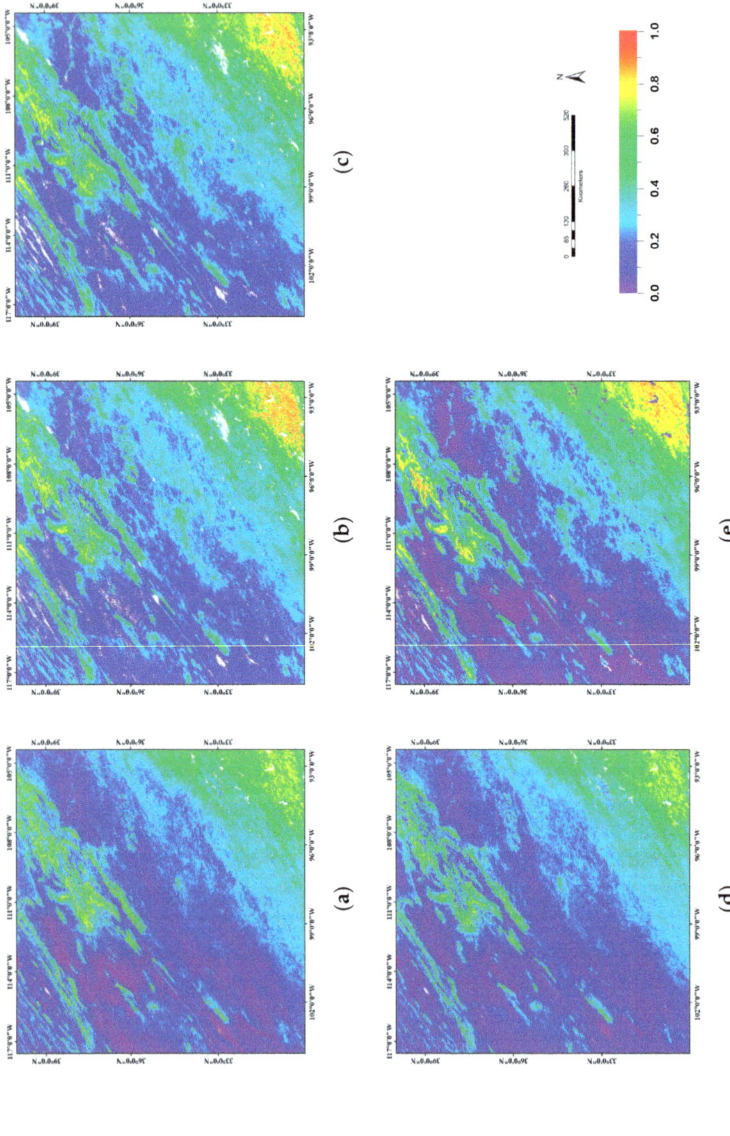

Figure 5. Maps of (**a**) VIIRS; (**b**) MODC5; (**c**) MODC6; (**d**) GLASS; and (**e**) GEOV1 FAPAR products for tile h09v05 on day 185, 2013. White color denotes water, and areas masked in grey correspond to pixels where these FAPAR products did not provide valid FAPAR values.

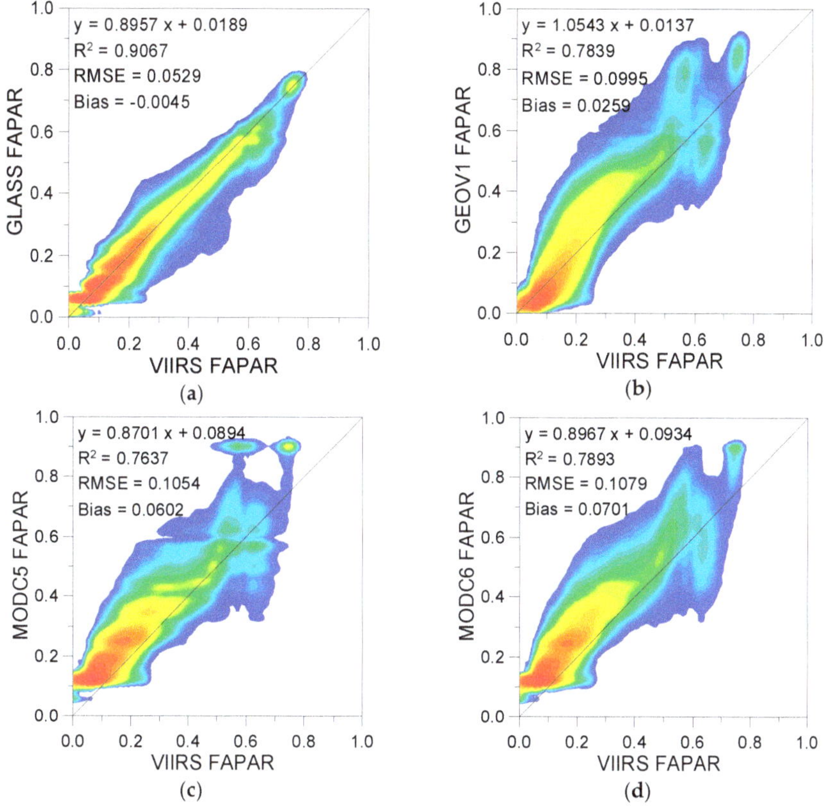

Figure 6. Density scatterplots between the VIIRS FAPAR values and the GLASS (**a**); GEOV1 (**b**); MODC5 (**c**); MODC6 (**d**) FAPAR values for tile h09v05 on day 185, 2013. Only collocated FAPAR values among these FAPAR products are included.

4.2.2. Temporal Analysis

Temporal consistency among the VIIRS, GLASS, MODC5, and MODC6 LAI and FAPAR products was evaluated over the selected sites. Figure 7 shows temporal profiles of the differences between GLASS, MODC5, MODC6 LAI/FAPAR products and VIIRS LAI/FAPAR products, and the similarities between the series of VIIRS LAI/FAPAR values and GLASS, MODC5, and MODC6 LAI/FAPAR values were shown in Table 3.

Figure 7a shows the temporal trajectories of LAI/FAPAR differences for the Konza site with grass and cereal crop biome type. There is an excellent agreement between VIIRS and GLASS LAI values during the whole year, with maximum differences of only 0.26 LAI units. The MODC6 LAI values were larger than VIIRS LAI values only from Julian days 153 to 185, and the differences between MODC5 and VIIRS LAI values were negative during the whole year. Just like VIIRS LAI values, the GLASS FAPAR values were higher than those of VIIRS only from Julian days 153 to 209. However, most of the differences between MODC5, MODC6 and VIIRS FAPAR values were positive for this year. The similarities between the series of VIIRS and GLASS LAI/FAPAR values (99.1311 and 98.2412, respectively) are slightly higher than those between the series of VIIRS and MODC5, MODC6 LAI/FAPAR values at this site.

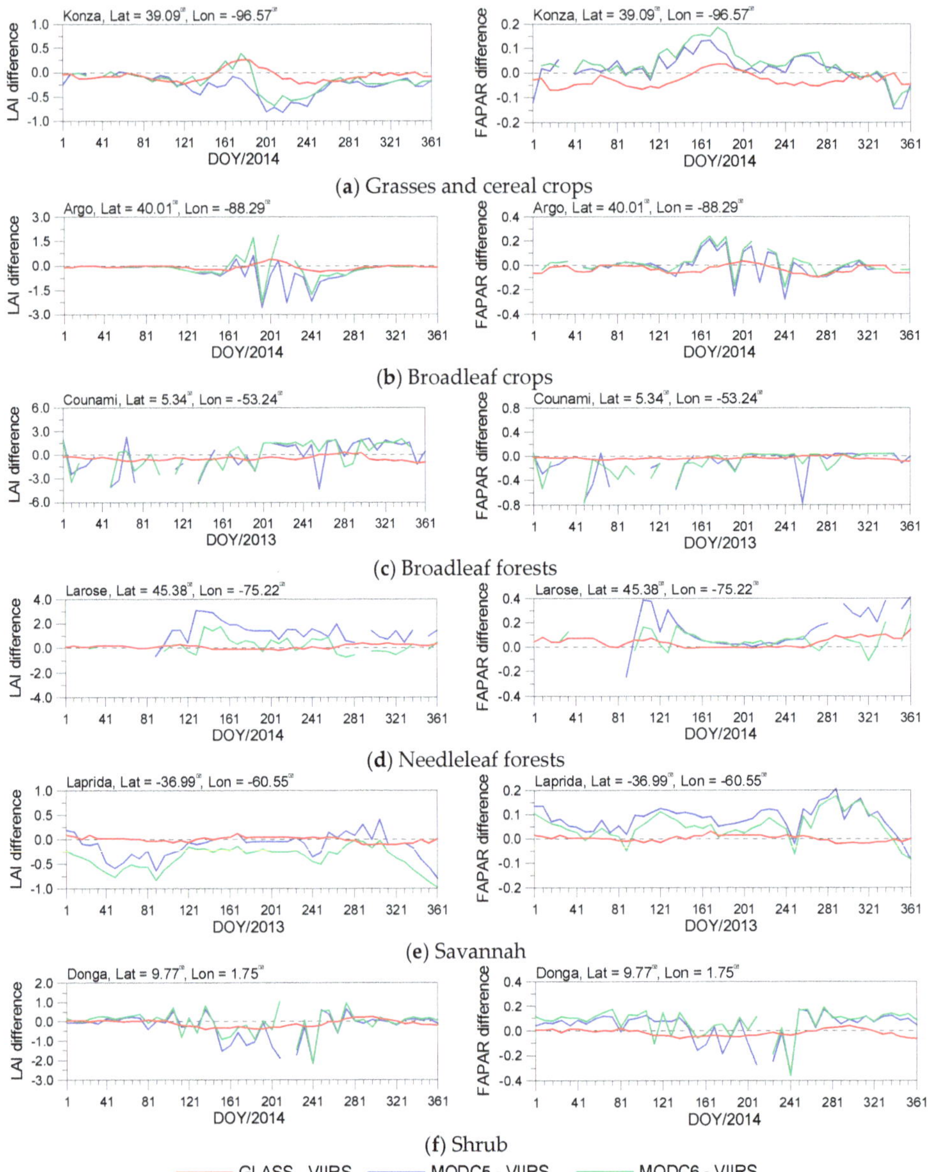

Figure 7. Temporal profiles of the differences between GLASS, MODC5, MODC6 LAI/FAPAR products and VIIRS LAI/FAPAR products at several sites with different vegetation types. The temporal profiles of LAI differences are shown in the left panel, and the temporal profiles of FAPAR differences are shown in the right panel. (**a**) Grasses and cereal crops; (**b**) Broadleaf crops; (**c**) Broadleaf forests; (**d**) Needleleaf forests; (**e**) Savannah; (**f**) Shrub.

The temporal profiles of LAI/FAPAR differences for the Argo site are shown in Figure 7b. The biome type for this site is broadleaf crops. During the nongrowing season, the LAI differences were approximately 0, which indicate that the VIIRS LAI values achieved very good agreement with the GLASS, MODC5 and MODC6 LAI values. During the growing season, the VIIRS LAI values were in good agreement with GLASS LAI values, with maximum differences of only 0.41 LAI units, but the difference between the MODC6 and VIIRS LAI value at day 209 reached 1.86 LAI units. At days 193, 217 and 241, the MODC5 and MODC6 LAI values were between 2.2 and 2.6 LAI units lower than the VIIRS LAI values, which stems partly from underestimations of the MODC5 and MODC6 LAI values caused by the residual cloud in the MODIS surface reflectance. The VIIRS FAPAR values were in good agreement with the GLASS FAPAR values at this site. The MODC5 and MODC6 FAPAR values, except for those contaminated by residual cloud, were between 0.1 and 0.2 units higher than the VIIRS FAPAR values during the vegetation growing season.

The temporal profiles of LAI and FAPAR differences for the Counami site with broadleaf forest biome type are shown in Figure 7c. Some MODC5 and MODC6 LAI/FAPAR values are missing due to residual cloud contamination in MODIS surface reflectance data. The MODC5 and MODC6 LAI values, except for those contaminated by residual cloud, were higher than those of VIIRS (up to 2.1 LAI unit) after day 201 for this year. At this site, the VIIRS LAI values were larger than those of GLASS (up to 1.0 LAI unit) and the VIIRS LAI values were higher than those of GLASS (up to 0.08 units) except for Julian days 265 to 297. The similarities between the series of VIIRS and GLASS LAI/FAPAR products at the Counami site are less than 50% due to the fact that the number of good observations was very limited in tropical rain forests. Nevertheless, the similarities between the series of VIIRS and GLASS LAI/FAPAR products are still larger than those between the series of VIIRS and MODC5, MODC6 LAI/FAPAR products.

Figure 7d shows the temporal trajectories of LAI and FAPAR differences for the Larose site with needleleaf forest biome type. The differences demonstrate that the MODC6 LAI/FAPAR values were in better agreement with the VIIRS LAI/FAPAR values than with the MODC5 LAI/FAPAR values. The MODC5 LAI values were between 1.0 and 3.1 LAI units higher than those of VIIRS especially during the vegetation growing season. The MODC5 and MODC6 FAPAR values were in good agreement with the VIIRS FAPAR values during the vegetation growing season. However, large discrepancies between these FAPAR products could be observed during the nongrowing season, when the MODC5 FAPAR values were between 0.2 and 0.4 units higher than those of VIIRS.

The temporal profiles of LAI and FAPAR differences for the Laprida sites with savannah biome type are shown in Figure 7e. The differences between the GLASS and VIIRS LAI values were lower than 0.11 LAI units and the differences between the GLASS and VIIRS FAPAR values were less than 0.03 units for the entire year, which demonstrated that VIIRS LAI/FAPAR values achieved very good agreement with those of GLASS at this site. The VIIRS LAI values were higher than those of MODC5 and MODC6 (up to 1.0 LAI units), whereas the VIIRS FAPAR values were lower than the MODC5 and MODC6 FAPAR values (up to 0.2 units) throughout the entire year.

The temporal trajectories of LAI and FAPAR differences for the Donga site with shrub biome type in 2014 are shown in Figure 7f. The VIIRS LAI values achieved good agreement with those of MODC5 and MODC6 during the nongrowing season. The MODC5 and MODC6 FAPAR values, except for those contaminated by residual cloud, were higher than the VIIRS FAPAR values (up to 0.2 units) during the whole year. The similarities between the series of VIIRS and GLASS LAI/FAPAR products were higher than 98%, which indicate that the VIIRS LAI/FAPAR values agree well with those of GLASS.

The similarities between the series of these LAI/FAPAR products in Table 3 demonstrate that the VIIRS and GLASS LAI/FAPAR products have the best agreement among these LAI/FAPAR products. The similarities between the series of the VIIRS and GLASS LAI/FAPAR values for each site are larger than those between the series of the VIIRS and MODC5, MODC6 LAI/FAPAR values. The similarities for LAI and surface reflectance in Table 3 also demonstrate that reconstruction of surface reflectance using the VIRR method and the proposed GRNN approach in this study are an important factor to ensure the high agreement between the VIIRS and GLASS LAI/FAPAR values. One the one hand, at the Argo and Konza sites, the series of the VIIRS and GLASS LAI/FAPAR values have large similarities when the similarities between the series of the reconstructed VIIRS and

MODIS surface reflectance data are large. At the Counami site, the similarities between the series of the reconstructed VIIRS and MODIS surface reflectance for the red, NIR and SWIR bands are lower than those at other sites (no more than 41%), and the similarities between the series of the VIIRS and GLASS LAI/FAPAR values at this site are also lower than those at other sites (no more than 50%). Therefore, reconstruction of surface reflectance using the VIRR method is an important factor that contributes to the high agreement between the VIIRS and GLASS LAI/FAPAR values. On the other hand, the similarity between the GLASS and VIIRS LAI time series can still reach 49.92% although the similarities between the series of the reconstructed VIIRS and MODIS surface reflectance data especially in the red and SWIR bands are small at the Counami site. What is more, the similarities between the GLASS and VIIRS LAI/FAPAR time series are larger than the similarities between the series of the reconstructed VIIRS and MODIS surface reflectance data for each site. Therefore, another important factor to ensure the high agreement between the VIIRS and GLASS LAI/FAPAR values is the proposed GRNN approach in this study which uses the same GRNN structure as the GLASS LAI retrieval method and uses GLASS LAI values as the LAI training data of the GRNNs for VIIRS LAI retrieval.

5. Conclusions

This paper contributes to the research community for data continuity of LAI/FAPAR (leaf area index/fraction of absorbed photosynthetically active radiation) across Advanced Very High Resolution Radiometer (AVHRR), Moderate Resolution Imaging Spectroradiometer (MODIS), and Visible Infrared Imaging Radiometer Suite (VIIRS) sensors. To generate Global LAnd Surface Satellite (GLASS) LAI and FAPAR products after the era of the MODIS sensor, the GLASS LAI/FAPAR retrieval algorithms were adapted in this study to retrieve LAI and FAPAR values from VIIRS surface reflectance time-series data. After reprocessing of the VIIRS surface reflectance to remove remaining effects of cloud contamination and other factors, a database generated from the GLASS LAI product and the reprocessed VIIRS surface reflectance for all Benchmark Land Multisite Analysis and Intercomparison of Products (BELMANIP) sites was used to train general regression neural networks (GRNNs). The reprocessed VIIRS surface reflectance data from an entire year were entered into the trained GRNNs to estimate the one-year LAI values, which were then used to calculate FAPAR values.

A cross-comparison indicates that the VIIRS LAI/FAPAR values were generally consistent with the GLASS, MODIS and GEOV1 LAI/FAPAR values in their spatial patterns. The VIIRS and GLASS LAI/FAPAR products have the best agreement (R^2 = 0.8972 and RMSE = 0.3054; and R^2 = 0.9067 and RMSE = 0.0529, respectively) among these LAI/FAPAR products. The high agreement between the VIIRS and GLASS LAI/FAPAR values is partly attributed to the proposed GRNN approach which uses the same GRNN structure as the GLASS LAI retrieval method and uses GLASS LAI values as the LAI training data of the GRNNs for VIIRS LAI retrieval. Reconstruction of surface reflectance using the VIRR method is another factor to ensure the high agreement between the VIIRS and GLASS LAI/FAPAR values. Therefore, the method proposed in this study can be used to generate longer time-series GLASS LAI and FAPAR products from VIIRS surface reflectance data.

However, the proposed GRNN approach requires that the input of the GRNNs is one year of surface reflectance without missing values. Therefore, the VIRR method must be used to fill the missing surface reflectance values before deriving LAI/FPAR. In addition, validation of VIIRS LAI and FAPAR products is now limited by the scarcity of LAI/FAPAR ground measurements. In the near future, the authors hope to carry out more extensive validation and analysis of the LAI and FAPAR values derived from VIIRS reflectance data by searching for ground measurements with which to perform direct validation.

Acknowledgments: AA121201 under the "State Program for High-Tech Research and Development (863 program)" and the National Natural Science Foundation of China under grant nos. 41171264 and 41331173. The authors would like to thank the anonymous reviewers for their valuable comments.

Author Contributions: Shunlin Liang and Zhiqiang Xiao conceived the study. Zhiqiang Xiao performed the data analysis and wrote the paper. Shunlin Liang, Tongtong Wang and Bo Jiang reviewed and edited the manuscript. All authors read and approved the manuscript.

Conflicts of Interest: The authors declare no conflict of interest.

References

1. Watson, D.J. Comparative physiological studies in the growth of field crops. I. Variation in net assimilation rate and leaf area between species and varieties, and within and between years. *Ann. Bot.* **1947**, *11*, 41–76.
2. Monteith, J.L. Solar radiation and productivity in tropical ecosystems. *J. Appl. Ecol.* **1972**, *9*, 747–766. [CrossRef]
3. GCOS. *Satellite-Based Products for Climate, Supplemental Details to the Satellite Based Component of the Implementation Plan for the Global Observing System for Climate in Support of the UNFCCC*; Report GCOS-107 (WMO/TD No. 1338); World Meteorological Organization (WMO): Geneva, Switzerland, 2006.
4. Liang, S. Recent developments in estimating land surface biogeophysical variables from optical remote sensing. *Prog. Phys. Geogr.* **2007**, *31*, 501–516. [CrossRef]
5. Wang, F.; Huang, J.; Tang, Y.; Wang, X. New vegetation index and its application in estimating leaf area index of rice. *Rice Sci.* **2007**, *14*, 195–203. [CrossRef]
6. Kimes, D.S.; Knyazikhin, Y.; Privette, J.L.; Abuelgasim, A.A.; Gao, F. Inversion methods for physically-based models. *Remote Sens. Rev.* **2000**, *18*, 381–439. [CrossRef]
7. Gascon, F.; Gastellu-Etchegorry, J.P.; Lefevre-Fonollosa, M.J.; Dufrene, E. Retrieval of forest biophysical variables by inverting a 3-D radiative transfer model and using high and very high resolution imagery. *Int. J. Remote Sens.* **2004**, *25*, 5601–5616. [CrossRef]
8. Xiao, Z.; Liang, S.; Wang, J.; Song, J.; Wu, X. A temporally integrated inversion method for estimating leaf area index from MODIS data. *IEEE Trans. Geosci. Remote Sens.* **2009**, *47*, 2536–2545. [CrossRef]
9. Myneni, R.B.; Hoffman, S.; Knyazikhin, Y.; Privette, J.L.; Glassy, J.; Tian, Y.; Wang, Y.; Song, X.; Zhang, Y.; Smith, G.R.; *et al.* Global products of vegetation leaf area and fraction absorbed PAR from year one of MODIS data. *Remote Sens. Environ.* **2002**, *83*, 214–231. [CrossRef]
10. Baret, F.; Hagolle, O.; Geiger, B.; Bicheron, P.; Miras, B.; Huc, M.; Berthelot, B.; Nino, F.; Weiss, M.; Samain, O.; *et al.* LAI, fAPAR and fCover CYCLOPES global products derived from VEGETATION. Part 1: Principles of the algorithm. *Remote Sens. Environ.* **2007**, *110*, 275–286. [CrossRef]
11. Chen, J.M.; Pavlic, G.; Brown, L.; Cihlar, J.; Leblanc, S.G.; White, P.; Hall, R.J.; Peddle, D.; King, D.; Trofymow, J.A.; *et al.* Validation of Canada–wide leaf area index maps using ground measurements and high and moderate resolution satellite imagery. *Remote Sens. Environ.* **2002**, *80*, 165–184. [CrossRef]
12. Hu, J.; Tan, B.; Shabanov, N.; Crean, K.A.; Martonchik, J.V.; Diner, D.J. Performance of the MISR LAI and FPAR algorithm: A case study in Africa. *Remote Sens. Environ.* **2003**, *88*, 324–340. [CrossRef]
13. Knyazikhin, Y.; Martonchik, J.V.; Myneni, R.B.; Diner, D.J.; Running, S.W. Synergistic algorithm for estimating vegetation canopy leaf area index and fraction of absorbed photosynthetically active radiation from MODIS and MISR data. *J. Geophys. Res.* **1998**, *103*, 32257–32275. [CrossRef]
14. Gobron, N.; Mélin, F.; Pinty, B.; Verstraete, M.M.; Widlowski, J.L.; Bucini, G. A global vegetation index for SeaWiFS: Design and applications. In *Remote Sensing and Climate Modeling: Synergies and Limitations*; Beniston, M., Verstraete, M.M., Eds.; Springer: Dordrecht, The Netherlands, 2001; pp. 5–21.
15. Gobron, N.; Pinty, B.; Verstraete, M.M.; Govaerts, Y. The MERIS Global Vegetation Index (MGVI): Description and preliminary application. *Int. J. Remote Sens.* **1999**, *20*, 1917–1927. [CrossRef]
16. Garrigues, S.; Lacaze, R.; Baret, F.; Morisette, J.T.; Weiss, M.; Nickeson, J.E.; Fernandes, R.; Plummer, S.; Shabanov, N.V.; Myneni, R.B.; *et al.* Validation and intercomparison of global Leaf Area Index products derived from remote sensing data. *J. Geophys. Res.* **2008**, *113*. [CrossRef]
17. Camacho, F.; Cernicharo, J.; Lacaze, R.; Baret, F.; Weiss, M. GEOV1: LAI, FAPAR essential climate variables and FCOVER global time series capitalizing over existing products. Part 2: Validation and intercomparison with reference products. *Remote Sens. Environ.* **2013**, *137*, 310–329. [CrossRef]
18. Liang, S.L.; Zhao, X.; Liu, S.H.; Yuan, W.P.; Cheng, X.; Xiao, Z.Q.; Zhang, X.T.; Liu, Q.; Cheng, J.; Tang, H.R.; *et al.* A long-term Global LAnd Surface Satellite (GLASS) data-set for environmental studies. *Int. J. Digit. Earth* **2013**, *6*, 5–33. [CrossRef]

19. Liang, S.; Zhang, X.; Xiao, Z.; Cheng, J.; Liu, Q.; Zhao, X. *Global LAnd Surface Satellite (GLASS) Products: Algorithms, Validation and Analysis*; Springer: New York, NY, USA, 2014.
20. Xiao, Z.; Liang, S.; Wang, J.; Xiang, Y.; Zhao, X.; Song, J. Long time-series global land surface satellite (GLASS) leaf area index product derived from MODIS and AVHRR data. *IEEE Trans. Geosci. Remote Sens.* **2015**, under review.
21. Xiao, Z.Q.; Liang, S.L.; Sun, R.; Wang, J.D.; Jiang, B. Estimating the fraction of absorbed photosynthetically active radiation from the MODIS data based GLASS leaf area index product. *Remote Sens. Environ.* **2015**, *171*, 105–117. [CrossRef]
22. Murphy, R.E.; Barnes, W.L.; Lyapustin, A.I.; Privette, J.; Welsch, C.; DeLuccia, F.; Swenson, H.; Schueler, C.F.; Ardanuy, P.E.; Kealy, P.S.M. Using VIIRS to provide data continuity with MODIS. In Proceedings of the IEEE 2001 International Geoscience and Remote Sensing Symposium (IGARSS '01), Sydney, Australian, 9–13 July 2001; Volume 3, pp. 1212–1214.
23. Lewis, J.M.; Martin, D.W.; Rabin, R.M.; Moosmüller, H. Suomi: Pragmatic visionary. *Bull. Am. Meteorol. Soc.* **2010**, *91*, 559–577. [CrossRef]
24. Vermote, E.; Justice, C.; Csiszar, I. Early evaluation of the VIIRS calibration, cloud mask and surface reflectance Earth data records. *Remote Sens. Environ.* **2014**, *148*, 134–145. [CrossRef]
25. The LAADS Web. Available online: http://ladsweb.nascom.nasa.gov/data/search.html (accessed on 15 January 2016).
26. The GLASS LAI at Beijing Normal University. Available online: http://www.bnu-datacenter.com/en (accessed on 15 January 2016).
27. The GLASS LAI at the Global Land Cover Facility. Available online: http://glcf.umd.edu (accessed on 15 January 2016).
28. Pedelty, J.; Devadiga, S.; Masuoka, E.; Brown, M.; Pinzon, J.; Tucker, C.; Vermote, E.; Prince, S.; Nagol, J.; Justice, C.; *et al.* Generating a long-term land data record from the AVHRR and MODIS instruments. In Proceedings of the 2007 IEEE International Geoscience and Remote Sensing Symposium, Barcelona, Spain, 23–28 July 2007; pp. 1021–1025.
29. Baret, F.; Weiss, M.; Lacaze, R.; Camacho, F.; Makhmara, H.; Pacholcyzk, P.; Smets, B. GEOV1: LAI and FAPAR essential climate variables and FCOVER global time series capitalizing over existing products. Part1: Principles of development and production. *Remote Sens. Environ.* **2013**, *137*, 299–309. [CrossRef]
30. The VALERI Validation Data. Available online: http://w3.avignon.inra.fr/valeri/ (accessed on 15 January 2016).
31. Cohen, W.B.; Maiersperger, T.K.; Turner, D.P.; Ritts, W.D.; Pflugmacher, D.; Kennedy, R.E.; Kirschbaum, A.; Running, S.W.; Costa, M.; Gower, S.T. MODIS land cover and LAI collection 4 product quality across nine sites in the western hemisphere. *IEEE Trans. Geosci. Remote Sens.* **2006**, *44*, 1843–1858. [CrossRef]
32. Yang, W.; Tan, B.; Huang, D.; Rautiainen, M.; Shabanov, N.V.; Wang, Y.; Myneni, R.B. MODIS leaf area index products: From validation to algorithm improvement. *IEEE Trans. Geosci. Remote Sens.* **2006**, *44*, 1885–1898. [CrossRef]
33. MODIS Collection 6 (C6) LAI/FPAR Product User's Guide. Available online: https://lpdaac.usgs.gov/dataset_discovery/modis/modis_products_table/mcd15a2h_v006 (accessed on 15 January 2016).
34. The GEOV1 LAI and FAPAR Product. Available online: http://www.geoland2.eu/ (accessed on 15 January 2016).
35. Document for VIIRS Surface Reflectance (SR) Intermediate Product (IP) Software. Available online: http://npp.gsfc.nasa.gov/documents.html (accessed on 28 February 2016).
36. Xiao, Z.; Liang, S.; Wang, T.; Liu, Q. Reconstruction of satellite-retrieved land-surface reflectance based on temporally-continuous vegetation indices. *Remote Sens.* **2015**, *7*, 9844–9864. [CrossRef]
37. Garcia, D. Robust smoothing of gridded data in one and higher dimensions with missing values. *Comput. Stat. Data Anal.* **2010**, *54*, 1167–1178. [CrossRef] [PubMed]
38. Baret, F.; Morisette, J.T.; Fernandes, R.A.; Champeaux, J.L.; Myneni, R.B.; Chen, J.; Plummer, S.; Weiss, M.; Bacour, C.; Garrigues, S.; *et al.* Evaluation of the representativeness of networks of sites for the global validation and intercomparison of land biophysical products: Proposition of the CEOS-BELMANIP. *IEEE Trans. Geosci. Remote Sens.* **2006**, *44*, 1794–1803. [CrossRef]
39. The BELMANIP Sites. Available online: http://postel.obs-mip.fr/?-Documents,53- (accessed on 15 January 2016).
40. Specht, D.F. A general regression neural network. *IEEE Trans. Neural Netw.* **1991**, *2*, 568–576. [CrossRef] [PubMed]

41. Campbell, S.G.; Norman, J.M. *An Introduction to Environmental Biophysics*, 2nd ed.; Springer-Verlag: New York, NY, USA, 1998.
42. He, L.; Chen, J.M.; Pisek, J.; Schaaf, C.B.; Strahler, A.H. Global clumping index map derived from the MODIS BRDF product. *Remote Sens. Environ.* **2012**, *119*, 118–130. [CrossRef]
43. Willmott, C.J. On the validation of modes. *Phys. Geogr.* **1981**, *2*, 184–194.
44. Zhang, X. Reconstruction of a complete global time series of daily vegetation index trajectory from long-term AVHRR data. *Remote Sens. Environ.* **2015**, *156*, 457–472. [CrossRef]
45. National Mapping Division, U.S. Geological Survey, U.S. Department of the Interior. *GCTP General Cartographic Transformation Package: Software Documentation*; National Mapping Program Technical Instructions: Reston, VA, USA, 1993.

Article

Quality Assessment of S-NPP VIIRS Land Surface Temperature Product

Yuling Liu [1,*], Yunyue Yu [2], Peng Yu [1], Frank M. Göttsche [3] and Isabel F. Trigo [4]

1 Earth System Science Interdisciplinary Center at University of Maryland, College Park, MD 20740, USA; peng.yu@noaa.gov
2 Center for Satellite Applications and Research, NOAA/NESDIS, College Park, MD 20740, USA; yunyue.yu@noaa.gov
3 Karlsruhe Institute of Technology (KIT), Eggenstein-Leopoldshafen 76344, Germany; frank.goettsche@kit.edu
4 IPMA, Instituto Português do Mar e da Atmosfera, Lisbon 1749-077, Portugal; isabel.trigo@ipma.pt
* Author to whom correspondence should be addressed; yuling.liu@noaa.gov; Tel.: +1-301-683-2573.

Academic Editors: Changyong Cao, Richard Müller and Prasad S. Thenkabail
Received: 24 July 2015; Accepted: 14 September 2015; Published: 21 September 2015

Abstract: The VIIRS Land Surface Temperature (LST) Environmental Data Record (EDR) has reached validated (V1 stage) maturity in December 2014. This study compares VIIRS v1 LST with the ground in situ observations and with heritage LST product from MODIS Aqua and AATSR. Comparisons against U.S. SURFRAD ground observations indicate a similar accuracy among VIIRS, MODIS and AATSR LST, in which VIIRS LST presents an overall accuracy of −0.41 K and precision of 2.35 K. The result over arid regions in Africa suggests that VIIRS and MODIS underestimate the LST about 1.57 K and 2.97 K, respectively. The cross comparison indicates an overall close LST estimation between VIIRS and MODIS. In addition, a statistical method is used to quantify the VIIRS LST retrieval uncertainty taking into account the uncertainty from the surface type input. Some issues have been found as follows: (1) Cloud contamination, particularly the cloud detection error over a snow/ice surface, shows significant impacts on LST validation; (2) Performance of the VIIRS LST algorithm is strongly dependent on a correct classification of the surface type; (3) The VIIRS LST quality can be degraded when significant brightness temperature difference between the two split window channels is observed; (4) Surface type dependent algorithm exhibits deficiency in correcting the large emissivity variations within a surface type.

Keywords: VIIRS LST EDR; split window algorithm; surface type dependency; LST uncertainty

---

1. Introduction

Land surface temperature (LST) is a critical parameter in the weather and climate system controlling surface heat and water exchange with the atmosphere [1]. It has been used in many applications, including weather forecasting [2,3], irrigation and water resource management particularly agricultural drought forecasting [4,5], and urban heat island monitoring [6]. Remote sensing in the thermal infrared (TIR) provides a unique resource of obtaining LST information at the regional and global scales [7]. Satellite LSTs have been routinely produced from geostationary and polar-orbiting satellites. As of 2014, more than 30 years of global satellite LST data had been accumulated, which is an important component of the Climate Data Records (CDRs).

Many algorithms have been developed for LST retrieval including single and multi-channel algorithms, e.g., [8–15]. Among these, regression algorithms based on the split window (SW) technique have been the most widely used due to their simplicity, effectiveness and robustness. The SW algorithm utilizes the differential atmospheric absorption in two adjacent channels within the thermal infrared atmospheric windows, generally centered at about 11 μm and 12 μm. Many efforts have been made since the late 1980s to extend the SW method, initially developed for sea surface temperature estimates, to retrieve the LST. With modifications to treat the spatio-temporal and spectral variations of the Land Surface Emissivity (LSE), the large difference between the LST and the air temperature, the total column water vapor (WV) in the atmosphere, and the viewing zenith angle (VZA), a variety of SW algorithms for LST retrieval have been developed [7,16]. Many operational LST products have been generated using SW algorithms such as Advanced Very High Resolution Radiometer (AVHRR), Advanced Along-Track Scanning Radiometer (AATSR) [17], Moderate Resolution Imaging Spectroradiometer (MODIS) [11], and Spinning Enhanced Visible and Infrared Imager (SEVIRI) [18,19].

The SW approach has been applied to the Visible Infrared Imaging Radiometer Suite (VIIRS) instrument, a primary sensor onboard the Suomi National Polar-orbiting Partnership (S-NPP) satellite for measuring earth surface parameters. VIIRS was designed to improve upon the capabilities of the operational AVHRR and provide observation continuity with MODIS [20]. LST is one of Environmental Data Records (EDRs) measured by VIIRS, with a moderate spatial resolution of 750 m at nadir during the satellite over-pass times at both day and night. Unlike the SW algorithm for MODIS LST production, coefficients of the SW LST algorithm for VIIRS LST production are surface type dependent, *i.e.*, they depend on the 17 International Geosphere-Biosphere Programme (IGBP) types. Validation of such surface type dependent LST algorithm is particularly needed.

As a quality control (QC) procedure, the Joint Polar Satellite System (JPSS) products including the EDRs are managed through a series of maturity status review processes (*i.e.*, beta, provisional validated V1, validated V2, and validated V3 stages). It is also designed to ensure that scientists, meteorologists and other specialists are prepared and able to utilize data through the JPSS program. The VIIRS LST product completed the validated V1 stage maturity review in December 2014. The V1 stage is a critical milestone in the JPSS EDRs production; it is defined as "using a limited set of samples, the algorithm output is shown to meet the threshold performance attributes identified in the JPSS level 1 requirements".

Some researchers have evaluated the VIIRS LST product during its beta maturity stage [7,21,22]. In this study, we present evaluations of the VIIRS LST data since its provisional stage from April 2014 when a newly calibrated algorithm coefficients set was implemented. The evaluations are based on comparisons of the VIIRS LST data with the ground station observations and with the heritage LST products from MODIS Aqua collection 5 and AATSR. Furthermore, a method is presented to quantify the uncertainty of LST derived from the surface type dependent algorithm. In addition, a proxy method, consisting of the use of the VIIRS LST algorithm to MODIS observations, is adopted for the interpretation of the cross comparison between the VIIRS and MODIS LSTs. The outline of the paper is as follows. Section 2 gives a detailed description of the data sets including the satellite data and ground in situ data as well as the data quality control procedures. Section 3 provides a description of the methodology used in this study. The validation results are presented in Section 4. The limitation of the VIIRS LST algorithm and the uncertainty caused by the surface type input are analyzed and discussed in Section 5. Finally, the concluding remarks are provided in Section 6.

2. Data

Multiple data sets were used in this study: VIIRS LST data, which are to be assessed; ground observations from the SURFace RADiation budget observing network (SURFRAD) and from the validation station "Gobabeb" (Namibia) operated by Karlsruhe Institute of Technology (KIT) are used as reference data for temperature based (T-based) validation; MODIS LST and AATSR LST are reference data used in cross validation. In addition, the radiative transfer simulation database is used for the theoretical analysis of the LST uncertainty caused by the surface type input.

2.1. VIIRS LST EDR

Two algorithms coexist in the VIIRS LST software package: SW and dual split window algorithms (DSW). The VIIRS EDR has been operationally generated using SW algorithm since 11 August 2012. The VIIRS moderate resolution channels M15 and M16 centered at 10.76 μm and 12.01 μm, respectively, are utilized in the LST algorithm.

$$LST_i = a_0(i) + a_1(i)T_{15} + a_2(i)(T_{15} - T_{16}) + a_3(i)(\sec\theta - 1) + a_4(i)(T_{15} - T_{16})^2 \quad (1)$$

Where $a_j(i)$ are algorithm coefficients derived from regression analyses; index i denotes 17 land surface types defined by the International Geosphere-Biosphere Programme (IGBP); T_{15} and T_{16} are corresponding brightness temperatures of M_{15} and M_{16} bands; θ is satellite viewing zenith angle. Physically, brightness temperature T_{15} in the SW algorithm is utilized as primary estimate of the land surface temperature; the brightness temperature difference ($T_{15} - T_{16}$), with its first order and second order, are applied for atmospheric corrections; and the satellite viewing zenith angle is applied further for atmospheric path correction. Algorithm coefficients $\{a_j\}$ are clarified by daytime and nighttime atmospheric conditions.

VIIRS LST data is archived and distributed by NOAA Comprehensive Large Array-data Stewardship System (CLASS) [23]. Although VIIRS cryoradiator opened doors on 19 January 2012, 1 February 2012 is chosen as the start date in this study considering the signal at very early stage might be unstable. For T-based validation, VIIRS LST data is from the CLASS subset data provided by NASA's Land Product Evaluation and Analysis Tool Element (LPEATE); for cross-comparison, the granule data is from CLASS. All VIIRS LST is reprocessed by using the up to date LUT. The local calculation has been verified with operational product and the difference is found within the floating calculation error.

2.2. Reference Data

2.2.1. MODIS LST Product

The time for VIIRS overpass equator is at local time 1:30 am/pm which is same as MODIS Aqua so the MODIS Aqua LST product MYD11_L2 (collection 5), equivalent to VIIRS LST EDR, is selected as a reference to evaluate the performance of the VIIRS LST. The generalized split window algorithm [11] is used to derive LST value from brightness temperature measurements in MODIS band 31 and band 32, centered at 11.02 μm and 12.03 μm, respectively.

$$LST = A_0 + (A_1 + A_2\frac{1-\varepsilon}{\varepsilon} + A_3\frac{\Delta\varepsilon}{\varepsilon^2})\frac{(T_{11} + T_{12})}{2} + (A_4 + A_5\frac{1-\varepsilon}{\varepsilon} + A_6\frac{\Delta\varepsilon}{\varepsilon^2})\frac{(T_{11} - T_{12})}{2} \quad (2)$$

where A_i (I = 0–6) are algorithm coefficients depending on viewing zenith angle, surface air temperature and water vapor content. ε and Δε are the mean and difference of the surface emissivity in band 31 and band 32. The coefficients are derived from regression analysis for a LST value ranging from Tair −16 K to Tair +16 K for C5 LST product [24].

2.2.2. AATSR LST Product

The basic form of the Advanced Along Track Scanning Radiometer (AATSR) LST algorithm expresses the LST as a linear combination of the two brightness temperatures T_{11} and T_{12}. The weak non-linearity by permitting the temperature difference to vary with a power n is introduced into the algorithm [25].

$$LST = a_{f,i,pw} + b_{f,i}(T_{11} - T_{12})^n + (b_{f,i} + c_{f,i})T_{12} \quad (3)$$

where $n = 1/\cos(\theta/m)$, θ is satellite zenith angle and m is a variable parameter controlling the dependence on view angle. $a_{f,i,pw} = d(\sec(\theta) - 1)pw + fa_{v,i} + (1 - f)a_{s,i}$, f is vegetation fraction and pw is total precipitable water $b_{f,i} = fb_{v,i} + (1 - f)b_{s,i}$, $c_{f,i} = fc_{v,i} + (1 - f)c_{s,i}$.

These coefficients are determined by regression over simulation dataset for 14 land cover classes (I = 1–14). The parameters d and m are empirically determined using the radiative transfer simulations for regions where validation data are available [25]. $m = 5$ and $d = 0.4$ are used in [26] for LST derivation. The available AATSR LST data covers the time period from 2002 to 2012. Therefore the data from 1 February 2012 to 8 April 2012 are used in this study.

2.2.3. SURFRAD Ground Observations

The SURFRAD stations provide high quality *in situ* measurements of surface upwelling and downwelling long wave radiations along with other meteorological parameters [27–29]. The pyranometer is usually mounted on a 10 meter high tower in each SURFRAD site, facing downward to measure the surface upwelling radiation. The spatial representativeness is about 70 m × 70 m [21]. Observations from SURFRAD stations have been widely used for evaluating satellite-based estimates of surface radiation, for validating hydrology, weather prediction, climate models and satellite LST products from ASTER, GOES and MODIS [1,15,30–33]. In this study, SURFRAD observations from February 2012 to April 2015 over 7 sites as shown in Table 1 are used for validation of the VIIRS LST retrieval.

Table 1. Geo-Location and surface type of the seven SURFRAD stations.

No.	Site Location	Station Acronyms	Lat(N)/Lon(W)	Surface Type
1	Bondville, IL	BON	40.05/88.37	Crop Land
2	Fort Peck, MT	FPT	48.31/105.10	Grass Land
3	Goodwin Creek, MS	GWN	34.25/89.87	Grassland
4	Table Mountain, CO	TBL	40.13/105.24	Grass/Crop Land
5	Desert Rock, NV	DRA	36.63/116.02	Shrub Land
6	Pennsylvania State University, PA	PSU	40.72/77.93	Mixed Forest
7	Sioux Falls, SD	SFX	43.73/97.49	Cropland

The *in situ* surface skin temperature, *Ts*, is estimated using the following equation

$$T_s = \left((R^{\uparrow} - (1 - \varepsilon)R^{\downarrow})/\sigma\varepsilon\right)^{1/4} \tag{4}$$

where $R^{\uparrow}$ and $R^{\downarrow}$ are upwelling and downwelling long wave fluxes respectively, ε is the surface broadband emissivity, and σ is Stefan-Boltzmann constant *i.e.*, 5.67051×10^{-8} W·m^{-2}·K^{-4}. ε is estimated from a spectral to broadband relationship [34] as shown in Equation (5).

$$\varepsilon = 0.2122 \times \varepsilon_{29} + 0.3859 \times \varepsilon_{31} + 0.4029 \times \varepsilon_{32} \tag{5}$$

where, ε_{31} and ε_{32} are narrowband emissivity of the MODIS bands 29, 31, and 32 centered at 8.52 μm, 11.03 μm and 12.02 μm. Instead of using the fixed emissivity value for each site, the monthly emissivity from the Global Infrared Land Surface Emissivity database [35] is used for broadband emissivity calculation to better characterize the emissivity change over sites.

The accuracy of the LST estimated by Equation (4) depends on the accuracy of the upwelling and downwelling radiation and the broadband emissivity. The accuracy of the pyrgeometer is claimed to be about 9 W·m^{-2} [27]. The broadband emissivity matches well the ground measurements, with a standard deviation of 0.0085 and a bias of 0.0015 [33].

2.2.4. Ground Observation at Gobabeb, Namibia

KIt isIt is LST validation station at Gobabeb (latitude 23.55°S, longitude 15.05°E, 406 m a.s.l.) is located in the hyper-arid climate of the Namibia Desert on large (several thousand km^2) and highly homogeneous gravel plains (Figure 1). The gravel plain consists mainly of gravel and sand (about 75%) and spatially well distributed desiccated grass (about 25%), but there are also some smaller wadis and rock outcrops. This site is well characterized and is dedicated to LST validation [18,36,37].

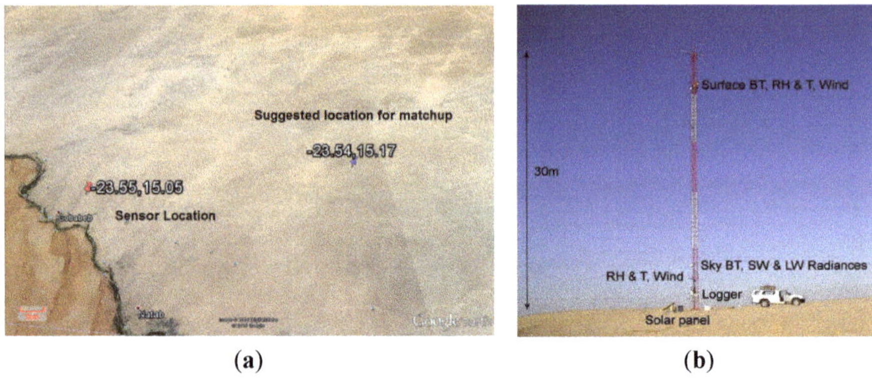

(a) **(b)**

Figure 1. (**a**) is Geographic landscape in Gobabeb station in Namibia and (**b**) is the instrumentation for LST measurement: two radiometers measure the surface-leaving radiance (9.6–11.5 µm) from the gravel plain, which is highly homogenous over at least 2500 km^2. A third radiometer measures sky radiance.

The station instruments are mounted at several heights of 30 m high wind profiling tower. The main instrument for the *in situ* determination of LST at KIt isIt is validation stations is the precision radiometer "KT15.85 IIP" produced by Heitronics GmbH, Wiesbaden, Germany. KT15.85 IIP radiometers measure thermal infra-red radiance between 9.6 µm and 11.5 µm, have a temperature resolution of 0.03 K and an accuracy of ±0.3 K over the relevant temperature range [36]. The KT15.85 IIP has a drift of less than 0.01% per month: the high stability is achieved by linking the radiance measurements via beam-chopping (a differential method) to internal reference temperature measurements and was confirmed by a long-term parallel run with the self-calibrating radiometer "RotRad" from CSIRO, which is continuously stabilized with 2 blackbodies [38]. Due to the KT-15 IIP's narrow spectral response function and the small distance between the radiometers and the surface, atmospheric attenuation of the surface-leaving TIR radiation is negligible. However, the measurements of the surface-observing KT-15 IIPs contain radiance emitted by the surface (*i.e.*, the target signal) as well as reflected downward IR radiance from the atmosphere, which needs to be corrected for [36]. Therefore, at each station an additional KT-15.85 IIP measures downward longwave IR radiance from the atmosphere at 53° VZA: measurements under that specific zenith angle are directly related to downward hemispherical radiance [39] so that no ancillary data for deriving ground truth LST are needed.

Brightness temperatures from the surface pointing radiometers are converted to radiances; these are corrected for reflected downwelling radiance using Satellite Application Facility on Land Surface Analysis [36] emissivity and measured downwelling radiance. LST is obtained from these corrected surface leaving radiances. Ground data, corresponding to observations taken every minute throughout the whole year of 2012, are used for VIIRS LST validation.

2.3. Quality Control Procedures

Quality control of the both ground data and satellite data is critical to reliable results. The following procedures are performed in this study.

(1) Ground data quality control

Two procedures are used for ground data quality control: the first takes into account the quality flag (QF) included in the data, e.g., in SURFRAD data set, a QF of zero indicates that the corresponding data point is good, having passed all QC checks; the other is the temporal variation test by checking the Standard Deviation (STD) of ground observations in 30 min temporal interval centered at the observation time. In our practice, it is observed that LST can be varied significantly in 30 min time scale. We therefore have tested using different threshold values to filter out noise data and found that 1.5 STD is a reasonable threshold to remove noisy data and to still keeping reasonable time-varying LST data.

(2) Satellite data quality control

The main purpose for satellite data quality control is to reduce the impact from the cloud contamination and suboptimal atmospheric conditions. Therefore, the QF for cloud condition in LST product is used to constrain the cloud to be confidently clear. In addition, the spatial variation test, *i.e.*, the STD of the 3 by 3 pixel brightness temperature of the channel at 11 μm centered at the matchup pixel, is applied as an additional cloud filter [15]. The STD should be small over thermal homogeneous surfaces unless there is cirrus or cloud cover. The spatial variation test is widely used in LST validation studies. For example, Li *et al.* [30] used the neighboring 5 × 5 box for MODIS LST validation using 10 year SURFRAD data. In this study, the threshold is set as 1.5 K, although it may be slightly higher (e.g., 1.75 K) for sites like Boulder [33]. We intend to include all angle measurements in the validation so the LST data quality flag is not applied as it includes the viewing zenith angle restriction.

(3) Match up process

For T-based validation, the spatially closest pixel to the site is used for the matchup and temporally the ground site observations match with the granule start time. Because the subset VIIRS LST data provided by LPEATE is aggregated to ~5 min, which is same to MODIS v5 LST data, the temporal difference between satellite observations and ground measurements is up to 5 min. Different from MODIS and VIIRS, the overpass time of AATSR LST is calculated at pixel level so that the AATSR LST data is concurrent to the ground measurements. To evaluate the relative agreement between VIIRS and MODIS LST products, Simultaneous Nadir Overpasses (SNOs) tool [40] is used to search for the "just-miss" scenes for both satellite. The SNO tool provides the service for three regions: polar area (north and south), low latitude area and continental US area. The SNO criteria are set as 10 min temporal difference and 250 km nadir distance. More than 100 SNOs are chosen in the cross comparison of VIIRS and MODIS Aqua granules acquired from 2012 to early 2014.

3. LST Assessment Methodology

There are many challenges in evaluation of the satellite LST products. Among these, one should bear in mind that: *in situ* LST measurements are usually performed within a very small area that may not represent well the measurement from satellite sensor, that is of the much larger pixel footprint; temporal variability of LST, particularly during the daytime, requires a stable *in situ* measurement and tolerance of very short time difference to the satellite observation; high quality *in situ* measurements are extremely limited, reducing the statistical significance of the results, and limiting the seasonal and global representativeness.

Three approaches are widely used for LST products validation: T-based method, radiance based (R-based) method and the cross-satellite comparison method. Obtaining reference (or "truth") LST value is the key. In the T-based and the cross-satellite approach, the reference LST is the ground measurement and the cross-satellite measurement, respectively. While in the R-base method, numerical radiative transfer model is applied to calculate the reference LST using the satellite sensed top of atmosphere brightness temperatures. This method, however, requires accurate atmospheric profile and surface emissivity information which are hard to obtain. In this study, we used the T-based method and cross-satellite comparison methods to assess the VIIRS LST quality.

3.1. T-Based Validation Method

The T-based method is a direct comparison analysis of ground measurements of LST and the corresponding satellite estimates. It is based on the assumption that the ground LST measurements would represent fairly well the satellite LSTs. Obviously, such assumption may be problematic in some ground sites where thermal homogeneity is a serious issue. Some field campaigns have been conducted to validate LST products over surfaces such as lake, grassland and rice fields [24,26,41–45]. However, the *in situ* data collected from field campaigns are very limited and costly. Researchers studied various methodologies to characterize/calibrate traditional ground station data for the T-based LST validation [46–48]. In this study, we use ground observations of surface leaving longwave radiation to estimate *in situ* LST such as from the SURFRAD [1,31,33,49]. The T-based method is limited by the spatial variability of LSTs, especially during the daytime [50].

3.2. Cross Satellite Comparison Method

The cross comparison with heritage satellite LST product is widely used for satellite LST evaluation [29,51,52]. However, this is not an absolute validation unless one of the satellite products has been independently validated [21]. As the VIIRS is expected to replace MODIS in the future, the cross comparison to the MODIS LST will provide the evaluation of the VIIRS LST retrieval performance with respect to characterization of the differences, *i.e.*, spatial pattern, systematic error budget, which may reflect the algorithm difference, limitations and error sources. Since the launch of MODIS Terra in December 1999, the MODIS LST data has been widely used and evaluated by many individual and institutional investigators [40,43,53].

LST product from AATSR is also used in this study. For the nadir view of AATSR, the instantaneous field of view (IFOV) is 1 km $\times$ 1 km at the center of swath, which is also the case for MODIS observations close to nadir. One of the special features of the AATSR instrument is its use of a conical scan to give a dual-view of the Earth's surface. AATSR LST product has also been available over a decade and many studies have been conducted to assess and validate this product [9,26,54]. The validation result indicates that AATSR LST data has a fairly good accuracy, e.g., an average error of -0.9 K with a STD of 0.9 K was reported by Coll *et al.* [26] using ground measurements from experimental site in an area of rice crops.

3.3. VIIRS LST Uncertainty to Input Imprecision

The need for a correct characterization of the uncertainty associated with satellite retrievals is becoming increasingly recognized. The VIIRS LST algorithm (see Section 2.1 for further details) uses as explicit inputs the TOA brightness temperatures (split-window channels M15 and M16) and the pixel view zenith angle. Other variables, such as the pixel land cover classification and solar zenith angle are implicit inputs, since they are used to select the correct set of coefficients to be used in the retrieval (see Equation (1)). The surface type misclassification implies the use of inaccurate emissivity for the split window bands, which are built up according to emissivity and land cover type data to represent different IGBP surfaces in the regression process therefore affecting the algorithm coefficients retrieval. The overall typing accuracy for the 17 land cover types is expected to be 70 percent at moderate spatial resolution [55], which means that about 30% of pixels might be misclassified as other surface types. Several studies have pointed surface emissivity as one of the most relevant sources of LST, and here we will consider in particular the expected uncertainty in VIIRS LST, which can be attributed to land cover/emissivity misclassification, as summarized by Equations (6) and (7).

$$S_i{}^2 = \sum_{j=1}^{17} (P_{ij} \times \sigma^2(\varepsilon_{ij})) \quad (6)$$

$$S_{sf}{}^2 = \sum_{i=1}^{17} \frac{S_i{}^2 \times N_i}{N_{total}} \quad (7)$$

where p_{ij} is the probability of (mis-)classification of surface type i (I = 1, 2 . . . 17) to be j (j = 1, 2 . . . 17).

Table 2. Surface type distribution over land.

No.	IGBP Land Surface Type	Percentage
1	Evergreen Needle Leaf Forests	1.91
2	Evergreen Broadleaf Forests	9.25
3	Deciduous Needle leaf Forests	1.17
4	Deciduous Broadleaf Forests	0.78
5	Mixed Forests	5.85
6	Closed Shrub Lands	0.06
7	Open Shrub Lands	15.18
8	Woody Savannahs	8.16
9	Savannahs	8.13
10	Grasslands	8.45
11	Permanent Wetlands	0.87
12	Croplands	7.25
13	Urban build-up	0.39
14	Croplands/Natural Vegetation Mosaics	4.11
15	Snow ice	10.46
16	Barren	12.54
17	Water Bodies	5.45

ε_{ij} is the LST difference between LST calculated with the equation for surface type i and with the equation for surface type j for each pixel with i surface type; $\sigma^2(\varepsilon_{ij})$ is the respective error variance. $S_i{}^2$ represents the error variance associated to the pixel land cover classification, for each IGBP type i under either day or night condition. S_{sf}^2 represents the error variance for all IGBP types and all day/night conditions. N_i represents the number of samples for surface type IGBP i. N_{total} represents the total number of samples for all cases. P_{ij} is obtained from the class composition of commission errors for VIIRS surface type quality assessment (Damien Sulla-Menashe, VIIRS ST V1 Quality Assessment 2 April 2014). The type of water bodies is not included in above commission error; therefore we exclude water bodies in the uncertainty analysis. N_i/N_{total} actually represents the proportion of each surface cover over land. The global surface type distribution over land (Table 2) based on statistical analysis of a whole year of surface type input is used in the theoretical analysis.

4. Results

4.1. Comparison with SURFRAD Data

Figure 2 shows overall comparisons of the VIIRS LSTs (a) and the MODIS LSTs (b) against the SURFRAD LSTs. The number of VIIRS matchups is twice that of MODIS matchups: On the one hand, this is due to a better coverage of VIIRS compared to MODIS; on the other hand, it is a result of different cloud flag definition. In the MODIS LST product, the cloud free pixels affected by nearby clouds are excluded, which is not the case for the VIIRS LST product. It presents that accuracy and precision of the VIIRS LSTs are −0.41 K and 2.35 K, respectively, which is better than that of the MODIS LSTs (*i.e.*, −1.36 K and 2.50 K, respectively). Note that a better accuracy/precision of the VIIRS LSTs are at nighttime (−0.24 K/1.97 K) compared to that at daytime (−0.71 K/2.86 K). The better nighttime performance is expected because the thermal heterogeneity is usually higher during daytime and the atmospheric water vapor is less and the land surface behaves almost homogeneously at night [7]. This result is consistent with the results of other studies [21,33,47,56].

Note also that overall the VIIRS LSTs are 0.95 K warmer than the MODIS LSTs.

A similar comparison of the VIIRS LSTs and the AATSR LSTs against the SURFRAD LSTs is shown in Figure 3. In which, bias and of the VIIRS LSTs are −0.78 K and 2.34 K, respectively, which is comparable to that of the AATSR LSTs (*i.e.*, −0.20 K and 2.42 K, respectively). Note that VIIRS LST is on average 0.5 K colder than AATSR LST.

Note that circled matchups of the VIIRS LSTs in Figures 2 and 3 are significantly lower than the ground measurements. These are suspicious cloud contaminated data since temperature of the cloud top is mostly lower than the land surface during the time. Although the VIIRS QF from the cloud mask product and additional cloud filtering have been utilized, it appears insufficient for the validation purpose. It is found that all matchups in the circle are with snow/ice cover, which suggests a degradation of the cloud detection over bright surfaces. Four types of misclassification have been found for snow/ice identification with the cloud mask including the multi-layered cloud misclassified as snow [57]. Cloud leakage has been reported by EDR groups and snow/ice/cloud differentiation has been listed as the major issue for further improvement [58,59]. Besides, the snow/ice EDR only provides temporal snow for daytime, which leads to the incorrect surface type used in the VIIRS LST retrieval at night. Therefore, the VIIRS nighttime LST is likely degraded by misuse of the surface cover information. In order to solve this problem, the nighttime snow/ice detection was introduced to the operational product on 22 May 2014. However, it cannot help the analysis of past data prior to that date.

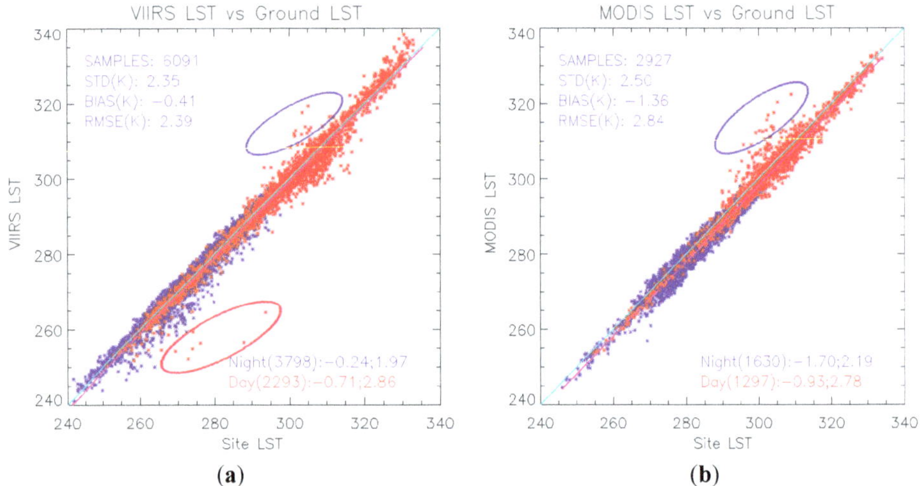

Figure 2. Scatter plots of the VIIRS LSTs (**a**) and MODIS LSTs (**b**) against the SURFRAD LSTs compared in the period from February 2012 to April 2015. Overall accuracy and precision of the satellite LSTs referring the SURFRAD LSTs are noted, as well as the daytime and nighttime cases. Some VIIRS LST plots are circled as suspicious cloud contaminated plots (red).

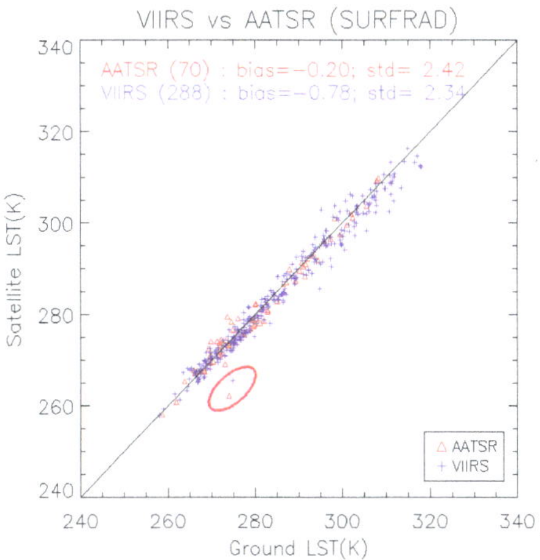

Figure 3. Scatter plots of the VIIRS LSTs (blue) and the AATSR LSTs (red) against the SURFRAD LSTs compared in the period from 1 February 2012 to 8 April 2012. Overall accuracy and precision of the satellite LSTs referring the SURFRAD LSTs are noted. Some VIIRS LST plots are circled as suspicious cloud contaminated plots.

Difference between the VIIRS LSTs and SURFRAD LSTs also demonstrates a strong seasonal variation. As shown in Table 3, the best agreement occurs in fall with a bias of −0.23 K and STD as of 1.82 K and the worst agreement in spring with a bias of −0.57 K and STD as of 2.56 K; the seasonal pattern is more significant at daytime than at nighttime. The seasonal variation is also reported in MODIS LST validation [30]. Two most relevant error sources in LST retrieval are atmospheric water vapor absorption and the surface emissivity uncertainty. Significant decrease of the atmospheric transmittance at 11 and 12 µm with increase of the water vapor introduces significant error in the split window algorithm when the surface temperature is high [60], which is the fundamental reason of the worsen performance in spring and summer compared to that in fall and winter. Besides, the large discrepancies in spring are also attributed to the cloud contamination over snow/ice surface (those matchups in the red circle of Figure 2 happened in spring) and considerably warm LST retrievals about 6–10 K greater than ground observations over Bondville station in late spring to early summer. As shown in blue circle of Figure 2, the feature is found in both VIIRS LST and MODIS LST validation results against SURFRAD observations. In their MODIS LST validation study, Li *et al.* [30] compared 10 years 16-day average NDVI and daily emissivity datasets from the MODIS observations and found that this feature might be caused by anomalous NDVI-emissivity relationship, *i.e.*, emissivity does not change accordingly with the NDVI change during the time period. Guillevic *et al.* [21] mentioned that validation results obtained for stations surrounded by croplands present strong seasonal dependency: station observations may be closer/deviate more from the temperature of surrounding fields, according to crop maturity. As such, VIIRS LST tends to be much lower than that of Bondville station when plants in surrounding fields (corn and soybeans) are well developed in summer and significantly higher after the harvest.

Table 3. Seasonal variation from the validation using SURFRAD measurements.

Season	Samples	Overall		Day		Night	
		Bias	STD	Bias	STD	Bias	STD
Spring	1549	−0.57	2.55	−0.58	3.16	−0.56	2.13
Summer	1433	−0.12	2.46	−0.90	3.70	0.26	1.40
Fall	1734	−0.23	1.82	−0.46	1.97	−0.07	1.70
Winter	1372	−0.72	2.21	−0.85	1.80	−0.63	2.44

In our dataset, however, we mainly observed higher LST at early growth stage from May to June but not after the harvest; and this feature is not obvious from other sites with cropland cover such as over the SXF site. Therefore, other impacts should be investigated. Some matchups with higher LST estimation are listed in Table 4. It mainly happens at local time 1–3 p.m. When crops are short, they will have little shade on ground. From the angles shown in columns of STZ (Satellite Zenith Angle), STAZ (Satellite Azimuth Angle), SOZ (Solar Zenith Angle) and SOAZ (Solar Azimuth Angle), the sun illuminates crops with a certain angle from 17 deg to 36 deg centering at about 20 deg, and the satellite views the soil surface and vegetation stem rather than the canopy.

Radiometer of the station, however, is always looking down at nadir measuring the surface upwelling radiance from the vegetation canopy. Such observation difference might be another reason leading to lower ground observations than satellite LST estimation.

Table 4. Details about the match ups over Bondville site with higher satellite LST retrievals than ground measurements.

Viirs_lst	Surfrad_lst	BT15	BT16	Date	Time	STZ	STAZ	SOZ	SOAZ
312.19	304.77	306.05	302.51	2013140	1905	33.39	−98.15	26.34	−133.20
313.65	305.98	308.04	304.88	2013163	1835	14.57	78.10	19.43	−146.77
311.39	301.43	306.27	303.76	2013164	1815	39.15	74.33	17.73	−159.42
313.91	304.41	307.21	303.14	2013169	1825	32.07	75.08	17.95	−155.56
310.84	303.62	304.40	300.88	2013170	1805	50.70	72.09	16.85	−169.68
313.14	305.44	308.69	306.60	2014128	1855	6.24	−100.44	27.15	−142.34
312.03	301.27	308.06	306.73	2014130	1815	44.90	73.11	23.26	−161.37
308.33	300.28	302.04	299.51	2014130	2000	67.60	−90.80	35.98	−118.65
309.05	302.28	305.02	303.27	2014144	1855	6.25	−97.24	24.05	−137.94
311.75	303.35	307.29	305.41	2014151	1825	38.72	74.25	19.61	−154.49
309.26	303.50	304.43	302.02	2014154	1910	25.26	−99.96	24.39	−130.55
306.46	299.39	303.54	302.82	2014165	1900	15.91	−99.07	22.50	−132.85

To characterize the spatial representativeness of the ground site LST, ASTER LST product is used by aggregating 90 m ASTER pixels to form 1 km pixels centered at each station [7,21]. In this study, google earth image is used for visual check of the surface heterogeneity. The SURFRAD sites DRA and FPK appear more homogeneous than other sites. The quality of validation results over relatively heterogeneous sites depend on the satellite footprint, geolocation accuracy, surface type accuracy as well as the emissivity settings of ground LST calculation. The discrepancies between VIIRS LST and ground measurements are analyzed over each site and associated surface types (Table 5). All sites except DRA and GWN present seasonal snow cover. The validation results are strongly impacted by cloud contamination in FPK and SXF sites for snow cover. The analysis result suggests land cover discrepancies between sites and satellite footprints. For example, FPK site and surroundings are located within grassland areas; however 90 out of 637 matchup pixels are classified as crop/vegetation mosaic, which results in a relatively large error. Similarly PSU site is with cropland cover on site; however, 17 matchups pixels are classified as deciduous broadleaf forests, which also causes a significant error. For the DRA site, although there were 149 matchups misclassified as barren surface, a good agreement between the satellite observations and ground measurements is obtained. This is possibly because that the emissivity setting for barren surface (emissivity pair of 0.965 and 0.97 for VIIRS band 15 and 16, respectively) is close to the bushy surface of the site. Furthermore, there would also be considerable bush shading at the site so that no obvious underestimation observed. It is also noted that the remaining 97 matchups are classified as closed shrubland at DRA site. According to the IGBP surface type definitions, shrub canopy cover is greater than 60% and 10%–60% for closed shrubland type and open shrubland type, respectively. Therefore surface type over the DRA site might change depending on the green and dry season.

Table 5. Discrepancies between VIIRS LST and Ground LST over site and associated surface types.

Site	Surface Types	Samples Number	Overall			Nighttime			Daytime		
			Bias	Std	Rmse	Bias	Std	Rmse	Bias	Std	Rmse
BON	Cropland	768	−0.42	2.92	2.95	−0.48	2.05	2.10	−0.27	4.33	4.33
BON	Snow/ice	39	0.12	1.34	1.33	0.12	0.83	0.80	0.12	1.50	1.48
DRA	Closed Shrublands	97	−0.96	1.42	1.71	−1.32	0.84	1.56	−0.45	1.88	1.91
DRA	open shrublands	1128	−0.18	1.57	1.58	−0.58	0.88	1.05	0.26	2.00	2.01
DRA	Barren	149	−0.23	1.55	1.56	−1.04	0.75	1.28	0.87	1.67	1.88
FPK	Grass	491	−0.19	1.84	1.85	0.07	1.63	1.63	−0.70	2.12	2.23
FPK	Crop/vegetation Mosaic	90	−1.13	2.61	2.83	−1.70	2.86	3.31	−0.08	1.69	1.67
FPK	Snow/ice	56	−3.16	5.57	6.36	-	-	-	−3.16	5.57	6.36
GWN	Woody Savannahs	390	0.06	2.69	2.69	1.39	1.75	2.23	−2.10	2.56	3.30
GWN	Crop/vegetation Mosaic	487	−0.18	2.52	2.52	1.28	1.61	2.06	−2.20	2.11	3.05
PSU	Deciduous broadleaf forests	21	−0.85	2.52	2.60	−0.48	2.55	2.51	−1.77	2.39	2.80
PSU	Grass	157	−0.28	1.85	1.86	−0.21	1.93	1.93	−0.37	1.75	1.77
PSU	Cropland	35	−1.16	2.20	2.46	−1.21	2.38	2.63	−0.91	1.04	1.31
PSU	Crop/vegetation Mosaic	406	−0.15	2.51	2.51	−0.19	2.56	2.56	0.00	2.34	2.32
PSU	Snow/ice	105	−1.30	3.10	3.35	−2.29	3.67	4.29	−0.72	2.56	2.64
SXF	Cropland	762	−0.44	2.33	2.37	−0.13	2.07	2.07	−1.08	2.69	2.90
SXF	Snow/ice	119	−1.91	3.64	4.10	−1.94	1.94	2.72	−1.90	4.10	4.50
TBL	Grass	749	−0.68	1.81	1.94	−0.70	1.59	1.74	−0.63	2.35	2.43
TBL	Snow/ice	41	−1.36	1.80	2.24	−2.43	0.80	2.54	−1.06	1.90	2.14

4.2. Comparison with Data from Gobabeb, Namibia

The same QC control procedure as for the SURFRAD sites is implemented and the validation results are shown in Figure 4. Gobabeb *in situ* LST are also used to validate MODIS Aqua LST (collection 5), which are used as a reference.

The results show that the VIIRS and MODIS algorithms underestimate *in situ* LST with a bias of 1.57 K and 2.97 K, respectively, whereas they achieve similar precisions of 2.06 K for VIIRS and 1.92 K for MODIS. As shown in Figure 1, the location used for the comparison is about 13 km east of the validation station, where the gravel plain is highly homogeneous over large areas [37]. Using additional *in situ* measurements across the gravel plains, Göttsche *et al.* [36] demonstrated that the surface conditions at Gobabeb station are highly representative of the gravel plains and that there is an excellent match between the operational SEVIRI LST retrieved by EUMETSAT's Land Surface Analysis-Satellite Application Facility (LSA-SAF) and Gobabeb station LST [18], with typical monthly biases of less than 1.0 K and rms errors of about 1.0 K to 1.5 K.

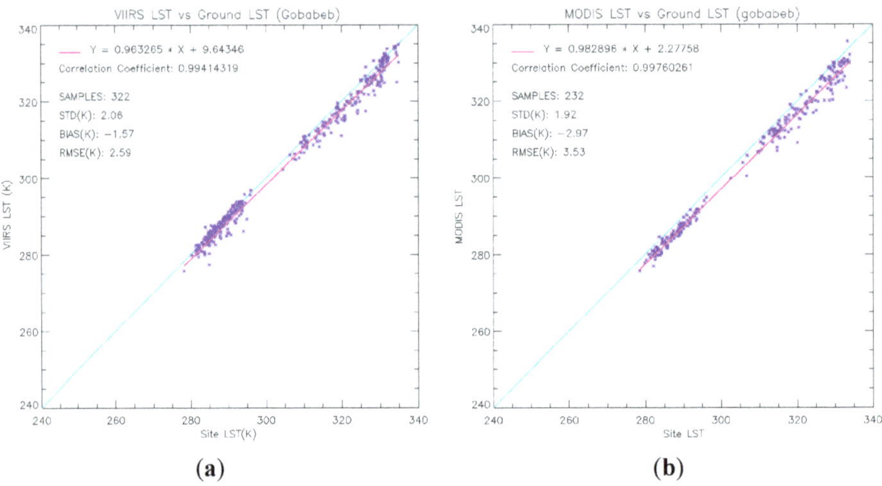

Figure 4. Validation result against the data in Gobabeb, Namibia in 2012: VIIRS LST (**a**) and MODIS LST V5 (**b**).

Wan [54] clearly described the underestimation of MODIS v5 LST over bare soil sites. Three possible sources are considered for the large LST error: (1) The original split window algorithm does not well cover the wide range of LSTs; (2) the large errors in surface emissivity values in bands 31 and 32 estimated from land-cover types; (3) effect of dust aerosols that has not been considered in the R-based validation. Emissivity adjustment model for bare soil pixel as well as a new set of split window algorithm coefficients is incorporated into the day/night algorithm, resulting C6 level 2 LST products. Above reasons (1) and (2) are also applicable for the large error in VIIRS LST. Reason 3 is not investigated in this study.

Besides, the misclassification of surface type over Gobabeb site is observed, with 4 matchup pixels being classified as evergreen broadleaf forest. These lead to large LST errors and were therefore removed from the validation results. Again this demonstrates the impact of surface type misclassification on LST estimates. It is necessary to understand the LST uncertainty associated with the surface type input. Section 5 presents this analysis in detail.

4.3. Cross Comparison with MODIS Aqua LST

The cross comparison of the VIIRS and MODIS LST product is conducted at granule level using the SNO service. As described in the quality control section, the temporal difference is restricted to 10 min. Over 100 scenes are chosen covering each month of one year in continental US, low latitude and polar area representing low, middle to high latitudes climate. The overall comparison results as shown in Figure 5, indicate that MODIS LST and VIIRS LST produce a consistent measurement with a bias of 0.77 K and STD of 1.97 K (VIIRS minus MODIS).

Considering that the cloud residue and surface cover difference within two satellite footprints have strong impact on the cross comparison, the spatial variation test is applied to both MODIS and VIIRS LSTs, which results in the exclusion of two third of match-up pixels as shown in Figure 5b. The viewing angle difference screening is further applied based on Figure 5b. Therefore Figure 5c, representing the cross comparison results with cloud and VZA screening, shows a bias of 0.7 K and STD of 1.13 K. In order to check if the discrepancy is due to the LST retrieval algorithm, a proxy like method is used for VIIRS LST calculation, *i.e.*, using MODIS sensor data records (BT and geometry information) as input for VIIRS LST retrieval. This way the impact of difference in the sensor data records can be excluded. The result shows a bias of 0. 5K and STD of 0.7 K (Figure 5d).

We also apply the above proxy like procedure to generate multiple daily global data, which leads to similar results. For example, a bias of 0.13 K and STD of 0.72 K is obtained for the global proxy comparison on 22 April 2014; a bias of 0.5 K and STD of 0.55 K is obtained on 19 December 2014. This exercise demonstrates that the algorithm difference is rather small in terms of uncertainty; more significant uncertainties are due to the sensor characterization, thermal heterogeneity of the land surface, temporal difference, and the angular anisotropy of land surface emissivity and temperature. A positive bias of the order of 0.5 K is found between VIIRS and MODIS LST. The comparison of VIIRS and MODIS LST over surface types is summarized in Table 6. It is noted that the comparison result does not cover all surface types, e.g., snow/ice in nighttime and closed shrubland in daytime, and the number of match-ups in each surface type varies significantly from 2 to 552,550. The surface types with less than 500 samples are excluded from the following discussion due to lack of statistical representativeness. In the daytime, the LST difference ranges from 0.2 K to 3.3 K for absolute bias and from 0.57 K to 1.49 K for STD; in the nighttime, the LST difference ranges from 0.06 K to 2.11 K for absolute bias and from 0.93 K to 1.53 K for STD. Large discrepancies are found over open shrubland, savannahs and barren soil, for which the difference is up to 3.3 K. The comparison results are restrained by availability of SNOs and accuracy of the VIIRS LST surface type information.

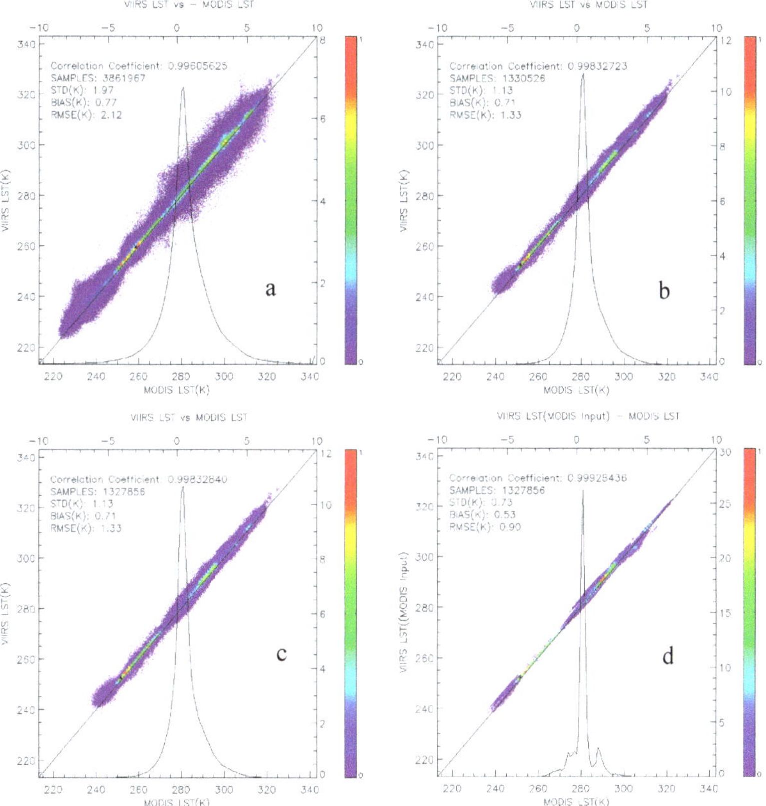

Figure 5. Cross-comparison results between VIIRS and AQUA for the whole period and area under analysis. (**a**) all comparison results under cloud clear condition ; (**b**) based on a, spatial variation tests are added ; (**c**) based on b, angle difference is added ; (**d**) based on c, VIIRS LST is calculated using MODIS data as input and then compare to MODIS LST.

However, we also observed some very large discrepancies in the cross satellite comparison of VIIRS and MODIS LST products. For example, difference over 10 K was observed on 28 December 2013 over Australia, where VIIRS overpasses at UTC 04:17 and MODIS Aqua overpasses at UTC 04:45. Therefore, the temporal separation is about 30 min. The corresponding granules are used as a case study to investigate the problem.

Table 6. Cross comparison of VIIRS LST and MODIS LST over surface type. This result is corresponding to Figure 5c. *i.e.*, the data has been filtered to include only LST with the angle difference within 10 degrees and possible cloud contamination excluded. The overpass includes areas in low latitude, high latitude and US.

Surface Type	Night			Day		
	Bias	STD	Samples	Bias	STD	Samples
Evergreen Needleleaf Forest	−0.36	1.15	31	0.24	1.01	12
Evergreen broadleaf Forest	−0.06	0.93	11110	0.20	0.92	40085
Deciduous Needleleaf Forest	1.70	2.09	104	−1.41	0.28	2
Deciduous Broadleaf Forest	0.50	0.93	1947	−0.46	0.97	1871
Mixed Forest	−0.10	1.28	5666	−0.72	1.19	218
Closed Shrublands	1.60	0.97	858	-	-	-
Open Shrublands	2.11	1.34	166680	−0.37	1.43	1097
Woody Savannahs	0.15	1.13	124278	0.46	1.36	7728
Savannahs	0.76	1.03	145338	3.34	1.01	505
Grasslands	0.46	1.24	51831	−0.33	1.35	259
Wetlands	0.61	1.34	4371	1.72	1.13	340
Croplands	0.21	1.23	26030	−0.32	1.06	11583
Urban	0.60	1.22	769	0.38	1.40	52
Natural Vegetation Mosaics	0.44	1.04	53593	1.14	1.49	2551
Snow/ice	-	-	-	0.47	0.57	552550
Barren	2.04	1.24	1222	1.08	1.12	111549
Water	1.40	1.53	3073	−0.19	1.00	553

The overall granule comparison of VIIRS and MODIS LST (Figure 6a) shows that VIIRS LST is statistically 2 K warmer than MODIS LST and the maximum difference is over 10 K. To examine whether these highest discrepancies are caused by the sensor input, we examine the brightness temperature at 11 μm and the BT difference of the two split windows; the results are displayed in Figure 6b,c, respectively.

It is found that the VIIRS BT_{15} is on average 1 K colder than MODIS BT_{31} (Figure 6b) but the VIIRS BT difference between two split windows is statistically 1 K higher than that in MODIS (Figure 6c), which is considered as the main cause for the large LST discrepancy as the impact of BT difference on VIIRS LST is not linear but quadratic growth. For verification, we calculate VIIRS LST using the same proxy method in generating Figure 5d and then compare with MODIS LST. The LST discrepancy becomes much smaller as shown in (Figure 6d) with a bias of 0.01 K. Therefore, it once again suggests that the algorithm difference is not the main cause for the large discrepancies. However, VIIRS LST might overcorrect the atmospheric absorption under very high BT difference condition, which results in the LST degradation for the particular case.

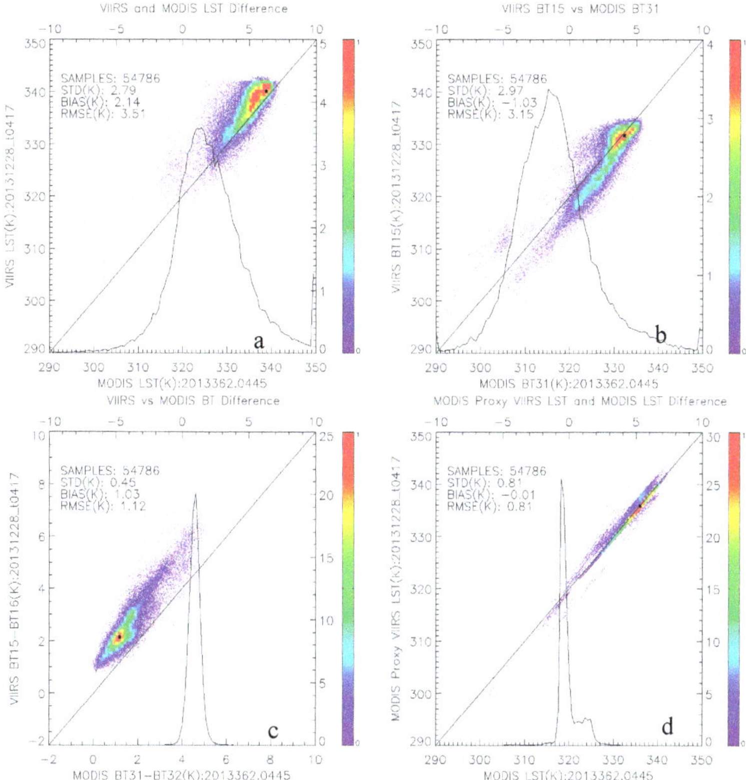

Figure 6. Cross-comparison results between VIIRS and AQUA of the case study on 28 December 2013. (**a**) Overall comparison results under cloud clear condition; (**b**) Brightness temperature comparison of VIIRS band 15 and MODIS Aqua band; (**c**) the BT difference comparison between VIIRS (BT15-BT16) and MODIS (BT31-BT32); (**d**) 31 based on a, VIIRS LST is calculated using MODIS data as input and then compare to MODIS LST

5. Discussion

5.1. Impact from the Non-Linear Term

Note that split window algorithm applied for the VIIRS LST retrieval is a heritage of the sea surface temperature (SST) retrieval; it is a linearization approach of the radiative transfer equation. Over water, surface emissivity difference between the two split window channels (centered at about 11 and 12 μm wavelengths, respectively) is ignorable so that brightness temperature (BT) difference between the split window channels represents well the atmospheric absorption. Over land, however, the BT difference includes the emissivity difference as well as the atmospheric absorption. In particularly, the VIIRS LST algorithm applies a linear and a quadratic term (Equation (1)), $(T_{11} - T_{12})$ and $(T_{11} - T_{12})^2$, for correcting atmospheric absorption. Emissivity difference between the two channels may introduce significant error. The non-linear SW algorithm is proposed primarily to improve the accuracy of LST retrieval [61]. Many similar forms of nonlinear SW algorithms have been developed in the literature.

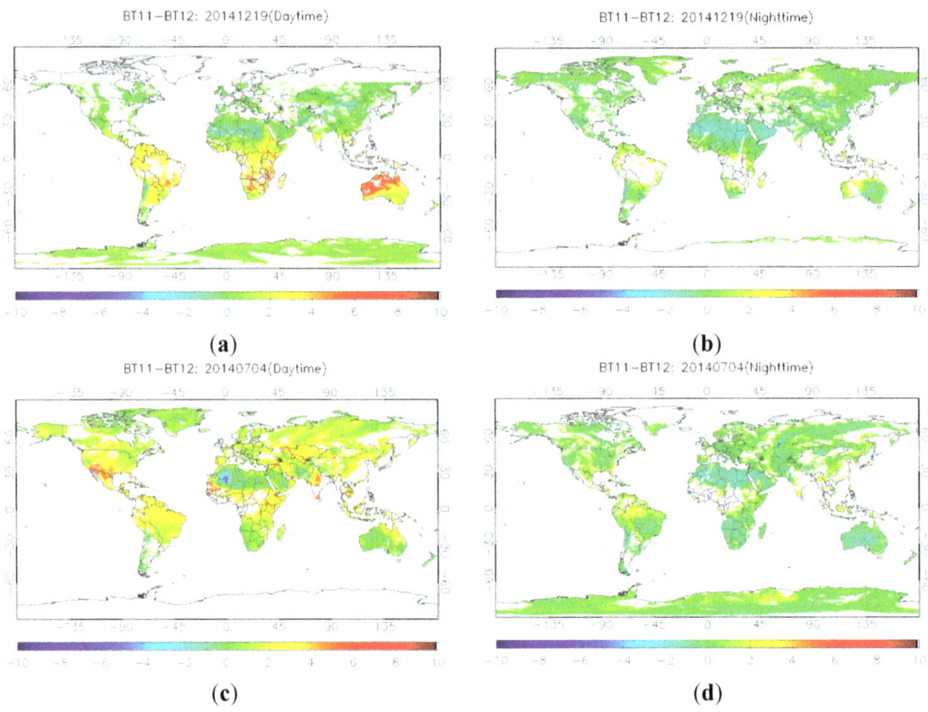

Figure 7. Global BT difference distribution map for 19 December 2014 at daytime (**a**) and nighttime (**b**); 4 July 2014 at daytime (**c**) and nighttime (**d**).

In our simulation database, the BT difference is within −1 K to 3 K for most cases but the real orbit data may go far beyond this range. Figure 7 shows the daily global BT distribution on 19 December 2014 and 4 July 2014, representing the hemisphere summer and winter. The global BT difference distribution presents an obvious regional feature, high BT difference centers in the low latitude area and low BT difference centers in middle and high latitude area with a seasonal variation. The BT difference distribution also varies under day and night condition, more homogeneous at nighttime compared to those at daytime. It is mostly less than 4 K at nighttime even in low latitude areas but close or even higher than 10 K at daytime particularly over Australia and South Africa. Refining the algorithm by extending the representativeness of the simulation database used for algorithm regression does not provide a promising improvement; it may improve over some areas but likely degrade over some other areas due to large variation at global scale. A specific coefficient set is recommended in this case to counter for the large variation within a surface type.

5.2. Uncertainty due to Error of Surface Type

Note also that the VIIRS LST algorithm is surface type dependent and therefore the impact of error in surface type input is a concern. Theoretical analysis of the algorithm uncertainty due to the surface type has been described in Section 3.3. Figure 8 shows that the overall LST uncertainty caused by surface type misclassification is 0.73 K, more specifically, 0.83 K for nighttime and 0.61 K for daytime. The impact varies significantly with different surface types and day/night conditions. The most significant impact is found over closed shrubland in which an LST uncertainty of 1 K is found at daytime and 1.9 K at nighttime. Though the accuracy over snow/ice surface type is as high as 91%, the 9% misclassification causes 1 K LST uncertainty. Therefore, the impact of surface type accuracy on LST uncertainty is closely related to the emission characteristics differences between the correct surface type and misclassified surface types. It is noted that the daytime uncertainty is smaller than nighttime uncertainty from the theoretical analysis. The combination of atmospheric conditions and SW coefficients turn night-time LST estimates more sensitive to errors in surface type classification than daytime retrievals.

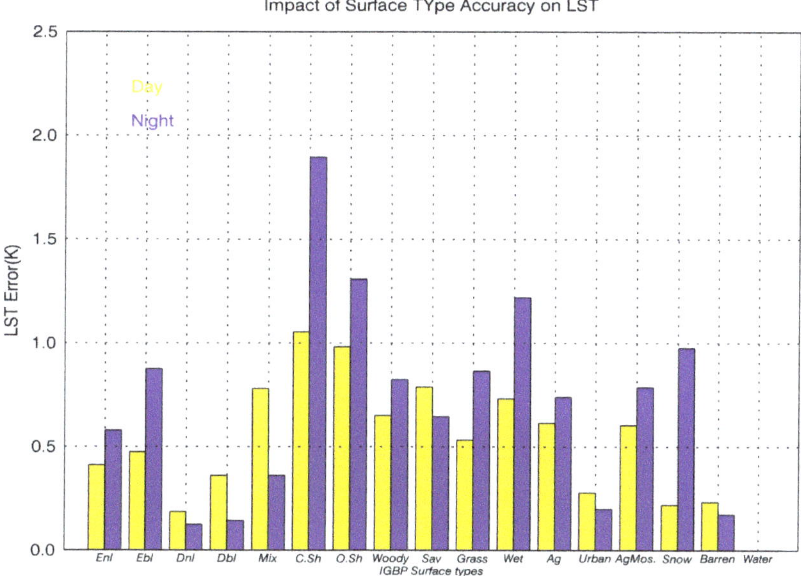

Figure 8. LST uncertainty associated with the uncertainty in surface type classification. These values are estimated using the simulation dataset for all surface types and day/night conditions.

In addition to the theoretical estimation, a set of global daily VIIRS data on 22 October 2014 for both daytime and nighttime are used as a real data case study. The blue line in Figure 9 represents the surface type accuracy from 0 (zero percent) to 1 (100 percent). The red line represents the LST uncertainty introduced by the surface type misclassification. As expected, the higher the surface type accuracy is, the smaller the uncertainty of the LST retrieval. Impact at daytime is more significant than that at nighttime. The surface type misclassification causes overall LST errors of 1.2 K, with 1.5 K and 0.8 K for daytime and nighttime respectively, which is larger than the theoretical results based on the simulation data. The maximum impact of 3 K is found at daytime over open shrubland and 1.4 K at nighttime over permanent wetland. LST performance is severely affected over closed shrub land, open shrub land, woody savannahs and croplands/Natural vegetation mosaics at daytime. Night-time LST uncertainties are generally smaller than those obtained for daytime, and tend to be higher over permanent wetland, close shrubland, open shrubland and evergreen broadleaf forest. Furthermore, some problems are observed from the surface type accuracy table, for instance closed shrub land, all classified as evergreen needle leaf forests which is quite distinct to shrub land in terms of surface emission characteristics, causing large LST errors in this case.

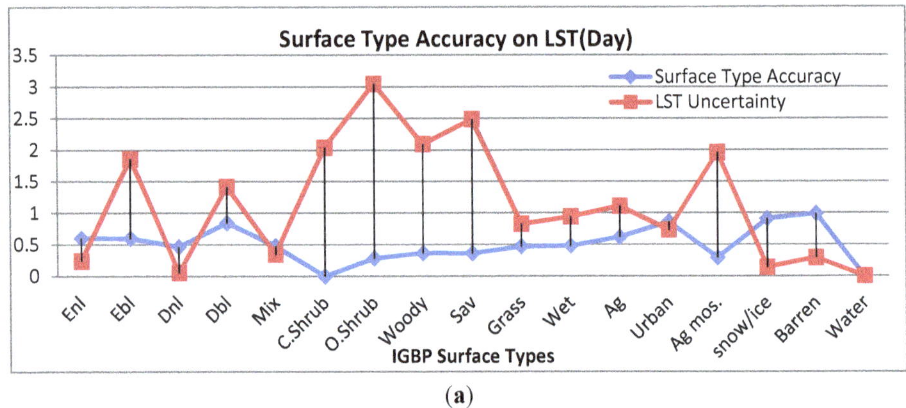

(a)

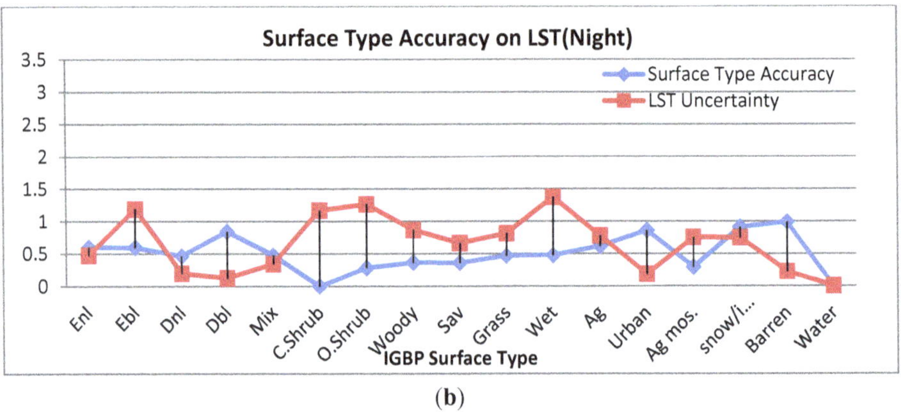

(b)

Figure 9. Impact of surface type accuracy (blue line, ranging from 0 to 1) on LST uncertainty (red line, in K) for daytime (**a**) and nighttime (**b**).

6. Conclusions

Two validation approaches are used in this study, namely the T-based method to compare the VIIRS LST retrieval with ground LST measurements and the comparison with different satellite products, e.g., MODIS Aqua LST product and AATSR LST. These two methods are complementary to each other and both are useful to quantify and characterize the accuracy of VIIRS LST product and help refine the LST retrieval algorithms. The comparisons with ground measurements in SURFRAD indicate that the VIIRS LST EDR yields a reasonable accuracy with an average bias of −0.41 K and RMSE of 2.38 K for the seven sites. The accuracy at nighttime is better than that in daytime. In addition, the VIIRS LST quality shows significant seasonality with a better performance in fall and winter than that in spring and summer. The comparisons with ground measurements in Gobabeb, Namibia, indicate that VIIRS and MODIS underestimate the LST over desert area by 1.57 K and 2.97 K, respectively.

The cloud contamination and surface type misclassification are found to have a great impact on the validation result. Additional cloud screening is strongly recommended in the LST validation and applications. The surface type misclassification under given accuracy introduces a 0.7 K uncertainty in LST when using the MODTRAN simulation database, lower than that from the real case (1.2 K or so). Currently, VIIRS LST is the only product using surface type dependent algorithm among the existing operational LST products. It strongly depends on the accurate classification of surface cover at satellite footprint. The consistency of the surface type product, surface type mixed pixel and misuse of the surface type information will definitely affect and limit the performance of the VIIRS LST retrieval. During the time of this study, surface type EDR is updated quarterly but now has changed to an annual product, which will potentially have a negative impact on surface type dependent LST retrieval. As shown in Table 4, there is more than one surface classification for matchups over DRA, FPK, GWN, and PSU sites in addition to the temporal snow/ice cover. It might be either misclassification or the natural variation of the surface cover along the year. The impact on LST validation varies depending on how distinct/ close the emission characteristics are among those surface types. With the annual surface type product, the LST algorithm should be able to account for significant emissivity variability which might go beyond the range for a typical IGBP surface type. It is certainly a big challenge for the current VIIRS LST algorithm. A possible solution will be to explicitly employ emissivity values as the parameters of the split window algorithm. Many algorithms of this type have been proposed, for instance, the LST algorithms for MODIS [11], for GOES [47] as well as for the future GOES-R [1]. Additional refinement is needed to improve the emissivity explicit algorithm performance over arid and semi-arid regions, e.g., dynamic surface emissivity setting [21] and separate coefficient set for this surface type [54].

The cross comparison with existing satellite data at granule level, e.g., MODIS LST product, indicates that VIIRS LST and MODIS LST are in very good agreement with over 100 SNO scenes. However, large discrepancies are found when the BT difference between the two split window channels is very large. This usually indicates a hot and humid weather condition and a coupled effect of significant spectral emissivity difference between split window channels. This high BT difference is mostly found at daytime and spatially distributed in low latitude area. This problem is not observed in the ground validation. The analysis results using MODIS sensor data as input for VIIRS LST retrieval clearly show that the large difference between VIIRS LST and MODIS LST for this particular situation is attributed to the sensor data difference, particularly the split window BT difference. Special attention needs to be paid to the usage of the quadratic term in LST retrieval, e.g., the contribution of this term has to be restricted so that it won't cause the overcorrection of the atmospheric effect, thus cause the overestimate of the LST. Possible solutions include the stratification of the retrieval by water vapor or a separate coefficient set particularly developed for very high BT difference situation.

Acknowledgments: This study was supported by NOAA grant NA09NES4400006 (Cooperative Institute for Climate and Satellites-CICS) at the University of Maryland/ESSIC. The manuscript contents are solely the opinions of the authors and do not constitute a statement of policy, decision, or position on behalf of NOAA or the U. S. Government.

Author Contributions: Yuling Liu designed the research, collected satellite data and implemented the data analysis as well as wrote the manuscript. Yunyue Yu guided the whole study including research contents, methodology *etc.* and has the greatest contribution on the revisions of the manuscript. Peng Yu provided some tools and ideas for uncertainty and data result analysis. Frank-M. Göttsche provided the *in situ* LST measurements over Africa and revised the manuscript with a focus on the part of temperature based validation. Isabel F. Trigo contributed to the interpretation of the results. All the authors worked on the revisions of the manuscript.

Conflicts of Interest: The authors declare no conflict of interest.

References

1. Yu, Y.; Tarpley, D.; Privette, J.L.; Flynn, L.E.; Xu, H.; Chen, M.; Vinnikov, K.Y.; Sun, D.; Tian, Y. Validation of GOES-R satellite land surface temperature algorithm using SURFRAD ground measurements and statistical estimates of error properties. *IEEE Trans. Geosci. Remote Sens.* **2012**, *50*, 704–713.
2. Meng, C.L.; Li, Z.-L.; Zhan, X.; Shi, J.C.; Liu, C.Y. Land surface temperature data assimilation and its impact on evapotranspiration estimates from the common land model. *Water Resour. Res.* **2009**, *45*, W02421. [CrossRef]
3. Zheng, W.; Wei, H.; Wang, Z.; Zeng, X.; Meng, J.; Ek, M.; Mitchell, K.; Derber, J. Improvement of daytime land surface skin temperature over arid regions in the NCEP GFS model and its impact on satellite data assimilation. *J. Geophys. Res. Atmos.* **2012**, *117*, D06117. [CrossRef]

4. Anderson, M.C.; Kustas, W.P.; Norman, J.M.; Hain, C.R.; Mecikalski, J.R.; Schultz, L.; Gonzalez-Dugo, M.P.; Cammalleri, C.; d'Urso, G.; Pimstein, A.; *et al.* Mapping daily evapotranspiration at field to continental scales using geostationary and polar orbiting satellite imagery. *Hydro. Earth Sys. Sci.* **2011**, *15*, 223–239. [CrossRef]
5. Anderson, M.C.; Allen, R.G.; Morse, A.; Kustas, W.P. Use of Landsat thermal imagery in monitoring evapotranspiration and managing water resources. *Remote Sens. Environ.* **2012**, *122*, 50–65. [CrossRef]
6. Rajasekar, U.; Weng, Q. Urban heat island monitoring and analysis by data mining of MODIS imageries. *ISPRS J. Photogramm. Remote Sens.* **2009**, *64*, 86–96. [CrossRef]
7. Li, H.; Sun, D.; Yu, Y.; Wang, H.; Liu, Y.; Liu, Q.; Du, Y.; Wang, H.; Cao, B. Evaluation of the VIIRS and MODIS LST products in an arid area of Northwest China. *Remote Sens. Environ.* **2014**, *142*, 111–121. [CrossRef]
8. Jiménez-Muñoz, J.C.; Sobrino, J.A. A generalized single-channel method for retrieving land surface temperature from remote sensing data. *J. Geophys. Res.* **2003**, *108*, 4688–4695. [CrossRef]
9. Prata, A.J. Land surface temperatures derived from the AVHRR and the ATSR. 1, Theory. *J. Geophys. Res.* **1993**, *98*, 689–16702.
10. Coll, C.; Caselles, V.; Sobrino, J.A.; Valor, E. On the atmospheric dependence of the split-window equation for land surface temperature. *Int. J. Remote Sens.* **1994**, *15*, 105–122. [CrossRef]
11. Wan, Z.; Dozier, J. A generalized split-window algorithm for retrieving land-surface temperature from space. *IEEE Trans. Geosci. Remote Sens.* **1996**, *34*, 892–905.
12. Gillespie, A.R.; Rokugawa, S.; Hook, S.J.; Matsunaga, T.; Kahle, A.B. *Temperature/Emissivity Separation Algorithm Theoretical Basis Document*, version 2.4; NASA/GSFC: Greenbelt, MD, USA, 1996.
13. Yu, Y.; Tarpley, D.; Privette, J.; Goldberg, M.; Raja, M.; Vinnikov, K.; Xu, H. Developing algorithm for operational GOES-R land surface temperature product. *IEEE Trans. Geosci. Remote Sens.* **2009**, *47*, 936–951.
14. Sun, D.; Pinker, R.T. Estimation of land surface temperature from a Geostationary Operational Environmental Satellite (GOES-8). *J. Geophys. Res.* **2003**, *108*. [CrossRef]
15. Gillespie, A.R.; Rokugawa, S.; Matsunaga, T.; Cothern, J.S.; Hook, S.; Kahle, A.B. A temperature and emissivity separation algorithm for Advanced Spaceborne Thermal Emission and Reflection Radiometer (ASTER) images. *IEEE Trans. Geosci. Remote Sens.* **1998**, *36*, 1113–1126. [CrossRef]
16. Dash, P.; Göttsche, F.-M.; Olesen, F.-S.; Fischer, H. Land surface temperature and emissivity estimation from passive sensor data: Theory and practice—Current trends. *Int. J. Remote Sens.* **2002**, *23*, 2563–2594. [CrossRef]
17. Coll, C.; Valor, E.; Galve, J.M.; Mira, M.; Bisquert, M.; García-Santos, V.; Caselles, E.; Caselles, V. Long-term accuracy assessment of land surface temperatures derived from the advanced along-track scanning radiometer. *Remote Sens. Environ.* **2012**, *116*, 211–225. [CrossRef]
18. Freitas, S.C.; Trigo, I.F.; Bioucas-Dias, J.M.; Goettsche, F.-M. Quantifying the uncertainty of land surface temperature retrievals from SEVIRI/Meteosat. *IEEE Trans. Geosci. Remote Sens.* **2010**, *48*, 523–534. [CrossRef]
19. Niclòs, R.; Galve, J.M.; Valiente, J.A.; Estrela, M.J.; Coll, C. Accuracy assessment of land surface temperature retrievals from MSG2-SEVIRI data. *Remote Sens. Environ.* **2011**, *2011*, 2126–2140. [CrossRef]
20. Hulley, G.C.; Hook, S.J. Intercomparison of versions 4, 4.1 and 5 of the MODIS land surface temperature and emissivity products and validation with laboratory measurements of sand samples from the Namib Desert, Namibia. *Remote Sens. Environ.* **2009**, *133*, 1313–1318. [CrossRef]
21. Guillevic, P.C.; Biard, C.J.; Hulley, G.C.; Privette, J.L.; Hook, S.J.; Olioso, A.; Göttsche, F.M.; Radocinski, R.; Román, M.O.; Yu, Y.; Csiszar, I. Validation of land surface temperature products derived from the Visible Infrared Imaging Radiometer Suite (VIIRS) using ground-based and heritage satellite measurements. *Remote Sens. Environ.* **2014**, *154*, 19–37. [CrossRef]
22. Justice, C.O.; Román, M.O.; Csiszar, I.; Vermote, E.F.; Wolfe, R.; Hook, S.J.; Friedl, M.; Wang, Z.; Schaaf, C.; Miura, T.; *et al.* Land and cryosphere products from Suomi NPP VIIRS: Overview and status. *J. Geophys. Res.* **2013**, *118*, 9753–9765. [CrossRef] [PubMed]
23. CLASS. Available online: http://www.nsof.class.noaa.gov (accessed on 14 September 2015).
24. Wan, Z. New refinements and validation of the collection-6 MODIS land-surface temperature/emissivity product. *Remote Sens. Environ.* **2014**, *140*, 36–45. [CrossRef]
25. Prata, A.J. Land Surface Temperature Measurement from Space: AATSR Algorithm Theoretical Basis Document. Available online: https://earth.esa.int/c/document_library/et_file?folderId=13019&name=DLFE-660.pdf (accessed on 15 December 2014).

26. Coll, C.; Caselles, V.; Galve, J.M.; Valor, E.; Niclòs, R.; Sánchez, J.M.; Rivas, R. Ground measurements for the validation of land surface temperatures derived from AATSR and MODIS data. *Remote Sens. Environ.* **2005**, *97*, 288–300. [CrossRef]
27. Augustine, J.A.; DeLuisi, J.J.; Long, C.N. SURFRAD—A national surface radiation budget network for atmospheric research. *Bull. Am. Meteorol. Soc.* **2000**, *81*, 2341–2357. [CrossRef]
28. Augustine, J.A.; Hodges, G.B.; Cornwall, C.R.; Michalsky, J.J.; Medina, C.I. An update on SURFRAD—The GCOS surface radiation budget network for the continental United States. *J. Atmos. Ocean. Technol.* **2005**, *22*, 1460–1472. [CrossRef]
29. Hulley, G.C.; Simon, J.H. The North American ASTER land surface emissivity database (NAALSED), version 2.0. *Remote Sens. Environ.* **2009**, *13*, 1967–1975. [CrossRef]
30. Li, S.; Yu, Y.; Sun, D.; Tarpley, D.; Zhan, X.; Chiu, L. Evaluation of 10 year AQUA/MODIS land surface temperature with SURFRAD observations. *Int. J. Remote Sens.* **2014**, *35*, 830–856. [CrossRef]
31. Liu, Y.; Yu, Y.; Tarpley, D.; Wang, X.; Wang, Z. Initial assessment of Suomi NPP VIIRS Land Surface Temperature (LST) algorithms ballroom G. In Presented at the AMS 93rd Annual Meeting, Austin, TX, USA, 5–10 January 2013.
32. Salibury, J.W.; D'Aria, D.M. Emissivity of Terrestrial Materials in the 8–14 μm atmospheric window. *Remote Sens. Environ.* **1992**, *42*, 83–106. [CrossRef]
33. Wang, K.; Liang, S. Evaluation of ASTER and MODIS land surface temperature and emissivity products using long-term surface longwave radiation observations at SURFRAD sites. *Remote Sens. Environ.* **2009**, *113*, 1556–1565. [CrossRef]
34. Wang, K.; Wan, Z.; Wang, P.; Sparrow, M.; Liu, J.; Zhou, X.; Haginoya, S. Estimation of surface long wave radiation and broadband emissivity using MODIS land surface temperature/emissivity product. *J. Geophys. Res.* **2005**, *110*, D11109. [CrossRef]
35. Seemann, S.W.; Borbas, E.E.; Knuteson, R.O.; Stephenson, G.R.; Huang, H.-L. Development of a global infrared land surface emissivity database for application to clear sky sounding retrievals from multi-spectral satellite radiance measurements. *J. Appl. Meteor. Climatol.* **2008**, *47*, 108–123. [CrossRef]
36. Göttsche, F.-M.; Olesen, F.-S.; Bork-Unkelbach, A. Validation of land surface temperature derived from MSG/SEVIRI with *in situ* measurements at Gobabeb, Namibia. *Int. J. Remote Sens.* **2013**, *34*, 3069–3083. [CrossRef]
37. Göttsche, F.-M.; Hulley, G.C. Validation of six satellite-retrieved land surface emissivity products over two land cover types in a hyper-arid region. *Remote Sens. Environ.* **2012**, *124*, 149–158. [CrossRef]
38. Kabsch, E.; Olesen, F.; Prata, F. Initial results of the land surface temperature (lST) validation with the Evora, Portugal ground-truth station measurements. *Int. J. Remote Sens.* **2008**, *29*, 5329–5345. [CrossRef]
39. Kondratyev, K.Y. *Radiation in the Atmosphere*; Academic Press: New York, NY, USA, 1969.
40. Simultaneous Nadir Overpasses (SNOs) Tool. Available online: http://ncc.nesdis.noaa.gov/SNOPredictions (accessed on 14 September 2015).
41. Theocharous, E.; Usadi, E.; Fox, N.P. *CEOS Comparison of IR Brightness Temperature Measurements in Support of Satellite Validation. Part I: Laboratory and Ocean Surface Temperature Comparison of Radiation Thermometers*; National Physical Laboratory: Teddington, UK, 2010.
42. Wan, Z.; Zhang, Y.; Zhang, Q.; Li, Z. Validation of the land-surface temperature products retrieved from Terra Moderate Resolution Imaging Spectroradiometer data. *Remote Sens. Environ.* **2002**, *83*, 163–180. [CrossRef]
43. Wan, Z.; Zhang, Y.; Zhang, Q.; Li, Z. Quality assessment and validation of the MODIS global land surface temperature. *Int. J. Remote Sens.* **2004**, *25*, 261–274. [CrossRef]
44. Coll, C.; Wan, Z.; Galve, J.M. Temperature-based and radiance-based validations of the V5 MODIS land surface temperature product. *J. Geophys. Res.* **2009**, *114*, D20102. [CrossRef]
45. Hook, S.J.; Vaughan, R.G.; Tonooka, H.; Schladow, S.G. Absolute radiometric in-flight validation of mid infrared and thermal infrared data from ASTER and MODIS on the Terra spacecraft using the Lake Tahoe, CA/NV, USA, automated validation site. *IEEE Trans. Geosci. Remote Sens.* **2007**, *45*, 1798–1807. [CrossRef]
46. Guillevic, P.; Privette, J.; Coudert, B.; Palecki, M.A.; Demarty, J.; Ottlé, C.; Augustine, J.A. Land Surface Temperature product validation using NOAA's surface climate observation networks-scaling methodology for the Visible Infrared Imager Radiometer Suite (VIIRS). *Remote Sens. Environ.* **2012**, *124*, 282–298. [CrossRef]
47. Hale, R.C.; Gallo, K.P.; Tarpley, D.; Yu, Y. Characterization of *in situ* locations for calibration/validation of satellite-derived land surface temperature data. *Remote Sens. Lett.* **2011**, *2*, 41–50. [CrossRef]

48. Ermida, S.L.; Trigo, I.F.; DaCamara, C.C.; Göttsche, F.M.; Olesen, F.S.; Hulley, G. Validation of remotely sensed surface temperature over an oakwood landscape-The problem of viewing and illumination geometries. *Remote Sens. Environ.* **2014**, *148*, 16–27. [CrossRef]
49. Sun, D.; Yu, Y.; Fang, L.; Liu, Y. Toward an operational land surface temperature algorithm for GOES. *J. Appl. Meteor. Clim.* **2013**, *52*, 1974–1986. [CrossRef]
50. Wan, Z.; Li, Z.-L. Radiance-based validation of the V5 MODIS land-surface temperature product. *Int. J. Remote Sens.* **2008**, *29*, 5373–5395. [CrossRef]
51. Jacob, F.; Petitcolin, F.; Schmugge, T.; Vermote, E.; French, A.; Ogawa, K. Comparison of land surface emissivity and radiometric temperature derived from MODIS and ASTER sensors. *Remote Sens. Environ.* **2004**, *90*, 137–152. [CrossRef]
52. Trigo, I.F.; Monteiro, I.T.; Olesen, F.; Kabsch, E. An assessment of remotely sensed land surface temperature. *J. Geophys. Res.* **2008**, *113*, D17108. [CrossRef]
53. Wan, Z.; Li, Z. A physics-based algorithm for land-surface emissivity and temperature from EOS/MODIS data. *IEEE Trans. Geosci. Remote Sens.* **1997**, *35*, 980–996.
54. Soliman, A.; Duguay, C.; Saunders, W.; Hachem, S.P.-A. Land surface temperature from MODIS and AATSR: Product development and intercomparison. *Remote Sens.* **2012**, *4*, 3833–3856. [CrossRef]
55. VIIRS Surface Type EDR ATBD. Available online: http://www.star.nesdis.noaa.gov/jpss/documents/ATBD/D0001-M01-S01-024_JPSS_ATBD_VIIRS-Surface-Type_A.pdf (accessed on 15 September 2015).
56. Yu, Y.; Privette, J.L.; Pinheiro, A.C. Analysis of the NPOESS VIIRS land surface temperature algorithm using MODIS data. *IEEE Trans. Geosci. Remote Sens.* **2005**, *43*, 2340–2350.
57. Hutchison, K.D.; Iisager, B.D.; Mahoney, R.L. Enhanced snow and ice identification with the VIIRS cloud mask algorithm. *Remote Sens. Lett.* **2013**, *4*, 929–936. [CrossRef]
58. Kopp, T.; Heidinger, A.; Thomas, W. VIIRS Cloud Mask (VCM) Provisional Status. Available online: http://www.star.nesdis.noaa.gov/jpss/documents/meetings/2013/EDRProvReview/VCM_provisional_brief_Jan2013.pdf (accessed on 10 June 2015).
59. Kopp, T.J.; Thomas, W.; Heidinger, A.K.; Botambekov, D.; Frey, R.A.; Hutchison, K.D.; Iisager, B.D.; Brueske, K.; Reed, B. The VIIRS Cloud Mask: Progress in the first year of S-NPP toward a common cloud detection scheme. *J. Geophys. Res. Atmos.* **2014**, *119*, 2441–2456. [CrossRef]
60. VIIRS LST ATBD. Available online: http://www.star.nesdis.noaa.gov/jpss/documents/ATBD/D0001-M01-S01-022_JPSS_ATBD_VIIRS-LST_A.pdf (accessed on 15 September 2015).
61. Li, Z.; Tang, B.; Wu, H.; Ren, H.; Yan, G.; Wan, Z.; Trigo, I.F.; Sobrino, J.A. Satellite-derived land surface temperature: current status and perspectives. *Remote Sens. Environ.* **2013**, *131*, 14–37. [CrossRef]

Article

Validation of S-NPP VIIRS Sea Surface Temperature Retrieved from NAVO

Qianguang Tu [1,2], Delu Pan [1,2] and Zengzhou Hao [2,*]

1 Department of Earth Sciences, Zhejiang University, Hangzhou 310027, China; thotho@163.com (Q.T.); pandelu@sio.org.cn (D.P.)

2 State Key Laboratory of Satellite Ocean Environment Dynamics, Second Institute of Oceanography, State Oceanic Administration, Hangzhou 310012, China

* Correspondence: hzyx80@sio.org.cn; Tel./Fax: +86-571-8196-3121

Academic Editors: Changyong Cao, Xiaofeng Li and Prasad S. Thenkabail

Received: 15 October 2015; Accepted: 7 December 2015; Published: 18 December 2015

Abstract: The validation of sea surface temperature (SST) retrieved from the new sensor Visible Infrared Imaging Radiometer Suite (VIIRS) onboard the Suomi National Polar-Orbiting Partnership (S-NPP) Satellite is essential for the interpretation, use, and improvement of the new generation SST product. In this study, the magnitude and characteristics of uncertainties in S-NPP VIIRS SST produced by the Naval Oceanographic Office (NAVO) are investigated. The NAVO S-NPP VIIRS SST and eight types of quality-controlled *in situ* SST from the National Oceanic and Atmospheric Administration *in situ* Quality Monitor (iQuam) are condensed into a Taylor diagram. Considering these comparisons and their spatial coverage, the NAVO S-NPP VIIRS SST is then validated using collocated drifters measured SST via a three-way error analysis which also includes SST derived from Moderate Resolution Imaging Spectro-radiometer (MODIS) onboard AQUA. The analysis shows that the NAVO S-NPP VIIRS SST is of high accuracy, which lies between the drifters measured SST and AQUA MODIS SST. The histogram of NAVO S-NPP VIIRS SST root-mean-square error (RMSE) shows normality in the range of 0–0.6 °C with a median of ~0.31 °C. Global distribution of NAVO VIIRS SST shows pronounced warm biases up to 0.5 °C in the Southern Hemisphere at high latitudes with respect to the drifters measured SST, while near-zero biases are observed in AQUA MODIS. It means that these biases may be caused by the NAVO S-NPP VIIRS SST retrieval algorithm rather than the nature of the SST. The reasons and correction for this bias need to be further studied.

Keywords: SST; S-NPP VIIRS; three-way error analysis; Taylor diagram; NAVO

---

1. Introduction

Sea Surface Temperature (SST; see Table 1 for a list of abbreviations used in this paper) is a fundamental variable at the ocean-atmosphere interface [1]. It affects the complex interactions between atmosphere and ocean at a variety of scales. Thus, SST datasets with high quality are needed for many applications, such as operational monitoring, numerical weather, and ocean forecasting, climate change research, and so on. SST is collected routinely from *in situ* measurements, such as ships, moored and drifting buoys. They are usually taken as ground truth while limited in time and space coverage. Nowadays, continuous SST retrieved from satellites is increasingly used. However, the infrared satellite measurements are often contaminated by clouds and the microwave satellite observations are unavailable in rainfall, near sea ice or land, *etc.* Additionally, they are measured from different sensors and platform types, it is necessary to assess their biases and errors carefully before further use [2].

Table 1. List of abbreviations.

Abbreviations	Full Name
AVHRR	Advanced Very High Resolution Radiometer
EUMETSAT	European Organisation for the Exploitation of Meteorological Satellites
GHRSST	Group for High Resolution SST
GTS	Global Telecommunication System
HDF	Hierarchical Data Format
HR-Drifter	high resolution SST drifters
iQuam	*in situ* Quality Monitor
JPL	Jet Propulsion Laboratory
MODIS	Moderate Resolution Imaging Spectro-radiometer
NASA	National Aeronautics and Space Administration
NAVO	Naval Oceanographic Office
netCDF	network Common Data Form
NOAA	National Oceanic and Atmospheric Administration
OBPG	Ocean Biology Processing
OISST	Optimum-Interpolation Sea Surface Temperature
OSI-SAF	EUMETSAT Ocean and Sea Ice Satellite Application Facility
OSTIA	Operational SST and Sea Ice Analysis
RSMAS	Rosenstiel School of Marine and Atmospheric Science
SST	Sea Surface Temperature
S-NPP	Suomi National Polar-orbiting Partnership
VIIRS	Visible Infrared Imaging Radiometer Suite

To provide gap-free SST for various applications, many interpolated datasets, such as Optimum-Interpolation Sea Surface Temperature (OISST), the Operational SST and Sea Ice Analysis (OSTIA), and so on, are made by various research groups [3,4] incorporating as more available data from satellites and *in situ* measurements as possible. However, these products do not include SST from the new sensor of the Visible Infrared Imaging Radiometer Suite (VIIRS). It is a primary sensor onboard the Suomi National Polar-orbiting Partnership (S-NPP) satellite which launched on 28 October 2011 and achieved provisional maturity status by early 2013 [5]. VIIRS started a new era of moderate-resolution imaging capabilities as a successive sensor of Advanced Very High Resolution Radiometer (AVHRR) and Moderate Resolution Imaging Spectro-radiometer (MODIS). High-quality global SST is critically needed at the present stage. First, there is an increasing concern with the aging of the MODIS because they have far exceeded their original retirement ages. Second, the comparatively long records of AVHRR SST have been already accumulated (over 30 years), while the latest and final AVHRR onboard National Oceanic and Atmospheric Administration (NOAA)-19 for the afternoon orbit was launched on 6 February 2009.

Many works on sensor measurements calibration and satellite derived products validation have been done in order to verify the performance of S-NPP VIIRS [5–10]. They showed that the VIIRS has been working very well since launched after a number of issues were resolved. The radiometric uncertainty for the reflective solar bands is generally believed to be comparable to that of MODIS within 2% in reflectance, while an agreement on the order of 0.1 K with AVHRR and other existing references for the sea surface temperature bands has been reached. In this paper, we focus on the performances of SST derived from S-NPP VIIRS. The existing operational SST algorithms developed by several government organizations and institutes, such as NOAA, Naval Oceanographic Office (NAVO), the Rosenstiel School of Marine and Atmospheric Science (RSMAS), and Ocean and Sea Ice Satellite Application Facility (OSI-SAF) from the European Organization for the Exploitation of Meteorological Satellites(EUMETSAT), have been evaluated for VIIRS SST retrieval [11]. To keep consistency for assimilation into analyses and models, the SST is retrieved from NAVO, which provides operational processing of SST retrievals for AVHRR since 1993. Initial comparison between buoys and VIIRS SST processing at NAVO show a high root mean square error and a warm bias for both daytime and nighttime [12]. With the improvements of the sensor calibrating and cloud screening, as well as the increasing number of matchups allowing better algorithm coefficients to be regressed, the NAVO S-NPP VIIRS SST became operational at the end of January 2013 [13].

In this paper, we particularly focus on the comparison between global *in situ* and NAVO S-NPP VIIRS SST. Previous works have shown the difficulty in comparing different sources of SST with their specific characteristics in global and regional scale [2,14,15]. To assess the accuracy of satellite SST, it is important to utilize as many *in situ* measurements as possible for validation, especially for regions of sparse data. The Taylor diagram is first used to demonstrate the differences of the RMSE, correlation and standard deviation between the NAVO S-NPP VIIRS SST and eight types of *in situ* measurements, and then select drifter SST as reference to validate the VIIRS SST. Then, a three-way error analysis in conjunction with the drifter and AQUA MODIS SSTs are employed to evaluate the performance of VIIRS SST. The results can be beneficial for other applications, such as the data interpretation, improvement of sensor design, SST retrieval algorithm and merging the VIIRS SST with other available data sources, *etc.*

This work is organized as follows: the NAVO S-NPP VIIRS, AQUA MODIS and *in situ* SST used in this work are described in Section 2. A strategy for matching the VIIRS SST and *in situ* SST and the distribution of the matchups is outlined in Section 3. Then an initial comparison between the VIIRS SST and various types of *in situ* SST is performed in Section 4. The root mean square error of VIIRS SST is estimated via three-way analyses in Section 5. Conclusions are presented in Section 6.

2. Data

2.1. NAVO S-NPP VIIRS SST

The VIIRS is first on the S-NPP satellite which launched on 28 October 2011 as a part of the Joint Polar Satellite System (JPSS) program [5]. S-NPP is in a sun-synchronous near-polar orbit at 824 km with a swath-width of ~3060 km. As a result, VIIRS provides complete coverage of the globe twice daily without inter-orbit gaps: once in the early afternoon (ascending overpass with a local time node of approximately 13:25) and once in the early morning (descending overpass). The VIIRS provides 22 spectral bands coverage from 0.4–12.5 μm (five imagery bands at ~375 m nadir resolution and 17 moderate resolution bands at ~750 m at nadir). S-NPP VIIRS SST processing relies on five moderate resolution bands (center at 0.67 μm, 0.87 μm, 3.7 μm, 10.7 μm and 12.0 μm) for cloud screening and SST computations. The S-NPP VIIRS is a subsequent sensor of AVHRR and MODIS for SST. The Non-Linear SST (NLSST) algorithm which has been used for MODIS and AVHRR was improved and designated for VIIRS at NAVO [13]. The processing is done in every "target arrays" which is a small granule with 10 × 6 pixel-windows. Cloud-screened and SST retrieved is done on the 2 × 2 pixel unit array, which resulted in the spatial resolution of ~1.5 km.

To avoid the bright reflective of solar radiation, the mid-infrared channel cannot be used in the daytime algorithm as follows:

$$SST_{day} = a_0 + a_1 T_{11} + a_2(T_{11} - T_{12})T_{sfc} + a_3(T_{11} - T_{12})(\sec(\theta) - 1) + a_4(T_{11} - T_{12}) \quad (1)$$

where a_0, a_1, a_2, a_3, a_4 are coefficients derived by regression analysis, T_{11} and T_{12} are the brightness temperatures at 11 μm (VIIRS band M15) and 12 μm (VIIRS band M16), respectively. T_{sfc} is a first guess SST that scales the coefficient multiplying the brightness temperature difference between T_{11} and T_{12}. θ is the sensor zenith angle in radians and this term compensates for the increasing path length when the scan is away from nadir.

The mid-infrared channel is included in the nighttime algorithm as follow:

$$SST_{night} = b_0 + b_1 T_{11} + b_2(T_{3.7} - T_{12})T_{sfc} + b_3(\sec(\theta) - 1) + b_4(T_{3.7} - T_{12}) \quad (2)$$

where b_0, b_1, b_2, b_3, b_4 are coefficients derived by regression analysis, $T_{3.7}$ is the measured brightness temperature at the mid-infrared channel 3.7 μm (VIIRS band M12). Although the VIIRS measures the skin temperature with infrared sensor, they are regressed to a nominal depth 1 m. There are ~1012 VIIRS granules (~85 s for each granule) every day can be received in near real-time at NAVO.

The level 2 NAVO VIIRS SST is integrated to the Group for High-Resolution SST (GHRSST) and quality flags ranging from zero to five are appended to the data follow the GHRSST Data Processing Specification (GHRSST-DPS) [16]. Only the best quality data with flag five are used for validation. They are available at ftp.nodc.noaa.gov/pub/data.nodc/ghrsst/L2P/VIIRS_NPP/NAVO [17].

2.2. MODIS SST

The MODIS is on the AQUA satellite which launched in 2002. The AQUA platform is in a sun-synchronous near-polar orbit at altitude of ~705 km with a local time ascending node of ~13:30. The MODIS measure $\pm 55^\circ$ from nadir, yielding a swath-width of 2330 km and imaging the entire earth every one to two days at three different spatial resolutions (two bands at 250 m, five bands at 500 m, and 29 bands at 1000 m). It provides 36 channels coverage from 0.4 μm to 14.4 μm. Specifically, the 3.959, 4.05, 11, and 12 μm channels are used for the SST retrieved. The MODIS SST is integrated to the GHRSST through a joint collaboration between the National Aeronautics and Space Administration (NASA) Jet Propulsion Laboratory (JPL), Ocean Biology Processing Group (OBPG), and the RSMAS [18]. The daily MODIS ocean products, such as the Level 1, geolocation, cloud mask products, and higher level geophysical product are processed by OBPG in Hierarchical Data Format (HDF). SST retrieval algorithm, error statistics and quality flagging is in the charge of RSMAS (for more details see [19]). The MODIS 11 μm SST algorithm [20] also uses the nonlinear SST (NLSST) algorithm [21]. Algorithm coefficients are derived via collocations to *in situ* observations and continuously verified by RSMAS. The coefficients are tuned to three segments depending on the brightness temperature difference (BTD, in °C) between the 11 and 12 μm channels: $\mathrm{BTD} \leqslant 0.5$, $0.5 \leqslant \mathrm{BTD} \leqslant 0.9$ and $\mathrm{BTD} \geqslant 0.9$. During nighttime, the mid-infrared algorithm SST (retrieved from the mid-infrared bands near 4 μm) is used for the baseline SST, while the weekly Reynolds OISST product is bilinear interpolated to the pixel location for the daytime baseline. JPL acquires MODIS level 2 SST from the OBPG in near real-time and reformats them to the GHRSST specification with complete metadata, such as quality flags, in network Common Data Form (netCDF). The MODIS SST is considered to be measurements of the skin layer SST. Approximately 288+ five minute observation granules with ~1 km resolution per day can be achieved from GHRSST via ftp.nodc.noaa.gov/pub/data.nodc/ghrsst/L2P/MODIS_A/JPL [22].

2.3. In Situ Data

In situ subsurface SSTs from five independent data sources obtained from NOAA *in situ* Quality Monitor (iQuam [23,24]) are used to generate matchups with the satellite data. In addition to four *in situ* data types (drifters, ships, tropical, and coastal moorings) from the single source, Global Telecommunication System (GTS), four more platform types have been added, including ARGO floats, high-resolution SST drifters (HR-Drifter), coral reef watch (CRW) buoys, and Integrated Marine Observing System (IMOS) track ships. The ships and drifters from GTS are flagged as GTS-Ship and GTS-Drifter to distinguish with the IMOS ships and HR-Drifter, respectively. The tropical moorings (T-mooring) are the moored buoys located at the tropical oceans and the coastal moorings (C-mooring) are the moored buoys located at the coastal areas. The original sampling rate of moored buoys is usually ~10 min and they are usually smoothed at hourly intervals (0000, 0100, 0200 ... 2200, 2300 UTC) to minimize the noise. Drifting buoys observations are also processed into hourly intervals. Some ships take hourly observations, but most take four observations a day at the synoptic reporting hours (0000, 0600, 1200. and 1800 UTC). Argo is an internationally-coordinated program aimed at seeding the global ocean with 3000 profiling floats, which measure temperature and salinity of the upper 2000 m of the ocean [25]. Although Argo floats are designed to shut off at 5 m from the surface in order to avoid contamination/degradation of salinity sensors by pollutants (bio-fouling) at the sea surface and preserve stability, most of them still measure temperature and salinity between 3 and 5 m. They may prove to be a useful global reference for stability as a longer time series accumulates. HR-Drifter is more accurate than the GTS-Drifter. A second decimal place was added in the HR-Drifter temperature in response to high accuracy requests from GHRSST. IMOS ships are from Integrated

Marine Observing System (IMOS). CRW is from Coral Reel Watch/NOAA. Since the *in situ* SSTs sensors are mounted onboard different platforms, maintained by different countries and agencies, the quality of *in situ* SSTs is often suboptimal and non-uniform. Buoys remain unattended in a hostile environment for years, and ship records are subject to human errors. Additional errors occur during data transmission (to satellite and back to ground), processing and distribution via GTS. The quality assurance for all of the *in situ* data thus has been done through the NOAA iQuam on a daily basis. A quality flag ranging from zero to five is also appended to the data follows the GHRSST-DPS. Also the best quality data with flag five are used in this study.

3. NAVO S-NPP VIIRS SST—*In Situ* SST Matchups

3.1. Collocation Criteria

The comparisons are conducted through using extensive collocations between the NAVO S-NPP VIIRS SST and *in situ* SST during 2014. For each *in situ* observation, all the near-simultaneous VIIRS observations (within ± 1 h) within a radius of 0.05° are selected. Since the *in situ* observations are organized into daily files, they are collocated with the VIIRS SST from the previous, current, and following day, to ensure accurate collocation at the beginning and end of the daily *in situ* files. The strict temporal criterion is to limit the influence of possible diurnal warming effects in the surface ocean, which can even reach several degrees at low wind and high insolation conditions [26]. Finally the collocated VIIRS SST is taken an average of the selected values, if and only if, the standard deviation within the collocated window was not exceeding the threshold applied. The standard deviation restriction is an additional filter to reduce the impact from the cloud contamination and suboptimal atmospheric conditions. In this study, the threshold is set as 1 °C.

3.2. Distributions of Matchups

A total number of 90,773 matchups between the VIIRS SSTs and *in situ* SSTs are obtained during 2014. The number of each type of matchups for daytime and nighttime are summarized in Table 2. The distribution of the matchups is shown in Figure 1. The GTS-Drifter provides a more complete global coverage (Figure 1a). However, their sampling is still sparse and non-uniform in some areas. The fewest of VIIRS-drifting buoy pairs are found in the tropical oceans and the shallow marginal seas. The tropical oceans are characterized by persistent cloud cover and divergent surface currents. The shallow marginal seas with complex topography and human factors lead to few drifting buoys. Figure 1b shows the coverage of other types of matchups. The VIIRS and GTS-Ship matchups are sparse and concentrated around major ship routes in the North Pacific and North Atlantic. There are fewer matchups in the southern hemisphere due to far fewer ships. Additionally, sampling rate of ships is low (as mention in Section 2.3) and the S-NPP VIIRS overpass at fixed local time, which coincide to the ships only in certain limited regions, also lead to fewer matchups. The sampling rate of moored buoys is high, while they are only available in the tropics, along coasts and some other limited areas. Some Argo observations are seen mainly at the deep open oceans. Dense observations of HR-Drifter are seen at north Atlantic and south Indian Ocean. IMOS ships are mainly surrounding the Australian regions and even the Southern Ocean. The VIIRS-CRW pairs are mainly located around the Australian coastal and the coral islands in the open oceans. Although each type of *in situ* data has some limitations, they seem to be complement to analyze the global errors of the VIIRS SST fully and uniformly.

Table 2. Supplementary statistical information corresponds to Figure 2.

Type	Time	Nums	Bias (°C)	Centered RMS (°C)	σ_{VIIRS} (°C)	σ_{inSitu} (°C)
GTS-Ship	Day	1816	0.05	0.92	8.42	8.55
	Night	3092	0.061	0.78	7.64	7.69
GTS-Drifter	Day	18,093	−0.07	0.39	8.44	8.62
	Night	28,295	−0.06	0.32	7.86	7.87
T-Mooring	Day	755	−0.19	0.54	3.06	3.07
	Night	734	0.09	0.28	3.30	3.28
C-Mooring	Day	4468	0.13	0.54	7.25	7.41
	Night	6488	0.17	0.47	7.38	7.45
Argo	Day	246	0.11	0.54	8.08	8.21
	Night	286	0.01	0.38	7.56	7.56
HR-Drifter	Day	2706	−0.09	0.36	8.04	8.15
	Night	4204	−0.08	0.38	7.73	7.72
IMOS-Ship	Day	8560	0.14	0.58	5.14	5.06
	Night	7583	0.11	0.45	4.02	4.01
CRW	Day	1228	0.07	0.50	1.61	1.59
	Night	2151	0.11	0.42	1.32	1.30

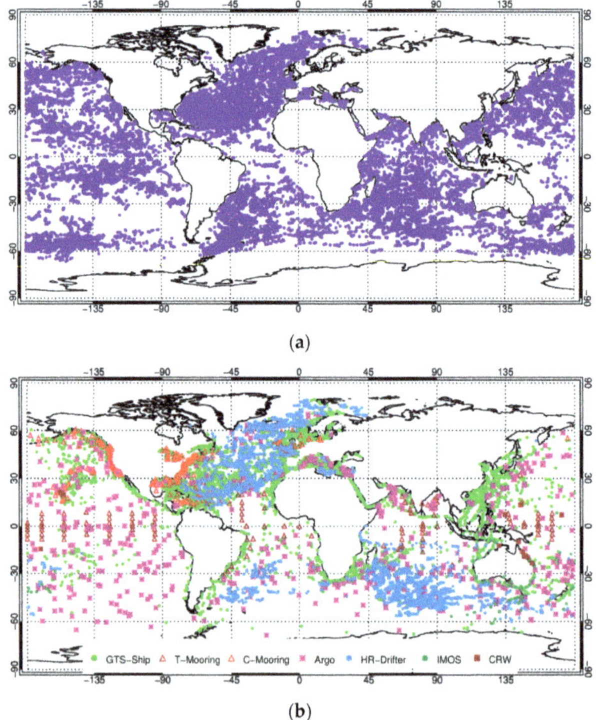

Figure 1. Distribution of NAVO S-NPP VIIRS-*in situ* matchups during 2014: (**a**) GTS-Drifter (Conventional drifters from GTS) measurements; (**b**) other types of measurements (GTS-Ship: conventional ships from GTS, T-Mooring: Tropical Moorings, C-Mooring: Coastal Moorings, Argo: Argo floats, HR-Drifter: High resolution drifters, IMOS: IMOS ships, CRW: Coral Reef Watch).

4. Initial Comparison between the NAVO S-NPP VIIRS SST and *in Situ* SST

Since various types of *in situ* SST originate from different countries and agencies for different purpose, it is necessary to identify significant deviations in their suitability for satellite validation first. The Taylor diagram [27] is used to present an overall comparison between different types of fields according to their correlation, centered root-mean squared (RMS) difference, and standard deviations. The correlation coefficient is used to quantify the similarity of different types of data. However, it is impossible to determine the range of variation using only the correlation. The centered RMS is then used to quantify the differences between different types of fields. Moreover, the standard deviations are also needed to give more complete properties for each type of fields. Figure 2 shows an overall comparison between the VIIRS SST and eight types of *in situ* data in Taylor diagram. The matchups are compartmentalized into daytime and nighttime to examine their difference. The angle to the horizontal axis indicates the correlation. In general, all types of *in situ* SST have very high correlation (larger than 0.99) with the VIIRS SST except the CRW. The standard deviations of *in situ* SST are normalized to the VIIRS SST. The VIIRS SST is, thus, located on the horizontal axis with a standard deviation of 1.0. The ratio of standard deviations is the distance from the origin of coordinate. Take the daytime GTS-ship and IMOS ships SST for example, their correlation with the VIIRS SST is similar, but the standard deviation of IMOS is lower than the VIIRS SST, by contrast, the standard deviation of GTS-ship SST is higher than the VIIRS SST. The normalized centered RMS for each type of *in situ* SST is the distance to the VIIRS SST. The GTS-Drifter and HR-Drifter present the lowest centered RMS errors and highest correlations. It shows a higher clustering around the VIIRS SST during nighttime. It means that the nighttime difference between the *in situ* SST and the VIIRS SST is a little smaller than the daytime.

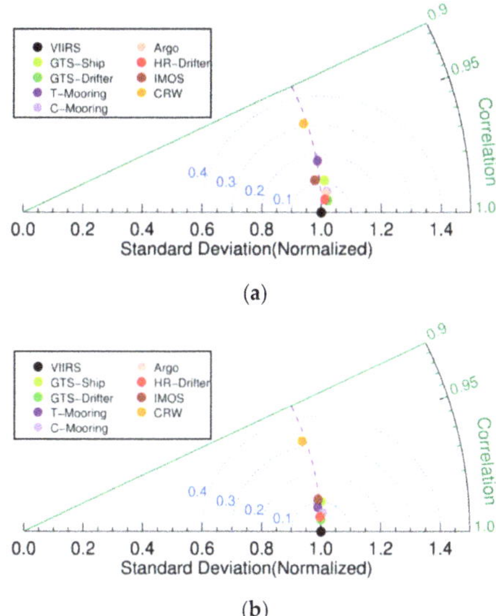

Figure 2. Comparisons between the NAVO S-NPP VIIRS SST and different types of *in situ* SST in Taylor diagram: (**a**) daytime; and (**b**) nighttime. Blue dot lines represent the normalized centered RMS error.

The values of centered RMS error and the standard deviation are listed in Table 2. Additionally, the number of matchups and the mean bias (VIIRS-*in situ*) are also provided to complement the information, since the overall bias has been removed in Taylor diagrams. The overall root-mean-square error (RMSE) can be derived from the square root of the sum of the squares of bias and centered RMS.

The difference is mainly coming from the observation errors by different sensors and the depth at which they measured. Donlon *et al.* [28] pointed out that the variability of vertical thermal structure at the upper ocean(~10 m) layer is complex, depending on the ocean mixing and air-sea exchange. Since the NAVO S-NPP VIIRS SST are regressed to a nominal depth 1 m, the mooring usually has a thermometer at ~3 m depth, drifter buoys are placed at ~0.2 m depth, ships and Argo measures several meters [29], the difference in near surface temperature vertical gradients should not be neglected. It results in a small positive bias for the NAVO S-NPP VIIRS SST with respect to the *in situ* SST which at greater depths, except the daytime tropical mooring. It should be caused by the NAVO S-NPP VIIRS SST retrieval algorithm. For the daytime algorithm, only two infrared channels M15 and M16, which are sensitive to columnar water vapor content, are used to derive the SST. In contrast, for the nighttime algorithm, an additional mid-infrared channel M12 is available and especially useful in the high water vapor conditions. The tropical moorings locate in the tropical regions where the water vapor is high. Thus, the high water vapor would affect the radiance in infrared channels and lead to large bias in daytime. A very small negative bias with RMSE less than 0.4 °C for both daytime and nighttime is found between the NAVO S-NPP VIIRS SST and GTS-Drifter and HR-drifter measured SST. Additionally, although a series of strict time and space collocation criteria is adopted in this study, the measurement at different time of day may also contribute to the bias. The centered RMS error of ship SST is higher than other *in situ* SST when compared with the NAVO S-NPP VIIRS SST.

5. Uncertainties in the VIIRS SST

The aim of validation in this study is to assess the magnitude and characteristics of uncertainties in the NAVO N-SPP VIIRS SST. Considering this, the overall RMSE of *in situ* measured SST should be small and the distributions of matchups should be uniform in space and time. Therefore, only the GTS-Drifter and HR-drifter (collectively called drifter) matchups are used to validate the NAVO N-SPP VIIRS SST according to the distributions showed in Section 3 and overall error statistics obtained in Section 4. The method of three-way error analysis enables the calculation of standard deviation of error on each observation from the collocations of three different types of observation. The matchups of the NAVO N-SPP VIIRS SST and drifter-measured SST are then collocated to the AQUA MODIS SST under the same match criteria.

Figure 3 shows the global distribution of mean biases in 5° × 5° boxes for the NAVO N-SPP VIIRS SST and the AQUA MODIS SST with respect to the drifter-measured SST during 2014. The biases are small and uniform in the majority of the global ocean. However, pronounced warm biases in the NAVO N-SPP VIIRS SST even up to 0.5 °C are observed in the Southern Hemisphere at high latitudes. Near-zero biases are observed in MODIS SST in these regions. It means that these biases may be caused by the NAVO N-SPP VIIRS SST retrieval algorithm rather than the nature of the SST.

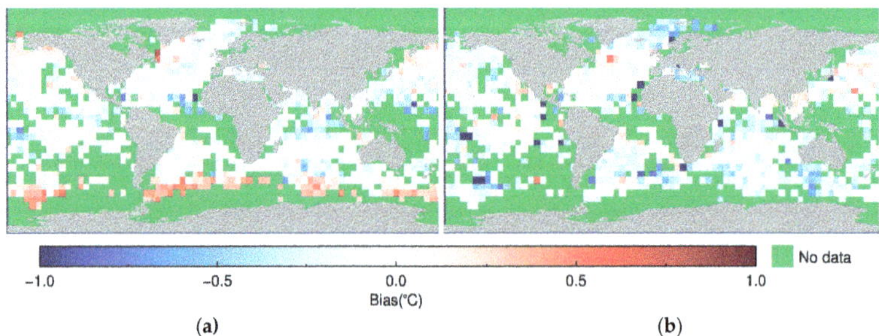

(a) (b)

Figure 3. Global distribution of mean biases in 5° × 5° boxes for (**a**) the NAVO N-SPP VIIRS SST and (**b**) the AQUA MODIS SST with respect to the drifter measured SST during 2014.

Three-way error analysis was firstly proposed to validate the wind speeds [30]. O'Carroll *et al.* [31] made a specific description for this method and applied it to calculate the standard deviation of error on two satellite sensors and drifting buoy SST. Similar analysis has been done to evaluate *in situ* data for satellite calibration and validation [15]. If the random errors σ are uncorrelated then they can be derived as follows:

$$\begin{aligned} \sigma_D &= \sqrt{0.5(V_{DV} + V_{DM} - V_{VM})} \\ \sigma_V &= \sqrt{0.5(V_{DV} + V_{VM} - V_{DM})} \\ \sigma_M &= \sqrt{0.5(V_{DM} + V_{VM} - V_{DV})} \end{aligned} \quad (3)$$

where *D*, *V*, and *M* indicate the observation types for Drifter, the NAVO N-SPP VIIRS, and the AQUA MODIS, respectively. V_{DV} is the variance of the difference between Drifter and the VIIRS, V_{DM} is the variance of the difference between Drifter and MODIS, V_{VM} is the variance of the difference between the VIIRS and MODIS.

Following O'Carroll *et al.* [31] and Xu *et al.* [15], the three-way error analysis is applied in this study to estimate the root mean square error in the NAVO N-SPP VIIRS SST, the drifter-measured SST and the AQUA MODIS SST. The equations have been solved in each 5° × 5° bin, and the results are shown in a form of three maps in Figure 4a,b respectively. In several bins, σ_V, σ_D and σ_M may be not exists when $V_{DV} + V_{DM} - V_{VM} < 0$, $V_{DV} + V_{VM} - V_{DM} < 0$ and $V_{VM} + V_{DM} - V_{DV} < 0$, respectively. It is likely because errors exist in the data or there is a violation of the non-correlation assumption and the respective boxes in Figure 4 are rendered as black color. For the drifter, σ_D values are the smallest and most uniform in the three types of observations. MODIS has the largest and most complex spatial structure of random error. There are large σ_M and σ_V in the Southern Hemisphere at high latitudes, which appear to follow and exist around the Antarctic circumpolar front. Large gradients exist in these regions. The large σ values may be due to real geophysical difference between a point measurement and a spatial average in high spatial varying regions. The diurnal variation usually more prominent in the skin layer which MODIS measures and the mismatch of time may lead to larger difference. Thus, the σ_V values lie between the σ_D and σ_M.

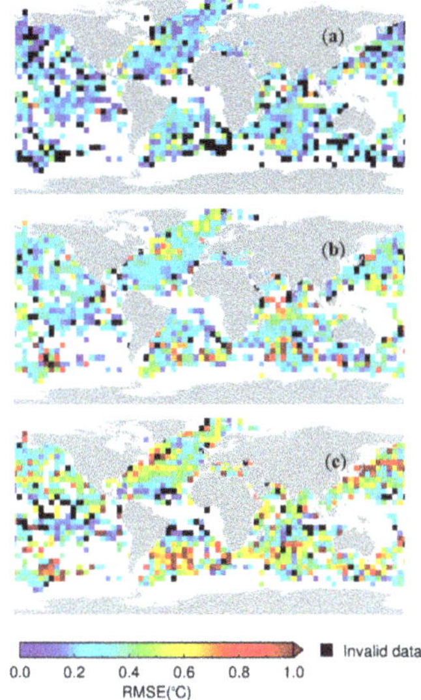

Figure 4. Global maps of root mean square errors (°C) in 5° × 5° resolution for (**a**) drifter measured SST; (**b**) NAVO S-NPP VIIRS SST; and (**c**) AQUA MODIS SST derived from three-way error analysis.

Figure 5 additionally plots histograms of σ_V, σ_D, and σ_M from Figure 4 in 0.05 °C bin width. As expected, the drifting buoy measurements have the smallest error. The value of σ_D varies from 0 to 0.8 °C with a median of ~0.24 °C. This estimate is in good agreement with the estimates of 0.23 °C by O'Carroll *et al.* [31] and 0.26 °C by Xu *et al.* [15]. For VIIRS, the RMSE range from 0 to 1 °C with a median of ~0.31 °C. For MODIS, σ_M is ~0.43 °C, but the histogram has a long tail extending out to 1.4 °C. The σ_M lies between previous investigations 0.38 °C [32] and 0.49 °C [33].

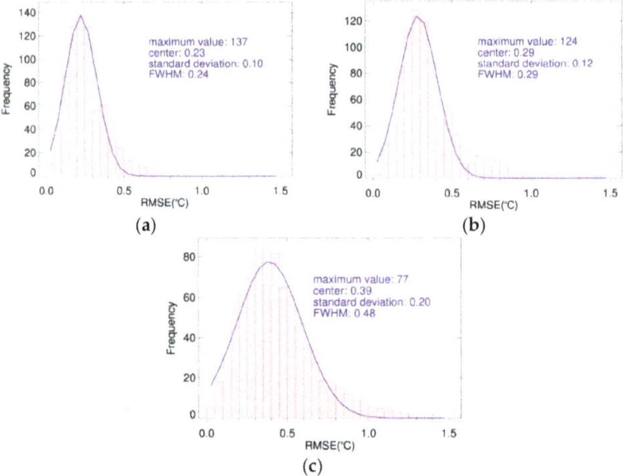

Figure 5. Histograms of root mean square errors in 0.05 °C bin width for (**a**) drifter measured SST; (**b**) NAVO S-NPP VIIRS SST and (**c**) AQUA MODIS SST, corresponding to Figure 4. The fitting of the Gaussian function to the RMSE is the blue line, the maximum value, center, standard deviation, and full-width-half-maximum (FWHM) of the Gaussian function are output on the plot as well.

Ideally, the distribution of random error should be normal. The Gaussian fitting shows normal in the range of 0–0.5 °C, 0–0.6 °C, and 0–0.8 °C for drifter measured SST, NAVO S-NPP VIIRS SST, and AQUA MODIS SST, respectively. However, the long tail on the right side does not fitting well in all the three type of observations. On the one hand, this can mainly result from a variety of causes that relate to how the SST is being made by the individual measurements or retrieval. For infrared remote sensing, the errors may result from the characteristics of the radiometer, and how well the measurements are calibrated, and the atmospheric correction algorithm. On the other hand, the validation techniques are also prone to error. The space-time mismatch leads to errors because of the nature of the SST variable and the temporal-spatial sampling difference in individual observations. Additionally, some of the drifting buoys may be used to derive both the VIIRS SST and MODIS SST retrieval algorithm coefficients, thus they are not completely uncorrelated and the RMSE estimated here may not be fully accurate by using three-way error analysis. The distributed characteristics of the RMSE should be considered for several analysis techniques, as in data assimilation and the merging of the satellite and *in situ* datasets.

6. Conclusions

This study was devoted to the analysis of the magnitude and characteristics of uncertainties in the NAVO S-NPP VIIRS SST. Eight types of *in situ* SST from five independent sources were collocated to the VIIRS SST within ±1 h and within ±0.05° of latitude and longitude. The distribution of the matchups shows that the drifters provide densest and most complete global coverage and other types of *in situ* data are only available in some limited areas.

An overall comparison results are performed in Taylor diagrams. They show that all types of *in situ* SST have very high correlation with the VIIRS SST except the CRW. The GTS-Drifter and HR-Drifter present very small negative mean bias and lowest center RMS errors less than 0.4 °C for both daytime and nighttime. For other types of *in situ* observations, ships measured SST have the largest center RMS errors 0.78 °C and 0.92 °C for nighttime and daytime, respectively, and positive mean bias exist. The bias may be due to the vertical gradients in near surface temperature. The NAVO VIIRS SST is at a nominal 1 m depth, which is deeper than the drifter measured SST and shallower than other *in situ* SST. However, a highest negative bias exists in the daytime tropical mooring. It may be due to the water vapor influence of daytime NAVO S-NPP VIIRS SST retrieval algorithm or time mismatch within 1 h when diurnal warming happening, or both.

Based on the overall comparison, drifters measured SSTs are selected to further investigate the uncertainty in the NAVO S-NPP VIIRS SST. Another SSTs retrieved from AQUA MODIS are used to help the analysis. The three-way error analysis shows that the errors are smallest and most uniform in drifter while largest and with most complex spatial structure in AQUA MODIS SST. The NAVO S-NPP VIIRS SST lie between the drifters measured SST and AQUA MODIS SST with high accuracy, and at a median RMSE of~0.31 °C ranging from 0–1 °C. The distribution of errors shows normality except for the long tail part on the right side and this should be considered when the SST are merged with other observations or assimilated into models. Global distribution of NAVO S-NPP VIIRS SST minus drifters measured SSTs shows pronounced warm biases up to 0.5 °C in the Southern Hemisphere at high latitudes, while near-zero biases are observed in AQUA MODIS SST minus drifters measured SSTs. It means that these biases may be caused by NAVO S-NPP VIIRS SST retrieval algorithm rather than the nature of the SST. The reasons and correction for this bias need to be further studied.

Acknowledgments: This study was supported by the Project of State Key Laboratory of Satellite Ocean Environment Dynamics, Second Institute of Oceanography (No. SOEDZZ1515) and the National Natural Science Foundation of China (No. 41476157).VIIRS and MODIS SST are available at GHRSST. The *in situ* data are obtained from iQuam/NOAA.

Author Contributions: Qianguang Tu designed the research, collected satellite and *in situ* data and implemented the data analysis as well as wrote the manuscript. Delu Pan and Zengzhou Hao guided the whole study including research contents, methodology *etc.* All the authors worked on the revisions of the manuscript.

Conflicts of Interest: The authors declare no conflict of interest.

References

1. Donlon, C.J.; Casey, K.S.; Robinson, I.S.; Gentemann, C.L.; Reynolds, R.W.; Barton, I.; Arino, O.; Stark, J.; Rayner, N.; Leborgne, P.; *et al.* The godae high-resolution sea surface temperature pilot project. *Oceanography* **2009**, *22*, 34–45. [CrossRef]
2. Castro, S.L.; Wick, G.A.; Jackson, D.L.; Emery, W.J. Error characterization of infrared and microwave satellite sea surface temperature products for merging and analysis. *J. Geophys. Res.-Oceans* **2008**, *113*. [CrossRef]
3. Reynolds, R.W.; Smith, T.M.; Liu, C.; Chelton, D.B.; Casey, K.S.; Schlax, M.G. Daily high-resolution-blended analyses for sea surface temperature. *J. Clim.* **2007**, *20*, 5473–5496. [CrossRef]
4. Donlon, C.J.; Martin, M.; Stark, J.; Roberts-Jones, J.; Fiedler, E.; Wimmer, W. The operational sea surface temperature and sea ice analysis (ostia) system. *Remote Sens. Environ.* **2012**, *116*, 140–158. [CrossRef]
5. Cao, C.; Xiong, J.; Blonski, S.; Liu, Q.; Uprety, S.; Shao, X.; Bai, Y.; Weng, F. Suomi npp viirs sensor data record verification, validation, and long-term performance monitoring. *J. Geophys. Res.-Atmos.* **2013**, *118*. [CrossRef]
6. Cao, C.; de Luccia, F.J.; Xiong, X.; Wolfe, R.; Weng, F. Early on-orbit performance of the visible infrared imaging radiometer suite onboard the suomi national polar-orbiting partnership (s-npp) satellite. *IEEE Trans. Geosci. Remote Sens.* **2014**, *52*, 1142–1156. [CrossRef]
7. Uprety, S.; Cao, C.Y. Suomi npp viirs reflective solar band on-orbit radiometric stability and accuracy assessment using desert and antarctica dome c sites. *Remote Sens. Environ.* **2015**, *166*, 106–115. [CrossRef]
8. Lei, N.; Wang, Z.P.; Xiong, X.X. On-orbit radiometric calibration of suomi npp viirs reflective solar bands through observations of a sunlit solar diffuser panel. *IEEE Trans. Geosci. Remote Sens.* **2015**, *53*, 5983–5990. [CrossRef]
9. Liang, X.; Ignatov, A. Avhrr, MODIS, and VIIRS radiometric stability and consistency in sst bands. *J. Geophys. Res.-Oceans* **2013**, *118*, 3161–3171. [CrossRef]
10. Liu, Y.; Yu, Y.; Yu, P.; Göttsche, F.M.; Trigo, I.F. Quality assessment of s-npp viirs land surface temperature product. *Remote Sens.* **2015**, *7*, 12215–12241. [CrossRef]

11. Petrenko, B.; Ignatov, A.; Kihai, Y.; Stroup, J.; Dash, P. Evaluation and selection of sst regression algorithms for jpss viirs. *J. Geophys. Res.: Atmos.* **2014**, *119*, 4580–4599. [CrossRef]
12. McKenzie, B.; May, D.; Cayula, J.-F.; Willis, K. Initial results of npp viirs sst processing at navoceano. *Ocean Sens. Monit. IV* **2012**, *8372*. [CrossRef]
13. Cayula, J.-F.P.; May, D.A.; McKenzie, B.D.; Willis, K.D. Viirs-derived sst at the naval oceanographic office: From evaluation to operation. *Ocean Sens. Monitor. V* **2013**, *8724*. [CrossRef]
14. Emery, W.J.; Baldwin, D.J.; Schlussel, P.; Reynolds, R.W. Accuracy of *in situ* sea surface temperatures used to calibrate infrared satellite measurements. *J. Geophys. Res.-Oceans* **2001**, *106*, 2387–2405. [CrossRef]
15. Xu, F.; Ignatov, A. Evaluation of *in situ* sea surface temperatures for use in the calibration and validation of satellite retrievals. *J. Geophys. Res. Oceans* **2010**, *115*. [CrossRef]
16. Donlon, C.; Robinson, I.; Casey, K.S.; Vazquez-Cuervo, J.; Armstrong, E.; Arino, O.; Gentemann, C.; May, D.; LeBorgne, P.; Piolle, J.; *et al.* The global ocean data assimilation experiment high-resolution sea surface temperature pilot project. *Bull. Am. Meteorol. Soc.* **2007**, *88*, 1197–1213. [CrossRef]
17. S-NPP VIIRS L2P Sea Surface Temperature from NAVO. Available online: ftp.nodc.noaa.gov/pub.data.nodc/ghrsst/L2P/VIIRS_NPP/NAVO (accessed on 23 August 2015).
18. Brown, O.B.; Minnett, P.J.; Evans, R.; Kearns, E.; Kilpatrick, K.; Kumar, A.; Sikorski, R.; Závody, A. *Modis Infrared Sea Surface Temperature Algorithm Algorithm Theoretical Basis Document Version 2.0*; University of Miami: Coral Gables, FL, USA, 1999.
19. OceanColor documents. Available online: http://oceancolor.gsfc.nasa.gov/DOCS/modis_sst/ (accessed on 20 November 2012).
20. Franz, B. Implementation of sst processing within the obpg. *Último Acceso* **2006**, *4*, 2014.
21. Walton, C.C. Nonlinear multichannel algorithms for estimating sea surface temperature with AVHRR satellite data. *J. Appl. Meteorol.* **1988**, *27*, 115–124. [CrossRef]
22. AQUA MODIS L2P Sea Surface Temperature from JPL. Available online: ftp.nodc.noaa.gov/pub/data.nodc/ghrsst/L2P/MODIS_A/JPL (accessed on 23 August 2015).
23. Xu, F.; Ignatov, A. *In situ* SST quality monitor (i quam). *J. Atmos. Ocean. Technol.* **2014**, *31*, 164–180. [CrossRef]
24. Ignatov, A.; Xu, F. *In situ* SST quality monitor: From iquam version 1 to version 2. In Proceedings of the 14th GHRSST Meeting, Woods Hole, MA, USA, 17 June 2013.
25. Roemmich, D.; Johnson, G.C.; Riser, S.; Davis, R.; Gilson, J.; Owens, W.B.; Garzoli, S.L.; Schmid, C.; Ignaszewski, M. The argo program observing the global ocean with profiling floats. *Oceanography* **2009**, *22*, 34–43. [CrossRef]
26. Kawai, Y.; Wada, A. Diurnal sea surface temperature variation and its impact on the atmosphere and ocean: A review. *J. Oceanogr.* **2007**, *63*, 721–744. [CrossRef]
27. Taylor, K.E. Summarizing multiple aspects of model performance in a single diagram. *J. Geophys. Res.-Atmos.* **2001**, *106*, 7183–7192. [CrossRef]
28. Donlon, C.J.; Minnett, P.J.; Gentemann, C.; Nightingale, T.J.; Barton, I.J.; Ward, B.; Murray, M.J. Toward improved validation of satellite sea surface skin temperature measurements for climate research. *J. Clim.* **2002**, *15*, 353–369. [CrossRef]
29. Sybrandy, A.; Niiler, P.; Martin, C.; Scuba, W.; Charpentier, E.; Meldrum, D. *Global Drifter Programme Drifter Design Reference*; DBCP Report; Data Buoy Cooperation Panel: Paris, France, 2009.
30. Stoffelen, A. Toward the true near-surface wind speed: Error modeling and calibration using triple collocation. *J. Geophys. Res.-Oceans* **1998**, *103*, 7755–7766. [CrossRef]
31. O'Carroll, A.G.; Eyre, J.R.; Saunders, R.W. Three-way error analysis between aatsr, amsr-e, and *in situ* sea surface temperature observations. *J. Atmos. Ocean. Technol.* **2008**, *25*, 1197–1207. [CrossRef]
32. Gentemann, C.L. Three way validation of MODIS and AMSR-E sea surface temperatures. *J. Geophys. Res.: Oceans* **2014**, *119*, 2583–2598. [CrossRef]
33. Minnett, P.J. The validation of sea surface temperature retrievals from spaceborne infrared radiometers. In *Oceanography from Space*; Springer: Berlin, Germany, 2010; pp. 229–247.

Article

The Potential of Autonomous Ship-Borne Hyperspectral Radiometers for the Validation of Ocean Color Radiometry Data

Vittorio E. Brando [1,2,3,*], Jenny L. Lovell [1], Edward A. King [1], David Boadle [4], Roger Scott [1] and Thomas Schroeder [1]

1 Commonwealth Scientific and Industrial Research Organisation (CSIRO), Oceans and Atmosphere, Hobart 7001, Australia; Jenny.Lovell@csiro.au (J.L.L.); Edward.King@csiro.au (E.A.K.); Roger.Scott@csiro.au (R.S.); Thomas.Schroeder@csiro.au (T.S.)
2 National Research Council of Italy, Institute for Electromagnetic Sensing of the Environment (CNR-IREA), Milano 20133, Italy
3 Currently at National Research Council of Italy, Institute of Atmospheric Sciences and Climate (CNR-ISAC), ROME 00133, Italy
4 Commonwealth Scientific and Industrial Research Organisation (CSIRO), Land and Water, Townsville 4811, Australia; David.Boadle@csiro.au
* Correspondence: vittorio.brando@csiro.au; Tel.: +61-3-6232-5041

Academic Editors: Changyong Cao, Xiaofeng Li and Prasad S. Thenkabail
Received: 29 November 2015; Accepted: 4 February 2016; Published: 16 February 2016

Abstract: Calibration and validation of satellite observations are essential and on-going tasks to ensure compliance with mission accuracy requirements. An automated above water hyperspectral radiometer significantly augmented Australia's ability to contribute to global and regional ocean color validation and algorithm design activities. The hyperspectral data can be re-sampled for comparison with current and future sensor wavebands. The continuous spectral acquisition along the ship track enables spatial resampling to match satellite footprint. This study reports spectral comparisons of the radiometer data with Visible Infrared Imaging Radiometer Suite (VIIRS) and Moderate Resolution Imaging Spectroradiometer (MODIS)-Aqua for contrasting water types in tropical waters off northern Australia based on the standard NIR atmospheric correction implemented in SeaDAS. Consistent match-ups are shown for transects of up to 50 km over a range of reflectance values. The MODIS and VIIRS satellite reflectance data consistently underestimated the *in situ* spectra in the blue with a bias relative to the "dynamic above water radiance and irradiance collector" (DALEC) at 443 nm ranging from 9.8×10^{-4} to 3.1×10^{-3} sr^{-1}. Automated acquisition has produced good quality data under standard operating and maintenance procedures. A sensitivity analysis explored the effects of some assumptions in the data reduction methods, indicating the need for a comprehensive investigation and quantification of each source of uncertainty in the estimate of the DALEC reflectances. Deployment on a Research Vessel provides the potential for the radiometric data to be combined with other sampling and observational activities to contribute to algorithm development in the wider bio-optical research community.

Keywords: ocean color radiometry; reflectance; ship-borne radiometry; validation

---

1. Introduction

Satellite sensors allow the observation of ocean color radiometry (OCR) that can be interpreted to monitor water constituents and quality [1,2]. The spectral remote sensing reflectance (R_{rs} (sr^{-1})) determined from top-of-atmosphere radiance is the primary ocean color product used for the generation of all higher-level products. Calibration and validation of OCR observations is a necessary and on-going requirement, to produce consistent time series of data and to ensure that processing methods accurately account for atmospheric and other environmental effects [1–4].

To determine R_{rs}, *in situ* radiometric measurements are performed with above- and in-water optical radiometers from fixed offshore platforms [5,6], moored buoys [7,8] or ships [9–11]. The above- and in-water approaches to determine R_{rs} rely on radiometric measurements analysed with different underlying assumptions: (i) in-water radiometry enables the determination of immediately below surface radiometric quantities based on the extrapolation of subsurface continuous or fixed-depth profiles of radiometric quantities [5,7,12]; (ii) above-water systems operate with non-nadir viewing geometry and can be corrected for the skylight reflected into the field-of-view by the air-sea interface [5,6,12,13].

Permanent fixed platforms and moorings supported by dedicated calibration and maintenance regimes provide the best quality observations for long-term calibration data, but ships can provide a suitable platform to collect spatially diverse ocean color data for global and regional validation purposes [2,4,14]. Targeted voyages with manually operated above-water radiometers can provide comprehensive datasets of a region at a given time [10,14–16], while autonomous above-water radiometers can provide more extensive spatial coverage if the near-surface effects such as sun glint, platform shadowing and spray are minimized by maintaining a suitable viewing geometry and by implementing adequate quality control and processing procedures [11,17–20].

As part of Australia's Integrated Marine Observing System (IMOS), an autonomous ship based system was commissioned to provide an above-water hyperspectral radiometry data-stream from Australian waters. The "Dynamic above water radiance and irradiance collector" (DALEC) was mounted on research vessels to capture data during daylight hours over multi-day voyages with minimal procedures for set-up, shutdown and maintenance. The hyperspectral data obtained from this instrument can be re-sampled for use in validation of a variety of satellite sensors. The instrument was initially deployed on the Australian Marine National Facility vessel, RV Southern Surveyor and collected data during 9 voyages between July 2011 and September 2012. This deployment resulted in refinement of the operating procedures and maintenance schedule requirements prior to deployment on the Australian Institute of Marine Science (AIMS) vessel, RV Solander.

To demonstrate the potential of the DALEC measurements for OCR validation, this study presents the comparison between OCR satellite imagery and autonomous ship based above-water hyperspectral radiometry for contrasting environments in tropical Australian waters. OCR data were acquired by the Aqua Moderate Resolution Imaging Spectroradiometer (MODIS) and the Suomi National Polar-Orbiting Partnership (SNPP) Visible Infrared Imaging Radiometer Suite (VIIRS). The next section presents the DALEC instrument deployment and processing procedure. Then the following section provides details on the study sites in north-western Australia, the MODIS and the VIIRS data processing and matchup analysis procedures. Results are presented in Section 4 and discussed in Section 5.

2. Ship Borne Above-Water Radiometry

2.1. Theoretical Background

In above-water radiometry the measurement geometry is defined by the solar zenith angle (θ_z), the sea-viewing zenith angle (θ_v), the sky-viewing zenith angle (θ_s), and the relative azimuth angle (ϕ) away from the solar plane (Figure 1). Measurements of the total upwelling sea surface radiance $L_u(\theta_v, \phi, \lambda)$ consist of the water-leaving radiance just above the sea surface $L_w(\theta_v, \phi, \lambda)$ as well the surface-reflected radiance $L_{sr}(\theta_v, \phi, \lambda)$ that includes direct (sun glint) and diffuse (reflected background sky) contributions [5,6,13,21]. The sea surface radiance reflection coefficient $\rho(\theta_v, \phi, \theta_z, \lambda)$ can be used to estimate $L_{sr}(\theta_v, \phi, \lambda)$ from sky radiance $L_{sky}(\theta_s, \phi, \lambda)$ measurements: $L_{sr} = \rho L_{sky}$.

Hence, to determine R_{rs} three (near-) simultaneous above-water measurements are needed, $L_u(\theta_v, \phi, \lambda)$, $L_{sky}(\theta_s, \phi, \lambda)$, and the total downwelling irradiance $E_d(\lambda)$:

$$R_{rs} \equiv \frac{L_w}{E_d} = \frac{L_u - L_{sr}}{E_d} = \frac{L_u - \rho L_{sky}}{E_d} - \epsilon \quad (1)$$

where $\epsilon(\lambda)$ is a residual sunglint and skyglint term. As the measured L_{sky} is typically an order of magnitude or more greater than L_w, using the correct value for ρ is critical for the estimation of R_{rs} [5,13,22]. In clear skies for a theoretical level surface, ρ equals the Fresnel reflectance, which is a function only of θ_v. Otherwise ρ is strongly dependent on sky conditions, viewing geometry, sea state, and to a lesser extent, wavelength [13]. To minimize the near-surface effects such as sun glint, platform shadowing and sea spray, angles of $\theta_v = 40°$ and $90° < \phi < 135°$ (ideally 135°) are considered suitable for ship-borne above-water radiometry [5,6,9,11,13,21,23]. Under these optimized viewing angles, ρ varies between 2.5%–8% at 550 nm with varying sea surface roughness and cloud cover [9,11,13,24].

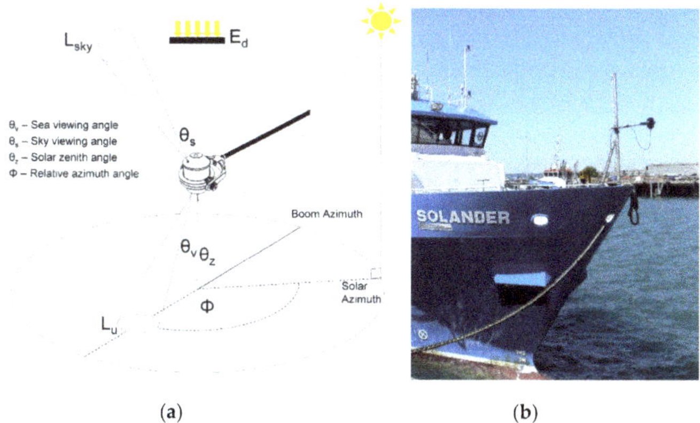

Figure 1. The "dynamic above water radiance and irradiance collector" (DALEC) hyperspectral radiometer: (**a**) Instrument schematics and measurement geometry (courtesy of *in situ* Marine Optics) (**b**) instrument mounted on RV Solander.

Several approaches have been developed to estimate ρ (and ε) as a function of measurement geometry, wind speed, cloud cover and wavelengths, even if in most cases ρ and ε are assumed to be spectrally invariant. Based on numerical simulations using Cox–Munk sea surfaces [25] and unpolarized ray tracing, Mobley [13] tabulated ρ (M99 hereafter) for clear skies as a function of Sun zenith angle, viewing direction and wind speed (W). Based on experimental results, Toole, Siegel, Menzies, Neumann and Smith [9] proposed a set of empirical values for ρ and ε for broad classes of cloud cover and wind speed. Following interpolation of the M99 tabulated ρ, Ruddick [10] modeled ρ dependence on sea state as a function of W and based on experimental results proposed a theoretical spectral shape for R_{rs} in the NIR (the "similarity spectrum") that can be used to estimate a spectrally invariant ε. Lee, Ahn, Mobley and Arnone [22] proposed a spectral optimization scheme based on a function of spectral inherent optical properties to estimate spectrally variant ρ and a spectrally invariant ε. Simis and Olsson [11] proposed the "fingerprint method", a spectral optimization minimizing the propagation of atmospheric absorption features to R_{rs} in clear to overcast skies. Garaba, Voss, Wollschlager and Zielinski [17] proposed a multi-model average of four approaches to retrieve the best approximation of R_{rs}. Recently Mobley [24] proposed a revised ρ table (M15 hereafter) computed using elevation- and slope-resolving surfaces and polarized ray tracing.

2.2. The DALEC Instrument

The DALEC is a hyperspectral radiometer developed in Australia by "*In situ* Marine Optics" (IMO hereafter) to measure $L_u(\theta_v, \phi, \lambda)$, $L_{sky}(\theta_s, \phi, \lambda)$, and $E_d(\lambda)$ in a simultaneous fashion during autonomous ship-based deployment (Figure 1). The DALEC sensor head contains three compact hyperspectral spectroradiometers (Carl Zeiss Monolithic Miniature Spectrometers), as well as a GPS and pitch and roll sensors, and is designed to be mounted on a boom positioned over the water, typically off the ship's bow (Figure 1). A deck unit contains a data logger, batteries, and a charging circuitry.

Each spectroradiometer records 200 channels (400–1050 nm) with spectral resolution of 10 nm, spaced at ~3.3 nm intervals. The Zeiss spectroradiometers in the DALEC measure in the 305–1140 nm range, but the system includes a UV filter with the dual purpose of guarding against second order effects in the diffraction grating and providing a wavelength range for measurement of the spectrometer dark current during data collection [26].

Following Mobley [13] recommendations, radiance channel viewing angles are fixed to 40° off nadir (L_u) and zenith (L_{sky}) when the sensor is held level (Figure 1). The L_u and L_{sky} sensors have a 5° field of view and the $E_d(\lambda)$ sensor has a cosine-like response. A passive gimbal mount with adjustable damping stabilizes the instrument while the ship is in motion to ensure consistent measurement geometry. An embedded compass, GPS and motor control adjust ϕ during data collection. To avoid viewing the ship, the DALEC automatically seeks the "ideal" measurement geometry within user-defined boom-relative limits (*i.e.*, $\phi = 135°$, or at least $\phi > 90°$). Pitch and roll sensors record data for quality control purposes.

2.3. Deployment on Board RV Solander

The DALEC instrument has been deployed since July 2013 on the RV Solander, operated by the AIMS across Australia's tropical waters (Figure 1). Custom fabricated mounting assemblies provide a robust rapid deployment and retrieval system for the DALEC on the forestay of RV Solander. This platform for observations enables controlled deployment and retrieval as well as quick servicing and cleaning. The instrument is deployed approximately two meters clear of the RV Solander's foredeck protruding one meter forward of the bow at approximately 7 meters above the sea surface (Figure 1). This provides an uninterrupted azimuthal sea viewing angle of ~270° and a sampling footprint surface for L_u of ~0.5 m^2. Furthermore it reduces the effects of sea spray from the ship's bow.

The DALEC is deployed opportunistically on the RV Solander on research voyages where staff can oversee the operation and daily maintenance of the instrument. The instrument is operated daily from 9AM to 4PM with data collection at approximately 1 Hz. Under normal operating conditions the instrument is left in place on the deployment boom and retrieved only for routine maintenance. The deployment is reviewed when the sea state reaches 2 m rising and when the sea state reaches 2–3 m the sensor head is retrieved and the boom is stowed against the forestay. The DALEC setup on the RV Solander has approximately 10 h of battery autonomy. The instrument is operated in the daylight and then retrieved in the vessel's dry lab for cleaning and battery charging. The three spectroradiometers in the DALEC are factory-calibrated annually against a NIST traceable lamp.

2.4. Data Management and Analysis

The radiometric data stream from the DALEC can be collected in real time through a dedicated WiFi network, however in the current installation data are internally logged then downloaded and archived at the end of the day. At the end of each research voyage, radiometric calibration (Level 1B) is applied using IMO's proprietary processing software (DALECproc v3.45). The radiometric calibrated data is then exported as CF compliant NetCDF and uploaded onto the IMOS Ocean Portal [27,28] where the data is freely available to the oceanographic community.

In this study, the radiometric calibrated DALEC spectra were filtered by applying thresholds on ancillary quality control parameters including pitch (<3°) and roll (<3°), sun zenith angle (<80°) and ship geometry (view angle > 60° from bow) [29]. As the wavelength factory calibrations of each Zeiss spectrometer differ slightly, to achieve a coherent wavelength alignment prior to reflectance processing, the spectra from the three sensors were re-sampled to 2 nm spacing with a spline interpolation [10]. To calculate R_{rs} with Equation (1) from the simultaneous above water measurements, the appropriate value for $\rho\,(\theta_v = 40°, \phi, \theta_s, W)$ was retrieved from the M15 tabulated ρ using an interpolated look-up-table approach. As the wind speed is not logged on RV Solander, a constant value of W = 3 ms^{-1} was used for data processing, consistent with the assimilated wind field analyses for the selected dates distributed by the Australian Community Climate and Earth-System Simulator (ACCESS) Numerical Weather Prediction system [30].

To correct for instantaneous sunglint effects captured in the ~0.5 m^2 sampling footprint surface for L_u, ϵ values were estimated by iteratively adjusting R_{rs} to the Ruddick [10] "similarity spectrum" in the 700–800 nm spectral range. To minimize the perturbing effects of surface roughness in the estimate of R_{rs}, Hooker, Lazin, Zibordi and McLean [5] suggested aggressively filtering out the higher measured L_u values. Hence, for the operational processing of autonomous above water radiometry Zibordi *et al.* [31] selected the lowest 20% of the L_u spectra to retrieve L_w. In this study we used spectra in the 5–25 percentile range of R_{rs} at 443 nm in the aggregation period (or at least 5 spectra) to eliminate the effects of potential extreme outliers. Temporal aggregation of the DALEC spectra was performed over a period of time equivalent to approximately one kilometer of transit (~3 minutes at 10 knots) to calculate the average and standard deviation to be used for further analyses.

3. Study Sites and OCR Imagery

In this study, DALEC data was compared to MODIS and VIIRS imagery to assess the potential of the continuous ship-borne radiometry measurements for OCR validation. Five clear sky dates were selected among the DALEC deployments on RV Solander carried out in April–August 2015 in Northwestern Australian waters (Figure 2, Table 1).

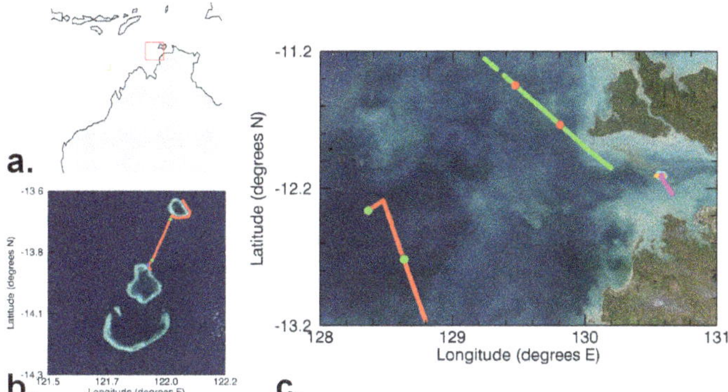

Figure 2. Study site. (**a**) Northern Australia location map for DALEC data used in this study, the green box indicates the position of image (**b**) while the red box indicates the position of image (**c**). True color images are from Moderate Resolution Imaging Spectroradiometer (MODIS) acquired on 12 April 2015 at 05:54 UTC (**b**) and on 24 May 2015 at 04:25 UTC (**c**). Contrasting dots overlaying the red and green transects identify segments of DALEC data used in matchups, as described in Section 4 and Table 1.

Table 1. Satellite overpass date and times and DALEC time intervals reported in Universal Time (UTC).

Site	Date	VIIRS Overpass Time	MODIS Overpass Time	DALEC Intervals
Scott Reef	12 April 2015	05:54	05:50	04:30–05:30
Beagle Gulf	24 May 2015	04:25	04:45	03:34–03:58; 05:25–06:44
Beagle Gulf	6 August 2015	04:37	05:25	04:35–04:52
Timor Sea	7 August 2015	04:18	04:30	02:30–04:48
Timor Sea	14 August 2015	05:26	04:35	02:37–07:12

3.1. Scott Reef

Scott Reef is an isolated coral reef system that rises steeply from north-western Australia's continental shelf. The waters around the reef are far removed from the influences of continental Australia and are dominated by the Indonesian through-flow, a relatively oligotrophic, warm low salinity current [32]. The reef experiences high tidal ranges that alternately expose and mostly submerge the reef structure leading to large internal waves at semi-diurnal frequencies [33]. These waves can bring nutrients from below the thermocline into the euphotic zone. DALEC data were acquired in the deep channel between the northern and southern atolls (namely, Seringapatam Reef and Northern Scott Reef) of the reef system (Figure 2b). The transect spans approximately 30 km and was acquired on 12 April 2015 (Table 1). All dates and times reported hereafter are expressed in Coordinated Universal Time (UTC), the notation UTC will be dropped for readability.

3.2. Beagle Gulf and Timor Sea

Beagle Gulf is the body of water into which Darwin Harbour opens. It is open to the west to the Timor Sea. The shallow waters in this area are influenced by strong tidal flows and high bottom stress resulting in intense vertical mixing. The water is turbid and may include entrainment of nutrients from deeper waters [34,35]. DALEC data were acquired on two voyages within this region. The first, on 24 May 2015 recorded data near the IMOS Darwin mooring located in the Beagle Gulf (12°24.0'S, 130°46.1'N) [36] (cyan track in Figure 2c) and the return transit to Darwin (magenta track in Figure 2c). During a second voyage in August 2015, DALEC data were acquired in Beagle Gulf (orange track in Figure 2c) on 6 August and further offshore in the Timor Sea on 7 August (green track in Figure 2c) and 14 August (red track in Figure 2c).

3.3. Satellite Data Processing and Matchup Analysis

VIIRS and MODIS data for the five selected dates (Table 1) were processed consistently using SeaDAS 7.2 [37]. The MODIS L0/PDS files acquired from NASA or local receiving stations were processed to L2 using SeaDAS with default calibration settings. The central wavelengths of the MODIS OC bands used in this study are 412, 443, 488, 531, 547, 667, 678 nm and the bandwidths range from 10–15 nm. The VIIRS data were downloaded from NASA as calibrated L1/SDR products and processed from L1 to L2 using SeaDAS with vicarious calibration set to match NASA R2014.0.1 reprocessing [38,39]. The central wavelengths of the VIIRS OC bands used in this study are 410, 443, 486, 551, 671 nm and the bandwidths are approximately 20 nm. For both sensors, R_{rs} was retrieved using the standard NIR atmospheric correction implemented in SeaDAS [40,41]; the out-of-band correction for water-leaving radiances was not applied.

To retain the satellite pixels at their natural resolution without averaging or sampling, ungridded MODIS and VIIRS R_{rs} data in swath format was used in the match-up analysis with the DALEC R_{rs} spectra. For comparison of spectra at time of overpass, DALEC spectra were aggregated over a period of time equivalent to approximately one kilometer in ship transit distance and used to calculate the mean and standard deviation at each wavelength. The closest pixel (in longitude-latitude space) to the mean position of the DALEC was identified in the satellite data swath and a 3×3 pixel subset was extracted from which we calculated the mean and standard deviation. To extend the range of comparisons, transects of DALEC data spatially aggregated to approximately 1 km scale were compared with satellite pixels. The pixels chosen for match-up were the closest pixel (in longitude-latitude space) to the mean position of the DALEC in each aggregation period.

As the 10 nm native spectral resolution of the DALEC radiometers is similar to the bandwidth of the OCR sensors, DALEC R_{rs} values were extracted for the wavelength closest to the band center of each satellite waveband, *i.e.*, the spectral response functions were not applied to the hyperspectral data. The comparison of R_{rs} values obtained from the DALEC and the satellite data is presented and summarized through the statistical metrics of the Mean Absolute Percent Difference (MAPD), the Root

Mean Squared Difference (RMSD), the correlation coefficient (R) and the bias (see Appendix). MAPD, RMSD, and bias used DALEC as the independent variable. These matchup summary statistics were calculated for each spectral band as well as aggregated over wavelengths.

4. Results and Discussion

This study presents the comparison between R_{rs} from VIIRS and MODIS ocean color imagery and the DALEC autonomous ship-borne above-water hyperspectral radiometry for contrasting environments in tropical Australian waters during April–August, 2015.

4.1. Scott Reef Transect

On 12 April 2015, at the time of the VIIRS overpass (05:54) and MODIS overpass (05:50), RV Solander was very close to the reef (blue dot in Figure 2b), so DALEC data from a South to North transect between the two sections of the reef spanning the time range 04:30–05:30 (green dots in Figure 2b) was compared to the satellite R_{rs}. DALEC data were aggregated spatially to approximately 1km scale for comparison with the closest satellite pixel. The time periods of DALEC aggregation were approximately 3 minutes.

Figure 3 illustrates the processing steps outlined in Section 2.4 to derive and aggregate R_{rs} from the simultaneous above water measurements. In this example, DALEC spectra (193 spectra) were extracted from a 1 km segment of the transect acquired at Scott Reef, at approximately 4 km from the first green dot in Figure 2b. The coefficients of variation (CV = standard deviation/average) at selected wavelengths for the DALEC spectra shown in Figure 3 are summarized in Table 2.

In this example, the L_u (Figure 3a) spectra exhibit larger variability than the associated E_d and L_{sky} (Figure 3b) spectra (e.g., CV for L_u (443) and L_{sky} (443) were 1.79% and 0.62%, respectively, Table 2). The variability in the individual L_w (Figure 3c) observations (instantaneous L_w) is then transferred to the instantaneous R_{rs} (Figure 3d) (e.g., CV at 443 nm were 2.18% and 2.10% respectively). In the red and NIR spectral range the variability in the instantaneous L_w and R_{rs} are one to two orders of magnitude larger than in the blue green regions due to the small absolute signal (e.g., CV at 671 nm and 745 nm were ~36% and 320%, respectively). In Figure 3e, the removal of ϵ by iteratively adjusting R_{rs} to the Ruddick [10] "similarity spectrum" leads to tighter spectra in the red and NIR spectral range while the variability in the blue and green spectral region is still retained (e.g., CV at 443 nm at 745 nm were 2.12% and 30%, respectively). By selecting the instantaneous R_{rs} in the 5%–25% range, the variability is then reduced in the blue and green spectral range, while in the red and NIR region it remains similar (e.g., CV at 443 nm at 745 nm were 0.87% and 28.3%, respectively).

Figure 4 reports the comparison between the DALEC aggregated R_{rs} and the VIIRS and MODIS R_{rs}. The example spectrum in Figure 4a is the result of the processing example in Figure 3f and is located approximately 4 km north of the first green dot in Figure 2b. R_{rs} values in blue, green and red bands for the one hour time period are shown in Figure 4b. The DALEC R_{rs} values at 443 nm decreased from ~0.010 to 0.067 sr^{-1} along the South to North transect, while R_{rs} values at 551 nm were constant at ~0.017 sr^{-1}. Overall there was a good agreement between DALEC and both satellite sensors for the clear, blue waters at this location (the spectrally aggregated R^2 was 0.985 and 0.993 for VIIRS and MODIS respectively, Table 3), but both satellite sensors underestimated DALEC reflectance in the blue bands (Figure 4a,b). Scatter plots of R_{rs} comparison of DALEC spectra with VIIRS and MODIS (Figure 4c,d) and the spectrally resolved summary statistics (Figure 5) show that the bias was almost nil for wavelengths above 500 nm, whilst there was a negative bias in the blue bands ranging −0.002 to −0.001 sr^{-1}. Moreover, VIIRS R_{rs} were closer to DALEC at 410 nm than MODIS at 412 nm (MAPD of 9.8% and 14.5% respectively), but the opposite behavior was observed at 443 nm (MAPD for VIIRS and MODIS of 22.2% and 17.7% respectively).

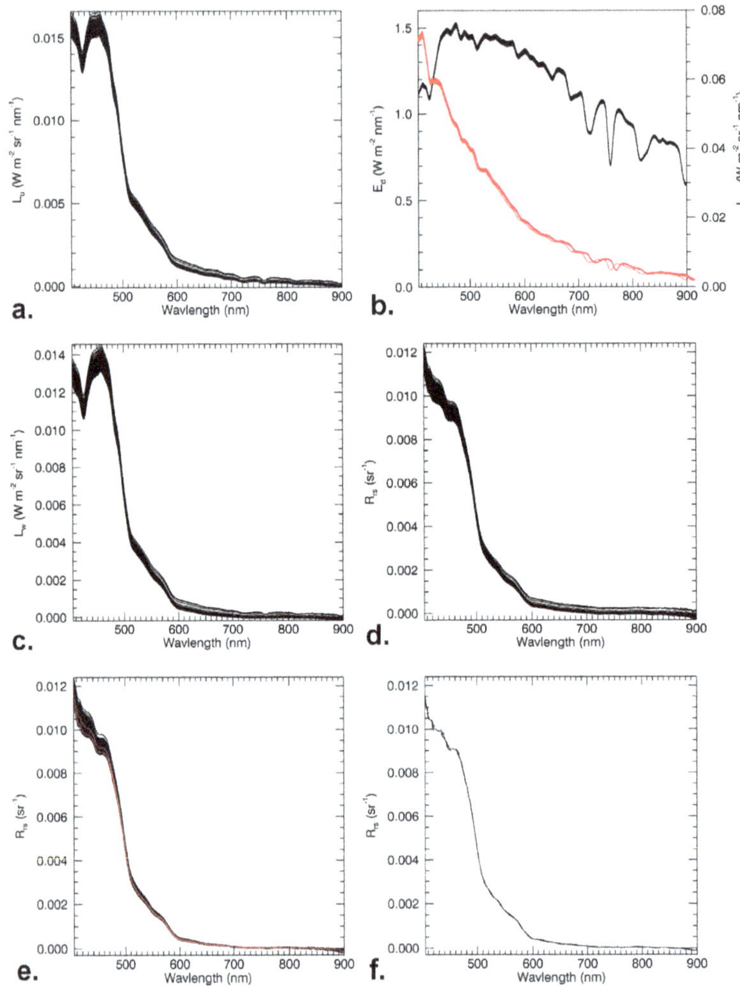

Figure 3. Example of DALEC processing sequence at Scott Reef on 12 April 2015 for 193 spectra spanning approximately 1 km. (**a**) L_u; (**b**) E_d and L_{sky} on two axes (L_{sky} in red on the right axis); (**c**) L_w; (**d**) instantaneous R_{rs}; (**e**) instantaneous R_{rs} after similarity spectrum correction with 5th and 25th percentile spectra indicated in red; (**f**) average and standard deviation of the aggregated R_{rs} (*i.e.*, of the 5–25 percentile range of the spectra).

Table 2. Coefficients of variation (CV = standard deviation/average) for all radiometric quantities for the DALEC spectra shown in Figure 3.

	CV at 443 nm	CV at 551 nm	CV at 671 nm	CV at 745 nm
L_u	1.79%	2.34%	7.94%	13.60%
E_d	0.33%	0.34%	0.35%	0.34%
L_{sky}	0.62%	0.72%	0.86%	0.90%
L_w	2.18%	3.36%	35.70%	315.00%
R_{rs}	2.10%	3.38%	36.20%	329.00%
$R_{rs} = R_{rs} - \epsilon$	2.12%	2.25%	7.36%	30.00%
$R_{rs} = R_{rs} - \epsilon$; (5%–25% iles)	0.87%	1.77%	7.36%	28.30%

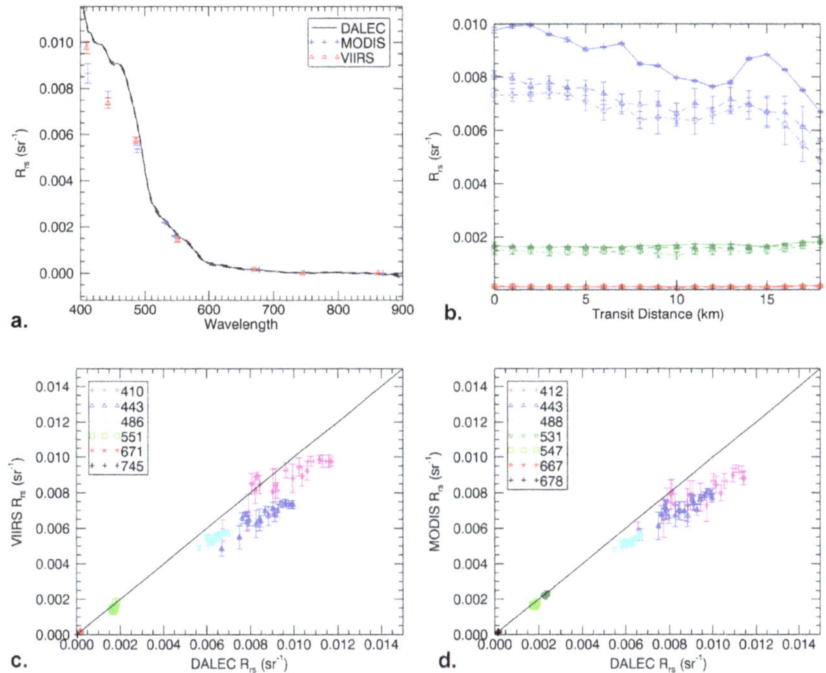

Figure 4. Spectral comparison at 1km scale between DALEC, VIIRS and MODIS acquired at Scott Reef on 12 April 2015. (**a**) DALEC, VIIRS and MODIS mean and standard deviation (DALEC standard deviation shown as dashed lines); (**b**) Transect plot of DALEC (+), VIIRS (triangle) and MODIS (diamond) at three spectral bands: the blue (443 nm), green (DALEC and VIIRS, 551 nm; MODIS 547 nm) and red (DALEC and VIIRS, 671 nm; MODIS 667 nm); (**c**) DALEC and VIIRS data; (**d**) DALEC and MODIS data. Black line is 1:1 in both (**c**) and (**d**). In all cases, error bars on DALEC data represent the standard deviation over the aggregation period and error bars on satellite data indicate the standard deviation over a 3x3 neighborhood of pixels.

Table 3. Spectrally aggregated match-up summary statistics. N is the number of pixels used in the statistical calculations.

	Date	MAPD	RMSD	Bias	R^2	N
VIIRS	12 April 2015	13.87	9.56×10^{-4}	-5.72×10^{-4}	0.985	19
	24 May 2015	15.64	1.81×10^{-3}	-7.78×10^{-4}	0.958	29
	7 August 2015	15.33	1.27×10^{-3}	4.97×10^{-4}	0.977	45
	14 August 2015	17.57	1.01×10^{-3}	-5.19×10^{-4}	0.916	54
	All dates	16.04	1.27×10^{-3}	-2.68×10^{-4}	0.961	147
MODIS	12 April 2015	12.05	8.36×10^{-4}	-4.71×10^{-4}	0.993	19
	24 May 2015	17.56	3.49×10^{-3}	-1.24×10^{-3}	0.911	28
	7 August 2015	15.40	1.68×10^{-3}	5.89×10^{-4}	0.960	49
	14 August 2015	16.17	6.87×10^{-4}	-3.26×10^{-4}	0.951	55
	All dates	15.67	1.86×10^{-3}	-2.18×10^{-4}	0.916	153

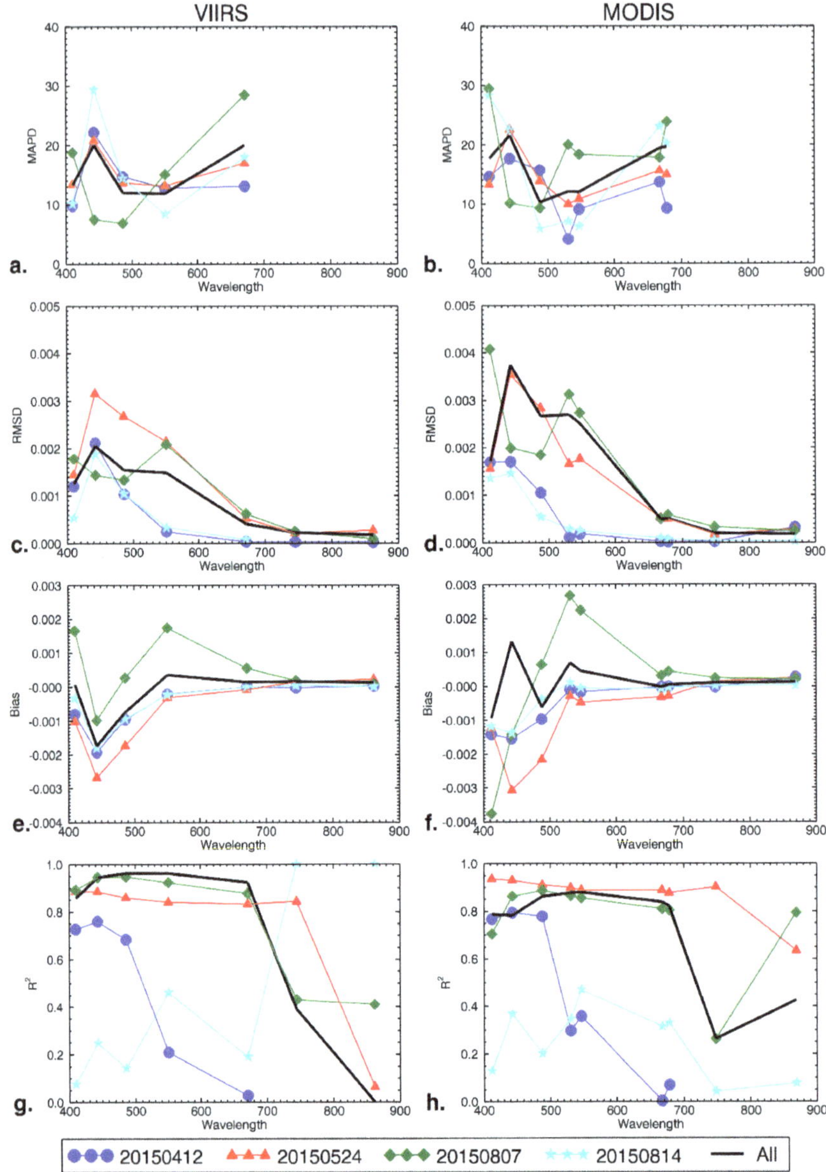

Figure 5. Spectrally resolved match-up summary statistics. (**a**) VIIRS MAPD; (**b**) MODIS MAPD; (**c**) VIIRS RMSD; (**d**) MODIS RMSD; (**e**) VIIRS bias; (**f**) MODIS bias; (**g**) VIIRS R^2; (**h**) MODIS R^2. MAPD has not been included for the NIR channels on as the very small signal magnitudes result in spurious values of relative difference. The very small range of data values in NIR at Scott Reef (12 April 2015) produced spurious correlation results, so these have not been plotted.

4.2. Beagle Gulf Mooring

DALEC data on 24 May 2015 and 6 August 2015 sample the same stretch of water near the IMOS Darwin mooring located in the Beagle Gulf. Satellite match-ups for 24 May (Figure 6a) used a 24 min aggregation of DALEC spectra (03:34–03:58) within 1 h of the MODIS overpass (04:45) and within 30 min of VIIRS overpass (04:25). The DALEC acquisitions on this transect terminated at 03:58. The satellite overpass times on 6 August were 04:37 for VIIRS and 05:25 for MODIS. DALEC data were aggregated over 17 min from 04:35 to 04:52 (Figure 6b). A second aggregation period closer to the MODIS overpass time produced very similar results (not shown) for MODIS with a slightly poorer matchup with VIIRS.

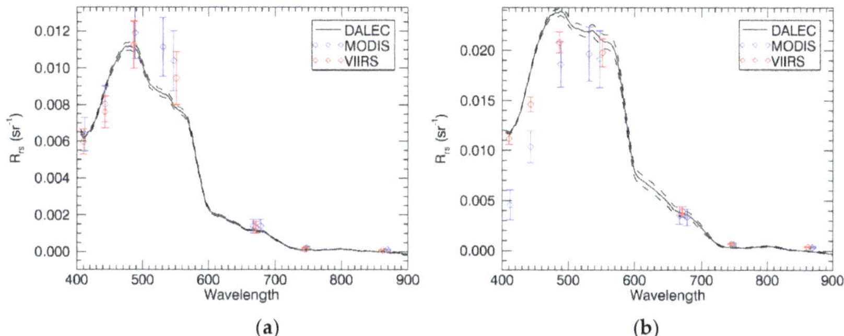

Figure 6. Spectral comparison between DALEC, MODIS and VIIRS near the Integrated Marine Observing System (IMOS) Darwin mooring located in in the Beagle Gulf. (**a**) 24 May 2015; (**b**) 6 August 2015. DALEC mean and standard deviation (dashed lines) over the aggregation period are shown. Error bars on satellite data indicate the standard deviation over a 3x3 neighborhood of pixels.

The spectra peaked in the green and did not vary significantly over the DALEC aggregation times. The spectrum on 6 August had a broader peak and overall higher intensity than the 24 May 2015 (R_{rs} at 490 nm of 0.0113 and 0.0244 sr^{-1} for 6 August 2015 and 24 May 2015 respectively), due to a higher amount of particulate matter suspended in the water column. Error bars on the satellite data are for a 3 × 3 neighborhood of pixels. This is a larger area than the DALEC was sampling over the time period and likely represents spatial variability rather than true uncertainty. The satellite data do not capture the magnitude of the reflectance on 6 August, but VIIRS reflectance in the blue is closer to DALEC than is MODIS.

4.3. Beagle Gulf and Timor Sea Transects

Transect data were acquired over 3 dates in different parts of the Timor Sea (Figure 2c, Table 1). The first, on 24 May 2015 is en-route through the Beagle Gulf from the IMOS Darwin mooring to Darwin. The second, on 7 August approaches within approximately 9 km of land (Tiwi Islands) while the third, on 14 August is well off-shore in the Timor Sea. The segments of DALEC transects on 7 and 14 August included in this analysis are indicated by the red and green dots respectively in Figure 2c. All spectra acquired along these transects peaked near 490 nm: R_{rs} at 490 nm was ~ 0.025 sr^{-1} in the spectra acquired on 24 May 2015 and 7 August 2015, while for those acquired on 14 August 2015 R_{rs} at 443 and 490 nm was ~0.007 sr^{-1} (Figure 7). There was a good agreement between DALEC and both satellite sensors for the green waters on these transects (the spectrally aggregated R^2 ranged 0.91 to 0.97 for VIIRS and MODIS, Table 3). Consistent with the Scott Reef transect, the bias between DALEC spectra and VIIRS and MODIS was almost nil for wavelengths above 530 nm in the spectra acquired on 24 May 2015 and 14 August 2015 (Figure 5, Figure 7), with a negative bias in the blue bands ranging −0.002 to −0.001 sr^{-1} for 14 August 2015 and reaching ~−0.003 sr^{-1} at 443 nm for 24 May 2015. The spectra acquired on 7 August 2015 showed significant bias in all the spectral range, being the only transect of this study with a positive bias at 550 nm.

The spectral MAPD ranged 10%–30% for MODIS and VIIRS and had similar spectral shape to the Scott Reef transect for the spectra acquired on 24 May 2015 and 14 August 2015, while on 7 August

2015 the MAPD at 410 nm was higher than at 443 nm. The larger scatter in MODIS on 7 August 2015 partially reflects a difference in flagging between VIIRS and MODIS. The data points that significantly exceed DALEC reflectances in the green on 7 August are the furthest in time from the satellite overpass and the closest to land. Some of these differences may be explained by the short term variability in dissolved and particulate matter in the water column due to wind- and tide-driven resuspension, as well as aerosol variability due to the land mass proximity.

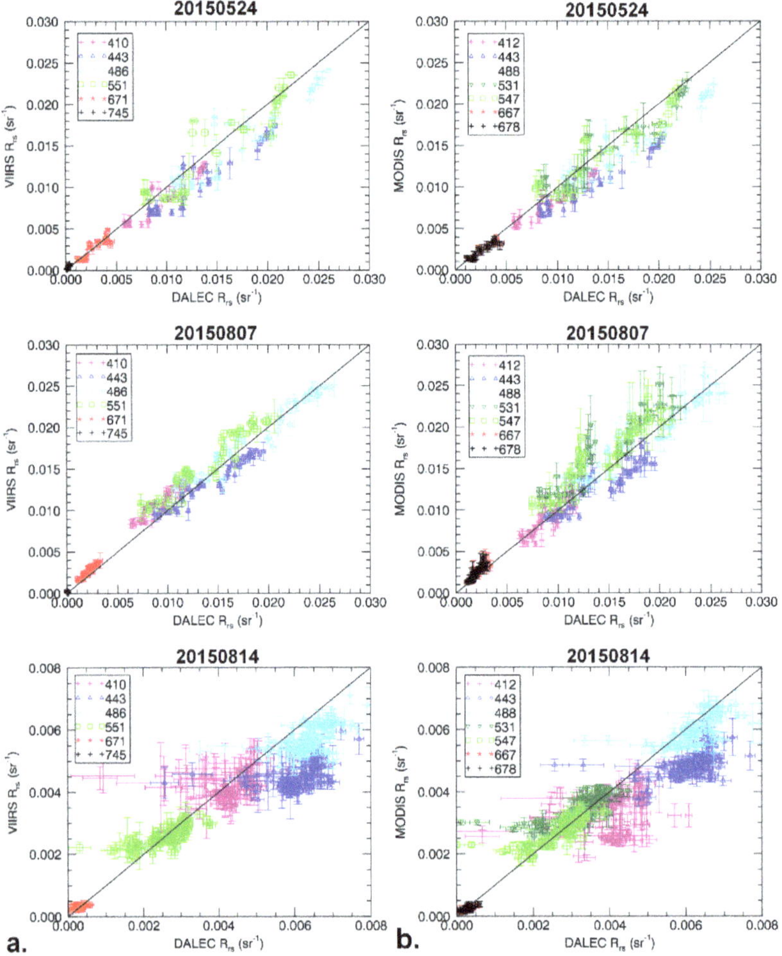

Figure 7. Spectral comparison in the Timor Sea for the dates shown and transects illustrated in Figure 2c, (**a**) VIIRS data; (**b**) MODIS data. DALEC data are aggregated to approximately 1km scale. Error bars on DALEC data represent the standard deviation over the aggregation period and error bars on satellite data indicate the standard deviation over a 3x3 neighborhood of pixels.

4.4. Sensitivity to DALEC Processing Parameters.

To evaluate the sensitivity of the DALEC R_{rs} to data processing parameters detailed in Section 2.4, instantaneous R_{rs} were calculated with varying W and ρ tables from all the 77255 sets of simultaneous above water measurements collected in the five dates. As wind speed is not logged on RV Solander, a constant W = 3 ms^{-1} was used in all analyses shown above to select $\rho\,(\theta_v = 40^\circ, \phi, \theta_s, W)$ from the

Mobley [24] tabulated ρ. To evaluate the effect of this assumption, all instantaneous R_{rs} were also calculated with W = 1 ms^{-1} and W = 5 ms^{-1} for comparison. Figure 8a shows that the effect of the variations of wind speed over the likely range during the observations is small (MAPD and bias of 1.6% and 1.8×10^{-4} at 443nm, respectively in the sample of 77255 R_{rs} spectra). Thus, the assumption of constant wind speed is not likely to significantly affect the results of the matchup analyses as the bias between DALEC and OCR R_{rs} was one order of magnitude larger than the effects of wind speed (e.g., bias at 443 nm was -3.0×10^{-3} to -1.20×10^{-3} sr^{-1} for MODIS and VIIRS *vs.* 1.8×10^{-4} sr^{-1}, Figure 5).

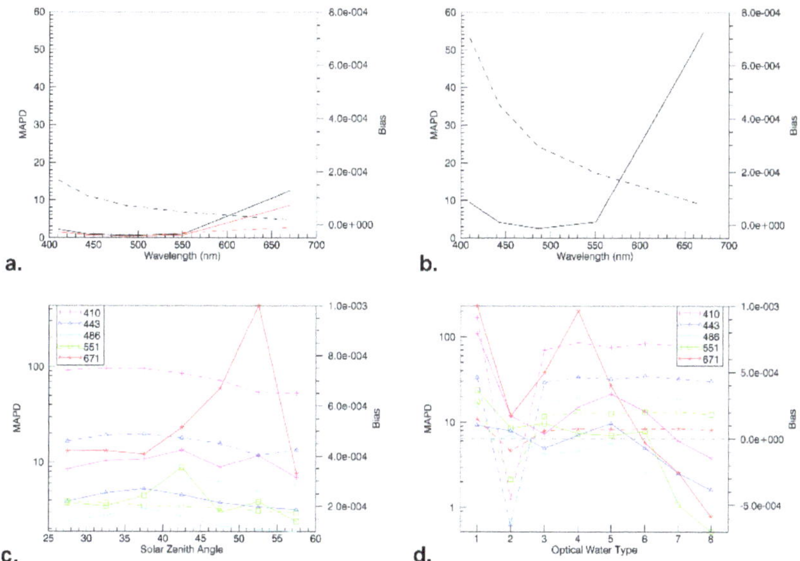

Figure 8. Sensitivity to processing parameters. (**a**) Mean Absolute Percent Difference (MAPD) and bias between R_{rs} calculated with wind speed of 1 ms^{-1} (black lines) and 5 ms^{-1} (red lines) relative to R_{rs} calculated with wind speed of 3 ms^{-1}; (**b**) MAPD and bias between R_{rs} calculated with M15 and M99; (**c**) MAPD and bias between R_{rs} calculated with M15 and M99 classified by sun zenith angle; (**d**) MAPD and bias between R_{rs} calculated with M15 and M99 classified by optical water types (OWTs). In all cases MAPD is shown as solid lines and left axis while Bias is shown as dashed lines and right axis; c. and d. use log scale for MAPD.

To evaluate the effect of the recently proposed M15 ρ table, R_{rs} was also calculated using the M99 ρ table for comparison. As shown in Figure 8b, R_{rs} calculated with M15 was lower than M99 at all wavelengths (e.g., the bias at 443 nm and 671 nm was 4.5×10^{-4} and 7.4×10^{-5} sr^{-1} respectively). These differences are similar in magnitude and sign to the bias between DALEC and OCR R_{rs} (e.g., bias *vs.* MODIS and VIIRS at 443 nm ranged from 1.5×10^{-4} to 1.5×10^{-3} sr^{-1}, Figure 5). The bias between DALEC R_{rs} calculated with M99 and the OCR R_{rs} would have ranged from ~5×10^{-4} to ~2×10^{-3} sr^{-1} (data not shown). Thus the adoption of the M15 ρ table instead of the M99 table effectively leads to significant reduction of the bias between DALEC and OCR R_{rs}.

To further analyze the influence of the ρ table, the data were collated according to sun zenith angles (Figure 8c, Table 4) and classification into optical water type (OWT) [42] (Figure 8d, Table 4). The eight OWTs are based on R_{rs} mean spectra sequentially numbered along a general trend of increasing optical complexity, where OWTs 1–3 represent the "blue-water" oceanic environments and OWTs 4–8 cover the optically complex and turbid shelf waters [42]. The graphs are dominated by a peak in the 671 nm data due to a small subset of data with very low signal at this wavelength.

In addition, there are very few spectra (51 out of 77255) classified in OWT2. Disregarding these points, the bias shows weak variation with sun zenith angle, tending to be largest in the 30–40 degree range. There was no clear dependence of bias on OWT, while the MAPD was smaller for OWTs 6–8 due to the higher signal in optically complex waters. The observed bias dependence on sun zenith angles may be due to observation geometry with respect to the ship heading, sun elevation, and relative azimuth between radiometer and sun, leading to water volume bidirectional components in the DALEC R_{rs}. Hence, in future work a bidirectional effect correction scheme should be implemented in the DALEC processing chain [4,6].

Table 4. Number of spectra *versus* sun zenith angle and OWT.

Sun Zenith Angle	25–30	30–35	35–40	40–45	45–50	50–55	55–60	
Number	15414	15289	8994	6717	7410	6517	6463	
OWT	1	2	3	4	5	6	7	8
Number	380	51	9705	17907	44	6219	42417	532

5. Conclusions

In this study transects of DALEC data spatially aggregated to approximately 1 km scale have been matched with VIIRS and MODIS satellite pixels over contrasting water types. Between 150 and 200 instantaneous R_{rs} spectra were collected for each km along the track of the RV Solander voyages. This spatially dense sampling enabled data reduction approaches based on aggressive data filtering (retaining only 20% of the data), and the final averaging was based on over 20–30 instantaneous R_{rs} spectra. This enabled comparison of 20 to 50 aggregated spectra per single OCR date, instead of the one or two spectra per scene that would have been attained with manually operated instruments. Thus automated radiometric data collection enhanced the number of matchups by one or two orders of magnitude.

The data agree very well for all wavelengths above 530 nm over a range of reflectance values (RMSD < 0.002 sr^{-1}). However, the satellite reflectance data are consistently below DALEC values in the blue bands by 7.5%–29%. MODIS and VIIRS data are in close agreement although the VIIRS reflectances at 410 nm tend to be higher (and thus closer to the DALEC R_{rs}) than MODIS at 412 nm (MAPD of 9.8%–19% and 13%–29% respectively). These findings were consistent with the general negative bias in the MODIS and VIIRS R_{rs} retrievals relative to *in situ* measurements found by NASA for the R2014.0.1 reprocessing [38,43]. However, as only 5 dates were considered in this study, these results cannot be considered conclusive for tropical Australian waters.

The differences in the blue region between the DALEC R_{rs} and the satellite R_{rs} may be due to issues with the absolute calibration in the OCR data, the atmospheric correction method leading to over correction of the atmospheric signal in the OCR data, or in inaccuracies in the DALEC R_{rs}. As MODIS and VIIRS were processed consistently, it is not possible in this study to assess the absolute calibration in the OCR data, or atmospheric correction methods. Some of the differences observed at 412, 443 and 488 nm may be attributable to the processing method used to calculate R_{rs} from the simultaneous above water measurements. The sensitivity analysis showed that the adoption of the M15 instead of the M99 tabulated ρ values almost halved the biases in the blue bands between DALEC and OCR R_{rs}. In this study both ρ and ϵ were assumed spectrally constant: the M15 tabulated ρ used in the study are based on ray tracing simulations at 550 nm. Some of the differences observed in the blue region between the DALEC R_{rs} and the satellite R_{rs} may be further reduced by using data processing methods based on spectrally variant ρ (e.g., Lee, Ahn, Mobley and Arnone [22]). This study did not include an investigation and quantification of each source of uncertainty in the estimate of the DALEC R_{rs}, such analysis of the uncertainty budget would include the effects on uncertainty of the system calibration, the superstructure perturbations, viewing geometries, distance of the instrument footprint from the hull, estimate of W, ρ and ϵ, non-cosine response of E_d and further environmental perturbation such as wave effects [4,6,12].

The DALEC sensors provide calibrated data in the range 400–1050 nm. Whilst this is a useful range for comparison with satellite data, the lack of data in the 350–400 nm range limits the methods of processing above-water radiometry measurements to retrieve R_{rs}. For example, the fingerprint method of Simis and Olsson [11] uses spectral features between 375 nm and 800 nm to determine ρ. This method is very attractive as it allows derivation of ρ from within the data itself, without the need for ancillary information (e.g., wind speed, cloud conditions) and is relatively insensitive to illumination geometry as long as the observations are within a reasonable relative azimuth range. The method of Kutser, *et al.* [44], that subtracts a power function to remove the effects of sky and sun glint by estimating a spectrally variant ϵ (λ) in waters highly absorbing in the blue spectral region, also requires data at 350–380 nm. The Zeiss spectroradiometers in the DALEC acquire data in the 305–1140 nm range, but the system currently includes a UV filter blocking the incoming signal up to 400 nm. Hence it would be advantageous for the DALEC measurement spectral range to be extended in the UV, with a suitable alternative filter allowing for dark current measurements, while still providing useful data in the 350–400 nm spectral range.

Automated instrumentation such as the DALEC requires standard operating procedures (SOP) to be adhered to in order to collect reliable data that can be used with confidence. The initial installation of the DALEC system on the RV Southern Surveyor provided some good quality data during cruises shortly after installation by expert operators. However, lack of attention to operating requirements resulted in the sensor losing track of its geometric settings and producing spurious azimuth readings. This experience has resulted in refined SOP and maintenance regimes in the current installation on RV Solander. The success of the deployment depends on good interactions with the onboard crew to ensure the daily maintenance and quality assurance. In addition, protocols must ensure that data are processed within a short time after completion of the voyage so that issues that arise during the processing can be addressed and if necessary procedures amended to ensure best quality data acquisition on future voyages.

In this study we have demonstrated that autonomous optical observations that maintain suitable viewing geometry yield a high number of matchups per scene. Measurements undertaken with autonomous radiometers deployed on a number of vessels would have the potential to provide a network of validation observations covering a diversity of water types and environmental conditions. To ensure consistency of data products across installations in the network, data collection would be based on standardized and fully characterized instruments and measurement procedures [31,45]. Systematic inconsistencies resulting from data processing, data reduction and quality control variations would be reduced by using a centralized single-source processor [14,15,31,45].

The operation of an automated above-water hyperspectral radiometer has significantly augmented Australia's ability to contribute to the validation of global and regional OCR algorithms for Australian waters. Automated radiometric data collection has been shown to produce good quality data providing suitable operating and maintenance procedures are adhered to. The hyperspectral data acquisition of the DALEC sensor allows filtering to match any available OCR sensor between 400–900 nm. Thus it has the potential for use in calibration and validation of satellite sensors now and into the future as well as for comparison of data between other ship-based sensors observing above- or below-water radiances. Furthermore, combining the DALEC data-stream with other bio-optical and bio-geochemical sampling from the same platform will result in a rich data source for design and testing of optical retrieval algorithms in collaboration with the wider biological oceanographic community.

Acknowledgments: This study was funded by IMOS and CSIRO. IMOS is supported by the Australian Government through the National Collaborative Research Infrastructure Strategy and the Super Science Initiative. VB was also supported by CNR, the RITMARE Flagship Project and the European Union (FP7-427 People Co-funding of Regional, National and International Programmes, GA n. 600407). The authors wish to acknowledge contributions of M. Slivkoff and W. Klonowski (*In-situ* Marine Optics) for technical advice in instrument operation and data processing; R Keen, P. Daniel, D. McKenzie, L. Woodward, R. Palmer, D. Mills, B. Muir, N. Thapar for technical and logistical support of instrumentation on the RV Southern Surveyor; L. Besnard and S. Mancini for publication of data through IMOS; K. Suber for image production; P. Daniel, M. Furnas, J. Benthusen for technical, logistical and contract support associated with the RV Solander deployment; the master and crew of RV Southern Surveyor and RV Solander; the Ocean Biology Processing Group at NASA GSFC for development, support and distribution of the SeaDAS software and for distributing the MODIS and VIIRS data. The constructive comments by T. Malthus, and C. Giardino, as well as four anonymous reviewers improved earlier versions of this manuscript.

Author Contributions: The general conception of this work was developed by Vittorio E. Brando, Jenny L. Lovell and Thomas Schroeder. Vittorio E. Brando and Jenny L. Lovell wrote the manuscript, Edward A. King and Thomas Schroeder contributed to the interpretation of the results. Vittorio E. Brando, David Boadle and Thomas Schroeder designed and carried out DALEC installations, operations, logistics and data management. Edward A. King and Roger Scott processed the satellite imagery. Jenny L. Lovell and Vittorio E. Brando performed the data analysis. All co-authors provided critical comments to the manuscript.

Conflicts of Interest: The authors declare no conflict of interest.

Appendix

The statistical measures used in this paper are described by the following equations where in each case x is the DALEC reflectance, y is the satellite-derived reflectance and N is the number of samples (pixels).

$$MAPD = \frac{100}{N}\sum\frac{|y-x|}{x} \tag{A1}$$

$$Bias = \frac{1}{N}\sum(y-x) \tag{A2}$$

$$RMSD = \sqrt{\frac{1}{N}(x-y)^2} \tag{A3}$$

$$R^2 = \left[\frac{\sum xy - \frac{(\sum x)(\sum y)}{N}}{\sqrt{\left(\sum x^2 - \frac{(\sum x)^2}{N}\right)\left(\sum y^2 - \frac{(\sum y)^2}{N}\right)}}\right]^2 \tag{A4}$$

References

1. McClain, C.R.; Feldman, G.C.; Hooker, S.B. An overview of the SeaWiFs project and strategies for producing a climate research quality global ocean bio-optical time series. *Deep-Sea Res. Pt. II* **2004**, *51*, 5–42. [CrossRef]
2. McClain, C.R. A decade of satellite ocean color observations. *Annu Rev. Mar. Sci* **2009**, *1*, 19–42. [CrossRef] [PubMed]
3. Bailey, S.W.; Hooker, S.B.; Antoine, D.; Franz, B.A.; Werdell, P.J. Sources and assumptions for the vicarious calibration of ocean color satellite observations. *Appl. Opt.* **2008**, *47*, 2035–2045. [CrossRef] [PubMed]
4. Zibordi, G.; Melin, F.; Voss, K.J.; Johnson, B.C.; Franz, B.A.; Kwiatkowska, E.; Huot, J.P.; Wang, M.H.; Antoine, D. System vicarious calibration for ocean color climate change applications: Requirements for *in situ* data. *Remote Sens. Environ.* **2015**, *159*, 361–369. [CrossRef]
5. Hooker, S.B.; Lazin, G.; Zibordi, G.; McLean, S. An evaluation of above- and in-water methods for determining water-leaving radiances. *J. Atmos Ocean. Tech.* **2002**, *19*, 486–515. [CrossRef]
6. Zibordi, G.; Melin, F.; Hooker, S.B.; D'Alimonte, D.; Holbert, B. An autonomous above-water system for the validation of ocean color radiance data. *IEEE Trans. Geosci. Remote Sens.* **2004**, *42*, 401–415. [CrossRef]
7. Clark, D.K.; Gordon, H.R.; Voss, K.J.; Ge, Y.; Broenkow, W.; Trees, C. Validation of atmospheric correction over the oceans. *J. Geophys Res.-Atmos* **1997**, *102*, 17209–17217. [CrossRef]
8. Antoine, D.; Ortenzio, F.; Hooker, S.B.; Becu, G.; Gentili, B.; Tailliez, D.; Scott, A.J. Assessment of uncertainty in the ocean reflectance determined by three satellite ocean color sensors (MERIS, SeaWiFs and MODIS-a) at an offshore site in the mediterranean sea (boussole project). *J. Geophys Res.-Ocean.* **2008**, *113*. [CrossRef]

9. Toole, D.A.; Siegel, D.A.; Menzies, D.W.; Neumann, M.J.; Smith, R.C. Remote-sensing reflectance determinations in the coastal ocean environment: Impact of instrumental characteristics and environmental variability. *Appl. Opt.* **2000**, *39*, 456–469. [CrossRef] [PubMed]
10. Ruddick, K.; Cauwer, V.; Park, Y.J.; Moore, G. Seaborne measurements of near infrared water-leaving reflectance: The similarity spectrum for turbid waters. *Limnol. Oceanogr.* **2006**, *51*, 1167–1179. [CrossRef]
11. Simis, S.G.H.; Olsson, J. Unattended processing of shipborne hyperspectral reflectance measurements. *Remote Sens. Environ.* **2013**, *135*, 202–212. [CrossRef]
12. Zibordi, G.; Ruddick, K.; Ansko, I.; Moore, G.; Kratzer, S.; Icely, J.; Reinart, A. *In situ* determination of the remote sensing reflectance: An inter-comparison. *Ocean. Sci* **2012**, *8*, 567–586. [CrossRef]
13. Mobley, C.D. Estimation of the remote-sensing reflectance from above-surface measurements. *Appl. Opt.* **1999**, *38*, 7442–7455. [CrossRef] [PubMed]
14. Bailey, S.W.; Werdell, P.J. A multi-sensor approach for the on-orbit validation of ocean color satellite data products. *Remote Sens Environ.* **2006**, *102*, 12–23. [CrossRef]
15. Werdell, P.J.; Bailey, S.W. An improved *in-situ* bio-optical data set for ocean color algorithm development and satellite data product validation. *Remote Sens. Environ.* **2005**, *98*, 122–140. [CrossRef]
16. Nechad, B.; Ruddick, K.; Schroeder, T.; Oubelkheir, K.; Blondeau-Patissier, D.; Cherukuru, N.; Brando, V.; Dekker, A.; Clementson, L.; Banks, A.C.; *et al.* Coastcolour round robin data sets: A database to evaluate the performance of algorithms for the retrieval of water quality parameters in coastal waters. *Earth Syst. Sci. Data* **2015**, *7*, 319–348. [CrossRef]
17. Garaba, S.P.; Voss, D.; Wollschlager, J.; Zielinski, O. Modern approaches to shipborne ocean color remote sensing. *Appl. Opt.* **2015**, *54*, 3602–3612. [CrossRef]
18. Garaba, S.P.; Zielinski, O. Methods in reducing surface reflected glint for shipborne above-water remote sensing. *J. Eur Opt. Soc.-Rapid* **2013**, *8*. [CrossRef]
19. Martinez-Vicente, V.; Simis, S.G.H.; Alegre, R.; Land, P.E.; Groom, S.B. Above-water reflectance for the evaluation of adjacency effects in earth observation data: Initial results and methods comparison for near-coastal waters in the western channel, UK. *J. Eur Opt. Soc.-Rapid* **2013**, *8*. [CrossRef]
20. McKinna, L.I.W.; Furnas, M.J.; Ridd, P.V. A simple, binary classification algorithm for the detection of trichodesmium SPP. Within the great barrier reef using MODIS imagery. *Limnol. Oceanogr.: Methods* **2011**, *9*, 50–66.
21. Fougnie, B.; Frouin, R.; Lecomte, P.; Deschamps, P.Y. Reduction of skylight reflection effects in the above-water measurement of diffuse marine reflectance. *Appl. Opt.* **1999**, *38*, 3844–3856. [CrossRef] [PubMed]
22. Lee, Z.P.; Ahn, Y.H.; Mobley, C.; Arnone, R. Removal of surface-reflected light for the measurement of remote-sensing reflectance from an above-surface platform. *Opt. Express* **2010**, *18*, 26313–26324. [CrossRef] [PubMed]
23. Hommersom, A.; Kratzer, S.; Laanen, M.; Ansko, I.; Ligi, M.; Bresciani, M.; Giardino, C.; Beltran-Abaunza, J.M.; Moore, G.; Wernand, M.; *et al.* Intercomparison in the field between the new wisp-3 and other radiometers (trios Ramses, ASD FieldSpec, and TACCS). *J. Appl. Remote Sens.* **2012**, *6*, 063615–063616. [CrossRef]
24. Mobley, C.D. Polarized reflectance and transmittance properties of windblown sea surfaces. *Appl. Opt.* **2015**, *54*, 4828–4849. [CrossRef] [PubMed]
25. Cox, C.; Munk, W. Measurement of the roughness of the sea sur- face from photographs of the sun's glitter. *J. Opt. Soc. Am.* **1954**, *44*, 838–850. [CrossRef]
26. Slivkoff, M. *In-situ* Marine Optics, Dalec measurements in the UV. Personal Communication, 2013.
27. IMOS. Imos Ocean Portal. Direct Access to Above Water Radiometry Data. Available online: http://data.aodn.org.au/IMOS/opendap/SRS/OC/radiometer/VMQ9273_Solander/2015/ (accessed on 28 January 2016).
28. About the Imos Ocean Portal. Available online: http://imos.aodn.org.au/webportal (accessed on 4 February 2016).
29. Slivkoff, M. *In-situ* Marine Optics, Advice on dalec data processing. Personal Communication, 2011.
30. Access NWP Data Information. Available online: http://www.bom.gov.au/nwp/doc/access/NWPData.shtml (accessed on 11 April 2015).

31. Zibordi, G.; Holben, B.; Slutsker, I.; Giles, D.; D'Alimonte, D.; Melin, F.; Berthon, J.F.; Vandemark, D.; Feng, H.; Schuster, G.; *et al.* Aeronet-OC: A network for the validation of ocean color primary products. *J. Atmos Ocean. Tech.* **2009**, *26*, 1634–1651. [CrossRef]
32. Gilmour, J.; Smith, L.; Cook, K.; Pincock, S. *Discovering Scott Reef: 20 Years of Exploration and Research*; Australian Institute of Marine Science: Townsville, Queensland, Australia, 2013.
33. Wolanski, E.; Deleersnijder, E. Island-generated internal waves at scott reef, Western Australia. *Cont. Shelf Res.* **1998**, *18*, 1649–1666. [CrossRef]
34. Condie, S.A. Modeling seasonal circulation, upwelling and tidal mixing in the Arafura and Timor seas. *Cont. Shelf Res.* **2011**, *31*, 1427–1436. [CrossRef]
35. Blondeau-Patissier, D.; Schroeder, T.; Brando, V.E.; Maier, S.W.; Dekker, A.G.; Phinn, S. Esa-meris 10-year mission reveals contrasting phytoplankton bloom dynamics in two tropical regions of northern Australia. *Remote Sens.* **2014**, *6*, 2963–2988. [CrossRef]
36. Lynch, T.P.; Morello, E.B.; Evans, K.; Richardson, A.J.; Steinberg, C.R.; Roughan, M.; Thompson, P.; Middleton, J.F.; Feng, M.; Sherrington, R.B.; *et al.* Imos national reference stations: A continental scaled physical, chemical and biological coastal observing system. *PLoS ONE* **2014**, *9*, E113652. [CrossRef] [PubMed]
37. Seadas General Description. Available online: http://seadas.gsfc.nasa.gov (accessed on 4 February 2016).
38. VIIRS Ocean Color Reprocessing 2014. Available online: http://oceancolor.gsfc.nasa.gov/cms/reprocessing/OCReproc20140VN.html (accessed on 4 February 2016).
39. Eplee, R.E.; Turpie, K.R.; Meister, G.; Patt, F.S.; Franz, B.A.; Bailey, S.W. On-orbit calibration of the suomi national polar-orbiting partnership visible infrared imaging radiometer suite for ocean color applications. *Appl. Opt.* **2015**, *54*, 1984–2006. [CrossRef] [PubMed]
40. Gordon, H.R. Atmospheric correction of ocean color imagery in the earth observing system era. *J. Geophys. Res.* **1997**, *102*, 17081–17106. [CrossRef]
41. Gordon, H.R.; Wang, M.H. Retrieval of water-leaving radiance and aerosol optical-thickness over the oceans with SeaWiFs—A preliminary algorithm. *Appl. Opt.* **1994**, *33*, 443–452. [CrossRef] [PubMed]
42. Moore, T.S.; Campbell, J.W.; Dowell, M.D. A class-based approach to characterizing and mapping the uncertainty of the modis ocean chlorophyll product. *Remote Sens Environ.* **2009**, *113*, 2424–2430. [CrossRef]
43. Modis-Aqua Ocean Color Reprocessing 2014. Available online: http://oceancolor.gsfc.nasa.gov/cms/reprocessing/OCReproc20140MA.html (accessed on 4 February 2016).
44. Kutser, T.; Vahtmae, E.; Paavel, B.; Kauer, T. Removing glint effects from field radiometry data measured in optically complex coastal and inland waters. *Remote Sens Environ.* **2013**, *133*, 85–89. [CrossRef]
45. Zibordi, G.; Melin, F.; Berthon, J.F. A time-series of above-water radiometric measurements for coastal water monitoring and remote sensing product validation. *IEEE Trans.Geosci. Remote Sens.* **2006**, *3*, 120–124. [CrossRef]

Article

Validation of the Suomi NPP VIIRS Ice Surface Temperature Environmental Data Record

Yinghui Liu [1,*], Jeffrey Key [2], Mark Tschudi [3], Richard Dworak [1], Robert Mahoney [4] and Daniel Baldwin [3]

1 Cooperative Institute for Meteorological Satellite Studies, University of Wisconsin-Madison, 1225 West Dayton St., Madison, WI 53706, USA; rdworak@ssec.wisc.edu

2 NOAA/NESDIS, 1225 West Dayton St., Madison, WI 53706, USA; jkey@ssec.wisc.edu

3 Colorado Center for Astrodynamics Research, University of Colorado, Boulder, CO 80309, USA; mark.tschudi@colorado.edu (M.T.); daniel.baldwin@colorado.edu (D.B.)

4 Northrop Grumman Aerospace Systems, Redondo Beach, CA 90278, USA; robert.mahoney@ngc.com

* Correspondence: yinghuil@ssec.wisc.edu; Tel.: +1-608-890-1893; Fax: +1-608-262-5974

Academic Editors: Changyong Cao, Magaly Koch and Prasad Thenkabail
Received: 28 October 2015; Accepted: 11 December 2015; Published: 18 December 2015

Abstract: Continuous monitoring of the surface temperature is critical to understanding and forecasting Arctic climate change; as surface temperature integrates changes in the surface energy budget. The sea-ice surface temperature (IST) has been measured with optical and thermal infrared sensors for many years. With the IST Environmental Data Record (EDR) available from the Visible Infrared Imaging Radiometer Suite (VIIRS) onboard the Suomi National Polar-orbiting Partnership (NPP) and future Joint Polar Satellite System (JPSS) satellites; we can continue to monitor and investigate Arctic climate change. This work examines the quality of the VIIRS IST EDR. Validation is performed through comparisons with multiple datasets; including NASA IceBridge measurements; air temperature from Arctic drifting ice buoys; Moderate Resolution Imaging Spectroradiometer (MODIS) IST; MODIS IST simultaneous nadir overpass (SNO); and surface air temperature from the National Centers for Environmental Prediction/National Center for Atmospheric Research (NCEP/NCAR) reanalysis. Results show biases of −0.34; −0.12; 0.16; −3.20; and −3.41 K compared to an aircraft-mounted downward-looking pyrometer; MODIS; MODIS SNO; drifting buoy; and NCEP/NCAR reanalysis; respectively; root-mean-square errors of 0.98; 1.02; 0.95; 4.89; and 6.94 K; and root-mean-square errors with the bias removed of 0.92; 1.01; 0.94; 3.70; and 6.04 K. Based on the IceBridge and MODIS results; the VIIRS IST uncertainty (RMSE) meets or exceeds the JPSS system requirement of 1.0 K. The product can therefore be considered useful for meteorological and climatological applications.

Keywords: Suomi NPP; ice surface temperature; Environmental Data Record; calibration and validation

1. Introduction

The Arctic has been warming at a greater rate than anywhere else on Earth, a trend that is projected to continue over the next century [1]. Continuous measurements of the surface temperature are therefore important to the understanding, monitoring, and forecasting Arctic climate change. The surface temperature of sea ice, hereafter referred to as the "ice surface temperature (IST)", is the controlling factor for sea ice growth and melt. Furthermore, the strength of low-level atmospheric temperature inversions depends on the IST. IST and precipitable water are interdependent [2], as both the lower tropospheric temperature structure and water vapor content determine the downwelling longwave radiation. Consequently, changes in sea ice characteristics, temperature inversion strength, precipitable water, and associated cloud cover all feed back to the surface temperature. Therefore, accurate and spatially robust surface temperature measurements are required to better understand feedbacks within the Arctic climate system and to better quantify the degree to which the Arctic is warming at a greater rate than the global average, known as "Arctic amplification" [3,4].

Satellite imagers with longwave infrared bands have been used to measure the ice surface temperature under clear conditions for many years. Key and Haefliger [5] presented an algorithm for estimating IST using split-window thermal channels at 11 and 12 μm of the Advanced Very High Resolution Radiometer (AVHRR) sensors. Key *et al.* [6] presented some improvements to the Key and Haefliger [5] procedure. The method was applied to a multi-decadal time series of AVHRR to study recent climate change in the Polar Regions [7–11]. Hall *et al.* [12] adopted the Key *et al.* [6] method for use with the Moderate-Resolution Imaging Spectroradiometer (MODIS) onboard the Terra and Aqua satellites. Their validation yielded a root mean square error (RMSE) of 1.2 K when compared to near-surface air temperature. Methods have also been developed for the estimation of land surface temperature from remotely sensed satellite-based thermal infrared data [13]. The accuracy of those retrievals has been assessed in a variety of validation studies [13–16].

IST is now produced with the Visible Infrared Imager Radiometer Suite (VIIRS) onboard the Suomi National Polar-orbiting Partnership (NPP) satellite. It is one of the official VIIRS products, or Environmental Data Records (EDR). VIIRS has some advantages over its heritage sensors [17] in that the split window bands at 11 and 12 μm in "moderate" resolution have higher spatial resolution than both AVHRR and MODIS, with approximately 1.6 km for VIIRS at the edge of the scan (750 m at nadir) compared to 4.8 km for MODIS (1 km at nadir). VIIRS also has a wider swath than MODIS, approximately 3000 km *versus* 2330 km. Figure 1 gives an example of a daily composite of the VIIRS IST EDR on 1 March 2015. It shows the lowest IST north of the Canadian Archipelago and Greenland, higher temperatures north of Alaska, and the warmest areas around the Kara Sea and Bering Strait.

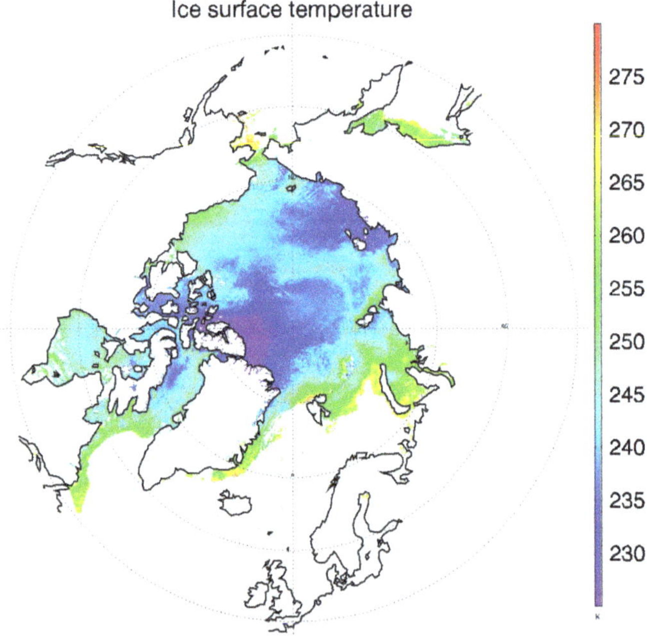

Figure 1. Ice surface temperature (IST) composite of the Visible Infrared Imaging Radiometer Suite (VIIRS) IST Environmental Data Record (EDR) from all overpasses over the Arctic on 1 March 2015.

This study presents the validation of the VIIRS IST EDR with multiple datasets, including aircraft measurements, 2 m air temperature from Arctic drifting buoys, MODIS IST, and surface air temperatures from an atmospheric reanalysis product. This work significantly extends the preliminary validation study of Key *et al.* [18] by employing additional IST product inter-comparisons, thereby providing a robust evaluation of the operational VIIRS IST product.

2. Data and Methods

The VIIRS IST EDR provides clear-sky surface skin temperature retrievals at VIIRS moderate resolution (750 m) for both day and night over snow and ice covered water surfaces. The VIIRS IST EDR algorithm is based on the split-window technique of Yu *et al.* [19]. This statistical regression method is implemented as:

$$\mathrm{IST} = a_0 + a_1 T_{11} + a_2 (T_{11} - T_{12}) + a_3 (\sec(z) - 1) \tag{1}$$

where T_{11} and T_{12} are the VIIRS brightness temperature at 10.8 μm and 12.0 μm in moderate resolution bands M15 and M16 respectively, z is the satellite zenith angle, and a_0, a_1, a_2, and a_3 are regression coefficients.

Different sets of coefficients are used for daytime and nighttime retrievals. Day is defined by solar zenith angles (SZA) less than 85 degrees. Retrievals are done only under clear sky conditions, which are identified with the VIIRS cloud mask [20]. The brightness temperature difference term in Equation (1) is for the correction of atmospheric water vapor absorption; the term that includes the satellite zenith angle accounts for the atmosphere path length. In this study, VIIRS IST EDR data produced by the Interface Data Processing Segment (IDPS) over the period August 2012 to July 2015 are used.

The algorithm described above is similar to the ice surface temperature algorithm first developed by Key and Haefliger [5] and later revised by Key *et al.* [6]. This approach is also similar to those used for sea surface temperature (SST) estimation. Other related algorithms are described by Comiso [21] and Lindsay and Rothrock [22].

The IDPS operational IST product [23] is based on prelaunch regression coefficients [24] derived from Global Synthetic Data [25,26] that includes the effects of VIIRS sensor relative spectral response and sensor noise, based on VIIRS F1 prelaunch, sensor characterization test measurements. At-sensor, clear sky, top-of-atmosphere (TOA) synthetic radiances for the VIIRS 10.8 μm and 12.0 μm IR bands were computed using the MODerate resolution atmospheric TRANsmission (MODTRAN) [27] radiative transfer model with inputs derived from physical properties based on NCEP Global Forecast System (GFS) snow/ice mask, surface skin-temperature and atmospheric data. VIIRS sensor effects were incorporated into the simulated radiances based on a VIIRS sensor model [28].

2.1. Validation Datasets

The datasets used to validate the VIIRS IST are summarized in Table 1. IST data sets from MODIS, specifically the Terra MODIS (MOD29) and Aqua MODIS (MYD29) Sea Ice Extent 5-Min L2 Swath 1 km datasets [29,30] from August 2012 to July 2015 are used to compare with the VIIIRS IST EDR. These data sets are in swath format at 1 km resolution at nadir for both daytime and nighttime under clear sky conditions, with swath coverage of 2330 km (cross track) by 2030 km (along track). To collocate the VIIRS IST EDR and MODIS IST, every overpass over the Arctic and Antarctic of the VIIRS IST EDR and MODIS IST are re-gridded to a 1 km Equal-Area Earth Grid (EASE-Grid), with 9025 by 9025 grid cells extending from 48.4 to 90°N over the Arctic and 8025 by 8025 grid cells extending from 53.2 to 90°S over the Antarctic. Only grids cells with VIIRS-MODIS observation time differences of less than 5 min, and indicated as cloud free from both VIIRS and MODIS cloud masks, are included in the matchup comparison.

Table 1. Validation datasets used in this study.

Validation Dataset	Parameter	Spatial Resolution	Spatial Coverage	Temporal Coverage
NASA IceBridge KT-19 IR Surface Temperature	Snow/ice temperature	15 × 15 m	Arctic and Antarctic	Arctic: 2012–2014 Antarctic: 2012–2013
MODIS Ice Surface Temperature	Snow/ice temperature	1 km	Arctic and Antarctic	August 2012–July 2015
MODIS simultaneous nadir overpass	Snow/ice temperature	0.05 degree longitude by 0.05 degree latitude	Arctic	March 2013–April 2014
Arctic drifting buoy	2 m air temperature	Point observations	Arctic	August 2012–June 2014
NCEP/NCAR reanalysis	Air temperature at 0.995 sigma level	2.5 × 2.5 degree latitude/longitude	Arctic and Antarctic	August 2012–July 2015

An additional comparison of VIIRS and MYD29 IST values based on S-NPP and Aqua Simultaneous Nadir Overpass (SNO) orbit swaths within a 5 min time window covering the Arctic is also performed. The matchup IST values for the SNO comparisons are binned and averaged to a 1/20th degree equal angle, Climate Model Grid (CMG). The SNO IST comparison matchups include only pixels indicated as confidently clear by the VIIRS cloud mask with satellite view zenith angles less than 45 degrees and ice concentrations greater than or equal to 95%, as estimated by the VIIRS Ice Concentration Intermediate Product (IP) [31]. For this comparison, 20 SNO orbit swaths from March 2013 to April 2014 are used, thus capturing a relatively broad range of temperatures.

IST measured with an airborne Heitronics KT-19.85 Series II infrared radiation pyrometer (KT-19) on a National Aeronautics and Space Administration (NASA) P3 during the NASA Operation IceBridge campaigns [32] is used to validate the VIIRS IST EDR. One of the NASA IceBridge mission goals is to better understand the processes that connect Polar Regions with the global climate system [33]. All KT-19 observations in the Antarctic in 2012 and 2013, and in the Arctic in 2014, are used to validate collocated VIIRS IST EDR. All available KT-19 temperature measurements that are within 375 m of the center of the VIIRS pixel are averaged. Only KT-19 temperature samples with standard deviations less than 1 K are used in matchup comparisons in order to eliminate cases with small-scale IST outliers that VIIRS would not be able to resolve. In addition, the closest KT-19 point to the VIIRS pixel must be within 15 min and 100 meters to be considered. Finally, a rigorous quality control is done to make sure that the VIIRS IST value is not contaminated by cloud.

If a daylight (SZA less than 85°) scene is used, a false-color image is created with VIIRS bands M5, M7, and M10 (0.67, 0.85, and 1.6 μm). These images are visually inspected to confirm that the daytime scene is clear sky. For the few available scenes that are night (SZA 90° or more), a visual inspection of the M15 and M16 (10.8, 12.0 μm) images is done. It should be noted that no cases with SZA between 80° and 90° are used in the statistics because of the large increase in random error observed in ice and cloud products for such low-sun conditions.

The NCEP/NCAR Reanalysis dataset [34] provides surface air temperature at the 0.995 sigma level from 1948 to the present. The daily surface air temperature data is available at 00Z, 06Z, 12Z, and 18Z with a spatial resolution of 2.5 × 2.5 degrees latitude/longitude. The following approach is used to match the NCEP/NCAR reanalysis surface temperature with VIIRST IST EDR. For each VIIRS IST EDR overpass, the IST values in the 1 km EASE-Grid projection are remapped to a 2.5 × 2.5 degree latitude/longitude grid. If the ratio of available remapped samples to the maximum possible sample number in a grid cell is larger than a threshold of 0.9, the mean of all the remapped VIIRS IST EDRs is assigned to that grid cell. Using the median of all the remapped VIIRS IST EDRs produces the same results, as shown in the following section.

Surface air temperatures (SAT) measured by drifting ice buoys from the International Arctic Buoy Programme (IABP) [35] are also used to assess the accuracy of the VIIRS IST EDR. Unfortunately, issues such as solar heating, frost and snow can lead to inaccurate SAT buoy measurements. Sensors with the most accurate measurements are the 2 m, shielded thermistors (personal communication, I. Rigor, 2015). In this study, SAT measurements from four drifting buoys from August 2012 to June 2014 are used, with buoy IDs 300025010128510, 300025010125530, 300025010123530, and 300234011045700, which are all ice mass balance buoys. More details of these buoys are available at http://iabp.apl.washington.edu/maps_daily_table.html. To collocate the buoy SAT observations and the VIIRS IST EDR, both are remapped to 1 km EASE-Grid, and only cases of collocated buoy SAT measurements and the VIIRS IST EDR within 2 h of each other are included.

2.2. Definition of Statistics

In the following sections three quantities are used to quantify the differences between two datasets. The bias, $\bar{\epsilon}$, is the average difference between observations in two datasets:

$$\bar{\epsilon} = \frac{1}{n}\sum_{i=1}^{n} e_i \tag{2}$$

where e_i is the difference between individual observations in the two samples—the "errors"—and n is the sample size. The root mean square error (RMSE) is the square root of the mean squared difference between two datasets:

$$RMSE = \sqrt{\frac{1}{n-1}\sum_{i=1}^{n} e_i^2} \tag{3}$$

The RMSE with the bias removed is the square root of the average squared deviation of the errors from the mean error, which is the standard deviation of the errors:

$$RMSE_{nobias} = \sqrt{\frac{1}{n-1}\sum_{i=1}^{n} (e_i - \bar{\epsilon})^2} \tag{4}$$

For the JPSS program, the absolute value of the bias in Equation (2) is termed the measurement accuracy. RMSE in Equation (3) is the measurement uncertainty. The RMSE with the bias removed, or the standard deviation, in Equation (4) is the measurement precision. If the bias is zero, RMSE and $\text{RMSE}_{\text{nobias}}$ are equal.

3. IST Validation Results

3.1. Comparison with MODIS IST

The MODIS IST product has been validated with surface air temperatures from weather stations and from buoy air temperatures, with a reported RMSE in the range of 1.2–1.3 K [12]. The MODIS IST product has been used in climate-related studies [29,30]. Comparisons of VIIRS IST EDR with MODIS IST can therefore demonstrate both consistency and accuracy. Histograms of the differences between VIIRS IST and MODIS IST match-ups are shown in Figure 2 for all cases, and in Figure 3 for cases with MODIS IST in the ranges 213–230 K, 230–240 K, 240–250 K, 250–260 K, 260–270 K, and 270–275 K. For over 2 billion match-up pairs, the VIIRS IST EDR shows a bias of −0.12 K with MODIS IST, and the $\text{RMSE}_{\text{nobias}}$ of 1.01 K. The majority of the IST differences have absolute values less than 1 K. This shows that in general the VIIRS IST EDR can be considered to be as accurate as the MODIS IST product. The good agreement can be attributed to similar split-window retrieval approach of both products, and good quality clear/cloudy identification using both the VIIRS and MODIS cloud masks.

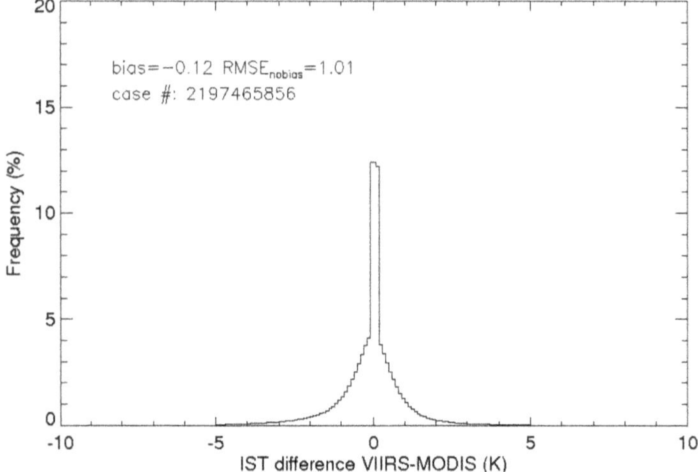

Figure 2. Histogram of ice surface temperature differences between Suomi National Polar-orbiting Partnership (S-NPP) VIIRS and MODIS (Aqua and Terra) in the Arctic and Antarctic for all cases from August 2012 to July 2015.

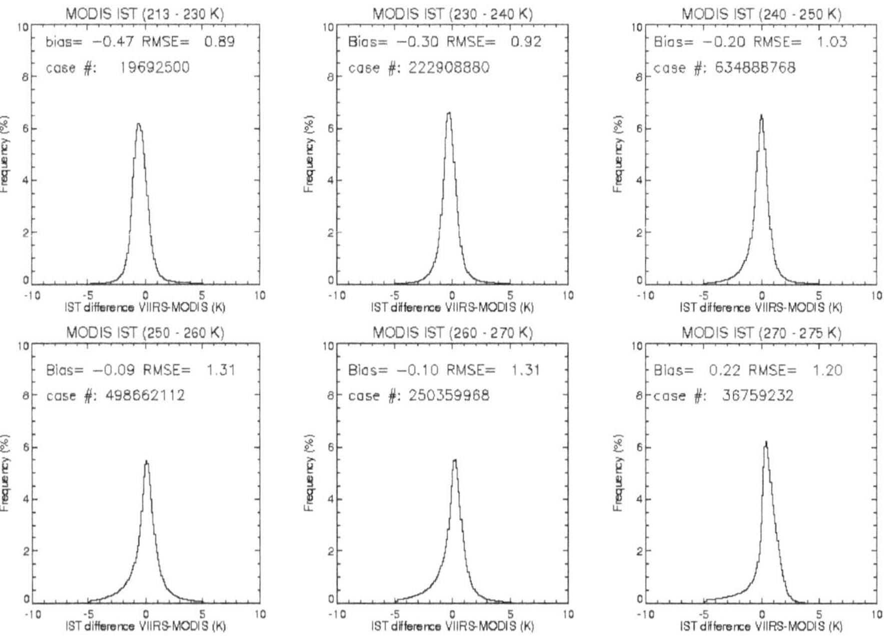

Figure 3. Histogram of ice surface temperature differences between NPP VIIRS and MODIS (Aqua and Terra) in the Arctic and Antarctic from August 2012 to July 2015 for cases with MODIS IST in the ranges 213–230 K, 230–240 K, 240–250 K, 250–260 K, 260–270 K, and 270–275 K. Measurement bias and RMSE$_{nobias}$ are indicated for each bin.

For the temperature bin with the MODIS IST between 270 and 275 K the bias is positive (0.22 K). The bias becomes negative (−0.10 K) for temperatures below 270 K, and becomes more negative (−0.47) for temperatures less than 230 K. The $RMSE_{nobias}$ values are relatively stable, with smaller values for lower temperatures. RMSE is close to $RMSE_{nobias}$ because of the small bias, with RMSE of 1.02 K for all cases, and 1.00 K, 0.96 K, 1.05 K, 1.31 K, 1.31 K, and 1.22 K for MODIS IST in the ranges 213–230 K, 230–240 K, 240–250 K, 250–260 K, 260–270 K, and 270–275 K. The bias characteristics can be attributed to the differences between how the retrieval coefficients are derived for VIIRS and MODIS ISTs. For the VIIRS IST EDR, different sets of coefficients are used for daytime and nighttime. For MODIS, different retrieval coefficients are derived for three temperature ranges of the 11 μm brightness temperature: less than 240 K, 240–260 K, and greater than 260 K. Thus, the MODIS IST retrievals may benefit from this approach, and have a similar performance for all three temperature intervals. The VIIRS IST EDR thus tends to have close agreement with MODIS IST for the middle temperature range, and have a higher bias when temperatures are low or high.

Comparisons of VIIRS IST EDR and MODIS IST for the Arctic only are shown in Figures 4 and 5. The bias and $RMSE_{nobias}$ for all cases, and for different temperature bins, have similar values and characteristics as those from all cases in the Antarctic. The RMSE is close to $RMSE_{nobias}$ because of the small bias, with RMSE of 0.98 K for all cases, and 1.04 K, 0.88 K, 0.95 K, 1.25 K, 1.31 K, and 1.18 K for MODIS IST in the ranges 213–230 K, 230–240 K, 240–250 K, 250–260 K, 260–270 K, and 270–275 K. The more rigorous matchups of the VIIRS IST EDR and MODIS IST with SNO for the Arctic yield an absolute bias of 0.16 K, $RMSE_{nobias}$ of 0.94 K, and RMSE of 0.95 Ks. (Figure 6).

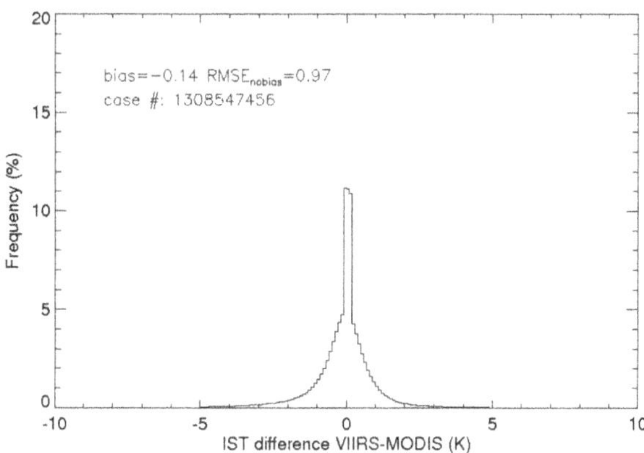

Figure 4. Histogram of ice surface temperature differences between NPP VIIRS and MODIS (Aqua and Terra) in the Arctic only for all cases from August 2012 to July 2015.

3.2. *Validation with KT-19 IST*

Though the inter-comparison of the VIIRS IST and the MODIS IST provides valuable information about the VIIRS IST EDR quality, the uncertainty of each needs to be determined with suitable "truth" data. The uncertainty in the MODIS IST has previously been evaluated with limited *in situ* validation [36]. The validation of the VIIRS IST EDR with KT-19 IST measurements provides the most accurate assessment of VIIRS IST EDR quality. The KT-19 sensor has been shown to accurately measure sea ice lead (fracture) temperatures of for leads 40 m wide or larger. With the rigorous procedures to collocate the KT-19 measurements with the VIIRS IST EDR, and with the elimination of the cloud contamination as detailed above, the comparison results give a high confidence assessment of the VIIRS IST EDR. With over 200 pairs of collocated VIIRS IST EDR and KT-19 IST over the Arctic and the Antarctic, the VIIRS minus KT-19 bias is −0.34 K, with an RMSE of 0.92 K (0.98 K) without (with) the bias included (Figure 7). With the bias removed, this RMSE is smaller than the 1.2–1.3 K value for the MODIS surface air temperature comparison in [12]. However, it should be noted that the IceBridge KT-19 measurements are concentrated in the springtime, when the surface temperatures are in the moderate range and the VIIRS IST EDR performs best. More measurements from other seasons, when the IST is near the melting point or is very low, are desirable.

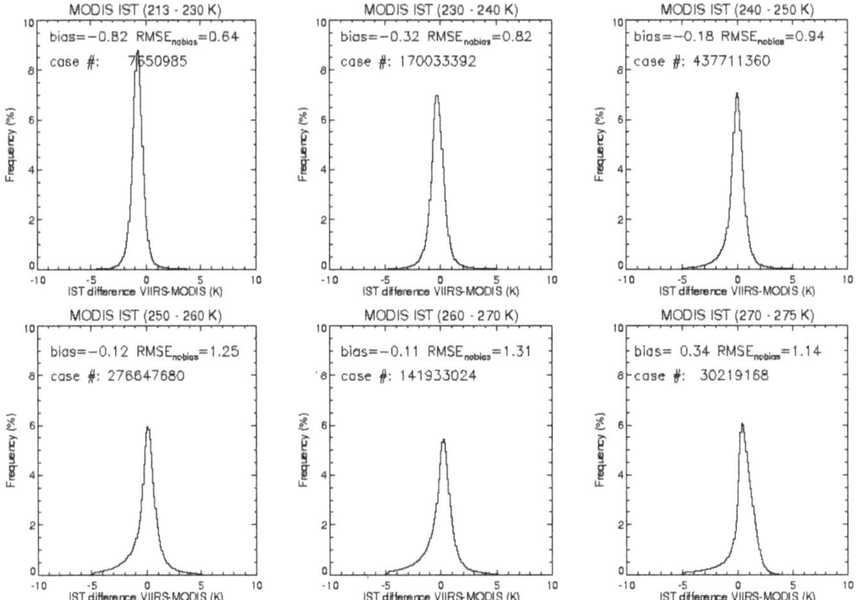

Figure 5. Histogram of ice surface temperature differences between NPP VIIRS and MODIS (Aqua and Terra) in the Arctic only from August 2012 to July 2015 for cases with MODIS ice surface temperature in the ranges 213–230 K, 230–240 K, 240–250 K, 250–260 K, 260–270 K, and 270–275 K. Measurement bias and $RMSE_{nobias}$ are indicated for each bin.

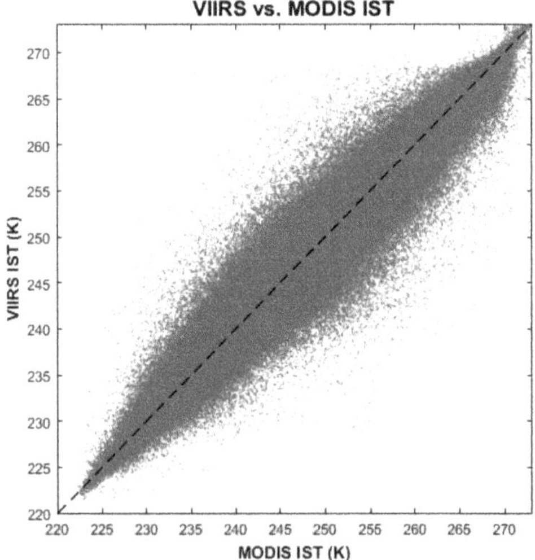

Figure 6. Scatter plot of IST from 20 NPP VIIRS and MODIS Aqua simultaneous nadir overpasses (SNO) over the Arctic from March 2013 to April 2014, with an overall bias of 0.16 K, $RMSE_{nobias}$ of 0.94 K, and RMSE of 0.95 K.

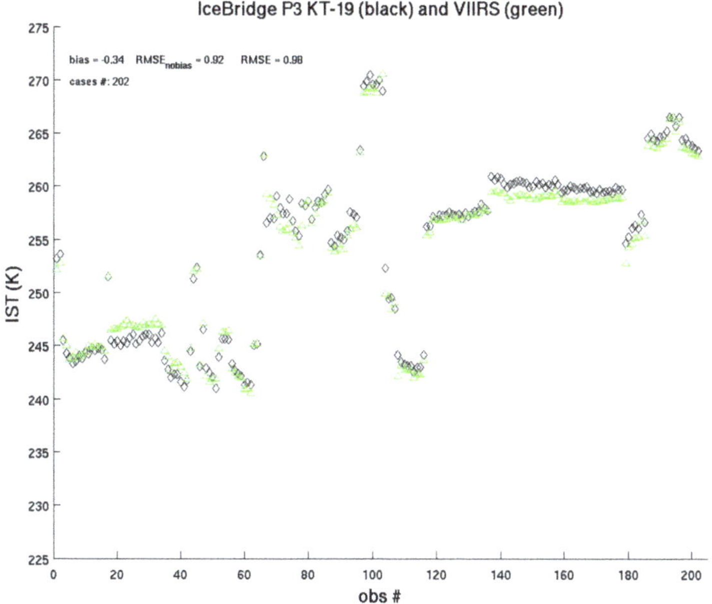

Figure 7. VIIRS IST (green) and KT-19 IST (black) for all coincident IceBridge flights with cloud-free observations over the Arctic (March–May 2014) and Antarctic (October–November 2012–2013).

3.3. Comparison with Drifting Buoy 2 m Air Temperature

The VIIRS IST EDR was also compared to the 2 m air temperature measured by drifting ice buoys in the Arctic (Figure 8). The results show a bias of −3.2 K, $RMSE_{nobias}$ of 3.7, and RMSE of 4.9 K, which means the VIIRS IST EDR is lower (colder) than the 2 m air temperature by over 3 K. The bias absolute values increase with increasing VIIRS IST.

3.4. Comparison with NCEP/NCAR Surface Air Temperature

Surface air temperature is assimilated in most of reanalyses and is also an output parameter [34,37]. However, the surface skin temperature over land and sea ice is not assimilated. The VIIRS IST EDR was compared to the surface air temperature from the NCEP/NCAR reanalysis [34]. Note that an additional comparison between the VIIRS IST and the NCEP-DOE Reanalysis 2 dataset [38] was also performed, but in our study the differences between the VIIRS IST and the NCEP/NCAR reanalysis were smaller than those from the VIIRS IST and NCEP-DOE comparisons. The results of the NCEP/NCAR comparison (Figures 9 and 10) show an overall bias of −3.41 K and an $RMSE_{nobias}$ of 6.04 K, so the VIIRS IST EDR is 3.41 K lower (colder) than the surface air temperature from the NCEP/NCAR reanalysis. On average, the $RMSE_{nobias}$ is larger than that of the drifting ice buoy comparison. The VIIRS IST has a minimum cold bias and $RMSE_{nobias}$ when VIIRS IST is between 270 and 275 K, opposite to the results from the comparison with drifting ice buoys. This cold bias generally increases with decreasing IST except for the IST range between 260 and 270 K. The maximum cold bias and $RMSE_{nobias}$ occur for IST between 213 and 230 K. The RMSE is 6.94 K for all cases, and 9.34 K, 7.35 K, 6.83 K, 6.87 K, 6.25 K, and 3.01 K for NCEP/NCAR surface air temperature in the ranges 213–230 K, 230–240 K, 240–250 K, 250–260 K, 260–270 K, and 270–275 K.

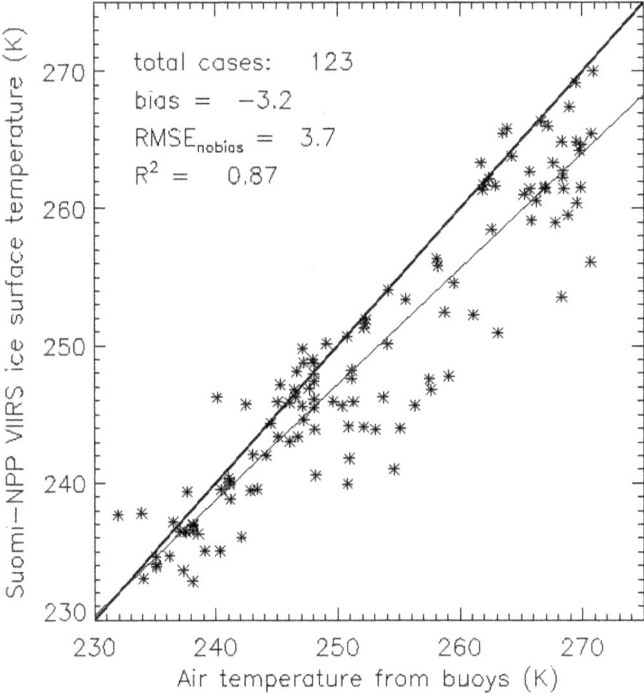

Figure 8. Scatterplot of surface air temperature from Arctic drifting ice buoys and VIIRS IST from August 2012 to June 2014. The thick line is the 1-to-1 ratio line; the thin line as the linear regression.

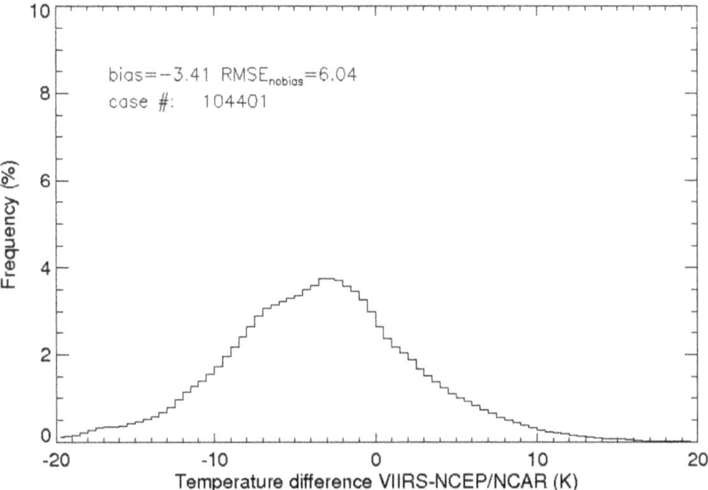

Figure 9. Histogram of ice surface temperature differences between VIIRS IST and National Centers for Environmental Prediction/National Center for Atmospheric Research (NCEP/NCAR) surface air temperature in the Arctic for all cases from August 2012 to July 2015.

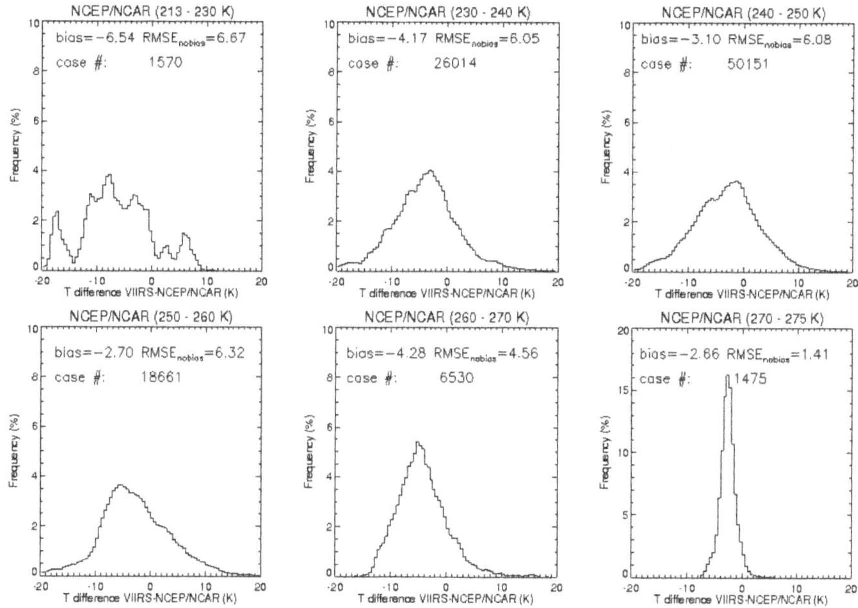

Figure 10. Histogram of ice surface temperature differences between VIIRS IST and NCEP/NCAR surface air temperature in the Arctic from August 2012 to July 2015 for cases with NCEP/NCAR surface air temperature in the ranges 213–230 K, 230–240 K, 240–250 K, 250–260 K, 260–270 K, and 270–275 K. Measurement bias and $RMSE_{nobias}$ are indicated for each bin.

Radionov *et al.* [39] showed that the average difference between the snow surface temperature and the 2 m air temperature is −0.7 K (surface is colder) during the June–August warm season, and −0.6 K in cold season from October to May. The annual mean difference in their measurements was −0.1 K. The largest differences are between −4.5 and −6.9 K from November to May, usually under clear sky conditions, and +2.1 to +5.5 in the warmer part of the year. Given that the VIIRS IST EDR is for clear-sky conditions only, differences larger than the averages reported in the Radionov *et al.*, study would be expected. Nevertheless, with thousands of cases in the comparisons of VIIRS, buoy, and reanalysis temperatures, a VIIRS cold bias of over 3 K may not be solely the result of air-skin temperature differences.

The large VIIRS IST cold bias, or reanalysis and buoy warm bias, may also be attributable to differences in the spatial coverage of VIIRS, drifting buoys, and the NCEP/NCAR reanalysis, and therefore the heterogeneity of the observed IST field. Due to the coarser resolution, the reanalysis is unable to resolve some of the complex IST features. The reanalysis could be smoothing out extreme values caused by features such as leads and polynyas that create small-scale temperature variations. A factor that could also explain the bias with drifting buoys is that some drifting buoys experience heating from sunlight, and as a result their temperatures may be biased high (warm). This would also cause a warm bias in the reanalysis, as buoy temperatures and pressures are assimilated [34]. Another third possibility is that cloud contamination, especially from high cirrus clouds, results in a low retrieved VIIRS IST.

4. Summary and Conclusions

Monitoring the surface temperature is important for climate studies globally, but it deserves special attention in the Polar Regions because of sparse surface observations, an extreme temperature range, and the unique surface and atmospheric characteristics. The VIIRS sensor onboard the Suomi

NPP and future Joint Polar Satellite System (JPSS) satellites provides an IST product with higher spatial resolution and wider coverage than heritage sensors such as AVHRR and MODIS with excellent radiometric quality. This study provided an assessment of the VIIRS IST EDR accuracy through comparisons with *in situ* and aircraft measurements, another satellite product, and a climate reanalysis.

Validation with KT-19 IST measurements from NASA Operation IceBridge flights over the Arctic and Antarctic in 2012–2014 yields a VIIRS IST cold bias of −0.34 K and an $RMSE_{nobias}$ of 0.92 K. An intercomparison with MODIS IST from 2012 to 2015 gives a VIIRS IST bias of −0.12 K and $RMSE_{nobias}$ of 1.01 K. The bias and $RMSE_{nobias}$ values show a dependence on temperature, with a positive bias for the warmer part of the range (270–275 K) and negative biases for lower temperatures. However, the bias and $RMSE_{nobias}$ values are relatively small in all temperature bins. A more rigorous simultaneous nadir overpass comparison shows an absolute bias of 0.16 K and an $RMSE_{nobias}$ of 0.94 K.

Comparisons of the VIIRS IST with 2 m air temperatures show a larger cold bias. Validation with drifting ice buoy 2 m air temperature yields a VIIRS IST bias of −3.2 K and $RMSE_{nobias}$ of 3.7 K. Comparisons to NCEP/NCAR Reanalysis surface air temperature produce a similar bias of −3.41 K and $RMSE_{nobias}$ of 6.04 K.

These validation and evaluation results demonstrate the high quality of the VIIRS IST EDR as indicated by the absolute bias and RMSE values for the IceBridge KT-19 measurement comparisons, where the KT-19 provide the best "truth" data for skin temperature. The results also show that the science quality of the VIIRS IST is similar to that of the MODIS IST. Based on the IceBridge and MODIS results, the VIIRS IST uncertainty (RMSE) meets or exceeds the JPSS system requirement of 1.0 K. The product can therefore be considered useful for meteorological and climatological applications. Furthermore, while we are not aware of any numerical weather prediction centers or climate reanalysis projects that assimilate satellite-derived ice surface temperature, the demonstrated accuracy of the IST product and the relatively large difference found in the comparison with the reanalysis surface temperatures provide motivation for research into its use in numerical models.

Acknowledgments: This work was supported by the Joint Polar Satellite System (JPSS) Program Office. MODIS ice surface temperature data is from the NASA Earth Observing System Data and Information System (EOSDIS) and the National Snow and Ice Data Center DAAC. NCEP Reanalysis data was provided by NOAA/OAR/ESRL PSD, Boulder, Colorado, USA, from their website at http://www.esrl.noaa.gov/psd/. The drifting buoy observations are from the International Arctic Buoy Programme, University of Washington at http://iabp.apl.washington.edu/index.html. KT-19 data is from the National Snow and Ice Data Center at http://nsidc.org/data/iakst1b. The VIIRS ice surface temperature data is from the JPSS Government Resources for Algorithm Verification, Independent Test and Evaluation (GRAVITE) system. The views, opinions, and findings contained in this report are those of the author(s) and should not be construed as an official National Oceanic and Atmospheric Administration or U.S. Government position, policy, or decision.

Author Contributions: Yinghui Liu designed the research, collected most of the data, carried out most of the data analysis, and wrote the manuscript. Jeffrey Key advised the whole study from the research method to the implementation, and has the greatest contribution on the revisions of the manuscript. Mark Tschudi introduced the KT-19 data, and did the initial analysis with Daniel Baldwin. Richard Dworak collected all the KT-19 data, and did the comparison of KT-19 with VIIRS IST EDR. Robert Mahoney did the VIIRS and MODIS Simultaneous Nadir Overpass analysis. All the authors contributed to the revisions of the manuscript.

Conflicts of Interest: The authors declare no conflict of interest.

References

1. IPCC. *Climate Change 2014: Impacts, Adaptation, and Vulnerability. Part B: Regional Aspects. Contribution of Working Group II to the Fifth Assessment Report of the Intergovernmental Panel on Climate Change*; Barros, V.R., Field, C.B., Dokken, D.J., Mastrandrea, M.D., Mach, K.J., Bilir, T.E., Chatterjee, M., Ebi, K.L., Estrada, Y.O., Genova, R.C., *et al*, Eds.; Cambridge University Press: Cambridge, UK; New York, NY, USA, 2014; p. 688.
2. Francis, J.A.; White, D.M.; Cassano, J.J.; Gutowski, W.J., Jr.; Hinzman, L.D.; Holland, M.M.; Steele, M.A.; Voeroesmarty, C.J. An arctic hydrologic system in transition: Feedbacks and impacts on terrestrial, marine, and human life. *J. Geophys. Res. Biogeosci.* **2009**, *114*. [CrossRef]

3. Solomon, S.; Qin, D.; Manning, M.; Chen, Z.; Marquis, M.; Averyt, K.B.; Tignor, M.; Miller, H.L.; IPCC. Summary for policymakers. In *Climate Change 2007: The Physical Science Basis*; Technical Report, Contribution of Working Group I, II and III to the Fourth Assessment Report of the Intergovernmental Panel on Climate Change; Cambridge University Press: Cambridge, UK; New York, NY, USA, 2007; p. 22.
4. Holland, M.M.; Bitz, C.M. Polar amplification of climate change in coupled models. *Clim. Dyn.* **2003**, *21*, 221–232. [CrossRef]
5. Key, J.; Haefliger, M. Arctic ice surface-temperature retrieval from AVHRR thermal channels. *J. Geophys. Res. Atmos.* **1992**, *97*, 5885–5893. [CrossRef]
6. Key, J.R.; Collins, J.B.; Fowler, C.; Stone, R.S. High-latitude surface temperature estimates from thermal satellite data. *Remote Sens. Environ.* **1997**, *61*, 302–309. [CrossRef]
7. Wang, X.; Key, J.R. Recent trends in arctic surface, cloud, and radiation properties from space. *Science* **2003**, *299*, 1725–1728. [CrossRef] [PubMed]
8. Wang, X.; Key, J.R. Arctic surface, cloud, and radiation properties based on the AVHRR polar pathfinder dataset. Part I: Spatial and temporal characteristics. *J. Clim.* **2005**, *18*, 2558–2574. [CrossRef]
9. Wang, X.; Key, J.R. Arctic surface, cloud, and radiation properties based on the AVHRR polar pathfinder dataset. Part II: Recent trends. *J. Clim.* **2005**, *18*, 2575–2593. [CrossRef]
10. Liu, Y.; Key, J.; Wang, X. Influence of changes in sea ice concentration and cloud cover on recent arctic surface temperature trends. *Geophys. Res. Lett.* **2009**, *36*. [CrossRef]
11. Liu, Y.; Key, J.; Wang, X. The influence of changes in cloud cover on recent surface temperature trends in the arctic. *J. Clim.* **2008**, *21*, 705–715. [CrossRef]
12. Hall, D.K.; Key, J.R.; Casey, K.A.; Riggs, G.A.; Cavalieri, D.J. Sea ice surface temperature product from MODIS. *IEEE Trans. Geosci. Remote Sens.* **2004**, *42*, 1076–1087. [CrossRef]
13. Duan, S.-B.; Li, Z.-L. Intercomparison of operational land surface temperature products derived from MSG-SEVIRI and Terra/Aqua-MODIS data. *IEEE J. Sel. Top. Appl. Earth Observ. Remote Sens.* **2015**, *8*, 4163–4170. [CrossRef]
14. Guillevic, P.C.; Privette, J.L.; Coudert, B.; Palecki, M.A.; Demarty, J.; Ottle, C.; Augustine, J.A. Land surface temperature product validation using NOAA's surface climate observation networks-scaling methodology for the visible infrared imager radiometer suite (VIIRS). *Remote Sens. Environ.* **2012**, *124*, 282–298. [CrossRef]
15. Wan, Z.; Li, Z.L. Radiance-based validation of the v5 MODIS land-surface temperature product. *Int. J. Remote Sens.* **2008**, *29*, 5373–5395. [CrossRef]
16. Li, Z.L.; Tang, B.H.; Wu, H.; Ren, H.; Yan, G.; Wan, Z.; Trigo, I.F.; Sobrino, J.A. Satellite-derived land surface temperature: Current status and perspectives. *Remote Sens. Environ.* **2013**, *131*, 14–37. [CrossRef]
17. Cao, C.; Xiong, J.; Blonski, S.; Liu, Q.; Uprety, S.; Xi, S.; Yan, B.; Weng, F. Suomi NPP VIIRS sensor data record verification, validation, and long-term performance monitoring. *J. Geophys. Res. Atmos.* **2013**, *118*, 664–678. [CrossRef]
18. Key, J.R.; Mahoney, R.; Liu, Y.; Romanov, P.; Tschudi, M.; Appel, I.; Maslanik, J.; Baldwin, D.; Wang, X.; Meade, P. Snow and ice products from Suomi NPP VIIRS. *J. Geophys. Res. Atmos.* **2013**, *118*, 12816–12830. [CrossRef]
19. Yu, Y.; Rothrock, D.A.; Lindsay, R.W. Accuracy of sea-ice temperature derived from the advanced very high-resolution radiometer. *J. Geophys. Res.: Oceans* **1995**, *100*, 4525–4532. [CrossRef]
20. Hutchison, K.D.; Roskovensky, J.K.; Jackson, J.M.; Heidinger, A.K.; Kopp, T.J.; Pavolonis, M.J.; Frey, R. Automated cloud detection and classification of data collected by the visible infrared imager radiometer suite (VIIRS). *Int. J. Remote Sens.* **2005**, *26*, 4681–4706. [CrossRef]
21. Comiso, J.C. Surface temperatures in the polar-regions from Nimbus-7 temperature humidity infrared radiometer. *J. Geophys. Res.: Oceans* **1994**, *99*, 5181–5200. [CrossRef]
22. Lindsay, R.W.; Rothrock, D.A. Arctic sea-ice surface-temperature from AVHRR. *J. Clim.* **1994**, *7*, 174–183. [CrossRef]
23. Baker, N. Joint Polar Satellite System (JPSS) VIIRS Ice Surface Temperature Algorithm Theoretical Basis Document Rev-A. Available online: http://npp.gsfc.nasa.gov/sciencedocs/2015-06/ 474-00052_ATBD-VIIRS-IST_A.pdf (accessed on 1 May 2013).

24. Ip, J.; Hauss, B. Pre-launch performance assessment of the VIIRS ice surface temperature algorithm. In Proceedings of the Fifth Annual Symposium on Future Operational Environmental Satellite Systems-NPOESS and GOES-R; 16th Conference on Satellite Meteorology and Oceanography; 89th Annual Meeting of the American Meteorological Society 89th, Phoenix, AZ, USA, 11–15 January 2009.
25. Baker, N. Joint Polar Satellite System (JPSS) NPP Operational Algorithm Proxy and Synthetic Test Data Description. Available online: http://www.star.nesdis.noaa.gov/jpss/documents/ OAD/GSFC_474-00096_JPSS_NPP_Operational_Algorithm_Proxy_and_Synthetic_Test_Data__Alt._doc._no._D45702_.pdf (accessed on 3 January 2012).
26. Shoucri, M.; Hauss, B. Everest: An end-to-end simulation for assessing the performance of weather data products produced by environmental satellite systems. *Proc. SPIE* **2009**, *7458*, 74580G.
27. Berk, A.; Anderson, G.P.; Acharya, P.K.; Bernstein, L.S.; Muratov, L.; Lee, J.; Fox, M.; Adler-Golden, S.M.; Chetwynd, J.H.; Hoke, M.L. MODTRAN5: 2006 Update. *Proc. SPIE* **2006**, *6233*, 62331F.
28. Mills, S. Simulation and test of the VIIRS Sensor Data Record (SDR) algorithm for NPOESS. In Proceedings of the Third Symposium on Future National Operational Environmental Satellites; 87th AMS Annual Meeting, San Antonio, TX, USA, 13–18 January 2007.
29. Hall, D.K.; Riggs, G. *MODIS/Terra Sea Ice Extent 5-Min L2 Swath 1km, Version 6*; NASA National Snow and Ice Data Center Distributed Active Archive Center: Boulder, CO, USA, 2015.
30. Hall, D.K.; Riggs, G.A. *MODIS/Aqua Sea Ice Extent 5-Min L2 Swath 1 km, Version 6*; NASA National Snow and Ice Data Center Distributed Active Archive Center: Boulder, CO, USA, 2015.
31. Baker, N. Joint Polar Satellite System (JPSS) Operational Algorithm Description (OAD) for VIIRS Sea Ice Concentration Intermediate Product (IP) Software. Available online: http://npp.gsfc.nasa.gov/sciencedocs/2015-06/474-00094_OAD-VIIRS-SIC-IP_B.pdf (accessed on 3 June 2013).
32. Krabill, W.B.; Buzay, E. *IceBridge KT-19 IR Surface Temperature, Version 1. (2012–2015)*; Updated 2014; NASA National Snow and Ice Data Center Distributed Active Archive Center: Boulder, CO, USA, 2012.
33. Studinger, M.; Koenig, L.; Martin, S.; Sonntag, J. Operation IceBridge: Using instrumented aircraft to bridge the observational gap between icesat and ICESat-2. In Proceedings of 2010 IEEE International, Geoscience and Remote Sensing Symposium (IGARSS), Honolulu, HI, USA; 2010; pp. 1918–1919.
34. Kalnay, E.; Kanamitsu, M.; Kistler, R.; Collins, W.; Deaven, D.; Gandin, L.; Iredell, M.; Saha, S.; White, G.; Woollen, J.; *et al.* The NCEP/NCAR 40-year reanalysis project. *Bull. Am. Meteorol. Soc.* **1996**, *77*, 437–471. [CrossRef]
35. Rigor, I.G.; Colony, R.L.; Martin, S. Variations in surface air temperature observations in the arctic, 1979–1997. *J. Clim.* **2000**, *13*, 896–914. [CrossRef]
36. Scambos, T.A.; Haran, T.M.; Massom, R. Validation of AVHRR and MODIS ice surface temperature products using *in situ* radiometers. *Ann. Glaciol.* **2006**, *44*, 345–351. [CrossRef]
37. Dee, D.P.; Uppala, S.M.; Simmons, A.J.; Berrisford, P.; Poli, P.; Kobayashi, S.; Andrae, U.; Balmaseda, M.A.; Balsamo, G.; Bauer, P.; *et al.* The ERA-Interim reanalysis: Configuration and performance of the data assimilation system. *Q. J. R. Meteorol. Soc.* **2011**, *137*, 553–597. [CrossRef]
38. Kanamitsu, M.; Ebisuzaki, W.; Woollen, J.; Yang, S.K.; Hnilo, J.J.; Fiorino, M.; Potter, G.L. NCEP-DOE AMIP-II reanalysis (R-2). *Bull. Am. Meteorol. Soc.* **2002**, *83*, 1631–1643. [CrossRef]
39. Radionov, V.F.; Bryazgin, N.N.; Alexandrov, E.I. *The Snow Cover of the Arctic Basin*; Technical Report, APL-UW TR 9701; Applied Physics Laboratory, University of Washington: Seattle, WA, USA, 1997.

Letter

An Investigation of a Novel Cross-Calibration Method of FY-3C/VIRR against NPP/VIIRS in the Dunhuang Test Site

Caixia Gao *,†, Yongguang Zhao *,†, Chuanrong Li, Lingling Ma, Ning Wang, Yonggang Qian and Lu Ren

Key Laboratory of Quantitative Remote Sensing Information Technology, Academy of Opto-Electronics, Chinese Academy of Sciences, Beijing 100094, China; crli@aoe.ac.cn (C.L.); llma@aoe.ac.cn (L.M.); wangning@aoe.ac.cn (N.W.); qianyg@aoe.ac.cn (Y.Q.); renlu@aoe.ac.cn (L.R.)

* Correspondence: caixiagao2010@hotmail.com (C.G.); zyggg22@163.com (Y.Z.); Tel.: +86-10-8217-8645 (C.G.)

† These authors contributed equally to this work.

Academic Editors: Changyong Cao, Richard Müller and Prasad S. Thenkabail

Received: 30 November 2015; Accepted: 14 January 2016; Published: 21 January 2016

Abstract: Radiometric cross-calibration of Earth observation sensors is an effective approach to evaluate instrument calibration performance, identify and diagnose calibration anomalies, and quantify the consistency of measurements from different sensors. In this study a novel cross-calibration method is proposed, taking into account the spectral and viewing angle differences adequately; the method is applied to the FY-3C/Visible Infrared Radiometer (VIRR), taking the Suomi National Polar-Orbiting Partnership (NPP)/Visible Infrared Imaging Radiometer Suite (VIIRS) as a reference. The results show that the relative difference between the two sets increases from January to May 2014, and becomes lower for the data on 24 July, 11 September, and 16 September, within approximately 10%. This phenomenon is caused by the updating of the calibration coefficients in the VIRR datasets with results from a vicarious method on June 2014. After performing an approximate estimation of the uncertainty, it is demonstrated that this calibration has a total uncertainty of 5.5%–6.0%, which is mainly from the uncertainty of the Bidirectional Reflectance Distribution Function model.

Keywords: cross-calibration; spectral adjustment; bidirectional reflectance distribution function model; FY-3C/VIRR; NPP/VIIRS

1. Introduction

To make full use of the ever-increasing EO satellite systems, radiometric calibration, especially post-launch calibration, is of critical importance for the various imaging sensor systems, because the performance of sensors is subject to change during launch and the subsequent exposure to the space environment [1–3]. On-board calibrators, reference to lamp sources and/or solar illumination or lunar illumination, and approaches using Earth scenes imaged in-flight are effective for operational calibration and monitoring of the performance of sensors. However, due to power, weight, and space restrictions, some satellites are not well characterized by on-board calibrators, especially in the solar reflective spectral region [4]; thus, the calibration approach using Earth scenes is significant.

Earth surfaces with suitable characteristics have long been used as benchmark or test sites to verify the post-launch radiometric calibration performance of satellite sensors. The associated methodologies are often referred to as vicarious calibration or cross-calibration [5]. Although vicarious calibration has been shown to be successful, its accuracy depends on the coincident surface measurements of test site, climate, and weather conditions, and this calibration technique is very complex, laborious, and expensive. The cross-calibration approach uses terrestrial targets to transfer radiometric calibration between satellite sensors without coincident surface measurements; it has been explored for evaluating instrument calibration performance, identifying and diagnosing calibration anomalies [5]. Therefore, a number of national and international efforts have been made [6] on cross-calibration. Notable among them is the Global Space-based Inter-Calibration System (GSICS) initiated in 2005 by the World Meteorological Organization (WMO) and the Coordination Group for Meteorological Satellites (CGMS). GSICS aims to produce consistent and well-calibrated measurements from a variety of the international operational meteorological satellites, such as the serial satellites of Geostationary Operational Environmental Satellite system (GOES), Meteosat, FY, *etc.* [7].

The Visible Infrared Radiometer (VIRR) is one of the key instruments onboard the Chinese meteorological sun-synchronous satellite FY-3C, which was successfully launched on September 2013. Similar to the Advanced Very High Resolution Radiometer (AVHRR) and Multi-channel Visible and Infrared Scanning (MVIRS), VIRR has no on-board calibration system for the solar reflective channels, and its post-launch calibration depends on cross-calibration and vicarious calibration. However, because the vicarious calibration is performed once a year due to the cost and measurement conditions, cross-calibration becomes a valuable approach to characterize the performance of VIRR. Xu *et al.* have performed cross-calibration using the Moderate Resolution Imaging Spectroradiometer (MODIS) with the simultaneous nadir overpass (SNO) method [8]. However, the SNO method requires simultaneous nadir observation of the two sensors over the same target and has strict thresholds on the solar zenith and view angles. These requirements would result in less eligible image pairs, thereby reducing the frequency of cross-calibration. Therefore, this study preliminarily proposes a novel method for the cross-calibration between the sensors without strict viewing angle and spectral consistency, such that the sensors' performance could be monitored more frequently. This method is applied to the VIRR data in the solar reflective band with different viewing angles, referring to the well-calibrated Visible Infrared Imaging Radiometer Suite (VIIRS) onboard the Suomi National Polar-Orbiting Partnership (Suomi-NPP) spacecraft. The calibration accuracy of VIIRS is approximately 2% in most solar reflective bands [9].

Section 2 gives a detailed description of the methodology, including data description and the calibration approach. Section 3 shows the cross-calibration results of VIIRS and VIRR. Section 4 gives the uncertainty analysis, and some discussions are presented in Section 5. Conclusions are drawn in Section 6.

2. Methodology

2.1. Dataset Description

The Dunhuang test site is in the Gobi desert, approximately 35 km west of the city of Dunhuang. The test site was an operational radiometric calibration and validation site for Chinese satellite sensors in 2001, located on the eastern edge of the Kumutage Penniform Desert in Gansu province, Southwest China. The whole target area for vicarious calibration is situated on a stabilized alluvial fan, 30 km $\times$ 30 km in size [10]. The atmosphere is dry, clean, and typically has low levels of aerosol loading, making it beneficial for the calibration experiments; the site was chosen as one of the Committee on Earth Observation Satellites (CEOS) calibration and validation test sites.

The VIRR onboard FY-3C is a heritage instrument from the Multispectral Visible Infrared Scanning Radiometer (MVISR) onboard FY-1C and D. It provides images in 10 spectral bands between 0.44 and 12.5 μm, with a spatial resolution of 1.1 km at nadir, and it includes five visible-near infrared bands, two shortwave bands, and three middle and thermal infrared bands; its data records could be used for vegetation and ocean colour monitoring. VIIRS collects radiometric and imagery data in 22 spectral bands within the visible and infrared region ranging from 0.4 to 12.5 μm, including 16 moderate-resolution (750-m pixels) and five imagery resolution (375-m pixels) bands, plus one panchromatic "Day-Night Band". The VIIRS spectral data are calibrated and geolocated in ground processing to generate Sensor Data Records (SDRs) [11,12]. The solar reflective bands of VIIRS, covering similar wavelength range as MODIS, are also calibrated by the solar diffuser (SD) and lunar observations, with a calibration accuracy of approximately 2% in most bands. Thus, bands M3, M4, I1, and I2 are used to cross-calibrate the corresponding VIRR bands 8, 9, 1, and 2; the characteristics of VIRR and VIIRS visible-near infrared channels are shown in Table 1 and Figure 1.

In addition, because the two sensors onboard different platforms have different overpassing times, only clear-sky scenes observed by both sensors are employed during 2014, so that the atmospheric effect resulting from different acquisition time could be reduced as much as possible, and no temporal matching is considered. In this study, a total of 11 scenes of VIRR images over Dunhuang site are acquired and used as the data source for cross-calibration with the corresponding VIIRS calibrated and geolocated SDR data. The dates and viewing geometries of these scenes are shown in Table 2. Note that some of the pairs have large differences in viewing direction between the two sensors, and this difference could incur larger calibration errors without an accurate Bi-directional Reflectance Distribution Function (BRDF) correction. As an example, Figure 2 shows the image of VIIRS band I2 (see Figure 2a) and VIRR band 2 (see Figure 2b) over the Dunhuang test site on 16 September 2014.

Table 1. The characteristics of VIRR and VIIRS visible-near infrared channels.

	Band	Centre Wavelength (μm)	Spectral Range (μm)	Spatial Resolution at Nadir (m)
FY-3C/VIRR	1	0.630	0.58–0.68	1100
	2	0.865	0.84–0.89	1100
	8	0.505	0.48–0.53	1100
	9	0.555	0.53–0.58	1100
NPP/VIIRS	I1	0.640	0.60–0.68	375
	I2	0.865	0.85–0.88	375
	M3	0.488	0.478–0.498	750
	M4	0.555	0.545–0.565	750

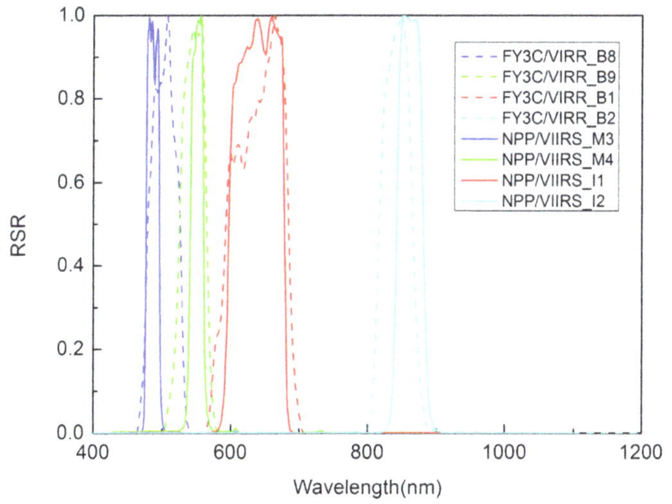

Figure 1. Spectral response functions of VIIRS and VIRR.

Table 2. The dates and viewing geometries of VIIRS and VIRR scenes.

Date	VIIRS				VIRR			
	SZA	SAA	VZA	VAA	SZA	SAA	VZA	VAA
8 January 2014	62.68	−173.55	37.31	74.33	64.50	162.11	21.07	−77.11
13 January 2014	62.03	−172.32	29.87	75.32	63.48	163.14	30.24	−75.71
24 January 2014	59.56	−174.62	37.35	74.32	61.78	160.39	23.50	−76.73
29 January 2014	58.40	−173.02	29.88	75.31	60.19	161.54	32.36	−75.38
13 March 2014	43.43	−171.67	37.61	74.22	45.60	157.12	29.94	−75.79
14 March 2014	42.73	−178.36	54.18	71.07	46.77	150.94	2.28	−87.08
24 March 2014	39.05	−171.91	43.72	73.23	41.62	154.53	22.92	−76.86
6 May 2014	24.28	−165.10	48.87	72.26	26.86	147.57	18.94	−77.53
24 July 2014	21.72	−156.56	29.69	75.37	22.27	152.56	47.02	−72.89
11 September 2014	36.16	−168.11	49.32	72.22	37.19	159.92	30.21	−75.77
16 September 2014	38.40	−165.41	43.84	73.27	38.56	163.97	37.89	−74.55

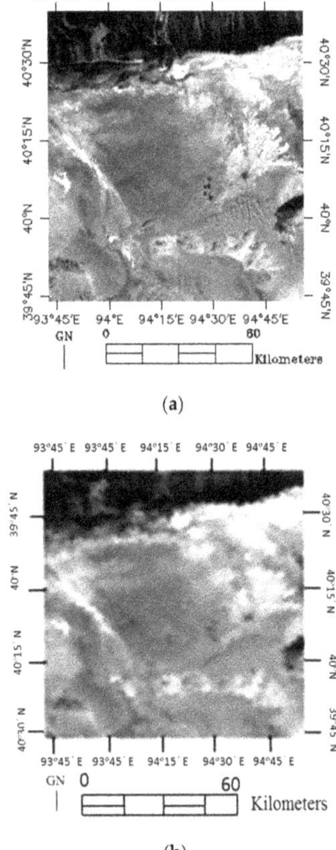

Figure 2. The images of VIIRS band I2 (**a**) and VIRR band 2 (**b**) over the Dunhuang test site on 16 September 2014.

2.2. Cross-Calibration Approach

The cross-calibration method involves comparison of the radiance/reflectance measured by the calibrated sensor with that measured by a well-calibrated sensor as a reference. This exercise can be reduced to spatiotemporal coincidences, *i.e.*, acquisition by both sensors at the same time and with the same viewing geometries (for example, the SNO method). Nevertheless, such coincidences are not very frequent when comparing two sensors with different orbits, altitudes, cycles, and local equatorial crossing time. In these conditions, the reference is, in general, not acquired at the same time for exactly the same spectral range and for the same viewing geometry [13]. For this reason, some corrections must be applied to take into account these aspects.

To alleviate the impact of viewing geometry on the cross-calibration, the Bidirectional Reflectance Distribution Function (BRDF) characteristics of Dunhuang site measured in 2011 with SVC HR1024 spectrometer are used. Measurements were acquired with the viewing zenith angle scanning from 0° to 70° with a step of 14°, and the relative azimuth angles between the sun and viewing direction varied from 0° to 180°, with a step of 30°. Seven datasets of hemispherical scanning measurements were used with the solar zenith angle ranging from 29° to 52°. Examples of the multi-angle Bidirectional Reflectance Factor (BRF) measurements corresponding to VIIRS bands M3, M4, I1, and I2 are shown in Figure 3. The figure shows a general increasing trend towards the backward scattering direction for these bands. In this study, first, with the atmospheric parameters at the VIIRS scenes' acquisition time [the water vapor content (WVC) extracted from the National Centers for Environmental Prediction (NCEP) reanalysis data, the ozone content extracted from the product of Ozone Monitoring Instrument (OMI) onboard the Aura satellite, and the assumed visibility of 40 km (see Table 3), the surface reflectance of Dunhuang site is derived from VIIRS reflectance at the top of atmosphere (TOA) using the radiative transfer model 6S. Next, the angular effect of surface reflectance is corrected with the BRDF model proposed by Roujean *et al.* [14], fitted with the multi-angle BRF measurements, and the surface reflectance along with VIRR viewing direction could be acquired. Subsequently, based on the atmospheric parameters at the VIRR scene acquisition times, the corresponding TOA reflectance is simulated using the 6S model. In this study, the default solar model in the 6S radiative transfer model is used to characterize the solar irradiance [15].

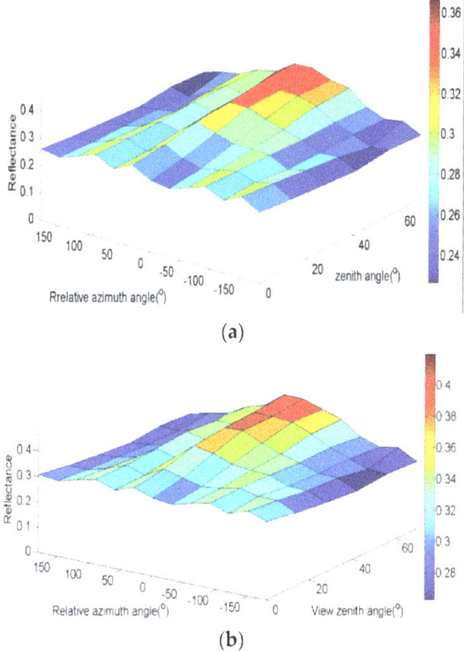

(a)

(b)

Figure 3. *Cont.*

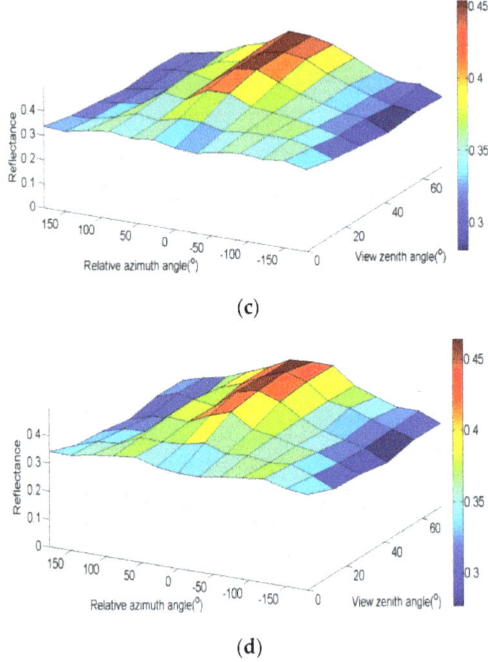

Figure 3. Measured Bidirectional Reflectance Factor of the Dunhuang site. (**a**) VIIRS M3; (**b**) VIIRS M4; (**c**) VIIRS I1; and (**d**) VIIRS I2.

Table 3. Atmosphere parameters at VIIRS and VIRR acquisition times.

Date	WVC at VIIRS Scenes Acquisition Time ($g \cdot cm^{-2}$)	WVC at VIRR Scenes Acquisition Time ($g \cdot cm^{-2}$)	Ozone (atm-cm)	Visibility (km)
8 January 2014	0.260	0.231	0.334	
13 January 2014	0.269	0.171	0.335	
24 January 2014	0.369	0.464	0.332	
29 January 2014	0.461	0.526	0.327	
13 March 2014	0.298	0.272	0.341	
14 March 2014	0.635	0.619	0.341	40.0
24 March 2014	0.551	0.524	0.347	
6 May 2014	0.751	0.754	0.280	
24 July 2014	1.234	1.198	0.280	
11 September 2014	0.848	0.844	0.297	
16 September 2014	0.397	0.318	0.297	

Different applications and technological developments in Earth observation necessarily require different spectral coverage [16]. Thus, spectral bands differ significantly among sensors, even for bands designed to observe at the same region of the electromagnetic spectrum; as a result, these sensors yield fundamentally different measurements that are not directly comparable. To remove the effect of spectral characteristics on cross-calibration, the spectral band adjustment factor k in a given spectral band i is calculated as a ratio of the TOA reflectance from two sensors with a simulation method. In this method, a series of TOA reflectance for VIIRS and VIRR sensors in a given spectral bands i are calculated using the 6S model. The band surface reflectances, varying from 0.1 to 0.5 with a step of 0.1, are used as inputs to drive the 6S model, together with the selected atmospheric states, the sensor's spectral response function, and the illumination and viewing geometries. Subsequently, the corresponding spectral band adjustment factor k_i is fitted by linear regression method.

After alleviating the impact of viewing geometry and spectral characteristics, the measured TOA reflectance of VIRR in a given band i could be evaluated with the simulated VIRR reflectance values. The flowchart is shown in Figure 4:

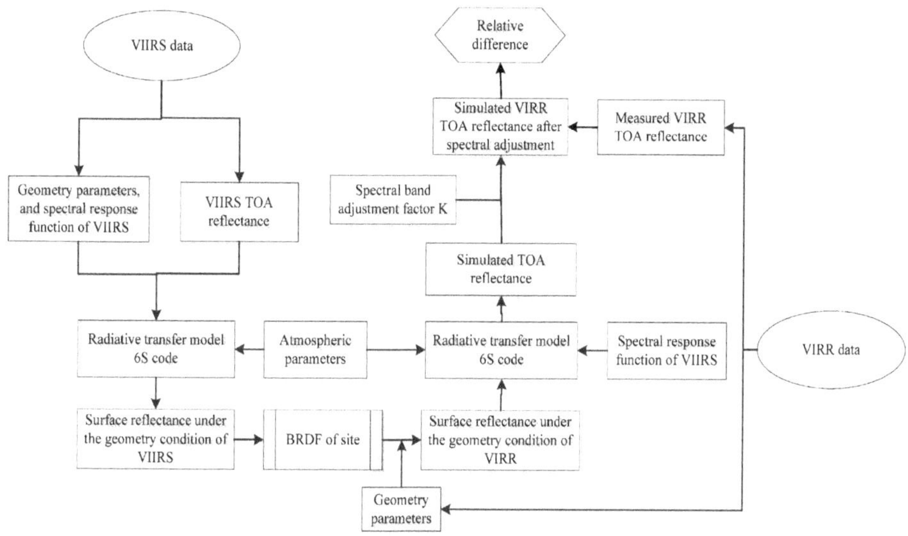

Figure 4. The flowchart of the cross-calibration procedure.

3. Results

To perform this cross-calibration, a common calibration area over the Dunhuang site is selected with geographic location of the top-left corner (40.25°N, 94.25°E) and the bottom-right corner (40.1°N, 94.5°E) of the area. This area has a coverage of approximately 2400 pixels and 1100 pixels in the VIIRS images at a spatial resolution of 350 m and 750 m, respectively, and 725 pixels in the VIRR images. Table 4 depicts its mean digital number (DN) values and standard deviation.

Table 4. Mean DN values and its standard deviation.

Date	Mean Value				STD Value			
	B8	B9	B1	B2	B8	B9	B1	B2
8 January 2014	140.95	145.17	83.62	87.08	1.92	2.51	1.89	2.57
13 January 2014	144.78	149.34	86.13	90.15	2.06	2.75	2.02	2.72
24 January 2014	153.59	160.44	92.99	95.53	2.25	2.95	2.21	2.87
29 January 2014	169.06	175.29	99.84	101.87	2.82	3.07	2.15	2.60
13 March 2014	215.26	232.47	136.25	145.70	3.59	4.68	3.40	4.66
14 March 2014	222.88	240.54	140.89	149.22	3.42	4.51	3.39	4.53
24 March 2014	230.96	251.25	147.83	156.00	3.25	4.14	3.08	4.06
6 May 2014	265.87	293.62	175.70	185.71	4.79	6.24	5.05	7.20
24 July 2014	249.63	269.75	156.51	155.94	3.72	4.65	3.50	4.88
11 September 2014	219.52	239.85	140.87	142.15	5.22	6.96	4.85	6.25
16 September 2014	218.79	237.21	139.20	144.28	4.36	5.65	4.00	5.35

With the aid of the atmospheric parameters (see Table 3), the atmospheric correction of VIIRS is performed for the aforementioned VIIRS datasets via 6S code, and then the derived surface reflectance are corrected with the BRDF model. The scatterplot of the VIIRS TOA reflectance and the derived surface reflectance (see Figure 5), and the scatter plot of the derived surface reflectance and the BRDF corrected values under the geometry condition of VIRR (see Figure 6) are shown below. Note that the atmosphere effect on the VIIRS M3 band is higher than for the other three bands, and the differences in Figure 6 denote the sole influence of the BRDF correction with positive or negative values, which depend on differences between the viewing geometries of the two sensors. With the aforementioned method, the spectral band adjustment factor for each of the image pairs is calculated. Figure 7 shows the spectral band adjustment factor on 24 July 2014. With the aid of the spectral band adjustment factor, surface reflectance and corresponding atmospheric parameters, the simulated VIRR TOA reflectance is acquired via 6S (see Figure 8) and is compared with the measured ones (see Figure 9). Note that the relative difference between the two sets increases from January to May 2014, with a maximum of more than 45%, and the relative differences are lower for the data on 24 July, 11, 16 September, approximately within 10%. The lower relative differences would result from the fact that VIRR has been calibrated using the vicarious method on June 2014, and the calibration coefficients in the datasets were updated. In contrast, the pre-launch calibration coefficient was adopted for these datasets before June 2014. Table 5 shows a comparison between the pre-launch calibration coefficient (gain) of the VIRR sensor and the post-launch one on June, 2014. Note that a large variation is presented on the gain values of the VIRR sensor.

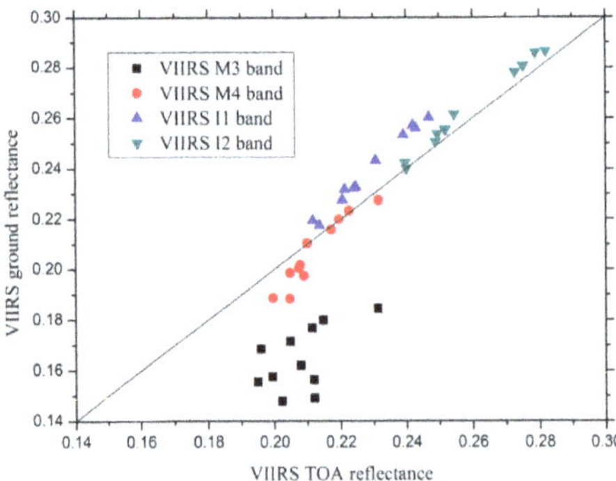

Figure 5. Scatterplot of the VIIRS TOA reflectance and its surface reflectance after atmospheric correction.

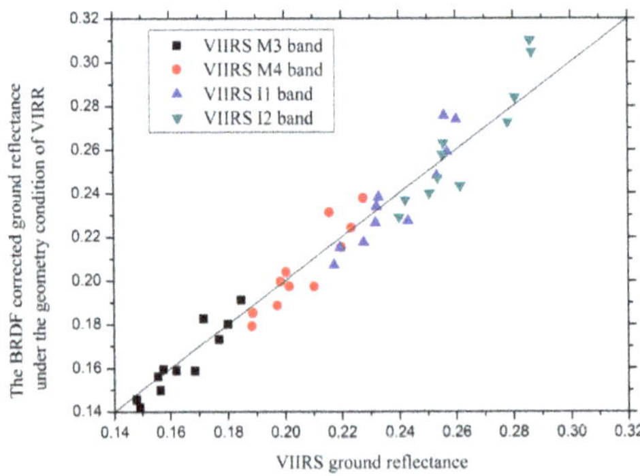

Figure 6. Scatterplot of the VIIRS surface reflectance and the BRDF corrected reflectance values under the geometry condition of VIRR.

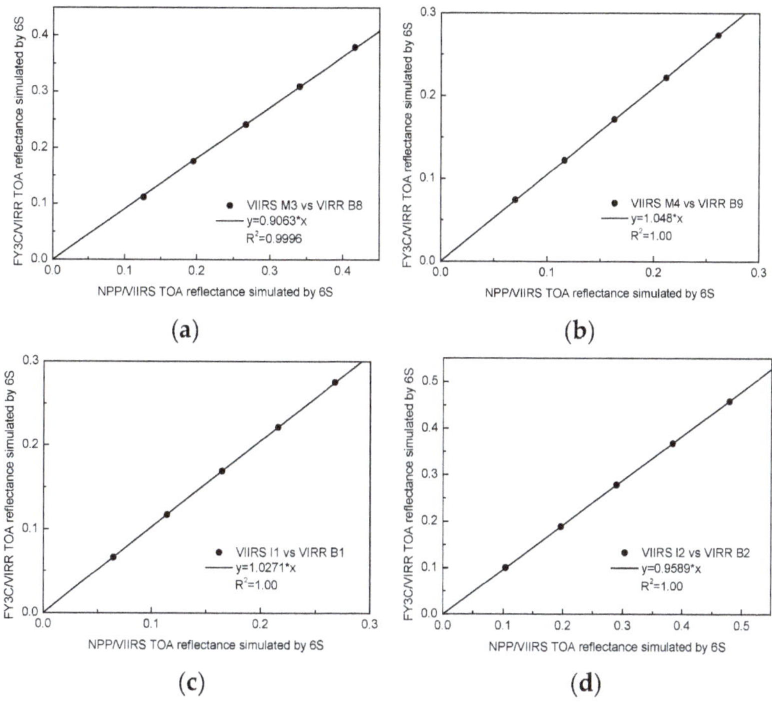

Figure 7. Spectral band adjustment factor between VIIRS and VIRR on 24 July 2014. (**a**) VIIRS M3 *vs.* VIRR B8; (**b**) VIIRS M4 *vs.* VIRR B9; (**c**) VIIRS I1 *vs.* VIRR B1; (**d**) VIIRS I2 *vs.* VIRR B2.

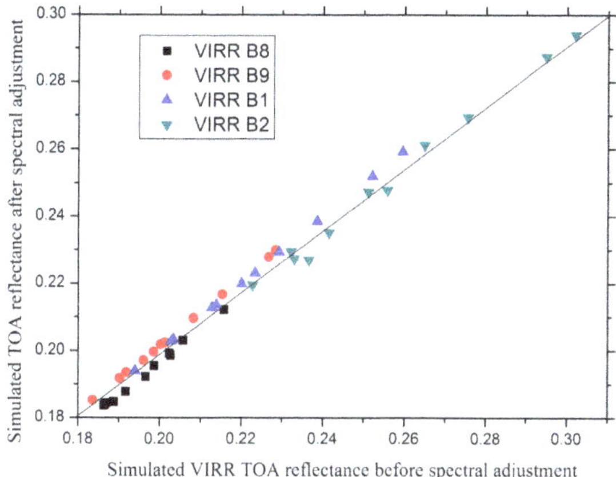

Figure 8. Scatterplot of the simulated VIRR TOA reflectance before spectral adjustment and the corresponding values after spectral adjustment.

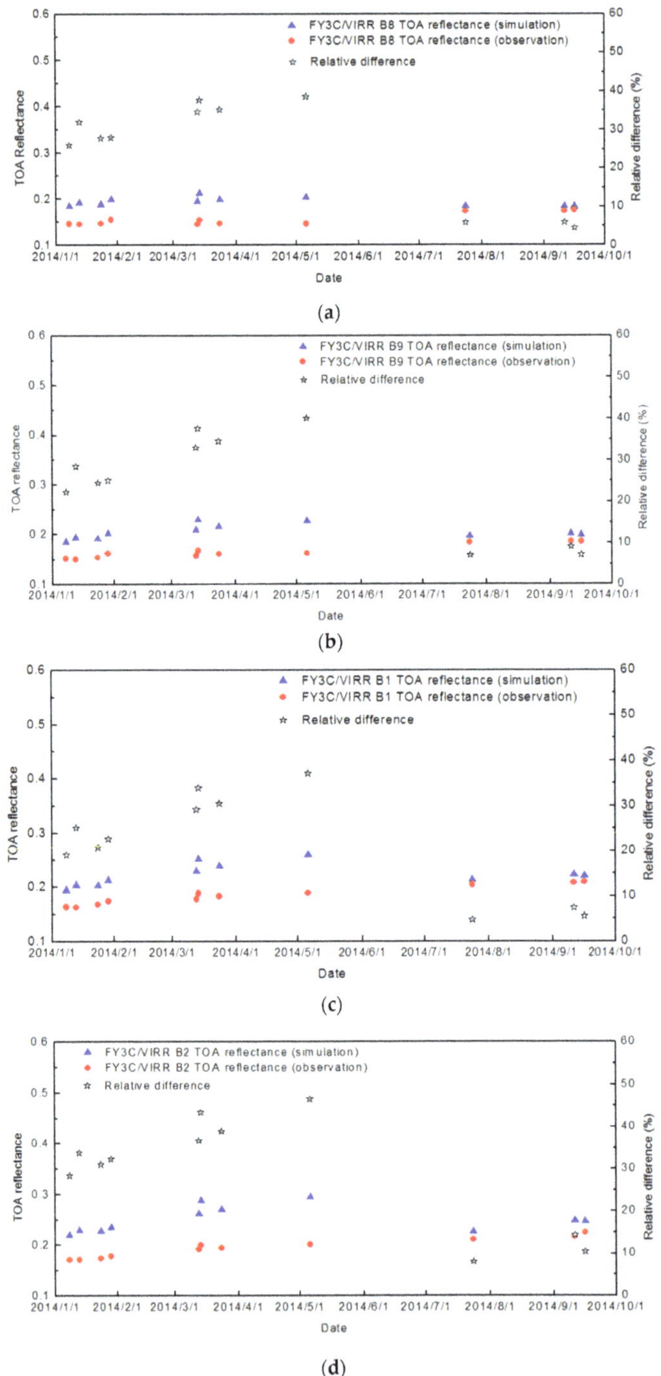

Figure 9. Scatterplot of the simulated VIRR TOA reflectance values after the spectral adjustment and the observed values. (**a**) VIRR B8; (**b**) VIRR B9; (**c**) VIRR B1; and (**d**) VIRR B2.

Table 5. The comparison between the pre-launch calibration coefficient (gain) of the VIRR sensor and the post-launch calibration coefficient (gain).

VIRR Band	Pre-Launch	Post-Launch (on June 2014)	Relative Difference
1	0.10118	0.12549	24.03%
2	0.10126	0.1297	28.09%
8	0.05061	0.06502	28.47%
9	0.05063	0.06455	27.49%

4. Uncertainty Analysis

Although the view geometry and spectral discrepancies between two sensors are reduced as much as possible in this study via BRDF correction and spectral adjustment, the accuracy of cross-calibration is still affected by several factors, such as the uncertainty of the BRDF model, atmosphere parameters, (WVC, visibility, ozone content, and aerosol model, *etc.*), image registration, 6S model, *etc.* Since not all the factors could be quantified, a rough estimation is performed in terms of the following factors in this study:

1. The uncertainty caused by the VIIRS calibration ($\sigma_{1,i}$ (band number i = 8, 9, 1, and 2)): the uncertainty of the VIIRS calibration is approximately 2% (see the second row in Table 6).
2. The uncertainty that is caused by the 6S model ($\sigma_{2,i}$ (band number i = 8, 9, 1, and 2)): according to the error transfer theory, the error of the 6S model (σ_{6s}), which is estimated to be approximately 2%, caused by this model error ($\sigma_{2,i} = \sqrt{2}\sigma_{6s}$) is approximately 2.83% (see the third row in Table 6).
3. The uncertainty caused by the BRDF effect ($\sigma_{3,i}$(band number i = 8, 9, 1, and 2)): the uncertainty of the fitted BRDF model proposed by Roujean is approximately 5%, resulting in an uncertainty within 4.5% of the simulated VIRR TOA radiance (see the fourth row in Table 6).
4. The uncertainty that is caused by the atmospheric parameters
   - In this study, the aerosol type is assumed as the continental aerosol model, which would create an error in the TOA reflectance simulation of VIRR because the Dunhuang test site has a mixed aerosol type of the continental and desert models. To analyze the effect of aerosol type on the cross-calibration ($\sigma_{4,i}$ (band number i = 8, 9, 1, and 2)), spectral adjustment factors and a new group of VIRR TOA reflectance for a given band i is generated when the desert aerosol model is used, and the relative differences are also computed (see the fifth row in Table 6). Similarly, the visibility is changed by ±10 km to analyze the uncertainty caused by aerosol type ($\sigma_{5,i}$ (band number i = 8, 9, 1, and 2), see the sixth row in Table 6). The results demonstrate that the uncertainty resulting from the aerosol type and the visibility are all within 1%.
   - The WVC and ozone contents are also important parameters in this cross-calibration approach. To investigate their effects, similar to the analysis method of visibility, it could be found that the 20% uncertainty in WVC and 10% uncertainty in ozone content would cause an uncertainty within 0.5% ($\sigma_{6,i}$ (band number i = 8, 9, 1, and 2)) and 0.2% ($\sigma_{7,i}$ (band number i = 8, 9, 1, and 2)), respectively, to the cross-calibration (see the seventh and eighth rows in Table 6).
5. Image co-registration error ($\sigma_{8,i}$ (band number i = 8, 9, 1, and 2)): the relative location offset of the cross-calibration test site in two images is inevitable, thereby affecting the accuracy of the cross-calibration. In this study, a sliding window method is used [17] to estimate that the image co-registration error would cause an uncertainty of approximately 0.3%–0.5% (see the ninth row in Table 6).

From these analyses, the total uncertainty for spectral band i (σ_i, i = 8, 9, 1, and 2) in the cross-calibration could be estimated with $\sigma_i = \sqrt{\sum_{j=1}^{8}\sigma_{j,i}^2}$ (j is the index of uncertainty terms); note that the uncertainty in the calibration for the four spectral bands is 5.5%–6.0%.

Table 6. Uncertainty analysis results.

Source	FY/VIRR B8	FY/VIRR B9	FY/VIRR B1	FY/VIRR B2
VIIRS calibration accuracy	2.00%	2.00%	2.00%	2.00%
6S model	2.83%	2.83%	2.83%	2.83%
BRDF	4.13%	4.33%	4.56%	4.42%
Aerosol type	0.51%	0.95%	0.97%	1.07%
Visibility	0.65%	0.56%	0.50%	0.32%
WVC	0.01%	0.01%	0.02%	0.41%
Ozone content	0.16%	0.18%	0.16%	0.01%
Image co-registration	0.31%	0.36%	0.44%	0.57%
Total uncertainty	5.47%	5.67%	5.85%	5.77%

5. Discussion

With the development of remote sensing technology and the urgent need for its quantitative application, much more accurate and higher-frequency radiometric calibration is required to monitor sensor performance. FY-3C/VIRR was launched in 2013 without an onboard calibrator, so that its calibration mainly depends on the various calibrations and cross-calibration. Due to the limitation of SNO cross-calibration method on the solar zenith and view angles, a new cross-calibration method is proposed that has great application potential. In this method, to eliminate the spectral difference between two sensors, the spectral adjustment factor is calculated with a linear regression method for reducing its low sensitivity to the variation of surface reflectance. Furthermore, to account for their viewing geometries differences between two sensors, a BRDF model is constructed with the measured BRF over a uniform area of Dunhuang site. After these corrections, the simulated VIRR TOA reflectance is acquired and is inter-compared with the measured one. Compared to the SNO method, the cross-calibration method can be performed without strict thresholds on view angles, and thus the performance of VIRR could be monitored more frequently with VIIRS, MODIS, *etc.* In addition, the method also could be applied to other sensors. However, it can be found that the accuracy of this method strongly depends on the accuracy of BRDF model, so the image pairs with very large viewing geometries would induce much higher uncertainties on the calibration. Therefore, in the future, further studies on the BRDF model, more accurate BRDF measurements, and the validation of the method over different surface types and synchronized ground measurements are required.

6. Conclusions

In this study, NPP/VIIRS was used as a reference sensor to cross-calibrate the FY-3C/VIRR with a proposed method to monitor its radiometric performance, taking into account the discrepancies in the geometries and spectral coverage between the two sensors. The results preliminary demonstrate that there exists a large difference between VIRR and VIIRS; this might be partly caused by the degradation of VIRR. The relative difference between the two datasets increases from January to May 2014, and is lower for the data on 24 July, 11 September, and 16 September, approximately within 10%. The phenomenon is because the calibration coefficients in the VIRR datasets were updated in June 2014, with results from the vicarious method. Furthermore, through an approximate error analysis, it is found that the total uncertainty for the cross-calibration is 5.5%–6.0%. Among the various factors, the uncertainties of atmosphere parameters have little effect on the accuracy of the cross-calibration; the accuracy of BRDF was found to be the main source affecting the calibration accuracy. Thus, a more accurate BRDF model is required to promote this method in the future for long-term application.

Acknowledgments: The work has been supported by the National High Technology Research and Development Program of China 863 program (2013AA122102) and by the National Natural Science Foundation of China (41301387). The authors thank the China Centre for Resources Satellite Data and Application for providing BRF measurement over the Dunhuang site, and the China Meteorological Administration for providing FY-3C satellite data. Thanks are also given to the anonymous reviewers.

Author Contributions: Caixia Gao and Chuanrong Li conceived and designed the experiments; Yongguang Zhao performed the experiments; Lingling Ma and Ning Wang analysed the data; Yonggang Qian contributed to the data processing; Lu Ren download the satellite data; Caixia Gao wrote the paper.

Conflicts of Interest: The authors declare no conflict of interest.

References

1. Teillet, P.M.; Fedosejevs, G.; Thome, K.J.; Barker, J.L. Impacts of spectral band difference effects on radiometric cross-calibration between satellite sensors in the solar-reflective spectral domain. *Remote Sens. Environ.* **2007**, *110*, 393–409. [CrossRef]
2. Kuusk, A.; Kuusk, J.; Lang, M.; Lükk, T. Vicarious calibration of the PROBA/CHRIS imaging spectrometer. *Photogramm. J. Finl.* **2010**, *22*, 46–59.
3. Gao, C.X.; Jiang, X.G.; Li, X.B.; Li, X.H. The cross-calibration of CBERS-02B/CCD visible-near infrared channels with Terra/MODIS channels. *Int. J. Remote Sens.* **2013**, *34*, 3688–3698. [CrossRef]
4. Ham, S.H.; Sohn, B.J. Assessment of the calibration performance of satellite visible channels using cloud targets: Application to Meteosat-8/9 and MTSAT-1R. *Atmos. Chem. Phys.* **2010**, *10*, 11131–11149. [CrossRef]
5. Teillet, P.M.; Barsi, J.A.; Chander, G.; Thome, K.J. Prime candidate earth targets for the post-launch radiometric calibration of space-based optical imaging instruments, optical engineering applications. *Int. Soc. Opt. Photonics* **2007**. [CrossRef]
6. Chander, G.; Hewison, T.J.; Fox, N.; Wu, X.; Xiong, X.; Blackwell, W.J. Overview of intercalibration of satellite instruments. *IEEE Trans. Geosci. Remote Sens.* **2013**, *51*, 1056–1080. [CrossRef]
7. Yu, F.F.; Wu, X.Q.; Goldberg, M. Recent operational status of GSICS GEO-LEO and GEO-GEO inter-calibrations at NOAA/NESDIS. In Proceedings of the IEEE International Geoscience and Remote Sensing Symposium, Vancouver, BC, Canada, 24–29 July 2011; pp. 989–992.
8. Xu, N.; Chen, L.; Wu, R.H.; Hu, X.Q.; Sun, L.; Zhang, P. In-flight intercalibration of FY-3C visible channels with AQUA MODIS. *Proc. SPIE* **2014**. [CrossRef]
9. VIIRS SDR Science Team. Visible Infrared Imaging Radiometer Suite (VIIRS) Sensor Data Record (SDR) Error Budget, 19 December 2013. Available online: http://www.star.nesdis.noaa.gov/smcd/spb/nsun/snpp/VIIRS/VIIRS_SDR_Error_Budget-2013.pdf (accessed on 16 October 2015).
10. USGS. Remote Sensing Technologies. Available online: http://calval.cr.usgs.gov/rst-resources/sites_catalog/radiometric-sites/dunhuang (accessed on 7 October 2015).
11. Cao, C.; Shao, X.; Uprety, S. Detecting light outages after severe storms using the S-NPP/VIIRS day/night band radiances. *IEEE Trans. Geosci. Remote Sens. Lett.* **2013**, *10*, 1582–1586. [CrossRef]
12. Cao, C.; de Luccia, F.J.; Xiong, X.X.; Wolfe, R.; Weng, F. Early on-orbit performance of the visible infrared imaging radiometer suite onboard the suomi national polar-orbiting Partnership (S-NPP) satellite. *IEEE Trans. Geosci. Remote Sens.* **2014**, *52*, 1142–1156. [CrossRef]
13. Lacherade, S.; Fougnie, B.; Henry, P.; Gamet, P. Cross calibration over desert sites: Description, methodology, and operational implementation. *IEEE Trans. Geosci. Remote Sens.* **2013**, *51*, 1098–1113. [CrossRef]
14. Roujean, J.L.; Leroy, M.; Deschamps, P.Y. A bi-directional reflectance model of the Earth's surface for the correction of remote sensing data. *J. Geophys. Res.* **1992**, *97*, 20455–20468. [CrossRef]
15. Neckel, H.; Labs, D. The solar radiation between 3300 and 12500. *Sol. Phys.* **1984**, *90*, 205–258. [CrossRef]
16. Chander, G.; Mishra, N.; Helder, D.L.; Aaron, D.; Choi, T.; Angal, A.; Xiong, X. Use of EO-1 Hyperion data to calculate spectral band adjustment factors (SBAF) between the L7 ETM+ and Terra MODIS sensors. In Proceedings of the IEEE International Geoscience and Remote Sensing Symposium (IGARSS), Honolulu, HI, USA, 25–30 July 2010; pp. 1667–1670.
17. Wang, Z.; Xiao, P.F.; Gu, X.F.; Feng, X.Z.; Li, X.Y.; Gao, H.L.; Li, H.; Lin, J.T.; Zhang, X.L. Uncertainty analysis of cross-calibration for HJ-1 CCD camera. *Sci. China* **2013**, *56*, 713–723. [CrossRef]

Article

Improved VIIRS and MODIS SST Imagery

Irina Gladkova [1,2,*], Alexander Ignatov [3], Fazlul Shahriar [1,4], Yury Kihai [2], Don Hillger [5] and Boris Petrenko [2]

1 City College of New York, NOAA/CREST, 138th St, New York, NY 10031, USA; fshahriar@gmail.com
2 Global Science and Technology, Inc., Greenbelt, MD 20770, USA; yury.kihai@noaa.gov (Y.K.); boris.petrenko@noaa.gov (B.P.)
3 NOAA STAR, NCWCP, 5830 University Research Court, College Park, MD 20740, USA; alex.ignatov@noaa.gov
4 Graduate Center, City University of New York, 365 Fifth Avenue, New York, NY 10016, USA
5 NOAA STAR, Regional and Mesoscale Meteorology Branch (RAMMB), Fort Collins, CO 80523, USA; don.hillger@noaa.gov
* Correspondence: gladkova@cs.ccny.cuny.edu; Tel.: +1-212-650-7002; Fax: +1-212-650-7003

Academic Editors: Changyong Cao, Dongdong Wang and Prasad S. Thenkabail
Received: 4 November 2015; Accepted: 11 January 2016; Published: 21 January 2016

Abstract: Moderate Resolution Imaging Spectroradiometers (MODIS) and Visible Infrared Imaging Radiometer Suite (VIIRS) radiometers, flown onboard Terra/Aqua and Suomi National Polar-orbiting Partnership (S-NPP)/Joint Polar Satellite System (JPSS) satellites, are capable of providing superior sea surface temperature (SST) imagery. However, the swath data of these multi-detector sensors are subject to several artifacts including bow-tie distortions and striping, and require special pre-processing steps. VIIRS additionally does two irreversible data reduction steps onboard: pixel aggregation (to reduce resolution changes across the swath) and pixel deletion, which complicate both bow-tie correction and destriping. While destriping was addressed elsewhere, this paper describes an algorithm, adopted in the National Oceanic and Atmospheric Administration (NOAA) Advanced Clear-Sky Processor for Oceans (ACSPO) SST system, to minimize the bow-tie artifacts in the SST imagery and facilitate application of the pattern recognition algorithms for improved separation of ocean from cloud and mapping fine SST structure, especially in the dynamic, coastal and high-latitude regions of the ocean. The algorithm is based on a computationally fast re-sampling procedure that ensures a continuity of corresponding latitude and longitude arrays. Potentially, Level 1.5 products may be generated to benefit a wide range of MODIS and VIIRS users in land, ocean, cryosphere, and atmosphere remote sensing.

Keywords: VIIRS; MODIS; imagery; bow-tie; aggregation; deletion; SST

---

1. Introduction

More than a dozen Advanced Very High Resolution Radiometers (AVHRRs) onboard National Oceanic and Atmospheric Administration (NOAA) satellites have been in operational use since 1978. The AVHRR onboard NOAA-19 (launched in February 2009) continues functioning well as of this writing, in addition to two AVHRRs acquired by the European Organization for the Exploitation of Meteorological Satellites (EUMETSAT) for use onboard their Metop satellites. Both Metop-A (launched in October 2006) and Metop-B (launched in September 2012) work well as of today and the remaining Metop-C is scheduled for launch in 2018. With life expectancy of six years, the Metop AVHRRs will likely be in orbit through at least the mid-2020s.

The Visible Infrared Imaging Radiometer Suite (VIIRS) is a new generation US imager, developed to succeed the AVHRR in NOAA operations. The first VIIRS sensor was launched on 28 October 2011 onboard the Suomi National Polar-orbiting Partnership (S-NPP). Four more instruments are lined up to fly onboard the Joint Polar Satellite System (JPSS) satellites, J-1 to J-4 from 2017–2026. NOAA and EUMETSAT entered into a Joint Polar System agreement, by which the S-NPP/JPSS VIIRS instruments cover the afternoon (1:30 P.M./A.M.) orbit, whereas the Metop AVHRRs cover the mid-morning (9:30 A.M./P.M.) orbit.

The NOAA VIIRS sensor builds upon the NASA Moderate Resolution Imaging Spectroradiometers (MODIS) flown onboard the two Earth Observation System (EOS) Terra (10:30 A.M./P.M. orbit) and Aqua (1:30 P.M./A.M. orbit) satellites since December 1999 and May 2002, respectively. Both VIIRS and MODIS carry a comprehensive set of spectral bands, and take measurements in a wide swath to provide (near) global daily coverage, with high spatial resolution and low radiometric noise. Although the design and performance of both sensors can support accurate monitoring of atmosphere, ocean, land and cryosphere from space, VIIRS improves upon all performance metrics [1,2].

Nevertheless, both VIIRS and MODIS are affected by two major imagery artifacts: striping and bow-tie distortions. Both have multi-detector push-broom design and double side rotating mirror, introduced to improve spatial resolution and reduce radiometric noise. The VIIRS imagery is further affected by two irreversible processing steps applied onboard–pixel deletion and aggregation, which are applied to reduce the data volume prior to its down-linking to the ground. These two artifacts should be corrected before VIIRS imagery can be used for visual analysis or downstream processing.

In particular, the NOAA sea surface temperature (SST) retrieval system, Advanced Clear-Sky Processor for Oceans (ACSPO), uses some uniformity filters (within an $n \times n$ pixel spatial window) for clear-sky identification [3], and a more comprehensive pattern recognition algorithm was recently proposed [4]. Continuous, non-distorted imagery in a swath projection is a prerequisite to these algorithms. As a first step, the destriping algorithm proposed in [5] was implemented in ACSPO [6]. This paper describes simple and computationally fast methods for approximating the values deleted onboard, and for resampling VIIRS imagery in order to correct for bow-tie distortions and deletions in ACSPO Level 2 product, while preserving the originally-observed data and associated geo-locations nearly intact. In fact, the original geo values will be slightly adjusted (only the longitudes in the overlapping portions of the scans, and only at the sub-pixel level) so that the corresponding SST imagery is physically continuous (note that typically, the imagery continuity is only met after remapping, in the Level 3 and higher processing level products). This is a very important step in the direction of advanced image processing at the swath level, to facilitate improvements of the cloud mask, or perform any other processing requiring spatial continuity (e.g., computation of the gradient field). The proposed resampling should also improve other applications that use spatial information at the L2 processing, such as ocean color, cloud properties *etc*.

The paper is organized as follows: Section 2 describes VIIRS scan geometry. The proposed approach for VIIRS is documented in Section 3, and its application to MODIS is discussed in Section 4. Section 5 discusses results of application of the proposed algorithm with ACSPO. The alternative VIIRS imagery option, the Ground Track Mercator (GTM) projection, is considered in Section 6. Section 7 shows the potential of the proposed re-sampling approach for some additional L2 products and Section 8 summarizes and concludes the paper.

2. VIIRS Scan Geometry and Imagery

With the unprecedented advance of the remote sensing technologies, the number of data users is growing, while the satellite instruments are becoming more sophisticated. Most users and even satellite data producers may be unfamiliar with the lower-level data processing, which requires knowledge of the instrument design and specifics of onboard processing. Thus, there is a growing need for intermediate user-friendly data products such as Level 1.5 [7], in which some of the instrument-specific data issues have been mitigated.

As input, the ACSPO processing uses VIIRS L1b data (also called sensor data records, SDR [8]) in the original swath projection. Aside from striping, there are three separate issues found in VIIRS SDRs: (1) onboard aggregation; (2) bow-tie distortions; and (3) onboard deletions in the bow-tie regions. These issues are related but distinct and should not be confused. The bow-tie distortions for instance are also found on MODIS, but the onboard aggregation and deletion are unique to the VIIRS. The geometry of these procedures has to be understood and its effects on imagery mitigated, before L2 processing relying on spatial context and patterns in the imagery is possible. In what follows, we briefly describe VIIRS scan geometry and then discuss each of the above issues and their impact on ACSPO L2 product in more details.

2.1. Scan Geometry

The VIIRS Rotating Telescope Assembly (RTA) sweeps in a cross-track direction a 112.56° Earth view sector corresponding to the view zenith angle (VZA) range of $\pm\sim70°$ on the ground and a swath of ~3040 km in the cross-scan direction. In the along-scan direction, the 16 detectors cover a strip of approximately 11.9 km at nadir and 25.9 km wide at the end of each scan. This increase is attributed to the scan geometry and Earth's shape, resulting in a panoramic "bow-tie" segment. A schematic of the VIIRS half-scan projection on the Earth surface from nadir to swath edge is shown in Figure 1 and the scanning footprints for three consecutive VIIRS scans are shown in Figure 2 (*cf.* Figure 2.2-12, p. 20 in [9] for comparison).

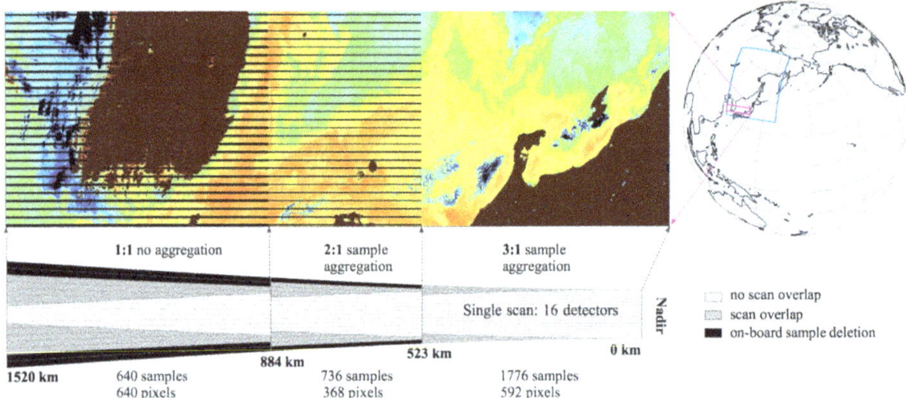

Figure 1. Top-left: Example of bow-tie deletions when the Visible Infrared Imaging Radiometer Suite (VIIRS) sea surface temperature (SST) image is displayed in the original swath projection. Deleted pixels are rendered in black and the land is shown in brown. **Top-right**: Location of the 10 min Advanced Clear-Sky Processor for Oceans (ACSPO) granule (18 October 2015 UTC) is shown by blue rectangle and its portion, displayed in the top-left, is shown in magenta. **Bottom**: Schematic representation of the left half of a single scan, showing bow-tie distortions, on-board deletions and aggregation for a single-gain M-band.

The neighboring whiskbrooms do not overlap at nadir but start overlapping at scan angles greater than approximately 19°. The overlap increases with scan angle and reaches ~12 km at the swath edge. However, in a swath projection, the overlapping pixels from the neighboring scans are appearing in the order they were acquired onboard, rather than according to their position on the Earth's surface (*cf.* Figure 2). As a result, the L1 imagery appears distorted. Examples of these distortions (referred to herein as the "bow-tie distortions") will be shown and analyzed in the following sections. Note that L1 imagery from both MODIS and VIIRS is subject to bow-tie distortions, although to a different degree.

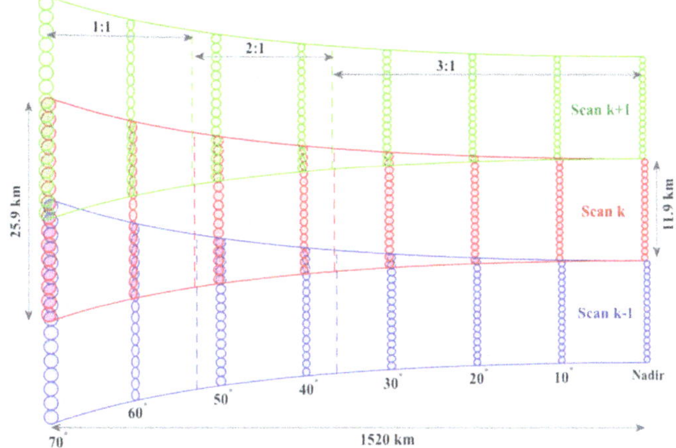

Figure 2. Schematic representation of VIIRS sampling for three consecutive whiskbroom (rendered in blue, red and green respectively), for several selected VZAs from 0° (nadir) to 70° (approximately representing the left edge of the scan) in 10° increments. The two axis of each ellipse represent the horizontal and vertical sampling intervals, respectively. The vertical dashed lines show positions where the aggregation factor changes.

2.2. Aggregation

The constant angular resolution of the sensor field-of-view results in an increasing pixel "footprint" projected onto the Earth for the view directions away from the nadir. For instance, the AVHRR Full Resolution Area Coverage (FRAC) and MODIS pixels grow from ~1 km at nadir to ~5–7 km, in a cross-track direction. The VIIRS instrument is subject to the same resolution degradation problem. The decision was made to employ an onboard aggregation algorithm, to reduce the variation of the pixel size across swath. At nadir, three ~0.24 km footprints are aggregated to form a single VIIRS "pixel" with a size of ~0.74 km. The aggregation changes from 3 × 1 to 2 × 1 at the scan angle ~31.589° and to 1 × 1 at 44.680° as shown in Figure 3 (*cf.* Figure 4 on p. 9 in [8] for comparison). The discontinuities in the pixel size due to the change in the aggregation scheme can cause artifacts during re-projection or resampling and should be treated properly.

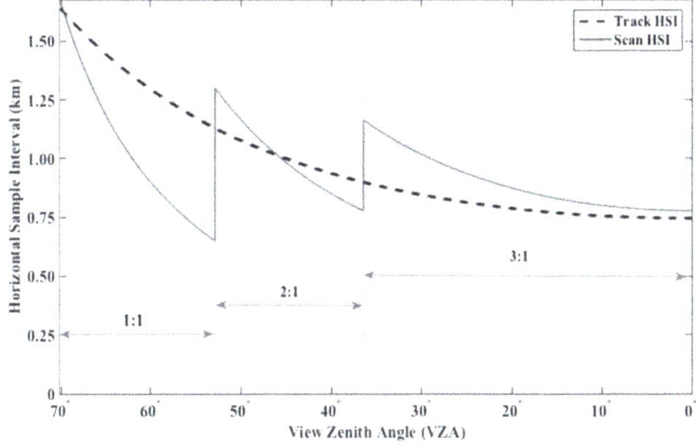

Figure 3. The sampling interval as a function of VZA, in along-track and along-scan directions.

2.3. Bow-Tie Distortion and Their Effect on ACSPO L2 Product

The bow-tie effect can be effectively removed by re-projecting the swath image onto a map. Therefore, users of Level 3 and higher level products never have to worry about the bow-ties. At NOAA, a limited number of VIIRS bands are processed into the imagery product, by mapping them to a Ground-Track Mercator (GTM) projection [10]. (Note that NOAA position its imagery product as an L2, although the mapping itself is an essential element of L3 processing.) The quasi-equidistant GTM grid is laid out in such a way that the columns are aligned along the ground track and the rows are positioned perpendicular to the columns. The NOAA imagery product is not subject to bow-tie artifacts and is primarily used for visual inspection of the VIIRS data. The utility of the GTM approach for SST applications is discussed later in Section 6.

As of this writing, a majority of VIIRS L2 products at NOAA are produced from L1 data on a pixel-by-pixel basis, and reported in the same swath projection. Such "in-pixel" L2 algorithms are not sensitive to the bow-tie distortions. However, some L2 algorithms do use spatial information. For instance, the ACSPO clear-sky mask employs spatial filters in which local statistics such as mean, range and standard deviation are computed over a sliding spatial window surrounding the pixel [3]. The bow-tie effects are currently ignored in ACSPO, because the re-projection is computationally expensive and there were no fast off-the-shelf algorithms easily available that preserve both the array dimensions and the corresponding geo-information. Recently, another algorithm, which uses data from the ambient pixels, was tested to smooth out the atmospheric correction term in the regression SST equations [11]. Most importantly, improvements to clear-sky mask are being explored based on patterns in the SST imagery [4]. Proper implementation of all these techniques assumes spatial continuity, which is not satisfied in the original imagery, especially at the swath edges. De-bowtizing of the L1 and/or L2 imagery is needed to improve the clear-sky mask and SST retrievals. The proposed algorithm is discussed in Section 3.

2.4. Onboard Deletions

Figure 2 shows that consecutive whiskbrooms overlap, progressively more so away from nadir, resulting in "duplicate" data. On MODIS, all data are kept and transmitted to the ground. On VIIRS, however, the decision was made to delete onboard the radiances measured in the overlapping portions of the scan, in order to reduce the data volume to be transmitted to the ground. The data in orange and magenta in Figure 1 show pixels deleted onboard the S-NPP. Simple estimates suggest that this onboard data deletion reduces the data volume by ~12.9%. When the VIIRS Raw Data Record (RDR; L0) and corresponding sensor data record (SDR; L1b) are created on the ground, the radiances in the deleted pixels are populated with fill values (whereas the corresponding geographical coordinates and angles are calculated and written to the RDR and SDR data files). These missing data on VIIRS complicate de-bowtizing and should be filled in (*i.e.*, MODIS-like VIIRS product should be first created).

A standard way to approximate a missing value is interpolation. The simplest interpolation scheme is a nearest neighbor (NN) approach with variations on what is considered to be a "neighbor". The best way to use a NN approach is to rely on the geographical neighbor (rather than on the closest row/column in a swath projection). Using geo-neighboring properly is intuitively straightforward, but may be computationally inefficient if the whole image has to be re-mapped first. To speed up the processing, one can remap only the values in the spatial proximity of the deleted pixels, but that again may not be fast since it requires finding the nearest neighbors for every missing value before proceeding with the estimations. Resampling of the whole image can be computationally very cheap, as will be shown later in this paper. Therefore, we first resample the L2 imagery and then perform approximation using one of the 2D spatial interpolation schemes as described in Section 3.3.

3. Re-Sampling: Requirements and Approach

To summarize the science-driven motivations, instrument-imposed specifics and real-time processing needs, our objective is a resampling that satisfies the following major requirements A_I–A_{III}:

A_I: Ensures spatial continuity of the imagery;
A_{II}: Provides minimal deviation from the original swath geo sampling grid;
A_{III}: Is computationally fast and appropriate for real time L2 processing.

Intuitively, what is minimally required is just an unfolding procedure that reorganizes the footprints according to their geolocation (rather than in the order they were acquired by the instrument and reported in the RDR/SDR swath data files). Note that the problem of bow tie distortions in MODIS has long been recognized (e.g., [12–16]). The proposed solution has been remapping, whereas our objective here is to preserve the original swath projection for L2 ACSPO SST product, deferring the mapping to L3 and higher processing levels.

The impact of the proposed resampling approach on SST imagery can be quickly previewed from Figure 4. A small crop from a typical SST image near the end of the swath and the corresponding image of the latitude (as it appears in the SDR geolocation file) are shown in Figure 4a,b. The same latitude values, but re-ordered according to our unfolding scheme (described in Section 3.1) are shown in Figure 4c and the corresponding re-sampled SSTs in Figure 4d. The bow-ties, obviously present in the original L2 SST product (Figure 4a), are "unfolded" along with the corresponding geolocation. Re-ordering according to predefined order is computationally cheap, meets the requirements A_{II} and A_{III} and almost meets the requirement A_I (see Section 3.2 for details). The last component of the resampling procedure is the adjustment of the longitude values in the bow-tie regions (described in the Section 3.3), which satisfies the A_I and fulfills all three requirements, A_I–A_{III}.

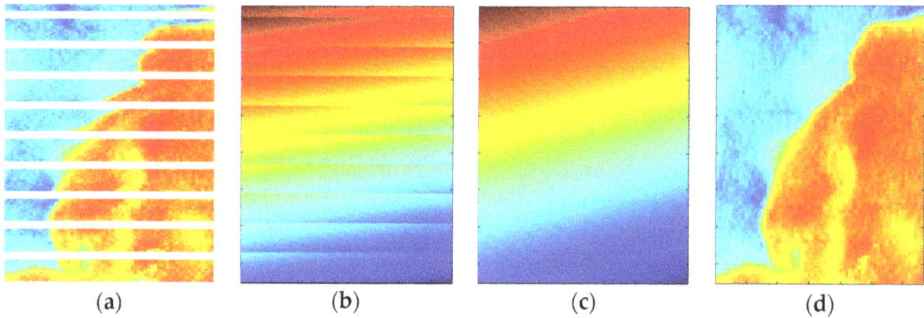

Figure 4. (**a**) Original VIIRS SST with bow-tie distortions and deletions (note the jumps and repeats along the SST thermal fronts, and that onboard deleted pixels are shown in white); (**b**) original latitudes; (**c**) unfolded latitudes; (**d**) corresponding resampled SSTs.

3.1. Unfolding

Unfolding is basically the re-ordering according to some pattern related to the geometry of the instrument's swath projection and the footprint locations of the scan on the Earth surface. This pattern is specific to the particular instrument and can be estimated using sorting procedures on a per-column basis for individual scan and then statistically determined based on the large set of scan-based re-ordering patterns. There are two types of VIIRS geolocation files–ellipsoid based (GMODO) and terrain corrected (GMTCO). Over ocean, the two are very close although there may be some differences due to geoid variations and in the coastal areas. In the ACSPO, the terrain-corrected GMTCO is used. However, the unfolding of the bow-tie distortions is performed in conjunction with the near elliptical Earth's shape, which requires using the ellipsoid geolocation file, GMODO. Figure 5a,b demonstrates the effect of unfolding for three consecutive scans. The detectors are shown with 16 distinct colors, ranging from yellow (for detector 1) to blue (for detector 16). The center of the swath (the nadir) has no overlap, so the original order remains the same. In Figure 5 this corresponds to a monotonic color change within individual scans from yellow to blue.

The unfolding pattern corresponding to Figure 5a,b is illustrated in Figure 5c, for one scan S_k. Only the left half-scan is shown in Figure 5c; to extend the unfolding procedure to the right side, the table should be reflected symmetrically with respect to the nadir. The rows in the Figure 5c correspond to VIIRS detectors and the columns define the column ranges with identical re-ordering pattern.

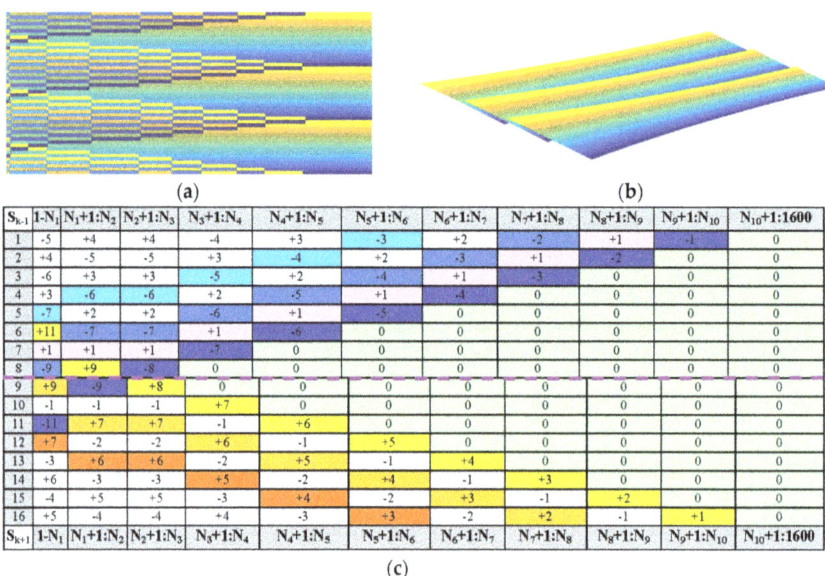

S_{k-1}	$1\text{-}N_1$	$N_1+1{:}N_2$	$N_2+1{:}N_3$	$N_3+1{:}N_4$	$N_4+1{:}N_5$	$N_5+1{:}N_6$	$N_6+1{:}N_7$	$N_7+1{:}N_8$	$N_8+1{:}N_9$	$N_9+1{:}N_{10}$	$N_{10}+1{:}1600$
1	-5	+4	+4	-4	+3	-3	+2	-2	+1	-1	0
2	+4	-5	-5	+3	-4	+2	-3	+1	-2	0	0
3	-6	+3	+3	-5	+2	-4	+1	-3	0	0	0
4	+3	-6	-6	+2	-5	+1	-4	0	0	0	0
5	-7	+2	+2	-6	+1	-5	0	0	0	0	0
6	+11	-7	-7	+1	-6	0	0	0	0	0	0
7	+1	+1	+1	-7	0	0	0	0	0	0	0
8	-9	+9	-8	0	0	0	0	0	0	0	0
9	+9	-9	+8	0	0	0	0	0	0	0	0
10	-1	-1	-1	+7	0	0	0	0	0	0	0
11	-11	+7	+7	-1	+6	0	0	0	0	0	0
12	+7	-2	-2	+6	-1	+5	0	0	0	0	0
13	-3	+6	+6	-2	+5	-1	+4	0	0	0	0
14	+6	-3	-3	+5	-2	+4	-1	+3	0	0	0
15	-4	+5	+5	-3	+4	-2	+3	-1	+2	0	0
16	+5	-4	-4	+4	-3	+3	-2	+2	-1	+1	0
S_{k+1}	$1\text{-}N_1$	$N_1+1{:}N_2$	$N_2+1{:}N_3$	$N_3+1{:}N_4$	$N_4+1{:}N_5$	$N_5+1{:}N_6$	$N_6+1{:}N_7$	$N_7+1{:}N_8$	$N_8+1{:}N_9$	$N_9+1{:}N_{10}$	$N_{10}+1{:}1600$

Figure 5. Three consecutive VIIRS whiskbrooms for the left half of the scan (from the edge of the swath to the nadir): (**a**) in swath projection (along with sorting patterns); (**b**) in the mapped projection. The detectors are shown with distinct colors ranging from yellow (detector 1) to blue (detector 16); and (**c**) reordering scheme corresponding to the left half of the VIIRS swath. The top and bottom halves of the reordering table are displaced intentionally, to emphasize that the corresponding sorting patterns may be different (see text for details).

Numbers in each cell represent the amount of shift for each of 16 rows in the scan S_k during the reordering. Positive values correspond to an increase of the row index and negatives correspond to the decrease of the row index. Zeros represent no change (no re-ordering). Some of the cells are color coded to facilitate the visual perception of the proposed re-ordering.

The dark blue cells, appearing together as a "ribbon" running across Figure 5c, correspond to a propagation of the last detector of the scan S_{k-1}, cutting through the scan S_k. The corresponding (negative) shifts, increasing in magnitude toward the end of the swath as the bow-tie overlaps increases, indicate the position of the last detector of the scan S_{k-1}. The lighter blue ribbon (2 cells up from the blue ribbon) corresponds to the 15nd detector of the scan S_{k-1}. These two ribbons are separated by a "pink ribbon" with a + 1 shift corresponding to the pixels in the same central scan, S_k (which in many cases, take the spaces on the blue ribbon). Another ribbon two more lines above, shown in cyan, stands for the 14th detector in the S_{k-1}. Corresponding propagation of the 1st–3rd detectors from the next scan, S_{k+1}, are shown in yellow-to-orange colors. In this particular case, the pattern for the propagation from S_{k+1} into S_k are symmetric to that of S_{k-1} into S_k, but generally speaking, this may not be always the case for the VIIRS.

3.2. Adjustments to the Generic Break Points

There are a total of 21 reordering zones determined by 20 break points, N_i, corresponding to column indexes where the reordering pattern changes. The central zone centered on the nadir (where no reordering is needed; columns 1600 and 1601) is sandwiched between N_{10} and N_{11}. The break points are the positions where the grid lines of S_k intersect with the grid lines of S_{k-1} and S_{k+1}. The upper (detectors 1 through 8) and the lower (detectors 9 through 16) parts of the table can have different break points, as the middle scan S_k can have different overlaps with the neighboring scans, S_{k-1} and S_{k+1}, caused by a slight variation in the sub-satellite track. Our analyses of VIIRS data suggest that such scan-to-scan variations of the sub-satellite track are small but sufficient to affect the reordering positions derived as one static set. The reordering pattern is thus general, but the break points N_i's need per-scan adjustments.

The adjustment of the initial position of an individual break point, N_k, follows the general numerical approach for finding the zero of a function, given a good initial approximation. The goal function is the distance between the elliptical latitude/longitude pair of the last detector of S_{k-1} at the current break point and the geo-location of the first eight detectors of S_k (for the upper part of the scan), and the first detector of S_k and the last eight detectors of S_{k+1} (for the lower part of the scan). Our implementation is using haversine formula for orthodromic distance. The schematic of the iterative procedure for break point adjustment is shown in Figure 6, where the crossing of the grid line of the 16th detector of S_k (green line) and the grid line of the 4th detector of S_{k+1} (blue line) corresponds to the actual 4th break points that needs to be computed based on the initial approximation marked by the dashed dotted line in Figure 6b.

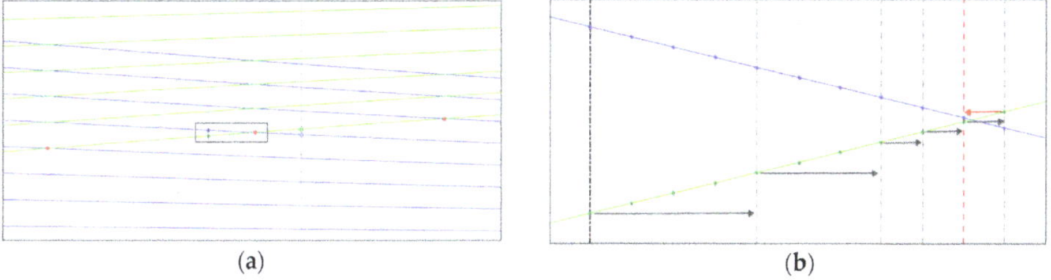

Figure 6. (**a**) Intersection of grid lines of S_k (green) and S_{k+1} (blue). Break points are shown with red asterisks; (**b**) Iterative adjustment procedure: black dash-dotted line shows the column with the initial break point approximation; gray dashed lines correspond to intermediate steps of the iterative adjustment procedure with arrows indicating the direction in which adjustments occur, pointing from the current position to the next approximation. The red dashed line is the final position of the break point, which is the approximation before last in the iterative sequence, which continues until the difference between the latitude values changes the sign.

3.3. Adjusting Longitude Values

The unfolding procedure described in Sections 3.1 and 3.2 would almost meet requirements A_I–A_{III}, if it would not be for the relative displacement of the consecutive scans, caused by Earth rotation. As the VIIRS instrument completes its S_k'th scan and gets ready for S_{k+1}'th one, the Earth has rotated by a certain amount, depending on the latitude and on the instrument scan rate. This causes the grid displacements between consecutive scans by the amount of this rotation (*cf.* Figure 7a for an example of displacement amounts between three consecutive scans at the end of the swath).

What really matters for the continuity assumption A_I, is the relative grid displacement with respect to the grid width. This ratio varies from the nadir to the end of the swath as shown in Figure 3. The largest relative displacement is at the nadir for the scan that sweeps through the equator. For the

VIIRS instrument, this displacement can exceed the pixel width (*i.e.*, ratio > 1), since the nominal VIIRS scan rate of 1.7864 s/scan and the speed of the Earth rotation at the equator is 2*pi*6380/(24*3600) km/s, resulting in the shift of ~0.828 km between the scans, which is larger than the 0.75 km pixel width at the nadir. However, there is no bow tie effect at the nadir, and reordering is done for columns 1:1307 and 1893:3200, where the pixels grow larger than 0.75 km. The relative displacement as a function of latitude and the pixel location in the swath (for the left half the swath) is shown in Figure 7b. The orange-to-red values of the 2D relative displacement surface indicate the cases, where the relative scan displacement exceeds one pixel. This happens in the low latitudes (tropics) around the nadir position and at the beginning of each aggregation zone. The shifts at the end of the swath are at a sub-pixel level (in agreement with the example shown in Figure 7a).

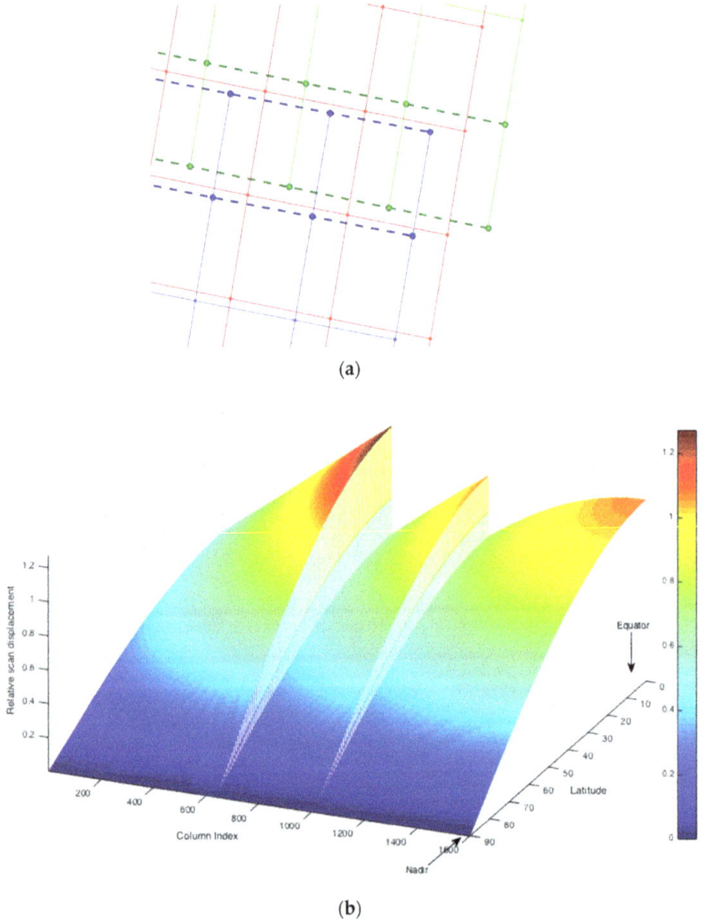

Figure 7. (**a**) The geo-locations of three consecutive VIIRS scans near the end of the swath are shown in green, red and blue. Portions of the scans corresponding to onboard deletions are marked by dashed line; (**b**) Relative scan displacement as a function of latitude and scan position.

Figure 8a shows the grid overlap between scans S_k and S_{k+1}, around the transition from the 3:1 to the 2:1 aggregation zone. It is small but not negligible. The order of grid pixels-as they appear in a particular column of latitude and longitude arrays after unfolding-is traversed with magenta arrows.

The zigzagging magenta path is caused by Earth rotation between consecutive scans, and the need to keep the swath projection non-displaced from one scan line to another. As long as the zigzagging is within the footprint boundaries (*i.e.*, the gridlines do not cross), the continuity (requirement A_I) is satisfied. The zigzagging exceeding the grid cell can affect the quality of the reordered SST imagery. It is especially noticeable along ocean thermal fronts or other features on the surface with large gradients.

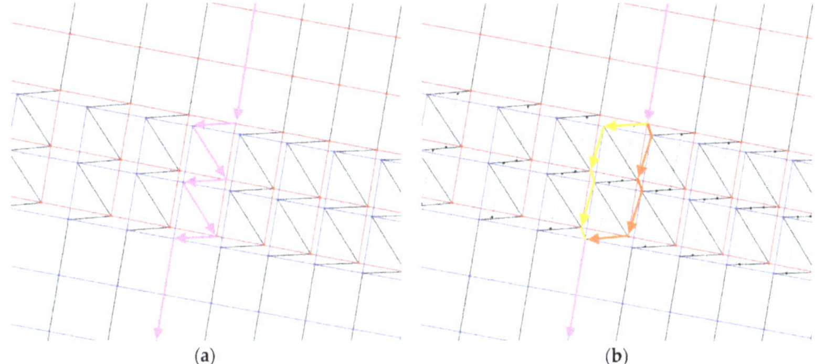

Figure 8. (**a**) Geo-locations around the transition from the 3:1 to the 2:1 sample aggregation scheme; magenta path connects interleaved footprints from different scans; (**b**) two alternate geo reordering schemes, involving column shifts, marked by yellow and orange paths.

One possible alternative to preserve the values of the original geo-grid and to mitigate the zigzagging effect is to additionally reorder columns. An example of such possible reordering is shown in Figure 8b with two alternative orders (marked by yellow and orange paths, respectively). Both paths would require a change of the original 3200 column setting to allow for the shifts persistently present between consecutive scans. A more attractive approach was deemed to (slightly) adjust the longitudes, which also ensures the spatial continuity while preserving the swath width intact. An example of this adjustment is shown in Figure 8b, with the new grid lines marked by dotted lines and new grid points by black dots. Note that only the longitudes are (slightly) changed, while the latitudes are reordered but not modified. The grid overlaps for scans S_{k-1}, S_k, and S_{k+1} around the transition from the 2:1 to the 1:1 sample aggregation scheme are shown in Figure 9.

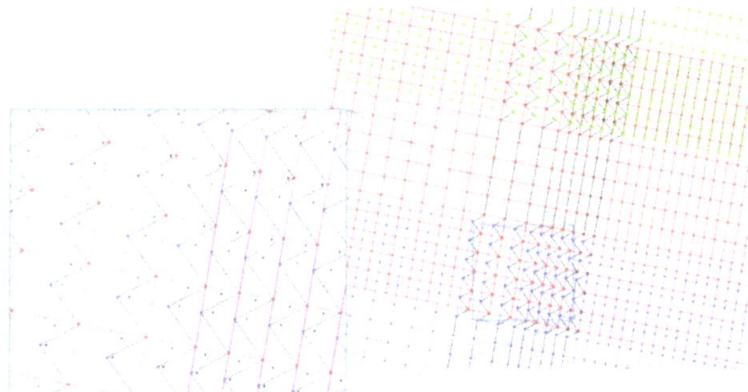

Figure 9. Geo-locations around the transition from the 2:1 to the 1:1 sample aggregation scheme. Unfolding order for this region corresponds to the N_5 +1:N_6 column of Figure 5c with only six middle detectors, 6 through 11, outside the bow-tie region. The insert on the left shows detailed overlap of the S_k and S_{k+1} grids. Black zigzagging line represents reordered path and black dotted line corresponds to proposed grid with adjusted longitude values.

The overlap due to bow-ties is significant and the unfolding order for this region corresponds to the N_5+1:N_6 region of Figure 5c. The latitude and longitude values for one column in this region zoomed into two scan's overlap are plotted in Figure 10 (top plot). The blue line in the latitude plot corresponds to the original scan order. The black dotted line (top plot of Figure 10) shows the latitude values after reordering, which makes the plot monotonic. Reordered longitudes, shown in a black solid line in Figure 10 (bottom plot) have a zigzagging pattern, as expected due to the described relative scan displacement. Simple (and computationally fast) 1D per-column interpolation, preserving the longitudes that have not been reordered (*cf.* cells with "0" entries in Figure 5c), results in the monotonic longitude values shown in magenta. Adjustments to the longitudes are only done for the reordered pixels. The corresponding modified grid is shown in the insert of Figure 9, where the gridlines are marked by black dotted lines and black dots correspond to new (longitude-adjusted) geo-locations.

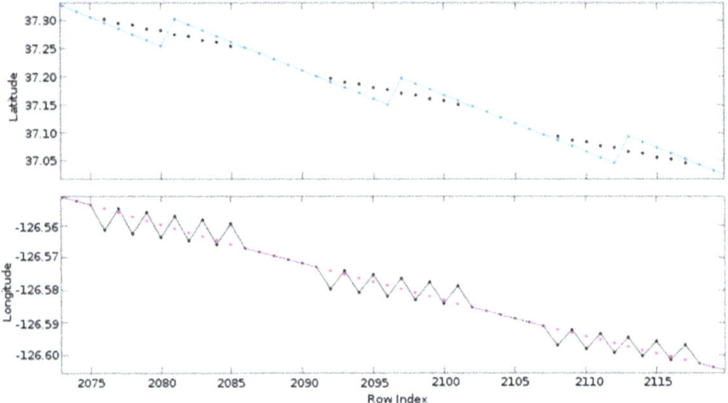

Figure 10. Top: Latitude values for the middle column of the grid portion shown in Figure 9. Blue line represents original latitude values and black line corresponds to latitudes reordered according to Figure 5c. **Bottom**: Longitude values for the same column. Black line corresponds to longitudes reordered according to Figure 5c. Zigzagging effect caused by Earth rotation is present at the bow-tie regions after reordering. Magenta line, representing adjusted longitudes, is monotonic. The adjustments are performed only at the overlapping portions of the consecutive scans.

Note that some (but not all) of the pixels whose longitudes are adjusted, correspond to the deleted onboard data (orange and magenta pixels of Figure 1), making changes to their geolocation easier to justify. For the resulting de-bowtized imagery to be used for image-processing SST applications, the missing radiances in the deleted pixels should be evaluated.

3.4. *Approximation of the Deleted Values*

With all of the reordering work done, the estimating of the missing values is a straightforward task: in the resampled imagery, the index-based neighbors are now also geo-neighbors, which makes distance-based weighted averaging a simple and well-justifiable option. For each onboard deleted value, we use the four closest (±1 row/column) neighbors and compute the Gaussian-weighted average with standard deviation proportional to the corresponding vertical footprint size. There might be some flagged values among those neighbors due to the present deletion/reordering scenario, but a closer look at Figure 5c reveals that there are always some non-deleted entries among the selected four horizontal/vertical neighbors. There might be cases when all four neighbors are flagged for some other reason (not relevant to onboard deletion). All of the flagged values are excluded from the averaging procedure, and when all four neighbors are flagged, then the deleted pixel is marked as invalid.

An example of resampled brightness temperature BT at 12 μm with estimated values in the deleted zones is shown in Figure 11c. For comparison, the original BT at 12 μm is given in Figure 11a. The presented crop is selected from the 10 min ACSPO granule acquired on 1 December 2015 at 21:40 UTC. "Repeats" typical for bow-tie distortions, which were apparent in Figure 11a, are not present in the unfolded reordered image shown in Figure 11b. The re-ordered image in Figure 11b also reveals the variation of the break-points, discussed in the Section 3.2. A zigzagging pattern, especially noticeable along thermal fronts, is also noticeable in the re-ordered image. Corrections to the longitude values, discussed in Section 3.3, undo the zigzagging artifacts caused by the shifts, as can be seen from resampled image in Figure 11c. The image shown in Figure 11c now meets all the A_I-A_{III} requirements above, and ready for all planned SST imagery-based enhancements at NOAA.

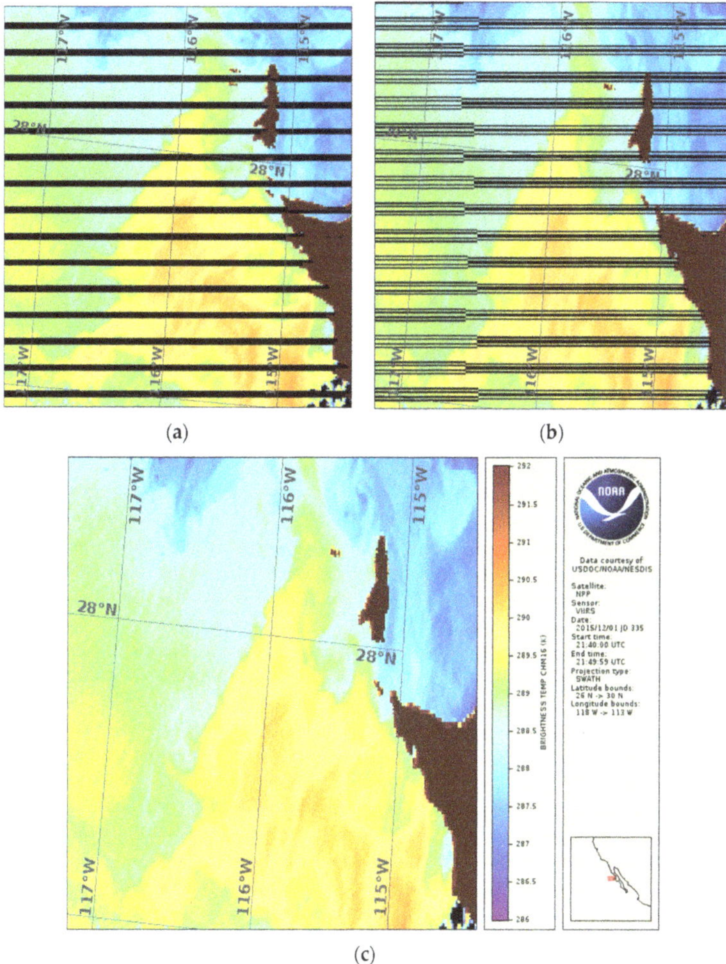

Figure 11. BT at 12 μm: (**a**) Original; (**b**) Reordered; (**c**) Resampled with missing values filled.

4. Resampling for MODIS

MODIS instrument has fewer detectors (10 *vs.* 16), coarser pixel resolution (1 *vs.* 0.74 km at nadir; ~5 *vs.* 1.5 at swath edge), faster scan rate (1.478 *vs.* 1.786 s per scan), narrower swath (2330 *vs.* 3040 km).

Otherwise, it is similar to VIIRS instrument: in particular, it is also subject to bow-tie distortions. A schematic of the MODIS half-scan projection on the Earth surface from nadir to swath edge is shown in Figure 12a [10]. The unfolding algorithm described in Section 3 can be directly applicable to MODIS, except its implementation is simpler, as there are no pixel aggregations or onboard deletions. Also, there is no need for the recalculation of the original longitudes, because the scan-to-scan displacement due to Earth rotation (~0.685 km in worst case scenario, at the nadir and at the equator) is smaller due to faster scan rate, and the displacement is always smaller than the pixel size. The reordering table for MODIS is given in Figure 12b. Note also that the break point adjustments are not needed for MODIS, because the displacements between consecutive scans are much more regular for both Aqua and Terra satellites than for the S-NPP. This is fortuitous because in MODIS L1b data, only terrain corrected geolocation is reported and the ellipsoid based geo locations (not rectified for terrain) are not readily available.

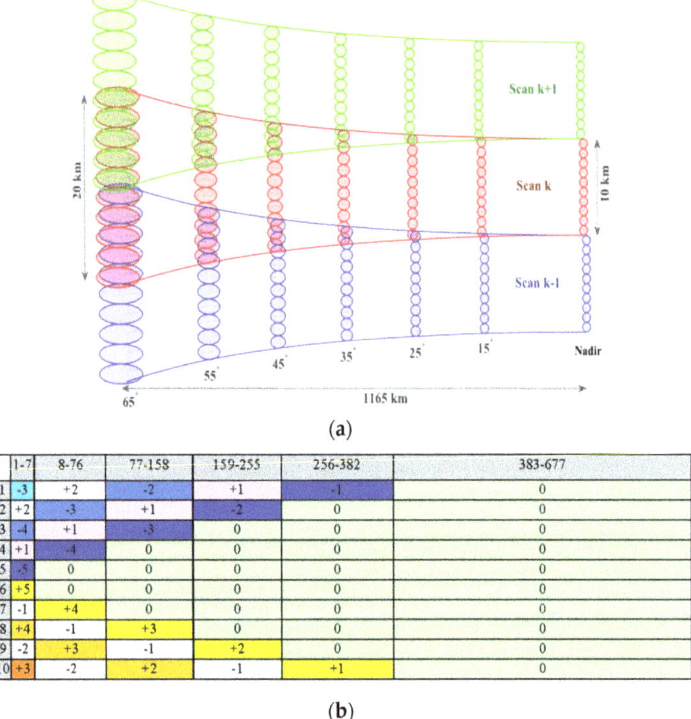

(**a**)

	1-7	8-76	77-158	159-255	256-382	383-677
1	-3	+2	-2	+1	-1	0
2	+2	-3	+1	-2	0	0
3	-4	+1	-3	0	0	0
4	+1	-4	0	0	0	0
5	-5	0	0	0	0	0
6	+5	0	0	0	0	0
7	-1	+4	0	0	0	0
8	+4	-1	+3	0	0	0
9	-2	+3	-1	+2	0	0
10	+3	-2	+2	-1	+1	0

(**b**)

Figure 12. (**a**) Schematic representation of Moderate Resolution Imaging Spectroradiometer (MODIS) sampling for three consecutive whiskbrooms (rendered in blue, red and green respectively), for several selected VZAs from 0° (nadir) to 65° (approximately representing the left edge of the scan) in 10° increments. The two axis of each ellipse represent the horizontal and vertical sampling intervals, respectively; (**b**) The reordering scheme for the left half of the MODIS swath.

5. Results and Evaluation

The impact of the proposed resampling on the performance of the current ACSPO clear-sky mask (ACSM; [3]) was initially evaluated and reported here. Recall that two ACSM tests—the "SST filter" and the "spatial uniformity check"—employ information in spatial windows surrounding the pixel [3]. In the resampled data, the performance of these two tests is expected to improve, *i.e.*, to reduce the number of the corresponding "false alarms", and the number of clear-sky ocean pixels, CN, and the corresponding clear-sky fraction, CF (defined as a ratio of clear-sky to the total number of ocean pixels), to increase.

To verify this expectation, global ACSPO products have been generated from Aqua and Terra MODIS (each containing 288 5-min granules) and S-NPP VIIRS (144 10-min granules), for one full day of global data (18 October 2015). Two ACSPO runs were performed: with the original SDR data and with the resampled SDRs. In what follows, the results of the runs are analyzed, with emphasis on the effect of resampling. Cross-platform differences (which may result from different spatial resolution, radiometric quality, and different overpass times) are relatively small and not analyzed here.

Number of clear-sky ocean pixels, CN, and the corresponding clear-sky fraction, CF, are shown in Table 1. For all sensors, the CN derived from the original SDRs is smaller than from the resampled SDRs. For the VIIRS, the increments are much larger than for MODIS. Recall that they include 12.9% of VIIRS pixels which have been deleted onboard (see Section 2.2). In contrast, the corresponding resampled VIIRS CFs (right Table) are only slightly larger. For VIIRS, two mutually offsetting mechanisms are responsible for the change in the CF. First, the deleted pixels are filled in from the four neighbors, and there is a high probability that at least one of them is cloudy. As a result, the filled pixels are expected to have a "cloudy" bias, which should lead to a decrease in the CF. On the other hand, the improved spatial uniformity due to resampling is expected to result in the improved ACSM, and larger CF. As Table 1 suggests, the improvements in the SST imagery (and consequently, in the ACSM) outweigh the "cloudy" bias in the filled pixels, resulting in a net absolute increase of 0.2% in the CF (average between day and night). In a relative sense, this delta is equivalent to an increase in the clear-sky sample by 1%, globally. For MODIS, there are no deleted pixels and no corresponding offsetting mechanism, and therefore a larger improvement is expected. Table 1 shows that indeed, the absolute increment between Terra and Aqua, day and night, is from 0.4% to 0.6% (approximately 2%–3% increase in the clear-sample).

Table 1. Top: Number of clear-sky ocean pixels, CN (millions). Down: Corresponding clear-sky fraction, CF (percent to total number of ocean pixels observed by the corresponding sensor). Data are from one full global day on 18 October 2015.

	Day		Night	
	Original	Resampled	Original	Resampled
S-NPP VIIRS	129.0	148.0	112.9	131.7
Aqua MODIS	48.1	49.0	40.9	42.2
Terra MODIS	47.3	48.3	44.4	45.8
	Day		Night	
	Original	Resampled	Original	Resampled
S-NPP VIIRS	21.00	21.10	18.87	19.20
Aqua MODIS	21.46	21.87	18.91	19.47
Terra MODIS	21.74	22.17	20.57	21.21

Figure 13 (left panels) additionally shows the CF (day and night combined) as a function of view zenith angle (VZA). For both sensors, the CF is largest around nadir (~22%–24%) and drops off to ~10%–15% at the swath edges. The resampled data always have a comparable or larger CF, in all bins. The relative differences between the two grey curves are shown on the right. The improvement is insignificant around nadir, and progressively increases towards swath edges, where it reaches from 6% to 8% for the VIIRS and from 9% to 12% for MODIS. The increment is a little larger at night, likely due to the use of reflectances during the daytime, which may be more subject to artifacts in visible imagery and result in higher screening rate.

Additional analyses (not shown here) suggest that the global statistics of the newly added SSTs are comparable to, or better than those derived from the original SDRs. Recall that a distinctive feature of the ACSPO SST product, compared to other partners' SST retrievals systems, is that retrievals are made in the full sensor swath. Increase by up to 8%–12% in the valid SST data rate at the slant view geometries (which recall are least populated with the retrievals and characterized by degraded observational conditions) is instrumental to provide more complete global coverage by the ACSPO SST product. Note that this assessment was done with the current ACSPO clear-sky mask, which minimally uses spatial information. The improvement is expected to be more significant with implementing the pattern recognition techniques [4], at which time the potential of the improved SST imagery will be more fully realized.

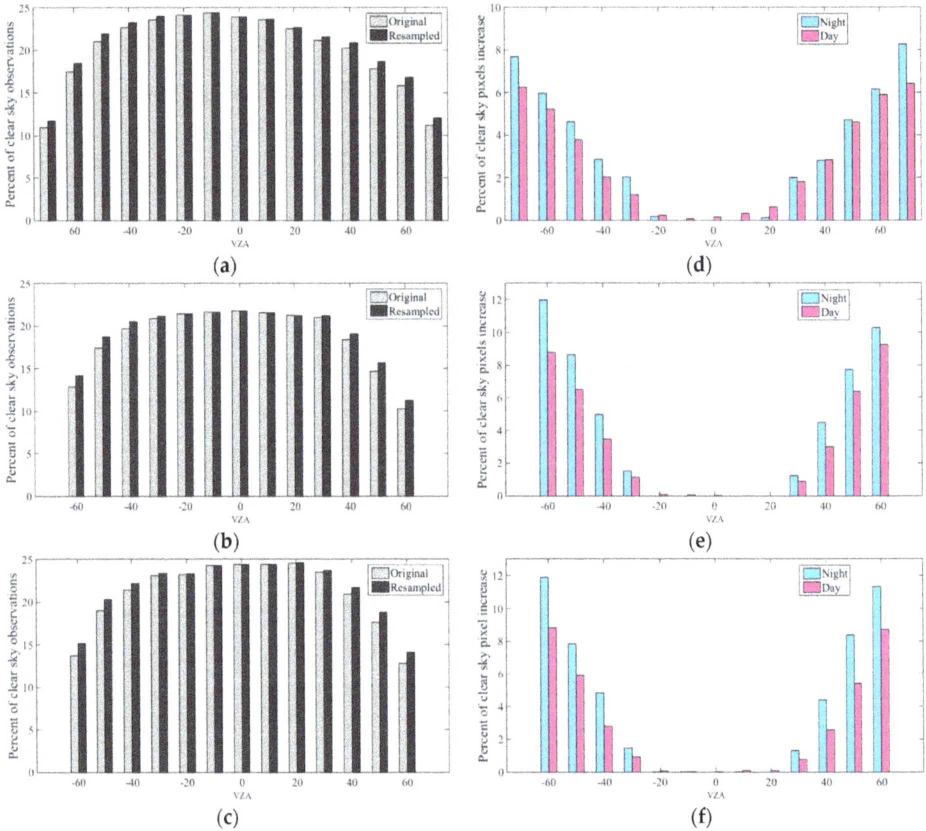

Figure 13. Number of clear sky observations for one day (18 October 2015) of global ACSPO SST data as a function of view zenith angle (VZA) for (**a**) S-NPP VIIRS; (**b**) Aqua MODIS; (**c**) Terra MODIS. Original data are shown in light gray and resampled in dark gray. Day and night data are combined together. Corresponding percent increase for (**d**) S-NPP VIIRS; (**e**) Aqua MODIS and (**f**) Terra MODIS (separated by night and day).

6. Comparison with Ground Track Mercator (GTM) Projection

As mentioned earlier in the paper, NOAA generates a GTM VIIRS imagery product, which is free of bow-tie artifacts including the stripes of missing data caused by onboard deletions. The GTM algorithm also suppresses the discontinuities at the switch from one aggregation scheme to another.

The measured digital counts (DC) are carried unaltered from the original row-column locations in the swath projection to the closest GTM grid position using nearest neighbor (NN) approach.

Overall, the GTM algorithm successfully removes the most obvious imagery issues and improves the VIIRS imagery. However, the NN implementation may introduce some other artifacts, which may not be as obvious from the visual perception, but nevertheless problematic for image processing that requires spatial information. A small, zoomed-in patch from the edge of the VIIRS swath is shown in Figure 14a. The original geo locations (dots) and corresponding grids (lines) are shown in blue, with the pixels deleted onboard rendered in magenta. The GTM data are shown in black (again, dots representing centers, and lines the corresponding grids). Red dashed arrows pointing to the GTM black dots indicate the DCs copied from the original swath data to the GTM locations. As can be seen from Figure 14a, some of the original DCs are duplicated (actually, one swath point may be reused up to four times). This will imply bias in computations of horizontal, vertical or both components of the (numerical) gradient. Such duplications will also bias the statistics computed within small spatial windows.

The replication patterns do change depending on the location within the scan, but the distribution of number of duplicates within the scan can be estimated by simply counting the number of identical DCs in the horizontal (identical DC values at *(i,j)* and *(i,j + 1)* location) and vertical (identical values at *(i,j)* and *(i + 1,j)* locations). Percentage of locations with the same values to the right (horizontal duplicates) and to below (vertical duplicates) are shown in Figure 14b. Toward the end of the scan, the percent is as large as 50% for both, vertical and horizontal count. As intuitively expected, the distribution of the horizontal duplicates closely follows the discontinuities inherited from the three different aggregation schemes. A more thorough analysis of GTM projection, its artifacts and impact on other products is beyond the scope of this paper. Our goal is improved clear-sky mask for SST and calculation of ocean thermal fronts (gradients). To that end, the re-sampling procedure proposed here is deemed a more fit-for-purpose imagery improvement technique compared to the GTM projection.

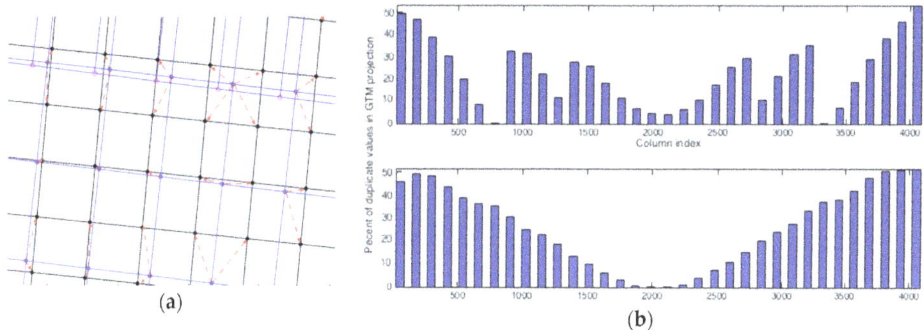

Figure 14. (**a**) Small crop from the end of the swath showing both grids: original swath (blue) and GTM (black). Pixels deleted onboard (magenta) would decrease the number of duplications in the GTM projection; (**b**) Distribution of GTM duplicates: (**top**) horizontal and (**bottom**) vertical. The bar graph uses 121 columns in each bin, except for the last bin.

7. Potential Benefits of Resampling for other Level 2 Products

The resampling procedure described here is a general algorithm that is specific to the instrument rather than to the derived L2 product. Although it was motivated by, and originally developed for the SST applications, it can be successfully used for other L2 products and applications as well. One such example is presented below, to illustrate its potential for ice analysis.

Figure 15 shows a 400 × 400 pixel crop taken at the end of the VIIRS swath for the 20:27:42.7 and 20:29:08.1 UTC passes on 1 September 2015 over Arctic Ocean. The crop in Figure 15 includes

Qikiqtaaluk Region (Nunavut, Canada) showing a part of the Northwestern Passage, Axel Heiberg, and Ellesmere Islands containing mountainous and glaciated terrain.

The White Glacier located on the Axel Heiberg Island shows the ice remaining here after the hot summer of 2015. This glacier has the longest continuous mass balance record of all high Arctic glaciers, and has been shrinking for the past two decades. There are multiple ice floats in the Sverdrup Channel, as well as a single ice float just above Eureka. The fine-scale details captured by the VIIRS instrument are preserved during the proposed resampling procedure. The difference between original M4 (0.55 μm) and M10 (1.6 μm) bands is shown in Figure 15a and its resampled counterpart in Figure 15b. The M4–M10 is the key feature widely employed for snow and ice applications. The ice patterns are much more pronounced and form more continuous patterns in the difference image.

The difference between terrain-corrected and ellipsoid longitudes in the original scan order (from GMTCO and GMODO VIIRS geo files) is shown in Figure 16a, and after resampling in Figure 16b. The elevation patterns attributed to Earth terrain are clearly seen in the de-bowtized resampled imagery, Note that resampled values of GMTCO are reported in ACSPO files, and Figure 16b suggests that that not only the SST imagery is approved but also the corresponding geolocation.

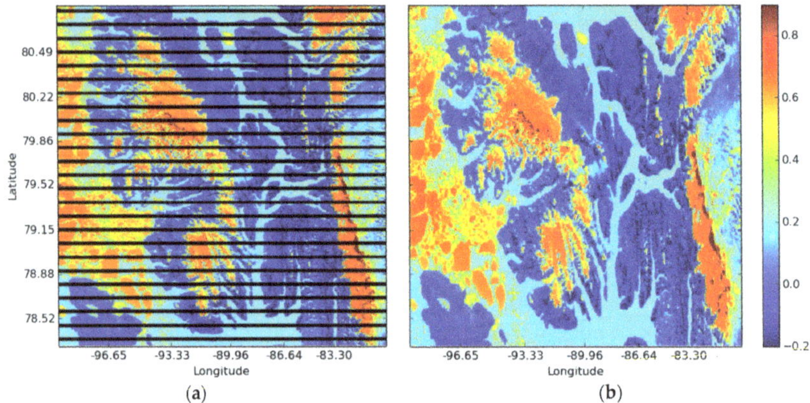

Figure 15. (**a**) Difference between original M4 (0.55 μm) and M10 (1.6 μm) bands; (**b**) difference between resampled M4 and M10.

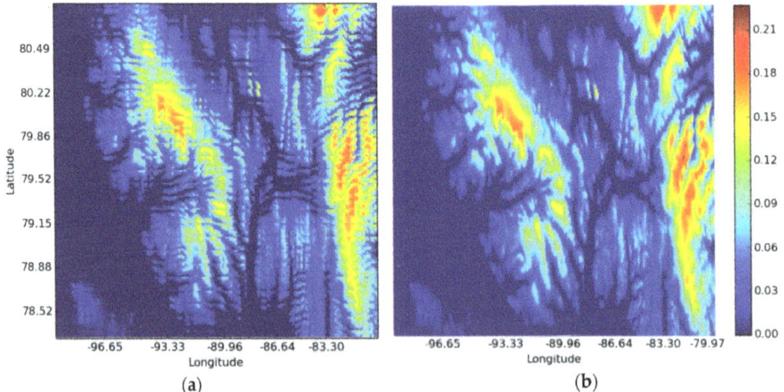

Figure 16. (**a**) Difference between terrain-corrected and ellipsoid longitudes in the original scan order (from VIIRS Moderate Bands SDR Terrain Corrected Geolocation (GMTCO) and Original (elliptical) Geolocation (GMODO) files); (**b**) De-bowtized difference after re-sampling.

8. Conclusions

Today VIIRS, along with its predecessor MODIS, remain the best suite-for-purpose US polar SST sensors in space. Their superior performance, including the high spatial resolution and low radiometric noise, is achieved by using push-broom technology, when multiple detectors are used to simultaneously collect measurements from several scan lines. The flipside of the multi-detector sensor design is striping and bow-tie distortions in the satellite imagery. Special pre-processing to minimize these instrumental artifacts and ensure spatial continuity of satellite imagery is critically important for users interested in satellite imagery, and in processing spatial patterns using the machine learning algorithms [4]. The NOAA JPSS SST Team has implemented a destriping algorithm in the NOAA SST ACSPO version 2.40, which became operational in May 2015. Work is underway to implement the bow-tie correction algorithm documented in this paper for the VIIRS and MODIS data in ACSPO version 2.50. The objective of these two releases is to set the stage for implementing the pattern recognition improvements described in [4] in ACSPO version 2.60.

Note that NASA which was responsible for the MODIS sensor onboard EOS Terra and Aqua satellites has opted to not correct for the effects of striping and bow-tie distortions. The intention was to keep the MODIS radiances and geolocation in Level 1 and 2 products intact (as measured). However, the mission of the NOAA National Environmental Satellite Data and Information Service (NESDIS) is serving its users data which are easy to display, use and understand. This requires minimization of the specifics of individual measurements, and emphasizing sensor data which are as artifact-free as possible. To that end, the EUMETSAT (a European counterpart of NESDIS, whose objective, similarly to NESDIS, is serving artifact-free data to the Numerical Weather Prediction customers and other users of satellite data) has generated a Level 1.5 data from the Spinning Enhanced Visible and Infra-Red Imager (SEVIRI) onboard the Meteosat Second Generation (MSG) satellite [7]. In SEVIRI Level 1.5 data, differences in detector responses are equalized, nonlinearity of the sensor response is corrected for, and the satellite imagery is converted into a standard reference projection (including the co-registration between channels) [7]. Therefore, radiances and geolocation on SEVIRI Level 1.5 are (slightly) changed from those measured onboard, but these adjustments proved extremely beneficial to the users of the SEVIRI data. Those science and sensor-analysis oriented users, who are interested in the original sensor data "as measured", can always resort to a Level 0 or 1 data.

From the Advanced Himawari Imager (AHI) onboard the Japanese new generation geostationary Himawari-8 satellite, the Japanese Meteorological Agency (JMA) produces a Level 1 product, using a processing philosophy similar to the one adopted in SEVIRI Level 1.5. The Advanced Baseline Imager (ABI) to be launched onboard the new generation GOES-R series will also report a geometrically corrected, radiometrically-equalized imagery. Discussions are also underway at NOAA to generate a VIIRS Level 1.5 product, which will be identical in the data format to the current VIIRS SDRs (L1b), but will include bow-tie corrected (filled-in and re-sampled, and potentially destriped) imagery. The JPSS SST Team strongly supports such intermediate product, as it will reduce the burden on the SST processing. We firmly believe that this processing should be done upstream, and it will benefit other JPSS L2 products, not only SST. Should NOAA decide to pursue the L1.5 route for the VIIRS, it is strongly recommended that the original L0 (RDR) and L1b (SDR) data are fully available and easily accessible to the users.

It has been suggested that the unfolded product is good for visual analysis, usable for inter-pixel tests (e.g., uniformity) and for SST evaluation, but may not be appropriate for any serious estimations of global climatic parameters or statistics, and therefore it should not be used as a data source for generating Level 2 and 3 products. In the next version of ACSPO product (v2.50), the resampled L2 imagery will be reported. We initially planned to fix the bow-tie distortions, and fill in deleted pixels, as an intermediate step and then restore the original VIIRS projection. However, many users of VIIRS data have issues with these artifacts, and therefore the decision was made to preserve the resampled imagery in the output files. Recall that for the vast majority of pixels, the geo-location will not change (only reordered to satisfy the requirement of local monotonicity). The percent of pixels, where geo-location has to be adjusted (by less than the footprint width) to preserve the alignment between different scan lines is very small. All VIIRS pixels filled in the bow-tie deletion areas, will be flagged as such, so it will be up to L3/4 producers whether or not to use those. However, all users of ACSPO data relying on SST imagery, will greatly benefit from the improved appearance of the product.

One can expect that the future satellite sensors will be as (or even more) complex as the current VIIRS and MODIS to ensure high resolution, low noise measurements in multiple spectral bands. Therefore the striping and bow-tie artifacts will likely continue to be present. Satellite data processing experts should therefore consider users' needs in simple, intuitive, and easy to understand and use sensor data. Generating a Level 1.5, and keeping it along with the raw data records (RDR; Level 0) or SDR (Level 1b) data in the NOAA archives, may be a practical compromise. We also strongly recommend avoiding the current onboard deletion of VIIRS pixels in bow tie areas, which degrades the data and further complicates the imagery improvements.

Acknowledgments: This work is conducted under the JPSS SST project funded by the JPSS Program Office, and the Ocean Remote Sensing Program funded by NOAA. We thank NOAA SDR Team (Changyong Cao), STAR JPSS Lead Lihang Zhou and JPSS Program Scientist Mitch Goldberg for supporting the SST re-sampling efforts. The views, opinions, and findings contained in this paper are those of the authors and should not be construed as an official NOAA or US Government position, policy, or decision.

Author Contributions: I. Gladkova developed the resampling algorithm as a supplementary pre-processing step for VIIRS SST image analysis. Her PhD student, F. Shahriar, has implemented the algorithm in python and C++. He hosts C++ implementation on github. The resampled imagery was initially converted back to the original L1b projection. A. Ignatov suggested to output the resampled SST and BT imagery to the ACSPO L2 files, extend the algorithm to MODIS, and generate L1.5 VIIRS and MODIS products with the corrected radiance imagery. He has actively participated in the discussions and evaluation, and wrote abstract, introduction and conclusion sections. Y. Kihai participated in the algorithm development, scrutinized the algorithm's outputs and provided valuable feedback on its performance. He performed global evaluations and helped to generate VIIRS and MODIS instrument-related figures. D. Hillger participated in discussion and helped with GTM comparisons, providing JPSS Imagery team feedback and perspective to the resampling algorithm and applications. B. Petrenko analyzed the effects of resampling on the SST statistics, with a focus on angular dependencies. All co-authors have reviewed and edited the entire manuscript.

Conflicts of Interest: The authors declare no conflict of interest.

References

1. Hillger, D.; Kopp, T.; Lee, T.; Lindsey, D.; Seaman, C.; Miller, S.; Solbrig, J.; Kidder, S.; Bachmeier, S.; Jasmin, T.; *et al.* First-light imagery from Suomi NPP VIIRS. *BAMS* **2013**, *94*, 1019–1029. [CrossRef]
2. Hillger, D.; Seaman, C.; Liang, C.; Miller, S.; Lindsey, D.; Kopp, T. Suomi NPP VIIRS imagery evaluation. *JGR* **2014**, *119*, 6440–6455. [CrossRef]
3. Petrenko, B.; Ignatov, A.; Kihai, Y.; Heidinger, A. Clear-sky mask for the Advanced Clear-Sky Processor for Oceans. *JTech* **2010**, *27*, 1609–1623. [CrossRef]
4. Gladkova, I.; Kihai, Y.; Ignatov, A.; Shahriar, F.; Petrenko, B. SST Pattern Test in ACSPO clear-sky mask for VIIRS. *Remote Sens. Environ.* **2015**, *160*, 87–98. [CrossRef]
5. Bouali, M.; Ignatov, A. Adaptive reduction of striping for improved sea surface temperature imagery from Suomi National Polar-Orbiting Partnership (S-NPP) Visible Infrared Imaging Radiometer Suite (VIIRS). *JTech* **2014**, *31*, 150–163.
6. Mikelsons, K.; Ignatov, A.; Bouali, M.; Kihai, Y. A fast and robust implementation of the adaptive destriping algorithm for SNPP VIIRS and Terra/Aqua MODIS SST. *Proc. SPIE* **2015**. [CrossRef]
7. Schmetz, J.; Pili, P.; Tjemkes, S.; Just, D.; Kerkman, J.; Rota, S.; Ratier, A. An Introduction to Meteosat Second Generation (MSG). *BAMS* **2002**, *83*, 977–992. [CrossRef]
8. Cao, C.; Xiong, X.J.; Wolfe, R.; DeLuccia, F.; Liu, Q.M.; Blonski, S.; Lin, G.; Nishihama, M.; Pogorzala, W.; Oudrari, H.; *et al. VIIRS Sensor Data Record (SDR) User's Guide, Version 1.2*; Technical Report; U.S. Department of Commerce: Washington, DC, USA, 2013.
9. JPSS VIIRS Geolocation ATBD, NASA GSFC JPSS CMO, JPSS Ground Project, 474–00053. 31 July 2011; p. 144. Available online: http://www.star.nesdis.noaa.gov/jpss/documents/ATBD/D0001-M01-S01-004_JPSS_ATBD_VIIRS-Geolocation.pdf (accessed on 12 January 2016).

10. JPSS VIIRS Imagery ATBD, NASA GSFC JPSS CMO, JPSS Ground Project, 474–00031, Rev. B. 15 January 2014; p. 45. Available online: http://www.star.nesdis.noaa.gov/jpss/documents/ATBD/D0001-M01-S01-008_JPSS_ATBD_VIIRS-Imagery_B.pdf (accessed on 12 January 2016).
11. Petrenko, B.; Ignatov, A.; Kihai, Y. Suppressing the noise in SST retrieved from satellite infrared measurements by smoothing the differential terms in regression equations. *Proc. SPIE* **2015**. [CrossRef]
12. Nishihama, M.; Wolfe, R.; Solomon, D.; Patt, F.; Blanchette, J.; Fleig, A. MODIS Level 1A Earth Location Algorithm Theoretical Basis Document (ATBD) Version 3.0. GSFC, SDST-092. p. 147. Available online: http://modis.gsfc.nasa.gov/data/atbd/atbd_mod28_v3.pdf (accessed on 12/01/2016).
13. MODIS Reprojection Tool Swath User's Manual, Dec 2010. Release 2.2. Land Processes DAAC USGS Center for Earth Resources Observation and Science (EROS). Available online: https://lpdaac.usgs.gov/sites/default/files/public/MRTSwath_Users_Manual_2.2_Dec2010.pdf (accessed on 12 January 2016).
14. Wolfe, R.; Roy, D.; Vermote, E. MODIS land data storage, gridding, and compositing methodology: Level 2 grid. *IEEE TGRS* **1998**, *36*, 1324–1338. [CrossRef]
15. Khlopenkov, K.; Trishchenko, A. Implementation and evaluation of concurrent gradient search method for reprojection of MODIS Level 1B imagery. *IEEE TGRS* **2008**, *46*, 2016–2027. [CrossRef]
16. Dwyer, J.; Schmidt, G. The MODIS reprojection tool. In *Earth Science Satellite Remote Sensing*; Springer-Verlag: New York, NY, USA, 2006; Volume 2, pp. 162–177.

Article

Comparison between the Suomi-NPP Day-Night Band and DMSP-OLS for Correlating Socio-Economic Variables at the Provincial Level in China

Xin Jing [1,2], Xi Shao [1], Changyong Cao [3], Xiaodong Fu [4] and Lei Yan [2,*]

1. Department of Astronomy, University of Maryland, College Park, MD 20742, USA; xjing@pku.edu.cn (X.J.); xshao@umd.edu (X.S.)
2. School of Earth & Space Sciences, Yaogan Bldg, Peking University, No.5 Yiheyuan Road, Haidian District, Beijing 100871, China
3. NOAA/NESDIS/STAR, College Park, MD 20742, USA; changyong.cao@noaa.gov
4. School of Economics, Mingde Bldg, Renmin University of China, Beijing 100872, China; rdfxd@126.com

* Correspondence: lyan@pku.edu.cn; Tel.: +86-010-62759765

Academic Editors: Changyong Cao, Richard Müller and Prasad S. Thenkabail
Received: 27 November 2015; Accepted: 21 December 2015; Published: 25 December 2015

Abstract: Nighttime light imagery offers a unique view of the Earth's surface. In the past, the nighttime light data collected by the DMSP-OLS sensors have been used as an efficient means to correlate regional and global socio-economic activities. With the launch of the Suomi National Polar-orbiting Partnership (Suomi-NPP) satellite in 2011, the day-night band (DNB) of the Visible Infrared Imaging Radiometer Suite (VIIRS) onboard represents a major advancement in nighttime imaging capabilities, because it surpasses its predecessor DMSP-OLS in radiometric accuracy, spatial resolution and geometric quality. In this paper, four variables (total night light, light area, average night light and log average night light) are extracted from nighttime radiance data observed by the VIIRS-DNB composite in 2013 and nighttime digital number (DN) data from the DMSP-OLS stable dataset in 2012, respectively, and correlated with 12 socio-economic parameters at the provincial level in mainland China during the corresponding period. Background noise of DNB composite data is removed using either a masking method or an optimal threshold method. In general, the correlation of these socio-economic data with the total night light and light area of VIIRS-DNB composite data is better than with the DMSP-OLS stable data. The correlations between total night light of denoised DNB composite data and built-up area, gross regional product (GRP) and power consumption are higher than 0.9 and so are the correlations between the light area of denoised DNB composite data and city and town population, built-up area, GRP, power consumption and waste water discharge. However, the correlations of socio-economic data with the average night light and log average night light of VIIRS-DNB composite data are not as good as with the DMSP-OLS stable data. To quantitatively analyze the reasons for the correlation difference, a cubic regression method is developed to correct the saturation effect of the DMSP stable data, and we artificially convert the pixel value of the DNB composite into six bits to match the DMSP stable data format. The correlation results between the processed data and socio-economic data show that the effects of saturation and quantization are two of the reasons for the correlation difference. Additionally, on this basis, we estimate the total night light ratio between saturation-corrected DMSP stable data and finite quantization DNB composite data, and it is found that the ratio is ~11.28 ± 4.02 for China. Therefore, it appears that a different acquisition time is the other reason for the correlation difference.

Keywords: nighttime light; socio-economic statistics; visible infrared imaging radiometer suite (VIIRS); day-night band; DMSP-OLS

1. Introduction

Remote sensing of the environment provides great opportunities to understand links between human and nature and global socio-economic changes. With rapid advances in remote sensing technology and its applications, it becomes increasingly more desirable to use remote sensing data to study and monitor the socio-economic environment. Nighttime light imagery stands distinctly against various remote sensing data sources, as it offers a unique view of the Earth's surface in the light of human activities. Nocturnal lighting becomes one of the hallmarks of modern development and provides a unique attribute for identifying the presence of development or human activity that can be sensed remotely. The presence of lighting across the globe is mostly due to some form of human activity, such as human settlements, shipping fleets, gas flaring or fire associated with swidden agriculture [1,2].

Satellite sensors, such as OLS on DMSP, have been acquiring day/night images since the early 1970s for applications, such as military surveillance, population estimation, monitoring social-economic development and power consumption and providing weather- and climate-related data [3]. The DMSP-OLS sensor distinguishes itself from the rest of passive, optical remote sensing in that the data can be acquired at night and are sensitive to light sources down to a minimum detectable radiance of 10^{-9} W/cm^2-sr [4,5]. In essence, the radiance detected by the sensor, after masking out clouds using the OLS thermal infrared channel, are mostly man-made light sources, primarily from cities, but also from oil-field gas-flare burn off, biomass burning and shipping fleets [6].

In the past, the remote sensing of nighttime light with DMSP-OLS was actively studied and shown to be an accurate, economical and straightforward way of mapping the global distribution and density of developed areas, as well as population [2]. Night light imagery data were also used in mapping regional economic activity at the national and regional level. Welch [7] showed quantitative relationships between DMSP-OLS nocturnal lighting images of the United States and population, urban area and electric energy utilization patterns. Sutton [8] showed that the correlation between DMSP-OLS data and population density within the urban areas in the United States can be as high as 0.9. Elvidge *et al.* [9] found that light area estimated from the DMSP-OLS data is highly correlated with gross domestic product (GDP) and electric power consumption. However, significant outliers in the relation between light area and population indicate that it is difficult for DMSP-OLS stable light products to provide direct detection of rural population. Doll *et al.* [10] developed a method to correlate the light area of a city derived from DMSP-OLS data and statistic data of local socio-economic development to map global economic activity (GDP) and carbon dioxide emissions at the regional level. Elvidge *et al.* [5] developed a method to perform radiance calibration for the digital number (DN) data of DMSP-OLS. With this method, Doll *et al.* [11] derived a linear relationship between the intensity of light observed by DMSP-OLS and gross regional product (GRP) for a sub-set of countries within the European Union and U.S. They concluded that different countries have different relationships with total radiance based on their cultures. Sutton *et al.* [12] developed predictive relationships between observed changes in nighttime satellite images derived from the DMSP-OLS and changes in population and GDP. Letu *et al.* [13] demonstrated the estimation of electric power consumption from saturated nighttime DMSP-OLS imagery after correction for saturation effects. Additionally, recently, Li *et al.* [14] used 38 monthly DMSP-OLS System composites covering the period between January 2008 and February 2014 to analyze the response of nighttime light to the Syrian Crisis. The results indicate that the nighttime light experienced a sharp decline as the crisis broke out. Coscieme *et al.* [15] presented a method based on DMSP-OLS nighttime data that uses nocturnal light data as a proxy measure for the evolution of the non-renewable fraction of national emergy flow. Coscieme *et al.* [15] found a strict correlation between the intensity of lights and the non-renewable component of national energy flow for more than 100 countries.

One comprehensive study of correlating DMSP-OLS imagery data with multiple socio-economic variables was conducted by Lo [16]. The DMSP-OLS imagery data were acquired between March 1996 and January to February 1997. Lo [16] modeled three types of population parameters (population, non-agricultural population and population density) at the provincial, city and county level in China, respectively, by using four types of variables (light area, percent light area, light volume and pixel mean) derived from OLS imagery data. It was found that the DMSP-OLS nighttime data produced reasonably accurate estimates of non-agricultural population at both the county and city levels using the algometric growth model and the light area or light volume as input. The logarithmic form of the algometric growth model is $\log Population = \log a + b \log A$. Here, a is a coefficient, b is an exponent and A is the built-up area of the settlement. Both a and b are empirically determined. Non-agricultural population density was best estimated using percent light area in a linear regression model at the county level. Lo [16] concluded that the 1-km resolution DMSP-OLS nighttime light image has the potential to provide estimation of the total and urban population of a country from space. Furthermore, Lo [16] presented the relationship between DMSP-OLS imagery data and additional socio-economic parameters at the provincial level in China, such as household, energy consumption, electricity consumption, gross value of industrial output, per capita rural income and per capita urban income. The correlation results between these additional socio-economic parameters and variables derived from the DMSP image were less emphasized in Lo [16] as compared to those obtained with population data.

While the DMSP-OLS is remarkable for its detection of dim lighting, there have been some limitations in DMSP-OLS, such as low spatial resolution (2.7 km ground sample distance), low radiometric resolution (six bit), a saturation effect in bright regions, lack of on-board calibration, lack of systematic recording of in-flight gain changes and lack of multiple spectral bands for discriminating lighting types [2].

With the launch of the Suomi National Polar-orbiting Partnership (Suomi-NPP) satellite in October 2011, the day-night band (DNB) of the Visible Infrared Imaging Radiometer Suite (VIIRS) onboard represents a major advancement in nighttime imaging capabilities [17–21]. DNB serves primarily to provide imagery of clouds and other Earth features over illumination levels ranging from full sunlight to quarter moon. Other applications of using DNB, such as light outage detections during major storms, have been recently demonstrated [19]. The basic parameters for DNB specifications can be found in Table 1 [22] (Shao *et al.*, 2013). The DNB is a *de facto* radiometer, because it uses an onboard calibration system to generate the radiances for Earth observations, compared to the DMSP-OLS, which is an imager and has no onboard calibration. The DNB of the VIIRS sensor utilizes a backside-illuminated charge coupled device (CCD) focal plane array (FPA) for sensing of radiances spanning seven orders of magnitude in one panchromatic (0.5 to 0.9 μm) reflective solar band (RSB). In order to cover this extremely broad measurement range, the DNB employs four imaging arrays that comprise three gain stages. The low gain stage (LGS) gain values are determined by solar diffuser data. In operations, the medium and high gain stage values are determined by multiplying the LGS gains by the medium gain stage (MGS)/LGS and high gain stage (HGS)/LGS gain ratios, respectively [23]. The DNB relies on collocation with multispectral measurements on VIIRS and other Suomi-NPP sensors for accurate geolocation. The spatial resolution of the DNB is approximately 750 m across the entire swath. This is achieved by performing on-chip aggregation of the CCD detector elements that form pixels, which results in 32 aggregation zones through each half of the instrument swath on either side of nadir. The aggregation zones near the end of scan (EOS) have fewer pixels than the zones near nadir, as the footprint of a single CCD detector element on the ground is much larger at EOS. These improvements, coupled with the multispectral complementary information from other collocated VIIRS channels, enables the use of Suomi-NPP to pursue quantitative applications heretofore restricted to daytime measurements, a true paradigm shift in nighttime remote sensing capability.

Shi *et al.* [24] suggested that VIIRS data might be more indicative of demographics and economics than DMSP data at both the city and the province scales by statistically comparing the correlations between nighttime light brightness and socio-economic variables. Ma *et al.* [25] investigated correlations of DNB nighttime light radiance with GDP, population, electrical power consumption and paved road areas, and this work indicated that these parameters had a significantly positive linear relation with nighttime light radiance. The application of VIIRS DNB nighttime data, beyond correlating with socio-economic data [26–32], also can be used to detect social insurgency [33].

Table 1. Design specifications for Suomi-National Polar-orbiting Partnership (Suomi-NPP) VIIRS-day-night band (DNB). HGS, high gain stage; MGS, medium gain stage; LGS, low gain stage.

Spectral Band	0.5 to 0.9 μm
Relative Radiometric Gains	119,000:477:1 (HGS:MGS:LGS)
Dynamic Range	Lmax/Lmin = 6,700,000
Number of Bits in analog to digital (A/D)	14 bits (16,384 levels) for HGS; 13 bits (8192 levels) for MGS and LGS
Spatial Resolution	750 m
Aggregation	32 aggregation zones
Time Delay Integration (TDI)	1, 3 and 250 pixels for LGS, MGS and HGS, respectively
Number of Samples per Scan	4064

Recent work by Li *et al.* [26] compared the capabilities of using DNB and DMSP-OLS data to model the gross regional product (GRP) in China. One variable, total night light (*TNL*), is derived from DNB and DMSP imagery data to model GRP at the provincial and county level in China with a linear regression model. It was shown that the *TNL* derived from Suomi-NPP DNB exhibit R^2 values of 0.8699 and 0.8544 when correlating with the provincial and county GRP, respectively, which are significantly better than the correlative relationship between the *TNL* from DMSP-OLS F16 (0.6923) and F18 (0.7056) satellites and GRP. This demonstrated that the DNB nighttime light imagery has a stronger capability in modeling GRP than those of the DMSP-OLS data. However, the comparison between Suomi-NPP DNB and DMSP-OLS in correlating with regional socio-economic variables performed by Li *et al.* [26] is limited to correlating one socio-economic parameter, *i.e.*, GRP, with one light variable (total night light) derived from nighttime imagery data. However, Li *et al.* [26] only gave the three general potential factors (the saturation effect of DMSP-OLS in city centers, the different acquisition time between DNB and DMSP-OLS data and the onboard calibration system on NPP-VIIRS) that make DNB data more efficient than the OLS data in modeling the economy, but without any quantitative analysis. Factors that cause the difference between DNB and DMSP observations in correlating with GRP remain to be investigated.

In this paper, we focus on comparing the performance of imagery data of the DMSP-OLS stable data with that of Suomi-NPP DNB composite data in correlating with multiple regional socio-economic parameters in China. We developed methods to remove the background noises that are not related to economic activities in DNB data. Different from the work of Li *et al.* [26], we calculate correlations between four light variables derived from nighttime imagery data and multiple socio-economic parameters to assess the difference between the DMSP-OLS stable data and the DNB composite data in correlating with socio-economic parameters. In view of the significant differences between DMSP data and DNB, such as different data quantization, the saturation effect of the DMSP stable data and data acquisition time at night, *i.e.*, DNB at ~1:30 a.m. *versus* DMSP-OLS at ~9 p.m. Equator cross time, we use a cubic regression model to correct the saturation pixels of the DMSP stable data, artificially quantize the pixel value of the DNB composite data into six bit and estimate the ratio of total night radiance between saturation-corrected DMSP-OLS stable and finite quantization DNB composite data for China.

In the following sections, we first introduce the data and regional areas studied in our work. In Section 3, we first illustrate the variables of interest derived from nighttime light data. Then, the noise masking method (NMM) and the optimal threshold method (OTM) are presented for removing background noise of DNB composite data. The correlation results between variables from nighttime data and socio-economic parameters are given. Section 4 explores the factors that contribute to the correlation difference between the DNB composite data and the DMSP-OLS stable data. The conclusion is given in Section 5.

2. Data and Study Area

2.1. Nighttime Imagery Data

In this study, both DMSP-OLS stable data and Suomi-NPP VIIRS DNB composite data over mainland China are used. The VIIRS-DNB (Figure 1) composite data used in this study are the first global cloud-free composite of VIIRS nighttime lights and acquired from the Earth Observation Group (EOG) of NOAA's National Geophysical Data Center (NGDC).

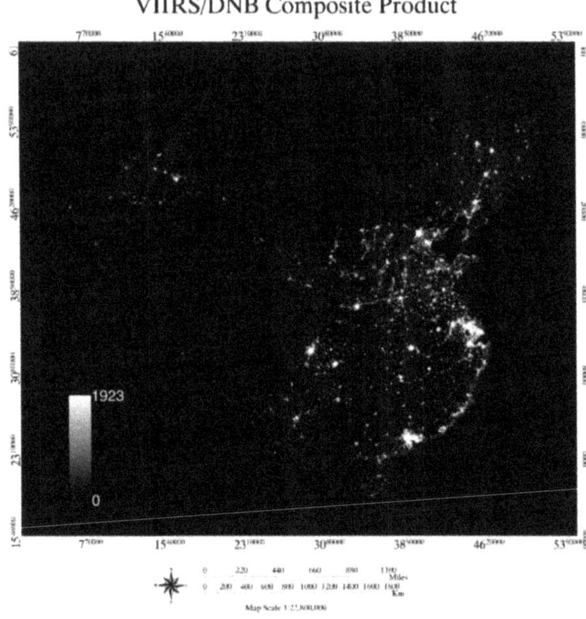

Figure 1. The VIIRS-day-night band (DNB) composite data of mainland China in 2013 (multiplying the pixel value by 10^{-9} gives radiance in units of W/cm^2-sr).

These global composite of VIIRS nighttime light data are assembled with observations collected at nights with zero moonlight during the periods of April and October in 2012 and January in 2013. Cloud screening was performed based on the detection of clouds using the observations from the M15 infrared thermal band of VIIRS. However, this data product has not been filtered to remove lights associated with fires, gas flares, volcanoes or aurora. Furthermore, the background noise has not been subtracted [34]. First, we used a mean algorithm to remove the abnormal lights that may be associated with fires and gas flares. Then, we introduced the noise masking method (NMM) and the optimal threshold method (OTM) to remove background noise in DNB composite data, and these methods are illustrated in Sections 3.2 and 3.3. 2012 composite data reveal significant geolocation issues, which were due to a DNB pointing error or systematic offset in computing the location of nadir. In the nighttime data, this resulted in a westward shift of entire scans, with off-nadir pixels more affected than pixels close to nadir. When making a composite (or average) product, the effect was an enlarged footprint in the track direction of the composite and an average radiance, which was not representative of the ground pixel, as it was an average of a close-by region depending on geolocation accuracy. NOAA/NGDC developed software to correct for these pointing errors, and these errors were fixed in the January 2013 composite. Therefore, we use January 2013 DNB composite data to compare to the DMSP-OLS stable product in this paper. January is not the burn season in China, so the biomass burning effect can be ignored in this month.

The DMSP stable data are also acquired from the NGDC/EOG. This dataset is cloud-free composites assembled using all of the archived smooth resolution data of DMSP-OLS that are available during calendar years. The products are in a spatial resolution of 30 arc seconds [35]. The composite products of DMSP data we use in this study were stable light products and acquired by the F18 satellite in 2012. In this data product, the background noise was identified and replaced with values of zero, and ephemeral events have been discarded, so that only lights from cities, towns and other sites with persistent lighting (including gas flares) remain [36]. All of the nighttime light imagery was re-projected using Albers conical equal-area projection with its original resolution. The details of nighttime light data are described in Table 2.

Table 2. Year and spatial resolution of the satellite imagery data used in this study and the year of the corresponding socio-economic data used to correlate with these imagery data. All of the satellite data are obtained from National Geophysical Data Center (NGDC)/Earth Observation Group (EOG).

Year	Imagery Data	Archived Composites	Resampled Imagery in this Study	Year of Socio-Economic Data to Correlate with
2013	VIIRS/DNB composite data	15 arc seconds	491 m	2013
2012	DMSP/F18 stable data	30 arc seconds	982 m	2012

In this paper, we assume that the night light distributions in mainland China during the 12-month composite in 2012 and the 1-month composite in 2013 are the same. Then, four types of variables, total night light (*TNL*), light area (*LA*), average night light (*ANL*) and log*ANL* of each administrative region, were derived from nighttime image. Details on these four variables can be found in Section 3.1.

2.2. Socio-Economic Data

In this study, socio-economic parameters chosen to be correlated with nighttime imagery data are acquired from China Statistical Yearbook for Regional Economy, China City Statistical Yearbook, and China Statistical Yearbook. In total, there are 12 socio-economic parameters chosen for correlating in this study. The abbreviations, sources and units of these parameters are described in Table 3.

Table 3. The abbreviations, sources and units of socio-economic parameters at the provincial level used in this study.

Socio-Economic Parameters	Abbreviations	Unit	Source
Total Population	TP	Ten thousands	a
City Population	CP	Ten thousands	d1
City and Town Population	CTP	Ten thousands	b
Household	HH		a
City Area	CA	Square km	c
Built-up Area	BUA	Square km	c
Gross Regional Product	GRP	One hundred million Yuan	a
GRP Per Capita	GRPPC	Yuan	d2
City and Town Per Capita Income	CTPCI	Yuan	b
Rural Per Capita Income	RPCI	Yuan	b
Power Consumption	PWC	A hundred million kilowatt hours	b
Waste Water Discharge	WWD	Ten thousand tons	a

a, China Statistical Yearbook; b, China Statistical Yearbook for Regional Economy; c, China City Statistical Yearbook; d1, calculated by multiplying city population density and city area; d2, calculated by dividing GRP by total population.

2.3. Region of Interest and Identification

To evaluate the correlations between socio-economic parameters and night light variables derived from DMSP-OLS and NPP-VIIRS imagery data, we focus on analyzing administrative regions at the provincial level in mainland China. Thirty-one provinces in mainland China are selected for the analysis. Boundary data of these 31 provinces at the scale of 1:4 million in ArcInfo format for year 2000 were obtained from the Data Sharing Infrastructure of Earth System Science [37]. The nighttime imagery data were registered to the corresponding provincial units using ENVI software. Figure 1 illustrates the composite observations by DNB of these 31 provincial regions of mainland China.

3. Methodology and Results

3.1. Variable Extraction from Nighttime Imagery Data

Both the DNB composite data and DMSP stable products are arranged in longitude/latitude format. In our analysis, nighttime image and provincial boundaries are projected using Albers equal-area conic projection available in ArcGIS, which is a conic, equal-area map projection that uses two standard parallels. With this projection, distortion is minimal between the standard parallels. This projection is best suited for land masses extending in an east-to-west orientation.

After applying projection, we extract the pixels and their values within each provincial boundary and derive four variables, total night light (*TNL*), light area (*LA*), average night light (*ANL*) and log*ANL*, which are defined as follows.

Following the approach of Lo [16] and Li *et al.* [26], the *TNL* indicates the total amount of light within a given administrative region and is closely related to the socio-economic activities in the region. It is calculated using the following formula:

$$TNL = \sum_{L_i > L_t} L_i \tag{1}$$

where L_t is the threshold value in mainland China and i is the i-th pixel with a pixel value $L > L_t$. For DMSP-OLS stable data, the pixel value is DN, and the L_t is zero. Additionally, for DNB composite data, the pixel value has a radiance unit of W/cm^2-sr; the L_t is zero when the data have been processed by the noise masking method and is the optimal threshold value when the data have been processed by the optimal threshold method. Each *TNL* value represents the sum of all pixel values larger than L_t in an administrative region. For each kind of nighttime imagery data, we can extract 31 *TNL* values for the 31 regions, respectively.

As Elvidge *et al.* [9] and Lo [16] suggested that light area estimated from the nighttime imagery data is highly correlated with socio-economic parameters, we also introduce the light area (*LA*) as a variable to be extracted from the imagery data, and it is calculated as:

$$LA = \sum_{L_i > L_t} A\left(L_i\right) \tag{2}$$

where A is area of the i-th pixel with pixel value $L > L_t$. In this case, the light area is the total area of an administrative region with a pixel value greater than the threshold value L_t. Similarly, 31 *LA* values can be extracted for each kind of nighttime imagery data.

The third variable we extract from the imagery data is average night light (*ANL*), and it is calculated as:

$$ANL_j = TNL_j / LA_j \tag{3}$$

where index j refers to the j-th administrative region and TNL_j and LA_j are the *TNL* and *LA* of region j, respectively, as defined in Equations (1) and (2). Intuitively, *ANL* can be correlated with per capita-type socio-economic parameters, such as per capita income.

Meanwhile, referencing to the work of Lo [16] and considering the division calculation of the variable *ANL*, we also introduce the fourth variable log*ANL* in this paper.

3.2. DNB Noise Filtering with the Noise Masking Method

While DMSP-OLS products are stable nighttime data and the background noise and ephemeral lights have been identified and replaced with values of zero [36], the DNB composite data acquired from NGDC have not been filtered to remove light signals associated with fires, gas flares, volcanoes or aurora. In calculating the *TNL*, *LA*, *ANL* and log*ANL*, the key is to remove the dark background noise and ephemeral noises, which are not related to socio-economic activities.

Before removing background noise of the DNB composite, we notice that there are a few outliers of the 2013 DNB composite data in northeast and western China. The outliers are probably caused by lights from the fires of oil or gas flares located in those areas. Since Beijing and Shanghai are the two most developed administrative regions in China, the pixel values of the other areas should not exceed those of the two regions theoretically [29]. The highest radiance of those two regions is 2.62×10^{-7} W/cm^2-sr, and other pixels whose radiance is larger than 2.625×10^{-7} W/cm^2-sr in the DNB composite data were smoothed by their eight neighbors.

After that, this preliminary corrected data also have background noise left. Making reference to the work of Lo [16] and Li *et al.* [26], we introduce two methods to remove these background noises.

Li *et al.* [26] developed a simple and approximate method for removing the background noise and ephemeral lights in DNB composite data through applying the mask generated from the 2010 DMSP-OLS stable data to 2012 DNB composite data. It was shown in Li *et al.* [26] that after applying this method, the correlation of *TNL* with regional GRP in China is largely increased. By using this method, Li *et al.* [26] made an assumption that the light areas in the years 2010 and 2012 are the same. During the time of their study, they can only acquire DMSP-OLS stable data in 2010, which is the closest year to 2012. In our work, the DMSP-OLS stable data in 2012 can be acquired, and we can generate a mask from the DMSP-OLS data in 2012, close to the year of DNB composite data. This method is referred as the noise masking method (NMM). Furthermore, we assume that the light areas in the years 2013 and 2012 are the same.

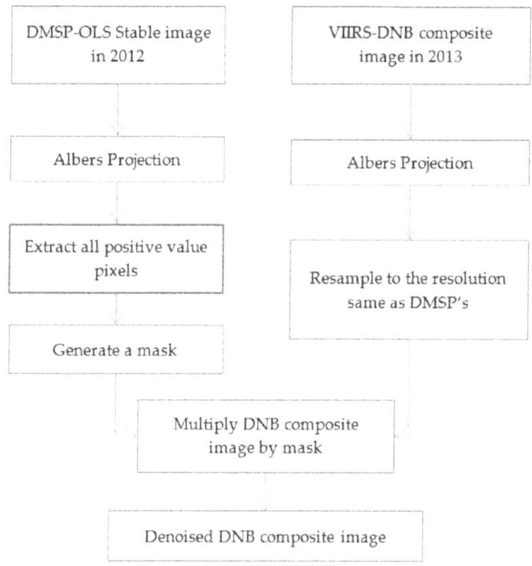

Figure 2. Flow chart of the noise masking method (NMM) in removing background noise for DNB composite data.

Figure 2 shows the flow chart of NMM. Different from the work of Li *et al.* [26], we use Albers projection in this work instead of the Lambert projection used in Li *et al.* [26]. The DNB composite data are resampled to the same resolution as that of DMSP-OLS stable light product, *i.e.*, 982 m. Then, we extract all of the pixels with a positive value (DN > 0) from 2012 DMSP stable data to generate a mask. The mask is applied to the DNB January 2013 composite imagery data. For pixels outside the mask, the pixel value of DNB data is set as NaN (not a number), and the pixel value is kept the same for pixels inside the mask.

3.3. DNB Noise Filtering with the Optimal Threshold Method

Although NMM can be effective at screening out background noise and ephemeral lights in DNB composite data, this method relies on the DMSP stable data to generate the mask. These masks might not be readily available for the DNB observations of interest and can be outdated as the new DNB observations are made available. Additionally, at the same time, NMM will exclude some small towns and road features that the DNB product is sensitive enough to pick up. Therefore, we introduce another method, the optimal threshold method (OTM), to remove the noises in DNB composite data. To determine the optimal threshold (L_T), we chose the correlation between *LA* and built-up area as the object function. The choice of this object function originates from the work by Lo [16] and Chen *et al.* [38], which showed that light intensity is closely related to the type of land use or land cover and depicts built-up area the best. The object function is defined as:

$$O(L_t) = \rho(LA(L > L_t), b) \tag{4}$$

where $\rho(LA(L > L_t), b)$ is the correlation coefficient between *LA* with a pixel value above the intensity threshold value L_t and built-up area *b*; L_t: intensity threshold value; LA $(L > L_t)$: light areas with a pixel value above the radiance threshold value L_t inside administrative regions; *b*: built-up areas of the regions in China that were acquired from China Yearbook.

The optimal threshold value L_T is therefore determined using:

$$O(L_T) = \max\left(O(L_t)|_{L_t=[L_{t1},\, L_{t2}]}\right) \tag{5}$$

so that the resulting correlation $O(L_T)$ between *LA* and built-up area reaches the maximum when $L_t = L_T$. Here, the light radiance threshold L_t varies from L_{t1} to L_{t2} to determine L_T. In our calculation, we use L_{t1} equal to 0 and L_{t2} equal to 10^{-7}.

Figure 3 illustrates the determination of the optimal threshold value L_T from DNB composite data. The X-axis is the radiance value varying from 0 to 10^{-7} W/cm^2-sr. The Y-axis is the correlation coefficient between *LA* and the built-up area of the provinces of interest. When the correlation coefficient reaches the maximum, the corresponding radiance value is determined as the optimal threshold value L_T. As can be seen from Figure 3, the optimal threshold value of the original DNB composite data is determined as $L_T = 2.15 \times 10^{-9}$ W/cm^2-sr.

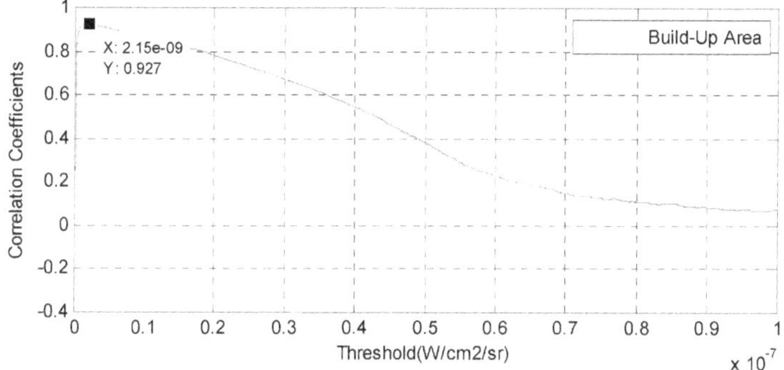

Figure 3. Correlation coefficient between built-up area and light area *vs.* different threshold values (see Equation (4)) for 31 provinces.

As we will show in Section 3.4.2, OTM is a more effective way than NMM to increase the correlation between *LA* and socio-economic parameters.

3.4. Correlation Results

We compute *TNL*, *LA*, *ANL* and log*ANL* using Equations (1) to (3) given above for DMSP-OLS stable data and VIIRS-DNB composite data, respectively. For DNB composite imagery data, we applied the NMM and OTM to remove the background noise in composite imagery. The relationships between these variables derived from nighttime imagery and socio-economic parameters are evaluated using the Pearson correlation coefficients.

3.4.1. Correlations with *TNL*

Table 4 and Figure 4 show the correlation results and scatter plots between *TNL* and socio-economic parameters, respectively. As shown in Table 4, *TNL* of DNB composite imagery derived with the NMM and OTM (hereinafter referred to as DNB NMM and DNB OTM, respectively) all have a better correlation with socio-economic parameters than that of *TNL*-derived from DNB composite imagery without noise filtering. This indicates the importance of noise removal in processing DNB composite data. *TNL* of DNB NMM has overall better correlation than that derived with OTM. With NMM, The *TNL* derived from the DNB composite image has correlation coefficients (ρ) with socio-economic parameters all above 0.7. In particular, the correlation of *TNL* with built-up area (BUA), GPR and power consumption (PWC) are the best, all above 0.9. The correlation of *TNL* with city population (CP), city and town population (CTP) and waste water discharge (WWD) are relatively in the middle, whose coefficients are 0.85, 0.84 and 0.86, respectively. The correlation of *TNL* with total population (TP), household (HH) and city area (CA) are relatively weak, but still strong in absolute value, whose coefficients are 0.70, 0.72 and 0.75, respectively.

Comparing the correlation of *TNL* of the DNB composite image with socio-economic parameters and that of DMSP-OLS stable data, we found that the *TNL*'s of DNB NMM and DNB OTM in general have a better correlation with most of the socio-economic parameters (except TP and HH) than the correlation derived with DMSP stable data. For *TNL* from DMSP-OLS stable data, the best correlation is with PWC, whose coefficient is 0.91; the correlations with TP, HH, BUA, GPR and WWD are in the range of 0.8; the correlations with CP, CTP and CA are relatively weak, but still strong in absolute value, whose coefficients are 0.74, 0.79 and 0.70, respectively.

Figure 4 shows the scatter plot of selected socio-economic parameters, such as BUA, GRP and PWC, *vs.* *TNL* from DNB NMM, DNB OTM and DMSP-OLS stable data, respectively. It can be seen that Guangdong has the highest BUA and GPR, and Jiangsu has the highest PWC in the year 2013; Guangdong also has the highest BUA, GPR and PWC in the year 2012. Meanwhile, Guangdong has the highest *TNL* as derived from both DNB NMM (2013) and DNB OTM (2013), and Shandong has the highest *TNL* as derived from DMSP-OLS (2012) stable data. Xizang has the lowest *TNL* as derived from both processed DNB composite data and DMSP stable data and has the lowest BUA, GPR and PWC. This suggests that Shandong, Jiangsu and Guangdong provinces are large administrative regions and more industrialized with more night light, which is consistent with the actual situation in China.

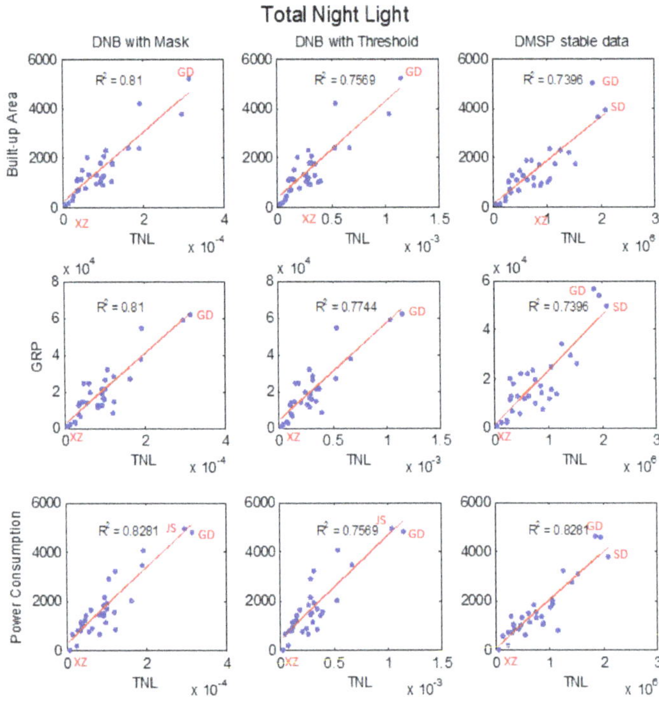

Figure 4. Scatter plots of total night light (*TNL*) *vs.* socio-economic parameters (1st row: built-up area; 2nd row: gross regional product (GRP); 3rd row: power consumption) together with the fitting curve (in red) from regression. *TNL* data used in the 1st, 2nd and 3rd column are derived from DNB composite data with NMM, the optimal threshold method (OTM) in 2013 and DMSP-OLS stable data in 2012, respectively. Red labels in the panel denote: GD, Guangdong; SD, Shandong; XZ, Xizang; JS, Jiangsu.

Table 4. Correlation coefficients between *TNL* and socio-economic parameters. *TNL*'s are derived from DNB composite data with or without noise filtering and DMSP-OLS F18 stable data, respectively.

	DNB *TNL* 2013			DMSP *TNL*
	Original	**NMM**	**OTM**	**F18 2012**
TP	0.43	0.70	0.63	0.80
CP	0.58	0.85	0.85	0.74
CTP	0.54	0.84	0.81	0.79
HH	0.45	0.72	0.66	0.81
CA	0.53	0.75	0.73	0.70
BUA	0.67	0.90	0.87	0.86
GRP	0.62	0.90	0.88	0.86
PWC	0.71	0.91	0.87	0.91
WWD	0.57	0.86	0.85	0.80

In summary, applying noise-filtering methods, *i.e.*, NMM or OTM, to DNB composite data helps improve the correlation between *TNL* and socio-economic parameters. The NMM produces the best performance in correlating *TNL* of DNB composite data with socio-economic parameters in comparison with another method. The correlation of *TNL* from DNB NMM is better than that of DMSP stable data for almost all of the parameters, except TP and HH, and that with PWC is comparable. Li *et al.* [26] studied the correlation only between *TNL* and GRP, and our results on this correlation are consistent with what they concluded.

3.4.2. Correlations with *LA*

Table 5 and Figure 5 show the correlation results and scatter plots between *LA* and socio-economic parameters, respectively.

As shown in Table 5, the *LA*'s derived from DNB NMM and DNB OTM all have significant better correlations with socio-economic parameters than that of *LA* derived from the DNB composite without noise filtering. The correlation coefficients of *LA* derived from DNB OTM are all above 0.77. In particular, the correlations of *LA* derived from DNB OTM with economic parameters (CTP, BUA, GRP, PWC and WWD) are among the best, all above 0.9. *LA* of DNB OTM has a much better correlation than that of DNB NMM, particularly in correlating with CP, CTP, CA, BUA, GRP, PWC and WWD. The best correlation of *LA* derived from DNB NMM has $\rho \sim 0.78$, *i.e.*, TP and HH, far less than $\rho > 0.9$ achieved for the correlation of *LA* using OTM with CTP, BUA, GRP, PWC and WWD. This illustrates that OTM is a more effective way than NMM to remove noisy background of DNB composite data and deriving socio-economic activity-related *LA* from the data. Therefore, using OTM can more effectively improve the correlative relationship between *LA* of DNB composite data and socio-economic parameters.

Table 5. Correlation coefficients between *LA* and socio-economic parameters. *LA*'s are derived from DNB composite data with or without noise filtering and DMSP-OLS F18 stable data, respectively.

	DNB 2013 *LA*			DMSP *LA*
	Original	NMM	OTM	F18 2012
TP	−0.27	0.78	0.78	0.78
CP	−0.33	0.58	0.86	0.59
CTP	−0.33	0.72	0.90	0.75
HH	−0.29	0.78	0.80	0.78
CA	−0.27	0.55	0.77	0.54
BUA	−0.23	0.74	0.93	0.73
GRP	−0.31	0.68	0.94	0.68
PWC	−0.16	0.75	0.95	0.75
WWD	−0.30	0.65	0.92	0.65

Since the administrative region of interest of DNB NMM data is generated from the DMSP-OLS mask, the correlations of *LA* using DNB NMM and the DMSP stable data are similar; only the correlations with CP, CTP, CA and BUA have small differences, which is because the years of the corresponding socio-economic data that DNB NMM and DMSP stable data correlated with are different. Therefore, the performances of the correlation of *LA* derived from both DNB NMM and DMSP-OLS stable data with socio-economic parameters are almost the same.

Figure 5 shows the scatter plot of selected socio-economic parameters, such as household, city and town population and power consumption *vs.* *LA* from DNB NMM, DNB OTM and DMSP-OLS stable data, respectively. From Figure 5, Guangdong and Shandong have the largest *LA* as derived from DNB OTM (2013) and DMSP-OLS stable data (2012), respectively. Xizang has the smallest *LA* as derived from both DNB denoised data and DMSP-OLS stable data. This relative ranking of *LA* among Guangdong, Shandong and Xizang is consistent with the ranking of *TNL*. Correspondingly, Shandong, Guangdong and Jiangsu have the highest household, city and town population and power consumption in the year 2013, respectively, Shandong has the highest HH, and Guangdong has the highest CTP and PWC in the year 2012. Still, Xizang province has the lowest built-up area, GPR and power consumption.

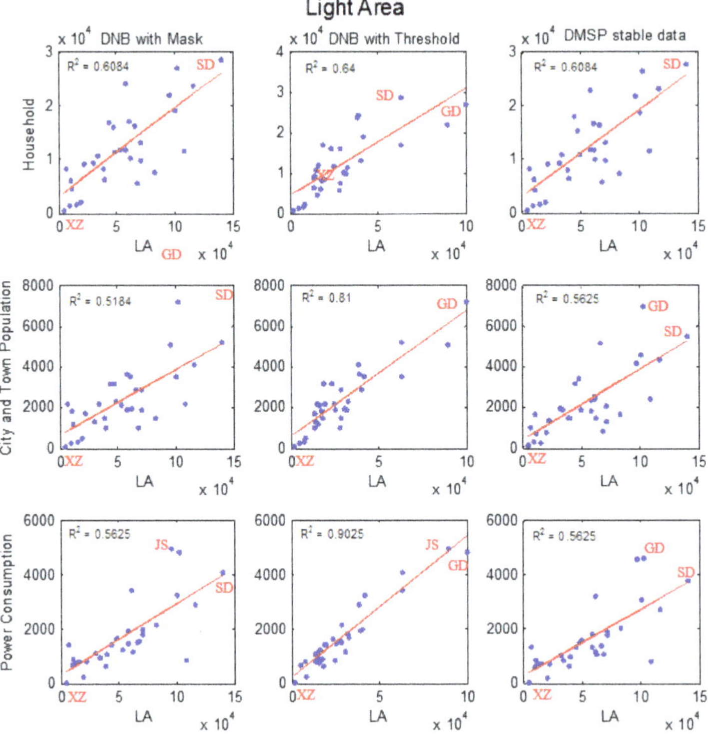

Figure 5. Scatter plots of light areas (*LA*'s) *vs.* socio-economic parameters (1st row: household; 2nd row: city and town population; 3rd row: power consumption) together with the fitting curve (in red) from regression. *LA* data used in the 1st, 2nd and 3rd column are derived from DNB composite with NMM and OTM in 2013 and DMSP-OLS stable data in 2012, respectively. Red labels in the panel denote: GD, Guangdong; SD, Shandong; XZ, Xizang; JS, Jiangsu.

From Tables 4 and 5 it can also be observed that using OTM in deriving *TNL* and *LA* from DNB composite data, the resulting correlations with BUA, GRP, PWC and WWD are all quite strong, *i.e.*, above 0.92 for *LA* and between 0.85 and 0.88 for *TNL*. This suggests that OTM is consistent in removing noise for calculating the correlation with both *LA* and *TNL* and is an effective method for filtering DNB data to model these socio-economic parameters. Meanwhile, based on the generation procedure of DNB NMM data, its performance for *LA* in correlating with the parameters is the same as DMSP stable data, but has better performance of *TNL* than DNB OTM data in correlating with the parameters, especially with BUA, GRP, PWC and WWD (economic parameters).

3.4.3. Correlations with *ANL* and log*ANL*

Tables 6 and 7 show the correlation results between socio-economic parameters and both *ANL* and log*ANL*, respectively. Figure 6 shows scatter plots between log*ANL* and multiple socio-economic parameters. Since the *ANL* is derived by dividing *TNL* with *LA*, it should be related to per-capita socio-economic parameters, such as per capita GRP and per capita income, in evaluating its correlation.

Table 6. Correlation coefficients between *ANL* and socio-economic parameters. *ANL*'s are derived from DNB composite data with or without noise filtering and DMSP-OLS stable data, respectively.

	DNB *ANL* 2013			DMSP *ANL*
	Original	**NMM**	**OTM**	**F18 2012**
GRPPC	0.69	0.71	0.67	0.84
CTPCI	0.78	0.81	0.77	0.88
RPCI	0.77	0.80	0.76	0.89

Table 7. Correlation coefficients between log*ANL* and socio-economic parameters. log*ANL*'s are derived from DNB composite data with or without noise filtering and DMSP-OLS stable data, respectively.

	DNB log*ANL* 2013			DMSP log*ANL*
	Original	**NMM**	**OTM**	**F18 2012**
GRPPC	0.83	0.84	0.75	0.89
CTPCI	0.84	0.88	0.79	0.90
RPCI	0.88	0.87	0.81	0.91

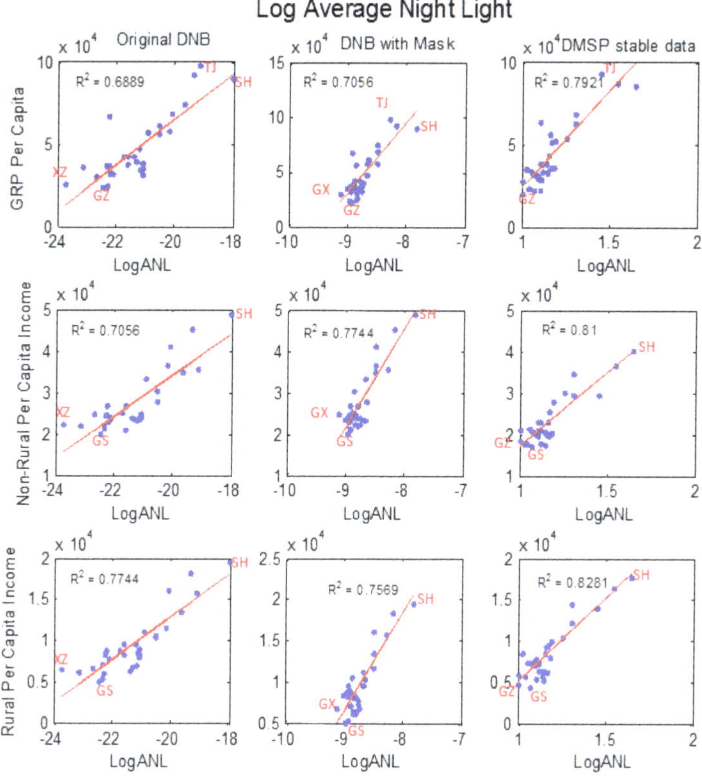

Figure 6. Scatter plots of log*ANL* *vs.* socio-economic parameters (1st row: GRPPC; 2nd row: CTPCI; 3rd row: RPCI) together with the fitting curve (in red) from regression. log*ANL* data used in the 1st, 2nd and 3rd column are derived from original DNB composite data in 2013, DNB NMM in 2013 and DMSP-OLS stable data in 2012, respectively. Red labels in the panel denote: GS, Gansu; GZ, Guizhou; GX, Guangxi; SH, Shanghai; TJ, Tianjin; XZ, Xizang.

Figure 6 shows the scatter plot of the selected socio-economic parameters, such as GRP per capita (GRPPC), CT per capita income (PCI) and RPCI *vs.* log*ANL* from original DNB composite data, DNB NMM and DMSP-OLS stable data, respectively. As mentioned above, *ANL* is derived by dividing *TNL* with *LA*, so the province with the highest *ANL* is different from that with the highest *TNL* and *LA*. Shanghai has the highest log*ANL* for both DNB composite data and DMSP-OLS stable data. Accordingly, Shanghai also has the highest CTPCI and RPCI over the two years of interest, and Tianjin has the highest GRPPC in both years. On the other hand, Xizang, Guangxi and Guizhou have the lowest log*ANL* for original DNB composite data, DNB NMM data and DMSP-OLS stable data, respectively. Guizhou has the lowest GRPPC, and Gansu has the lowest CTPCI and RPCI in both years. This indicates that Shanghai is more industrialized, has a higher living standard per capita and, therefore, more night light emission in *ANL*, which is indeed the scenario in China.

4. Analysis and Discussions

From Section 3.4, it can be seen that the correlation of socio-economic data with the *TNL* and *LA* of denoised DNB composite data is in general better than with DMSP-OLS stable data. The correlations with *ANL* and log*ANL* of DNB composite data are not as good as with DMSP-OLS stable data. Therefore, this section focuses on the cause analysis of the correlation difference with variable *TNL*. In this section, we analyze the reasons for the correlation difference between DMSP stable data and DNB NMM data, rather than DNB OTM data, because, since the DNB NMM is derived from DMSP stable data, we can pay more attention to the nighttime light intensity difference between DNB composite and DMSP stable data instead of the difference caused by noise removal methods. For the same reason, we will not analyze correlation difference with variables *LA*, *ANL* and log*ANL* in this paper.

To find out the factors that cause the correlation difference between DNB NMM and DMSP-OLS stable data, we here analyze the primary difference (Table 8 [5,35]) between DNB composite and DMSP-OLS stable data, such as the effects of the saturation and quantization of the pixel value and the *TNL* ratio between different nighttime datasets.

Table 8. Main differences between VIIRS-DNB and DMSP-OLS data.

Data	Spatial Resolution	Pixel Value	Saturation Effect	Quantization	Acquisition Time (Local Time)
VIIRS/DNB	750 m (Observed) 15 arc second (NGDC)	Radiance	NO	14 bit [39]	1:30 a.m.
DMSP-OLS	2700 m (Observed) 30 arc second (NGDC)	DN	YES	6 bit [5]	~9 p.m.

4.1. *Effect of Saturation*

DN data of the DMSP-OLS stable light image can be saturated at centers of city areas where nighttime light is strong [13]. At full spatial resolution (called "fine"), the OLS collects data with a normal pixel size of 0.56 km. Onboard averaging of five by five blocks of fine data produces smoothed data with a ground sampling distance (GSD) of 2.7 km [2]. In this case, saturated and non-saturated fine pixels get averaged together, so the resultant data appears non-saturated [40]. The stable products are made using all of the available archived DMSP-OLS smooth resolution data for calendar years and re-mapped with 1 km × 1 km spatial resolution, so that the sub-pixel saturation phenomenon is not as obvious as the smoothed data. Meanwhile, China is a developing country, and the percentages of saturation area ($area_{saturation}/area_{pixel\ value>0}$) in administrative regions are small (the largest three percentage regions are Beijing, Shanghai and Tianjin, whose saturation area percentages are 0.182, 0.048 and 0.023, respectively). Therefore, we determine that the sub-pixel saturation effect is negligible in this work.

Letu *et al.* [13] developed a correction method for the saturation light by using a cubic regression equation in the power supply areas in Japan, China and other countries in Asia. The correlation results between cumulative *DN*s and electric power consumption of each prefecture in China increases after the correction for the saturation DMSP stable light. In this paper, we follow the correction method of Letu *et al.* [13] to estimate the total DN values in the saturation areas and assess the effect of saturation on the correlation difference.

The cubic regression equation is based on the tendency of DN change in non-saturated areas to correct the saturation effect. The cubic regression equation is [13]:

$$DN_T = DN_{NS} + \int_A^B \left(ax^3 + bx^2 + cx + d\right) \tag{6}$$

where DN_T is the corrected total DN of the administrative region, DN_{NS} is the total DN of the non-saturation area and A and B are the lower and upper limits of the total number of pixels in the saturation area, respectively. Additionally, a, b, c, d are coefficients that were obtained from a four-dimensional simultaneous equation on the least-squares method.

The correlation results between socio-economic parameters and *TNL* derived from the saturation-corrected DMSP-OLS stable data are listed in Table 9, and even though the correlation difference from that of *TNL* derived with DMSP-OLS stable data, Δρ, differs at the third decimal point, all of the correlations with saturation corrected data have been improved. This indicates that this cubic regression saturation correction method is effective, and the saturation effect of DMSP stable data is one of the reasons for the correlation difference.

Table 9. Correlations between socio-economic parameters and *TNL* of DMSP stable data, *TNL* of saturation-corrected DMSP stable data, *TNL* of DNB NMM and *TNL* of DNB NMM after quantization. Δρ: difference in the comparison with the correlation derived from *TNL* of DMSP stable data and DNB NMM, respectively.

	TNL of DMSP Stable Data			*TNL* of DNB NMM Data		
	Original	Saturation Corrected		Original	Quantization	
			Δρ			Δρ
TP	0.801	0.808	0.007	0.695	0.684	−0.011
CP	0.738	0.747	0.009	0.850	0.860	0.010
CTP	0.789	0.797	0.008	0.841	0.844	0.003
HH	0.809	0.816	0.007	0.721	0.710	−0.011
CA	0.700	0.709	0.009	0.754	0.756	0.002
BUA	0.864	0.870	0.006	0.904	0.898	−0.006
GRP	0.856	0.865	0.009	0.899	0.910	0.011
PWC	0.911	0.917	0.006	0.906	0.909	0.003
WWD	0.799	0.808	0.009	0.861	0.877	0.016

Figure 7 shows the correction results for the saturation pixels in Beijing and Tianjin. The correction data vary considerably from the DN of the non-saturated area, and therefore, we could estimate the DN of the saturation areas.

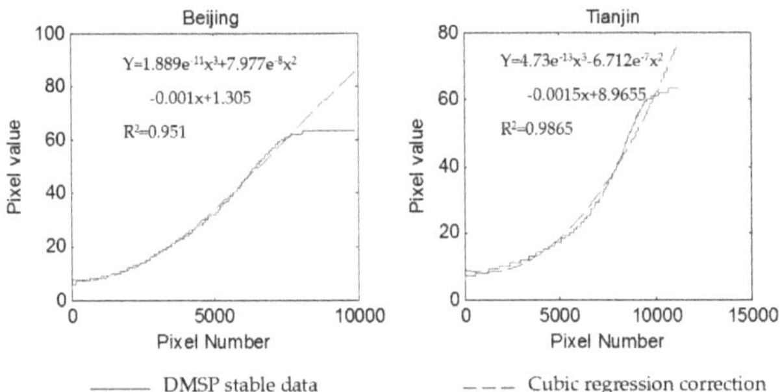

Figure 7. Correction of the saturation light by the cubic regression equation of Beijing and Tianjin, respectively.

4.2. Effect of Quantization

The radiometric signals observed by the DNB sensor are digitized using 14 bits for the HGS and 13 bits for the MGS and LGS. The fine quantization of HGS enhances the appearance of terrestrial light emissions, including faint city lights. By applying gain coefficients and offsets, raw data from DNB observation are converted into radiometric units, *i.e.*, W/cm^2-sr [19]. On the other hand, the pixel value of DMSP-OLS stable data obtained from NGDC is of a digital number (DN) in six-bit format with a value between zero and 63.

Fourteen bit and six bit are different in quantization steps. There are more gray levels for 14-bit data compared to six-bit DMSP stable data, so the different quantization might affect the correlation results. Since there is no absolute radiometric calibration for the DMSP-OLS observation in 2012, in this subsection, we show the relationship between the DN value and the radiance of DMSP-OLS stable data firstly. Based on this premise, to study the effect of finite quantization embedded in the DMSP stable data on the performance of correlations with socio-economic parameters, we artificially transform the radiance value of DNB NMM data into six-bit format to match the DN value of the DMSP-OLS stable data format and then compare the resulting correlations.

The DMSP-OLS stable data are acquired under operational conditions, and the gain is varied both along each scan line and as the satellite follows its polar orbit. However, the gain is not recorded in the data stream [40]. In this work, we assume that the gain of the operational stable light product, which is taken by sensors set at the variable, but highest level of gain [41], is fixed (*i.e.*, 55) [40].

The instrument gain of DMSP-OLS is a setting that determines how the detector converts the radiance into a digital number. The transform equation is [40,42]:

$$DN = DN_{max}\left(R/R_{sat}\right) \tag{7}$$

where DN is the digital number of the DMSP-OLS data and R is the corresponding radiance. DN_{max} is 63, and R_{sat} is the saturation radiance of the detector. Additionally, the following equation gives the relationship between gain setting (G) and saturation radiance:

$$\log_{10}R_{sat} \approx -\left(C + G\right)/20 \tag{8}$$

where C is a constant coefficient of the relationship between gain and saturation that can be acquired from a pre-launch calibration graph and is subject to change as the instrument degrades. The unit of R_{sat} is W/cm^2/sr. Even though the constant coefficient C for the DMSP-OLS F18 sensor is unreachable, based on the assumption mentioned above and Equations (7) and (8), we can get that the relationship between DN of a specific DMSP-OLS stable dataset, and its corresponding radiance value is linear.

After that, we should find the radiance range of DNB NMM data that will be quantized. We started with the hypothesis that the distribution of the composite night light in China is stable in the years 2012 and 2013, so that the brightness levels of DNB NMM data and DMSP stable data are the same. Therefore, if the DNB composite data have been artificially saturated, the saturated pixel percentage is the same as that of DMSP stable data in China mainland. Then, we arrange the pixels of DMSP stable data and DNB NMM data with a DN value in a gradually increasing order, determine the corresponding radiance value for the saturated DN of DNB composite data, *i.e.*, $L_{sat} = 3.606 \times 10^{-8}$ W/cm^2-sr, and set pixel values of DNB NMM data larger than L_{sat} equal to L_{sat}. Then, we perform an inverse transformation of Equation (7) so that the radiance data from DNB NMM data can be converted to six-bit format to match the DMSP stable data format.

Figure 8 shows the histograms of DNB NMM data transformed with finite quantization for Beijing and Tianjin. Table 9 lists the correlation coefficients between socio-economic parameters and *TNL* of DNB NMM data after quantization. From Table 9, we notice that, after quantization processing of DNB NMM, the correlations of CP, GRP and WWD have improved compared to the original DNB NMM correlations. Meanwhile, quantization processing makes the correlations with TP, HH and BUA worse. Therefore, the effect of quantization on the correlation difference is noticeable, but socio-economic parameter dependent.

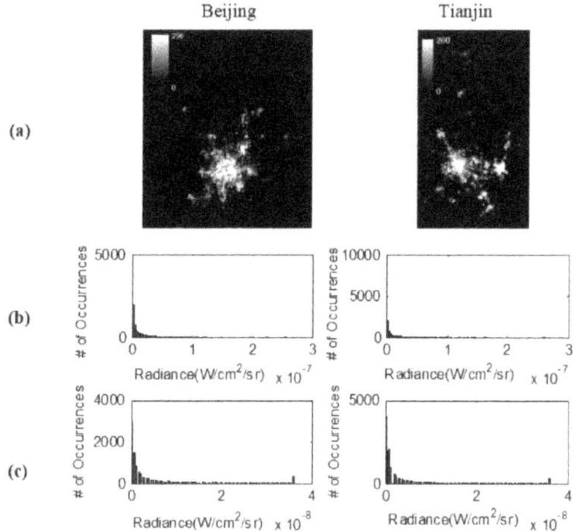

Figure 8. (**a**) DNB imagery derived with NMM for Beijing (left) and Tianjin (right); (**b**) histograms of DNB NMM data; (**c**) histograms of DNB NMM data transformed with finite quantization for Beijing (left) and Tianjin (right), respectively.

4.3. Fluctuation of the TNL Ratio

In this sub-section, we compare the difference in the *TNL* of saturation-corrected DMSP stable data and finite quantization DNB NMM data and finite quantization DNB NMM data for individual provinces. For comparison's sake, we calculate the *TNL* ratio using the pixel value of DNB NMM data instead of the radiance value, and multiplying the pixel value by 10^{-9} gives radiance in units of W/cm^2-sr.

The *TNL* ratio for 31 provinces derived from the saturation-corrected DMSP stable data in 2012 and quantized DNB NMM data in 2013 ranges from 3.4 to 24.9. The mean of the ratio is ~11.28 $\pm$ 4.02. Figure 9 shows the *TNL* ratio for 31 provinces. This indicates that the fluctuation of these ratios in Figure 9, other than saturation and quantization effects, is the other reason for the correlation difference between DNB composite data and DMSP-OLS stable data. From Figure 9, for provinces that have a large built-up area and are well industrialized, such as Beijing (2), Guangdong (6), Jiangsu (15),

Shanghai (23), Tianjin (27) and Zhejiang (31), their *TNL* ratios are relatively small. For provinces that are relatively underdeveloped, such as Gansu (5), Guangxi (7), Henan (12), Jiangxi (16), Neimenggu (19) Ningxia (22) and Yunnan (30), their *TNL* ratios are larger.

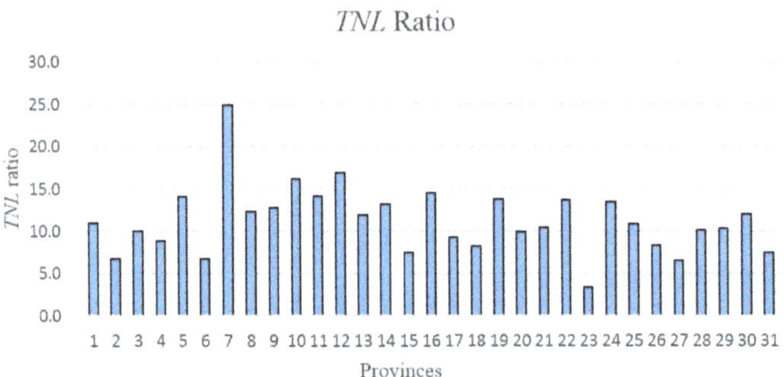

Figure 9. *TNL* ratio between saturation-corrected DMSP stable data and quantized DNB composite data for 31 provinces in China. Names corresponding to the indices of these provincial regions are listed in Table A1. Mean ratio = 11.28 and standard deviation of the ratio = 4.02.

DMSP satellites operate in Sun-synchronous orbits with nighttime overpassing at local time from 7 p.m. to 9 p.m. [2]. Additionally, the Suomi-NPP satellite was placed into Sun-synchronous orbit with local equatorial crossing times at ~1:30 a.m. during the nighttime [18]. Because of their different observation times, the characteristics of observed radiance data are quite different. At 1:30 a.m., people are asleep, and residential light sources and light emitted from vehicles are reduced, but the commercial and city infrastructure light sources are still on. Therefore, the reason for the phenomenon of the *TNL* ratio fluctuation is mainly because of the different data acquisition times. In the well-industrialized regions, commercial and city infrastructure lights are still on at midnight, so the light intensity changes are relatively small at night and midnight. On the contrary, the night light intensity has larger variation in the underdeveloped regions at night and midnight.

We note the different instantaneous field of views (IFOV) and spectral responses will affect *TNL*; but, for the *TNL* ratio, the IFOV and spectral response difference will be a constant factor, and the ratio fluctuation tendency will not change. However, still, these ratios only serve as a preliminary and rough estimate for the difference in the nighttime light emission at different night times, *i.e.*, at ~8 p.m. *vs.* at ~1:30 a.m., in different provincial regions. Other factors, such as saturation correction and finite quantization uncertainty, different data collection times in the year and sensor calibration, *etc.*, can certainly contribute to the overall uncertainty of the ratio estimation.

We use a nearest-neighbor model to resample DMSP stable data to the same spatial resolution as DNB composite data, and its *TNL* is four times the original DMSP stable data. This will not change the correlation results between DMSP stable data and statistics.

Therefore, the reasons that caused the correlation difference between nighttime data can be the effects of the saturation and quantization of DMSP stable data, and a different acquisition time is another reason for the correlation difference.

5. Conclusions

In this paper, we calculate the correlations between four variables (*TNL*, *LA*, *ANL* and log*ANL*) derived from nighttime light data and 12 socio-economic parameters at the provincial level in mainland China to compare the performance of Suomi-NPP VIIRS/DNB composite data and DMSP-OLS stable data in correlating with regional socio-economic parameters. The noise masking method and optimal threshold method have been used to remove the background noise of DNB composite data that is not related to economic activities before calculating the correlations.

From the correlation results, we can find that the OTM is effective at noise removal for both *TNL* and *LA* variables of DNB composite data, and the NMM is effective at noise removal for *TNL* of DNB composite data. Quantitatively, the correlations between *TNL* of DNB NMM and BUA, GRP and PWC are higher than 0.90. Additionally, for the *LA*, OTM can improve the correlations significantly. Correlations between *LA* of DNB OTM and CTP, BUA, GRP, PWC and WWD are higher than 0.9. For the *ANL* and log*ANL*, the processing of DNB composite data with NMM and OTM has little effect on the correlations in comparison to the original DNB composite data. All of the results demonstrate that OTM is consistent at removing noise and is an effective method for filtering DNB composite data to model these socio-economic parameters. In addition, from an application perspective, OTM is not bounded by time, but the NMM depends on the masks derived from DMSP stable data of the most recent years.

A comparison is also performed of the relationship between DNB composite and DMSP-OLS stable data with socio-economic parameters through correlation analysis. *TNL* and *LA* of DNB composite data have a better correlation than DMSP-OLS stable data in general. For *TNL*, DNB NMM has a better correlation with all of the socio-economic parameters (except TP and HH) than the correlation derived with DMSP/F18 stable data. For *LA*, DNB OTM has a better correlation with all of socio-economic parameters than the correlation derived with DMSP/F18 stable data. However, the correlation between *ANL*/log*ANL* and DNB composite data is not as good as DMSP stable data.

To analyze the factors contributing to the correlation difference between DNB composite data and DMSP stable data, we studied the effects of the differences in their saturation effect, quantization, spatial resolutions and the *TNL* ratios. A cubic regression method is developed to correct the saturation effect of DMSP stable data. Additionally, we artificially convert DNB composite data into a six-bit value to match the DMSP stable data format. The correlation results between the processed data and socio-economic data show that the effects of saturation and finite quantization are two reasons for the correlation difference. Additionally, on this basis, we estimate the *TNL* ratio between saturation-corrected DMSP stable data and finite quantization DNB composite data, and it is found that the ratio is ~11.28 $\pm$ 4.02 for China. Based on the characteristic of the *TNL* ratio fluctuation, the fluctuation tendency of the ratio is mainly due to different acquisition times: DMSP and Suomi-NPP satellites overpass at local time about 8 p.m. and 1:30 a.m., respectively. At 1:30 a.m., residential and vehicle light sources are reduced, but commercial light sources are left. The fluctuation tendency consists of the situation in which the night light intensity in more developed regions changes less at night and midnight (Figure 9). That means that the social economic parameters we considered, which have a good correlation with the VIIRS-DNB composite, are indicators of human-related activity. This does not mean that these activities cease when humans are asleep. This means that societal development, city infrastructure and, consequently, light emissions are all correlated.

Note that the ratio of *TNL* between DMSP-OLS stable and DNB composite data is a rough estimate and can be affected by other factors, such as saturation correction and finite quantization uncertainty, different data collection times in the year, sensor calibration, *etc.*

In this paper, the VIIRS DNB composite, like some eliminated ephemeral events in OTM, has no further removal, and some faint sources of VNIR emissions have been removed wrongly. These are the next steps of our work about the nighttime light.

The Suomi-NPP VIIRS/DNB is a major step forward from DMSP-OLS in its night-imaging capabilities. The advantages of the DNB sensor are clear: higher radiometric accuracy, finer spatial resolution and higher geometric quality. Additionally, and more importantly, the radiometric data are more reliable and inter-comparable due to the on-board calibration process and three-gain stage of DNB, which ensures no saturation effect at night. The comparison results in this paper confirm this and show that with DNB data, we can quantitatively determine the regional night light in radiance units and assess the correlation with socio-economic parameters. Additionally, our study demonstrates the promising aspects of applying well-calibrated VIIRS-DNB data to estimate long-term regional socio-economic development. With the effort from NOAA to improve the calibration of VIIRS DNB products, it is anticipated that the VIIRS nighttime lights will enable advances in more applications of nighttime imaging products.

Acknowledgments: This study is partially funded by the Joint Polar Satellite System (JPSS) program. The manuscript contents are solely the opinions of the authors and do not constitute a statement of policy, decision or position on behalf of NOAA or the U.S. government.

Author Contributions: Xin Jing, Xi Shao and Changyong Cao conceived of and designed the experiments. Xin Jing and Xi Shao performed the experiments. Xin Jing and Xi Shao analyzed the data. Xin Jing, Xiaodong Fu and Lei Yan contributed auxiliary data. Xin Jing and Xi Shao wrote the paper. Changyong Cao and Lei Yan edited and revised the paper.

Conflicts of Interest: The authors declare no conflict of interest.

Appendix A

Table A1. Socio-economic parameters during 2013 for 31 provinces in mainland China (units and abbreviations for these socio-economic parameters are listed in Table 3).

	Index Number	Abbreviations	TP	CP	CTP	HH	CA
Anhui	1	AH	6030	1380	2886	16,155	5852
Beijing	2	BJ	2115	1826	1825	6166	12,187
Chongqing	3	CQ	2970	1133	1733	9071	6134
Fujian	4	FJ	3774	1105	2293	11,340	4299
Gansu	5	GS	2582	568	1036	6280	1450
Guangdong	6	GD	10,644	4947	7212	27,050	16,136
Guangxi	7	GX	4719	942	2115	11,809	6104
Guizhou	8	GZ	3502	623	1325	9296	1828
Hainan	9	HN	895	246	472	2060	1265
Hebei	10	HB	7333	1608	3528	19,036	6478
Heilongjiang	11	HLJ	3835	1361	2201	11,554	2766
Henan	12	HN	9413	2321	4123	23,644	4658
Hubei	13	HB	5799	1841	3161	15,907	7349
Hunan	14	HN	6691	1430	3209	16,935	4312
Jiangsu	15	JS	7939	2884	5090	22,008	14,308
Jiangxi	16	JX	4522	960	2210	10,730	2113
Jilin	17	JL	2751	1127	1491	8206	3596
Liaoning	18	LN	4390	2324	2917	13,106	13,974
Neimenggu	19	NMG	2498	885	1466	7521	8356
Ningxia	20	NX	654	264	340	1684	2106
Qinghai	21	QH	578	164	280	1414	560
Shandong	22	SD	9733	2945	5232	28,638	21,635
shanghai	23	SH	2415	2415	2164	8377	6341
Shanxi	24	SX	3630	1057	1908	9903	2999
Shaanxi	25	SAX	3764	862	1931	10,203	1555
Sichuan	26	SC	8107	1866	3640	24,136	6433
Tianjin	27	TJ	1472	664	1207	4515	2334
Xinjiang	28	XJ	2264	706	1007	5634	1620
Xizang	29	XZ	312	62	74	636	339
Yunnan	30	YN	4687	806	1897	11,739	3337
Zhejiang	31	ZJ	5498	1998	3519	17,034	10,992
	BUA	GRP	GRPPC	CTPCI	RPCI	PWC	WWD
Anhui	1777	19039	31,574	25,006	8098	1528	266,234
Beijing	1306	19501	92,201	45,274	18,338	913	144,580
Chongqing	1115	12657	42,615	26,850	8332	813	142,535
Fujian	1263	21760	57,657	33,383	11,184	1701	259,098

Table A1. *Cont.*

	Index Number	Abbreviations	TP	CP	CTP	HH	CA
Gansu	727	6268	24,276	20,149	5108	1073	64,969
Guangdong	5232	62164	58,403	36,504	11,669	4830	862,471
Guangxi	1154	14378	30,468	25,029	6791	1238	225,303
Guizhou	695	8007	22,863	21,413	5434	1126	93,085
Hainan	296	3146	35,156	24,920	8343	232	36,156
Hebei	1787	28301	38,595	24,143	9102	3251	310,921
Heilongjiang	1758	14383	37,504	21,149	9634	845	153,090
Henan	2289	32156	34,161	23,687	8475	2899	412,582
Hubei	2007	24668	42,539	25,181	8867	1630	294,054
Hunan	1505	24502	36,619	24,643	8372	1423	307,227
Jiangsu	3810	59162	74,520	35,131	13,598	4957	594,359
Jiangxi	1151	14339	31,708	22,949	8782	947	207,138
Jilin	1344	12981	47,188	23,544	9621	654	117,703
Liaoning	2386	27078	61,680	27,905	10,523	2008	234,508
Neimenggu	1206	16832	67,383	26,978	8596	2182	106,920
Ningxia	421	2565	39,221	23,767	6931	811	38,528
Qinghai	157	2101	36,350	22,131	6196	676	21953
Shandong	4187	54684	56,184	30,628	10,620	4083	494,570
shanghai	999	21602	89,450	48,879	19,595	1411	222,963
Shanxi	1041	12602	34,717	24,014	7154	1832	138,030
Shaanxi	915	16045	42,628	24,109	6503	1152	132,169
Sichuan	2058	26261	32,393	23,894	7895	1949	307,648
Tianjin	747	14370	97,623	35,656	15,841	774	84,210
Xinjiang	1065	8360	36,927	22,388	7297	1540	100,720
Xizang	120	808	25,887	22,561	6578	31	5005
Yunnan	936	11721	25,007	24,698	6141	1460	156,583

References

1. Doll, C.N.H. *CIESIN Thematic Guide to Night-Time Lights Remote Sensing and Its Applications*; The Trustees of Columbia University: New York, NY, USA, 2008.
2. Elvidge, C.D.; Erwin, E.H.; Baugh, K.E.; Ziskin, D.; Tuttle, B.T.; Ghosh, T.; Sutton, P.C. Overview of DMSP nightime lights and future possibilities. In Proceedings of the 2009 Joint Urban; Remote Sensing Event, Shanghai, China, 20–22 May 2009; pp. 1–5.
3. Elvidge, C.D.; Cinzano, P.; Pettit, D.R.; Arvesen, J.; Sutton, P.; Small, C.; Nemani, R.; Longcore, T.; Rich, C.; Safran, J.; *et al.* The Nightsat mission concept. *Int. J. Remote Sens.* **2007**, *28*, 2645–2670. [CrossRef]
4. Elvidge, C.D.; Baugh, K.E.; Kihn, E.A.; Kroehl, H.W.; Davis, E.R. Mapping city lights with nighttime data from the DMSP Operational Linescan System. *Photogramm. Eng. Remote Sens.* **1997**, *63*, 727–734.
5. Elvidge, C.D.; Baugh, K.E.; Dietz, J.B.; Bland, T.; Sutton, P.C.; Kroehl, H.W. Radiance calibration of DMSP-OLS low-light imaging data of human settlements. *Remote Sens. Environ.* **1999**, *68*, 77–88. [CrossRef]
6. Croft, T.A. Nighttime images of the earth from space. *Sci. Am.* **1978**, *239*, 86–98. [CrossRef]
7. Welch, R. Monitoring urban population and energy utilization patterns from satellite data. *Remote Sens. Environ.* **1980**, *9*, 1–9. [CrossRef]
8. Sutton, P. Modeling population density with night-time satellite imagery and GIS. *Comput. Environ. Urban Syst.* **1997**, *21*, 227–244. [CrossRef]
9. Elvidge, C.D.; Baugh, K.E.; Kihn, E.A.; Kroehl, H.W.; Davis, E.R.; Davis, C.W. Relation between satellite observed visible-near infrared emissions, population, economic activity and electric power consumption. *Int. J. Remote Sens.* **1997**, *18*, 1373–1379. [CrossRef]
10. Doll, C.N.H.; Muller, J.P.; Elvidge, C.D. Night-time imagery as a tool for global mapping of socioeconomic parameters and greenhouse gas emissions. *Ambio* **2000**, *29*, 157–162. [CrossRef]
11. Doll, C.N.H.; Muller, J.P.; Morley, J.G. Mapping regional economic activity from night-time light satellite imagery. *Ecol. Econ.* **2006**, *57*, 75–92. [CrossRef]

12. Sutton, P.C.; Elvidge, C.D.; Ghosh, T. Estimation of gross domestic product at sub-national scales using nighttime satellite imagery. *Int. J. Ecol. Econ. Stat.* **2007**, *8*, 5–21.
13. Letu, H.; Hara, M.; Yagi, H.; Naoki, K.; Tana, G.; Nishio, F.; Shuhei, O. Estimating energy consumption from night-time DMPS/OLS imagery after correcting for saturation effects. *Int. J. Remote Sens.* **2010**, *31*, 4443–4458. [CrossRef]
14. Li, X.; Li, D. Can night-time light images play a role in evaluating the syrian crisis? *Int. J. Remote Sens.* **2014**, *35*, 6648–6661. [CrossRef]
15. Coscieme, L.; Pulselli, F.M.; Bastianoni, S.; Elvidge, C.D.; Anderson, S.; Sutton, P.C. A thermodynamic geography: Night-time satellite imagery as a proxy measure of emergy. *Ambio* **2013**, *43*, 1–11. [CrossRef] [PubMed]
16. Lo, C.P. Modeling the population of china using DMSP operational linescan system nighttime data. *Photogramm. Eng. Remote Sens.* **2001**, *67*, 1037–1048.
17. Lee, T.E.; Miller, S.D.; Turk, F.J.; Schueler, C.; Julian, R.; Deyo, S.; Dills, P.; Wang, S. The NPOESS VIIRS day/night visible sensor. *Bull. Am. Meteorol. Soc.* **2006**, *87*, 191–199. [CrossRef]
18. Miller, S.D.; Mills, S.P.; Elvidge, C.D.; Lindsey, D.T.; Lee, T.F.; Hawkins, J.D. Suomi satellite brings to light a unique frontier of nighttime environmental sensing capabilities. *Proc. Natl. Acad. Sci. USA* **2012**, *109*, 15706–15711. [CrossRef] [PubMed]
19. Cao, C.; Shao, X.; Uprety, S. Detecting light outages after severe storms using the S-NPP/VIIRS day/night band radiances. *IEEE Geosci. Remote Sens. Lett.* **2013**, *10*, 1582–1586. [CrossRef]
20. Cao, C.; de Luccia, F.J.; Xiong, X.; Wolfe, R.; Weng, F. Early on-orbit performance of the visible infrared imaging radiometer suite onboard the suomi national polar-orbiting partnership (S-NPP) satellite. *IEEE Trans. Geosci. Remote Sens.* **2014**, *52*, 1142–1156. [CrossRef]
21. Liao, L.B.; Stephanie, W.; Steve, M.; Bruce, H. Suomi NPP VIIRS day-night band on-orbit performance. *J. Geophys. Res. Atmos.* **2013**, *118*, 12705–12718. [CrossRef]
22. Shao, X.; Cao, C.; Uprety, S. Vicarious calibration of S-NPP/VIIRS day-night band. *Proc. SPIE* **2013**. [CrossRef]
23. Geis, J.; Florio, C.J.; Moyer, D.; Rausch, K.W.; de Luccia, F.J. VIIRS day-night band gain and offset determination and performance. *Proc. SPIE* **2012**, *8510*, 851012.
24. Shi, K.; Yu, B.; Huang, Y.; Hu, Y.; Yin, B.; Chen, Z.; Chen, L.; Wu, J. Evaluating the ability of NPP-VIIRS nighttime light data to estimate the gross domestic product and the electric power consumption of China at multiple scales: A comparison with DMSP-OLS data. *Remote Sens.* **2014**, *6*, 1705–1724. [CrossRef]
25. Ma, T.; Zhou, C.; Tao, P.; Haynie, S.; Fan, J. Responses of SUOMI-NPP VIIRS-derived nighttime lights to socioeconomic activity in China's cities. *Remote Sens. Lett.* **2014**, *5*, 165–174. [CrossRef]
26. Li, X.; Xu, H.; Chen, X.; Li, C. Potential of NPP-VIIRS nighttime light imagery for modeling the regional economy of China. *Remote Sens.* **2013**, *5*, 3057–3081. [CrossRef]
27. Chen, X.; Nordhaus, W. A test of the new VIIRS lights data set: Population and economic output in Africa. *Remote Sens.* **2015**, *7*, 4937–4947. [CrossRef]
28. Shi, K.; Huang, C.; Yu, B.; Yin, B.; Wu, Y.H.J. Evaluation of NPP-VIIRS night-time light composite data for extracting built-up urban areas. *Remote Sens. Lett.* **2014**, *5*, 358–366. [CrossRef]
29. Chen, Z.; Yu, B.; Hu, Y.; Huang, C.; Shi, K.; Wu, J. Estimating house vacancy rate in metropolitan areas using NPP-VIIRS nighttime light composite data. *IEEE J. Sel. Top. Appl. Earth Obs. Remote Sens.* **2015**, *8*, 2188–2197. [CrossRef]
30. Shi, K.; Yu, B.; Hu, Y.; Huang, C.; Chen, Y.; Huang, Y. Modeling and mapping total freight traffic in China using NPP-VIIRS nighttime light composite data. *Gisci. Remote Sens.* **2015**, *52*, 274–289. [CrossRef]
31. Yu, B.; Shi, K.; Hu, Y.; Huang, C.; Chen, Z.; Wu, J. Poverty evaluation using NPP-VIIRS nighttime light composite data at the county level in China. *IEEE J. Sel. Top. Appl. Earth Obs. Remote Sens.* **2015**, *8*, 1217–1229. [CrossRef]
32. Yu, B.; Shu, S.; Liu, H.; Song, W.; Wu, J.; Chen, L.W.Z. Object-based spatial cluster analysis of urban landscape pattern using nighttime light satellite images: A case study of China. *Int. J. Geogr. Inf. Sci.* **2014**, *28*, 2328–2355. [CrossRef]
33. Li, X.; Zhang, R.; Huang, C.; Li, D. Detecting 2014 northern IRAQ insurgency using night-time light imagery. *Int. J. Remote Sens.* **2015**, *36*, 3446–3458. [CrossRef]

34. NOAA. VIIRS Nighttime Lights-2012. Available online: http://ngdc.noaa.gov/eog/viirs/download_viirs_ntl.html (accessed on 18 October 2015).
35. NOAA. Version 4 DMSP-OLS Nighttime Lights Time Series. Available online: http://ngdc.noaa.gov/eog/dmsp/downloadV4composites.html (accessed on 18 October 2015).
36. NOAA. Version 4 DMSP-OLS Nighttime Lights Time Series. Available online: http://ngdc.noaa.gov/eog/gcv4_readme.txt (accessed on 18 October 2015).
37. Data Sharing Infrastructure of Earth System Science. Available online: http://www.geodata.cn/data/publisher.html (accessed on 18 October 2015).
38. Chen, J.; Zhuo, L.; Shi, P.J.; Toshiaki, I. The study on urbanization process in China based on DMSP/OLS data: Development of a light index for urbanization level estimation. *J. Remote Sens.* **2003**, *7*, 168–175.
39. Miller, S.D.; Turner, R.E. A dynamic lunar spectral irradiance data set for NPOESS/VIIRS day/night band nighttime environmental applications. *IEEE Trans. Geosci. Remote Sens.* **2009**, *47*, 2316–2329. [CrossRef]
40. Ziskin, D.; Baugh, K.; Feng, C.H.; Ghosh, T.; Elvidge, C.D. Methods used for the 2006 radiance lights. *Proc. Asia Pac. Adv. Netw.* **2010**, *30*, 131–142. [CrossRef]
41. Hsu, F.C. Global Radiance Calibrated Nighttime Lights Product Readme. Available online: http://ngdc.noaa.gov/eog/dmsp/radcal_readme.txt (accessed on 18 October 2015).
42. Kohiyama, M.; Hayashi, H.; Maki, N.; Higashida, M.; Kroehl, H.W.; Elvidge, C.D.; Hobson, V.R. Early damaged area estimation system using DMSP-OLS night-time imagery. *Int. J. Remote Sens.* **2004**, *25*, 2015–2036. [CrossRef]

MDPI AG
St. Alban-Anlage 66
4052 Basel, Switzerland
Tel. +41 61 683 77 34
Fax +41 61 302 89 18
http://www.mdpi.com

Remote Sensing Editorial Office
E-mail: remotesensing@mdpi.com
http://www.mdpi.com/journal/remotesensing

www.ingramcontent.com/pod-product-compliance
Lightning Source LLC
Chambersburg PA
CBHW051700210326
41597CB00032B/5322

* 9 7 8 3 0 3 8 4 2 3 1 8 8 *